Lecture Notes in Computer Science   1240

Edited by G. Goos, J. Hartmanis and J. van Leeuwen

Advisory Board: W. Brauer   D. Gries   J. Stoer

Springer-Verlag Berlin Heidelberg GmbH

José Mira   Roberto Moreno-Díaz
Joan Cabestany (Eds.)

Biological and Artificial Computation: From Neurosciene to Technology

International Work-Conference on Artificial
and Natural Neural Networks, IWANN'97
Lanzarote, Canary Islands, Spain
June 4-6, 1997
Proceedings

 Springer

Series Editors

Gerhard Goos, Karlsruhe University, Germany
Juris Hartmanis, Cornell University, NY, USA
Jan van Leeuwen, Utrecht University, The Netherlands

Volume Editors

José Mira
Departamento de Inteligencia Artificial
Universidad Nacional de Educación a Distancia
Senda del Rey s/n, E-28040 Madrid, Spain
E-mail: jmira@dia.uned.es

Roberto Moreno-Díaz
Centro Inter. de Investigacion en Ciencias de la Computacion
Universidad de las Palmas de Gran Canaria
Campus de Tafira, E-35017, Canary Islands, Spain
E-mail: roberto@grumpy.dis.ulpgc.es

Joan Cabestany
Departament d'Enginyeria Electronica, Universitat Politécnica de Catalunya
C/ Gran Capita s/n, E-08034 Barcelona, Spain
E-mail: cabestan@eel.upc.es

Cataloging-in-Publication data applied for

Die Deutsche Bibliothek - CIP-Einheitsaufnahme

Biological and artificial computation : from neuroscience to
neurotechnology ; proceedings / International Work Conference on
Artificial Neural Networks, IWANN '97, Lanzarote, Canary Islands,
Spain, June 4 - 6, 1997. José Mira ... (ed.).
  (Lecture notes in computer science ; Vol. 1240)
  ISBN 978-3-540-63047-0   ISBN 978-3-540-69074-0 (eBook)
  DOI 10.1007/978-3-540-69074-0

CR Subject Classification (1991): I.2, F.1.1, C.1.3, C.2.1, G.1.6, I.5.1, B.7.1,
J.1, J.2

ISSN 0302-9743
ISBN 978-3-540-63047-0

© Springer-Verlag Berlin Heidelberg 1997
Originally published by Springer-Verlag Berlin Heidelberg New York in 1997

Typesetting: Camera-ready by author
SPIN 10548822   06/3142 – 5 4 3 2 1 0   Printed on acid-free paper

Preface

Neural computation is considered here in the dual perspective of *analysis* (as reverse engineering) and *synthesis* (as direct engineering). As a science of *analysis*, neural computation seeks to help neurology, brain theory, and cognitive psychology in the understanding of the functioning of the nervous system by means of computational models of neurons, neural nets, and subcellular processes, with the possibility of using electronics and computers as a "laboratory" in which cognitive processes can be simulated and hypotheses proven without having to act directly upon living beings.

As a *direct* engineering (how can we build sub-symbolic intelligent machines?), neural computation seeks to complement the symbolic perspective of artificial intelligence (AI), using the biologically inspired models of distributed self-programming and self-organizing networks, to solve those non-algorithmic problems of function approximation and pattern classification having to do with changing and only partially known environments. Fault tolerance and dynamic reconfiguration are other basic advantages of neural nets.

In the sea of meetings, congresses, and workshops on ANNs, IWANN'97, the fourth International Work-Conference on Artificial Neural Networks, that took place in Lanzarote, Canary Islands (Spain), 4 - 6 June, 1997, focused on the three subjects that most worry us:

(1) The search for biologically inspired new models of local computation architectures and learning along with the organizational principles behind the complexity of intelligent behavior.

(2) The search for some methodological contributions in the analysis and design of knowledge-based ANNs, instead of "blind nets", and in the reduction of the knowledge level to the sub-symbolic implementation level.

(3) The cooperation with symbolic AI, with the integration of connectionist and symbolic processing in hybrid and multi-strategy approaches for perception, decision, and control tasks, as well as for case-based reasoning, concept formation, and learning.

To contribute to the formulation and partial solution of these global topics, IWANN'97 offered a brain-storming interdisciplinary forum in advanced neural computation for scientists and engineers from biology, neuroanatomy, computational neurophysiology, molecular biology, biophysics, mathematics, computer science, artificial intelligence, parallel computing, electronics, cognitive sciences, and all the concerned applied domains (sensory systems and signal processing, monitoring, diagnosis, classification and decision making, intelligent control and supervision, perceptual robotics and communication systems).

The papers presented here correspond to talks delivered at IWANN'97, organized by the Universidad Nacional de Educación a Distancia (UNED), Madrid, Universidad de Las Palmas de Gran Canaria, and Universidad Politécnica de

Catalunya, in cooperation with the Asociación Española de Redes Neuronales (AERN), IFIP Working Group in Neural Computer Systems, WG10.6, Spanish RIG IEEE Neural Networks Council, and the UK&RI Communication Chapter of IEEE.

Sponsorship has been obtained from the Spanish CICYT and DGICYT (MEC) and the organizing universities (UNED, Las Palmas, and Catalunya).

After the evaluation process, 142 papers were accepted for oral presentation or poster, according to the recommendations of reviewers and the author's preferences. The three extended papers corresponding to the invited speakers (DeFelipe, Eckhorn, and Ienne) have been included as introductions to the corresponding topics of neuroscience, neural modeling in perception, and implementation.

We would like to thank all the authors as well as all the members of the international program committee for their labor in the production and evaluation of the papers. Only by proceeding with this severe averaging of the external experts' reviews, could we be sure to maximize the originality, technical quality, and scientific relevance of this event. We also would like to mention the effort of the authors of rejected papers, mainly because they were immature proposals or topics not covered by IWANN.

Last but not least, the editors would like to thank Springer-Verlag, in particular Alfred Hofmann, for the continuous and excellent cooperative collaboration from the first IWANN in Granada (1991, LNCS 540), the successive meetings in Sitges (1993, LNCS 686) and Torremolinos (1995, LNCS 930), and now in Lanzarote.

The papers published in this volume present the current situation in natural and artificial neural nets, with a significant increase in the contributions related to the biological foundations of neural computation and the computational perspective of neuroscience. We have organized the papers in the following sections:

* *Biological Foundations of Neural Computation*
* *Formal Tools and Computational Models of Neurons and Neural Nets Architectures*
* *Plasticity Phenomena (Maturing, Learning and Memory)*
* *Complex Systems Dynamics*
* *Cognitive Science and IA*
* *Neural Nets Simulation, Emulation and Implementation*
* *Methodology for Data Analysis, Task Selection and Nets Design*
* *Neural Networks for Communications, Control and Robotics*

This book endeavors to summarize the state of the art in neural computation with a focus on biologically inspired models of the natural nervous system. The complexity of the nervous system is now accepted, and a significant part of the scientific community has returned to anatomy and physiology, rejecting the temptation to use models which are clearly insufficient to cope with this complexity. At the same time there is an increasing interest in the use of computational models of neural networks to improve our understanding of the functional organization of the brain. Finally, there is also evidence of a lack of formal tools enabling the hybridization of the symbolic and connectionistic perspectives of artificial intelligence in the common goal of making computational the knowledge of human experts in technical domains related with perception, communication, and control. All these developments, as reported in these proceeding, are needed in order to bring neuroscience and computation closer together. To recognize the disparity that exists between the richness and fineness of the nervous system and the crudeness we use in handling it is a good step forward.

Madrid, March 1997

J. Mira Mira

R. Moreno-Díaz

J. Cabestany Moncusi

Contents

1. Biological Foundations of Neural Computation

2. Formal Tools and Computational Models of Neurons and Neural Net Architectures

3. Plasticity Phenomena (Maturing, Learning and Memory)

4. Complex Systems Dynamics

5. Cognitive Science and AI

6. Neural Nets Simulation, Emulation and Implementation

7. Methodology for Data Analysis, Task Selection and Nets Design

8. Neural Networks for Perception

9. Neural Networks for Communications, Control and Robotics

Microcircuits in the Brain

Javier DeFelipe

Instituto Cajal (CSIC), Avenida Dr. Arce, 37, 28002-Madrid, Spain
E-mail: defelipe@cajal.csic.es

Abstract. There is an increasing interest in computational models of neural networks, based on real synaptic circuits, for investigating the functional organization of the brain. It seems obvious that the utility of these models would be greater as more detailed circuit diagrams become available. The present paper provides some general quantitative data on the synaptic organization of the cerebral cortex and summarizes certain basic features of cortical microcircuits which might be useful for computational models of the cerebral cortex.

Introduction

The brain is composed of a very large number of neurons (on the order of several billion) which are heavily interconnected through synapses in an ordered fashion, forming highly intricate neuronal networks of variable complexity, depending on brain structure and species. It is clear that the synaptic organization (microcircuitry) in different parts of the brain differs considerably among themselves and, therefore, data obtained in one structure are not necessarily applicable in another. Thus, synaptic organization must be examined separately in particular regions. The neocortex is the choice region of numerous theorists and experimentalists because of its direct involvement in many aspects of mammalian behavior and because it is the most human part of the nervous system; that is to say, it is the brain structure where those capabilities are localized, such as speech and thought, that distinguish humans from other mammals. However, there are a number of difficulties in studying the neocortex. Among the most important are: first, the fibers entering and leaving the neocortex do not form distinct intracortical pathways and very little is known about the characteristics of the information carried by these fibers; second, there is a high degree of complexity. For example, in one mm^3 of mouse cerebral cortex it has been estimated that there are approximately 90.000 neurons, $3km$ of axonal length, $300m$ of dendritic length and 700 million synapses [1]. Nevertheless, it should be noted that the cerebral cortex is not like the universe (to which it is often compared) with unknown limits containing an immeasurable number of elements, but it has well-defined limits and a finite number of elements. It is thought that the neuronal circuits of the human neocortex have reached their finest organization and highest level of complexity; but, fortunately, during mammalian evolution many aspects of neocortical organization are maintained, and data obtained in a given species can often be extrapolated to other species. This is particularly important when data can only be obtained with experimental approaches which, for obvious reasons, cannot be performed with humans. This fact, together with the recent development of powerful methods of characterizing cortical circuits, has allowed us to analyze in detail certain fundamental aspects of neocortex microorganization which are of extraordinary importance for better understanding the functioning of the cerebral cortex. In recent years, computational models of neural networks based on real cortical networks have become useful tools for studying certain aspects of the functional organization of the neocortex (see refs. 2,3). It appears that as more detailed circuit diagrams become available, the more we will learn with computer simulations about the role of each element of the circuit. Thus, it is likely that increased interaction will occur in the next few years between theorists interested in neural computation and experimentalists working on cortical microanatomy. In the present paper, I shall try to summarize using diagrams those features of microcircuits of the neocortex which, in my opinion, are essential. Since knowledge of cortical synaptic circuits is rather incomplete, these diagrams represent, of course, an oversimplification of real cortical microcircuits.

Some historical notes and general aspects of the organization of the neocortex

The detailed study of cerebral cortex organization began at the end of last century with Cajal's master descriptions of the structure of the cortex using the Golgi method. Before Cajal's discoveries, very little was known about the neuronal elements of the cortex or the possible relationships between intrinsic neurons and afferent fibers. This was due to the fact that the early methods of staining only allowed visualization of neuronal cell bodies and a tiny portion of their proximal processes, and some poorly-stained fibers which had unknown origins [4]. The Golgi method, however, permitted the visualization of individual neurons with all their parts: soma, dendrites and axon. Furthermore, the finest morphological details were readily observed in Golgi-stained cells, which led to the characterization and classification of neurons, and to the study of their possible connections (Fig.1).

The neocortex does not have a uniform structure: it has been divided into a series of cortical areas which are distinguished by their histological and functional characteristics. However, these cortical areas are not separated by any insulating morphological boundaries, but constitute a continuous thin structure whose neurons are arranged in horizontal layers (layers I-VI). In addition to neurons, the neocortex contains neuroglia, nerve fibers and blood vessels. The nerve fibers can be subdivided into extrinsic and intrinsic. The *extrinsic fibers* are represented by those originating outside the neocortex and include fibers from the thalamus, brain stem, basal forebrain, claustrum, amygdala, hypothalamus and hippocampus. Nevertheless, there are clear regional variations in the distribution and density of these fibers, as well as species differences. The *intrinsic fibers* are those originating in the neurons of the neocortex itself and can be further subdivided into short-range and long-range. The short-range fibers are distributed near, or at a short distance, from the parent neuron (local axons), while the long-range fibers are those coming from other cortical areas (commissural and ipsilateral corticocortical fibers). The terms "cortical inputs" or "afferent systems or fibers" refer to extrinsic and corticocortical fibers collectively, the thalamocortical and corticocortical fibers being the major cortical inputs.

There are two major classes of neurons: spiny neurons and smooth nonpyramidal neurons. Spiny neurons are cells whose dendritic surfaces are covered with spines and are represented by pyramidal cells and spiny stellate cells. *Pyramidal cells* constitute the most abundant and characteristic neuronal type of the cerebral cortex and are distinguished by the pyramidal shape of their cell bodies, which give rise to a prominent apical dendrite directed towards the cortical surface (Fig.1, *left*). They are the projection neurons of the cerebral cortex and are located in all layers (except layer I); but, in general, their projection sites differ depending on the layer where their cell bodies are located [5,6]. However, their descending axons give rise to a variety of axonal collaterals that engender profuse terminal axonal arborizations which make up one of the most important components of intracortical circuitry [7]. *Spiny stellate cells* represent a morphologically heterogeneous group of neurons located in layer IV of the primary sensory areas, whose dendritic arborization remains mostly within this layer. Some of them project to other cortical areas, but the majority are short-axon cells (interneurons) whose axons are distributed within layer IV, or in the layers above or below [8]. Since layer IV represents one of the main regions of thalamic axon terminals, spiny stellate cells have often been considered the principal target of thalamic input and, therefore, as key components in the initial steps of the processing of thalamic input. Electron microscope examination of synaptic connections between thalamic axons and identified neurons [6,9] has proved that spiny stellate cells are postsynaptic to thalamocortical axons, however many types of pyramidal and nonpyramidal cells also receive thalamocortical synapses in even a greater percentage than spiny stellate cells. *Smooth nonpyramidal neurons* are short-axon cells with smooth or sparsely spiny dendrites. They are found in all layers and it has been estimated that approximately 1 out of 4 or 5 of all neurons are smooth nonpyramidal cells (see below), but different proportions are found in different layers. They show a variety of morphologies and there are 10 basic types recognized on the basis of their patterns of axonal arborization and synaptic connectivity [10,11].

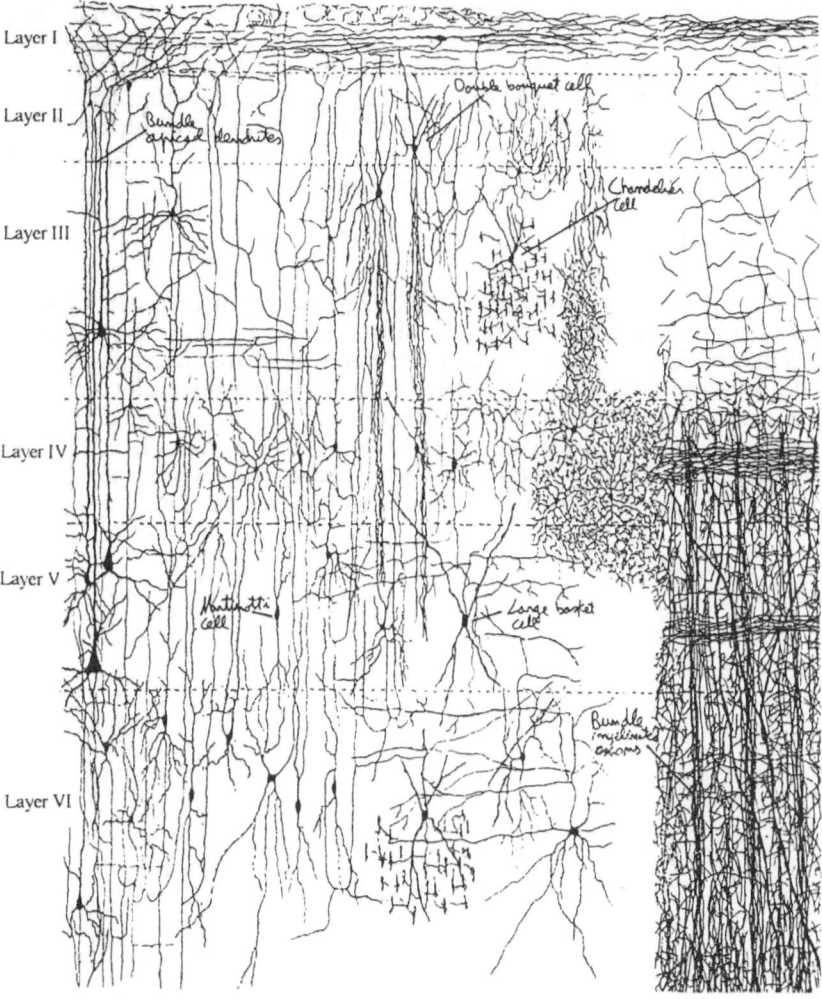

Fig. 1. Drawing showing the main types of cortical neurons found in Golgi studies and myelinated axons (*right*). This drawing has been modified and extended by the author of the present paper from a drawing originally presented by Bonne (*L'écorce cérébrale*, Rev. Gén. Histol. 2: 291-581, 1906) to illustrate the main types of cortical neurons described by Cajal.

At the electron microscope level, two main types of synapses are recognized, mostly on the basis of the morphology of the postsynaptic density: asymmetrical and symmetrical synapses. Asymmetrical synapses form between 75-95% of all synapses, whereas the remaining 5-25% are symmetrical synapses. The major sources of asymmetrical synapses are spiny neurons and the main cortical afferent fiber systems which are excitatory in function, while most axon terminals forming symmetrical synapses originate in the population of smooth nonpyramidal cells, and many of these cells and the axon terminals forming symmetrical synapses have been shown, using immunocytochemical techniques, to be immunoreactive for GABA (or its synthesizing enzyme, glutamic acid decarboxylase, GAD) which is the main inhibitory neurotransmitter of the cerebral cortex [6,7,12-15]. Thus, in general, asymmetrical synapses are considered to be excitatory and symmetrical synapses inhibitory and, consequently, excitatory synaptic circuits dominate over inhibitory ones. Finally, in sections stained for myelin (Fig. 1, *right*) there has been revealed in the deep half of the cortex the existence of numerous and regularly distributed small, long bundles of myelinated axons (radial fasciculi) and the presence of bands of myelinated axons running horizontally, mainly in layers I, IV and V. The origins of these myelinated axons are from both extrinsic and intrinsic fibers, pyramidal cells being the major source of myelinated axons in the radial fasciculi and in the horizontal bands in layers IV and V.

Basic circuit

As pointed out above, the introduction of the Golgi method permitted the study of possible connections between intrinsic neurons and afferent fibers. This led to the great advance of tracing the first circuit diagrams of the cerebral cortex, which were increasing in complexity as more data were available (Fig. 2). These circuits reached the highest level of complexity and refinement with Lorente de Nó [16], whose schematic circuit diagram of the cerebral cortex is still frequently used in text books. However, in most cases, the exact connections between neurons and afferent fibers could not be specified and it was unknown whether the connections between neurons were performed by one, two, several or many synapses.

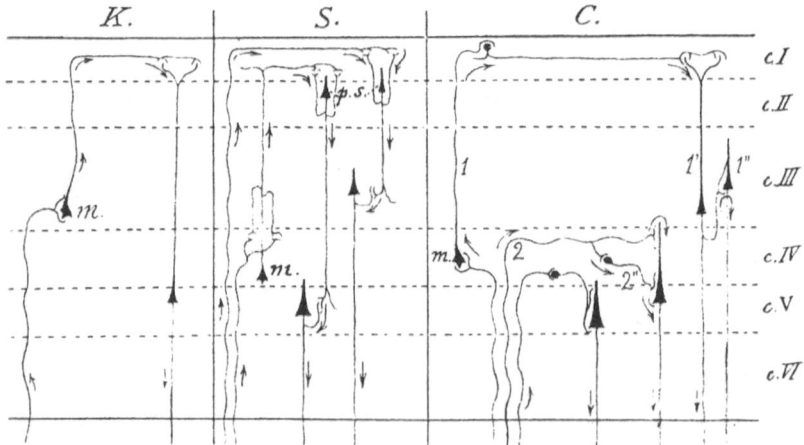

Fig. 2. Early schematic circuit diagrams of the cerebral cortex after Kölliker (*K*) Schaffer (*S*) and Cajal (*C*). *m*, Martinotti cells; *ps*, superficial polymorphic cells. From Bonne (*L'écorce cérébrale*, Rev. Gén. Histol. 2: 291-581, 1906).

The introduction of electron microscopy represented the next crucial step in the to study cortical circuits, Gray [17] and Colonnier [18] being the first to examine in detail the cerebral cortex; they also classified cortical synapses into either types I and II (Gray) or asymmetrical and symmetrical (Colonnier). Combining Wallerian degeneration with electron microscopy [19] permitted the identification of thalamocortical axon terminals and their synaptic connections [20]. After these studies, a number of researchers examined at the electron microscope level the synaptic connections of the main afferent fiber systems (principally thalamocortical synapses) (reviewed in [21]), but the types of postsynaptic neurons could not be identified with certainty. Furthermore, the connections between the different types of neurons were still highly speculative. However, Szentágothai [22,23] was able to integrate data from studies using the Golgi method and electron microscopy to make useful circuits diagrams. But it was the introduction of the combination of the Golgi method and electron microscopy [24,25], and especially the gold-toning technique of Fairén and colleagues [26], which allowed the accurate study of the ultrastructural characteristics and synaptic connections of Golgi-impregnated neurons. This technique led to the further development of combinations of a variety of techniques for electron microscopy (among which the most important are degeneration, enzyme histochemistry and immunocytochemistry) in the study of synaptic (afferent and efferent) connections and chemical characteristics of Golgi-impregnated neurons, of retrogradelly horseradish peroxidase (HRP)-labeled neurons and of physiologically characterized and intracellularly HRP-labeled neurons [6,21,27]. These approaches have been used in recent years by a number of researchers and have led to important advances in the knowledge of the chemical and synaptic organization of the cerebral cortex, allowing us to trace cortical circuit diagrams quite precisely [6,15]. In spite of this, there is still at present a considerable lack of data about synaptic circuits. For example, the particular characteristics of inputs to pyramidal and nonpyramidal cells identified by their physiological properties, chemical features, projection sites (in the case of pyramidal cells) or synaptic connections are still basically unknown. Therefore, it should be kept in mind that many characteristics of the main patterns of cortical synaptic connections, or the basic circuit, is largely incomplete. Nevertheless, certain important features of synaptic circuits are known that may be helpful in setting up the basis for drawing a "sufficiently" complete basic circuit in the near future. These features, which are based on data obtained from many researchers (reviewed in refs. 6,7,28), can be summarized as follows.

Pyramidal cells are the output neurons; therefore, as more data are available about synaptic inputs to pyramidal cells, the more will be known about intracortical processing of information. Synapses are found in dendrites, somata and axon initial segments of pyramidal cells, but the majority are found in dendrites (on the order of a few thousands), followed by the soma (a few hundred) and then the axon initial segment (a few dozen). However, there is a particular distribution of synapses over these neuronal surfaces. The soma and axon initial segment receive synapses only of the symmetrical type, whereas on dendrites both asymmetrical and symmetrical are found. The main source of symmetrical synapses is from inhibitory GABAergic interneurons, while for asymmetrical synapses the sources are the major cortical afferent fiber systems and the local axon of pyramidal cells, which are excitatory. Each dendritic spine receives one synapse which is asymmetrical, but some spines (10%) receive an additional synapse that is either asymmetrical or symmetrical. Although the density of spines is not uniform in all regions of the dendritic arbor, and the number of synapses on the dendrites, soma and axon initial segments is variable depending on the type of pyramidal cell, the synaptic values given below might be considered, in general, as approaches to mean values.

For dendrites, the total number, types and distribution of synapses are based on the following estimations and assumptions. There are 15 spines per $10\mu m$ of dendrite, the length of the dendrites being a few millimeters. The total number of synapses per $10\mu m$ of dendrite is 20, of which 85% are considered to be asymmetrical (n= 17) and 15% symmetrical (n= 3). The total number of synapses on spines is 16.5 (15 + 1.5, because 10% of spines receive 2 synapses). Of the symmetrical synapses, 31% are on spines (n= 0.93, which represent 4.65% of all synapses) and the rest (n= 2.07, which represent 10.35% of all synapses) on the dendritic shaft. Of the asymmetrical synapses, 15.57 are on spines (77.85% of all synapses), the remaining 1.43 being asymmetrical synapses (7.15% of all synapses) found on the dendritic shaft. The soma width varies approximately between $12\mu m$ and $60\mu m$ (giant pyramidal cells of Betz), with the surface area ranging from $300\mu m^2$ to $6400\mu m^2$. Two types of somata regarding synaptic density can be distinguished: one of high density (2 synapses per $10\mu m^2$) and the other of low density (1 synapse per $10\mu m^2$). The axon initial segments, which are approximately $25\mu m$ in length and $1\mu m$ in diameter, can also be subdivided into those with high synaptic density (approximately 25 synapses) and low synaptic density (approximately 3 synapses). These data can be useful for constructing a variety of pyramidal cell-like elements with realistic synaptic weights for computational models. For example, it is possible to construct a pyramidal cell-like element based on the morphological parameters found in a real pyramidal cell, say in a medium-size pyramidal cell with a soma of 20 μm in width and having $1000\mu m^2$ of somatic surface area and a total dendritic length of $2500\mu m$. This element would have the following synaptic values:

Number of synapses on dendrites: 2500 x 2= 5000
-dendritic shafts
asymmetrical: 5000 x 0.0715= 357.5
symmetrical: 5000 x 0.1035= 517.5
total= 875

-dendritic spines (n= 3750 spines)
asymmetrical: 5000 x 0.7785= 3892.5
symmetrical: 5000 x 0.0465= 232.5
total= 4125

Number of synapses on the soma= 200 symmetrical (with high density) or 100 symmetrical (with low density)

Number of synapses on the axon initial segment= 25 symmetrical (with high density) or 3 symmetrical (with low density)

TOTAL= 5225 maximum (4250 asymmetrical, 975 symmetrical); 5103 minimum (4250 asymmetrical, 853 symmetrical)

Many types of cortical GABAergic neurons innervate pyramidal cells. These GABA cells are the main source of symmetrical synapses on pyramidal cells and can be divided into four major groups (Fig. 3): *D cells*, which innervate only, or almost only, dendrites (for example, double bouquet cells and cells with axonal arcades); *DS cells* and *SD cells*, which form numerous synapses with both the dendrites and soma (for example, large and small basket cells), but with different degrees of preference, *DS* cells being those that innervate preferentially the dendrites and *SD* cells the soma; and *AIS cells*, which form numerous synapses with the axon initial segment, the chandelier cell being the only known representative neuron of this group to form synapses only with the axon initial segment. It has been shown (directly or indirectly) that all these neurons form and receive synapses from

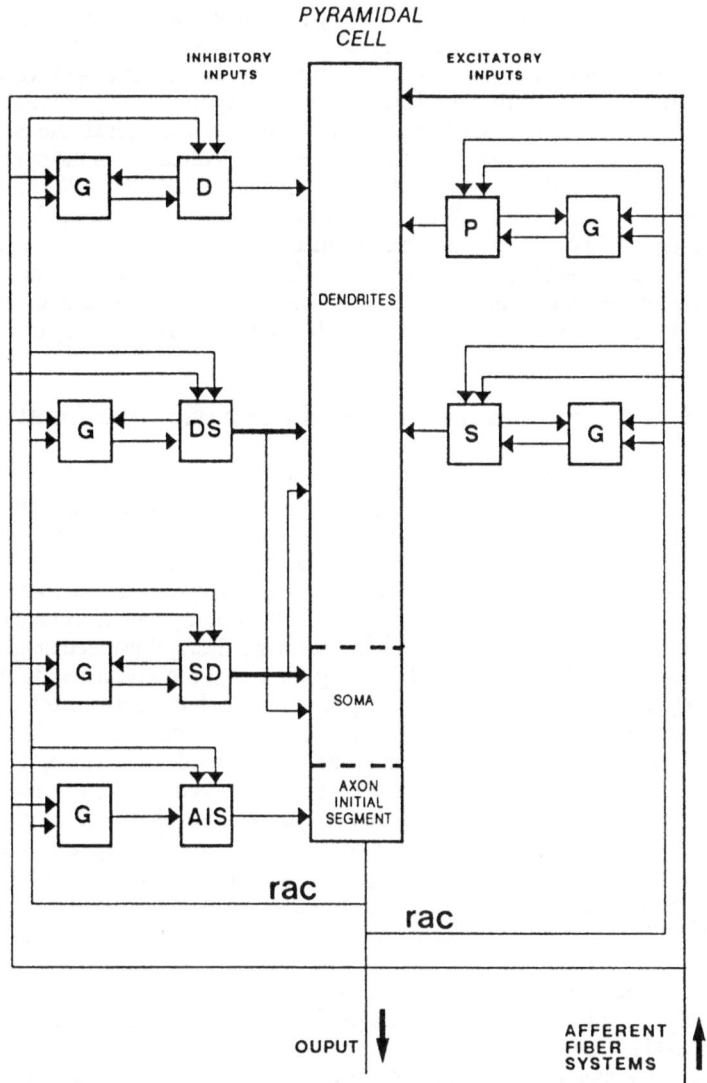

Fig. 3. Schematic diagram of the synaptic inputs and main patterns of local connections of the pyramidal cell. The pyramidal cell is the output neuron of the cerebral cortex and receives inhibitory inputs from a variety of GABAergic cells (*D*, *DS*, *SD*, *AIS*) that form synapses with all parts of the pyramidal cell (dendrites, soma and axon initial segment). The excitatory inputs terminate only on the dendrites and come from the various cortical afferent systems and the local axonal arborization of other pyramidal (*P*) and spiny stellate (*S*) cells. The local neurons that innervate the pyramidal cell receive inhibitory synapses from GABAergic cells (*G*) and excitatory synapses from the recurrent axon collaterals (*rac*) of the pyramidal cell and from afferent fibers. For further details, see "Basic circuit".

other GABAergic nonpyramidal cells (*G* in Fig. 3), except for *AIS* cells, which receive synapses from other nonpyramidal cells, but do not form synapses with them. As shown in Fig. 3, the excitatory inputs to pyramidal cells terminate on the dendrites and come from the local axonal arborization of other pyramidal (*P*) and spiny stellate (*S*) cells and from the various cortical afferent systems, which, in turn, form synapses with all the neurons that innervate the pyramidal cell. In addition, the recurrent axon collaterals of the pyramidal cell may form synapses with the same cells that innervate it (*D, DS, SD, AIS, P* and *S* cells) and with the GABAergic cells (*G*) that innervate those neurons. Since the number of synapses that a pyramidal cell receives is on the order of several thousand and, in general, the number of synapses made by a single axon originating from a given cortical or subcortical neuron is in tens or less, each single pyramidal cell receives synapses from many sources; in other words, it is involved in a great variety of circuits and, thus, probably in a variety of functions.

Columnar organization, modules and the basic functional unit

In Nissl-stained sections, it is observed (more clearly in some cortical areas than in others) that pyramidal cells form small vertical aggregates. However, this ordering is more obvious in sections stained for myelin, where the bundles of vertically-oriented myelinated fibers (radial fasciculi) (Fig. 1, *right*) indicate a regular, vertical arrangement of groups of pyramidal cells. These facts have been known since the first cyto- and myelo-architectonic studies. For example, it was Meynert at the end of the last century who discovered the radial fasciculi, the vertical arrangement of pyramidal cells being clearly illustrated in his drawings (see [4]). In addition, Cajal in his studies using the Golgi method often described and illustrated bundles of apical dendrites (see Fig. 6 in [29]), which is also a reflection of the vertical grouping of pyramidal cells (Fig. 1, *left*). However, Lorente de Nó [16] was the one who emphasized vertical organization, and the first to consider the cortex as formed by "*elementary units*" of operation. He considered the cortex as consisting of small cylinders "*composed of vertical chains of neurons*" that include all cortical layers, having a specific afferent fiber as an axis. He says: "*all the elements of the cortex are represented in it, and therefore it may be called an elementary unit, in which, theoretically, the whole process of the transmission of impulses from the afferent fibre to the efferent axon may be accomplished*". This columnar organization did not receive much attention until the physiological studies of Mountcastle [30,31], who provided experimental evidence for the columnar organization. This organization has been proven both anatomically and physiologically in a number of cortical areas [32,33], the works of Hubel and Wiesel in the visual cortex [34] giving the most solid support to this organization.

Mountcastle proposed that the neocortex is composed of replicated local neural circuits which are often called modules. Each module processes information from its input to its output, the basic modular unit being the *minicolumn*, which represents the smallest functional unit of cortical organization [32]. This minicolumn is formed by a vertically-oriented group of interconnected cells, which are contained in a vertical cylinder of tissue with a diameter of about $30\mu m$, and crossing all cortical layers, about 600 million being the number of minicolumns present in the human neocortex [32]. Based on the immunocytochemical studies of Hendry and colleagues [35] on the number and proportion of GABA immunoreactive neurons in the monkey neocortex, within this cylinder there would be approximately 35 GABAergic and 155 non-GABAergic neurons in the visual cortex, and 25 GABAergic and 70 non-GABAergic neurons in the rest of the cortical areas. The majority of non-GABAergic neurons are pyramidal cells and, as stated above, they are arranged in vertical groups in such a way that their apical dendrites and myelinated axons form bundles. The bundles of apical dendrites have been particularly studied by Peters and colleagues [36,37]. Recently, Peters and Yilmaz [38] have also examined in detail the spatial relationship between bundles of myelinated axons and apical dendrites in the monkey visual cortex and have proposed that the functional neuronal units of the visual cortex are pyramidal cell modules, each pyramidal cell module being contained within a vertical cylinder of cortical tissue of about $23 \mu m$ in diameter. However, pyramidal cells are not the only cortical cells that show a regular organization, since the axons of the so-called double bouquet cells, which are characterized by the vertical bundling of their axons that traverse several layers [11,39,40] (Fig. 1), form a regular microcolumnar system with a center-to-center spacing of $15-30 \mu m$ [41] (see below).

Finally, the terminal portions of the axons ("candles") of single chandelier cells impregnated by the Golgi method, or intracellularly-labeled with HRP, have been shown to be distributed in clusters [42-44] which innervate small groups of pyramidal cells, with many pyramidal cells located within the limits of their axonal arborization not being innervated [44]. This distribution has also been found when chandelier cell axon terminals are stained in large numbers using immunocytochemistry for the calcium-binding protein parvalbumin [45]. These immunoreactive chandelier cell axon terminals are also found to form small clusters, so that all pyramidal cells within the cluster appear to be innervated, but between the clusters many may not be [46]. Furthermore, axon initial segments of pyramidal cells receive a variable number of axoaxonic synapses and some of them seem not to be postsynaptic to chandelier cells [44,47]. Since pyramidal cells projecting to a common target are grouped in small clusters as well [48] these observations support the suggestion that chandelier cells may influence the activity of some clusters of output cells, but not others [44].

Local circuit

A schematic diagram of the main intracortical pathways and circuits could be as follows (Fig. 4). Pyramidal cells may be grouped in vertical cylinders ("vertical units") of about $30 \mu m$ in diameter and extending from layer II to VI. Their main axons are clustered together, forming a bundle of myelinated axons (radial fasciculi). According to Peters and Sethares [38], these bundles contain an average of 34 myelinated axons in the monkey visual cortex, which is a number that could be taken as an approach to the number of pyramidal cells that form the vertical units. In general, pyramidal cells in different layers have different morphological and chemical characteristics, project to different sites, and are involved in different synaptic connections, in such a way that the vertical unit could be subdivided into upper, middle and deep local subunits. The upper subunit would include pyramidal cells located in layer II and the upper part of layer III, the middle subunit pyramidal cells located in the lower part of layer III and layer IV, and the deep subunit pyramidal cells located in layers V and VI. Within each subunit, pyramidal cells may or may not be interconnected, but the pattern of synaptic connections in the neighboring upper, middle and deep subunits would be similar within a given area, although they may differ from other areas. Each subunit contains pyramidal cells that form vertical connections with other subunits located within the same unit and horizontal connections between adjacent vertical units, but not all pyramidal cells within a given subunit participate equally in these connections: some are involved mostly in vertical connections, while others mostly in horizontal connections [49-53]. Part of the inputs to the cells forming these subunits may be from the same or different afferent fibers, depending on the origin of these fibers. For example, the bulk of nonspecific thalamocortical axons, claustral and serotoninergic fibers, although terminating in different layers, are coincident in some layers [54] and, therefore, cells within the upper, middle and deep subunits receive partially overlapping and partially non-overlapping information. Thus, when a given afferent system is activated, two pyramidal cells located in different subunits within the same or different vertical units may be similarly affected, while activation of a different afferent system may affect mainly, or exclusively, only one of the two cells.

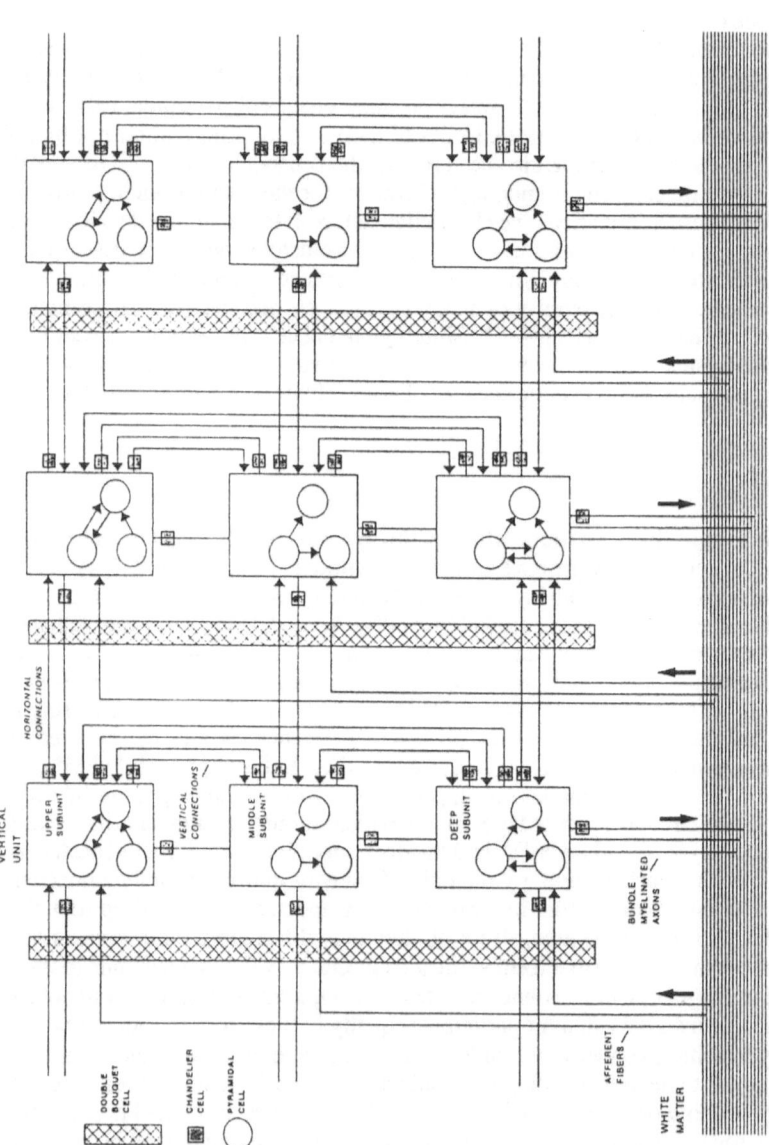

Fig. 4. Schematic diagram of the main intracortical pathways and circuits. The cerebral cortex is composed of multiple vertical units of information processing. Each vertical unit has been subdivided across the layers in upper, middle and deep subunits, and are composed of pyramidal cells. Within each subunit, pyramidal cells may or may not be interconnected, but the pattern of synaptic connections in the neighboring upper, middle and deep subunits would be similar within a given area, although they may differ from other areas. The only GABAaergic neurons represented are chandelier cells and double bouquet cells. For further details, see "Local circuit".

The pyramidal cells forming the upper, middle and deep subunits are innervated by a variety of GABAergic neurons (see above) which exert their effects either locally, vertically across layers, or horizontally within one or more layers, depending on the shape and distribution of their axonal arborization. Furthermore, the horizontal spread of the axonal arborization of all GABAergic neurons is greater than $30\mu m$, except for double bouquet cells, whose axon collaterals form vertical bundles of about $10\mu m$ in diameter [41]. Thus, each interneuron is involved in the GABAergic input to several subunits belonging either to the same or different vertical units. In the schematic diagram of Fig. 4, chandelier cells have been given special importance because they are considered to be one of the most powerful types of inhibitory neuron, and which probably play a crucial role in the control of the efferent activity of each subunit. Finally, double bouquet cells have also been considered as key components in this diagram, mainly for two reasons. First, double bouquet cells are thought to be one of the major sources of inhibitory axospinous synapses on pyramidal cells [7] and dendritic spines represent the major postsynaptic sites of pyramidal cells that receive excitatory synapses (see above). Second, double bouquet cells constitute a microcolumnar inhibitory system that could act on groups of pyramidal cells located in different subunits within vertical units and, therefore, these cells might serve as "minicolumnar enhancer elements" to increase discrimination between inputs from excitatory neurons to neighboring vertical units. In conclusion, the cerebral cortex might be envisaged as composed of multiple, small vertical units of information processing. Each vertical unit is subdivided across layers into different subunits of processing, these subunits being composed of pyramidal cells which are involved in a wide variety of functions.

Acknowledgments

The author is grateful to Ana María Sainz-Pardo for her help with the drawings shown in figures 3 and 4. This work was supported by FIS grant 96 /2134.

References

1. A. Schüz, G. Palm. *Density of neurons and synapses in the cerebral cortex of the mouse.* J. Comp. Neurol. **286**: 442-455, 1989.
2. O. V. Favorov, D.G. Kelly. *Minicolumnar organization within somatosensory cortical segregates: I. Development of afferent connections.* Cereb. Cortex **4**: 408-427, 1994.
3. O. V. Favorov, D.G. Kelly. *Minicolumnar organization within somatosensory cortical segregates: II. Emergent functional properties.* Cereb. Cortex **4**: 428-442, 1994.
4. E. G. Jones. *History of cortical cytology.* In: Cerebral cortex. Vol.1. Cellular components of the cerebral cortex, pp. 1-32. A. Peters, Jones, E.G (Eds.). New York: Plenum Press, 1984.
5. E. G. Jones. *Laminar distribution of cortical efferent cells.* In: Cerebral cortex. Vol.1. Cellular components of the cerebral cortex, pp. 521-553. A. Peters, E. G. Jones (Eds.). New York: Plenum Press, 1984.
6. E. L. White. *Cortical Circuits: Synaptic Organization of the Cerebral Cortex. Structure, Function and Theory.* Boston: Birkhäuser, 1989
7. J. DeFelipe, I. Fariñas. *The pyramidal neuron of the cerebral cortex: Morphological and chemical characteristics of the synaptic inputs.* Prog. Neurobiol. **39**: 563-607, 1992.
8. J. S. Lund. *Spiny stellate neurons.* In: Cerebral cortex. Vol.1. Cellular components of the cerebral cortex, pp. 255-308. A. Peters, E. G. Jones (Eds.). New York: Plenum Press, 1984.
9. T. F. Freund, K.A.C. Martin, I. Soltész, P. Somogyi, D. Whitteridge. *Arborisation pattern and postsynaptic targets of physiologically identified thalamocortical afferents in striate cortex of the macaque monkey.* J. Comp. Neurol **289**: 315-336, 1989.

10. E.G. Jones. *Varieties and distribution of non-pyramidal cells in the somatic sensory cortex of the squirrel monkey.* J. Comp. Neurol. **160**: 205-268, 1975.

11. A. Fairén, J. DeFelipe, J. Regidor. *Nonpyramidal neurons. General account.* In: Cerebral cortex. Vol.1. Cellular components of the cerebral cortex, pp. 201-253. A. Peters, E. G. Jones (Eds.). New York: Plenum Press, 1984.

12. C. R. Houser, J. E. Vaughn, S. H. C. Hendry, E. G. Jones, A. Peters. *GABA neurons in the cerebral cortex.* In: Cerebral cortex. Vol.2. Functional properties of cortical cells, pp. 63-89. E. G. Jones, A. Peters (Eds.). New York: Plenum Press, 1984.

13. A. Peters. *Synaptic specificity in the cerebral cortex.* In: Synaptic Fuction, pp. 373-397. G. M. Edelman, W. E.Gall, W. M. Cowan (Eds.). New York: John Wiley, 1987.

14. J. DeFelipe. *Neocortical neuronal diversity: chemical heterogeneity revealed by co-localization studies of classic neurotransmitters, neuropeptides, calcium binding proteins and cell surface molecules.* Cerebral Cortex **3**: 273-289, 1993.

15. E. G. Jones. *GABAergic neurons and their role in cortical plasticity in primates.* Cerebral Cortex. **3**: 361-372., 1993.

16. R. Lorente de Nó. *Architectonics and structure of the cerebral cortex.* In: Physiology of the nervous system, pp. 291-330. J. F. Fulton (Ed.). New York : Oxford University Press, 1938.

17. E.G. Gray. *Axo-somatic and axo-dendritic synapses of the cerebral cortex: An electron microscopic study.* J. Anat. **93**: 420-433, 1959.

18. M. Colonnier. *Synaptic patterns on different cell types in the different laminae of the cat visual cortex. An electron microscope study.* Brain Res. **9**: 268-287, 1968.

19. M. Colonnier. *Experimental degeneration in the cerebral cortex.* J. Anat. (Lond.) **98**: 47-53., 1964.

20. E. G. Jones. *An electron microscopic study of the termination of afferent fiber systems within the somatic sensory cortex of the cat.* J. Anat. (Lond.) **103**: 595-597, 1968.

21. E. L. White. *Thalamocortical synaptic relations: a review with emphasis on the projections of specific thalamic nuclei to the primary sensory areas of the neocortex.* Brain Res. Rev. **1**: 275-311, 1979.

22. J. Szentágothai. *The "module-concept" in cerebral cortex architecture.* Brain Res. **95**: 475-496, 1975.

23. J. Szentágothai.*The neuron network of the cerebral cortex: A functional interpretation.* Proc. R. Soc. London. Ser. B. **201**: 219-248, 1978.

24. W. K. Stell. *Correlation of retinal cytoarchitecture and ultrastructure in Golgi preparations.* Anat. Rec. **153**: 389-397, 1965.

25. T. W. Blackstad. *Mapping of experimental axon degeneration by electron microscopy of Golgi preparations.* Z. Zellforsch. **67**: 819-834, 1965.

26. A. Fairén, A.Peters, J. Saldanha. *A new procedure for examining Golgi impregnated neurons by light and electron microscopy.* J. Neurocytol. **6**: 311-337, 1977.

27. P. Somogyi. *Synaptic connections of neurones identified by Golgi impregnation: Characterization by immunocytochemical, enzyme histochemical, and degeneration methods.* J. Electron. Microsc. Tech. **15**: 332-351, 1990.

28. M. L. Feldman. *Morphology of the neocortical pyramidal neuron.* In: Cerebral cortex. Vol.1. Cellular components of the cerebral cortex, pp. 123-200. A. Peters, E. G. Jones (Eds.). New York: Plenum Press, 1984.

29. S. R. Cajal. *Estudios sobre la corteza cerebral humana III: Corteza acústica.* Rev. Trim. Micrográf. **5**: 129-183, 1900.

30. V. B. Mountcastle. *Modality and topographic properties of single neurons of cat's somatic sensory cortex.* J. Neurophysiol. **20**: 408-434, 1957.

31. T. P. S. Powell, V. B. Mountcastle. *Some aspects of the functional organization of the cortex of the postcentral gyrus of the monkey: A correlation of findings obtained in a*

single unit analysis with cytoarchitecture. Bull. Johns Hopkins Hosp. **105**: 133-162, 1959.

32. V. B. Mountcastle. *An organizing principle for cerebral function: The unit module and the distributed system.* In: The mindful brain, pp. 7-50. V. B. Mountcastle, G. M. Edelman (Eds.). Cambridge, MA: MIT press, 1978.

33. E.G. Jones. *The columnar basis of cortical circuitry.* In: The clinical neurosciences, W. D. Willis (Ed.), pp. 357-383. New York: Churchill Livingstone, 1983.

34. Hubel, D. H., and Wiesel, T. N. *Functional architecture of macaque monkey cortex.* Proc. R. Soc. Lond. B **198**: 1-59, 1977.

35. S. H. C. Hendry, H.D. Schwark, E.G. Jones, J. Yan. *Numbers and proportions of GABA-immunoreactive neurons in different areas of monkey cerebral cortex.* J. Neurosci. **7**: 1503-1519, 1987.

36. A. Peters, T.M. Walsh. *A study of the organization of apical dendrites in the somatic sensory cortex of the rat.* J. Comp. Neurol. **144**: 253-268, 1972.

37. A. Peters, C. Sethares. *Organization of pyramidal neurons in area 17 of the monkey visual cortex.* J. Comp. Neurol. **306**: 1-23, 1991.

38. A. Peters, C. Sethares. *Myelinated axons and the pyramidal cell modules in the monkey primary visual cortex.* J. Comp. Neurol. **365**: 232-255, 1996.

39. S. R. Cajal. *Estudios sobre la corteza cerebral humana I: Corteza visual.* Rev. Trim. Micrográf. Madrid **4**: 1-63, 1899.

40. P. Somogyi, A. Cowey. *Double bouquet cells.* In: Cerebral cortex. Vol.1. Cellular components of the cerebral cortex, pp. 337-360. A. Peters, E. G. Jones (Eds.). New York: Plenum Press, 1984.

41. J. DeFelipe, S. H. C. Hendry, T. Hashikawa, M. Molinari, E. G. Jones. *A microcolumnar structure of monkey cerebral cortex revealed by immunocytochemical studies of double bouquet cell axons.* Neuroscience **37**: 655-673, 1990.

42. A. Fairén, F. Valverde. *A specialized type of neuron in the visual cortex of cat : a Golgi and electron microscope study of chandelier cells.* J. Comp. Neurol. **194**: 761-779, 1980.

43. T. F. Freund, K. A. C. Martin, A. D. Smith, P. Somogyi. *Glutamate decarboxylase-immunoreactive terminals of Golgi-impregnated axoaxonic cells and of presumed basket cells in synaptic contact with pyramidal neurons of cat's visual cortex.* J.Comp. Neurol. **221**: 263-278, 1983.

44. J. DeFelipe, S. H. C. Hendry, E. G. Jones, D. Schmechel, D. *Variability in the terminations of GABAergic chandelier cell axons on initial segments of pyramidal cell axons in the monkey sensory-motor cortex.* J. Comp. Neurol. **231**: 364-384, 1985.

45. J. DeFelipe, S. H. C. Hendry, E. G. Jones. *Visualization of chandelier cell axons by parvalbumin immunoreactivity in monkey cerebral cortex.* Proc. Natl. Acad.Sci. USA **86**: 2093-2097, 1989.

46. M. R. del Río, J. DeFelipe. *A study of SMI 32-stained pyramidal cells, parvalbumin-immunoreactive chandelier cells and presumptive thalamocortical axons in the human temporal neocortex.* J. Comp. Neurol. **342**: 389-408, 1994.

47. I. Fariñas, J. DeFelipe. *Patterns of synaptic input on corticocortical and corticothalamic cells in the cat visual cortex. II. The axon initial segment.* J. Comp. Neurol. **304**: 70-77, 1991.

48. E. G. Jones, S. P. Wise. *Size, laminar and columnar distribution of efferent cells in the sensory-motor cortex of monkeys.* J. Comp. Neurol. **175**: 391-438, 1977.

49. C. D. Gilbert, T. N. Wiesel. *Clustered intrinsic connections in cat visual cortex.* J. Neurosci. **3**: 1116-1133, 1983.

50. K. A. C. Martin, D. Whitteridge. *Form, function, and intracortical projections of spiny neurons in the striate visual cortex of the cat.* J. Physiol. (Lond.) **353**: 463-504, 1984.

51. J. DeFelipe, M. Conley, E. G. Jones. *Long-range focal collateralization of axons arising from corticocortical cells in monkey sensory-motor cortex.* J. Neurosci. **6**: 3749-3766, 1986.
52. H. D. Schwark, E. G. Jones. *The distribution of intrinsic cortical axons in area 3b of cat primary somatosensory cortex.* Exp. Brain Res. **78**: 501-513, 1989.
53. H. Ojima, C. N. Honda, E. G. Jones. *Patterns of axon collateralization of identified supragranular pyramidal neurons in the cat auditory cortex.* Cerebral Cortex **1**: 80-94, 1991.
54. E. G. Jones. *Identification and classification of intrinsisc circuit elements in the neocortex.* In: Dynamic aspects of neocortical function, pp. 7-40. G. Edelman, W. M. Cowan, W. E. Gall (Eds), New York : John Wiley, 1985.

Some Reflections on the Relationships Between Neuroscience and Computation

J. Mira & A.E. Delgado

Dpto. de Inteligencia Artificial
Facultad de Ciencias. UNED
C/ Senda del Rey s/n. 28040 Madrid. SPAIN

Summary

We are still a long way from understanding the organization and functioning of the Nervous Systems. At the same time there is a large disparity between the richness and fineness of the nervous phenomenology and the crudeness of the point neuron simplistic models we use in its modeling. This distances neuroscience from computation in an almost irreversible fashion.

To contribute to the posing and partially solving of these problems we present in this paper a summary of our work that has been marked by two recurrent themes: (i) *the search for a methodology of the natural*, with the introduction of distinct levels and domain of description (ii) *the search for inspiration in neuroscience* seeking new ideas for a more realistic model of neural computation. The final purpose is to move computation a step closer to neuroscience and vice versa.

1. Introduction

The usual paradigm in terms of which the neural processes are interpreted is the computational one, in the sense of Kuhn [Kuhn, 1970]. In this paradigm we begin by differentiating between environment and system (anatomical compartment under consideration). The behavior of the system is then described in terms of a set of inputs ("stimuli" at this level), a set of outputs ("response messages" to the next connected compartments) and a set of computable processes (logic and/or integro-differential operators) that applied to these inputs and to some internal variables ("state"), modify these "state" and produces a set of outputs that replicate and predict the stimulus-induced neuronal responses in the real system.

This computational approach to neuroscience seeks to help neuroscientist to uncover the complex information processing and knowledge acquisition and transformation underlying the anatomy and physiology of the nervous systems. The final purpose is the understanding of the functioning of neurons, neural nets and sub-cellular processes in a sense similar to that of physics (a *"brain theory"*).

At the same time, these neural models has been used, since the historical time of biocybernetics and bionics (with the works of W. McCulloch and W. Pitts, N. Wiener, J. von Neumann, J. McCarthy, H. von Foerster, O.G. Selfridge, R. Ashby, D. McKay, A. Newell, M. Minsky, S. Papert, S.C. Kleene, M. Davis, A. Uttley, C. Shannon, U. Maturana, J. Lettvin and M. Arbib, to name but a few), as source of inspiration in engineering and computation [McCulloch, 1965], [Shannon & McCarthy, 1956]. The basic idea is that both, live beings and machines, can be understood using the same organizational and structural principles of communication (coding and information theory), control (motor planning), pattern recognition, self-organization, evolution, modularity and massive parallelism, associative memory, reliability (fault-tolerance) and plasticity (maturing, learning and self-programming) [Wiener, 1947], [Yovits & Cameron, 1960]. Consequently, we can get some profit in studying the biological solutions to those non algorithmic problems of *pattern recognition, decision,* and *planning* in changing and only partially known environments that had proved to be so hard for the symbolic perspective of Artificial Intelligence (AI).

Unfortunately, in the "fiorello" of neural computation [Rumelhart et al., 1986], both as modeling tool and as alternative or complement to the symbolic perspective of AI, the same simplistic models of the past were repeated. This distances computation from neuroscience in an almost irreversible fashion because the lack of methodology and the fact that the formal tools used are clearly insufficient. Strictly speaking, most of these models, what Segev call *"morpho-less" point neuron* [Segev, 1992], says nothing about the neural functioning of real brains. The deepest and more for-reaching aspects of neural tissue are practically ignored because neural modeling is "a priory" limited by the nature of the formal and conceptual tools used to describe processes. In modeling neurons we build mathematical and for logical formulations of experimental data and relations. This means that the properties and limitations of the mathematical operators are imposed on the final model.

At the same time, as plain computational models of modular, distributed and self-programming architectures in engineering, these old models of adders followed by sigmoids and supervised learning of the backpropagation type, are nearly exhausted, and consequently, some fresh air is also needed in the computational side of the river.

As a preliminary reflection on the relationship between computation and neuroscience, two problems are now clear to us, [Mira, 1995].

P1. The intrinsic complexity of the nervous systems of which we don't have the "original drawing" for the design recovery. This is a problem of "reverse neurophysiology" made worse by the lack of a methodology of the natural and theoretical developments to integrate the sea of experimental data and techniques available.

P2. The use of very poor models in neural computation for the local function and the associated "learning" processes (parameter adjustment in skeletal models). The panorama of analogical and logical formal tools usually available for neural modeling and, consequently, the neurophysiological data which serves as their basic, are not sufficient to describe, model and predict the most genuine aspects of the nervous system. This problem, again, is made worse by the lack of theory and methodology in the field artificial neural nets (ANN's). The mathematical foundations of this modular, parallel and self-programming computation are not still well established.

The purpose of this paper is to contribute to the posing of these two problems. In order to do this, we will first comment on some methodological aspects of computation that could be of interest in the interpretation of the neurophysiological findings and the planning of new experiments. Then, we move from neuroscience to computation in search of a different sort of inspiration. We need to know more about the operative languages (neural codes) associated with the different levels of description of neural nets, which will allow us to substitute neurons, dendro-dendritic circuits and synaptic contacts for more effective computational models in the context of the genuine functions of the nervous systems. Until we know this "more" about real biological computation, instead of looking after trivial models of analogic or logic nature or replications of the input-output correlations at the biophysical level, it could be of more interest to use the neurophysiological findings as a source of inspiration in search of some *"modes of computation"* apparently genuine of the nervous tissue in the sense of being capable of partially reproducing, not the details of implementation, but the logical organization, the wealth and the complexity of the emergent behavior of real neural nets. The final purpose is to move computation a step closer to neuroscience and vice versa.

2.   From Computation to Neuroscience

2.1.   Levels of description

In neural modeling, when we are faced with the complexity of the nervous system, it is useful to make use of a hierarchy of levels of description: *subcellular*, (physico-chemical processes), *cellular* (biochemical and electric), *organic*, and that of *global behaviour*. Each descriptive level is characterized by a phenomenology, a set of entities and relationships, as well as some organizational and structural principles of its own. From the neural modeling viewpoint, each compartment in each level is represented in terms of an input space and on output space, along with the transformation rules which link both spaces. These spaces are, always, representational, with a sign-significance structure which acquires value in the domain of the observer that interprets the experimental results. Each level has it own set of signals it understand but, when the description of the computation in the level is made by an external observer, as usual, we need always to state in a clear manner the particular *semantics* of these input and output spaces (sign-significance relationship) to account for previous processes experience and evolution. Also, only a specific set of questions are pertinent for each level.

Every level is partially closed to organization and structure. This closure brings about the stability of the level and its inclusion within a *hierarchy* in which each level is linked to the one below by means of a process of *reduction*. The link with the level above is carried out by means of emergency processes which, in order to be understood, require the *injection of knowledge* from and external observer (the neurobiologist) in order to add the semantic tables necessaries to raise the reduced semantics of the signal level to the entities of the new level. Otherwise, only irrelevant interpretations are obtained.

A frequent source or error in the interpretation of experimental data lies in the mixing of entities belonging to different levels and the attempt to explain data and processes pertinent only in a high semantic level, such as memory, purpose or feelings using only experimental findings and models that belongs to very low levels. Hence, evolution, learning and all the social dimensions of animal behaviour are neglected.

In computation, the proposal of three levels (knowledge, symbolic and physical) is due to [Newell, 1981] and [Marr, 1982] (figure 1). The models at the knowledge level (KL) are concerned with the natural language descriptions of a calculus, including the corresponding algorithm. The symbolic level (SL) is below and corresponds to the programs. That is to say, to the symbols, data structures and sequence of instructions using only the primitives of a formal language. Finally, the physical level (PL) has to do with the implementation that leads us from the program to the hardware, ending the design. These three levels of description are related, though not in a single, and causal manner.

In moving down to a lower level, (a problem of direct engineering), information is always lost because neither the machine which implements algorithm, nor the algorithm which resolves a problem are uniquely determined. Nevertheless, in computation we have pre-established and unequivocal laws to follow in this reduction process, from formal languages to machine level language (compilers & interpreters). Alternatively, in neuroscience we don't know the equivalent genetic laws.

In moving up, to the above level the algorithms and data structures of the program being executed at that moment cannot be deduced from the detailed knowledge of what is happening in the electronic circuits of a computer. Analogously, when we study the nervous systems we are not able to deduce the "algorithms" of our neural nets using only the detailed knowledge (at the biophysical level) of what is happening in some synapse of some neuron. This is a problem of reverse neurophysiology [Mira, 1995]. *"Given* a system described at physical level in terms of local behavior, with input signals and output responses to these specific signal, *find* (a) the set of functional specifications from which it originates, and (b) the additional levels of understanding concerning the potential information processing algorithms of which (a) is only one of the possible implementations".

It we remember the reduction processes used to understand computation, in particular the existence of an intermediate symbolic level and we move now *from computation to neuroscience*, it could be apparent the need of the following developments:

1) *A language for neurophysiological signals with codes* (spikes, slow potentials, formulations for adaptive thresholding, excitation-inhibition

process, dynamic bindings, synchronizations and oscillations and firing patterns generators).

2)   *A set of abstractions*, from the signals level (anatomical implementation independent) to the symbolic level. That is to say, abstractions of entities and relations from the anatomical and neurophysiological "details" and proposition of *"signal-independent"* models of information processing in neural nets.

3)   *A set of abstractions* of new entities and relations emergent from the symbolic to the knowledge level. These entities must be symbolism independent and, still, anatomical implementation independent.

2.2.   The observer Agent and the two domains of description

To understand the meaning of neural function it is essential to introduce the distinction between two domains: the level's *own domain* (OD) and *the domain of the external observer* (EOD) which interprets facts and relations at that level using models of "what could happens" as external knowledge (figure 1).

The figure of the observer and the differentiation between phenomenology and its description comes from physics and has been reintroduced and elaborated in the field of neuroscience by [Maturana 1975] and [Varela 1979] and in AI by [Mira and Delgado 1987, 1995]. The concept of the observer implicitly contains the idea of a *frame* of reference. When the neurologist observes and "measures" the behaviour of a neuron, two descriptions naturally arise in two different frames of reference with different semantics. The first is particular to the system being observed (spikes, slow potentials or biochemical descriptors in the physical level, for example). The second is particular to the observer who generally not only observe in a passive manner, but also interpret results (i.e. inject knowledge and attach meanings) and plans experiments in term of a model.

In the OD, processes are inseparable from processors. Things (correspondences between input and output signals) occur "as they must occur", following their own structural laws. In computation, at the physical level, inverters invert and counters count because they can do nothing else: this is what they have built for. A very different thing is the interpretation that the observer (EOD) gives these cross-correlations between signals. According to Maturana and Varela, what the observer calls "stimulus" (EOD) are here disturbances (OD), and the observed "responses" (EOD) are compensations (OD) for the disturbances according to the built-in mechanisms for adaptive stability. These laws are not explicit and has to be abstracted by the observer.

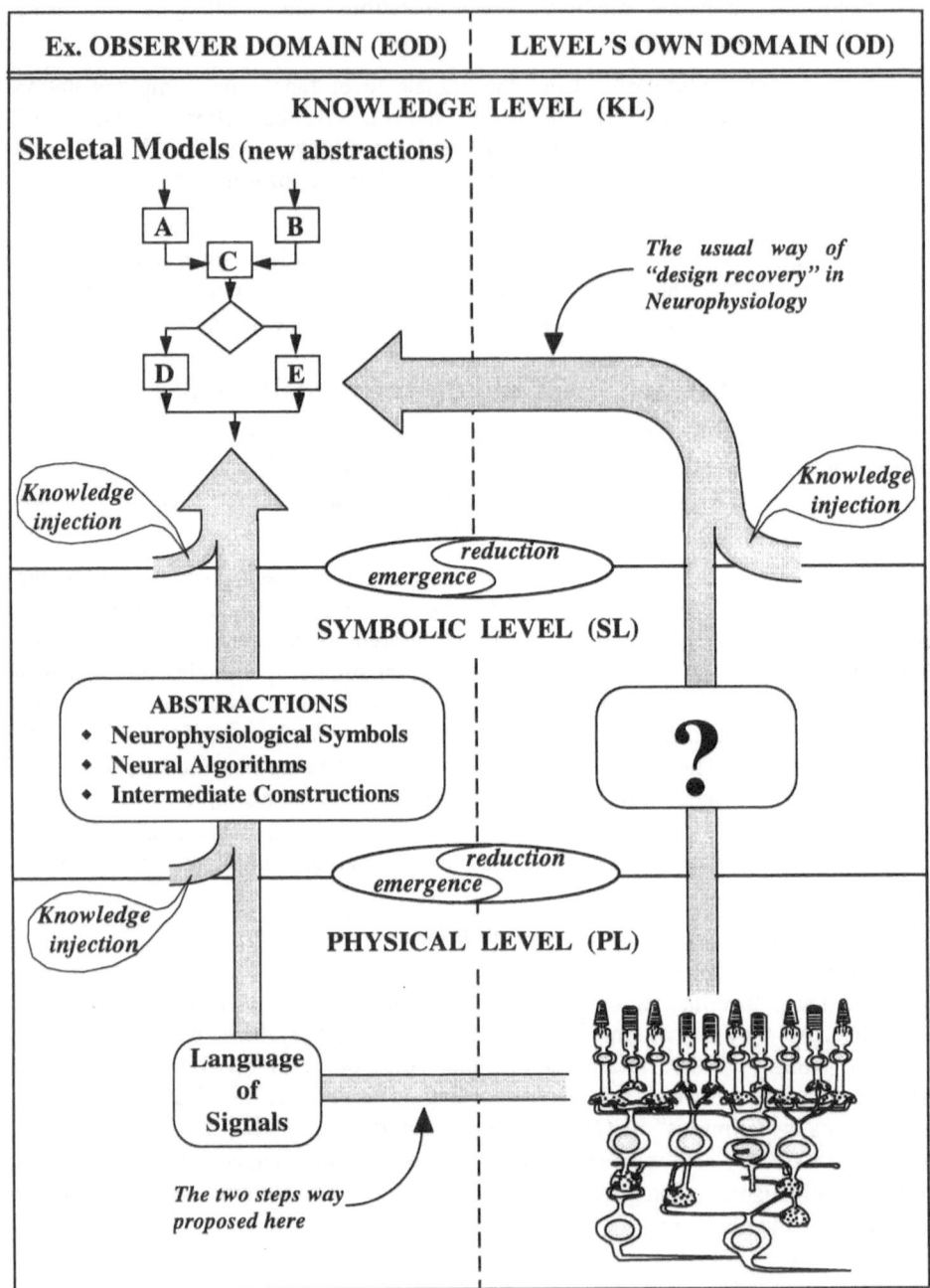

Figure 1. Levels (KL, SL & PL) and domains (EOD, OD) of description as introduced by Newell and Mar in computation and by Maturana and Varela in Biology. Note that the final model ("design recovered") is always at the KL in the EOD.

3.   From Neuroscience to Computation

We have seen some reflections on what neuroscience could "learn" from the levels and domains of description used in computation. The three proposed developments are: a *language for neurophysiological signals*, a *set of abstractions at the symbolic level* and new abstractions of *"skeletal models" in the knowledge level*. Now we consider here some modes of computation apparently genuines of the nervous tissue: anatomically engraved control structures for synaptic computation, local processes of activity accumulation, lateral interaction and reflex arches.

3.1.   Anatomically engraved control structures

Inside a neuron we have the diversity of contacts (dendrite-dendrite, dendrite-soma, ...) along with the specificity of local networks sufficient to carry out all the control structures (*"if ... then ..."*) necessaries to have any spatio-temporal program. In figure 2.a we present varions local circuits in the sense of [Rakic, 1975] and [Shepherd, 1978] that illustrate this point.

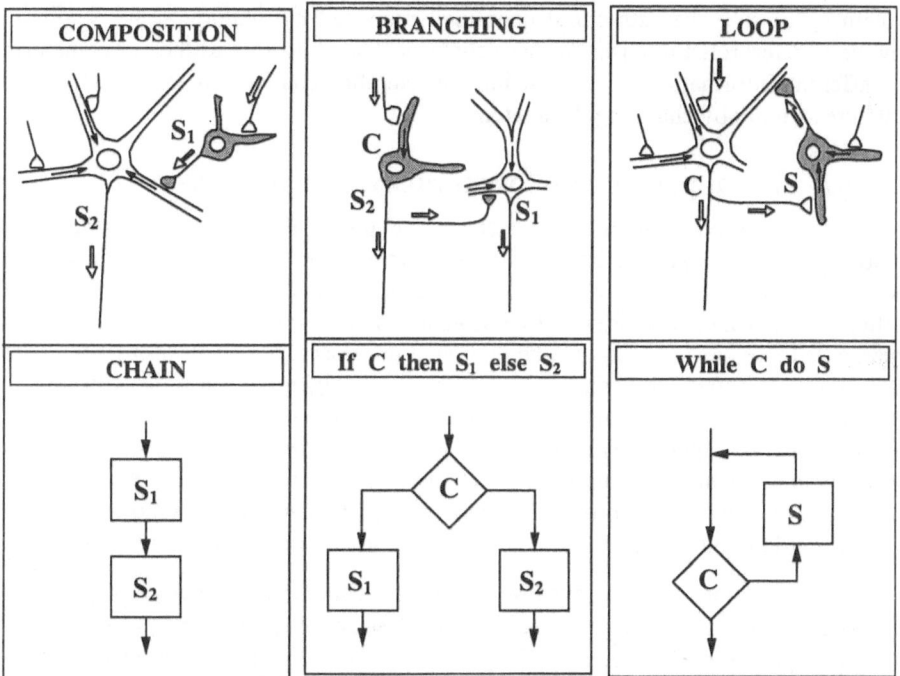

Figure 2.   In the upper part we show some of the local circuits of the olfactory bulb as proposed by Shepherd. Open profiles indicate presumed excitatory synaptic action; shaded profiles indicate presumed inhibitory action. The lower part of the figure is a suggestion of the corresponding control structure that could be executed on these local circuits. Since the configuration changes with the afferent information, these anatomical micro-programs are always under topographic re-programming.

These are circuits consisting of short-axon neurons interacting with local processes in a well-defined manner, basically by means of integration and association interactions. As Rakic and Shepherd indicate, it was Ramón y Cajal who in 1889 first suggested the importance of local circuits in which the entire process (afference, computation, response) is physically located in a very small volume.

In the lower part of figure 2 we suggest a possible interpretation of the local circuits shown in the upper part of the same figure, as control structures of anatomical programs. All the control structures found in most algorithmic languages (*"composite sequence S_1, S_2"*, *"if C then S_1 else S_2"*, *"while C do S"*) can be executed by a subset of the local circuits that we find in the dendritic field of a neuron [Mira & Delgado, 1986] and viceversa, all the dendro-dendritic circuits that we find in the organization of synapses and neurons intro circuits within the different regions of the nervous systems can, *probably*, be *synthesized* using *only* these control structures. This suggest that we have to look at the neuroanatomy very carefully because *in the anatomical structure we have the generic program of each neuron.* Since the configuration details of these contacts changes with the afferent information, learning is always taking place with the diversity and specificity of neurotransmitters and of the chemical processes of protein synthesis. The "structural coupling" proposed by Maturana and Varela can now be re-interpreted in terms of these changes. That is to say, as the net is maturing, the afferent information is introducing the suitable changes in the local controls structures simply by changing the anatomy.

3.2. Local processes of accumulation in synapses

We have seen how local circuits can be interpreted as control structures in anatomical microprograms. So, even before arriving at the level of the neuron and considering only local processes existing within its dendritic fields, there is already sufficient experimental evidence to formulate *an inferential model of neuron* with *data route* functions (convergent and divergent processes, passive transmission, lateral inhibition and reflex arches), *local computation* (algebraic addition, filtering and neurotransmitter specific functions), *memory* and *wave-forms generation*. All these functions operate repetitively in space and time in a single neuron and amply cover all the programming mechanisms used by computers. The neuron is too complex to be considered an "elementary processor". In fact, their behaviour looks more like a large information-integrating centre with, at least, as many processors as there are synapses.

Let us consider now the synapse as the new candidate to "elemental processor" [Shepherd, 1990], and look after an abstract description (signal independent) of the associated processes (the C, S_1 and S_2 of the previous section). Through the entire revision of functional circuits and the physiological actions of synapses within local circuits we find over and over again what we could call a "generic style of local computation" consisting of:

1. Some degree of autonomy around each synapse with a proper time scale.

2. A great variety of convergent processes with specific spatio-temporal input patterns.

3. A local function of accumulation (non linear spatial and temporal integration) of these spatio-temporal coincidences with modulation and changes in threshold (adaptation).

4. Constant "forgetfulness" and attenuated transmission of this state of local activity.

3.3. Lateral inhibition and reflex arches

Local computation in synapses could be equivalent to the NAND gate in digital electronics. That is to say, the basic operator from which any other function can be synthesized through the repeated use of this operator with the adequate connection schemes. Among all these schemes of specific connectivity that appear in practically of the levels of integration and in all the nervous systems, both at the peripheral and central areas, we consider particularly relevants as *"skeletal models of computation"* to be abstracted away from the anatomical implementation "details", the *lateral inhibition* and the *reflex arches*.

Since Ernst Mach published his studies on the interdependence of neighboring elements in the retina and the Harline and Ratliff quantitative studies of the *limulus*, lateral inhibition has been considered as the structure which, probably, most recurs in the nervous system, and is simultaneously the skeletal scheme of computation on which a great part local computation can develop. Anatomically, it corresponds to schemes such as the one in figure 3.a in which there is a layer of local processors (neurons or synapses) of the same type (LP) and possessing local connectivity such that the unit response not only depends on its own inputs, but on the inputs and response of the neighbouring LP's. In general, the interaction is of the inhibitory type such that the activity of one LP diminishes when its neighbours are active. We can distinguish between recurrent and non-recurrent actions and, in both cases, between additive and multiplicative-divisive, according to the mathematical operation which best adjusts experimental results.

From the computational point of view, the function of lateral interaction schemes is to realize a partition of the input space into three region: *centre, periphery* and *excluded*. It does the same with feedback from the output space and in both cases carries out a local process (LP) over the central zone and another one over the peripheral zone. Subsequently, it compares the results of these processes and generates a local decision which enters its outputs space. This process was first described as a filter to detect spatio-temporal contrasts, and later, in the context of self-organizing nets and other inferential models as the *"winner takes all"* algorithm.

The interesting thing about these schemes of connectivity with inhibitory feedbacks is that they exist at practically every level of vertebrates and invertebrates, from the first stages of sensorial pathways (retina), to the most central layers of the cerebral cortex. The other interesting aspect is that the lateral interaction nets define computational families ("skeletal models" at the knowledge level, as in figure 1) which can be projected in their particular cases (analog, logic or inferential) when we select the synaptic *"operator"*.

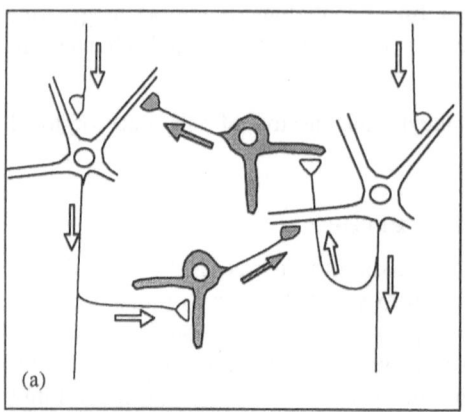 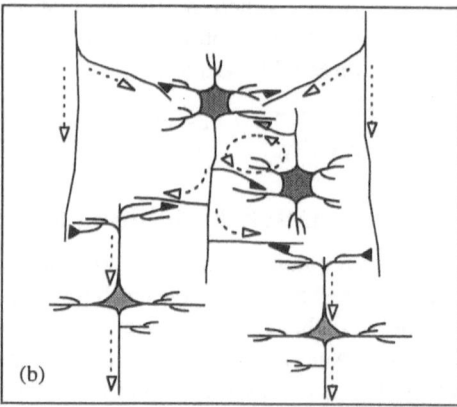

Figure 3.   Connectivity schemes. (a) Recurrent lateral inhibition between local processors (LP's). (b) Reflex arches with LP's of temporal association that open and close the association pathways according to the persistence of spatio-temporal coincidences. Extinction is always in course and dominates if there is no more reinforcement Observe that both schemes can be synthesized using only the control structures of figure 2.

Another one of the basic structures of the nervous system is the *reflex arch* which enables the spatio-temporal association of stimuli and is the support of associative functions and of some learning processes. The generic structure of the nets which carry out the spatio-temporal association of stimuli, the accumulation of persistent coincidences, and their subsequent extinction is shown in figure 3.b. Each input to a LP possesses a direct pathway to the output LP and another one to association-accumulation LP's. Each response LP can be triggered by afferences of the direct pathway or by afferences of the association pathway, as long as the accumulated excitation level surpasses a certain threshold value. Both the creation and the extinction of the association are considered adaptive responses of the organism to changes in the environment. They are formulated by means of a net of LP's which first accumulate the associations and later forgets them if there is no reinforcement. Both. Lateral interaction and reflex arch schemes can be synthesized combining only the three control structures of figure 2.

4.   Conclusions

We have examined in this paper the two points which most concern us in neural computation. a) the apparent lack of a methodology of the natural for the interpretation and integration of neurophysiological data in a similar manner as we do in computation (using three levels –knowledge, symbolic and physic– and two domains, observer and own) and b) the deeply-held conviction that the usual morpho-less point neuron models are not sufficient to describe and predict the genuine aspects of the nervous systems.

The current state of the first point has been presented in the section 2 ("from computation to neuroscience") and ends with a conjecture: We need a language for neurophysiological signals and two sets of abstractions from the signals level to the symbolic and knowledge levels.

The proposal concerning the second point is to turn our eyes once again to neuroanatomy to find an inexhaustible source of inspiration about new styles of computation.

As a final reflection on what can we do to bring computation closer to neuroscience and viceversa we suggest: (1) accept the complexity of real neurons, and look at it carefully; (2) place emphasis on methodology and (3) promote more *interdisciplinary teams* to reduce the disparity that exists between the phenomenological richness and fineness of the nervous systems and the crudeness we use in its handling.

5. Acknowledgment

We acknowledge the financial support of the Spanish CICYT, under project TIC 94-95 and of the CAM research department under project I+D 11-94.

6. References

[Kuhn, 1970] Kuhn, T.S. (1970). *"The Structure of Scientific Revolutions"*. The Univ. of Chicago Press, Chicago.

[Marr, 1982] Marr, D. (1982). *"Vision"*. Freeman, New York.

[Maturana, 1975] Maturana, H.R. (1975)."The Organization of the Living: A theory of the Living Organization". *Int. J. Man-Machine Studies*, 7, 313-332.

[McCulloch, 1965] McCulloch, W.S. (1965). *"Embodiments of Mind"*. The MIT Press. Cambridge, Mass.

[Mira & Delgado, 1986] Mira, J. Delgado, A.E. (1986). "On Some Fresh Air in Neural Modelling". *Cybernetics and Systems'86*. R. Trappl, ed. 311-318. D. Reidel Pub. Comp. Dordrecht.

[Mira & Delgado, 1987] Mira, J. and Delgado, A.E., (1987) *"Some Comments on the Antropocentric Viewpoint in the Neurocybernetic Methodology"*. Proc. of the Seventh International Congress of Cybernetics and Systems, Vol. 2, pp. 891-895. London (1987).

[Mira & Delgado, 1995] Mira, J. and Delgado, A.E. (1995). "Aspectos Metodológicos en IA". *Aspectos Básicos de la IA*. J. Mira et al. (Eds.) 53-84. Sanz y Torres. Madrid.

[Mira, 1995] Mira, J. (1995). "Reverse Neurophysiology: The "Embodiments of Mind" Revisited". *Brain Processes, Theories and Models*. R. Moreno-Díaz and J. Mira-Mira (eds.) 37-49. The MIT Press, Mass.

[Newell, 1981] Newell, A., "The Knowledge Level". *AI Magazine, summer* (1981) 1-20.

[Rakic, 1975] Rakic, P. (1975). *"Neuroscience Research Program Bulletin on L.C.N."*, 13, 3, 299-314. Boston.

[Rumelhart *et al*, 1986] Rumelhart, D.E., Hinton, G.E., and Williams, R.J. (1986). "Learning Internal Representations by Error Propagation". *Parallel Distributed*

Processing: Explorations in the Microstucture of Cognition, Rumelhart, D.E., and McClelland, J.L. (Eds.), Vol. 1: Foundations. 318-362, MIT Press, Cambridge, Mass.

[Segev, 1992] Segev, I. (1992). "Single Neurone Models: Oversimple, Complex and Reduced". *Trends in Neurosciences,* Vol. 15, No. 11, 414-421.

[Shannon & McCarthy eds., 1956] Shannon, C.E. and McCarthy, J. eds., (1956). *Automatas Studies,* Princeton University Press, N. Jersey.

[Shepherd, 1975] Shepher, G.M. (1975). *"Neuroscience Research Program Bulletin on L.C.N.",* 13, 3, 344-352. Boston.

[Shepherd, 1990] Shepherd G.M.(ed). (1990) *The Synaptic Organization of the Brain.* Oxford Univ. Press.

[Varela, 1979] Varela, F.J. (1979). *"Principles of Biological Autonomy".* North-Holland. New York.

[Wiener, 1947] Wiener, N. (1947). *"Cybernetics".* MIT Press and J. Wiley, New York.

[Yovits & Cameron, 1960] Yovits, M.C. and Cameron, S. (1960). *"Self-Organizing Systems",* Pergamon Press, Oxford.

Different Types of Temporal Correlations Obtained in Pairs of Thalamic Visual Neurons Suggest Different Functional Patterns of Connectivity

Casto Rivadulla and Javier Cudeiro

Dpto. Ciencias de La Salud I (E.U. Fisioterapia), Univ. de La Coruña and Unidad de Cirugía Experimental (Lab. Neurofisiología), Hospital Juan Canalejo, La Coruña, Spain.

ABSTRACT

The traditional view of the dorsal lateral geniculate nucleus is that of a mere relay nucleus, faithfully transmitting retinal visual information to the visual cortex. More recent studies have pointed to a more complex role in the processing of this information, with inputs from cortex and sub-cortical structures modulating the form and strength of the relayed visual information. Nevertheless, the vast majority of studies assume that this processing within the geniculate is compartmentalised such that there is no cross-talk between geniculate relay cells, each essentially carrying out a completely separate analysis of its extrageniculate input. Here we provide preliminary data suggesting a much greater level of communication between geniculate relay cells, with information shared over a modest local area, but between radically different cell types. We suggest that models of visual processing should include intrageniculate circuitry allowing passage of information between different streams (originally first thought to occur at the level of the cortex) to be a component of thalamic analysis of visual information.

INTRODUCTION

The dorsal lateral geniculate nucleus (dLGN) is the main thalamic relay for visual information en route to the visual cortex. The surface of the dLGN is topographically laid out such that cells have retinotopically organized receptive fields. It is further segregated into different laminae, whose input is dominated by one eye (Hubel and Wiesel, 1961). Furthermore, afferent fibres from retinal ganglion cells establish a highly ordered pattern of connections on target dLGN cells: X type ganglion cells make synaptic contacts on X dLGN cells and Y ganglion cells contact Y dLGN cells (Hubel and Wiesel, 1961; Hoffmann et al., 1972). This pattern of connectivity gives rise to parallel channels in the visual processing stream implicated in the analysis of different components of the visual scene (Stone et al., 1979; Sherman and Spear, 1982). However, the dLGN is much more than a relay nucleus. At this level incoming retinal information is influenced both by visual (feedback from visual cortex) and non-visual inputs (from subcortical nuclei: for a recent review see Sherman and Guillery, 1996) which critically modulate the excitability of the cells thereby regulating the amount and content of signal to be sent to the cortex. Interestingly, retinal afferents represent only 12% of total synaptic contacts on any relay neuron (Montero, 1991). Single unit activity of dLGN cells has been studied in great detail with respect to pharmacological properties, receptors types, neurotransmitters, spontaneous and visually elicited responses: studies of patterns of conectivity and individual neuron responses are legion. However, only a very few papers have focused on the interactions between geniculate relay cells. Amongst these, broad correlations between relay cells of different origins (in the order of tens of milliseconds) have been shown previously (Stevens and Gerstein, 1976, Sillito et al., 1994, Neuenschwander and Singer 1996), but little or no attention has been paid to tight correlations, of less than one millisecond separation. Recently, in a paper by Alonso et al. (1996), it was shown that neighbouring neurons with overlapping receptive fields of the same type, spike in a synchronized fashion with interspike intervals of around 1 ms, indicating a common input from divergent retinal afferents. It was suggested that these dLGN cells project to the same target cell in visual cortex, and thus this synchronization could reinforce the thalamic input to cortical simple cells.

In work we report here, we show that these type of very precise correlations are much more common that has been previously reported, and it could be generated not only by a common input (Alonso et al. 1996) but also by a direct excitatory connection between LGN cells of different types, allowing the system to integrate visual signals carried by different channels even at the thalamic level.

METHODS

Experiment were carried out on adult cats (2.5-3.5 Kg) anaestethized with halothane (0.1-5% in $N_2O : O_2$ 70-30%) and paralised with Pavulon (1mgr/kg/h). Rectal temperature, heart rate, end-tidal CO_2 , and EEG were continuously monitored throughout the experiment. Detailed experimental procedures for preparation and mantenaince of the cats has been described previously (Cudeiro and Sillito, 1996).

Extacellular recording of single-unit activity from pairs of cells was made in the A laminae of the dLGN by mean of pairs of electrodes with a tip separation of 1-2mm in the horizontal plane. The receptive fields of all neurons were within 12 degrees of the area centralis. A schematic diagram of the experiemntal design is shown in figure 1. Data were collected on a personal computer (Macintosh Power PC, 7100/80), and visual stimuli were presented on a computer screen (SONY Multiscan GDM-17") 57 cm in front of the animal with a refresh rate of 80 Hz. Spike times are stored with a 0.02 ms interval resolution and can be subsequentely processed with respect to any aspect of the stimulus variable used during data collection. Visual stimuli routinely consisted of full field sinusoidal drifting gratings with spatial and temporal frequencies qualitatively selected to evoke a strong response from both cells. Contrast ((Lmax -Lmin)/(Lmax+Lmin)) was 0.6 with a mean luminance of 14 cd/m^2. Activity from both cells was recorded simultaneously during periods of 10-20 min in two different conditions: spontaneous discharge and visually driven discharge. dLGN cells were clasified as X or Y type on the basis of standard criteria (for details see Cudeiro and Sillito, 1996), which include linearity of spatial summation, using phase-reversing sinusoidal gratings, receptive field size and presence or absence of shift effect.

Cross-correlograms were computed based on the technique described by Perkel et al. (1967b) with a resolution of 0.1 ms, in order to detect mono- and di-synaptic connections. In brief: a histogram is built of time differences (t) between each spike in an interval T in the first spike train and each spike in the other train within a time window, t. The first cell is arbitrarily considered the reference cell, and so bins to the left of zero, the negative times, represent times when the second cell fired before the reference cell. The centre (zero bin) shows the number of occurences the two cells firing simultaneously, and bins to the right indicate occasions when the reference cell fired before the second. All correlograms were shuffled in order to avoid artefacts time-locked to the stimulus (Perkel et al., 1967a). For quantitative analysis, a correlation index (the percentage of the spikes that accumulated in the peak above baseline, see Reid and Alonso, 1995) was taken as a measure of correlation strength, and the width of the peaks in the cross-correlograms were measured at half-height. The baseline was defined in our case as the total number of spikes between -2.5 and -1.0 ms, which is the maximum period obtained for a direct excitatory synaptic connection between two observed neighbouring neurons (Perkel et al., 1967b; Michalski et. al., 1983)

RESULTS

We have recorded 34 pairs of neurons in lamina A and A1 of the dLGN. 24 pairs were recorded from electrodes separated 1 mm and 10 from electrodes separated by 2 mm.

In the first group, with tip separations of 1mm, 54% (n=13) of the pairs showed a strong peak in their correlograms. In 2 out of these 24 pairs, we observed peaks of correlation centred on the zero bin, indicating the neurons often fired synchronously, suggesting they share a common input. This is illustrated in figure 2a. The histogram shows the correlation obtained between two Y cells of the same polarity (both ON centre). The spikes were collected over a 15 minute period in the absence of visual stimulation. The peak has a width of 0.2 ms and has a reasonable correlation strength (20%). Surprisingly, the peaks obtained in the cross-correlograms of the vast majority of our cells (13/24 pairs) appeared systematically displaced from zero (mean=0.51ms±0.03 SEM). A typical example is shown in figure 2b, where the correlation between two X cells, in this case of different polarity (ON and OFF), is illustrated. The peak with a 0.6 ms phase lag from zero, demonstrating a tight correlation between the dLGN cells (half-width, 0.7ms; strength 80%), and is suggestive of a direct excitatory synaptic connection between the two studied neurons. Here, as in figure 2a, the correlogram was computed with spikes recorded over 15 minutes in the absence of visual stimuli, thereby allowing us to avoid any superimposed stimulus-dependent correlation.

All recorded cells were also studied during visual stimulation, and when a positive correlation was found with spontaneous discharge, it always appeared in those spike trains obtained under visual stimulation. This is shown in figure 3 where two different examples are presented. In figure3a the upper histogram shows the correlation achieved between two cells of diffent type and polarity, X ON and Y OFF, over 10 minutes of visual stimulation (strength 100%), and the lower histogram illustrates the correlogram constructed from the spontaneous discharge of the same neurons (strength 50%). A similar example is shown in figure 3b, but in this case an ON Y and ON X cells are correlated. The strength of the correlation after 10 minutes of visual stimulation was 32%, and after 20 minutes of spikes collection in the spontaneous discharge condition was 46%. This sort of correlation has been found between different combinations of cell types (X / Y, ON / OFF), without any preference in our sample. It is interesting to note that the only two pairs which showed correlations suggestive of common input (peak centred on zero), were found between neurons of the same type and polarity.

In the second group of our sample, with electrode tip separation of 2mm (n=10), no correlations of any type were found in any of the recorded pairs.

DISCUSSION

In this paper we have used cross-correlation analysis to study types of interaction between pairs of neurons in cat dLGN. The results described here indicate that nearly half of the pairs recorded from electrodes separated by 1mm can show very tight correlated firing regardless of cell type and polarity, most often with a phase lag (mean=0.51 ms), with only 2 pairs out of 24 showing zero phase lag. Initially this seems to strand in contradition to previous work by Alonso et al. (1996) who showed that only 15% of dLGN studied by mean of electrodes 100-400μm apart, presented positive cross-correlograms, found maily in cells with identical characteristics and with overlapping receptive fields, which was interpreted by the authors as demostrative of a common input, as the correlations were centred on the zero bin. This discrepancy can be explained by taking into account the following: (i) the distance between our electrodes was 1mm. It is immediatly obvious that the receptive fields of the cells would be unlikely to overlap, explaining why we have found only two pairs suggestive of a common input. (ii) the proportion of synapses established by dLGN relay cells intrageniculate collaterals is thought to be very low (~0.5%, Montero, 1991), although morphological evidence for some interaction between relay cells does exist (Hamos et al., 1985). However, available anatomical data indicates that 10% of neighbouring geniculate cells receive input from any given ganglion cell (Hamos et al., 1987), suggesting that common input from retinal axons has to occur more often than direct excitatory connections between dLGN relay cells. This fits very well with Alonso et al. (1996) data, and the authors interpretation is that neurons which produce spikes within 1ms of each other are sharing a common input. However, this may be an overestimation of the real percentage of pairs actually centred on the zero, induced simply their by pooling of all data together.

Available work in the literature demonstrates that the existence of crosscorrelograms computed from cells located in the striate visual cortex of the cat and characterized by narrow, high peaks slightly shifted from the centre (mean 0.62 ms) are suggestive of direct synaptic connections between neurons (Michalski et al., 1983; Toyama et al., 1981). Our data seem to suggest that a high proportion of cells in the dLGN of the cat receive excitatory connections from neighbouring cells. This is in direct contradiction with the morphological data, yet even with a small sample such as we present here, it is hard to believe that with more data it fit the anatomical evidence. We can suggest two possible explanations. Firstly the available methods used to quantify the very complex synaptic contacts within the dLGN do not allow a very precise identification of terminals, and may underestimate intrageniculate contacts. Evidence for such problems clearly exist, for example about cortical input to the dLGN (see Montero, 1991 and Weber et al., 1989). Secondly, our classification methods are different . Based on the work in the visual cortex (Michalski et al., 1983; Toyama et al., 1981) and on the systematic localization of the peaks found in our correlograms around 0.5 ms, we have interpreted our data as direct connections between neurons and not as common input. But even if we adopt the other classification scheme, the percentage of neurons which would share the same retinal axon is very high in comparison to the data of Alonso et al. (1996).

In summary, we have presented here consistent and reliable data suggesting that the very precise correlations which have been described at the level of the dLGN, are much more common than previously reported, and can be explained in most of the cases by the presence of direct excitatory connection between dLGN cells of different types. From the functional point of view, this would permit the system to mix different aspects of the visual signal at a very early stage. It is well know that combination of inputs, for instance ON/OFF, is a characteristic of complex cells in the visual cortex (Hubel and Wiesel, 1959;1962) as a result of combination of inputs from other cortical cells. We are tempted to speculate that a similar type of integration already exists in the visual thalamus, for example with output contacting a cortical complex cell. Future experiments should shed some light on this hypothesis.

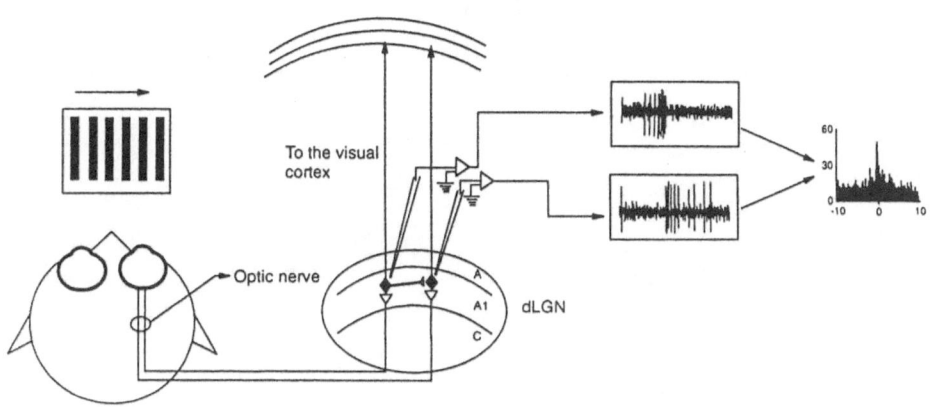

FIGURE 1

Diagrammatic representation of our experimental setup. A pair of electrodes is inserted into the A laminae of the dLGN for simultaneous recording of unitary extracellular activity (in the diagram, two cells (filled symbols) located in lamina A1). To the right we illustrate the crosscorrelogram analysis of the two spike trains.

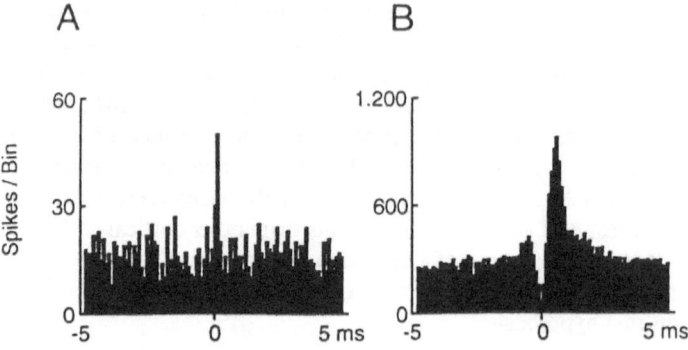

FIGURE 2

Here we show typical examples of the two types of correlations we have found. Part A shows a correlogram obtained from the background discharge of two dLGN cells. The peak is centred on zero. In part B we illustrate another correlogram also obtained from the resting discharge of two recorded neurons, but in this case the peak is shifted 0.6 ms to the right of zero. See text for further details.

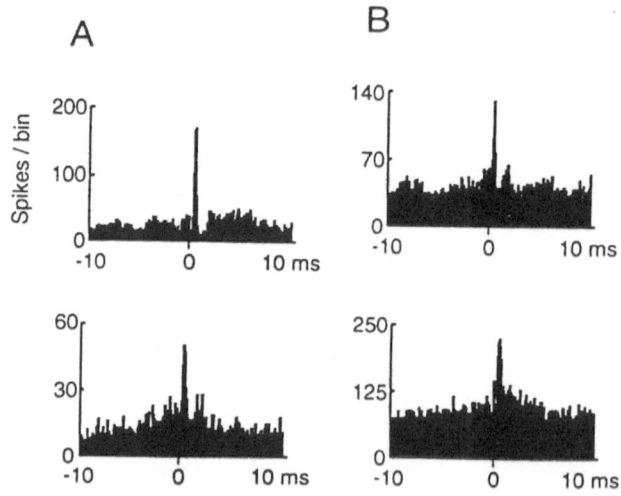

FIGURE 3

We illustrate in this figure two different examples (A and B) of positive crosscorrelations with the peak displaced from the zero bin. The upper panel of the picture show the correlograms obtained computing the spike trains from the cells resting discharge while the bottom panel presents the croscorrelograms obtained during visual stimulation. See text for details.

BIBLIOGRAPHY

Alonso, JM., Usrey, WM., Reid, RC.:
Precisely correlated firing in cells of the lateral geniculate nucleus. Nature 383 (1996) 815-819

Cudeiro, J., Sillito, AM.:
Spatial frequency tuning of orientation-discontinuity-sensitive corticofugal feedback to the cat lateral geniculate nucleus. J. Physiol. London. 490.2 (1996) 481-492

Hamos, JE., Van Horn, SC., Raczkowski, D., Uhlrich, D., Sherman, SM.:
Synaptic connectivity of a local circuit neurone in lateral geniculate nucleus of the cat. Nature 317 (1985) 618-621

Hamos, JE., Van Horn, SC., Raczkowski, D., Sherman, SM.:
Synaptic circuits involving an individual retinigeniculate axon in the cat. J. Comp. Neurol. 259 (1987) 165-192

Hoffmann, K.P., Stone, J., Sherman, SM.:
Relay of receptive-field properties in dorsal lateral geniculate nucleus of the cat. J. Neurophysiol. 37 (1972) 518-531

Hubel. DH., Wiesel, TN. :
receptive fields of single neurones in the cat's striate cortex J. Physiol. London. 148 (1959) 574-591

Hubel. DH., Wiesel, TN. :
Integrative action in the cat's lateral geniculate body. J. Physiol. London. 155 (1961) 385-398

Hubel. DH., Wiesel, TN. :
Receptive fields, binocular interaction and functional architecture in the cat's visual cortex. J. Physiol. London. 160 (1962) 106-154

Michalski, A., Gerstein, GL., Czarkowska, J., Tarnecki, R.:
Interactions between cat striate cortex neurons. Exp. Brain Res. 51 (1983) 97-107

Montero, VM.:
A quantitative study of synaptic contacts on interneurons and relay cells of the cat lateral geniculate nucleus. Exp. Brain Res. 86 (1991) 257-270

Neuenschwander, S., Singer, W.:
Long-range synchronization of oscillatory light responses in the cat retina and lateral geniculate nucleus. Nature 379 (1996) 728-733

Perkel, DH., Gerstein, GL., Moore, GP.:
Neuronal spike trains and stochastic point processes. I. The single spike train. Biophys. J. 7 (1967a) 391-418

Perkel, DH., Gerstein, GL., Moore, GP.:
Neuronal spike trains and stochastic point processes. I. Simultaneous spike trains. Biophys. J. 7 (1967b) 419-440

Reid, RC., Alonso, JM.:
Specificity of monosynaptic connections from thalamus to visual cortex. Nature 378 (1995) 281-284

Sherman, SM., Spear , PD. :
Organization of the visual pathways in normal and visually deprived cats. Physiol. Rev. 62 (1982) 738-855

Sherman, MS., Guillery, RW.:
Functional organization of thalamocortical relays. J. Neurophysiol. 76 (1996) 1367-1395.

Sillito, AM., Jones, HE., Gerstein, GL., West, DC.:
Feature-linked synchronization of thalamic relay cell firing induced by feedback from the visual cortex. Nature 369 (1994) 479-482

Stevens, JK., Gerstein, GL.:
Interactions between cat lateral geniculate neurons. J. Neurophysiol. 39 (1976) 239-256

Stone, J., Dreher, B., Leventhal, A.:
Hierarchical and parallel mechanisms in the organization of visual cortex. Brain Res. Revs. 1 (1979) 345-394

Toyama, K., Kimura, M., Tanaka, K.:
Cross-correlation analysis of interneuronal connectivity in cat visual cortex. J. Neurophysiol. 46 (1981) 191-201

Development of *on-off* and *off-on* Receptive Fields Using a Semistochastic Model

E. M. Muro,[1] P. Isasi,[1] M. A. Andrade,[2] and F. Morán[3]

[1] Grupo de Vida Artificial. Departamento de Informática.
Universidad Carlos III de Madrid. Butarque 15, 28911 Leganés, Spain.
email:emuro@gaia.uc3m.es isasi@gaia.uc3m.es
[2] European Bioinformatics Institute, Minxton, Cambridge CB10, 1SD, UK.
email: andrade@ebi.ac.uk
[3] Grupo de Biofísica. Departamento de Bioquímica y Biología Molecular.
Universidad Complutense de Madrid. Ciudad Universitaria, 28040 Madrid, Spain.
email: fmoran@solea.quim.ucm.es

Abstract. A model for ontogenetic development of receptive fields in the visual nervous system is presented. The model uses a semistochastic approach where random uncorrelated activity is generated in the input layer and propagated through the network. The evolution of the synaptic connections between two neurons are assumed to be a function of their activity, with two interpretations of the Hebb's rule: (a) the synaptic weight is modified proportional to the product of the activity of the two connected neurons; and (b) proportional to the statistical correlation of their activity. Both models explain the origin of either *on-off* and *off-on* receptive fields with symetric and non symetric forms. These results agree with previous models based on deterministic equations. The approach presented here has two main advantages. Firstly the lower computer time that allows the study of more complex architectures. And secondly, the possibility of the extension of this model to cover more complex behavior, for instance, the inclusion of time delay in the transmition of the activity between layers.

keywords: neural networks, self-organization, receptive fields, unsupervised learning, semistochastic models.

1 Introduction

The nervous visual system of most higher mammals is formed and organized during the prenatal period in a process dependent of the spontaneous activity inner to the visual system [3]. This process gives raise to a highly structured distribution of neurons and connections, needed for the complex selective properties that appear in the adult visual system. One of the characteristics of the primary visual cortex is the organization in columns selective to the orientation of the stimulus. The origin of this selectivity is based on the formation of receptive fields with different shapes and orientation [9].

Apparently, the information contained in the genome is not sufficcient to establish the adult functional state of the system. However, several neurophysiological models using neural networks have shown that this ontogenetic development is based on self-organization processes driven by quite simple rules, that can be equally codified in the individuals of the system (the neurons) [11], [6], [7], [8] and [10].

In previous works [1] and [2] we have developed a neural network model that account for the ontogenetic development of occularity domains and variable-sized receptive fields. These models are based on the principles of cortical self-organization early proposed by von der Malsburg [4]. The final or mature state of the connectivity between two neural layers in the visual system is calculated as the solution of a set of differential equations for given initial conditions. Thus, starting from a desordered state, given by connection (weight) matrixes with random values, the system evolves *deterministically* towards the final ordered solution that accounts for a given functional state.

In this paper a modification of our previous model is presented. Using the same approach in the calculation of the evolution of the connecting weights, we have introduced here a semistochastic approach. The input activity is randomly generated and propagated through the network. In this way the activity of any couple of connected neurons can be calculated numerically. The modification of the corresponding weight is then performed according to two interpretations of the Hebb's rule: (a) proportional to the product of the activities; and (b) proportional to the correlation. As described elsewere [2], hebbian and anti-hebbian learning is used for activating and inhibiting connections, respectively. The cycle of generation and propagation of activity and modification of the weights (unsupervised learning) are repeated until a convergence in the weight distribution is encountered. Then, the final connectivity structure is analized. As a first approximation we have shown that the application of either of the Hebb's rule interpretations leads to similar results.

Finally, the extension of this model to account for some other characteristics or global properties of the visual system is discussed.

2 Model

In this work we use a two layer architecture [2], which attempts to reproduce the global property evolution of a locally ruled system as observed in the visual nervous system. The input layer is constituted by n neurons, (see fig.1) and the output layer by m. N_i^a ($i = 1, \ldots, n$) represents the i input layer neuron and N_j^b ($j = 1, \ldots, m$) the j output layer neuron. We have used the same descriptive criterium in the rest of elements of our model. The two layers are connected using different unidirectional weighted connections (synaptic weights): 1) action activators ($\mathcal{W}$) between both layers and 2) lateral inhibitors in the output layer ($\mathcal{Q}$). Both connectors control the level of activity interchanged among different neurons. Inhibitors model the inhibiting action of the visual cortex interneurons. Since the activating/inhibiting properties of connectors cannot be interchanged, a neuron in the input layer can directly inhibit no neuron in the output layer. The strength of these unidirectional connections has been artificially represented by two matrixes $\mathcal{W}(t)$ and $\mathcal{Q}(t)$ with real positive weights:

$$\mathcal{W}_{ij}(t) > 0 \tag{1}$$

$$\mathcal{Q}_{kl}(t) > 0 \tag{2}$$

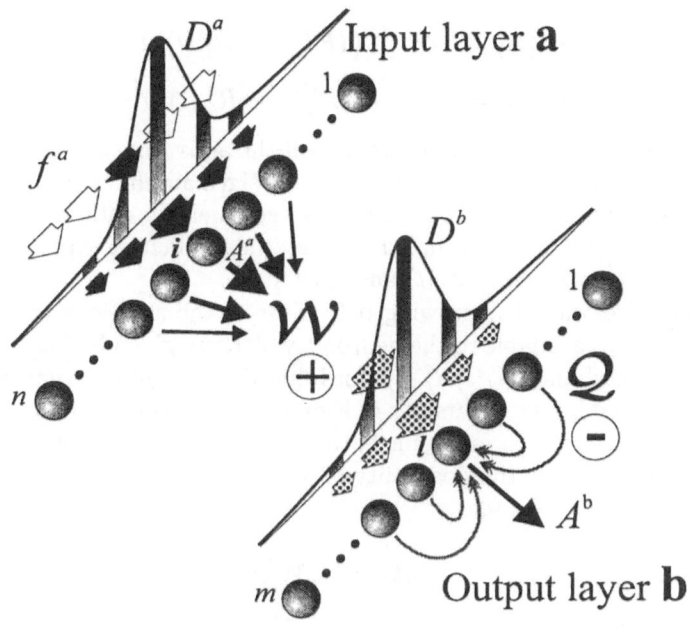

Fig. 1. Schematic representation of the architecture of the model.

The dynamics of the system can be described as follows. A random input activity pattern $f_h^a(t)$ $h \in [1, \ldots, n]$, both spatially and temporally uncorrelated, is introduced to the input layer. This activity is laterally propagated using a decreasing diffusion gaussian distribution D, establishing simple-neighbour relationships. The activity generated at each neuron is transmitted to the output layer through $\mathcal{W}(t)$, the matrix of activating connections, and diffused using another gaussian function. This activity is also received by the inhibiting interneurons. The output activity of each neuron in this layer is the balance between the input activity and the inhibiting action representing the interneuron dynamics. At a higher level, cooperative and competitive processes can be obtained through a self-organization process led by local rules, fixed diffusion functions and synaptic plasticity.

As a result of synaptic plasticity, the above referred weights are assumed to change stepwise and dynamically through the following evolution equations:

$$\mathcal{W}_{ij}(t+1) = \mathcal{W}_{ij}(t) + \Delta\mathcal{W}_{ij}(t) \tag{3}$$

$$\mathcal{Q}_{kl}(t+1) = \mathcal{Q}_{kl}(t) + \Delta\mathcal{Q}_{kl}(t) \tag{4}$$

According with a previous model described by [2] the increments can be calculated as:

$$\Delta\mathcal{W}_{ij}(t) = \beta\mathcal{W}_{ij}(t)\left(\left(F_{ij}^w(t) - R_{ij}^w(t)\right)\right) + \alpha \tag{5}$$

$$\Delta\mathcal{Q}_{kl}(t) = \beta\mathcal{Q}_{kl}(t)\left(\left(F_{kl}^q(t) - R_{kl}^q(t)\right)\right) + \alpha \tag{6}$$

where $F_{ij}^w(t)$ and $F_{kl}^q(t)$ respectively represent the activating and inhibiting growing synaptic factors, which control the rate of weigth amplification. These factors model the self-amplification properties observed when synaptic weights are modified [12]. Both equations are coupled through the implicit or explicit dependence of the growing factors with $\mathcal{W}_{ij}(t)$ and $\mathcal{Q}_{kl}(t)$.

The model restricts the growing behaviour of the solutions by limiting the synaptic resources available to the neurons, which are given by $\mathcal{W}_{ij}(t)R_{ij}^w(t)$ (for activating weights) and $\mathcal{Q}_{kl}(t)R_{kl}^q(t)$ (for inhibiting weights). Synaptic plasticity speed is controlled by the parameter β. Finally, the parameter α is a consequence of (1) and (2) in order to maintain both the activating and inhibiting connection properties avoiding the possibility of a property interchange. Following [2], restriction factors are set to:

$$R_{ij}^w(t)) = \gamma_w\mathcal{W}_{ij}^2(t) \tag{7}$$

$$R_{kl}^q(t)) = \gamma_q\mathcal{Q}_{kl}^2(t) \tag{8}$$

where γ_w and γ_q are the parameters controlling the resources available to neurons.

The process carried out by each neuron or processor depends on the kind of input activity pattern the system is fed with. Random positive real input patterns have been used to simulate spontaneous photoreceptors activities in the ontogenetic development of the visual nervous system. On the other hand, the development of self-organized connections depends on the way in which weights are initialized. We have set them to:

$$(\mathcal{W},\mathcal{Q})_{ij}(0) = m\left((1-b)r + \frac{b}{1+d_{ij}}\right) \tag{9}$$

where r is a random real number in $[0,1]$, m is the maximum value any weight can take, b is a real number in $[0,1]$ representing the retinotopic degree, and d_{ij} is the distance between the two neurons connected by $(\mathcal{W},\mathcal{Q})_{ij}(t)$.

In this work, two different kinds of dynamically changing growing factors are presented, both of them calculated using different implementations of hebbian [5] and anti-hebbian plasticity.

2.1 Activity-product learning model

As a first step we have used the usual implementation of Hebb's rule based on the product of the output activities of the neurons joined by the respective connection. In this way, at any time, the growing synaptic factors $F_{ij}^w(t)$ and $F_{il}^q(t)$ are calculated as:

$$F_{ij}^w(t) = \delta_w A_i^a(t) A_j^b(t) \tag{10}$$

$$F_{kl}^q(t) = \delta_q A_k^b(t) A_l^b(t) \tag{11}$$

where δ_w and δ_q are learning rates and $A_i^a(t)$ and $A_j^b(t)$ represent respectively N_i^a and N_j^b output activities.

At time t and in each first layer neuron N_h^a we set $f_h^a(t)$ equal to a randomly generated input in the interval $[0,1]$. The output activity of a first layer neuron is then calculated as the contribution of its own activity and those of the other neurons after being modulated (in a neighbouring sense) using a fixed diffusion function D^a, which decreases with the distance between neurons.

$$A_i^a(t) = \sum_{h=1}^{n} f_h^a(t) D_{hi}^a \tag{12}$$

The activity outputs produced by the first layer neurons are propagated through activating connections:

$$I_j^b(t) = \sum_{i=1}^{n} A_i^a(t) W_{ij}^a(t) \tag{13}$$

The resulting total input activity of any second-layer neuron is then calculated as the result of its own activity and the neighbours activities affected by another fixed lateral-diffusion function.

$$S_k^b(t) = \sum_{j=1}^{m} I_j^a(t) D_{jk}^b \tag{14}$$

Taking into account the second layer inhibiting interneuron connections, the output neuron activity is:

$$A_l^b(t) = S_l^b(t) - \sum_{k=1}^{m} S_k^b(t) Q_{kl}(t) \tag{15}$$

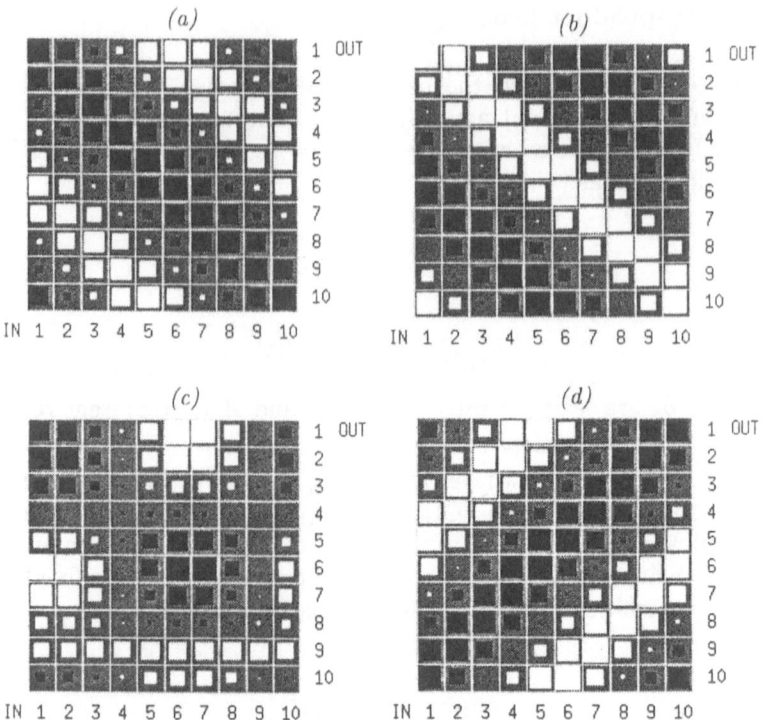

Fig. 2. Receptive fields obtained with the activity-product learning model. *(a)On-off* homogeneous receptive fields. *(b)Off-on* homogeneous receptive fields. *(c)On-off* non-homogeneous receptive fields. *(d)On-off* antiretinotopic receptive fields.

Results of the activity-product learning model. In order to analize the self-organization process results, an array E has been used with the element $E_{il}(t)$ representing the influence of the activation of the input activity $f_i^a(t)$ of neuron N_i^a on the output activity $A_l^b(t)$ of the second layer neuron N_l^b. If $E_{il}(t) > 0$, the first layer neuron (N_i^a) belongs to the *on* part of the receptive field of N_l^b neuron, and if $E_{il}(t) < 0$ it belongs to the *off* part of its receptive field. Therefore, $E_{il}(t)$ constitutes the N_l^b receptive field. After some lengthy straight-forward algebra, it is easily deduced that:

$$E_{il}(t) = \sum_{j=1}^{n} D_{ij}^a \sum_{k=1}^{m} \mathcal{W}_{jk}(t) \left(D_{kl}^b - \sum_{o=1}^{m} D_{ko}^b \mathcal{Q}_{ol}(t) \right) \qquad (16)$$

As a first approach, we have used one-dimensional input and output layers with 10 neurons. This simple topology reduces the complexity (number of connections, computing time, ...) of the problem. In order to avoid border effects, the low number of neurons is compensated by toroidal boundary conditions in any of the layer geometries.

Results are presented using Hinton's diagrams (see fig.2), where black and white colours are respectively related to the on and off parts of the receptive fields.

In fig.2, homogeneous *on-off* receptive fields have been obtained after 3.10^5 time cycles (a). In this case high retinotopic initiation bias has been used. If low retinopic initiation bias is used then *off-on* receptive fields are found (b). For some middle initial conditions an antiretinotopic homogeneous solutions are obtained (d). When the synaptic growth restriction is decreased some non symetric solutions can be obtained (c). This indicates the existence of different attractors depending on the initial conditions and controlling parameters. It can be deduced that the atraction basin of the symetric retinotopic solution diminishes when the competence term decreases. Then, for some initial conditions the system lay on other non symetric solutions. For long term simulations, most of the solutions evolve towards symetric solutions, either *on-off*, *off-on*, retinotopic, or anti-retinotopic.

2.2 Activity-correlation learning model

At time t_p, the synaptic growth factor is taken proportional to the statistical correlation between the different output activities of those neurons joint until t_p by the connections we are interested in.

$$F_{ij}^w(t_p) = \delta_w \, \rho \left(A_i^a(t_p), A_j^b(t_p) \right) \qquad (17)$$

$$F_{kl}^q(t_p) = \delta_q \, \rho \left(A_k^a(t_p), A_l^b(t_p) \right) \qquad (18)$$

where ρ is the correlation given by:

$$\rho(x,y) = \frac{cov(x,y)}{\sigma_x \sigma_y} = \frac{\overline{xy} - \overline{x}\,\overline{y}}{\sigma_x \sigma_y} \qquad (19)$$

where x and y are data sets and σ_x and σ_y their respective standard deviation. Therefore

$$\mathcal{A}_i^a(t_p) = [A_i^a(t_0), A_i^a(t_0 + \Delta t), \dots, A_i^a(t_0 + p\Delta t)] \qquad (20)$$

$$\mathcal{A}_j^b(t_p) = [A_j^b(t_0), A_j^b(t_0 + \Delta t), \dots, A_j^b(t_0 + p\Delta t)] \qquad (21)$$

Results of the activity-correlation learning model. Using similar strategies to those used in the activity-product growing model experiments, the results plotted in fig.3 have been obtained.

The different receptive fields shown in fig.3, have been obtained after the same number of time cycles, and similar solutions are found. Homogeneous receptive fields are obtained using a high retinotopic initiaton bias (a) and can be shifted in order to find antiretinotopic fiels decreasing the retinotopic initiation degree

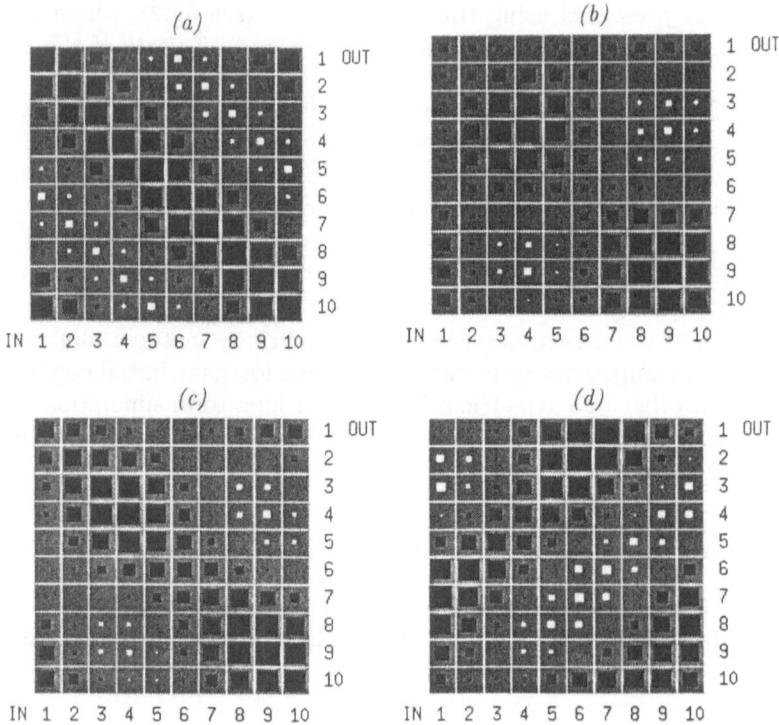

Fig. 3. Receptive fields obtained with the activity-correlation learning model. *(a)On-off* homogeneous receptive fields. *(b)On-off* non homogeneous receptive fields. *(c)On-off* non homogeneous receptive fields. *(d)Off-on* antiretinotopic receptive fields.

(d). As before, inhomogeneous receptive fields can also be found (b) and (c). If this inhomogenity could be found in a two or tree dimensional layers system, each receptive field would then be able to detect other important global properties such as orientation.

3 Conclusions

The results obtained (fig.2 and fig.3) agree with those found using the deterministic model proposed in earlier works [2]. In that case, statistic was used to correlate the input patterns, and as a result of this correlating process, deterministic differential equations were obtained. The two different self-organized models proposed here are semistochastic. Their development depends on a random input pattern generated at real time, in the same way that in natural systems. This activity patterns are then propagated through the network and the activity of the different neurons is calculated each time cycle. Since lower computational time is needed in these new models, future works are expected to be able to describe the self-organized dynamics of more complex global properties that can be detected by neurons of the visual cortex.

4 Acknowledgments

This work has been supported in part by grant No. BIO96-0895 from CICYT (Spain). We are grateful to D. Álvarez, R. Sánchez and M. A. Monge for useful discussions. E.M.Muro acknowledges support from the MEC-FPU program.

References

1. Andrade, M.A. & F. Morán. (1996) Structural study of the development of ocularity domains using a neural network model. *Biol. Cybern.* **74**: 243-254
2. Andrade, M.A. & F. Morán (1997) Receptive field map development by anti-hebbian learning. *Neural Networks* (in press.)
3. Frégnac, Y. & M. Imbert. 1984. Development of neuronal selectivity in primary visual cortex of cat. *Physiol. Rev.* **64**, 325-434
4. Haussler, A.F. & C. von der Malsburg. 1983. Development of retinotopic projections: an analytical treatment. *J. Theor. Biol.* **2**, 47-73
5. Hebb, D.O. 1949. *The Organization of Behaviour*, New York:Wiley, Introduction and Chapter 4,'The first stage of perception: growth of the assembly', pp.xi-xix, 60-78
6. Linsker, R. 1986a. From basic network principles to neural architecture: Emergence of spatial-opponent cells. *Proc. Natl. Acad. Sci. USA.* **83**, 7508-7512
7. Linsker, R. 1986b. From basic network principles to neural architecture: Emergence of orientation-selective cells. *Proc. Natl. Acad. Sci. USA.* **83**, 8390-8394
8. Linsker, R. 1986c. From basic network principles to neural architecture: Emergence of orientation columns. *Proc. Natl. Acad. Sci. USA.* **83**, 8779-8783
9. Orban, G.A. (1984). *Studies on Brain Function. Neuronal Operations in the Visual Cortex.* Berlin: Springer-Verlag
10. Miller, K.D. 1992. Development of orientation columns via competition between ON- and OFF-center inputs. *NeuroReport*, **3**, 73-76
11. von der Malsburg, C. 1973. Self-organization of orientation sensitive cells in the striate cortex. *Kybernetic.* **14**, 85-100
12. von der Malsburg, C. 1990. Network self-organization. *In An Introduction to Neural and Electronic Networks.* (S.F. Zornetzer, J.L. Davis, and C.Lau, eds.), San Diego, CA: Academic Press, 421-432

The Classification of Spatial, Chromatic, and Intensity Features of Simple Visual Stimuli by a Network of Retinal Ganglion Cells

Shuy Shoham (1), Remus Osan (1), Josef Ammermuller (2), Almut Branner (1), Eduardo Fernandez (3), Richard A. Normann (1,4).

(1) The Dept. of Bioengineering, Univ. of Utah, Salt Lake City, Utah, (2) Oldenburg Univ., Oldenburg, Germany, (3) Alicante Univ., Alicante, Spain, (4) corresponding author.

ABSTRACT

We are investigating the representation of simple visual objects by groups of retinal ganglion cells and are simultaneously recording the responses of ganglion cells in the isolated turtle retina with 15 out of an array of 100 penetrating microelectrodes. Stimulation is with circular spots of light of various intensities, diameters and colors. We have trained a three layer artificial neural network to estimate the stimulus parameters and have challenged it to classify the color, size and intensity of test stimuli. Individual ganglion cells are poor encoders of stimulus features, but the 15 cells in our sample allow one to classify intensity, color and spot diameter to within 0.6 log units, 61 nm, and 0.68 mm, respectively.

INTRODUCTION

There is considerable preprocessing of visual information that takes place by the vertebrate retina before it is sent to higher visual centers. The preprocessing begins in the photoreceptors and is elaborated by each of the neurons in the radial and lateral retinal pathways. Before the retinal visual signals reach the first synapse, they have already been substantially preprocessed. Each cone transduces the local light incident upon the retina into an electrical signal. The phototransduction process has an "automatic gain control" whereby cone gain (mV/photon) is set by the flux of light that strikes each cone. This gives the cone both high contrast sensitivity and good dynamic range over a very wide range of ambient illumination conditions (1). The mosaic of cones, and the different absorption spectra of the three types of photopigments found in the red, green and blue cones separates the incident colored light image into three superimposed primary images. Thus, cones perform a spatial, chromatic and intensity decomposition of the incident light stimulus.

The cones also begin the process of spatial band pass filtering. This spatial filtering is augmented by the triad synapse composed of the cones, horizontal cells and the bipolar cells. The horizontal cells have large receptive fields that provide an inhibitory surround for the bipolar cells. Light that excites cones that are presynaptic to a bipolar cell produces a direct response in the bipolar cell (the "center" response), while light some distance away excites cones that inhibit the direct response (the "surround" response) (2). This spatial antagonism, called lateral inhibition, produces a band pass spatial filtering of the visual image. By the time that the encoded visual image has reached the second order retinal neurons, it has lost information about the absolute light intensities striking the retina, as well as the very low spatial frequencies in the image.

These observations on retinal anatomy and physiology illustrate the complexities of retinal signal processing. By the time that the visual image has reached the ganglion cell level, it has experienced additional transformations. Each ganglion cell in this network is encoding many aspects of the visual image: color, space, intensity, temporality, etc. As a consequence, the identification of the color, intensity or shape of visual stimuli should be possible by examining the firing patterns of the group of ganglion cells that are stimulated by the visual stimuli. We have begun a systematic examination of this notion.

The variability of ganglion cell responses is expected to confound the process by which the parameters of the stimulus can be identified from a set of responses (3). Ganglion cells do not have a stationary transfer function but manifest considerable variability in their responses to repetition of simple stimuli. This variability can be extreme in some cells and is less significant in others.

Further, the study of the representations of complex stimuli by groups of ganglion cells has been limited by a lack of a suitable means to sample the simultaneous activity of large numbers of cells. Our understanding of the biophysics of the individual ganglion cells has been made possible by the use of single microelectrodes to record the responses of individual ganglion cells to simple patterns of light. How these cells interact in a network to encode all the features of a complex image can best be understood using multiple microelectrodes. The recent development of high density microelectrode arrays has provided the tools that can be used to begin the study of these network properties.

We have used a high density array of penetrating microelectrodes to investigate the representations of simple visual stimuli by a small network of ganglion cells. Our study has focused on two questions: how accurately can we classify simple visual stimuli from the patterns of firing of the groups of neurons sampled by our electrode array; and, to what degree does the variability in the ganglion cell responses complicate this classification process. We have used a three level perceptron neural network (4), trained with back propagation to classify the responses of 15 simultaneously recorded ganglion cells to circular spots of various colors, intensities and diameters. Our multielectrode data at this point must be regarded as being of a preliminary nature, and we are presenting our findings as "work in progress". We have found that while individual ganglion cells are quite poor classifiers of the general color, form, and intensity of visual stimuli, individual ganglion cells have response specializations: some are better classifiers of intensity, others are better at colors. However if the classification is performed using all fifteen recorded responses, the network is able to classify the intensity, color, and form of the stimulus to within 0.6 log units, 61 nm, and 0.62 mm, respectively. Correlations of 0.65-0.8 were observed between the estimated stimulus parameters and the actual parameters.

METHODS

Our extracellular ganglion cell recordings were made in the isolated superfused turtle retina preparation (5) using an array of 100, 1.5 mm long needles (The Utah Electrode Array). Our manufacturing techniques have been described elsewhere (6). The needles penetrate into the retina and record from groups of ganglion cells. They are built from silicon on a square grid with a 400 micron pitch. The distal 50 microns of each needle is metalized with platinum and forms an active electrode. The remaining parts of the silicon array are insulated with polyimide. In about ½ of the electrodes, we can use template generation and matching (7) to isolate single units from the multiunit responses typically recorded with the array.

In this study we have selected electrodes that had the highest signal-to-noise ratios, and have isolated 1 to 2 units from the multiunit responses with the expedient of high thresholds. We refer to these as "single units" throughout this paper. While stable recordings could be made in this preparation in retinas that had been isolated for over eight hours, the retinas we used in our experiments were typically limited to four hours post isolation. The retinas were isolated from the pigment epithelium, and mounted photoreceptor side down and perfused with a physiological solution (5). The light stimuli were delivered to the photoreceptor surface, and the electrode array was inserted into the retina from the ganglion cell side.

Light stimuli were produced from a tungsten lamp with wavelength selection by narrow band interference filters. Intensities were controlled with neutral density filters. Responses were amplified with a 16 channel 25,000 gain bandpass differential amplifier (low and high corner frequencies of 250 and 7500 Hz). Fifteen channels of data plus one stimulus channel were digitized with a commercial multiplexed A/D board and data was stored in a Pentium based computer.

The electrophysiological data was presented to a variety of neural network models for training and subsequent classification of the test data. The neural networks were three layer perceptron models, with 330 input nodes (15 neurons with 22 time bins for each neuron), a hidden layer with from 25 to 35 nodes, and three output nodes. Training was with back propagation with momentum and adaptive learning rates. The training and data sets were obtained by random selection of two subsets of the ganglion cell data: a training subset composed of 1046 sets of 15 simultaneously recorded responses (22 time bins long), and 523 test sets of 15 simultaneously recorded responses.

RESULTS

Response recordings: The goal of this experiment was to learn how well recordings from a network of ganglion cells could be used to predict the shape, color, and intensity of the visual stimulus that evoked the set of responses. Eight sets of 15 simultaneously recorded responses from 15 electrodes to progressively dimmer flashes of 633 nm, 3.4mm diameter, light are shown in Figure 1. The stimulus duration was 0.2 seconds, and it was followed by a 0.24 second period of darkness. The data is displayed as a gray level two dimensional plot. Time is plotted on the abscissa (in terms of 20 msec time "bins"), unit number is plotted on the ordinate, and the firing rate in spikes/sec (calculated from the number of spikes in a 20 msec time interval) is represented in gray levels. All of the units shown in this figure are of the "on - off" type. It is noted that while the individual responses share certain kinetics, each response is unique in terms of the number of action potentials and their relative temporal sequence during the response.

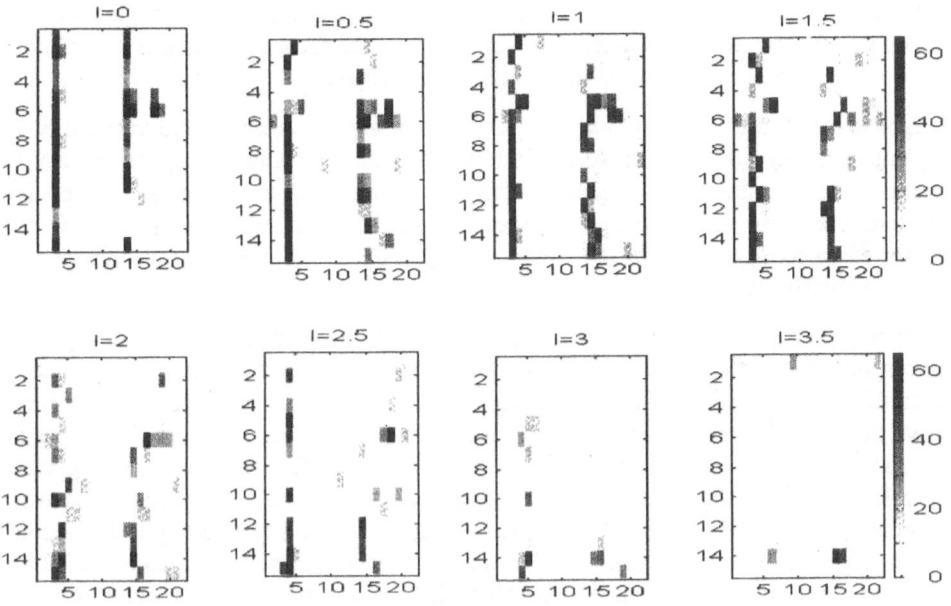

Figure 1-Dependence of 15 ganglion cell firings on stimulus intensity (brightest stimuli in upper left, dimmest in lower right). Time (represented in bin numbers) on abscissa, unit number on ordinate and firing rate in gray levels.

To illustrate how color and spot diameter are represented in ganglion cell firing patterns, we show in Figure 2, three sets of 15 simultaneously recorded ganglion cell responses to spots of light of various diameters and colors. Figure 2a shows a reference set of averaged responses to 633 nm, 3.4 mm diameter circular spots of light, of an intensity approximately 3 log units above threshold (defined as the intensity that evoked responses in 1/3 of the units). Figure 2b shows how the neuron set responded to 546 nm flashes, again, about 3 log above threshold. Finally, Figure 2c shows the dependence of the set of neuronal responses on spot diameter. Here, the spot diameter was decreased from 3.4 mm to 1.13 mm diameter. As expected, some units (those whose receptive fields were not well stimulated by the smaller spot) produced a smaller response, while others produced a slightly more vigorous response. These three figures and Figure 1 illustrate the point that when the neural response set is taken as a whole, there are clearly aspects of the units that seem to respond differentially to various parameters of the stimulus.

48

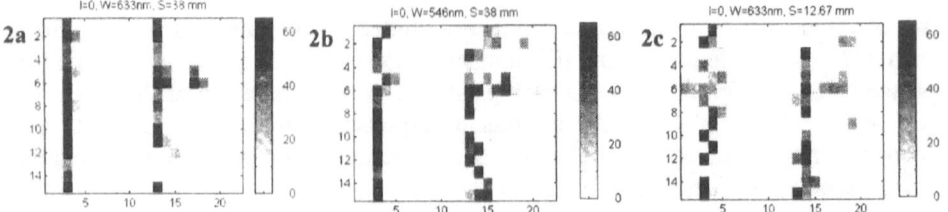

Figure 2-Effect of color (b)and spot size (c) on a reference set of 15 ganglion cell responses (a). Time (represented in bin numbers) on abscissa, unit number on ordinate and firing rate in gray levels.

Response Variability: Ganglion cell responses manifest considerable variability in sequential presentations of a fixed stimulus. The number and timing of the action potentials that make up the response vary from stimulus to stimulus. Thus, the differences seen in the responses of Figures 1 and 2 could reflect differences in the coding of various stimulus parameters, or they could simply reflect this nonspecific response variability.

An example of this variability is illustrated in Figure 3 where each panel is a plot of the responses of one of the 15 simultaneously recorded ganglion cells to a sequence of seven 633 nm, 3.4 mm diameter, 0.2 second long flashes that were about 3 log units above threshold. In each panel, time (reflected in 20 msec bin numbers) is plotted on the abscissa, and individual trial number is plotted on the ordinate. Firing rates (calculated from the 20 msec bins) are represented in gray level. These responses were recorded consecutively over a 3.5 second period, and were preceded by a 3.5 second sequence of flashes that were about 0.5 log units less intense. It is seen that some ganglion cells manifested considerable variability, while others had a more constant response. This figure illustrates the well known observation that ganglion cell responses to a repetitive stimulus vary substantially between cells and between responses. We conclude from these figures that while the intrinsic variability of responses from stimulus to stimulus could make discrimination of two very similar stimuli problematic, this will be less of a problem for classifying stimuli that are substantially different from each other.

Is Variability Global or Local? If the variability shown in Figure 3 were of a global nature, then the firing rates in all cells would wax and wane as a group and the changes in ganglion cell activity would not have a major effect on the encoding and our subsequent differential classification of the stimulus. However, if the variability were very local (i.e., some units increased their responses while adjacent units decreased their responses), the variability would make stimulus classification much more error prone.

Our results here are preliminary, and consist at this point of plots (not shown) of the total number of action potentials evoked by the stimulus in each of the fifteen cells for each of the seven successive presentations. This plot shows little consistent change in the number of spikes in each cell for each presentation. Some cells increased their firing while other decreased their firing. Our tentative conclusion is that the response variability is a local phenomenon. Clearly however, more work is needed before one can draw a significant conclusion.

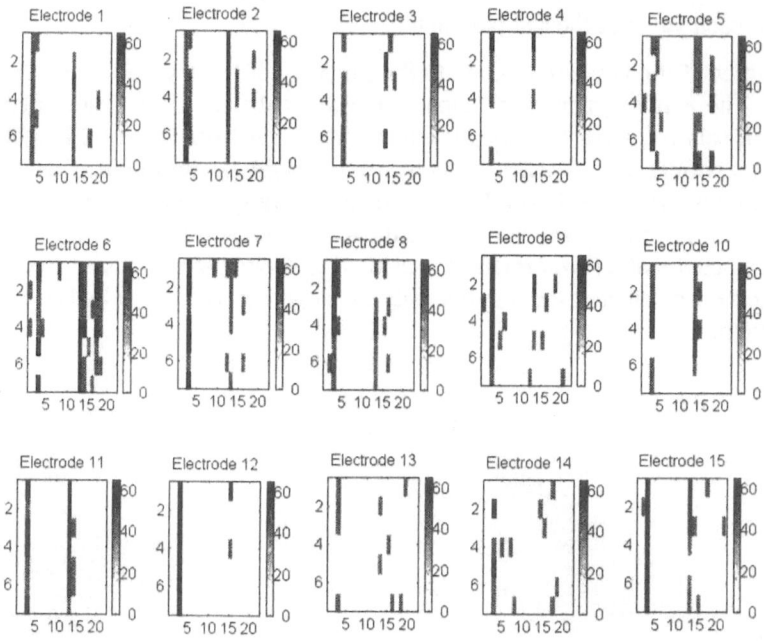

Figure 3-Responses of 15 ganglion cells to 7 sequential flashes of light. Each panel plots the responses of each cell to the 7 flashes. Time (represented in bin numbers) on abscissa, stimulus number on ordinate and firing rate in gray levels.

Optimization of the neural network: Neural network architectures were implemented that contained from 25 to 35 nodes in the hidden layer. The models were trained and tested on their ability to classify the stimuli that evoked the set of test responses. The number of training epochs were optimized to produce the highest correlations between test stimulus parameters and neural network classifications.

For the data used in this study, hidden layers containing from 25 to 35 nodes were found to only subtly affect the performance of the models. Thus, for most modeling, we focused our work on networks containing 25 nodes in the hidden layer. Neural network models were also trained only to specific time bins in our response set. As expected from the response kinetics (Figures 1-3), classifications of test data showed the highest correlations at time bins 3 and 13, the times of maximal neural activity.

Classifications using entire data set: The most challenging classification problem for our neural network model is to fully specify the color, intensity, and spot diameter of randomly selected test stimuli when the network has been trained using a subset of the entire data set. The classification capabilities of the optimized network architecture were tested by presenting the input nodes with fifteen simultaneously recorded responses from our test data set. The classifications were made over the full range of 8 intensities, 9 colors and 6 spot diameters used in our stimulus set. Performance of the network was judged with histograms of the number of classifications of the stimulus parameter at each value of the test parameter.

The capabilities of the network to classify the intensity of the test stimulus is given in Figure 4a. The 8 histograms were constructed for 8 different test intensities. The value of each column is the classification of the neural network, and its height is the number of classifications. Perfect classifications would be represented as 8 single columns, running from the lower left to the upper right. Figure 4a illustrates two important features. First, as the intensity of the test stimulus was lowered, the network generally predicted that a lower intensity was used to evoke the responses. Second, the network often miscalculated the test intensity as judged by the width and mean value of each histogram..

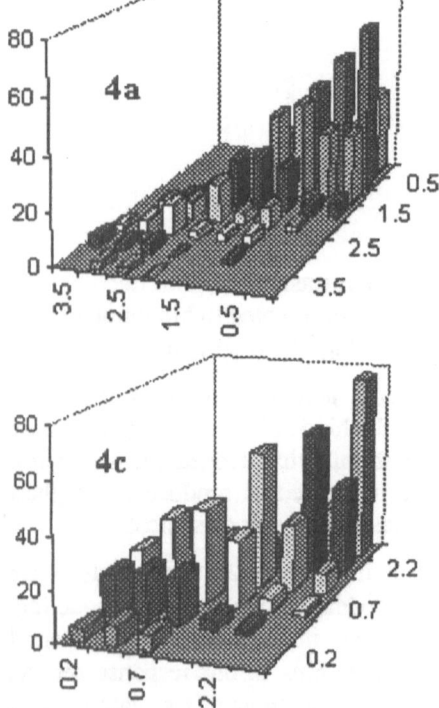

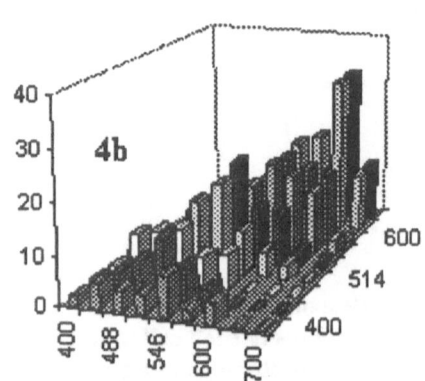

Figure 4- Histograms showing the classifications by the neural network of test flash intensity (a), color (b) and spot diameter (c). Firing rates are plotted on the y-axis. The known value of the stimulus parameter is plotted on the z-axis, and the neural network classification is plotted on the x-axis.

We have also used the network to classify the color and diameter of the test spots, and these results are shown in Figures 4b and 4c. As for intensities, the network often made misclassifications of the color and diameter of the stimuli, but generally, the histogram of the network responses were centered around the actual test values.

Figure 4 indicates that the neural network is able to classify the parameters of the test stimuli, but it does so with some errors. We have performed a linear regression between the test parameters predicted by the neural network and the actual parameters, and have used the correlation coefficients and standard deviation as a quantitative index of the classification performance of the neural network. These performance criteria are presented below.

	Intensity	Color	Spot Diameter
Correlation Coefficient	0.74	0.65	0.81
Standard Deviation	0.59 log units	60.7 nm	0.62 mm

Classification and training with variations in only one stimulus parameter: The classification task we have given the neural network is admittedly daunting. This is especially true when one considers that the dimensions of the training set contain inherent inconsistencies (bright 700nm flashes are indistinguishable from dimmer 633 nm flashes, both wavelengths only excite long wavelength cones to any significant extent). To make the classification task more tractable, we have restricted our training and test sets to contain only variations in one stimulus parameter. When the stimulus set only contains our largest diameter spots, and only 633 nm stimuli, the network classification was only slightly improved. Correlation coefficients for intensity classifications of 0.77 were obtained. Greater correlation coefficients were expected, but the reduced size of the training data set resulted in this value.

Classifications by individual neurons: We have used neural networks to learn how well individual ganglion cells are able to classify the color, intensity and diameter of the stimuli in our test data set. Fifteen separate neural networks were implemented and trained with our training set; each neural network was trained with data from a single unit. The number of training epochs for each network was optimized to provide the greatest correlations between the predictions for the test data sets. The correlation coefficients for all fifteen neural network classifications for color (top row) , intensity (middle row), and spot diameter (bottom row) are shown in Figure 5.

Not surprising, the individual neural networks performed poorly in the classification task. Also not surprising, some neurons were better classifiers of color than others (units 6 and 11), while others were better at classifying intensity units (5 and 13). As can be seen from Figure 5, none of the neurons were good classifiers of all three stimulus parameters: intensity, color, and spot diameter.

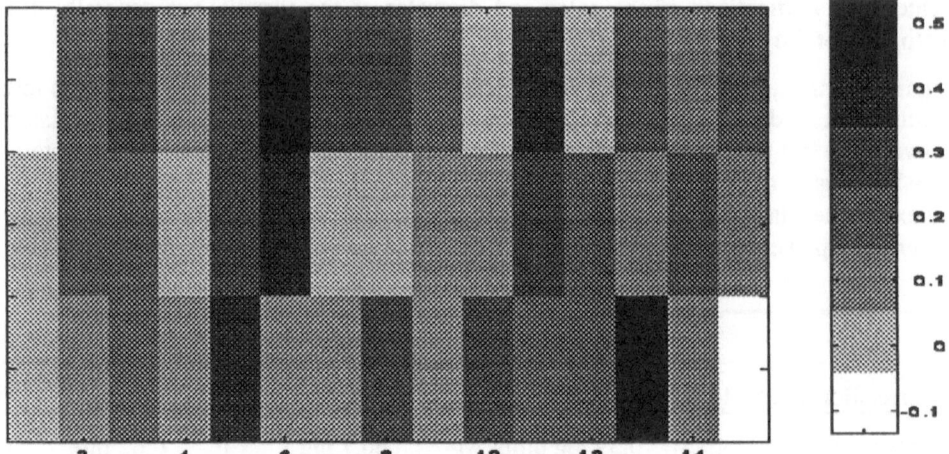

Figure 5-Correlation coefficients for classifications by each of all 15 units. Top, middle and bottom rows show results of single unit classifications of color, intensity, and sport diameter, respectively. Ordinate is unit number.

DISCUSSION

This study was motivated to verify the intuitive notion that since spatial, temporal, chromatic and intensity information about visual stimuli is encoded in the spatial/temporal firing pattern of networks of ganglion cells, reliable estimates of the parameters of visual stimuli could be made from the responses of subsets of ganglion cells. We have been able to show that a neural network is able to use the temporal firing pattern of fifteen simultaneously recorded ganglion cells to classify the color, size and intensity of a circular test spot with greater precision than was possible using individual single unit responses.

Unfortunately, the neural network performed this task with considerable error. Many of the errors were anticipated, and, as described earlier, were a simple consequence of the classification conflicts inherent in the original data set. Specifically, in some cases, we trained the network and challenged it to classify the color and size of spots that were of intensities below threshold (in this situation, the ganglion cells were "unaware" that any stimulation had occurred). In other cases, as seen in Figure 1, the responses to bright lights were often very similar: the network would have a particularly hard time distinguishing saturating intensities from supersaturating intensities. Finally, the network was expected to classify the spot size and intensities of short wavelength stimuli, a region of the spectrum where our incandescent lamp emitted minimal energy. Considering the number and size of these inherent "potholes" in the road that leads to accurate classification, it is not surprising that the neural network classified with less that total precision.

We have pointed out that we regard this study as "work in progress" and look forward to refining our physiological techniques so we can have greater confidence in our data.

The data set that was used in our neural networks, while large, should have been even larger to allow us to train the networks for stimulus variations in only one stimulus parameter. The isolated retina preparation also has another major difficulty; because the retina is stripped away from the pigment epithelium, photopigment regeneration is at best minimal and in some cases, nonexistent. This means that photosensitivity is poor initially, and gets poorer with each test stimulus and that test stimuli must be quite bright in order to evoke superthreshold responses.

In summary, the use of high density, silicon based microelectrode arrays is allowing us to begin to get a glimpse of how more complex visual stimuli are represented by the firing patterns of networks of ganglion cells. The use of neural networks to classify input stimuli allows one to better understand the extent of coding that occurs in selected groups of neurons, and how this encoding takes place, in terms of both space and time. The techniques described herein should provide us with new insights into noise and variability in ganglion cell responses, the stationarity of the coding process, and the role of coincident firing of groups of neurons in visual feature representation.

REFERENCES

(1) Normann, R.A. and Perlman, I.: The effects of background illumination on the photoresponses of red and green cones. J. Physiol., 286, 491-507, 1979.

(2) Werblin, F.S. and Dowling, J.E.: Organization of the retina of the mudpuppy, Necturus Maculosus. II. Intracellular recordings. J. Neurophysiol. 32, 339-355 (1969).

(3) Levine, M.W. :Variability in the maintained discharges of retinal ganglion cells. J. Opt. Soc. Am, 4, 2308-2320, (1987).

(4) Perceptron neural networks obtained from MATLAB Neural Network Toolbox.

(5) Perlman, I., Normann, R.A., Chandler, J. and Lipetz, L. :The effects of calcium ions on L-type horizontal cells in the isolated turtle retina. Visual Neuroscience, 4, 53-62, (1990).

(6) Jones, K.E., Campbell, P.K. and Normann, R.A. :A glass/silicon composite intracortical electrode array: Annals of Biomedical Engineering 20, 423-437 (1992).

(7) Nordhausen, C.T., Maynard, E.M. and Normann, R.A.: Single Unit Recording Capabilities of a 100 Microelectrode Array. Brain Research 726, 129-140 (1996).

Geometric Model of Orientation Tuning Dynamics in Striate Neurons

Igor A. Shevelev, Konstantin A. Saltykov and George A. Sharaev

Department of Sensory Systems, Institute of Higher Nervous Activity and Neurophysiology,

Russian Academy of Sciences, 5-a Butlerova St., 117865 Moscow, Russia

We simulated dynamics of orientation tuning (OT) of neurons in the cat primary visual cortex (VC) using a simple geometric model of receptive field (RF) with different dynamics of configuration and weight of excitatory and inhibitory subzones. The typical cases of OT dynamics were received in the model neurons that correspond to a similar behavior of neurons in the cat striate cortex. Parameters of the model were estimated that are critical for the main types of OT dynamics in the natural conditions. They were: the size and location of excitatory and inhibitory subzones in RF, their weight and type of dynamics. It was shown that the selective and acute OT of neurons may be formed by intracortical inhibition, while the dynamics of preferred orientation (PO) is connected only with dynamic reorganization of inhibitory subzone of RF.

INTRODUCTION

Mechanisms of the selective sensitivity of units in the cat and monkey visual cortex to orientation of bar or slit are up to now the subject of discussion [1-13]. Different models pay the leading role in orientation tuning either to excitatory [3,5,10,12,13], or to inhibitory [2,4-9,12-14] intracortical connections. Meanwhile, it is well known that RF of striate neurons includes both excitatory and inhibitory subzones, the relative location and some other parameters of which determine OT [5,7-10,14].

Study of orientation tuning of units in area 17 of the cat visual cortex with method of temporal slices [1,7-9,11,15-20] revealed dynamic changes of orientation tuning characteristics in all neurons studied. In 2/3 of neurons tuning successfully changed during some tens - some hundreds of ms after stimulus onset. Most typical was successive shift of PO along orientation range, that we named as scanning of the orientations [7,8,16-20]. It coincided with widening and then sharpening of OT. In contrast neurons with stable PO (1/3 of cases) demonstrated only the dynamic change of the width and selectivity of tuning that developed in counterphase to the same characteristics of the first group.

The main task of our study was to simulate different types of OT dynamics on a simple geometric model of RF, that was realized on PC, and to reveal features and characteristics of RF subzones that can be responsible for these dynamics.

METHODS

For our model we used known neurophysiological data on RF configuration and composition in the cat and monkey striate cortex [5,7,14,15,21]. Our modeled RF consisted from excitatory and inhibitory subzones (EZ and IZ). They were presented by the homogeneous rectangles with different relative localization, size and weight. Each zone was characterized by six parameters: length (1), width (2), x- and y-shift of zone center from the center of "flashing" stimulus (3 and 4), weight (5) and its sign (6) - excitatory (1) or inhibitory (0). Simulation of the RF configurations was limited by the reason of simplicity: we used minimal number of RF zones with configurations not far from the natural.

Parameters of the RF zones were determined for 10 successive moments, that is, we constructed the set of 10 successive RF maps. Stimulus ("light bar") was centered in RF excitatory subzone and we rotated it around the center in successive trials with a step of 22.5°. For each temporal step and stimulus orientation we calculated the weighted number of pixels of each sign (1 or 0) made active by the stimulus. Algebraic sum of signals from EZ and IZ was calculated for their overlapping parts.

Polar graphs of OT were plotted afterwards where stimulus orientation was shown as a radius-vector of corresponding orientation and its length corresponds to the difference between reaction of the EZ and IZ. Orientation tuning was characterized by: 1) preferred orientation (PO): angle of rotation (in degrees) of the "flashing" stimulus in relation to the horizontal taken as a 0°; 2) tuning width (F_l in °) measured at the level of 0.66 from OT maximum; 3) tuning selectivity (R): ratio of responses to the PO and the worst one; 4) quality of OT (F), that was directly proportional to R and inversely proportional to normalized ($F_l/180$) tuning width: $F=R \cdot 180/F_l$.

RESULTS

Dynamics of width and selectivity of orientation tuning

Simulation revealed dynamic changes of OT under variation of each of six parameters of RF. Fig. 1 shows RF with EZ (A) stimulated by a long "flashing" bar. In dynamics we introduced here typical [7,8,14,15,21] change of RF size with constant weight (B). This RF successively widened in dynamics (tacts 1-6), and then narrowed (tacts 6-10). It is seen that the polar graphs of OT (C) are changed in parallel with the RF size. Tuning is wide at the first moment due to full overlap of the stimulus of each orientation of all RF area. Then (2-nd tact) OT narrowed to 45°, widened up to 112-126° at the moments 3-7, and then extremely acuted (F_l narrowed up to 5°). The magnitude of the unit response (A in Fig. 1D) dynamically increased and then decreased according to dynamics of EZ. Orientation selectivity and detection quality are changed here oppositely to dynamics of F_l.

Stimulus width influences OT: its doubling (Fig. 1E) leads to modification of all of OT characteristics. Thus, F_l became wider and direction of OT dynamics became counterphasic to change of RF area and response magnitude (compare F_l dynamics in Fig. 1D and E).

Fig. 2 illustrates influence on OT of the stimulus shift from the RF center in x- and y-coordinates (A and D). In the first case at the initial moments the unit is relatively sharply tuned to horizontal orientation (B and C), but then (tacts 3-6) it begins to prefer orientations that are more and more close to vertical and tuning became wider. Then tuning returns to horizontal and again sharpens.

56

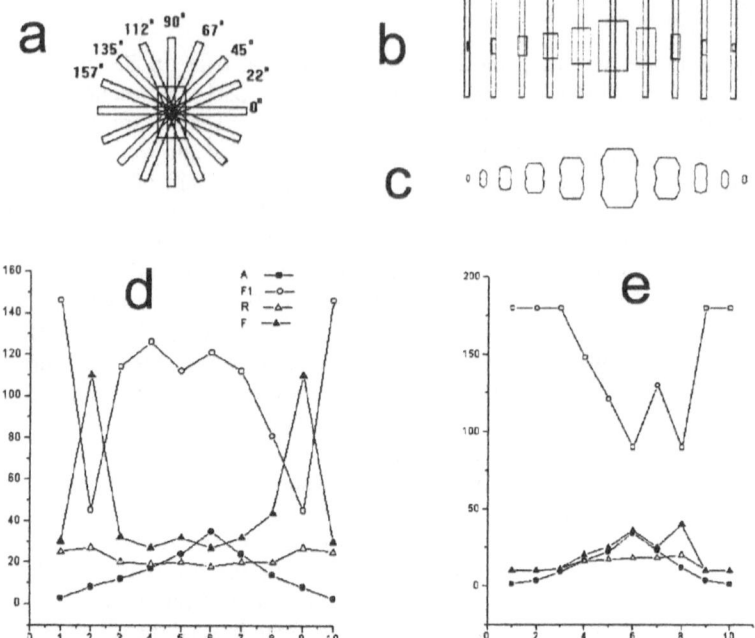

Fig. 1. Area dynamics of the receptive field (RF) excitatory zone (b) and of orientation tuning (OT) (c) in simulated neuron under stimulation by the centered slit of different orientation (a). (d) Dynamics of OT characteristics (Å - response magnitude, F_1 - tuning width, R - tuning selectivity and F - tuning quality). Abscissa - temporal tacts, ordinate - tuning characteristics. (e) the same, as in (d), but for the stimulus of doubled width.

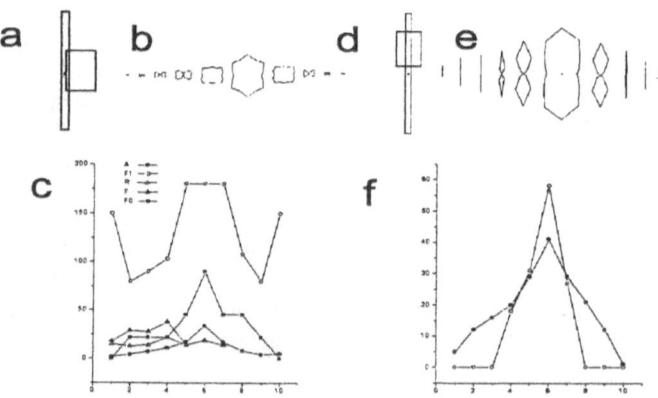

Fig. 2. Influence of the lateral (a) and longitudinal (d) stimulus shift in RF on the dynamics of OT (b and e, correspondingly) and on its characteristics (c and f). Other details as in Fig. 1.

Under vertical stimulus shift (Fig. 2 D) at the beginning and at the end of dynamics the unit seems to be an absolute detector of vertical (E, tacts 1-3 and 8-10). At the moments 4-7 OT widens from 18° to 58°, but with the same preferred orientation.

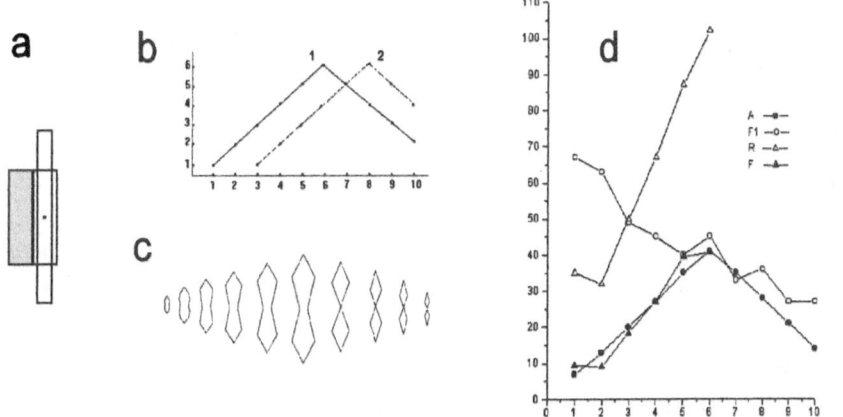

Fig. 3. Dynamics of OT (c) and its characteristics (d) for the simulated unit which RF (a) has an excitatory and side-inhibitory (stretched) zones. (b) Dynamics of weight of excitatory (1) and inhibitory (2) zones. Other details as in Fig. 1.

Simulation of RF with two parallel non-overlapping EZ and IZ of stable area (Fig. 3A) revealed the influence of the dynamics of zone weight (B) that changes the response magnitude (C) and OT characteristics (D). Before switching on of an inhibition OT is here relatively wide (67°), but successively acuted then (D) as a result of inhibitory suppression of the responses to a part of orientation range (0-45°). It is followed by an increase of orientation selectivity. Overlapping on 30% of the parallel EZ and IZ of RF improves OT width and selectivity. In this case the IZ step by step blocks the responses to all orientations except the vertical. If IZ weight is increased twice faster than of EZ, the OT sharpening comes even faster.

Dynamics of preferred orientation

Fig. 4 represents the case of dynamics of an area of end-stopping IZ of RF that changed its vertical dimension (A). As a result it approaches on the stable EZ with constant weight. As can be seen in the polar graphs of OT (B) in this case PO is shifted from 90° (1) up to 180° (10) traveling through intermediate orientation of 135-157°. Such dynamics of PO appears due to an inhibitory action of IZ that blocks the response to vertical and then gradually - to orientations more and more far from vertical up to the moment when the only permitted orientation is horizontal (0° or 180°). Thus the dynamic of end-stopping IZ of RF allows to simulate the successive dynamical rotation of PO in relatively simple and natural way. What about OT width, at the beginning it is small (OT is sharp), then it widens to became sharp again at the final moments (tacts 6-10).

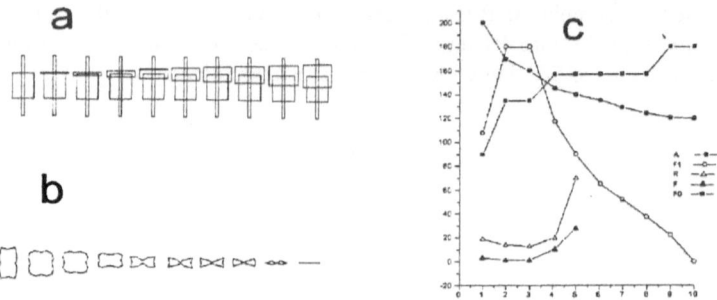

Fig. 4. Dynamics of OT (b) and its characteristics (c) in the RF (a) of the simulated neuron with excitatory zone and end-stopping inhibitory zone of increased area. F_o in (c) - PO of slit. Other details as in Fig. 1.

Fig. 5A illustrates the case of RF dynamics with the shift of localization of the IZ center. Both RF zones are here stable in weight and area, but IZ is vertically shifted about the center of EZ. At the first stage OT is wide (B, C), then (A, tacts 2-6) IZ approaches to EZ and overlaps it up to central part. That leads to inhibitory suppression of responses to some part of orientation range (B). The unit became sharply tuned to 67° instead of 90° and its selectivity strives for infinity, while width of OT decreases from 40° (2-nd tact) to 15° (5-th tact). When IZ overlapped near half of EZ the response became smaller (tacts 5-7) and OT widened up to 67° (tact 6-th). Further shift of the IZ leads to the PO shift on 90° (up to 112°) and to sharpening of OT.

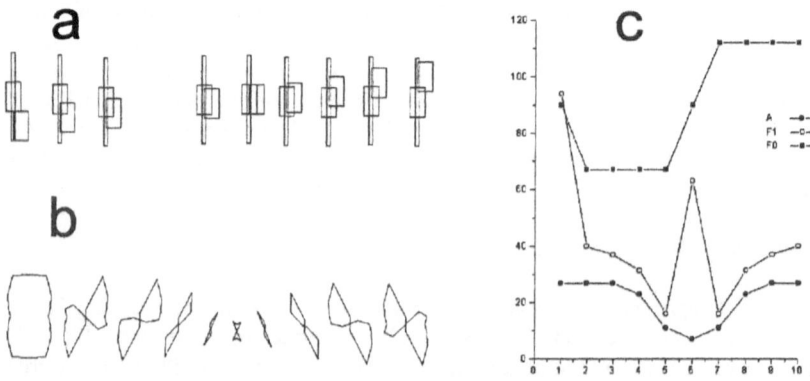

Fig. 5. Dynamics of OT (b) and its characteristics (c) in the RF (a) of the simulated neuron with excitatory zone and side-inhibitory zone which is shifting along the excitatory zone. Other details as in Fig. 1.

DISCUSSION

Simulation of the realistic geometrical configurations of RF revealed significant dependence of the output parameters of OT on small variations of the input characteristics of RF. Thus, dynamic widening of EZ leads to widening of OT without change in PO (Fig. 1). Stimulus centered in such RF with single EZ failed to reveal sharp orientation tuning (Fig. 2). This agrees with neurophysiological evidences of inhibitory formation and/or sharpening of OT in neurons of the visual cortex [4,6-9,13,20].

It was revealed that the change in the weight, area and localization of IZ of the simulated RF leads to a different type of OT dynamics. Thus, dynamic change in the weight of EZ and side-IZ of constant configuration is accompanied by the dynamics of OT width, selectivity and quality (Fig. 3): under inhibitory influences a unit became high-quality orientation detector with narrow OT and high selectivity.

Other type of OT dynamics is characterized by the successive shift of PO that was typical for 2/3 of the real striate neurons in the cat visual cortex [7,8,16-20]. Our simulation shows that this type of OT dynamics appears only under dynamic change in topography of IZ. In neurons with end-stopping IZ of RF this shift or rotation of PO may be based on widening of the IZ and its approach to the EZ. Contrary, RF with side-inhibition exhibits such shift only under change in coordinates of IZ (if it is shifted along the long axis of EZ). It is essential, however, that the change in the area of end-inhibitory zone can provide for the shift of PO of up to 90°, while dismission of the side-IZ can ensure only PO shift limited to 45° (Fig. 5). Dynamics of the width and selectivity of OT in this case are qualitatively similar to the change of OT under dynamics of weight of IZ (Fig. 3).

It must be stressed that simulated transformation of the IZ can have a neurophysiological basis. Thus, in striate and LGB neurons in cat the dynamic of EZ and IZ of RF was demonstrated: they dynamically widened and partially overlapped, but then narrowed again [7,12,14,15,21]. It was shown also the shift of IZ center in RF of cat visual neurons [9,15,20,21].

Two types of OT dynamics in real neurons [7,8,16-20] allow to suppose two types of their RF dynamics. Thus, neurons with dynamically stable PO [7,9,17-20] hypothetically change in time only the weight of EZ and IZ that leads only to variation of acuteness, selectivity and quality of OT. It is important that under local blockade of inhibition by microiontophoretical application of bicuculline [4,6,9] neurons of this type exhibit PO dynamics that indicate on an inhibitory stabilization of OT in natural conditions by a blocking of convergence from units with different tuning. Contrary to that type, units with changing of PO under normal condition [1,7-9,11,12,16-20] can undergo dynamic reorganization of the area and weight of RF zones.

It may be concluded that simulated dynamics of RF and OT is in accordance with known neurophysiological data and allows to predict some testable modifications of EZ and IZ of RF that can be the basis of the real types of OT dynamics in neurons of the cat visual cortex.

BIBLIOGRAPHY

1. *Celebrini S., Thorpe S., Trotter Y., Imbert M.* // Vis. Neurosci. 1993. V.10, P.811.
2. *Creutzfeldt O.D., Kuhnt V., Benevento L.A.* // Exp. Brain Res. 1974, V.21, P.251.
3. *Douglas R.J., Koch C., Mahowald M., Martin K.A., Suarez H.H.* // Science. 1995. V.269, P.981.

4. *Lazareva N.A., Eysel U.T., Shevelev I.A., Sharaev G.A.,* // Neurophysiol. 1995. V.27, N 1, P.54.

5. *Lund J.S., Wu Q., Hadingham P.T., Levitt, J.B.* // J. Anat. 1995. V.187, P.563.

6. *Pfleger B., Bonds A.B.* // Exp. Brain Res. 1995. V.104, P.81.

7. *Shevelev I.A.* Neurons of the visual cortex: Adaptivity and dynamics of receptive fields. Moscow: Nauka, 1984. 232 p.

8. *Shevelev I.A.* // Sensory Systems. 1994. V.8. N 3, P.245.

9. *Shevelev I.A., Eysel U.T., Lazareva N.A., Sharaev G.A.* // Neurophysiol. 1995. V.27, N 2, P.110.

10. *Somers D.C., Nelson S.B., Sur M.* // J. Neurosci. 1995. V.15, P.5448.

11. *Vogels R., Orban G.A.* // Exp. Brain Res. 1991. V.84, P.1.

12. *Volgushev M., Vidyasagar T.R., Pei X.* // Vis. Neurosci. 1995. V.12, P.621.

13. *Worgotter F., Koch C.* // J. Neurosci. 1991. V.11, P.1959.

14. *Volgushev M.À.* // Dokladi Acad.Nauk USSR. 1989. V. 309, N 6. P.1483.

15. *Podvigin N.F.* Dynamic properties of neuronal structures of the visual system. Leningrad: Nauka, 1979. 158 p.

16. *Shevelev I.A., Sharaev G.A.* // Dokladi Acad.Nauk USSR. 1981. V.256. P.1506.

17. *Shevelev I.A., Sharaev G.A.* // Neurophysiol. 1981. V.13, P.451.

18. *Shevelev I.A. Sharaev G.A.* // Neurosci. 1982. V.8S, P.192.

19. *Shevelev I.A., Sharaev G.A.* // Neurophysiol., 1985. V.17, P.35.

20. *Shevelev I.A., Sharaev G.A., Lazareva N.A., Novikova R.V., Tikhomirov A.S.* // Neurosci. 1993. V.56, P.865.

21. *Shevelev I.A., Volgushev M.A., Sharaev G.A.* // Neurosci. 1992. V.51, P.445.

Acknowledgment - The study was partly supported by the Russian Foundation for Basic Sciences (Grant # 96-04-48043).

Neuronal Circuitry in the Medial Cerebral Cortex of Lizards

J.A. Luis de la Iglesia and C. Lopez-Garcia

Neurobiología. Biología Celular. Facultad de Ciencias Biológicas
Universidad de Valencia. 46100.-Burjassot, Valencia, Spain
Phone and Fax: 34 6 386 4781, e-mail: lopezc@uv.es

Abstract
The medial cortex of lizards is a simple three-layered brain region displaying many characteristics which parallel the hippocampal fascia dentata of mammals. Its principal neurons form a morphologically diverse population, partly as a result of the prominent continuous growth of this nervous centre. By using the classical Golgi impregnation method we describe here the morphology of the principal neurons (8 types) and the short-axon interneurons (18 types) populating the medial cortex of Podarcis hispanica as well as the connections between them.

Histology of the medial cerebral cortex of lizards
The cerebral cortex of lizards is formed by four main cortical areas, the medial, dorsomedial, dorsal and lateral cortices (Fig. 1). All these areas share a simple cytoarchitectonic pattern, most neuronal cell bodies occurring closely packed in a conspicuous cell layer which lies between two poor-celled strata, the outer and the inner plexiform layers.

Most medial cortex neurons have their somata located in the cell layer, they are called "principal" or "projection" neurons because their axons project outside the medial cortex. They send zinc-enriched axons towards the juxtasomatic dendritic segments of the bipyramidal neurons of the dorsal and dorsomedial areas forming conspicuous synaptic fields comparable to the stratum lucidum of the hippocampal CA3 area of mammals (Lopez-Garcia, Martinez-Guijarro, 1988). These axons form large presynaptic boutons which accumulate zinc inside their synaptic vesicles and are glutamate immunoreactive thereby closely resembling the hippocampal mossy fibres (Martinez-Guijarro et al. 1991a). On the grounds of their morphology, connectivity, immunocytochemical and histochemical properties and late ontogenesis, medial cortex projection neurons closely resemble dentate granule cells.

The medial cortex principal neurons have their dendrites located aside the cell layer. Two dendritic "bouquets" located in the plexiform layers (bitufted neurons) receive the main axonal input which is highly segregated in horizontal laminae. Axons coming from the lateral olfactory cortex segregate in the external third of the outer plexiform layer ("lizard perforant path") (Martinez-Garcia et al. 1986). An intermediate sublamina of the outer plexiform layer is the recipient of axons coming from ipsilateral dorsal cortex bipyramidal neurons. In addition, axons coming from ipsi- and contra-lateral bipyramidal neurons of the dorsomedial cortex terminate in a distinct juxtasomatic lamina in the outer plexiform layer, thus resembling the commissural projection of the mammalian hippocampus which ends in the juxtagranular zone of the molecular layer. Moreover, in lizards, the commissural projection forms an additional lamina just beneath the cell layer where the projection neurons extend their basal dendrites (Lopez-Garcia et al. 1992). This infragranular dendritic sublayer does not exist in rodents but, surprisingly, it has a counterpart in primates.

The plexiform layers are made up of the principal cell dendrites intermingled with the incoming axons along with the radial glial scaffold and a reduced population of isolated cells, including solitary neurons and microglia cells. Short-axon neurons of the medial cortex plexiform layers, although scarce, are very heterogeneous (Luis de la Iglesia et al. 1994). Most of them are GABA-immunoreactive (Schwerdtfeger, Lopez-Garcia, 1986) and display diverse neuropeptide immunoreactivity, calcium binding protein immunoreactivity (Martínez-Guijarro et al. 1993) and even NADPH activity (Davila et al. 1995). The interneurons in the inner plexiform layer show similarities with those of the dentate hilus of the mammalian hippocampus (Lopez-Garcia et al. 1988a).

Late ontogenesis and regenerative potential of the lizard medial cortex

Perhaps the most important feature shared by the lizard medial cortex and the hippocampal fascia dentata is their common late ontogenesis. In rodents about 85 % of granule cells are generated in the first three postnatal weeks of life, a period which lasts longer in primates (Rakic, Nowakowski, 1981). Posteriorly, during juvenile and adult life a reduced neurogenetic potentiality still remains in the fascia dentata (Bayer et al. 1982). In the case of lizards, postnatal neurogenesis is a robust phenomenon which persists throughout the life span of individuals and contributes to the continuous growth of the neuronal population of the medial cortex (Lopez-Garcia et al. 1984; Lopez-Garcia et al. 1988b). In lizards, this neurogenetic activity varies seasonally and may be dramatically enhanced after experimental lesion. The neurotoxine 3-acetylpyridine may specifically kill about 95 % of medial cortex neurons causing a loss of spatio-temporal memory in treated animals which is surprisingly recovered after a few weeks (Font et al. 1989; Font et al. 1991). A few days after the lesion, remnant neuroblasts located in the medial cortex ependyma show a proliferation burst (Molowny et al. 1995). The newly generated immature neurons migrate trough the inner plexiform layer using the radial glia as a guided pathway until being recruited into the medial cortex cell layer. There they mature, grow dendrites and send zinc-enriched axons restoring the lesioned medial cortex (Lopez-Garcia, 1993).

The neurons of the lizard medial cortex shown with the Golgi method

The ancient silver chromate impregnation method of Golgi still is the best approach to analyse the neuronal population of a nervous centre. The Golgi-Colonnier impregnation procedure (Colonnier, 1964) was performed in a series of 60 brains of the lizard species *Podarcis hispanica*. A pool of 1174 well-impregnated neurons (Luis de la Iglesia, 1995) were observed in the light microscope and drawn using a camera lucida apparatus. Neuronal archetypes were defined considering intrinsic morphological characters (dendritic tree and axonal arbour patterns, dendritic spines, etc.) as well as location of the dendritic tree into the different synaptic fields (afferent laminae) and the projection of their axonal arbour (i.e., whether recurrent-collateral branches exist, etc.). Main data are summarised in Figure 2 and Table 1.

In short, principal neurons are projection neurons giving off descending axons with deep collateral branches provided with prominent zinc-rich-glutamate-immunoreactive axonal boutons whereas the main axonal branch reaches adjacent cortical areas and the bilateral septum. Five of them, "heavily spiny granular" (monotufted, medium-sized), "heavily spiny bitufted" (large), "spiny bitufted" (medium-sized), "sparsely spiny bitufted" (small) and "superficial multipolar" (small) were found in the cell layer, whereas the three others lay outside this layer and were regarded as ectopic types ("outer plexiform ectopic bitufted", "inner plexiform ectopic bitufted" and "inner plexiform monotufted"). Additional secondary criteria —soma position and shape—, allowed us to further classify bitufted neurons into three distinct subtypes each: "superficial-round", "intermediate-fusiform", and "deep-pyramidal". Moreover, a variety of small impregnated cells were observed; they probably represent newly generated immature neurons which have not yet completed their development.

Interneurons form a scarce but complex neuronal population distributed by all the medial cortex strata; they display GABA and diverse calcium binding protein or neuropeptide immunoreactivities, thus they likely are inhibitory local interneurons. According their location in the different laminae (place of specific axonal inputs) we have distinguished four main types of interneurons, which must be involved in feed forward inhibition of principal neurons (Fig. 3). Five types are in the outer plexiform layer, three types are in the cell layer and ten others in the inner plexiform layer (six in the juxtasomatic zone and four in the deep-zinc-enriched zone) (Luis de la Iglesia et al. 1994).

The first group of interneurons are those located in the outer plexiform layer. They are: short axon aspinous bipolar neuron (sarmentous neuron), short axon aspinous juxtasomatic neuron (coral neuron), short axon sparsely spinous multipolar neuron (stellate neuron), short axon sparsely spinous juxtasomatic multipolar neuron (deep stellate neuron), and sparsely spinous juxtasomatic horizontal neuron (couchant neuron). Most of them are gamma-aminobutyric acid (GABA) and parvalbumin immunoreactive, and are thus probably involved in medial cortex inhibition; moreover, a small fraction of them displayed beta-endorphin immunoreactivity. The distribution of these neuronal types is not uniform in the laminae of the outer plexiform layer. Sarmentous and stellate neurons overlap the axonal field projection coming from the dorsal cortex and the thalamus, whereas deep stellate and couchant neurons overlap ipsi-and contralateral dorsomedial projection fields as well as raphe serotoninergic and opioid immunoreactive axonal plexi. Thus, these neuronal types may be involved in the control of specific inputs to the medial cortex by feed-forward inhibition of principal neurons; nevertheless, feed-back inhibition may also occur regarding deep stellate neurons that extend deep dendrites to the zinc-rich bouton field

The second group of interneurons are those located in the cell layer. They are the "granuloid neurons" (ascending-axon spiny neurons located in the outer edge of the cell layer), "web-axon neurons" (sparsely spiny neurons with ascending-descending axons) and "deep-fusiform neurons" (ascending-axon aspinous neurons with fusiform radially oriented somata). These neurons are scarce; they are probably GABA and parvalbumin-immunoreactive and presumably participate in feed forward as well as in feed back inhibition of the principal projection cells of the lizard medial cortex.

The third group of interneurons are those located in the juxtasomatic lamina of the inner plexiform layer, (a zone comparable to the dentate subgranular zone of the hippocampus). They are the "smooth vertical neurons" (aspinous radially oriented neurons with ascending axons), "smooth horizontal neurons" (aspinous horizontally oriented neurons with ascending axons), "small radial neurons" (small fusiform radially oriented neurons with ascending axons), "large radial neurons" (large fusiform radially oriented neurons with ascending axons), "pyramidal-like radial neurons" (aspinous fusiform to pyramidal neurons with descending axon) and "spheroidal neurons" (sparsely spiny small multipolar neurons with short local axon). They are all probably GABA and parvalbumin-immunoreactive and involved in feed forward inhibition of principal medial cortex cells.

The fourth group of interneurons are those located in the deep inner plexiform layer (comparable to the hilus of the fascia dentata). They are the "giant-multipolar neurons" (large multipolar neurons with short local axons), "long-spined polymorphic neurons" (large and small polymorphic neurons with ascending axons), "periventricular neurons" (ascending-axon neurons with pyriform to fusiform vertical somata close to the ependyma) and "alveus-horizontal neurons" (bipolar horizontal neurons arranged among the myelinated alvear fibres with axons forming local plexuses in the inner plexiform layer). These neurons are GABA-immunoreactive and either neuropeptide (somatostatin-neuropeptide Y) or paravalbumin-immunoreactive. They seem to be involved in feed back or even occasionally in feed forward inhibition phenomena.

Inhibitory control of interneurons over principal neurons are summarised in Figure 3. The importance of every inhibitory circuit depends on the "weight" of the synaptic input which, on the other hand, depends on the number of synaptic contacts and the distribution of them (whether they are axo-dendritic, axo-somatic or axo-axonic). Golgi impregnation frequency of every neuronal type and posterior electron microscopic counts of synapses, combined with electrophysiological recordings, ought to be used for stimating the weight of every inhibitory circuit.

The architecture of the lizard medial cortex (as well as that of the hippocampal fascia dentata) is closely related with the olfactory input (from the olfactory cortices) trough the "perforant path" and with commissural input (from CA3 pyramidal neurons); both kinds of inputs presumably connect with all the principal neurons as expected for a primordial neuronal network (Kohonen, 1988). Perhaps the addition of control inputs (interneurons of the plexiform layers) may help the selection of a few principal neurons which result activated by that broad input, as occurs with "O'Keefe position neurons" (O'Keefe, Speakman, 1987) of the rat hippocampus.

Regarding the cerebral circuitry (Fig. 4),the lizard medial cortex and the hippocampal fascia dentata are centres located relatively far from both the sensorial and the motor centres. They appear as centres closely involved in making the "individual cognitive maps" and the management of the "spatial memory"; in addition, they both seem to "filter" codified information selecting what is important and discarding what is not.

References

Bayer SA, Yackel JW, Puri PS (1982) Neurons in the dentate gyrus granular layer substantially increase during juvenile and adult life. Science 216:890-892

Colonnier M (1964) The tangential organization of the visual cortex. J Comp Neurol 98:327-344

Davila JC, Megias M, Andreu MJ, Real MA, Guirado S (1995) NADPH diaphorase-positive neurons in the lizard hippocampus: a distinct subpopulation of GABAergic interneurons. Hippocampus 5:60-70

Font E, Garcia-Verdugo JM, Martinez-Guijarro FJ, Alcantara S, Lopez-Garcia C (1989) Neurobehavioral effects of 3-acetylpyridine in the lizard Podarcis hispanica. Eur J Neurosci 2S:148

Font E, García-Verdugo JM, Alcántara S, Lopez-Garcia C (1991) Neuron regeneration reverses 3-acetylpyridine-induced cell loss in the cerebral cortex of adult lizards. Brain Res 551:230-235

Kohonen T (1988) Self-organization and associative memory. Springer-Verlag, Berlin

Lopez-Garcia C, Tineo PL, Del-Corral J (1984) Increase of the neuron number in some cerebral cortical areas of a lizard, Podarcis hispanica, (Steind., 1870), during postnatal periods of life. J Hirnforsch 25:255-259

Lopez-Garcia C, Martinez-Guijarro FJ, Berbel P, Garcia-Verdugo JM (1988a) Long-spined polymorphic neurons of the medial cortex of lizards: a Golgi, Timm, and electron-microscopic study. J Comp Neurol 272:409-423

Lopez-Garcia C, Molowny A, Rodriguez-Serna R, Garcia-Verdugo JM, Martinez-Guijarro FJ (1988b) Postnatal development of neurons in the telencephalic cortex of lizards. In: Schwerdtfeger WK, Smeets W (eds) The forebrain of reptiles: Current concepts of structure and function. Karger, Basel, pp 122-130

Lopez-Garcia C, Molowny A, Martinez-Guijarro FJ, Blasco-Ibañez JM, Luis de la Iglesia JA, Bernabeu A, Garcia-Verdugo JM (1992) Lesion and regeneration in the medial cerebral cortex of lizards. Histol Histopath 7:725-746

Lopez-Garcia C (1993) Postnatal neurogenesis and regeneration in the lizard cerebral cortex. In: Cuello AC (ed) Neuronal cell death and repair. Elsevier Sci.Pub. pp 237-246

Lopez-Garcia C, Martinez-Guijarro FJ (1988) Neurons in the medial cortex give rise to Timm-positive boutons in the cerebral cortex of lizards. Brain Res 463:205-217

Luis de la Iglesia JA, Martinez-Guijarro FJ, Lopez-Garcia C (1994) Neurons of the medial cortex outer plexiform layer of the lizard *Podarcis hispanica*: Golgi and immunocytochemical studies. J Comp Neurol 341:184-203

Luis de la Iglesia JA (1995) Las neuronas del cortex medial de la lagartija Podarcis hispanica. Estudio con el metodo de Golgi. Doctoral Thesis, Univ. Valencia

Martinez-Garcia F, Amiguet M, Olucha F, Lopez-Garcia C (1986) Connections of the lateral cortex in the lizard Podarcis hispanica. Neurosci Lett 63:39-44

Martinez-Guijarro FJ, Soriano E, Del Rio JA, Lopez-Garcia C (1991) Zinc-positive boutons in the cerebral cortex of lizards show glutamate immunoreactivity. J Neurocytol 20:834-843

Martínez-Guijarro FJ, Soriano E, Del Río JA, Blasco-Ibáñez JM, Lopez-Garcia C (1993) Parvalbumin-containing neurons in the cerebral cortex of the lizard *Podarcis hispanica*: Morphology, ultrastructure, and coexistence with GABA, somatostatin, and neuropeptide Y. J Comp Neurol 336:447-467

Molowny A, Nacher J, Lopez-Garcia C (1995) Reactive neurogenesis during regeneration of the lesioned medial cerebral cortex of lizards. Neuroscience 68:823-836

O'Keefe J, Speakman A (1987) Single unit activity in the rat hippocampus during spatial memory task. Exp Brain Res 68:1-27

Rakic P, Nowakowski RS (1981) The time of origin of neurons in the hippocampal region of the rhesus monkey. J Comp Neurol 196:99-128

Schwerdtfeger WK, Lopez-Garcia C (1986) GABAergic neurons in the cerebral cortex of the brain of a lizard (Podarcis hispanica). Neurosci Lett 68:117-121

Table 1. Neuronal archetypes in the medial cerebral cortex of *Podarcis hispanica*

Layer	Type	Descriptive name		Short name	n	%
opl	i1	Short-axon, aspinous bipolar neuron		Sarmentous	15	1.28
opl/cl	i2	Short-axon, aspinous juxtasomatic neuron		Coral	11	0.94
opl	i3	Short-axon, sparsely spinous multipolar neuron		Stellate	21	1.79
opl/cl	i4	Short-axon, sparsely spinous juxtasomatic multipolar neuron		Deep-stellate	11	0.94
opl/cl	i5	Sparsely spinous juxtasomatic horizontal neuron		Couchant	6	0.51
cl	p1	Heavily spiny granular neuron		HSGN	53	4.51
cl	p2	Heavily spiny bitufted neuron		HSBN	171	14.57
cl	p2a	*idem*	*Superficial*	*Sup HSBN*	*46*	*3.92*
cl	p2b	*idem*	*Fusiform*	*Fus·HSBN*	*49*	*4.17*
cl	p2c	*idem*	*Pyramidal*	*Pyr HSBN*	*50*	*4.26*
cl	p2d	*idem*	*Quadrangular*	*Qua HSBN*	*28*	*2.39*
cl	p3	Spiny bitufted neuron		SBN	450	38.33
cl	p3a	*idem*	*Superficial*	*Sup SBN*	*137*	*11.67*
cl	p3b	*idem*	*Fusiform*	*Fus SBN*	*189*	*16.10*
cl	p3c	*idem*	*Pyramidal*	*Pyr·SBN*	*124*	*10.56*
cl	p4	Sparsely spiny bitufted neuron		SSPBN	130	11 07
cl	p4a	*idem*	*Superficial*	*Sup SSBN*	*57*	*4.86*
cl	p4b	*idem*	*Fusiform*	*Fus SSBN*	*48*	*4.09*
cl	p4c	*idem*	*Pyramidal*	*Pyr SSBN*	*25*	*2.13*
cl	p5	Superficial spiny multipolar neuron		Sup·SMN	9	0.77
cl	p6	Outer plexiform ectopic bitufted neuron		OPEBN	2	0.17
cl	p7	Inner plexiform ectopic bitufted neuron		IPEBN	29	2.47
cl	p8	Inner plexiform ectopic monotufted neuron		IPEMN	3	0.26
cl	cl-i1	Ascending-axon, Superficial Spiny Neuron		Granuloid	1	0.09
cl	cl-i2	Sparsely Spiny Neuron with Ascending-Descending Axon		Web-axon	2	0.17
cl	cl-i3	Ascending-axon, Aspinous Fusiform Vertical Neuron		Deep-fusiform	1	0.09
jip	i1	Ascending-axon, Aspinous Vertical Neuron		Smooth vertical	12	1.02
jip	i2	Ascending-axon, Aspinous Horizontal Neuron		Smoth horizontal	23	1.96
jip	i3	Ascending-axon, Small Fusiform Radial Neuron		Small radial	3	0.26
jip	i4	Ascending-axon, Large Fusiform Radial Neuron		Large radial	2	0.17
jip	i5	Descending-axon, Aspinous Pyramidal-Like Radial Neuron		Pyramidal-like radial	5	0.43
jip	i6	Short-axon, Sparsely Spinous Small Multipolar Neuron		Spheroid	17	1.45
dip	i7	Short-axon, Giant Multipolar Neuron		Giant-multipolar	21	1.79
dip	i8	Ascending-axon, Long-spined Polymorphic Neuron		Long-spined polymorphic	107	9.11
dip	i9	Ascending-axon, Periventricular Vertical Neuron		Periventricular	9	0.77
dip	i10	Short-axon, Alveus-horizontal Neuron		Alveus-horizontal	13	1.11

cl: cell layer; **dip:** deep zone of the inner plexiform layer; **i** (n): interneuron types; **p**(n): principal neurons; **jip:** juxtasomatic zone of the inner plexiform layer, **opl:** outer plexiform layer, **opl/cl:** outer margin of the cell layer; **n:** number of impregnated cells.

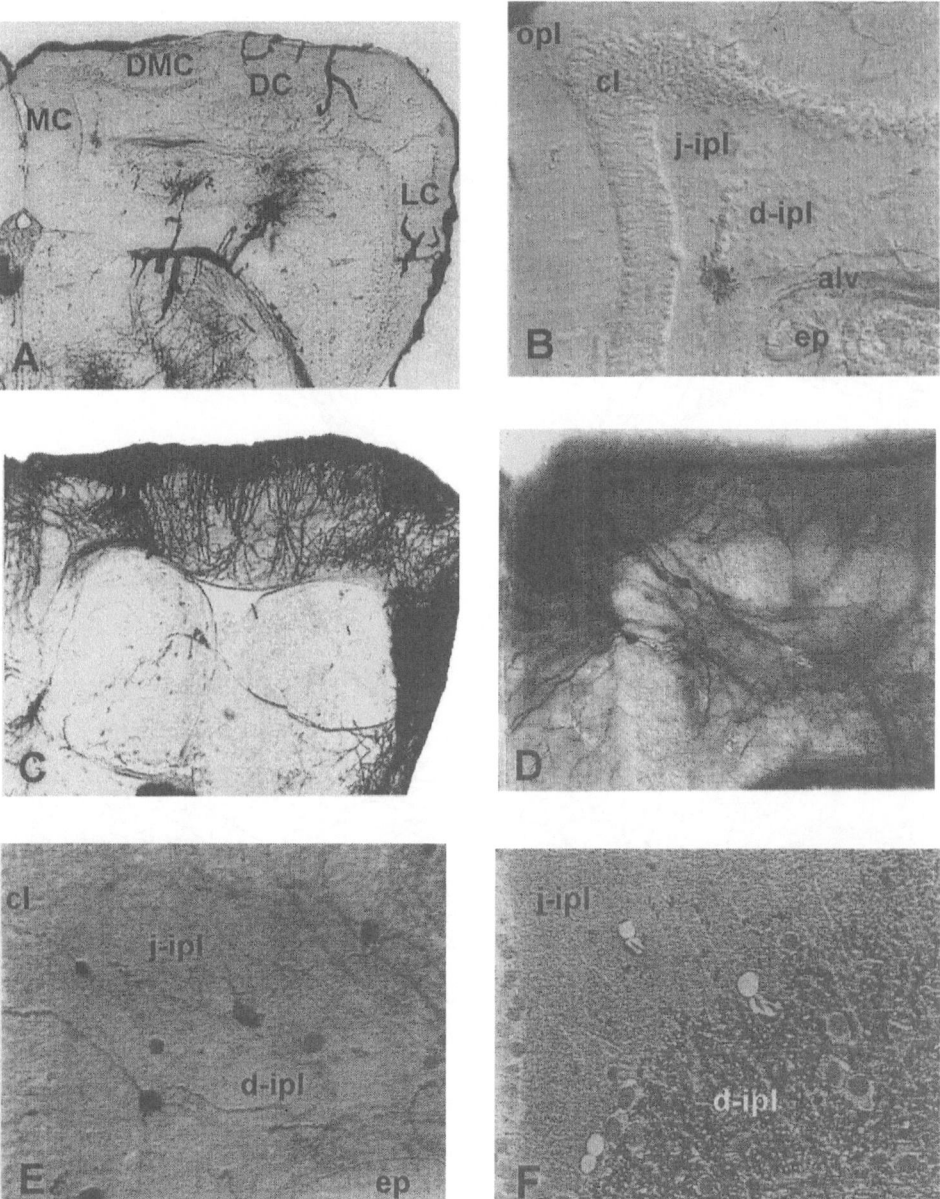

Figure 1.- A. Transversal section to a Golgi-impregnated brain in which cerebral cortex is almost completely devoid of impregnated cells. Cerebral cortical areas: **MC**, medial cortex; **DMC**, dorsomedial cortex; **DC**, dorsal cortex; **LC**, lateral cortex. **B.** Trilaminar arrangement of the medial cortex at a higher magnification. **C.** Typical aspect of the cerebral cortex of a well Golgi-impregnated brain. **D.** Detail of the medial cortex showing some principal projection neurons in the cell layer. **E.** Parvalbumin-immunostaining. **F.** Timm staining for vesicular-zinc detection, (**opl**, outer plexiform layer; **cl**, cell layer; **j-ipl**, juxtasomatic or basal dendritic zone of the inner plexiform layer; **d-ipl**, deep zone of the inner plexiform layer; **alv**, alveus; **ep**, ependyma).

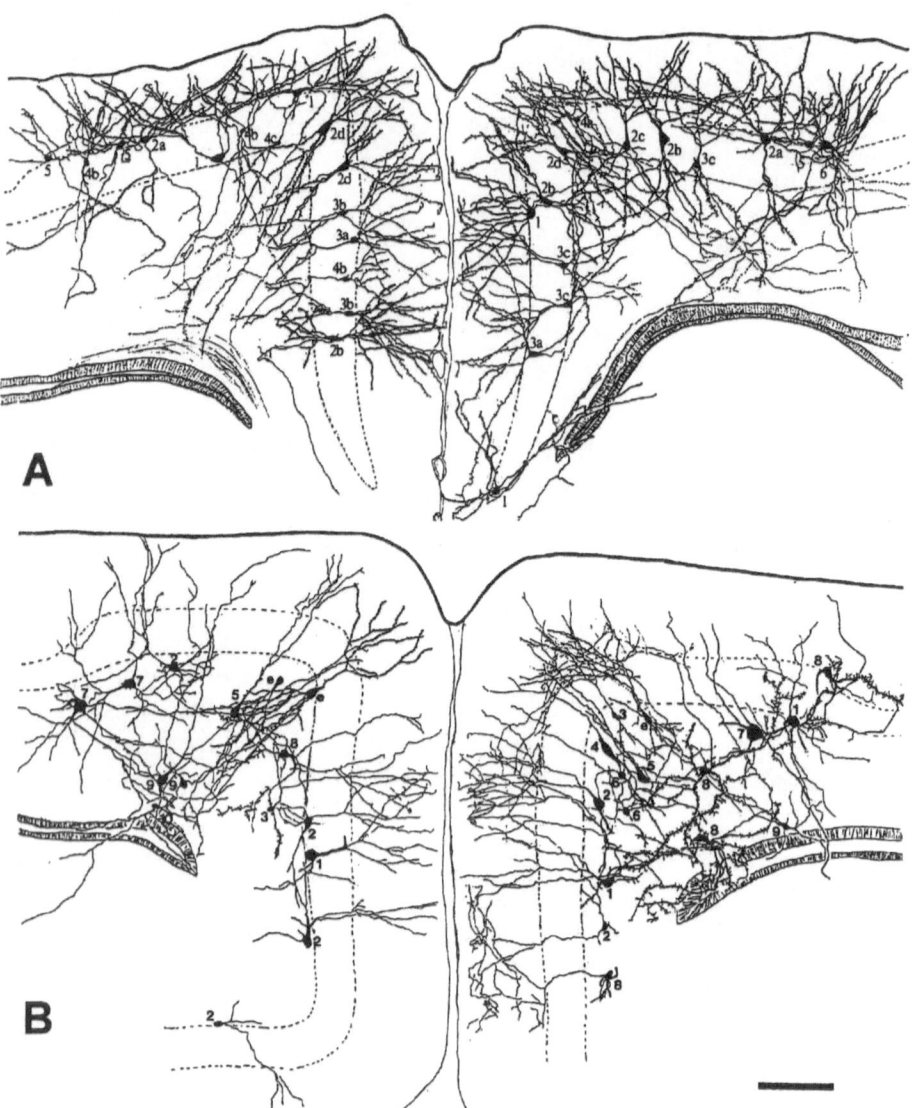

Figure 2.- A.Line drawing composition illustrating the different types/subtypes of principal projection neurons found in the medial cortex cell layer. Numerals 1-6 coincide with the ordinal names p2-p6 shown in Table 1. **B.** Idem, different interneuronal types found in the inner plexiform layer of the lizard medial cortex; some ectopic projection neurons located in the juxtasomatic zone of the inner plexiform layer are also represented (**e**) Numerals 1-10 coincide with those ordinal names i1-i10 (jip-dip layers) shown in Table 1. (Scale bar - 100 microns)

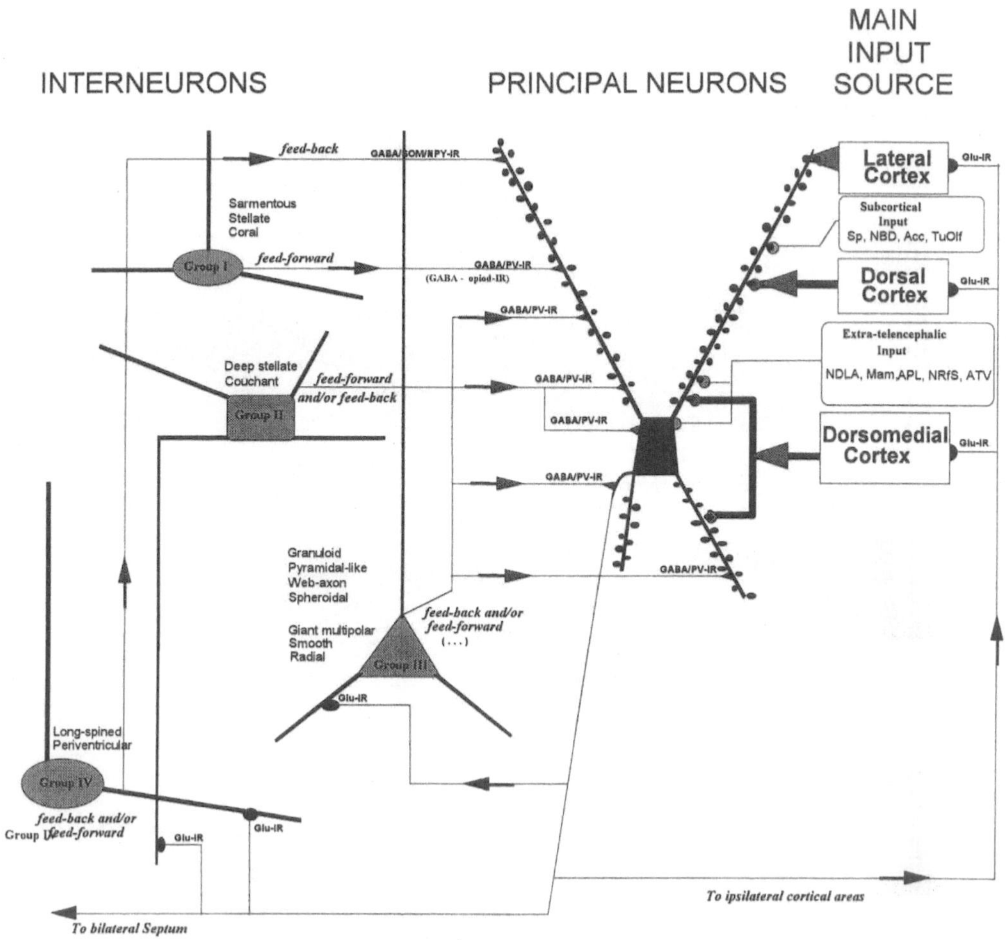

MAIN INPUT SOURCE

INTERNEURONS　　　　　**PRINCIPAL NEURONS**

feed-back　　GABA/SOM/NPY-IR

Sarmentous
Stellate
Coral

Group I　feed-forward　　　　GABA/PV-IR
(GABA - opiod-IR)

GABA/PV-IR

GABA/PV-IR

Deep stellate
Couchant　feed-forward
and/or feed-back

Group II

GABA/PV-IR

GABA/PV-IR

GABA/PV-IR

Granuloid
Pyramidal-like
Web-axon
Spheroidal

Giant multipolar
Smooth
Radial　　　feed-back and/or
feed-forward
(...)

Group III

GABA/PV-IR

Glu-IR

Long-spined
Periventricular

Group IV
feed-back and/or
Group I feed-forward

Glu-IR　Glu-IR

Lateral Cortex　Glu-IR

Subcortical Input
Sp, NBD, Acc, TuOlf

Dorsal Cortex　Glu-IR

Extra-telencephalic Input
NDLA, Mam, APL, NRfS, ATV

Dorsomedial Cortex　Glu-IR

To ipsilateral cortical areas

To bilateral Septum

Figure 3.- Scheme on the medial cortex neuronal circuitry. **Acc:** nucleus accumbens; **APL:** lateral preoptic area; **ATV:** ventral tegmental area; **NBD:** Broca diagonal band nucleus; **NDLA:** dorsolateral anterior thalamic nucleus; **NRfS:** superior raphe nucleus; **Mam:** mamilary nuclei; **Sp:** septum; **TuOlf:** tuberculum olfactorium. **GABA:** gamma amino butyric acid; **GABA/SOM/NPY-IR:** GABA, somatostatin and neuropeptide Y immunoreactive cells and boutons; **GABA/PV-IR:** GABA and parvalbumin immmunoreactive; **Glu-IR:** glutamate immunoreactive.

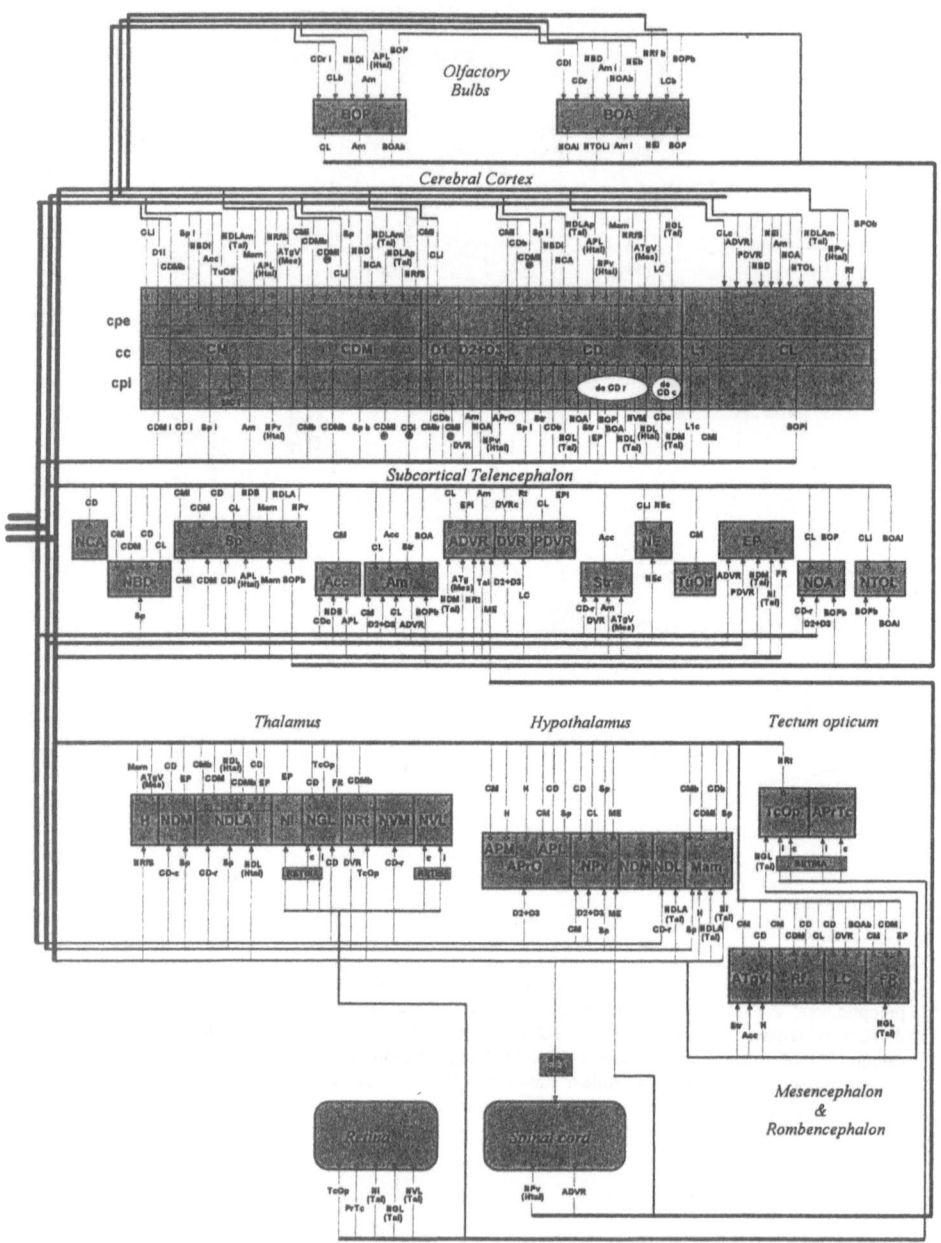

71

Figure 4.- Schematic wiring of the main axonal connections of the reptilian cerebral cortex and related structures. Small circles represent cell somata and arrows signal axonal projections. Colour codes main systems: **brown**: olfactory bulb connections, **red**: cerebral cortex connections, **pink**: subcortical connections, **blue**: extratelencephalic connections (thalamus, hypothalamus, tectum opticum and brain stem), **lihgt blue**: spinal cord connections, **green**: retinal connections.
Abbreviations: **Acc**, nucleus accumbens; **ADVR**, anterior dorso-ventricular ridge; **Am**, amygdaloid complex; **APL**, lateral preoptic area; **APM**, medial preoptic area ; **APrO**, preoptic area; **APrTc**, pretectal area; **ATgV**, ventral tegmental area; **b**, bilateral projection; **BOA**, accesory olfactory bulb ; **BOP**, principal olfactory bulb; **c**, contralateral projection; **cc**, cell layer; **CD**, dorsal cortex; **CDM**, dorsomedial cortex; **CL**, lateral cortex; **CM**, medial cortex; **cpe**, outer plexiform layer; **cpi**, inner plexiform layer; **D1**, medial subregion of the dorsal cortex; **D2+D3**, intermediate and lateral subregions of the dorsal cortex; **DVR**, dorsal ventricular ridge; **EP**, pallial thickening; **FR**, reticular formation; **H**, habenular nucleus; **i**, ipsilateral projection; **L1**, medial subregion of the lateral cortex; **LC**, locus coeruleus; **m**, pars magnocellularis; **Mam**, mamilary bodies; **Mes**, mesencephalon; **NBD**, Broca diagonal band nucleus; **NCA**, anterior commissure nucleus; **NDL**, dorsolateral thalamic nucleus; **NDLA**, anterior dorsolateral thalamic nucleus; **NDM(Htal)**, dorsomedial hypothalamic nucleus; **NDM(Tal)**, dorsomedial thalamic nucleus; **NE**, nucleus sphericus; **NGL**, lateral geniculate nucleus; **NI**, nucleus intercalatus; **NOA**, anterior olfactory nucleus; **NPv**, periventricular hypothalamic nucleus; **NRt**, nucleus rotundus; **NTOL**, lateral olfactory tract nucleus; **NVL**, ventrolateral thalamic nucleus; **NVM**, ventromedial thalamic nucleus; **p**, pars parvicellularis; **PDVR**, posterior dorsoventricular ridge; **Rf**, raphe nuclei; **Sp**, septum; **Str**, striatum; **TcOp**, tectum ᵔpticum; **TuOlf**, olfatory tuberculum.

Interactions Between Environmental and Hormonal Oscillations Induce Plastic Changes in a Simple Neuroendocrine Transducer

R. Alonso[1,2], I. López-Coviella[1,2], F. Hernández-Díaz[1], P. Abreu[1,2], E. Salido[2,3], and L. Tabares[4]

[1]Laboratory of Cellular Neurobiology, Department of Physiology and [2]Research Unit, and [3]Molecular Laboratory, University of La Laguna School of Medicine; [4]Department of Medical Physiology and Biophysics, University of Sevilla School of Medicine.

Abstract

Steroid hormones may affect, simultaneously, a wide variety of neuronal targets and influence the way neural networks interact and the way the brain reacts to the environment. Some of the neuronal effects of steroid hormones may be very fast and affect specific membrane conductances or second messenger cascades, while others may last a long time and exert profound influences on gene expression. The cross-talk between these forms of action may be crucial for the regulation of the way the brain develops and differentiates, changes with age, or remodels its synaptic circuitry during life. Here, we present some evidence of specific molecular changes brought about by gonadal steroids on a simple neuroendocrine transducer, the pineal gland.

Introduction

Analysis of the interactions between the endocrine and nervous system has greatly contributed to the understanding of molecular mechanisms involved in cell-to-cell communication. Central and peripheral nerve cells control the synthesis and release of different hormones along the endocrine system, linking environmental, behavioral and experience-dependent changes, as well as emotional states with overall body function. Conversely, peripheral circulating hormones feedback on the brain throughout the life cycle, and exert powerful effects on neural architecture and synaptic function (Fink, 1994; García-Segura et al., 1994; Naftolin et al., 1996), and induce significant modifications in mood, mental state, and memory (Fink et al., 1996). Adrenal steroids play significant roles in waking cognitive activity, probably through influencing the magnitude of long-term potentiation (LTP) in the hipoccampus, and mediate plastic changes induced by aging or stressful experiences (McEwen, 1996; Fuxe et al., 1996; de Kloet et al., 1996). Gonadal hormones influence neural plasticity in two ways: first, during critical periods of development, sex steroids promote cellular and molecular events which will determine sexual differentiation of the brain (Arnold and Gorski, 1984; García-Segura et al., 1994); second, during adult life, oscillations in their circulating levels affect a wide variety of neuronal phenomena, ranging from cyclic remodeling of synaptic circuitry to transynaptic modulation of neurotransmission (for review see   Alonso and López-Coviella. 1997).

Steroid hormone actions on neural plasticity

The molecular mechanisms used by steroid hormones to affect neural architecture and synaptic function are mediated by, at least, two types of receptors (Fig 1 a, b). Delayed, long-term to permanent, organizational steroid effects begin with the activation of intracellular receptors that can act as ligand-dependent transcription factors, and hence regulate neuronal gene expression (Beato, 1989; Truss and Beato. 1993). These effects present a wide range of latencies (from minutes to days), and may involve transcriptional modification of neurotransmitter synthesis and release. as well as long-lasting changes in postsynaptic receptor sensitivity. Some of the genomic effects of steroid hormones are not necessarily exerted directly on target neurons, but instead are mediated through actions on glial cells. which. in turn, produce certain neurotrophic factors that may regulate neuronal growth. survival. and plasticity (Singh et al., 1995).

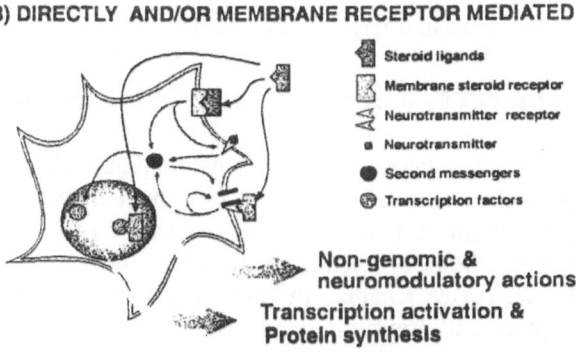

A) INTRACELULAR STEROID RECEPTOR MEDIATED

- Steroid ligands
- Intracellular steroid receptor
- Neurotransmitter receptor
- Neurotransmitter
- Steroid converting enzyme

Transcription activation & Protein synthesis

B) DIRECTLY AND/OR MEMBRANE RECEPTOR MEDIATED

- Steroid ligands
- Membrane steroid receptor
- Neurotransmitter receptor
- Neurotransmitter
- Second messengers
- Transcription factors

Non-genomic & neuromodulatory actions

Transcription activation & Protein synthesis

Fig. 1: Simplified molecular mechanisms of steroid action on nerve cells

In addition to genomic mechanisms, short- and intermediate-term effects of steroids may involve interaction with membrane receptors and direct modulation of ionic channel activity, coupling to guanine nucleotide-binding regulatory proteins (G-proteins), or second messenger systems (Orchinik and McEwen, 1994; Baulieu and Robel, 1995). Rapid and reversible effects of steroid hormones on neuronal excitability have been studied in several in vivo and in vitro experimental preparations involving synaptic actions mediated by excitatory, as well as inhibitory neurotransmitter aminoacids. Thus, several progesterone metabolites are able to directly modulate a chloride conductance by interacting with $GABA_A$ receptors in a stereospecific fashion (Lambert et al., 1996). On the other hand, 17β-estradiol, when applied locally at physiological concentrations, potentiates the response of glutamate receptors to specific agonists, or to the Schaffer collateral stimulation in CA1 hippocampal neurons via activation of second messenger cascades (Gu and Moss, 1996).

An emerging concept in understanding neuronal responses to steroid hormones is the interaction between their long-term and short-term effects. For example, the estradiol-induced potentiation of glutamate-mediated excitation of CA1 hippocampal neurons is further enhanced if cells are obtained from estrogen-primed animals, indicating that rapid, nongenomic actions of estradiol are influenced by long-term prior exposure to the hormone (Wong and Moss, 1992). These findings probably reflect cross-talk between genomic and non-genomic (often membrane-mediated) mechanisms operating within the same neuron, or even interacting with the same molecular targets. For example, estrogens are able to act on membrane receptors and activate the cyclic AMP (cAMP) cascade, thereby inducing a rapid phosphorilation of cAMP-response element-binding protein (CREB) in some brain regions (Gu et al., 1996; Zhou et al., 1996). The CREB protein is a nuclear transcription factor that, once phosphorilated, binds to cAMP-response element (CRE) on each DNA strands regulating neuronal gene expression (Sheng and Greenberg, 1990). Since CREB proteins may be convergent points for regulation of synaptic efficacy, which serve to couple short-term and long-term associative forms of synaptic plasticity (Carew, 1996; Dash et al., 1991; Emptage and Carew, 1993), the molecular substrates affected by steroid hormones on neurons may be the same as those involved in memory formation. This suggests that the study of gonadal and adrenal hormones actions on neuronal targets may constitute an useful experimental approach to investigate the molecular basis of neural information storing dynamics.

The mammalian pineal as a model of neuroendocrine integration

Neuroendocrine transduction at the mammalian pineal gland is an excellent model to study steroid effects on cellular signal transduction systems. This gland, located deep inside the brain in most mammals, is mainly controlled by a single presynaptic noradrenergic input acting on adrenoceptors present in the surface of pineal cells, named pinealocytes. Changes in light intensity and duration are detected by retinal photoreceptors and transmitted, via the retinohypothalamic projection (RHP), to the suprachiasmatic nucleus (SCN) in the hypothalamus, considered a major circadian rhythm generator or biological clock, with intrinsic oscillatory activity that is entrained to a 24-hour rhythm by light-dark cycles. From the SCN, circadian activity is further transmitted to the pineal gland through a multisynapuc pathway that relays in the paraventricular nucleus (PVN) of the hypothalamus, the intermediolateral cell column (IML) of the spinal cord, and the superior cervical ganglion (SCG) (Klein, 1993; Fig. 2), from where postganglionic sympathetic nerve terminals reach the pineal and release the neurotransmitter norepinephrine (NE). The rate of pineal noradrenergic transmission is accelerated during the night, thereby causing a nocturnal rise in the production of the pineal hormone melatonin (Reiter, 1991).

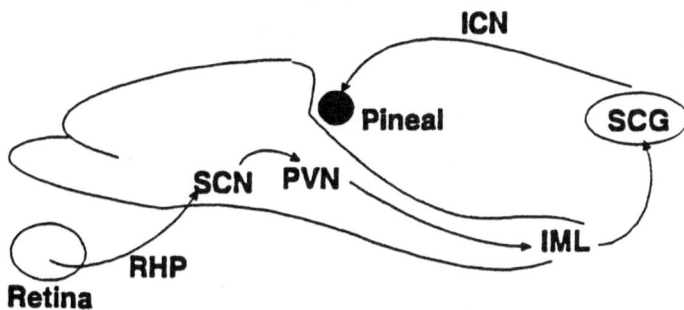

Fig. 2: Neural circuitry conecting retinal photoreceptors and pineal gland in mammals

In the rat, NE activates both β- and α_1-adrenergic receptors in the pinealocyte membrane, and initiates a series of events leading to stimulation of melatonin synthesis (Sugden, 1989) (Fig. 3). β-adrenoceptor activation elevates adenylate cyclase activity through a stimulatory Gs-protein, and induces a rapid increase in cAMP levels. Concurrent activation of α_1-adrenoceptors markedly potentiates accumulation of cAMP via a mechanism involving activation of phospholipase C (PLC), hydrolysis of phosphatidylinositol (PI), elevation of intracellular Ca^{2-}, and translocation and activation of protein kinase C (PKC) (Yu et al., 1993). Acting as a second messenger, intracellular cAMP activates a cAMP-dependent protein kinase (PKA), promotes mRNA transcription, and causes a transient rise in the activity of serotonin N-acetyltransferase (NAT) (Fajardo et al. 1992), the rate-limiting enzyme in melatonin production. The pineal gland, as other neuroendocrine tissues, contains an isoform of cAMP-response element modulator gene (CREM), named inducible cAMP early repressor (ICER), which acts as potent repressor of cAMP-induced gene transcription (Stehele et al., 1993). Interestingly, the expression of this gene displays circadian rhythmicity and is also under adrenergic control, which suggests that it could be responsible for the decline in NAT activity at the end of night period, ensuing several hours of adrenergic stimulation. All in all, the operation of this biological transduction system, with an output signal of elevated circulating levels of melatonin during night hours, probably encodes information about the status of the endogenous clock in terms of phase and amplitude (Arendt, 1995).

With the above system in mind, we and others have shown that changes in circulating levels of gonadal steroids also affect pineal melatonin synthesis and release (Alonso et al., 1993, 1996). In female rats, cyclic elevations in circulating levels of ovarian hormones, responsible for the positive feedback cascade that triggers a surge of luteinizing hormone releasing hormone (LHRH) and induces ovulation (Fink, 1988, 1994), also modulate pineal activity and reduce melatonin synthesis and release during the night of proestrus (Ozaki et al., 1978 ; Moujir et al., 1990). These steroidal effects are exerted both at the level of the postganglionic neurons sending noradrenergic fibers to the pineal, as well as on the sensitivity of adrenergic receptors in the pinealocytes (Alonso et al., 1993, 1995, 1996). Since, in the rat, melatonin has been shown to exert an inhibitory effect on LHRH release (Vanacek, 1991), the ovarian hormone-induced inhibition of pineal melatonin production may constitute part of an integrated mechanism aimed at lowering circulating melatonin levels at the time of ovulation.

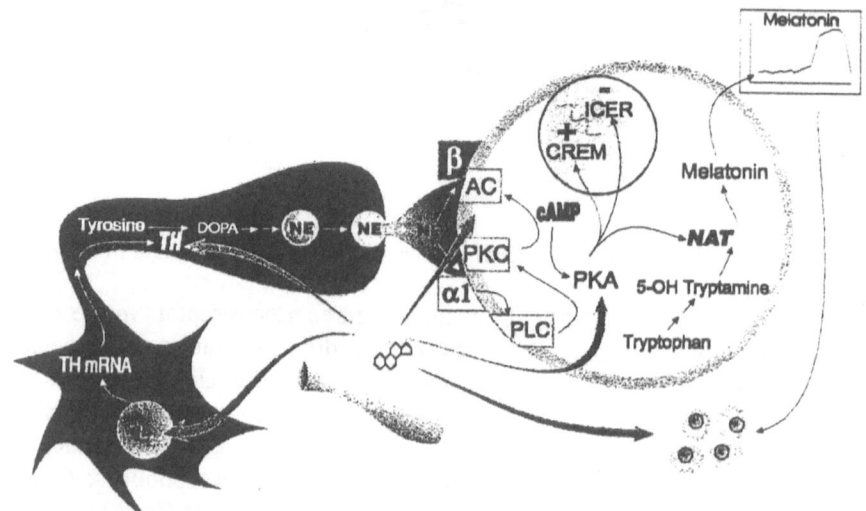

Fig. 3: Intracellular signals involved in regulation of pineal melatonin synthesis

Estradiol modulation of adrenoceptor sensitivity in rat pinealocytes

Treatment of ovariectomized rats with estradiol physiological doses (E_2) and/or Progesterone (Pg) blocks the elevation of pineal melatonin production induced by isoproterenol, a β-adrenergic agonist (Moujir et al., 1990). Therefore, we have examined the effects of adrenoceptor activation on cAMP accumulation in isolated pinealocytes from female rats under various states of hormonal manipulations. It was found that the expected β-adrenergic-induced increase of cAMP intracellular levels was significantly enhanced in pinealocytes from either ovariectomized rats, or from animals previously treated with specific ovarian hormone blockers. In addition, the potentiation effect of concurrent activation of β- and $α_1$-adrenergic receptors was further increased in rats lacking circulating ovarian hormones or pretreated with antihormones (Alonso et al., 1995).

To determine whether or not ovarian steroids directly modulate pinealocyte response to adrenoceptor activation, we have also studied the changes in second messenger levels in cultured pineal cells exposed to physiological concentrations of 17β-estradiol (Hernández-Díaz et al., unpublished). Immature (21 day-old) female rats were ovariectomized and left undisturbed for two days, time at which they were sacrificed thirty minutes before the onset of darkness (i.e., seeking a high population of active adrenoceptors), and had their pineals quickly removed and processed as described elsewhere (Alonso et al., 1995, 1996). Dispersed pinealocytes were incubated and exposed to 17β-estradiol (0.1 nM) or vehicle for 48 hours, prior to being stimulated for 15 minutes with adrenergic agonists (1 μM isoproterenol, ISO; or phenylephrine, PE). Our data show that the increase in cAMP or inositol triphosphate (IP_3) intracellular levels following activation of both β- and α_1-adrenergic receptors were reduced in estradiol-treated pinealocytes (Fig. 4).

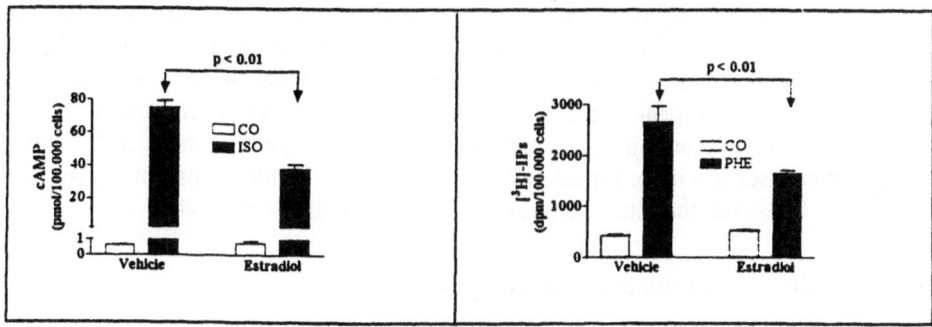

Fig. 4: Effect of exposure to 17β-estradiol on adrenergic-induced cAMP and IP_3 accumulation in pinealocytes from ovariectomized rats

Furthermore, we have also examined the changes in intracellular Ca^{2-} levels in response to α_1-adrenoceptor activation in single pineal cells under similar conditions, using the membrane permeable form of the Ca^{2+} indicator dye Fura-2 (Fura-2/AM; Marín et al., 1996). The application of PE to pinealocytes in culture elicited transitory Ca^{2+} signals consisting of a single (or double) spike(s) followed by a plateau (Fig. 5a). However, when pineal cells had been previously exposed to physiological concentrations of 17β-estradiol, α_1-adrenergic-induced Ca^{2-} spikes were significantly reduced, or even completely blocked (Fig. 5b).

These results indicate that exposure of female rat pinealocytes to physiological concentrations of estradiol (like those occurring during the proestrous stage of the rat estrous cycle), depresses their response to adrenoceptor activation. In the absence of plasma estradiol oscillations (i.e., at the basal estrogen circulating levels outside the proestrus), pinealocyte intracellular Ca^{2+} and cAMP concentrations rise by the concurrent activation of both α_1- and β-adrenoceptors (Sugden, 1989). The increase in cAMP appears to act at three different levels : a) inducing the transcription of NAT or some still unknown NAT regulator; b) stimulating NAT translation; and c) maintaining NAT in an active form. Estradiol could attenuate each one of the three

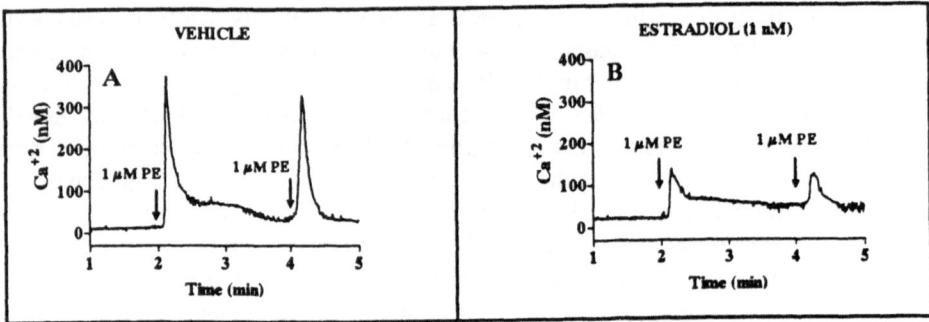

Fig. 5: Effect of exposure to 17β-estradiol on adrenergic-induced intracellular Ca²⁺ changes in pinealocytes from ovariectomized rats

responses by decreasing the accumulation of intracellular cAMP. Even though the minimal time required for estradiol effect. as well as its duration. remains to be determined. it appears to be an intermediate-term phenomenon. Since. in the rat. estradiol circulating levels peak and decline several hours before the nocturnal increase in the rate of pineal adrenergic transmission (Smith et al.. 1975 ; Abreu et al.. 1991). the decreased sensitivity of pineal cells to adrenergic stimulation probably lasts several hours following the return to basal levels of circulating estradiol.

Concluding remarks and future prospects

At the synaptic level. plasticity is expressed by changes in the amount of transmitter release or in the sensitivity of postsynaptic receptors for the transmitter. As a generalization. long-lasting plastic changes appear to require the activation of intracellular second messengers. changes in the phosphorilation state of certain proteins. or the transcription of new genes encoding the molecules involved in storage of neural information (Dash et al.. 1991 ; Emptage and Carew. 1993). However. while short- and long-term enhancement of synaptic efficacy is starting to be understood at the molecular level (Carew. 1996), the molecular basis of different forms of synaptic depression remain unknown. Therefore. the estradiol-induced inhibition of pinealocyte responsiveness to adrenergic stimulation may constitute an appropriate model to investigate the molecular interactions underneath cellular depression.

What may be the molecular targets for estradiol action on pinealocytes? From the various mechanisms mentioned here. estradiol could affect adrenoceptor sensitivity by. at least. two types of mechanism: a) interacting with the cell membrane and thus directly altering receptor responses; or b) activating nuclear estrogen receptors and. thereby. influencing the expression of some still unknown proteins which. in turn. could modify the phosphorilation state of other factors. In addition. since ICER is also under adrenergic regulation in the pineal cell (Stehle et al.. 1993). estrogens could also modulate pinealocyte adrenergic-induced cascades at other levels. as it is the case in other brain regions (Gu et al.. 1996; Zhou et al.. 1996).

Neurotransmitter-steroid hormone interactions, as those presented here for the pineal, probably represent a basic example of transynaptic steroid regulation present in other neuronal systems. It should be emphasized that, because of the wide distribution of putative steroid targets, sustained elevations of their circulating levels are able to act, simultaneously, on a wide variety of neuronal elements. This constitutes an effective mechanism by which a circulating molecule may be utilized in the coordination of different neural networks involved in the expression of a particular behavior, or in the regulation of several related neuroendocrine adaptations (Fink, 1994 ; Fink et al., 1996). Furthermore, steroids may exert both inhibitory and stimulatory effects on different neuronal systems, acting through several molecular mechanisms and having a wide range of latencies and duration. It should be pointed out that some of the brain regions that are susceptible of steroidal modulation are among those involved in cognitive functions and memory storing (Fink et al., 1996 ; McEwen, 1996). This perhaps may explain why some neurological diseases associated with impaired mental faculties during aging are more prevalent in females than in age-matched males (Bachman et al., 1992), at a time when age-related changes in the pattern of sex hormones secretion occur. Therefore, the use of simple and accessible biological models for hormonally-induced neural plasticity may constitute a worthwhile approach to the understanding the molecular bases of certain brain diseases.

Acknowledgments

Supported in part by DGCYT PM94-0590 to R.A., ULL 227-068/93 to P.A., and GAC 92/069 to I.L.-C. F.H.-D. held a predoctoral fellowship from GAC.

Bibliography

Alonso, R., Abreu, P., Fajardo, N. (1993). Steroid influences on pineal melatonin production. In Melatonin. Biosynthesis, Physiological Effects, and Clinical Applications, H.-S. Yu & R.J. Reiter (Eds), CRC Press, Boca Raton, pp. 73-105.

Alonso, R., Abreu, P., Fajardo, N., Hernández-Díaz, F., Díaz-Cruz, A., Hernández, G., Sánchez-Criado, J- (1995). Ovarian hormones regulate α_1- and β-adrenoceptor interactions in female rat pinealocytes. NeuroReport 6 : 345-348.

Alonso, R., Abreu, P., López-Coviella, I., Hernández, G., Fajardo, N., Hernández-Díaz, F., Díaz-Cruz, A., Hernández, A. (1996). Gonadal steroid modulation of neuroendocrine transduction : a transynaptic view. Cell. Mol. Neurobiol. 16 : 357-382.

Arendt, J. (1995). Melatonin and the Mammalian Pineal Gland, Chapman & Hall, London.

Alonso, R., López-Coviella, I. (1997). Gonadal steroid modulation of neural plasticity : the cellular basis of neuroendocrinology. Neurochem. Res. In press.

Arnold, A.P., Gorski, R.A. (1984). Gonadal steroid induction of structural sex differences in the central nervous system. Ann. Rev. Neurosci. 7 : 413-442.

Bachman, D.L., Wolf, P.A., Linn, R., Knoefel, J.E., Cobb, J., Belanger, A., D'Agostino, R.B., White, L.R. (1992). Prevalence of dementia and probably senile dementia of the Alzheimer type in the Framingham study. Neurology 42 : 115-119.

Baulieu, E.-E., Robel, P. (1995). Non-genomic mechanisms of action of steroid hormones. In Non-reproductive actions of sex steroids (Ciba Symposium 191), John Wiley & Sons, Chichester, pp. 24-42.

Beato, M. (1989). Gene regulation by steroid hormones. Cell 56 : 335-344.

Carew, T.J. (1996). Molecular enhancement of memory formation. Neuron 16 : 5-8.

Dash, P.K., Karl, K.A., Colicos, M.A., Prywes, R., Kandel, E.R. (1991). CAMP response element-binding protein is activated by Ca^{2+}/calmodulin- as well as cAMP-dependent protein kinase. Proc. Natl. Acad. Sci. USA 88 : 5061-5065.

De Kloet, E.R., Rots, N.Y., Cools, A.R. (1996). Brain-corticosteroid dialog : slow and persistent. Cell. Mol. Neurobiol. 16 : 345-356.

Emptage, N.J., Carew, T.J. (1993). Long-term synaptic facilitation in the absence of short-term facilitation in Aplysia neurons. Science 262 : 253-256.

Fajardo, N., Abreu, P. Alonso, R. (1992). Determination and kinetic properties of serotonin N-acetyltransferase in bovine pineal gland by using HPLC with fluorimetric detection. J. Pineal Res. 13 : 80-84.

Fink, G. (1994). Molecular principles from neuroendocrine models : steroid control of central neurotransmission. In Progress in Brain Research-Neuroscience : From the Molecular to the Cognitive, F Bloom (Ed), Elsevier, Amsterdam, pp. 139-147.

Fink, G., Sumner, B.E.H., Rosie, R., Grace, O., Quinn, J.P. (1996). Estrogen control of central neurotransmission : effect on mood, mental state, and memory. Cell. Mol. Neurobiol. 16 : 325-344.

Fuxe K., Díaz R., Cintra, A., Bhatnagar, M., Tinner, B., Gustafsson, J-K., Agnati, L.F. (1996). On the role of glucocorticoid receptors in brain plasticity. Cell. Mol. Neurobiol. 16 : 239-258.

García-Segura, L.M., Chowen, J.A., Párduz, A., Naftolin, N.F. (1994). Gonadal hormones as promoters of structural synaptic plasticity : cellular mechanisms. Progr. Neurobiol. 44 : 279-307.

Gu, Q., Moss, R.L. (1996). 17β-estradiol potentiates kainate-induced currents via activation of the cAMP cascade. J.Neurosci. 16 : 3620-3629.

Gu G., Rojo, A.A., Zee, M.C., Yu, J., Simerly, R.B. (1996). Hormonal regulation of CREB phosphorilation in the anteroventral periventricular nucleus. J. Neurosci. 16 : 3035-3044.

Klein, D.C. (1993). The mammalian melatonin rhythm-generating system. In Light and Biological Rhythms in Man, Neuroscience, L Weterberg (Ed), Pergamon Press, Oxford, pp. 55-72.

Lambert, J.J., Belelli, D., Hill-Venning, C., Callachan, H., Peters, J.A. (1996). Neurosteroid modulation of native and recombinant GABA$_A$ receptors. Cell. Mol. Neurobiol. 16 : 155-174.

Marín, A., Ureña, J., Tabares, L. (1996). Intracellular calcium release mediated by noradrenaline and acetylcholine in mammalian pineal cells. J. Pineal Res. 21 : 15-28.

McEwen, B.S. (1996). Gonadal and adrenal steroids regulate neurochemical and structural plasticity of the hippocampus via cellular mechanisms involving NMDA receptors. Cell. Mol. Neurobiol. 16 : 103-116.

Moujir, F., Bordón, R., Santana, C., Hernández, G., Abreu, P., Alonso, R. (1990). Ovarian steroids block the isoproterenol-induced elevation of pineal melatonin production in female rats. Neurosci. Lett. 119 : 12-14.

Naftolin, F., Leranth, C., Horvath, T.L., García-Segura, L.M. (1996). Potential neuronal mechanisms of estrogen actions in synaptogenesis and synaptic plasticity. Cell. Mol. Neurobiol. 16 : 213-224.

Orchinik, M., McEwen, B.S. (1995). Rapid steroid actions on the brain : a critique of genomic and non-genomic mechanisms. In M. Wehling (Ed), CRC Press, Boca Raton, pp 77-168.

Ozaki, Y., Wurtman, R.J., Alonso, R., Lynch, H.J. (1978). Melatonin secretion decreases during the proestrous stage of the rat estrous cycle. Proc. Natl. Acad. Sci. USA 75 : 531-534.

Reiter, R.J. Pineal gland. Interface between the photoperiodic environment and the endocrine system. TEM 2 : 13-29, 1991.

Sheng, M., Greenberg, M.E. (1990). The regulation and function of c-fos and other immediate early genes in the nervous system. Neuron 4 : 477-485.

Singh, M., Meyer, E.M., Simpkins, J.W. (1995). The effects of ovariectomy and estradiol replacement on brain-derived neurotrophic factor messenger ribonucleic acid expression in cortical and hippocampal brain regions of female Sprague-Dawley rats. Endocrinology 136 : 2320-2324.

Smith, M.S., Freeman, M.E., Neill, J.D. (1975). The control of progesterone secretion during the estrous cycle and early pseudopregnancy in the rat : prolactin, and gonadotrophin and steroid levels associated with rescue of the corpus luteum of pseudopregnancy. Endocrinology 96 : 219-226.

Stehle, J.H., Foulkes, N., Molina, C., Simonneaux, U., Pevet, P., Sassone-Corsi, P. Adrenergic signals direct rhythmic expression of transcriptional repressor CREM in the pineal gland. Nature 365 : 314-320, 1993.

Sugden, D. Melatonin biosynthesis in the mammalian pineal gland. Experientia 45 : 922-932, 1989.

Truss, M., Beato, B. (1993). Steroid hormone receptors : interaction with deoxyribonucleic acid and transcription factors. Endocr. Rev. 14 : 459-479.

Wong, M., Moss, R.L. (1992). Long-term and short-term electrophysiological effects of estrogen on the synaptic properties of hippocampal CA1 neurons. J. Neurosci. 12 : 3217-3225.

Yu, L., Schaad, N.C., Klein, D.C. (1993). Calcium potentiates cyclic AMP stimulation of pineal arylalkylamine N-acetyltransferase. J. Neurochem. 60 : 1436-1443, 1993.

Zhou Y., Watters, J.J., Dorsa, D.M. (1996). Estrogen rapidly induces the phosphorilation of the cAMP response element binding protein in rat brain. Endocrinology 137 : 2163-2166.

Current Source Density Analysis as a Tool to Constrain the Parameter Space in Hippocampal CA1 Neuron Models

Pablo Varona[1], José Manuel Ibarz[2], Juan Alberto Sigüenza[1], Oscar Herreras[2]

[1] Instituto de Ingeniería del Conocimiento, Dpto. de Ingeniería Informática.
Universidad Autónoma de Madrid, 28049 Madrid, Spain.
E-mail: varona@irene.iic.uam.es
[2] Dept. Investigación, Hospital Ramón y Cajal, 28034 Madrid, Spain.
E-mail: oscar.herreras@hrc.es

Abstract. We propose the use of Current Source Density (CSD) computer simulations as a useful technique to constrain the parameter space in compartmental models of hippocampal CA1 neurons. These simulations allow a direct comparison with physiological data from current source density analysis and straightforward testing of hypothesis.

1 Introduction

There are several experimental approaches to study the information processing in real neurons, and specifically the distribution of electrical properties along their morphology. Intracellular recordings can yield precise data of the electrical behavior at the region where the electrode is placed. Patch clamp techniques are used to place several electrodes in a single neuron in order to simultaneously record its activity at different locations. However, these techniques are at the moment open to artefacts and have not yet the desired spatial resolution. A method capable of yielding precise information about the distribution of electrical activity in neurons is current source-density (CSD) analysis [1]. CSD analysis calculates the sources and sinks generated at the extracellular medium by membrane currents from a population of neurons. In suitable structures such as the laminated CA1 region, this analysis provides detailed profiles of the electrical activity along the entire length of individual neuronal elements [2].

On the other hand, several computer models have been proposed to help the study of the information processing carried out by real neurons. Compartmental models [3] with enough resolution can be used to test hypothesis related to the generation and propagation of action potentials within a neuron. The computer implementation of these models gives insight into the role of different ionic channels and the effect of their distribution along the neuron morphology. However, it is a drawback for these models the lack of physiological data related to channel ion distribution and conductance densities. In general, there is not enough reliable experimental data to accurately delimit their huge parameter space. It is very common to find models that support opposite hypothesis differing only in the unknown values of the parameters.

Current source density analysis can also be simulated in computer models. This is a most useful technique since the neuron model parameters can be tuned up by directly comparing the results of experimental CSD and simulated CSD analysis. In this article we describe how this can be done for hippocampal CA1 neuron models.

2 Single neuron model

Each neuron in the population used for the CSD simulation was a 22-compartment unit. This simple neuron consisted of 1 soma, 4 basal, 9 apical and 8 axonal compartments –see table 1–.

The value of the transmembrane potential is calculated according to the spatially discretized cable equation:

$$C_{m,l}\frac{dV_l}{dt} = \frac{(E_{m_l} - V_l)}{R_{m,l}} + \frac{(V_{l+1} - V_l)}{r_{l,l+1}} + \frac{(V_{l-1} - V_l)}{r_{l,l-1}} + i_{ext} - i_{achans,l} \qquad (1)$$

where $C_{m,l}$ is the membrane capacitance for the l compartment, V_l is the membrane potential, E_{m_l} is the equilibrium potential, $R_{m,l}$ is the membrane resistance, $r_{l,l+1}$ is the coupling resistance between the lth and the $l + 1$th compartments, i_{ext} is an external stimulus current and $i_{achans,l}$ is the current flowing through the active ion channels.

We used 7 active ionic currents: Na, Ca and five K currents: delayed rectifier (DR), A-type transient, short-duration Ca and voltage-dependent, long-duration Ca-dependent, and small persistent muscarinic. For the axonal active compartments only Na and K_{DR} channels where modeled.

The conductance gate variables for the active ion channels are calculated using a Hodking-Huxley type formalism:

$$i_{achan} = g_{achan} \prod_j x_j^{p_j}(V_l - E_{eq}) \qquad (2)$$

$$\frac{dx_j}{dt} = \alpha_{x_j}(V_l)(1 - x_j) - \beta_{x_j}(V_l)x_j \qquad (3)$$

where g_{achan} is the ion channel maximum conductance, x_j are the channel gate variables, p_j represents the number of identical gates for variable x_j, and E_{eq} is the equilibrium potential for the ionic species.

The equations of the gate rate variables for the soma and the dendritic compartments $\alpha_{x_j}, \beta_{x_j}$ used in this model are described in [4]. For the unmielinized axonal compartments we used the gate rate equations described in [5].

In this paper we propose the use of experimental and simulated CSD analysis to constrain the values and distribution of the maximum conductances (g_{achan}'s) for the compartmental neuron model. Three sets of conductance density distributions proposed in the literature were used for the simulated CSD analysis: set 1 corresponds to a model of passive dendrites [4]; set 2 reproduces a hot spot in the axon hillock and small active conductances in the soma and in the dendrites [6]; and set 3 is a gradient-like distribution of active conductances through the soma and apical dendrites. The values of the electrotonic parameters and Na densities used for the active compartments in the three simulations are summarized in table 1.

Compartment	R_M ($\Omega \cdot m^2$)	R_A ($\Omega \cdot m$)	C_M ($F \cdot m^{-2}$)	l (μm)	d (μm)	g_{Na_1} ($S \cdot m^{-2}$)	g_{Na_2} ($S \cdot m^{-2}$)	g_{Na_3} ($S \cdot m^{-2}$)
b4	1.46	2.0	0.0153	50	2.2	-	-	-
b3	1.46	2.0	0.0153	50	2.4	-	-	-
b2	1.46	2.0	0.0153	50	2.6	-	-	-
b1	1.46	2.0	0.0153	50	2.8	-	-	-
soma	3.00	2.0	0.0153	50	4.0	1800	40	485
a1	3.00	2.0	0.0075	50	3.2	-	40	485
a2	3.00	2.0	0.0075	50	3.0	-	40	323
a3	1.46	2.0	0.0075	50	2.8	-	40	242
a4	1.46	2.0	0.0153	50	2.7	-	40	194
a5	1.46	2.0	0.0153	50	2.6	-	40	162
a6	1.46	2.0	0.0153	50	2.5	-	40	138
a7	1.46	2.0	0.0153	50	2.5	-	-	-
a8	1.46	2.0	0.0153	50	2.5	-	-	-
a9	1.46	2.0	0.0153	50	2.5	-	-	-
axon hillock	3.00	2.0	0.0075	10	4.0	323	30000	485
is1	0.10	1.0	0.0075	37	2.0	1000	1000	1000
is2	0.10	1.0	0.0010	37	1.0	1000	1000	1000
m1	50.0	1.0	0.0010	75	1.0	-	-	-
m2	50.0	1.0	0.0010	75	1.0	-	-	-
r1	0.10	1.0	0.0075	20	2.0	1000	1000	1000
m3	50.0	1.0	0.0010	200	1.0	-	-	-
r2	0.10	1.0	0.0075	20	2.0	1000	1000	1000

Table 1: Electrotonic parameters and Na conductance densities for the three CA1 neuron models. Compartments: b, basal dendrite; a, apical dendrite; is, initial segment; m, mielinized; and r, node of Ranvier.

3 CSD simulation

Transmembrane currents generate field potentials at the extracellular medium that can be recorded using electrodes. The value of the total field potential at any point is the linear superposition of the fields generated at each membrane portion from every cell in the region. The current source and sink densities (a scalar average of the transmembrane currents per unit volume) can be recovered from the measured spatial distribution of field potentials.

In order to build a CSD simulation, a total of 6144 neurons were distributed in a 3d array (64x24x4). Somas were placed $16\mu m$ apart in all three directions. The lattice has the same size as the estimated region of stimulation for the physiological experiment (see below). An array of 14 recording points ($50\mu m$ apart) was located at the center of the cell lattice parallel to the dendrosomatic (z) axis. Each recording point calculated the instantaneous field potential at the position where it was placed according to:

$$\Phi(t) = \frac{1}{4\pi\sigma} \sum_{i=1}^{cells} \sum_{j=1}^{compts} \frac{i_{m(ij)}(t)}{r_{ij}} \tag{4}$$

where σ is the extracellular conductance, r_{ij} is the distance from the recording point to the ij compartment, and $i_{m(ij)}$ is the total transmembrane current at that compartment. Equation (4) is valid for point current sources distributed in a linear noncapacitive and isotropic medium.

The current density and extracellular field potential are related by:

$$\nabla\sigma \cdot \nabla\Phi = -I_{CSD} \tag{5}$$

In a laminated structure such as CA1 where the cell population is assumed to be homogeneous in two of the three dimensions and a large number of neurons are synchronously excited, the equipotential surfaces are parallel to the horizontal $x - y$ plane. This approximation is widely used in the literature [7]. Thus, the overall extracellular currents flow only in the z direction and (5) becomes:

$$I_{CSD}(t) = -\sigma \frac{d^2\Phi}{dz^2} \tag{6}$$

This magnitude represents the transmembrane current per unit volume and can be compared with the transmembrane current at each compartment of the model neurons for a correct interpretation of CSD analysis.

4 Experimental CSD

The experimental CSD analysis was performed in the hippocampal CA1 region of anesthetized rats. Simultaneous action potential discharge in pyramidal cells was evoked by antidromic stimulation of efferent pyramidal cell fibers in the alveus. A stimulating electrode of the concentric type was used, activating a strip of about 400 μm width.

A single recording micropipette was lowered in successive 50 μm steps from the alveus down to 450 μm beneath the cell body layer. Constancy of response was assured by monitoring the waveform recorded by another stationary electrode. Additional preparation and recording technical details are explained elsewhere [2]. CSD profiles were calculated using equation (6) with a σ of 300 $S \cdot cm^{-1}$.

5 Results

Alvear stimulation evoked a propagating negative potential and current sink at the level of pyramidal cell somata which evolved to positive potential and smaller current sources and sinks as they moved into the apical and basal dendritic regions. Figure 1 shows the laminar field potentials and associated CSD profiles obtained in a typical physiological experiment.

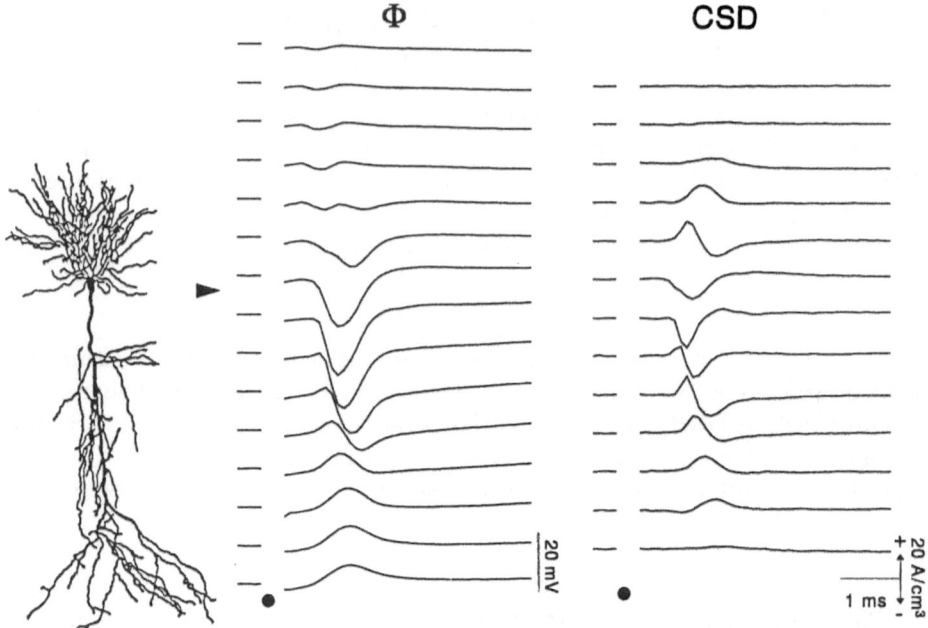

Fig. 1. Experimental laminar profiles of alvear-evoked antidromic field potentials (Φ) and the corresponding current source densities (**CSD**). Potentials were recorded at 50 μm intervals parallel to the dendrosomatic axis of the pyramidal cells. Profiles are displayed with a zero line to facilitate comparison of sink-source relationships. A scaled diagram of a CA1 pyramidal cell is shown on the left. Arrow indicates the soma level.

For the simulated CSD the stimulus consisted in a 1 nA–0.05 ms long pulse in the distal axonal compartment of all neurons. Compartmental transmembrane currents were calculated using GENESIS simulator [8]. Calculations of field potentials and CSDs were programmed in C code (a implementation of these routines is available for XSim Neural Network Simulator [9]).

Figure 2 shows the field potentials measured by the array of recording points for the three different conductance distributions. Note the larger span of negative potentials for Φ_3. The corresponding CSDs are shown in figure 3. As expected, they did not mirror the field potentials, i.e., the correct location and polarity of the sources and sinks is clearly revealed by the CSDs but not by the field potentials.

Figure 4 shows the transmembrane currents of apical, basal and soma compartments. Note that CSD profiles at the basal compartments differ from the transmembrane currents.

By inspection of figures 1, 2, and 3 we can see that the best range of con-

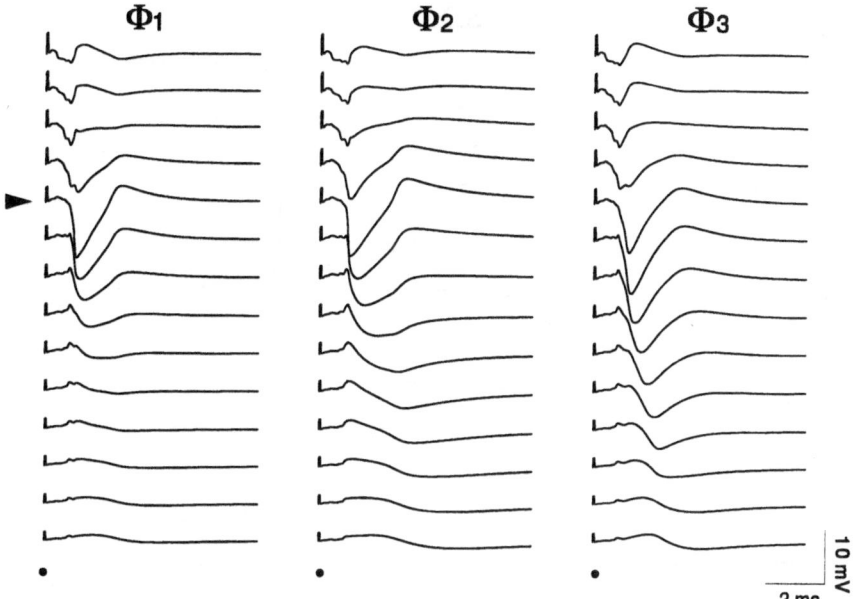

Fig. 2. Field potentials from three sets of simulations using different channel densities −see table 1−: Φ_1: active conductances placed in the soma with smaller values in the axon hillock, Φ_2: *hot spot* in the axon hillock and small conductances in the soma and apical dendrite. Φ_3: medium conductances in the soma with a small decreasing gradient of values through the apical dendrite.

ductance parameters that match the experimental data is that following a small gradient from the soma throughout the apical dendrite.

6 Discussion

Current source density simulations allowed comparisons of direct experimental measurements (field potentials and extracellular current density) with those obtained from model neurons all along its dendrosomatic axis. This technique can be used to constrain the parameter space of compartmental models as well as to correctly interpret the experimental CSD analysis. Particularly, the distributions of maximum conductances for the different ion channels throughout the neuron morphology can be estimated from these simulations. Even with the simplified compartmental CA1 model used here, it is already evident that the gradient−like distribution yields closer fields and CSD profiles to experimental data than passive or low dendritic conductance neuron models.

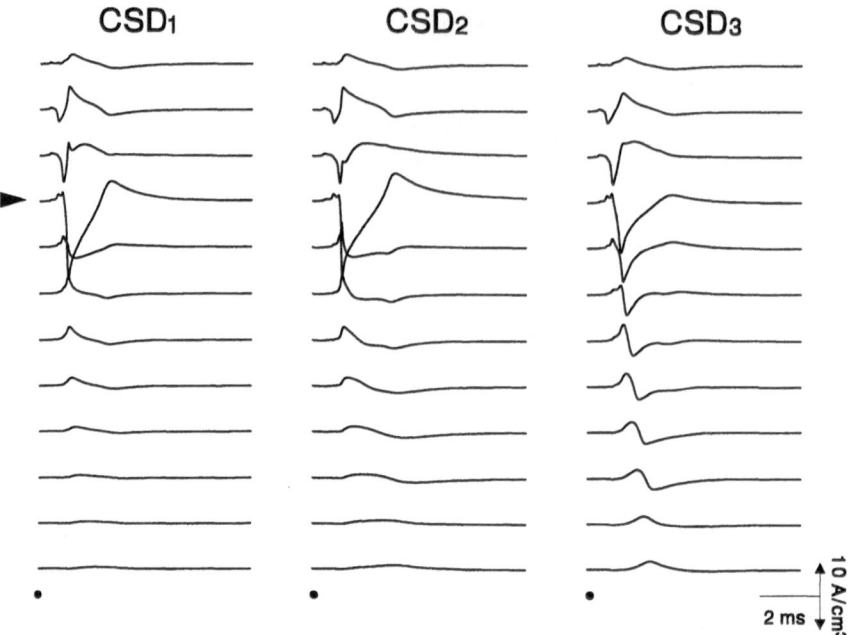

Fig. 3. Currents source densities calculated from the field potential profiles shown in figure 2.

Although model field potentials are more directly comparable to experimental data, it is obvious that current densities are much more accurate in terms of interpretation of sources and sinks. The more detailed the morphology with branched dendrites and realistic surface area, the more accurate the CSD computer analysis is to be expected. The implementation of activation delays, as pointed out by experimental results [10], can also make this simulations more realistic.

Additional interpretation hints can be obtained by the comparison of the compartmental transmembrane currents and simulated CSDs. For instance, although the three simulated field potentials were comparable around the somatic region, as were the CSDs –figures 2 and 3–, the individual I_m's –figure 4– showed that there was not current sink in the soma compartment for distribution 2. This occurred because active axonal compartments overlap the basal dendrite and their contribution to the field potentials and CSD profiles is mixed.

The progressive tune up of this simulated CSD will gradually provide a more restricted parameter space by delimiting two of the major missing parameters in modeling, namely, the spatial distribution and realistic maximum values for active conductances.

Compartmental models are often criticized on the grounds that they may be

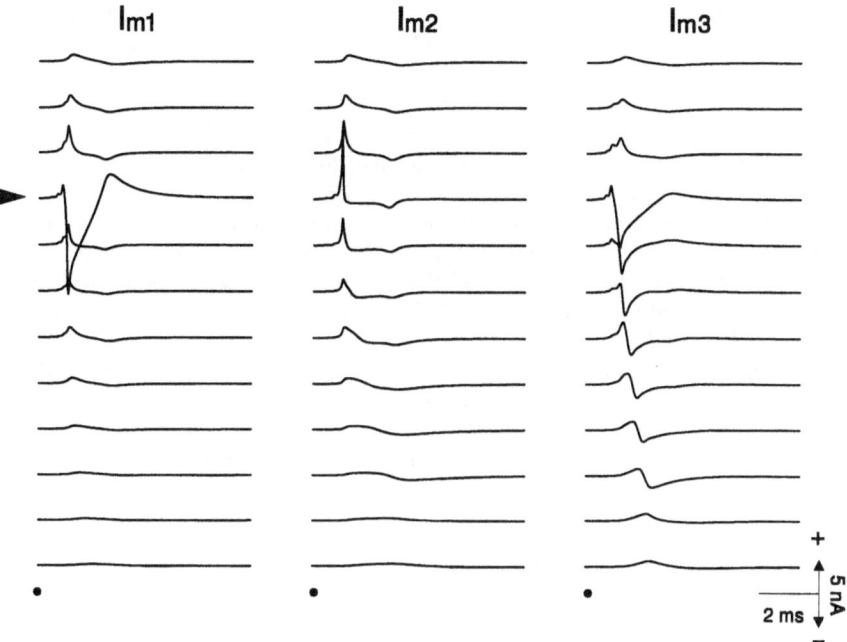

Fig. 4. Transmembrane compartmental currents for the three models with different channel densities.

capable of reproducing almost any behavior with different parameter settings. Although this is an exaggerated claim, there is a need for methods to evaluate compartmental models with respect to the available data (recently the use of bayesian methods of inference have been proposed for this task [11]).

The alternative method we propose here will ultimately lead to single global solutions for each specific set of three-dimensional field maps, and may prove as a useful device linking modeling and experimental research.

References

1. C. Nicholson, J. A. Freeman. *Theory of Current Source-Density Analysis and Determination of Conductivity Tensor for Anuran Cerebellum.* J. Neurophysiol. **38**: 356-368, 1975.
2. O. Herreras. *Propagating Dendritic Action Potential Mediates Synaptic Transmission in CA1 Pyramidal Cells In Situ.* J. Neurophysiol. **64**, No 5: 1429-1441, 1990.
3. C. Koch, I. Segev. Methods in Neuronal Modeling. From Synapses to Networks. MIT Press, 1989.
4. E. N. Warman, D. M. Durand, G. L. F. Yuen. *Reconstruction of Hippocampal CA1 Pyramidal Cell Electrophysiology by Computer Simulation.* J. Neurophysiol. **71**, No 6: 2033-2045, 1994.

5. R. D. Traub, R. K. S. Wong, R. Miles, H. Michelson. *A model of a CA3 hippocampal pyramidal neuron incorporating voltage-clamp data on intrinsic conductances.* J. Neurophysiol. **66**: 635-650, 1991.
6. Z. F. Mainen, J. Joerges, J. R. Huguenard, T. J. Sejnowski. *A Model of Spike Initiation in Neocortical Pyramidal Neurons* Neuron **15**: 1427-1439, 1995.
7. J. Stone, J. A. Freeman. *Synaptic organization of the pigeon's optic tectum: a Golgi and current source-density analysis.* Brain Res. **27**: 203-221, 1971.
8. J. M Bower, D. Beeman. *The Book of Genesis. Exploring Realistic Neural Models with the GEneral NEural SImulation System.* Springer-Verlag Telos, 1995.
9. P. Varona, J. A. Sigüenza. *Introducing XSim: A Neural Network Simulator That Incorporates Biological Parameters"* in LNCS 930 pp. 650-657 Mira-Sandoval (Eds.). Springer Verlag, 1995.
10. P. Andersen, T. V. P. Bliss, K. K. Skrede. *Unit analysis of hippocampal population spikes.* Exp. Brain Res. **13**: 208-221, 1971.
11. P. F. Baldi, M. C. Vanier, J. M. Bower. *On the use of Bayesian Methods for Evaluating Compartmental Neural Models.* Submitted to the Journal of Computational Neuroscience.

Spontaneous Activity of Hippocampal Cells in Various Physiological States

Nico Stollenwerk[+], Liset Menéndez de la Prida[*]
& Juan Vicente Sanchez-Andrés[*]

[+]Forschungszentrum Jülich, D-52425 Jülich, Germany

n.stollenwerk@kfa-juelich.de

[*]Departamento de Fisiología, Universidad de Alicante,
E-03080 Alicante, Spain

liset@juanvi.fisi.ua.es

Abstract

Time-series of interspike intervals from intracellular recordings in spontaneously firing hippocampal CA1 cells from rabbits are investigated. Slice preparations in three different experimental conditions are tested: *immature* from young rabbits up to seven postnatal days, *mature* from adult rabbits, and *bicuculline*-treated. also from adult rabbits. Bicuculline is a convulsion inducer through inhibition blockade.

Preparations from immature rabbits show the dynamical behavior of a Poissonian process, while those from mature animals are periodically firing with additional Gaussian noise. Bicuculline-treated slices from mature animals show again the same dynamics as from immature, namely Poissonian noise. In addition to the Poissonian processes in some immature and bicuculline preparations large proportions of bursts occur. We correct these bursts away and find better agreement with the null hypothesis of Poissonian processes. In some cases the frequent bursts even lead to signs of short time predictability, which is assumed to characterize deterministic chaos. However, we can also remove this predictability by burst correction.

To test for stochasticity we use the Kolomogorov-Smirnov test which is especially well suited for small data sets like in the present experiments. Explicit graphical analysis of the distribution function provides useful qualitative comparison. The predictability tests for nonlinearity are applied to original data and its Gauss-scaled surrogates to test for any linearly correlated process. Conclusions for the dynamical interplay between excitatory and inhibitory networks are drawn.

1   Introduction

In slice preparations we characterize three experimental states of the hippocampal network by its distribution of interspike intervals. Additionally we search for signs of determinism through a predictability test. The three states are immature from young rabbits up to 7 postnatal days, mature from adult rabbits, and with bicuculline preparated slices from adult rabbits. Bicuculline is an inhibition blocker, hence it generates epilepsy-like activity in the preparations.

In this contribution we present a detailed analysis of the distributions of intracellularly recorded interspike intervals of the different preparations: The immature state shows clear signs of an independent Poissonian process, where in some cases additionally occuring bursts change a part of the distribution as is shown in another study (Menéndez de la Prida, Stollenwerk, Sanchez-Andrés 1997), while the mature state shows periodic behavior with Gaussian noise. The predictability test is used to characterize the dissection into the exponentially distributed single spikes and the additional bursts. Finally, the bicuculline preparations indicate the same behavior as the immature.

In section two we describe briefly the theoretical background of the statistical analysis of interspike intervals based on the distribution function. The comparison of different null hypotheses is illustrated by two example data sets. Section three describes a more detailed analysis on an especially interesting data set from the *immature* slices. Section four presents the results of our analyses to all investigated data sets. Finally, after a summary we discuss implications of the present results as well for the direct neurobiological interpretation as for further investigations on modelling biological systems and artificial neural networks.

2   Interspike Intervals, Poissonian and Gaussian Processes

Experimentally, intracellular recordings were obtained in thin transversal hippocampal slices from New Zealand white rabbits in the CA1 hippocampal region (stratum pyramidale). The slices were prepared as described in Menéndez de la Prida, Bolea, Sanchez-Andrés (1996). From the recorded voltage measurements, interspike interval time series are obtained, i.e. the length x_i of time between spike i and spike $(i + 1)$ we call interspike interval (ISI).

For a first inspection of the ISI time series and its statistical and dynamical behavior see Fig. 1 a) with a typical time series of immature CA1 cells (data set *imm2*) and its scatter plot in b), and in Fig. 2 a) and b) the corresponding for a typical mature cell in the same hippocampal region (data set *mat2*).

The simplest stochastic process which can occur in interevent interval time series is the Poissonian process: Starting from the hypothesis that there is no change in probability to have an event for each time step Δt, the distribution of the time intervals between two events is just the exponential distribution (see e.g. van Kampen 1992) with probability distribution density

$$p_P(x) = a \cdot e^{-a \cdot x} \tag{1}$$

with a the constant probability p to have an event per time step Δt, hence $a = p/\Delta t$. This distribution is just described by one parameter a.

The next simple stochastic process is the independent Gaussian process, giving the Gaussian distribution for the ISIs (e.g. Deutsche Bundesbank 1991, ten-DM bill):

$$p_G(x) = \frac{1}{\sigma\sqrt{2\pi}} \; e^{-\frac{(x-\bar{x})^2}{2\sigma^2}} \tag{2}$$

hence described by two parameters, mean $\bar{x}$ and standard deviation σ.

To avoid the data hungry procedure of binning the observed data into classes to inspect the probability distribution $p(x)$ and perform statistical tests on this, we just rank order the data (original data set x_i and rank ordered x_j) and plot the data points x_j on the abscissa and the rank index j on the ordinate. This gives an estimator, which we call observed distribution function ODF, for the distribution function defined from the distribution $p(x)$ by

$$P(x_e) := \int_{-\infty}^{x_e} p(x) \cdot dx \quad . \tag{3}$$

For the Poissonian process the theoretical distribution function TDF is

$$P_P(x_e) = \int_0^{x_e} a e^{-ax} \cdot dx = 1 - e^{-a \cdot x_e} =: TDF(x_e) \quad . \tag{4}$$

The estimator ODF from the data points x_j can be writen as

$$ODF(x_e) := \frac{1}{N} \sum_{j=1}^{N} \Theta(x_e - x_j) \tag{5}$$

with the step function $\Theta(x)$ defined by

$$\Theta(x) := \begin{cases} 1 & for \quad x \geq 0 \\ 0 & for \quad x < 0 \end{cases} \quad .$$

In Fig. 1 c) we compare ODF from an immature slice preparation and TDF for the Poissonian process with parameter a fitted to the data set under investigation, here $imm2$. The relation between a and the data points x_i can be obtained from a *Maximum log-Likelihood Principle* with the result that a is simply the inverse of the mean of all data points.

The maximal absolute distance between ODF and TDF is the Kolmogorov-Smirnov test value (furtheron called KS-value), which has a well known approximate distribution function under the null hypothesis of independent realizations (Press *et al.* 1992):

$$P(KS) = 1 - 2 \sum_{j=1}^{\infty} (-1)^{j-1} \cdot e^{-2j^2 \cdot M \cdot KS} \tag{6}$$

with

$$M := \left(\sqrt{N} + 0.12 + \frac{0.11}{\sqrt{N}}\right)^2 \tag{7}$$

approximating well already for $N \geq 4$, with N the number of data points x_i. For $N \geq 100$ an even simpler expression is $M \approx N$. However, we use the original form of M to be prepared for small data sets.

For our data set $imm2$ with $N = 354$ data points and $a = 1.29~sec.^{-1}$ we obtain a KS-test value of $KS = 0.0554$ and $P(KS) = 0.78$. Hence in only 78% of the cases a purely random process would give a smaller distance KS between single N-points realization and theoretical distribution. But in more than 10% of the cases larger KS-values occur. A Poissonian process cannot be rejected as underlying process for this observed ISI time series.

For the Gaussian process the distribution function is just the celebrated error-function (adjusting the data to mean zero and unit variance) with its characteristic S-shape. No simple analytic expression can be given like for the exponential distribution, such that TDF for the Gaussian is only accessible numerically. In Fig. 2 c) for a mature preparation (data set $mat2$) the ODF and TDF for Gaussian distribution compare well. The KS-test gives $P(KS) = 0.957$ when tested for Gaussian, but complete rejection $(P(KS) > 1 - 10^{-10})$ for Poissonian. The complete results for all investigated data sets can be read off from Table 1 for the tests on Poissonian, and from Table 2 for the tests on Gaussian for the mature.

3 Burst Correction and its Effects on Distribution and Predictability

Although the data set $imm2$ confirms immediately the hypothesis of Poissonian processes (as do all four mature data sets for the hypothesis of Gaussian with about 5% confidence level, see Table 3), most immature preparations give observed distribution functions which deviate from TDF for Poissonian processes especially for small ISIs, as can be seen in data set $imm4$ in Fig. 3 for example. In Fig. 3 c) the deviations are obviously strongest for small ISIs, resulting in a well known two-humped distribution $p(x)$.

As a key question to such data sets it is often asked in the literature: Is there much more information present than for a pure Poissonian process, eventually is there even determinism present, which might be detected by nonlinearity tests (see e.g. Schiff et al. 1994(a)). In our data sets we have found only one clear candidate with some hints for short time predictability as a sign of underlying determinism, which can be separated statistically from first order correlated stochastic processes, see Fig. 5 a). The result is comparably weak as presented in Schiff et al. 1994(a) (we use the same prediction algorithm as they do, see Chang et al. 1994). Our case is the data set $imm4$, which is investigated in more detail in the following.

Starting from our observation of deviations between ODF and Poissonian TDF in $imm4$, which appear more in small ISI-values than in the larger ones,

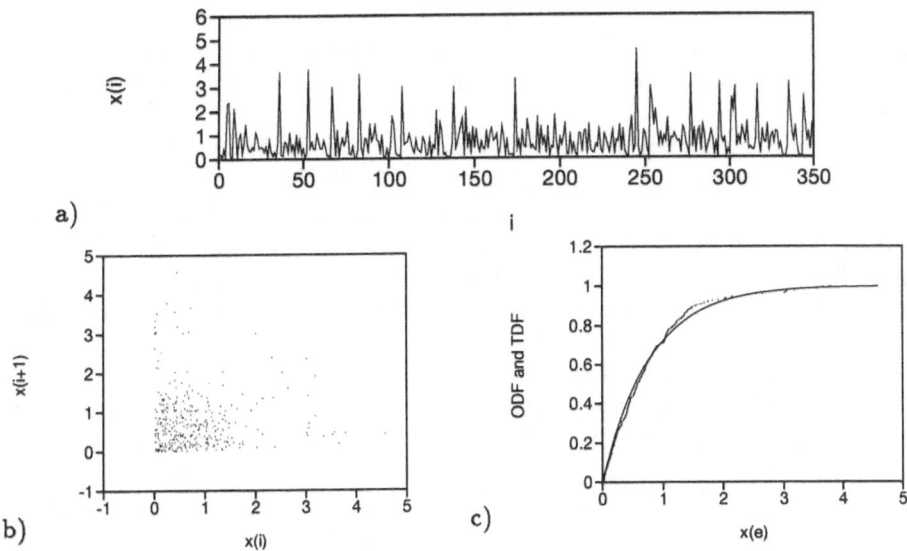

a)

b)

c)

Figure 1: *a) Time series and b) return plot of ISI data set imm2. c) Comparison of the observed distribution function ODF from data set imm2 (marked by points) with the theoretical distribution function TDF (full lines) obtained from our null hypotesis, the Poissonian process.*

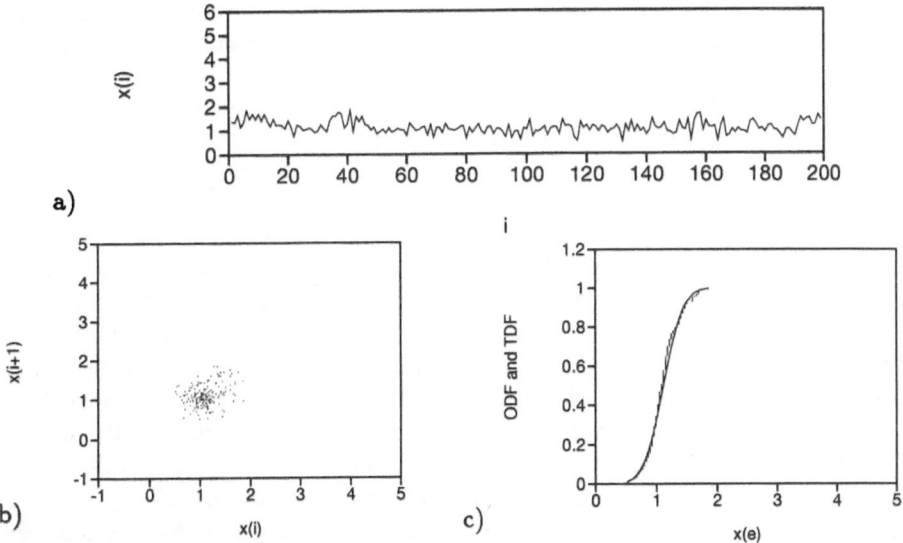

a)

b)

c)

Figure 2: *a), b) and c)show the same quantities as in Fig. 1, now for an example of the mature slices, here data set mat2. The ODF in this case is much better described by the S-shaped TDF of a Gaussian process as underlying hypothesis than with a Poissonian TDF, as was used in Fig.1 c).*

we propose a simple burst correction for removing a special type of small subsequential ISIs and see, what results for the remaining time series in terms of distribution and higher order correlations being characterized by the prediction test.

We correct the bursts in the following way: Two subsequent interspike intervals which are smaller than a given value δ, the *burst criterion parameter*, are considered as not existing. If the next ISI is still smaller than δ, this also vanishes etc.. As result of the algorithm a whole burst is replaced by a single spike. The underlying hypothesis is that according to the sum of inputs a cell fires. If this firing results in a single spike or a burst, is not considered. This procedure we call *burst correction*.

We test the burst corrected ISI-time series for different values δ against the null process of Poissonian, each with $P(KS)$, the sumed probability of maximal distances. The result is shown in Figure 4 a): The original data set *imm4* shows a test result of $P(KS) = 1$ up to 10 digits. For small burst corrections of $\delta = 0.1$ it is still higher than 0.98. Then for a wide range of δ the burst corrected time series agrees well with the exponential distribution, i.e. $P(KS) \leq 0.95$. A large minimum of $P(KS)$ in respect to δ appears around $\delta = 0.2$ *sec.*, where $P(KS) \leq 0.75$.

For $\delta = 0.2$ *sec.* we show the explicit effect of the burst correction on *imm4* in Figure 4 b). The comparison of the ODF of the burst corrected data set and its theoretical explanation TDF is much better than for the raw data (Fig. 3 c)). The result of the KS-test is $P(KS) = 0.7180$. Hence a good agreement between data and the null hypothesis of Poissonian is achieved. The effects of burst corrections to the other immature as well as the bicuculline preparations are given in Table 2 by its $P(KS)$, each for $\delta = 0.2$ *sec.*.

To investigate the effect of our burst correction procedure on possible further structure than being tested for in the distribution, we apply a test on predictability in the time series as used in Chang *et al.* (1994) and Schiff *et al.* (1994(a)). It is a simplified version of the previously investigated predictability tests using locally linear predictors for the future value of the actual time series point (Farmer, Sidorowich 1987, and Sugihara, May 1990). Again, the predictability can be removed completely by our burst correction. For detailed results see Menéndez de la Prida, Stollenwerk, Sanchez-Andrés (1997). Summarizing the results given there, no information is contained in the bursts corrected time series. Remains the question of possible information in the burst structure.

To address this question we investigate the interburst intervals (IBIs) to see if the information of a spiking neuron might be rather coded in the bursts than in single spikes. The resulting IBI time series is too short to reliably apply a prediction test ($N = 20$ in *imm4*). The investigation of the distribution however showed no statistically significant deviation from a Poissonian process. So the bursts are in its direct time sequence also not a good candidate for information coding on the investigated time scale.

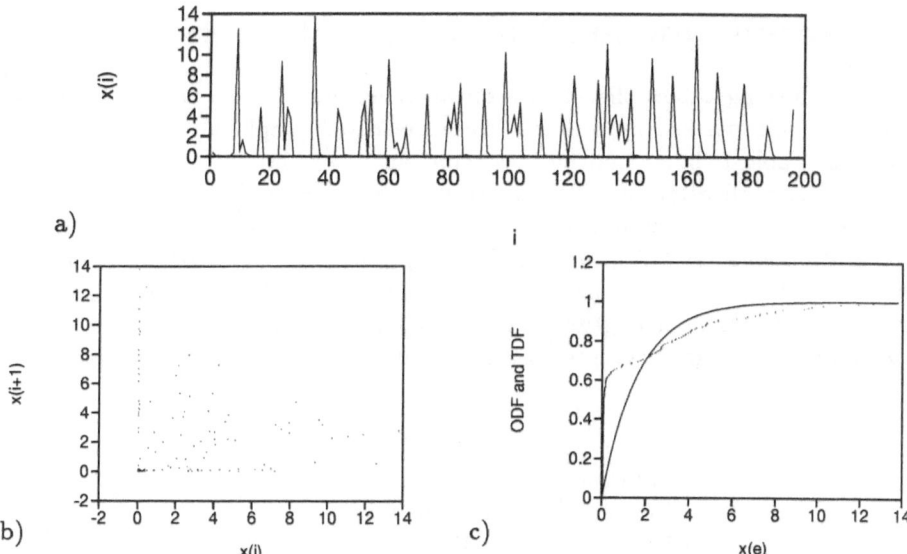

a)

b)

c)

Figure 3: *Data set imm4 presented in its different aspects a) time series, b) scatter plot and c) distribution function, as described in Fig.1. Most obvious are the differences between this data set imm4 and data set imm2 from Fig.1 in its deviations from the exponential distribution function in c), namely a higher proportion of small ISI than expected under the hypothesis of Poissonian processes.*

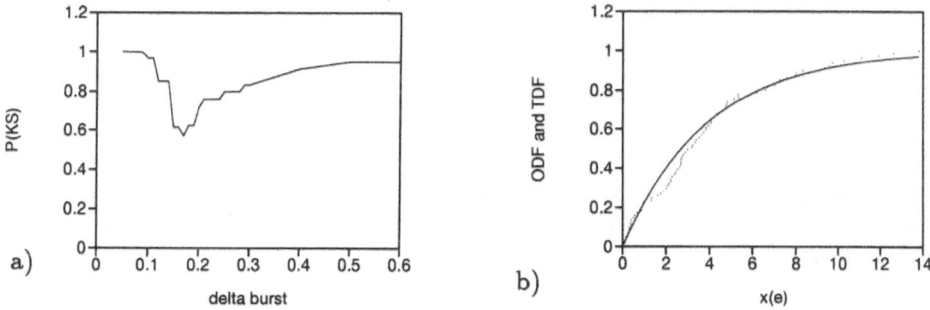

a)

b)

Figure 4: *a) The result of the KS-test of data set imm4, i.e. the value of the probability distribution function $P(KS)$ of the largest deviation between ODF and TDF, the KS-value, as a function of the correction for bursts with different values of the burst criterion parameter δ. b) Comparison of the observed distribution function ODF from the burst corrected data set imm4 with the theoretical distribution function TDF obtained from our null hypotesis, the Poissonian process. Within the statistical fluctuations the both distributions agree well with each other $(P(KS) = 0.7180)$.*

4 Results on the Different Data Sets

In the following tables the results on all investigated data sets are reported. The data sets marked by a star (*) are presented graphically in the previous sections. Table 1 shows that besides *imm2* and *bicu1* no data set is statistically accepted as realization of a Poissonian process inside a 1% confidence interval. The set *imm2* is accepted with more than 20% confidence and the bicuculline preparation *bicu1* at 3.2%, such that the hypothesis is supported after which the bicuculline leads to similar firing patterns as is obseved in immature slices. This hypothesis is further strengthened by the burst correction result on *bicu1*, where a Poissonian process is accepted with even more than 25% confidence (see Table 2).

data set	N	a	$P(KS)$
imm1	150	1.038	1
imm2*	354	1.289	0.78
imm3	269	0.997	0.999970
imm4*	196	0.610	1
imm5	347	1.629	0.9999960
bicu1	187	0.663	0.968
bicu2	240	1.125	0.999999940
mat1	139	-	0.999985
mat2	199	-	1
mat3	147	-	1
mat4	170	-	1

Table 1: *Statistical test results on Poissonian processes. For each data set the total number of observed interspike intervals N, the inverse mean a as the unique parameter for the exponential distribution, and the Kolmogorov-Smirnov-test result $P(KS)$ under null hypothesis of a Poissonian process are given. Since the mature preparations are better characterized by a Gaussian process, in these cases no values for a are denoted.*

In table 2 for the burst corrected ISI time series further three of the remaining four immature preparations are accepted by more than 1% as Poissonian processes. The effect is most drastically present in *imm4*, the only candidate with signs of short time predictability. A similarly strong effect is observed in *bicu1*. As turned out by inspection of the burst corrected *bicu2bc* neither a Poissonian nor a Gaussian null hypothesis can be accepted, but the distribution function seems to lie inbetween these two alternatives. Further experiments have to clarify this finding.

In the only immature case being rejected from the hypothesis of a bursting Poissonian process, *imm1*, there is a minimal $P(KS) = 0.99987$ at burst correction $\delta = 0.05$ with $a = 0.657$. We conclude our considerations on the burst correction process by investigating the distribution of the interburst intervals originated from the *imm4* data set, hence called *ibi4*. The hypothesis of a Poissonian process is well accepted by a nearly 40% confidence level.

Table 3 gives the results of tests on Gaussian distribution for the four mature slice preparations. In all cases the hypothesis is accepted by a 3.3% to 6.1% confidence level. The mean values indicate a firing frequency of about 1 Hz.

data set	N	a	P(KS)
imm1bc	82	-	0.9999936
imm2bc	-	-	-
imm3bc	209	0.785	0.962
imm4bc*	81	0.258	0.718
imm5bc	255	1.232	0.984
bicu1bc	164	0.585	0.716
bicu2bc	192	-	1
ibi4	20	0.0648	0.607

Table 2: *Test results on burst corrected Poissonian processes. Each burst correction with $\delta = 0.2$ sec.. The data set imm2 could be characterized well by a pure Poissonian process, hence no burst correction is needed in addition. The set bicu2bc has even under any burst correction parameter no acceptable P(KS).*

data set	N	$\overline{x}$	σ	P(KS)
mat1	139	0.872	0.574	0.9574
mat2*	199	1.117	0.263	0.9572
mat3	147	1.237	0.441	0.967
mat4	170	1.351	0.431	0.939

Table 3: *Test results on Gaussian processes. The characterizing parameters of the Gaussian distribution, mean $\overline{x}$ and standard deviation σ, are given besides N and $P(KS)$. In all four cases the null hypothesis can be accepted by confidence levels of 3.3% to 6.1%.*

5 Summary and Discussion

The present study shows during maturation of the hippocampal network clearly a shift from bursting Poissonian firing towards Gaussian fluctuating periodic dynamics in spontaneous activity. Epilepsy inducing convulsants can change the dynamics back to an immature like state by blocking the inhibitory network. This implies a functional role of the inhibitory network during maturation and possibly pathological epileptic states as underdevelopment of the inhibitory system. Previously stated findings of slight short time predictability, which we also observed, can be explained in our case by simply randomly occuring bursts, i.e. subsequently following short interspike intervals. The bicuculline preparations indicate the same behavior as the immature, which is a first indication that during maturation the inhibitory system is established to create periodically firing neuronal circuits.

Two hypotheses can be drawn from these findings: If learning through memorizing is a major goal of the developing hippocampus (as is widely stated in the literature, see e.g. Kandell, Schwartz, Jessel 1992 and Traub, Miles 1991), then the inhibitory system plays a key role in organizing periodic firing patterns as a basis of memory. The second hypothesis has clinical relevance: imperfect development of the inhibitory system during maturation might cause epileptiform activity in the hippocampus, which is a major source for mid brain epilepsy (Menéndez de la Prida, Bolea, Sanchez-Andrés 1996). Especially for epilepsies with lacking histological findings (e.g. in form of tumors) this might be an interesting explanation for certain types of epilepsy as a purely network generated property.

Up to now, the inhibitory system is very difficult to access in experiments, since directly correlated activation is only to be expected in the excitatory system. Only such correlations in excitation are being investigated frequently. A dynamical characterization, as presented here, can help to learn more about the whole interplay between inhibition and excitation. Furthermore, the direct access to the inhibiting interneurons is at present still technically limited.

In classical neural network models, inhibition is functionally restricted to damp in total the excitatory system, such that the excitation alone is often supposed to carry the information process. But as the inhibitory system has a drastic effect on the whole dynamics of the entire system, as becoming clear from the present study, then such networks of recurrent excitation and inhibition can perform a very wide range of dynamics including deterministic chaos (Pasemann 1995) and even might use the flexibility of chaotic systems in respect to tiny parameter changes (Stollenwerk, Pasemann 1996(a)). These are the direct implications to be drawn from this contribution for future research on artificial neural networks. Recently, extersive research is performed on the functional role of possibly chaotic dynamics in biological neural systems and artificial recurrent neural networks (Skarda, Freeman 1987, Varela 1989, Babloyantz, Lourenço 1994, Stollenwerk, Pasemann 1996(b)).

On the biological side, not the single interspike interval but rather the temporally changing mean firing rate should be the correct level of description in actual neurobiological systems, since biological neurons integrate over many single spikes. For a final decision on the appropriate time scales, preparations yielding longer time series and further experiments, e.g. periodic forcing as a simplest method of putting questions to the neuropreparation have to be performed. On first results on field potentials in hippocampal slices under periodic forcing see e.g. Hayashi, Ishizuka (1995). Pathological states can give possibilities for directly influencing the deterministic part of brain dynamics (Schiff *et al.* 1994(b) and related to it Christini, Collins 1995, and Pierson, Moss 1995).

6 Acknowledgments

We would like to express our thanks to F. Pasemann and F. Drepper (Jülich) for helpful discussions on various aspects of the presented results and thank S. Bolea, A. Vergara and S. Moya for their continuous collaboration. This work was supported by grant 96/2012 from *Fondo de Investigación Sanitaria.* L. M.P. was supported by fellowships from GV.

References

[1] Babloyantz, A. & Lourenço, C. (1994). Computation with chaos: A paradigm for cortical activity. *Proc. Natl. Acad Sci. USA* **91**, 9027–9031.

[2] Chang, T., Schiff, S.J., Sauer, T. Gossard, J.P. & Burke, R.E. (1994). Stochastic versus deterministic variability in simple neuronal circuits: I Monosynaptic spinal cord reflexes. *Biophysical J.* **67**, 671–683.

[3] Christini, D.J. & Collins, J.J. (1995). Controlling neuronal noise using chaos control. *Phys. Rev. Lett.* **75**, 2782–2795.

[4] Farmer, J.D. & Sidorowich, J.J. (1987). Predicting chaotic time series. *Phys. Rev. Lett.* **95**, 845–848.

[5] Hayashi, H. & Ishizuka, S. (1995). Chaotic responses of the hippocampal CA3 region to a mossy fiber stimulation in vitro. *Brain Research*, **686**, 194–206.

[6] van Kampen, N.G. (1992). Stochastic Processes in Physics and Chemistry. *North-Holland*, Amsterdam.

[7] Kandel, E.R., Schwartz, J.H. & Jessel, T.M. (1992). Essentials of Neural Sciences and Behavior. *Appleton & Lange.*

[8] Menendéz de la Prida, L., Bolea, S. & Sanchez-Andrés, J.V. (1996). Analytical characterization of spontaneaous activity evolution during hippocampal development in the rabbit. *Neurosci. Lett.* **218**, 1–3.

[9] Menéndez de la Prida, L., Stollenwerk, N., & Sanchez-Andrés, J.V. (1997). Evidence for nonlinearities in immature CA1 hippocampal neuronal activity. *submitted.*

[10] Pasemann, F. (1995). Neuromodules: A dynamical systems approach to brain modelling. in: *Supercomputing in Brain Research - From Tomography to Neural Networks* eds. H. Herrmann, E. Pöppel, D. Wolf, (World Scientific, Singapore), 331–347.

[11] Pierson, D. & Moss, F. (1995). Detecting periodic unstable points in noisy chaotic and limit cycle attractors with application to biology. *Phys. Rev. Lett.* **75**, 2124–2127.

[12] Press et al. (1992). Numerical Recipes in C. *Cambridge University Press*, Cambridge.

[13] Rapp, P.E., Albano, A.M., Zimmerman, I.D. & Jiménez-Montaño (1994). Phase-randomized surrogates can produce spurious identifications of non-random structure. *Phys. Lett.* **A 192**, 27–33.

[14] Schiff, S.J., Jerger, K., Chang, T., Sauer, T. & Aitken, P.G. (1994(a)). Stochastic versus deterministic variability in simple neuronal circuits: II Hippocampal slices. *Biophysical J.* **67**, 684–691.

[15] Schiff, S.J., Jerger, K., Doung, D.H., Chang, T., Spano, M.L. & Ditto, W.L. (1994(b)). Controlling chaos in the brain. *Nature*, **370**, 615–620.

[16] Skarda, C.A. & Freeman W.J. (1987). How brains make chaos in order to make sense of the world. *Behav. Brain Sci.* **10**, 161–195.

[17] Stollenwerk, N., & Pasemann, F. (1996(a)). Control Strategies for Chaotic Neuromodules. *International Journal of Bifurcation and Chaos*, **6**, 693–703.

[18] Stollenwerk, N., & Pasemann, F. (1996(b)). Switching in Self-Controlled Chaotic Neuromodules. *World Congress on Neural Networks, San Diego, California*, Sept. 15-18, 1996, (pp. 680–684). INNS Press and Lawrence Erlbaum, New Jersey.

[19] Sugihara, G. & May, R.M. (1990). Nonlinear forecasting as a way of distinguishing chaos from measurement errors in time series. *Nature* **344**, 734–741.

[20] Theiler, J., Eubank, S., Longtin, A., Galdrikian, B. & Farmer, J.D. (1992). Testing for nonlinearity in time series: The method of surrogate data. *Physica* **D 58**, 77–94.

[21] Traub, R.D. & Miles, R. (1991). Neuronal networks of the hippocampus. *Cambridge University Press*, Cambridge.

[22] Varela, F.J. (1988). Cognitive Science, a cartography of current ideas. (London).

Neural Network Model of Striatal Complex

Boris Aleksandrovsky[Ψ]*, Fernando Brücher*[*]*, Gary Lynch*[*]*, and Richard Granger*[*]
[Ψ]*Brightware, Inc., Novato, CA, U.S.A*
[*]Center for Neurobiology of Learning and Memory, UCI, Irvine, CA, U.S.A

Abstract

Basal ganglia bridges posterior and anterior neocortices, serving a variety of functions which are involved in execution and planning of complex motor and cognitive sequences.. We forward a motor production model which is based on known anatomical, physiological and behavioral studies of basal ganglia involvement in production of complex motor sequences in mammals (see [3] and [27] for review.) Learning in the model is based on the phenomenon of LTP [7] and LTD [42][34], both suspected substrates of memory, which have been observed in striatum. The model is illustrated with an example of adaptive production of handwriting via simulated robotic strokes.

1. Introduction

According to the classical view [3], basal ganglia consist of a number of segregated parallel circuits, combined output from all circuits project not only to the motor cortical areas, but to virtually all frontal and prefrontal lobe [26], prompting researchers to hypothesize involvement of basal ganglia into executive planning, cognitive and memory functions. Within each parallel circuit further subdivisions can be made; for instance those which segregates projections from different cortical motor areas [4]. Additionally striatum has well defined microarhitectural levels; alas striatosomes and matrix compartments have very different input/output connections [25]. It is evident that the emergent functionality can only be realistically modeled when all those levels are investigated. It is therefore a premise of the present paper to bring together different level of inquiry into a single modeling framework.

In this work we will only concern ourselves with the function of basal ganglia as an executive planning circuit for motor movements.. Those circuits are concentrated inside the basal ganglia motor circuit. The motor circuit receives somatotopically organized primary projections from the motor and somatosensory cortices, also some projections from supplementary motor area, arcuate premotor area, and other cortical regions in Brodman's areas 5 and 6. Somatotopic organization is maintained throughout the circuit. Cells in the motor circuit act upon ventral and central thalamic regions (VLo, VApc, VAmc, and CM in rodents) [11], which in its turn project to primary motor cortex [49]. Motor cortex then acts upon distal musculature via corticospinal tract.

2. Neurobiology of the Striatal Complex

Let us turn now to detailed description of anatomical and physiological organization of basal ganglia as is relevant to the model being described. Figure 1 below illustrates the discussions in this section:

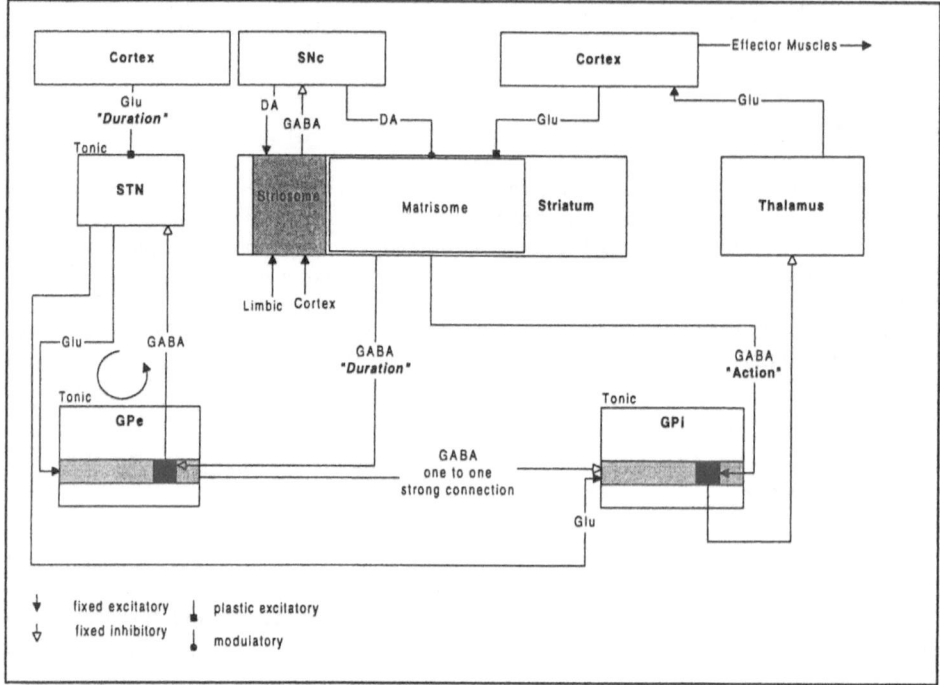

Figure 1. Model of basal ganglia.

2.1. Intrinsic Organization of the Striatum

The predominant cell type of the striatum is medium spiny neuron (later MSN), which accounts for 96% of all cell mass; each MSN has numerous collateral projections that contact neighboring cells in the proximal dendritic tree or soma, creating a strong lateral inhibition [46]. The remaining 4% of striatal cells are various types of interneurons [31]; function of which is mostly long-lusting modulatory [26] and is ignored in the present model for simplicity. Striatum contains two neurochemically different and architecturally segregated compartments, striatosomes and matrix (see [22] for review). In the latter zone smaller independent units called matrisomes have been identified. Striatosomes are vertical cylinder divisions intervened in the matrix across the extent of striatum, which receive afferents primarily from limbic structures and project to SNc. Matrisomes occupy approximately 80% of the area in striatum, receive excitation from cortex and project to both segments of the pallidum over both direct and indirect pathways. It is known that distinct populations of striatal cells project to direct (GPi/SNr) and indirect (GPe) pathways; it is also proposed that these cells are segregated in distinct matrisomes [46]. The boundaries of striatosomes and matrisomes are rather strict preventing them from sharing input or outputs with other units.

The inputs from cortex to striatum are random sparse, due to an 'en passant' type of connectivity. MSN's are the target of major projections from cortex and thalamus; and is also the only neuron type projecting out of the striatum. MSN also receives input from the amygdala and SNc dopaminergic neurons. As a rule, local afferents terminate on proximal dendrites and perykarion, while distant influences (cortex, thalamus, amygdala) project to distal dendritic field. Cortical afferents carry glutamate as their neurotransmitter, SNr projections contain dopamine, and local interactions are mediated by GABA, acetylcholine and substance P. One of the intriguing biophysical properties of MSN is the ability of sustaining activation for a relatively large time ("plateau depolarization" of [54] [53]). MSN's physiological characteristics is bi-phasic excitation pattern with two distinct states called "up" and "down". "Up" is a state of depolarization which can only be reached by integrated massive afferent excitation, but can be maintained for extremely long (1-5 sec) periods of time [53] [37] [36]. "Down" is a characteristic hyperpolarized state, which is the usual state in the absence of synchronic afferent excitation.

A number of generalized local circuit architectures are found throughout the extend of striatum. Lateral inhibition, facilitated through local axon collaterals, serves to promote resolution enhancement and integration, similar to those found in cortex. Dopaminergic projections from SNc can play both inhibitory and excitatory role; this is made possible by existence of two different types of dopamine receptors in striatum, D1 and D2 [24].

2.2. Intrinsic Organization of SNc

SNc is composed of closely interlocked clusters of dopamine neurons, which project widely throughout the extend of the striatum. Dopaminergic projection of SNc is considered the main factor in the reinforcement learning model of the striatal complex. The dopamine surge has been shown to be present when new stimuli or unexpected reward is experienced. SNc has two distinct projections paths: one for striatosomes and the other for matrix.

2.3. Intrinsic Organization of GPe

GPe is populated by relatively small number of large cells, with extensive discoid dendritic arborizations [44]. Those disks are oriented perpendicular to the upcoming striatal afferents, thus allowing for a high degree of convergence of striatal afferents (huge discoid dendritic trees have estimated 40,000 striatal synapses per cell). Striatal afferents terminating in GPe contain both GABA and enkephalin as their neurotransmitter. GPe intrinsic cells have a high level of spontaneous activity; the action of inhibitory striatal afferents is de-activation of this activity.

Most of the efferent projections from GPe, which are gabaergic inhibitory, go to STN; strong one to one intrapallidum (GPe→ GPi) connections exist as well. Since GPe discoid neurons exhibit high spontaneous activity, in the absence of afferent inhibition GPe tonically inhibits its projection targets - STN and GPi. In the presence of afferent inhibition, GPe projection neurons will disinhibit their efferent targets. Additionally GPe projects to GPi via collaterals of pallidosubthalamic neurons which are inhibitory [50]. This action produces shunting effect on the STN→GPi afferents. Physiological effect of this action is fine modulation of the time course of the thalamic action, since it would extend the time which is used for indirect pathway to counterbalance the direct pathway.

2.4. Intrinsic organization of GPi/SNr

Cytoarchitectonically GPi is homologous to GPe. Since dendritic arborization is so vast it is reasonable to assume that afferents from striatum, GPe and STN converge on the same cell population [44] [50]. In the absence of striatal afferent activity high spontaneous discharge rates of GPi neurons tonically inhibit the thalamus. When input from striatum arrives, a specific zone of GPi becomes disinhibited, which in its turn disinhibits the thalamus and its afferent targets (notable primary motor cortex) and the corresponding movement is initiated. The indirect pathway action through STN is of an excitatory nature and turns to counterbalance the influence of the direct pathway (as discussed below), so thalamus is being gradually deactivated.

2.5. Intrinsic organization of the STN

STN is comprised of medium-sized densely packed projection neurons, and a negligible number of interneurons) [14] . It receives massive afferent excitatory glutamatergic projections from broad areas of cerebral cortex (including prefrontal, cingulate, motor and somatosensory areas) [43] [35] [47]., and inhibitionary gabaergic from GPe, as well as differential projections from various brain stem nuclei, thalamus and striatum [47], which are ignored for the purposes of the presented model. Evidence [13] supports the existence of SNc→STN dopaminergic pathway; coupled with the evidence of modulatory effects of dopamine on the physiology of subthalamic neurons [10] this hints on the potential plasticity in STN by the mechanism not unlike that described in the striatum (refer to later sections 2.6 and 3).

STN's efferent projections are glutamatergic and excitatory to both GPe and GPi/SNr complex [38]. STN is tonically active, but GPe projection back to STN form a negative feedback circuit that keeps STN quiet in the absence of striatal input to GPe. Subthalamopallidal projection has been shown to be more potent than the corresponding striatopallidal projection [29] [45] due to the nature of the former to terminate more proximally to the cell body than the latter. STN→GPe projections are diffused all over external pallidal output target [12].

2.6. Mechanisms of synaptic plasticity in basal ganglia

Possible substrates of learning in striatum are based on phenomenon of LTP [7] [41] [40] and LTD [42] [34], which are suspected substrates of memory throughout the brain. Experiments which prove induction and maintenance of LTP in neostriatum were performed in nucleus accumbens [39] [48] [51]. Condition of LTP elicitation in striatum involve double tetanus simulation [20]. LTD in striatum was observed in different labs [52] [8], conditions for its induction in striatum are single tetanus simulation followed by an absence of subsequent activation. Realistic conditions for learning in basal ganglia might involve modulatory reward-reinforcement action of dopamine [9] (suggested by proximal position of dopaminergic afferents on striatal cells [46].) Presence of LTP and LTD in basal ganglia points out to the potential learning capabilities in the basal ganglia pathways. Possible site of this learning is hypothesized to be cortico-striatal, making possible reinforcement of goal-oriented motor activity. Another possible plastic site can be the cortico-subthalamic projection, with similar plasticity characteristics as cortico-striatal pathway, but is possibly modulated by different non-dopaminergic mechanisms.

3. Computational model of basal ganglia production of motor sequences

Basal ganglia receives commands from executive centers of the frontal lobe, and acts upon the thalamus, alternatively inhibiting and disinhibiting the corresponding thalamic nuclei. Cortical command (to initiate a sequence of movements) to striatum is distributed across multiple cortical areas - the information about the state of the muscles (motor cortex), environment (sensory cortices) and context of intended action (frontal cortex) should be present at the same time in order for a sequence of movements to be initiated [19]. In the current simulation cortical input is approximated to be a random vector taken from non-overlapping distributions for each distinct cortical command.

Balance and interplay between direct and indirect pathway which exert opposing influences on thalamus (direct pathway leads to excitation of thalamus, indirect to the inhibition) are at the core of the basal ganglia functionality. Basal ganglia direct pathway (cortex $\rightarrow$ striatum $\rightarrow$ GPi/SNr $\rightarrow$ thalamus $\rightarrow$ cortex) plays the role of the *execution* pathway. Striatal output cells send its efferent projections to the restricted patches of cells in GPi. This projection is, however, less potent then corresponding subthalamopallidal pathway, which converge on the same population of cells. Thus initially GPi is excited, thalamus inhibited and no muscular action is performed. However, in time negative feedback loops in STN $\rightarrow$ GPe pathway lead to the weakening of subthalamopallidal efferents, and striatopallidal influences win over the cells in GPi. This leads to gradual activation of the thalamus, and thus to gradual emergence of muscular action. When next sequential command arrives from cortex to start a different action, activity in indirect pathway prevails again, and current action is abruptly abandoned in favor of just arrived one.

Conjunction of activity in indirect pathway (cortex $\rightarrow$ striatum $\rightarrow$ GPe $\rightarrow$ STN $\rightarrow$ GPi/SNr $\rightarrow$ thalamus $\rightarrow$ $\rightarrow$ cortex) and the efferent copy of the cortrico-striatal excitation send over corticosubthalamic pathway [47] directly influence the amount of time muscular action is being executed; thus the indirect will be termed the *timing* circuit in the paper. Due to extensive axon arbotizations in STN cells, single cortical excitation drives STN to fast saturation once it reaches the firing threshold. The magnitude of cortical excitation controls directly the timing of the circuit; up to the maximum time constrained by the fixed depolarization time of the MSN in striatum. Striatal activation of GPe lead to its long-lasting inhibition [53] and subsequent disinhibition of pallidosubthalamic targets. This leads to activation of a very potent subthalamopallidal pathway. Initially the amount of activity there prevails over the opposing influence of striatopallidal pathway. However, due to negative feedback loops in STN → GPe pathway, GPe starts gradual re-activation. Correspondingly the amount of STN activation gradually diminishes, allowing for striatopallidal activity to overcome its influence. This leads to activation of the thalamus and muscular action.

The above discussion implies the existence of a neural mechanism which prevents more than a single action to be executed at a single time through local basal ganglia pathway (of course a number of unrelated actions can be executed simultaneously; however this involves action through different anatomically segregated circuits within basal ganglia). STN serves as such arbiter of actions by virtue of its projection to GPe being diffuse (as discussed in section 2.5), thus guaranteeing that only the first action (corresponding to striatal winner) follows through. Additional functional implication of extensive corticosubthalamic projection is hypothesized to be involvement of that projection into an *"emergency braking mechanism"*. Thus, in case peripheral feedback has arrived to the cortex indicating the action had been performed resulting in negative implications to the organism, an overwhelmingly strong excitation pattern is being send through STN all over GPi resulting in over-excitation of GPi and effectively in inhibition of all afferent thalamus.

Striatum has been implicated in a number of stereotypical automatic learned behaviors in both humans (handwriting, reaching, speech) and animals (rearing, grasping) [2] [1] [3]. All those behaviors involve automatic sequence of action in response for a single cortical command. Steps that follow receive sensory feedback, but no additional direct command is generated by the cortex if behavior is executed as planned. During the automatic execution of sequence cortex remains as an observer, with a single role of stopping the execution on catastrophic failure [1]. The role of the striatum thus becomes of recognizing of appropriate command generated by cortex and then executing the rest of the sequence based on somatosensory feedback and the current state of the striatum. The cortical feedback thus serves as a cue for subsequent steps in the sequence, and the current state of striatum (make possible by long depolarization of its cells) serves as a level of context which helps to chain actions in a single sequence. The end of the sequence is signaled by the absence of output from striatum, thus resulting in no feedback and therefore no cue for subsequent action.

3.1. Learning in the model

We hypothesize that reinforcement reward-associated learning occurs in striatum. Learning in the cortico-striatum pathway is mediated by dopamine reward, which is send to striatum by SNc [26]. Cortical command for generation of the motor action primes the population of striatum cells which win the intra-matrix competition. Winners fire and bring about certain motor action; if the results of that action warrants the reward, that reward signal is generated by SNc. SNc emits dopamine projections to the striatum which are broad and non-specific [30], thus the reward signal touches the cells which were prior cortically primed and the conditions for learning becomes satisfied [33].

There are two basic ways of weight modification considered, LTP and LTD. LTP increases the synaptic weights in proportion to the cell input; LTD correspondingly decreases it. Combining rules for LTP and LTD, it is possible to obtain one that modifies the weights to position them closer to the input vector X.

$$\Delta W = (alphaLTP-alphaLTD) * (X-W)$$

, where the learning constants alphaLTP and alphaLTD correspond to the modified step size of learning (see [33] for formal description of learning step and ceiling variations)., W is a current weight vector and X is the current input vector.

Issue can be taken on relative and absolute sizes of LTP and LTD learning constants. Reasonable small value of alpha is required for stability and eventual convergence. Values of learning constants are related to resolution capabilities of the network, but choosing a smaller value would significantly slow down training.

4. Computer simulations of the model

In this model we will concentrate on the action of a single matrisome, as an independent functional unit in a striatum. Model can thus be scaled up to provide for the function of a set of striatal patches by considering the function of conjunction of matriosomes. A matrisome consist of various WTA patches; each inhibition patch is comprised solely of cells projecting to one of the two output paths of striatum, direct or indirect pathways [46]. We consider these two types of patches to be in roughly equal number throughout a single matriosome unit.

The output of striatum is represented by a single binary vector, one part of which represents activation of cells which project to GPe, another represents activation of cells which project to GPi. Actions of those structures and associated direct and indirect will operationally transform striatal high level command vector into an action with *directional* and *timing* components.

The weights of the simulated striatal neurons are initialized with a uniform random distribution. Training proceeds with each successive step of the learning sequence presented in order; appropriate synaptic modification is performed as described in section 3.1. In case of catastrophic performance cortex will terminate the sequence using its projection to STN, and the training must resume from the very beginning. (this intuitively corresponds to the fact that the sequence will be later executed as a single atomic action, for modification of any components of it the whole sequence has to be re-learned.)

Dopamine reward signal is generated only in the case of "unexpected" enhancement in performance, when the outcome of the action is considerably better then the predicted outcome. The quantitative threshold is imposed on the judgment about the relative differences between the action and "expected" outcome; it serves as a parameter which controls the speed of learning; the finer the difference which produces the reward is, the faster is learning. During training each presentation of the input produces a small change in the weights, random in the direction of the input. A hard limit on the magnitude of weights is imposed to avoid out-of-range fluctuations. The 'unexpected' reward architecture and the random change of weights allows for gradual improvement toward minimizing the error. Obviously getting a reward in each step of the sequence is more efficient that a single reward at the end the sequence. If the reward signal were real valued instead of binary, its magnitude can be used to control the learning rate, gradually decreasing for finer and finer tuning; this however is not implemented in the current model for simplicity sake, and is left for future research.

The random change in the weights allows the implementation of a random search algorithm, that tries new weights and keeps the ones that produce a better performance. The random search algorithm is shown in figure 2. In the first step a small random change is introduced into the weights in the direction of the actual input. The change is stored in short term memory, i.e. a short term modification of synaptic efficacy. Then if the performance of the step with the new weights is better than predicted, the proposed change gets fixed, presumably by the action of the dopamine surge. The learning cycle continues reaching the minimum possible error asymptotically. To produce the random change in the weights in the direction of the input, is reasonable to assume the availability of the past inputs for circular training. If the inputs are orthogonal as suggested below, the changes for each step in the sequence are temporary isolated to separate synaptic sites. If the inputs for each step are not orthogonal the changes done to adjust one step will affect the others, resulting in sub-optimal performance. One simple way to avoid this problem is to orthogonalize the inputs. In relation to biology one can assume the inputs to be binary and the synaptic sites to have a refractory states lasting over the duration of the whole sequence. This effectively results in the fact that once a site is used is remains unmodified for the rest of the sequence. Because each input uses independent synaptic sites, one cell can store arbitrary long sequences, limited only be the number of synaptic sites on that cell.

Figure 3A below shows a possible way for a GP cell to convert a binary input from striatum to real valued activation; considering a single GP cell with random weights that receives several (1000) different striatum binary vectors of dimension 100, the resulting values are presented in increasing order. The cell is capable to work as the binary converter, with low and high saturation zones, and a linear zone in the middle. Figures 3B, 3C and 3D show the results of dynamic simulations developed to analyze the basal ganglia circuit. The figure(s) show GPe (heavy line), and STN (thin line), with a soma potential (dashed line) and firing frequency (solid line). At time $t=200$ the basal ganglia circuit receives input from cortex, which results in long depolarization of MSN's in STR; this in its turn inhibits GPe (the time that the MSN remain depolarized is considered constant in the simulation.) While GPe is inhibited its inhibitionary action on STN is lifted, allowing STN to slowly overcome the threshold and fire; once this happens excitatory axon collaterals within STN drive its activity to saturation. The brief but potent input of cortex can make the firing of STN occur much faster, thus gradually controlling the timing parameters of the circuit. Eventually STN returns to background activation, allowing the circuit to perform next step.

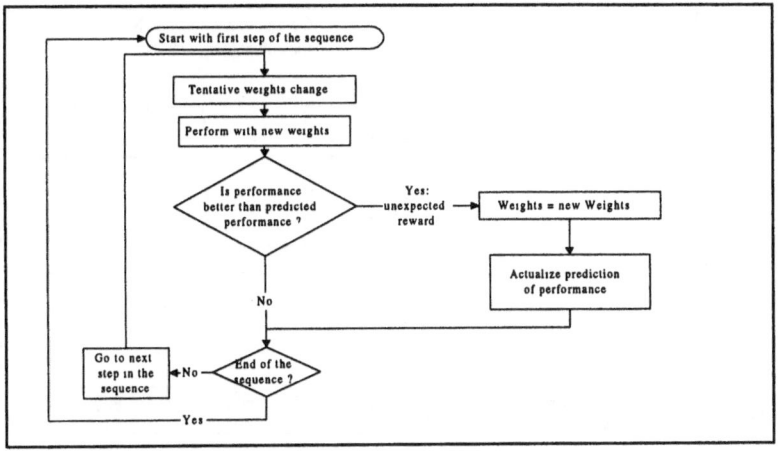

Figure 2. Training algorithm.

4. Experiments in handwriting synthesis

Performance and learning capabilities of a striatal network were illustrated on the example of adaptive production of handwriting via simulated robotic strokes. Samples of actual handwriting from NIST Special Dataset 2 [21] were presented to the network as training data. No effort was made to model a lower level muscle control (smoothing of movements, which is believed to be controlled by spinal cord and cerebellum), so all actions are represented by straight lines. The task was to generate a sequence of straight lines of variable orientation and length, which would most closely approximate the curvature of handwriting samples. To simplify the network only eight actions and eight times were considered to be in the output repertoire. The possible actions are movements from an origin point in directions separated by 45°. The times are mapped to the lengths of the lines in 8 increasing lengths (1 through 8 in a single unit increments). The sequence being trained is defined by a number of motor steps, each characterized by direction and time. The output in each step is represented as n-dimensional vector, one component of which represents direction and the other time. The input in the first step of the sequence is assumed to be a command send from the cortex, which contains the unique designator for the sequence to be run. The input in the following steps consist of somatosensory feedback, and "state variables" which indicate which step of the sequence is actually being executed (these latter components are necessary because is possible that the same somatosensory state will happen again within a sequence.)

Figure 4 depicts samples of sensory images of handwriting digits used in training and the corresponding images produced by the simulated basal ganglia network after 10, 50, 100 and 500 training cycles:

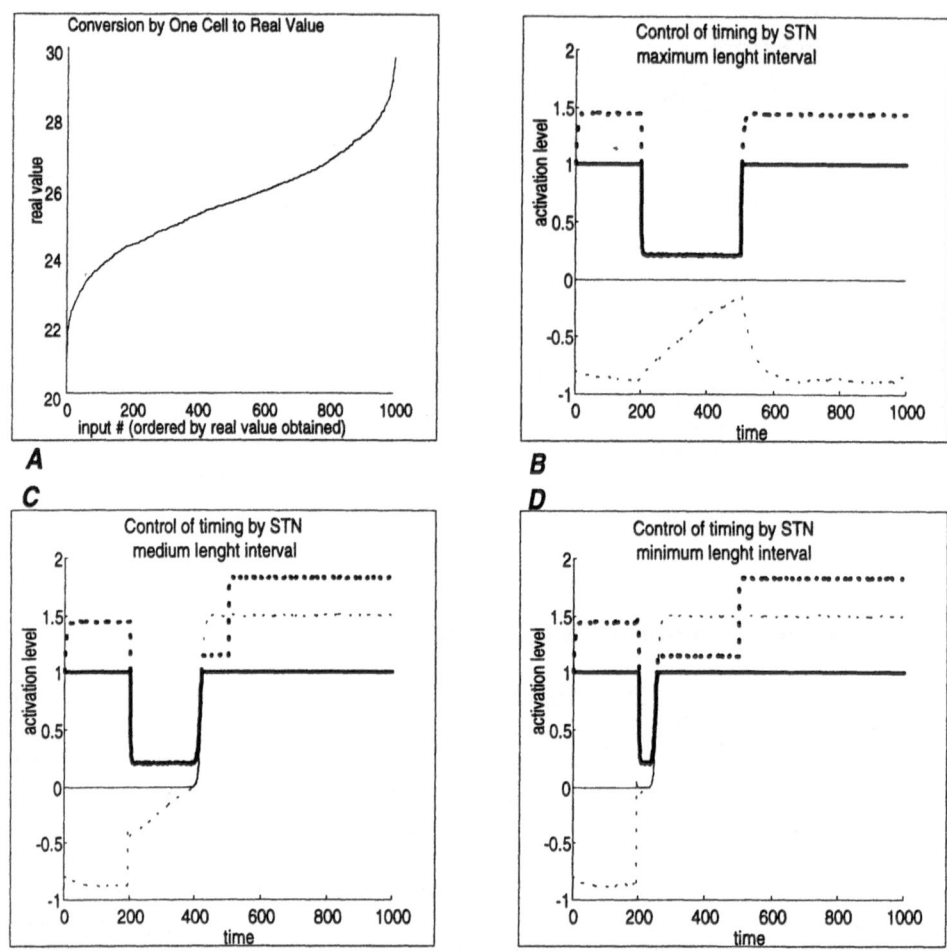

Figure 3. Parameters of the simulation of the model.

In addition to serving a good illustration of the works of the model of striatal complex described in this paper, the problem of handwriting generation has important practical and commercial applications [see 28] - from generating additional training data for pattern recognition machines, to generating personal advertisement content and expedition of interpersonal communications.

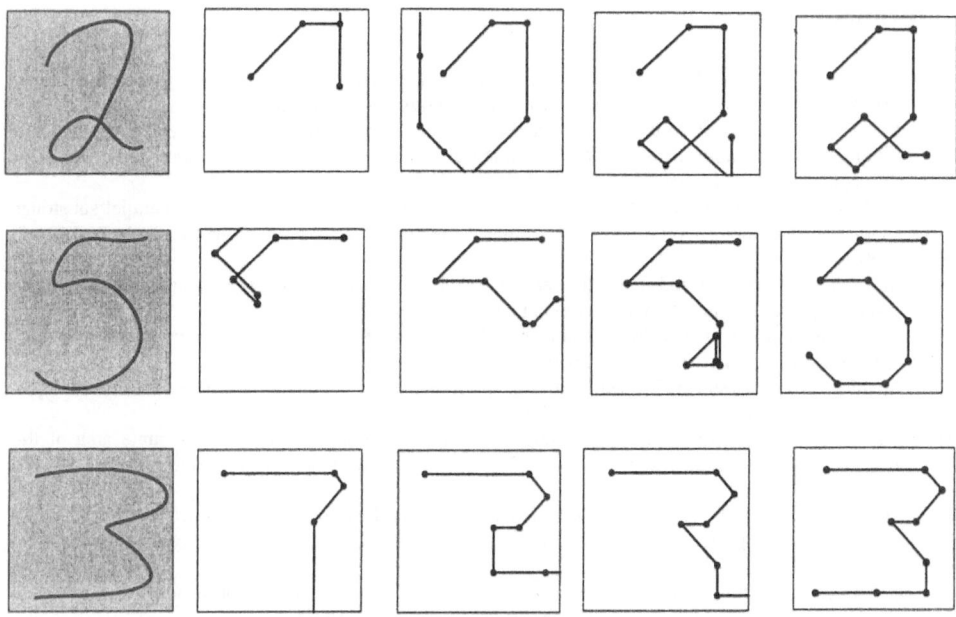

Figure 4: Illustration of Handwriting Synthesis performed by Striatal Network

5. Conclusions

We have presented in this paper realistic, highly biologically motivated model of striatal complex, which is based on the observed anatomy, physiology and biochemistry of the structures which comprise basal ganglia. We have shown preliminary results which are demonstrated to be promising to the potential application of resulting algorithms to image synthesis. Moving beyond simulation of a single muscle group invariably will suggest novel approaches to simulation of complex movements involving a whole body of the organism. Generalizing applications of basal ganglia circuitry to robotic posture, navigation and balancing will serve as an example of such application areas.

Glossary of anatomical terms

GP   - globus pallidum
GPe   - globus pallidum external segment
GPi   - globus pallidum internal segment
SNc   - substantia nigra parc compacta
SNr   - substantia nigra parc reticulata
STN   - subthalamic nucleus

114

References

[1]   Alexander , DeLong M. Microstimulation of the primate neostriatum. I. Physiological properties of striatal microexcitable zones. Journal of Neurophysiology, 1985 Jun, 53(6):1401-16.

[2]   Alexander, DeLong M. Microstimulation of the primate neostriatum. II. Somatotopic organization of striatal microexcitable zones and their relation to neuronal response properties. Journal of Neurophysiology, 1985 Jun, 53(6):1417-30.

[3]   Alexander, M. Crutcher, and M. DeLong. Basal ganglia-thalamocortical circuits: parallel substrates for motor, oculomotor, "prefrontal" and "limbic" functions. Progress in Brain Research, 85:119-46, 1990.

[4]   Alexander, M. DeLong, and P. Strick. Parallel organization of functionally segregated circuits linking basal ganglia and cortex. Annual Review of Neuroscience, 9:357--81, 1986.

[5]   Alexander, G. and Crutcher, M. Functional Architecture of Basal Ganglia Circuits: Neural Substrates of Parallel Processing. 1990. TINS 13(7):266-71.

[6]   Angulo, J. & McEwen, B. Molecular aspects of neuropeptide regulation and function in the corpus striatum and nucleus accumbens. Brain Research Reviews 19 (1994) 1-28.

[7]   Bliss and T. Lomo. Long-lasting potentiation of synaptic transmission in the dentate area of the anaesthetized rabbit following stimulation of the perforant path. Journal of Physiology, 232:331--356, 1973.

[8]   Calabresi, P. Maj, A. Pisani, N. Mercuri, and G. Bernardi. Long-term synaptic depression in the striatum: physiological and pharmacological characterization. Journal of Neuroscience, 12(11):4224--33, 1992.

[9]   Calabresi, P., Pisani, A., Mercuri, N. and Bernardi, G. The corticostriatal projection: from synaptic plasticity to dysfunctions of the basal ganglia. TINS (1996) 19, 19-24.

[10] Cambell G., Eckardt M., and Weight F. Dopaminergic mechanisms in subthalamic nucleus of rat: analysis using horseradish peroxidase and microiontophoresis. Brain Research, 333:261-270, 1985.

[11] Carpenter, K. Nakano, and R. Kim. Nigrothalamic projections in the monkey demonstrated by autoradiographic technics. Journal of Comparative Neurology, 165(4):401--15, February 1976.

[12] Carpenter M., Carleton S., Keller J., and Conte P. Connections of the subthalamic nucleus in the monkey. Brain Research, 1981 Nov 9, 224(1):1-29.

[13] Canteras N., Shammah-Lagnado S., Silva B., and Riccardo J. Afferent connections of the subthalamic nucleus: a combined retrograde and anterograde horseradish peroxidase study in the rat. Brain Research, 513:43-59, 1990.

[14] Chang H., Kita H., and Kitai S. The fine structure of the rat subthalamic nucleus: an electron microscope study. Journal of Comparative Neurology. 221: 113-123, 1983.

[15] Chesselet M. and Delfs J. Basal ganglia and movement disorders: an update. TINS. 19: 417-22, 1996.

[16] Dubois, B., Defontaines, B., Deweer, B., Malpani, C., and Pillon, B. Cognitive and Behavioral Changes in Patients with Focal Lesions of the Basal Ganglia. Advances in Neurology, Vol. 65. 1995. 29-41.

[17] Flaherty and A. Graybiel. Output architecture of the primate putamen. Journal of Neuroscience, 13(8):3222--37, 1993.

[18] Freund, J. Powell, and A. Smith. Tyrosine hydroxylase-immunoreacive boutons in synaptic contact with identified striatonigral neurons, with particular reference to dendritic spines. Neuroscience, 13(4):1189--1215, 1984.

[19] Fuster J. Frontal lobes. Current Opinion in Neurobiology 3:160-165, 1993.

[20] Garsia-Munoz, S. Young, and P. Groves. Presynaptic long-term changes in excitability of the corticostriatal pathway. Neuroreport, 3(4):357--60, 1992.

[21] Garris, M. Design and collection of a handwriting sample image database. Social Science Computer Review, Summer 1992, vol.10, (no.2):196-214.

[22] Gerfen. The neostriatal mosaic: multiple levels of compartmental organization in the basal ganglia. Annual Review of Neuroscience, 15:285--320, 1992.

[23] Goldman-Rakic. Cellular and circuit basis of working memory in prefrontal cortex of non-human primates. Progress in Brain Research, 85:325--35, 1990.

[24] Grace, A. The tonic/phasic model of dopamine system regulation: its relevance for understanding how stimulant abuse can alter basal ganglia function. Drug and Alcohol Dependence 37 (1995) 111-129.

[25] Graybiel A. Neurotransmitters and neuromodulators in the basal ganglia. Trends in Neurosciences, 7(13), 1990.

[26] Graybiel, A. et. al. The Basal Ganglia and Adaptive Motor Control, 1994, Science 265: 1826-31.
[27] Groenewegen, H. Berendse, J. Wolters, and A. Lohman. The anatomical relationship of the prefrontal cortex with the striatopallidal system, the thalamus and the amygdala: evidence for a parallel organization. Progress in Brain Research, 85:95--116, 1990.
[28] Guyon I. Handwriting synthesis from handwritten glyphs. International Workshop on Frontiers in Handwriting Recognition V, University of Essex, Great Britain, 1996.
[29] Hazrati L. and Parent A. Convergence of subthalamic and striatal efferents at pallidal level in primates: an anterograde double-labeling study with biocytin and PHA-L. Brain research. 569:226-340, 1992.
[30] Hedreen J. and DeLong M.. Organization of striatopallidal, striatonigral, and nigrostriatal projections in the macaque. Journal of Comparative Neurology, 304(4):569--95, 1991.
[31] Kawaguchi, Y., Wilson, C., Augood, S., & Emson, P. Striatal Interneurons: Chemical, Physiological and Morphological Characterization. 1995. TINS 18(12) 527:35.
[32] Kimura, M. Role of basal ganglia in behavioral learning. Neuroscience Research 22 (1995) 353-358.
[33] Kilborn, R. Granger, and G. Lynch. Effects of LTP on response selectivity of simulated cortical neurons. J. Cognitive Neuroscience, 1996. (in press).
[34] Kirkwood, S. Dudek, J. Gold, C. Aizenman, and M. Bear. Common forms of synaptic plasticity in the hippocampus and neocortex in vitro. Science, 260(5113):1518--21, 1993.
[35] Kitai S. and Deniau, J. Cortical inputs to the subthalamus: intracellular analysis. Brain research, 241: 411-415, 1981.
[36] Kita, T. Kita, and S. Kitai. Active membrane properties of rat neostriatal neurons in an in vitro slice preparation. Experimental Brain Research, 60(1):54--62, 1985.
[37] Kita, H. Kita, and S. Kitai. Passive electrical membrane properties of rat neostriatal neurons in an in vitro slice preparation. Brain Research, 300(1):129--39, 1984.
[38] Kitai S. and Kita H. Anatomy and physiology of the subthalamic nucleus: a driving force in the basal ganglia. In M.B. Carpenter and A. Jayaraman (Eds.) The Basal Ganglia II - Structure and Function: Current Concepts, Plenum Press, New York, pp. 357-373, 1987.
[39] Kombian and R. Malenka. Simultaneous LTP of non-NMDA- and LTD of NMDA-receptor-mediated responses in the nucleus accumbens. Nature, 368(6468):242--6, 1994.
[40] Larson and G. Lynch. Theta pattern stimulation and the induction of LTP: the sequence in which synapses are stimulated determines the degree to which they potentiate. Brain Research, 489:49--58, 1989.
[41] Larson, D. Wong, and G. Lynch. Patterned stimulation at the theta frequency is optimal for induction of long-term potentiation. Brain Research, 386:347--350, 1986.
[42] Malenka. Synaptic plasticity in the hippocampus: LTP and LTD. Cell, 78(4):535--8, 1994.
[43] Parent, A. Comparative neurobiology of the basal ganglia. New York : J. Wiley, 1986.
[44] Parent, A. Extrinsic connections of the basal ganglia. Trends in Neuroscience, 13(7), 1990.
[45] Parent, A and Hazrati, L. Anatomical aspects of information processing in primate basal ganglia, 1993, TINS 16(3):111-6
[46] Parent, A and Hazrati, L. Functional anatomy of the basal ganglia. I. The cortico-basal ganglia-thalamo-cortical loop. Brain Research Reviews 20(1995) 91-127.
[47] Parent, A and Hazrati, A. Functional anatomy of the basal ganglia. II. The place of subthalamic nucleus and external pallidum in basal ganglia. Brain Research Reviews 20(1995) 128-154.
[48] Pennartz, R. Ameerun, H. Groenewegen, and H. L. da Silva. Synaptic plasticity in an in vitro slice preparation of the rat nucleus accumbens. European Journal of Neuroscience, 5(27):107--17, February 1993.
[49] Schell and P. Strick. The origin of thalamic inputs to the arcuate premotor and supplementary motor areas. Journal of Neuroscience, 4(2):539--60, February 1984.
[50] Smith, T. Wichmann, and M. DeLong. The external pallidum and the subthalamic nucleus send convergent synaptic inputs onto the single neurones in the internal pallidal segment in monkey: anatomical organization and functional significance. The basal ganglia IV, Percheron (eds.). 1994.
[51] Uno and N. Ozawa. Long-term potentiation of the amygdalo-striatal synaptic transmission in the cource of development of amygdaloid kindling in cats. Neuroscience Research, 12(1):251--62, October 1991.
[52] Walsh. Depression of excitatory synaptic input in rat striatal neurons. Brain Research, 608(1):123--8, 1993.
[53] Wilson. Dendritic morphology, inward rectification, and the functional properties of neostrital neurons. in single neuron computation. McKenna, Davis and Zornetzer, eds. 1991.
[54] Wilson and P. Groves. Spontaneous firing patterns of identified spiny neurons in the rat neostriatum. Brain Research, 220(1):67--80, 1981.

Symmetry and Self-Organization
of the Oculomotor Neural Integrator

Thomas J. Anastasio

University of Illinois
Beckman Institute
405 N. Mathews Ave.
Urbana, IL 61801 USA
tstasio@uiuc.edu

Abstract

Modeling and experiment support the assertion that the oculomotor neural integrator works through positive feedback via reciprocal inhibition. Clinical studies reveal that the integrator is operational immediately after birth, suggesting that it develops prenatally. Using facts and computational techniques from dynamic systems theory, it can be shown that self-organization of the oculomotor neural integrator could occur through the construction during development of symmetrically connected clusters of neurons. This method is neurobiologically plausible and robust, and networks designed using it reproduce the dynamics observed for brainstem neurons that are part of the actual oculomotor neural integrator.

INTRODUCTION

In order to control the movement of the eye using the extraocular muscles, ocular motoneurons must have information related both to eye velocity and position (Robinson 1989). The eye velocity and position commands are needed to overcome the viscosity and elasticity of the eye, respectively. All of the premotor neurons in the oculomotor system, whether part of the vestibular, optokinetic, saccadic, pursuit, or vergence subsystems, carry a command proportional to eye velocity. The eye position command is generated by the oculomotor neural integrator, a neural circuit composed of neurons in the vestibular and prepositus nuclei in the brainstem, which receives eye velocity commands as input and temporally integrates them to produce eye position commands as output.

Modeling and experiment both indicate that temporal integration is produced by positive feedback within the brainstem neural networks that compose the integrator. The premotor neurons that carry eye velocity commands to the oculomotor neural integrator are spontaneously active, so the integrator must have some way of integrating the modulated activity but not the spontaneous carrier. The elegant solution to this problem, proposed by Cannon and associates (1983), takes advantage of the fact that the oculomotor system is arranged bilaterally and in push-pull. In their model of the integrator, the vestibular and prepositus neurons on either side of the brainstem reciprocally inhibit each other, thereby exerting net positive feedback on themselves. The reciprocal arrangement allows them to reject the spontaneous carrier rate of the premotor neurons but integrate the modulated, push-pull eye velocity signal. This solution requires that the reciprocal inhibitory connections are balanced bilaterally.

This integrator model was developed first for a two-neuron network, in which both neurons were modeled simply as linear, first-order, low-pass elements with membrane (resistor-capacitor) time constants (TAUmem) of 5ms. A low-pass element will act as an integrator for input signals at frequencies that are higher than its corner frequency. The corner frequency in radians/second is equal to the reciprocal of the time constant in seconds. The corner frequency of the individual neurons with TAUmem of 5ms would be 200 radians/second. This is much too high to be useful, considering that the frequency range of oculomotor velocity commands can extend to frequencies lower than 0.1 radians/second. Thus, the main goal of the positive feedback in the integrator network is to lengthen the apparent time constant of the neurons from TAUmem to some longer network time constant, TAUnet. Cannon and associates (1983) showed that, when the two neurons in the network inhibited each other with a connection of absolute value W, TAUnet would equal TAUmem/(1−W). Their analysis of the two-neuron integrator showed that the time constant of the integrator network (TAUnet) will get longer as the strength of the reciprocal inhibitory connections (W) gets closer to 1. The integrator network will become unstable for W > 1. Their analysis also showed that the gain of the low-pass network is proportional to its time constant, TAUnet.

Integration in the two-neuron network is sensitive to the precise value of W. To overcome this problem, Cannon and associates (1983) distributed the reciprocal inhibitory connections over a population of neurons so that the value of any individual connection was less critical. As in the two-neuron case, the connectivity in the distributed model was symmetrical in that the inputs sent out and received by any neuron in the network were the same. Thus, the network connectivity matrix equaled its transpose. The model predicted that removal of a subset of the connections should lower the apparent time constant of the combined output of the neural integrator. In the real brainstem, reciprocal inhibitory connections are carried over the commissures. Experiments in which subsets of brainstem commissural fibers were lesioned in fact produced sever integrator deficits, lending support to the model (Anastasio and Robinson 1991).

Recently, the question of how the integrator is constructed has been raised. Arnold and Robinson (1991) trained a neural network to be an integrator using an error-driven, supervised learning algorithm. They found that the network learned to have symmetrical connections, as in the previous model by Cannon and associates (1983). In fact, all of the reciprocal inhibitory connections made by the learning algorithm had approximately the same absolute value. The main problem with the learning hypothesis for integrator construction is that the neural integrator is operative immediately postnatally, before any learning through experience can take place (Weissman et al. 1989). Although postnatal modification of the integrator can occur through error-driven learning (Kapoula et al 1989), it appears that an operating neural integrator develops before birth by some other mechanism. The purpose of this study is to investigate a possible strategy for self-organization of the oculomotor neural integrator that does not rely on learning through experience, but might be programmed into some prenatal developmental process.

METHODS

The linear, two-neuron model of Cannon and associates (1983) was expanded to include larger numbers of neurons (from 4 up to 200). Various patterns of connectivity were set, either directly or through iterative processes, and the resulting networks were analyzed computationally using dynamic systems tools from MATLAB (The MathWorks, Inc.). These tools were used to extract the eigenvalues, and to determine poles and zeros and degrees of observability. They were also used to examine network dynamics through the computation of transfer functions and Bode plots.

RESULTS

Experimental and modeling work provides clues into the strategy might be used by the developing nervous system as it constructs the oculomotor neural integrator network. The neural integrator is composed of neurons in the vestibular and prepositus nuclei. Neurophysiological studies of these neurons (e.g. Shinoda and Yoshida 1974; Lopez-Barneo et al. 1979; Escudero et al. 1992) establish three crucially important facts. The first is that the responses of vestibular and prepositus neurons do not appear to oscillate. The second is that the phase lag of the responses of vestibular and prepositus neurons relative to the velocity signals they receive never exceeds 90 degrees, although it may bottom out at values between 0 and 90 degrees. This result indicates that the oculomotor neural integrator does not integrate its inputs more than once. Instead, it is limited to 1 order or less (a fractional order between 0 and 1) of integration (Anastasio 1994). The third important fact is that the responses of vestibular and prepositus neurons are stable in the sense that they do not grow without bound after they are stimulated but decay at various rates back to their spontaneous baseline.

The main clue from the modeling studies is that both the directly constructed (Cannon et al. 1983) and the learned (Arnold and Robinson 1991) integrator network models have symmetrical connectivity matrices. It was previously thought that this symmetry was necessary only to maintain network balance. However, system symmetry accomplishes much more that than in neural integrator networks. For example, it is a well established fact from linear algebra that, although an asymmetrical matrix may have complex eigenvalues, a symmetrical matrix can have only real eigenvalues (Noble and Daniel 1977). The implications of this for system dynamics are that the activities of neurons in integrator networks with symmetrical connectivity matrices will never oscillate. Therefore, a developmental process that promotes symmetry in connections within the integrator network will avoid not only imbalance but oscillation as well.

Simulation studies on symmetrically connected networks provide further insights. Symmetrical integrator networks were constructed in which each neuron on one side had reciprocal connections to every neuron on the other side. A simple, four-neuron example of such a symmetrical, completely connected network is shown in Figure 1 on the left. The phase lags of the neurons in these networks never exceeded 90 degrees, as observed neurophysiologically. An important characteristic of these symmetrical systems is that they had only 1 dominant eigenvalue or pole, and all subdominant poles were canceled by zeros. The implications of this are that the dynamics of neurons in symmetrically and completely connected integrator networks will be governed by only one time constant, and it will be equal to the opposite reciprocal of the dominant pole. When that pole is negative the integrator will be stable. Also, because the dominant pole can be brought arbitrarily close to 0, the time constant can be made arbitrarily long and the neurons will act as order-1 integrators over a wide frequency range. In contrast, asymmetrical networks, even those that are balanced and non-oscillatory (as in Figure 1 on the right), can have multiple, equal dominant poles, and/or substantial subdominant poles, that are not all canceled by zeros. The dynamics of neurons in those networks would be governed by multiple poles, producing integration of order greater than 1, with the consequence that their phase lags would increase to many multiples of 90 degrees. That behavior is never observed neurophysiologically.

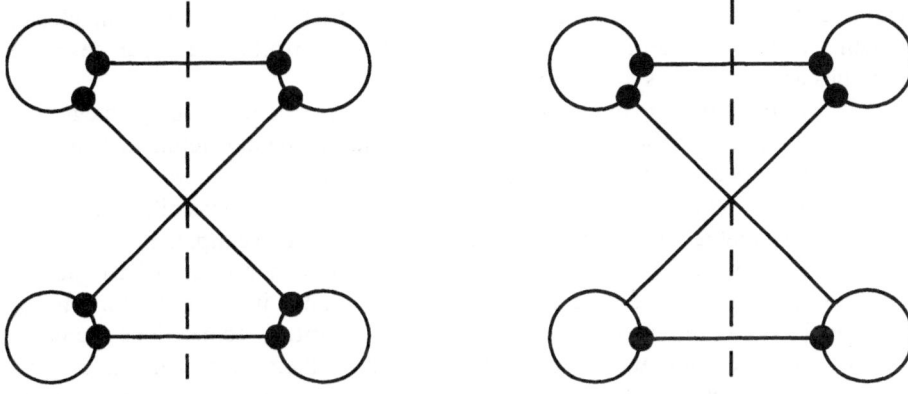

Figure 1. Two simple, four-neuron integrator networks that are similar in connectivity but show different dynamics. Both networks are arranged with two neurons on either side of a model brainstem. In the network on the left, each neuron on one side makes reciprocal, inhibitory connections with both neurons on the other side. The four reciprocally connected neurons in this network form a cluster in which the dynamics of each neuron is governed by a single pole, which causes them to act as order-1 integrators. In the network on the right, one reciprocally connected pair sends its outputs to a second reciprocally connected pair. Each pair acts as an order-1 integrator by itself, so the output of the first pair is integrated again by the second pair, resulting in order-2 integration. Neurons that compose the real oculomotor neural integrator could not be connected as shown on the right since they are not observed to integrate with an order greater than 1.

These theoretical facts and modeling results suggest that a neural integrator could be constructed simply by making symmetrical, reciprocal connections between brainstem neurons. To make self-organization even simpler, the developmental program could specify that all of the inhibitory connections between neurons in a reciprocally connected cluster must have the same absolute value.

That value will be inversely proportional to the number of neurons in the cluster, so the strength of the reciprocal connections can be small for large clusters. Stability of the dynamics of the cluster could be ensured by increasing the value of the connection weights incrementally and stopping as the network approaches marginal stability. This is equivalent to observing the dominant pole (Luenberger 1979). It is feasible that some other structure (such as the cerebellum, for example) could monitor the development of the integrator, because fully connected systems such as the proposed integrator clusters are completely observable. The implication of this is that the dominant pole of the cluster is observable in principle by observing the response of any neuron in the cluster. Simulation studies show that this iterative strategy very reliably produces neural integrator networks.

This developmental strategy is supported by another well established fact from linear algebra. Symmetrical matrices are well-conditioned, so that their eigenvalues are relatively insensitive to changes in connection strength values. The implication of this is that an integrator network would retain its properties, such as its lack of oscillatory behavior and limit to order-1 integration, even with departures from perfect symmetry in its connections. Therefore, the developmental process that promotes symmetry in connections need not be perfect. As long as the cluster is stable, a near-symmetrical integrator network will exhibit the same desirable properties as a perfectly symmetrical one.

Presumably, the integrator as a whole would contain many integrating clusters. To ensure no more than 1 order of integration, the developmental strategy would have to specify that neurons in an integrating cluster could not receive inputs from neurons in other integrating clusters. This suggests the possibility that some neurons, those that are not in integrating clusters, could receive inputs from multiple clusters. The analysis of the two-neuron integrator (Cannon et al. 1983) showed that both the gain and the time constant of the integrator increase proportionally, and sensitively, as the strength of the reciprocal inhibitory connections increases toward 1. Since clusters would vary in the strengths of their reciprocal connections, the receiver neurons (i.e. those not in clusters) would receive low-pass inputs in which the gains and time constants are proportional to each other, but vary from cluster to cluster. Over some range of frequencies, the sum of these inputs would approximate an integrated output with an order between 0 and 1 (Anastasio 1994). Networks constructed according to this strategy, with integrating clusters and separate receiver neurons, reproduce the range of dynamics actually observed for real vestibular and prepositus nuclei neurons, as shown in Figure 2.

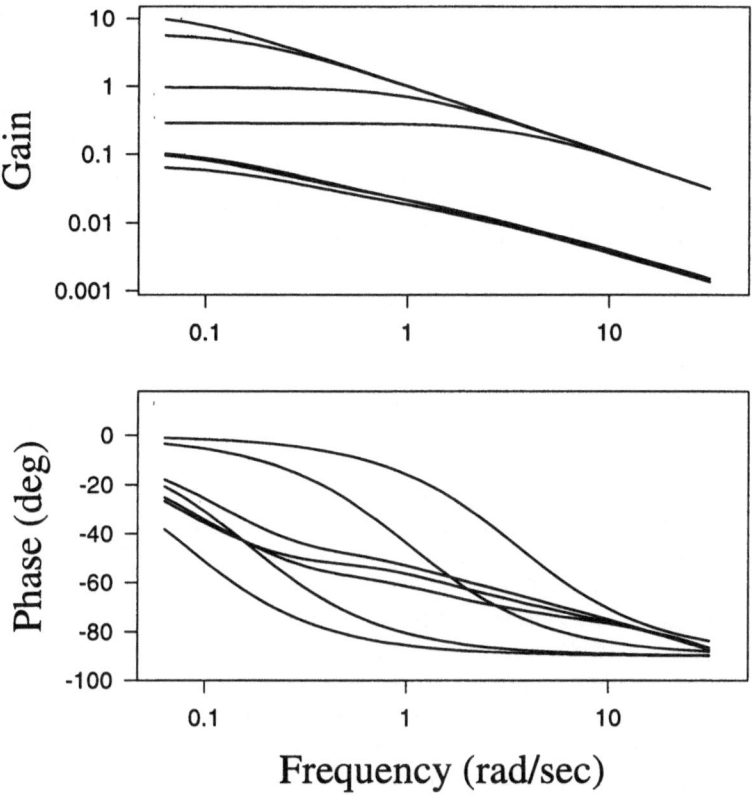

Figure 2. Bode plot of representative neurons from an integrator network constructed as described above, with integrating clusters of neurons and separate receiver neurons. Gain (or amplitude ratio computed as output/input) and phase lag (of the output relative to the input in degrees) is plotted as a function of input frequency in radians/second. The dynamics of the neurons are essentially those of low-pass elements, with gain decreasing and phase lag increasing with frequency. None of the neurons have gain slope more negative than −1 nor phase lag greater than 90 degrees, showing that none integrate with order greater than 1. For some of the neurons there are ranges of frequency in which gain slope is less negative than − 1 and phase lags by less than 90 degrees, indicating that in these regions the order of integration is less than 1. Dynamics similar to these are observed for the neurons that compose the real oculomotor neural integrator, such as those found in the vestibular and prepositus nuclei.

DISCUSSION

Developmental neurobiology has revealed a wealth of mechanisms (e.g. Purves and Lichtman 1985; Jacobson 1993) through which neurophysiologically operative structures could be constructed. The development of the oculomotor neural integrator could begin within a set of progenitor neurons, bilaterally arrayed in the brainstem, as they start to send out projections over the midline. An axonal projection that finds a target neuron on the other side could be retraced by a reciprocal projection. The establishment of reciprocal connections between two contralateral progenitor neurons could trigger the proliferation of those progenitors into an integrating cluster of neurons. The triggering event, mediated perhaps by mutual exchange of chemical messengers, could set such parameters as the number of progeny neurons and the strength of the reciprocal connections to be made between them (which are inversely related, see above). It could also establish common cell-surface receptor molecules for all neurons in the cluster on either side. In the presence of high concentrations of trophic factors, these cell-surface receptors would ensure that growing axons from one side of the cluster would find their correct targets on the other side, forming an approximately equal, complete, symmetrical set of reciprocal connections.

Cell-surface receptor molecules could guide the entire development of the oculomotor neural integrator, given the rule that a developing axon can make a connection only if it has the ability to recognize the cell-surface marker of the target neuron. Cell-surface receptors would be cluster-specific. Reciprocal connections within clusters would be assured because neurons on one side of the cluster would be able to recognize their counterparts on the other side. Connections between clusters would be prohibited because neurons in one cluster would lack the ability to recognize the cell-surface markers of neurons in other clusters. Progenitor neurons that do not initially establish reciprocal connections would differentiate into receiver neurons. They would not express a cluster-specific marker, nor would they be able to recognize any cluster-specific markers. However, they would express a non-specific marker that could be recognized by neurons in any cluster. Thus, receiver neurons could not project to neurons in any cluster, but could receive projections from neurons in every cluster. Both cluster and receiver neurons could project outside the integrator, to motoneurons, for example.

All of the developmental mechanisms sketched-out above are entirely plausible. Other developmental processes that were not discussed above, such as activity-dependent mechanisms, may also play a role in the development of the oculomotor neural integrator. The finding that a neural integrator with desirable properties can be constructed reliably and robustly, simply on the basis of symmetry in reciprocal connections, ushers in a whole host of possible strategies by which it could develop prenatally. The foregoing demonstrates that temporal integration by oculomotor neural networks does not have to be learned, but can be self-organized during development.

CONCLUSION

This study shows that the oculomotor neural integrator could self-organize during the development of the nervous system according to a simple and neurobiologically plausible strategy. That strategy involves nothing more than specifying that connections in reciprocally connected clusters of neurons should all have approximately the same strength, and that strength can take any value so long as the cluster is stable. It must further specify that neurons in one integrating cluster cannot receive inputs from neurons in another integrating cluster. However receiver neurons, which are not part of integrating clusters, can receive and sum the outputs of any number of neurons that are in integrating clusters. Neurons in artificial networks constructed using such a process of simulated development have dynamic properties similar to those that are observed for the neurons that compose the actual oculomotor neural integrator.

ACKNOWLEDGMENTS

The author expresses his thanks to Drs. Harry Hilton, Juraj Medanic, and Petros Voulgaris for many helpful discussions on the topics of linear algebra and systems theory. This work was supported by National Institutes of Health Grant PHS 1 R29 MH50577.

REFERENCES

Anastasio TJ (1994) The fractional-order dynamics of brainstem vestibulo-oculomotor neurons. Biol Cybern 75: 1-9

Anastasio TJ, Robinson DA (1991) Failure of the oculomotor neural integrator from a discrete midline lesion between the abducens nuclei in the monkey. Neurosci Lett 127: 82-86

Arnold DB, Robinson DA (1991) A learning network model of the neural integrator of the oculomotor system. Biol Cybern 64: 447-454

Cannon SC, Robinson DA, Shamma S (1983) A proposed neural network for the integrator of the oculomotor system. Biol Cybern 49: 127-136

Escudero M, de la Cruz RR, Delgado-Garcia JM (1992) A physiological study of vestibular and prepositus hypoglossi neurons projecting to the abducens nucleus in the alert cat. J Physiol 458:539-560

Jacobson M (1993) Developmental Neurobiology, 2nd edition, Plenum Press, New York

Kapoula Z, Optican LM, Robinson DA (1989) Visually induced plasticity of post-saccadic ocular drift in normal humans. J Neurophysiol 61: 879-891

Lopez-Barneo J, Darlot C, Berthoz A (1979) Functional role of the prepositus hypoglossi nucleus in the control of gaze. Prog Brain Res 50: 667-679

Luenberger DG (1979) Introduction to Dynamic Systems, John Wiley, New York

Noble B, Daniel JW (1977) Applied Linear Algebra, Prentice-Hall, Englewood Cliffs

Purvis D, Lichtman JW (1985) Principles of Neural Development, Sinauer, Sunderland

Robinson DA (1989) Control of eye movements. In: Brooks VB (ed) Handbook of Physiology, sect. 1: The nervous system, vol II, part 2. American Physiological Society, Bethesda, 1275-1320

Shinoda Y, Yoshida K (1974) Dynamic characteristics of responses to horizontal head angular acceleration in vestibuloocular pathway in the cat. J Neurophysiol 37: 653-673

Weissman BM, DiScenna AO, Leigh RJ (1989) Maturation of the vestibulo-ocular reflex in normal infants during the first two months of life. Neurology 39: 534-538

Quantal Neural Mechanisms Underlying Movement Execution and Motor Learning

J.M. Delgado-García, A. Gruart, J.A. Domingo and J.A. Trigo

Laboratorio de Neurociencia, Facultad de Biología, Universidad de Sevilla
Avda. Reina Mercedes, 6, Sevilla-41012, SPAIN
Phone: + 34-54-625007 Fax: + 34-54-612101
e-Mail: labneuro@obelix.cica.es

Abstract

The nictitating membrane/eyelid motor system appears to be a suitable experimental model for the study of neural processing underlying the genesis and control of spontaneous, reflex and learned motor acts. A short review of available data regarding the kinematic, time-domain, and frequency-domain properties of this motor system is presented. Lid movements seem to be generated by a ≈ 20 Hz oscillator in close relationship with central neural processes involved in attentive states. A quantal organization of this motor system is revealed mostly when considering the sensory-motor strategies related to the generation of new motor responses.

Tremor as a basic substrate for movement execution and acquisition

For many years, most medical textbooks have considered tremor as the exclusive result of some brain pathological condition. From this point of view, tremor is usually defined as a more or less rhythmic (and unwanted) movement of the whole body or of some body appendage. Tremor may be present at rest or during intentional movements, and may cover a wide range of oscillation frequencies and peak amplitudes, depending upon the specific brain malfunction. However, a very early description (Horsley and Schäfer, 1886) of motor effects produced by the electrical stimulation of the cerebral cortex correctly indicated that muscular response does not vary with the frequency of the stimulus, but maintains a rate of ≈ 10 contractions per second. Present understanding of brain function proposes that movement has a basically discontinuous nature and that tremor is the result of the oscillatory properties of involved neural circuits, and of the inertial and viscoelastic characteristics of executing muscle effectors (Llinás, 1991; Vallbo and Wessberg, 1993).

Motor behavior (walking, running, chewing) is rhythmic and is usually the result of the orderly repetition of some basic movements. However, when studied in controlled experiments, slow elbow, wrist, and finger simple movements also appear as discontinuous, that is, as composed of a succession of undulations very regular in both frequency and amplitude. Moreover, each body moving part has a dominant oscillation frequency which depends on the pulsatile activity of central neural elements controlling it, on the organized interplay of involved agonist and antagonist muscles, and on its biomechanical characteristics (Halliday and Redfearn, 1956; Marshall and Walsh, 1956; Brooks, 1974; LLinás, 1991; Vallbo and Wessberg, 1993).

Central neural oscillators thus play a crucial role in movement initiation and coordination. For example, fingers oscillate at a dominant frequency of 8-10 Hz (Vallbo and Wessberg, 1993) and it has been shown that finger movements are initiated at the moment of maximum angular velocity (momentum) during the oscillation, that is, at points with the greatest kinetic energy (Goodman and Kelso, 1983). On the other hand, each period of oscillation results from the organized succession of agonist and antagonist muscle actions, and the latter have been shown to be the result of the timed succession of appropriate neural motor commands (Llinás, 1991; Vallbo and Wessberg, 1993).

The discontinuous nature of movement execution becomes more evident during movements involving the learning of a new motor task. This is an indication of neural processing for a motor solution to the temporo-spatial task, as the same muscles are involved but they are now under the pressure of being coordinated with a different motor program. With training, coordination and dumping are improved by neural mechanisms, and a more precise and coordinated movement is performed.

The nictitating membrane/eyelid motor system as an experimental model for the study of reflex and learned motor responses

The study of brain function during movement and of the plastic properties of neural circuits involved in motor learning requires the election of a suitable experimental model. Among the main advantages of the nictitating membrane/eyelid motor system are the following: i) the eyelid is load-free, has an almost negligible mass and have no (functionally relevant) proprioceptors; ii) both inputs and outputs to the system can be controlled and quantified; eyelid movements can be precisely recorded with the magnetic field, search coil technique and eyelid-evoking stimuli (puffs of air, flashes of light, sounds) can be easily applied in a quantitative way (see Gruart et al., 1995); and iii) the electromyographic activity of the involved muscles can also be recorded and quantified. At the same time, there is a rapidly increasing amount of anatomical information about involved neural circuits and about the electrophysiological properties of representative neural populations that participate in the different types of eyelid response. Finally, the classical conditioning of the nictitating membrane/eyelid response has been used for more than 30 years to the study of neural plastic properties involved in the acquisition of new motor responses (see Gormezano et al., 1983 for references).

Kinematics and frequency-domain properties of eyelid movements

The kinematics, time-domain, and frequency-domain properties of eyelid movements have recently been described in conscious cats (Gruart *et al.*, 1995; Domingo *et al.*, 1996). In short, spontaneous blinks are produced by a fast (up to 2000 deg/sec) downward movement of the upper lid (the one usually recorded) followed by a slower lid movement in the upward direction. The profiles and kinetic properties of reflex blinks depend on the sensory modality of the evoking stimulus. Air puff-evoked blinks are fast downward lid movements, usually followed by a succession of smaller downward waves, their number depending upon the duration of the stimulus. The first fast downward lid movement has an almost fixed rise time of ≈ 25 ms and a very short latency (≈ 10-12 ms). The amplitude of this first downward lid movement depends on stimulus intensity but, as the duration of the movement is invariable, different amplitudes are achieved with changes in lid downward velocity. The late waves that follow the initial one have a mean duration of 40-50 ms and sometimes exceed the duration of the stimulus. Flash-evoked blinks have a longer latency (≈ 40-50 ms) and are usually composed of a single downward lid displacement followed by a slow upward movement. Tone-evoked blinks are also of a long latency, and rather small in amplitude, with an easy fatigability.

Frequency-domain analysis of reflexively-evoked eyelid motor responses with the fast Fourier transform suggests the existence of a 20-25 Hz oscillator underlying these reflex blink responses (Domingo *et al.*, 1996). This latter finding is very surprising as most described motor systems (finger, wrist, arm, etc.) present a dominant frequency < 10 Hz, and are supposedly controlled by the well-known intrinsic oscillatory properties of inferior olive neurons (Llinás, 1991). As proposed below, the eyelid motor system, because of its peculiar biomechanical properties, and because of its involvement with eye movements and attentive processes, needs to be controlled by a faster neural circuit, able to move the lid in synchrony with the 40-Hz coherent magnetic activity recorded in humans during attentive states (Llinás and Ribary, 1993). Synchronous 25-35 Hz oscillations have also been recorded in the somatosensory cortex of alert behaving cats (Bouyer *et al.*, 1987).

In the present paper we will present some data regarding the frequency-domain properties of conditioned eyelid responses and the role of identified facial motoneurons innervating the *orbicularis oculi* muscle. This is an interesting question for two reasons. First, given that conditioned responses are generated by neural circuits on the basis of selected sensory cues used as conditioned stimulus, the involvement of the above-mentioned 20-25 Hz oscillator can be determined along the acquisition of a new motor response. Second, firing properties of motoneurons innervating the muscle that directly produces the downward lid movement (i.e., the *orbicularis oculi* muscle) have to be modified until the motoneuron is able to fire during the conditioned stimulus-unconditioned stimulus interval in a progressive way in order to produce a conditioned response.

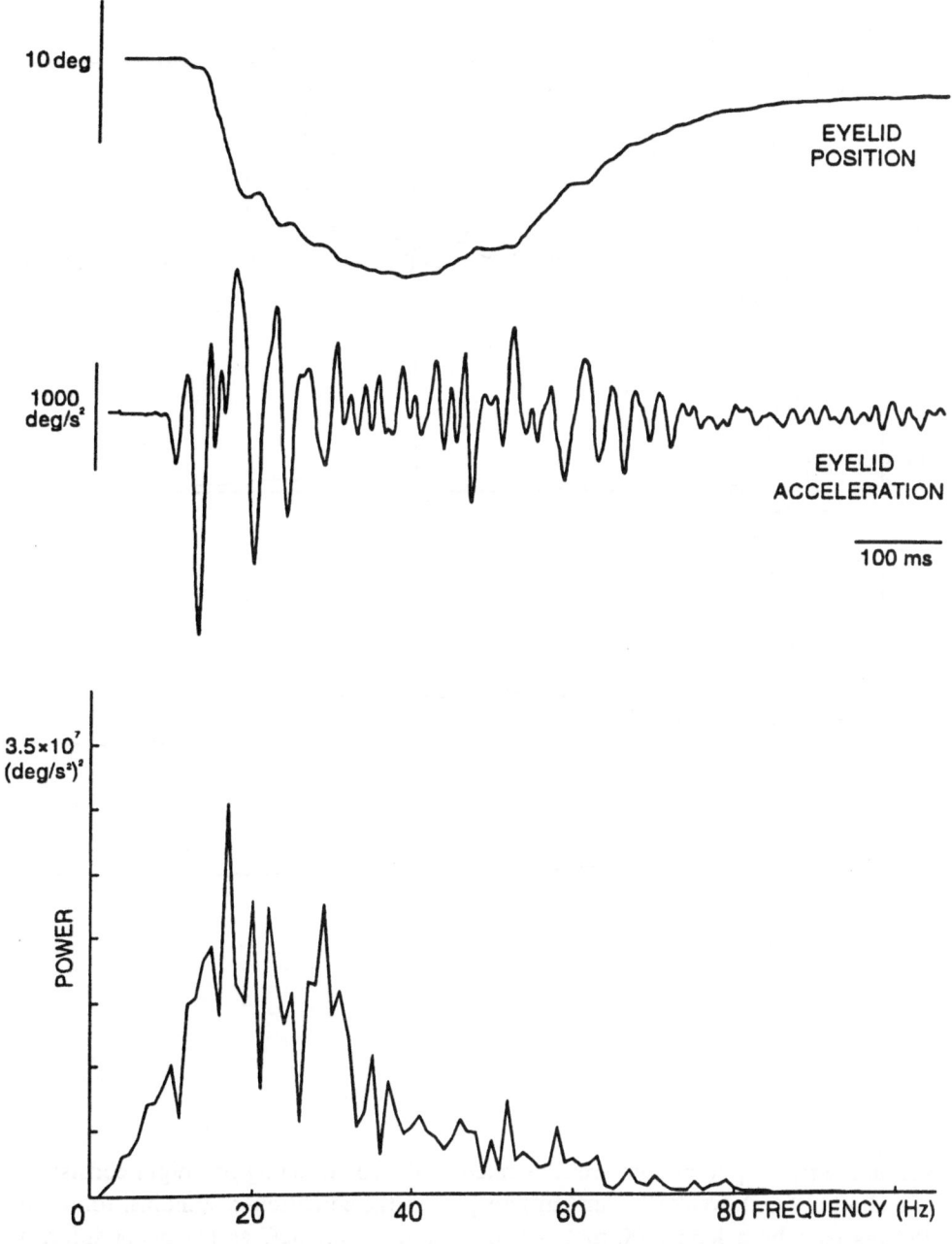

Figure 1. From top to bottom are illustrated: i) a conditioned eyelid response evoked by the sole presentation of the conditioning stimulus (a 600 Hz, 90 dB, 350 ms tone); ii) the acceleration profile of the lid movement; and iii) the power spectrum of the illustrated acceleration record. Calibrations for lid displacement and acceleration are indicated. The illustrated power spectrum (y axis) is expressed in acceleration units $(deg/s^2)^2$.

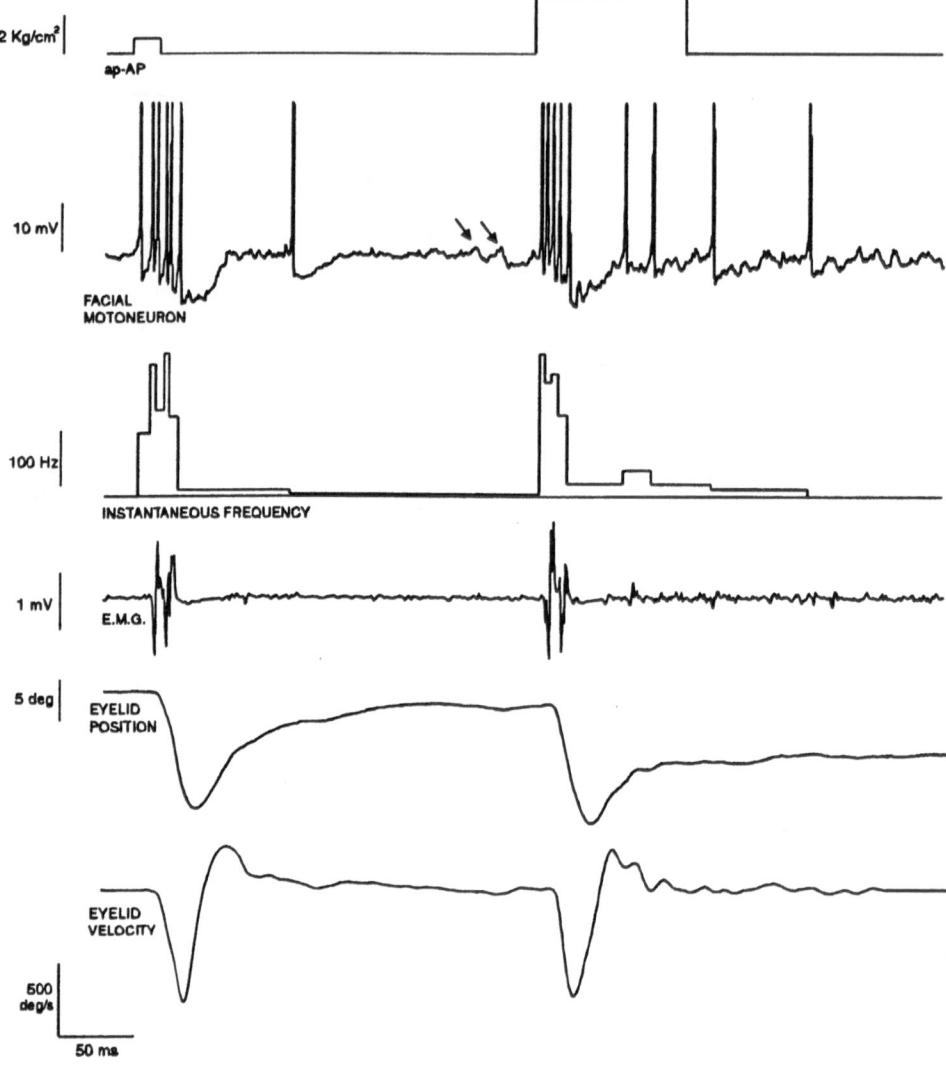

Figure 2. From top to bottom are illustrated: i) the conditioning paradigm consisting of a short (20 ms), weak (0.8 km/cm²) air puff as the conditioned stimulus, followed 250 ms later by a long (100 ms), strong (3 kg/cm²) air puff as the unconditioned stimulus; ii) the activity of a facial motoneuron recorded during the 2nd conditioning session; iii) the instantaneous frequency (spikes/s) of the recorded unit; iv) the electrical activity of the orbicularis oculi muscle; v and vi) the position and velocity of the evoked eyelid response. Note that a single spike only was produced at tⱼhe conditioned stimulus-unconditioned stimulus time interval. Note also the phasic nature of this motoneuron and the hyperpolarization period that follows every burst of action potentials. See text for further details.

Figure 1 illustrates an example of an eyelid conditioned response obtained in a well-conditioned animal with a *delayed* tone-air puff classical conditioning paradigm. The animal was presented with a 600 Hz, 90 dB, 350 ms tone as the conditioned stimulus, followed 250 ms from the beginning of the tone by a 3 kg/cm², 100 ms as the unconditioned stimulus. Thus, the beginning of the air puff was delayed in relation to the beginning of the tone, but the two stimuli finished simultaneously. The eyelid position record shown in Figure 1 corresponds to the conditioned blink evoked by the sole presentation of the conditioned stimulus. The acceleration profile of the evoked conditioned response presents evident oscillations with a dominant peak at $\approx$ 20 Hz. Accordingly, data presented in Figure 1 point to the fact that, as already described for reflex eyelid responses (Gruart *et al.*, 1995), conditioned blinks are generated by a neural oscillator of $\approx$ 20 Hz.

It should be pointed out that conditioned eyelid responses did not appear during the successive conditioning sessions having an all-or-nothing character. On the contrary, conditioned responses appeared for the very first time as a minimum quanta ($\approx$ 2-3 deg in amplitude and $\approx$ 50 ms in duration) of lid downward movement. This point is further illustrated in Figures 2 and 3.

Data shown in Figure 2 were obtained in another animal during a *trace* short, weak air puff/long, strong air puff classical conditioning paradigm. In this case a short (20 ms), weak (0.8 kg/cm²) air puff was presented to the ipsilateral cornea as the conditioned stimulus, followed 250 ms later by a long (100 ms), strong (3 kg/cm²) air puff as the unconditioned stimulus. Thus, no stimulus was present during the conditioned stimulus/unconditioned stimulus time interval. In this situation the conditioned response had to be generated by the central nervous system at that precise time interval without the simultaneous presence of any sensory cue. Records illustrated in Figure 2 were taken from the 2nd conditioning session. It should be noted that the single action potential produced by the recorded facial motoneuron and the minimum evoked eyelid movement during the unconditioned stimulus/conditioned stimulus interval represent a quantum of conditioned response and illustrate clearly the quantal nature of neural processes involved in the acquisition of new motor responses.

Data illustrated in Figure 3 were obtained in the same animal during the 5th conditioning session. In this case, the conditioned response was perfectly accomplished because facial motoneurons were able to fire at a low rate during the conditioned stimulus/unconditioned stimulus time interval.

Concluding remarks

Several hypotheses have previously been proposed regarding the site of motor learning, particularly in relation to classical conditioned eyelid responses. Unpublished data from our laboratory indicate that facial motoneurons are already able to fire

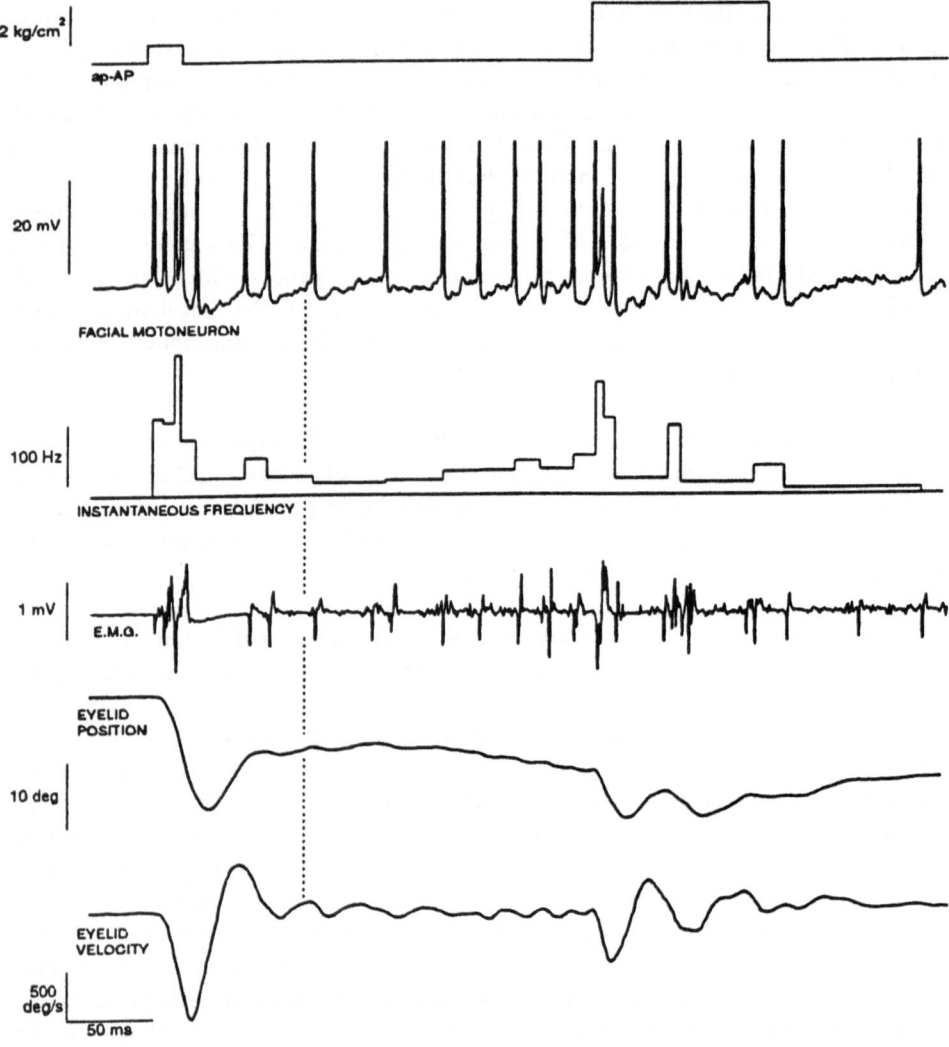

Figure 3. From top to bottom are illustrated: i) the conditioning paradigm consisting of a short (20 ms), weak (0.8 km/cm^2) air puff as the conditioned stimulus, followed 250 ms later by a long (100 ms), strong (3 kg/cm^2) as the unconditioned stimulus; ii) the activity of a facial motoneuron recorded during the 2nd conditioning session; iii) the instantaneous frequency (spikes/s) of the recorded unit; iv) the electrical activity of the orbicularis oculi muscle; v and vi) the position and velocity of the evoked eyelid response. Records were taken from the same animal and correspond to the 5th conditioning session. Note an increase in the activity of the recorded motoneuron during the time interval between the end of the conditioned stimulus and the beginning of the unconditioned one. The dotted line indicates that each action potential of the facial motoneuron is followed by a corresponding muscle potential and by a subsequent lid downward displacement

rhythmically at a dominant frequency of ≈ 20-25 spikes/s (see Figure 3) and that their intrinsic active membrane properties endow them with this rhythmic firing. However, other brain structures are able to fire with the same dominant frequency, mainly the pericruciate cortex (Aou et al., 1992) and the cerebellar interpositus nucleus (Gruart and Delgado-García, 1994). A main difference between these two brain structures is that motor cortex units seem to lead with their firing the beginning of each quantum of lid conditioned movement, while interpositus neurons lag the beginning of each downward lid wavy movement. In each case, the contribution of each neural structure to the learned blink is probably different and complementary.

In conclusion, eyelid movements seem to be generated by an ≈ 20-Hz oscillator. The oscillator probably has a distributed nature, because neuronal pools as different as facial motoneurons (Domingo et al., 1996), interpositus neurons (Gruart and Delgado-García, 1994), and motor cortex units (Aou et al., 1992) fire at this dominant frequency. On the other hand, the quantal release of learned movements suggests the presence of a neural step generator, able to be modified in both duration and amplitude according to the motor needs posed by the organization of the different sensory cues present at the external milieu. This proposed step generator will carry the needed neural plasticity, as the 20-Hz oscillator is simply released according to the available sensory-motor information.

Acknowledgements

This work has been supported by grants from the Spanish D.G.I.C.Y.T. (PB93-1175), C.I.C.Y.T. (SAF 96-0160) and Junta de Andalucía (PAI-3045). We would like to thank Mr. Roger Churchill for his help in the idition of the manuscript.

References

Aou, S., Woody, C.D. and Birt, D. Changes in the activity of units of the cat motor cortex with rapid conditioning and extinction of a compound eye blink movement. *J. Neurosci.*, 12: 549-559, 1992.

Bouyer, J.J., Montaron, F., Wahnee, J.M., Albert, M.P. and Rougeul, A. Anatomical localization of cortical beta rhythms in cat. *Neurosci.*, 22: 863-869, 1987.

Brooks, V.B. Some examples of programmed limb movements. *Brain Res.*, 71: 299-308, 1974.

Domingo, J.A., Gruart, A. and Delgado-García, J.M. A 20-Hz oscillator underlies the acquisition of conditioned eyelid responses. *26th Ann. Soc. Neurosci. Meeting*, Washington, 3: 1875, 1996.

Halliday, A.M. and Redfearn, J.W.T. An analysis of the frequencies of finger tremor in healthy subjects. *J. Physiol. (Lond.)*, 134: 600-611, 1956.

Horsley, V. and Schäfer, E.A. Experiments on the character of the muscular contractions which are evoked by excitation of the various parts of the motor track. *J. Physiol. (Lond.)*, 7: 96-110, 1886.

Gormezano, I., Kehoe, E.J. and Marshall, B.S. Twenty years of classical conditioning research with the rabbit. *Prog. Psychobiol. Physiol. Psychol.*, 10: 197-275, 1983.

Gruart, A., Blázquez, P. and Delgado-García, J.M. Kinematics of spontaneous, reflex, and conditioned eyelid movements in the alert cat. *J. Neurophysiol.*, 74: 226-248, 1995.

Gruart, A. and Delgado-García, J.M. Discharge of identified deep cerebellar nuclei neurons related to eye blinks in the alert cat. *Neurosci.*, 61: 665-681, 1994

Llinás, R.R. The noncontinuous nature of movement execution. In *Motor Control: Concepts and Issues*, Humphrey, D.R. and Freund, H.-J., eds. John Wiley and Sons, New York, pp. 223-242, 1991.

Llinás, R.R. and Ribary, U. Coherent 40-Hz oscillation characterizes dream state in humans. *Proc. Nat. Acad. Sci. (USA)*, 90: 2078-2081, 1993.

Marshall, J. and Walsh, E.G. Physiological tremor. *J. Neurol. Neurosurg. Psychiat.*, 19: 260-267, 1956.

Vallbo, Å.B. and Wessberg, J. Organization of motor output in slow finger movements in man. *J. Physiol. (Lond.)*, 469: 673-691, 1993.

A Model of Cerebellar Saccadic Motor Learning Using Qualitative Reasoning

J.L. Krichmar[1,2], G.A. Ascoli[1], L. Hunter[1,2,3] and J.L. Olds[1]

[1]Krasnow Institute for Advanced Study, George Mason University, Fairfax, VA
[2]Department of Computational Science and Informatics, George Mason University, Fairfax, VA
[3]National Library of Medicine, Bethesda, MD

ABSTRACT

We present a novel approach to modeling neural behavior using a "qualitative reasoning" algorithm. The Qualitative Reasoning Neuron (QRN) is capable of qualitatively reproducing single neuron behavior, but is computationally simple enough to use in large scale neural networks without loss of critical details. QRN simulations of a single Purkinje cell (~1600 compartments) show significant speedup over a recent GENESIS model. A large scale model of the cerebellar cortex (256 neurons, ~300,000 compartments) is used to simulate a saccadic eye movement task. The model reproduces *in vivo* Purkinje cell bursting patterns during saccades. We simulate rapid and gradual adaptation paradigms and show that error correction is possible when climbing fiber input is periodic and contains no error signal.

INTRODUCTION

Saccades are rapid eye movements to acquire a novel target accurately on the fovea. During saccades, the oculomotor system cannot receive eye position input from the visual system. Therefore, error adjustment must occur after movement is complete. Evidence suggests that the cerebellum is necessary for the execution of saccades and the subsequent correction of inaccurate saccades (1). In the cerebellar cortex, lobules VI and VII of the vermis region, known as the oculomotor vermis (OV), have been demonstrated to be associated with saccadic processing (2-4). Purkinje cell axons from the OV make inhibitory synapses onto the fastigial ocular region (FOR) in the deep cerebellar nuclei (5-7) and thus represent the output of this control system.

The OV, analogous to the rest of the cerebellar cortex, receives two types of inputs; climbing fibers and mossy fibers. Climbing fibers fire at a low rate and are thought to be either associated with error correction (8) or timing of movements (9). Mossy fiber input, active during saccadic movements, is thought to encode eye position, which arises from the dorsolateral pontine nucleus (7), a motor efference copy, which originates from the nucleus reticularis tegmenti pontis (NRTP) via the superior colliculus (2, 3, 10) and muscle proprioceptive information (11). Lesion studies and both recording and stimulating experiments have revealed a distinct pattern of firing in the FOR and OV (2-7). Thus, it appears that neurons on the ipsilateral side of the OV, with respect to movement, and the contralateral FOR fire before and during a saccade. In contrast, neurons on the contralateral OV and ipsilateral FOR fire at the end of a saccade. Cells in both the OV and FOR appear to have preferred directions with broad tuning widths (4, 12).

The OV appears to be responsible for adaptation (i.e. error correction) of saccades. Marr and Albus have hypothesized that the cerebellum is the locus of motor learning (13, 14). Subsequent to their theoretical exposition, long-term depression (LTD) of Purkinje cell synaptic responses was demonstrated by concomitant parallel fiber and climbing fiber activity (8). At the molecular level, Purkinje cell LTD appears to be due to the desensitization of AMPA receptors, near the parallel fiber-Purkinje cell synapse, to glutamate (15). There is much disagreement about the type of information conveyed by the climbing fiber into the cerebellum. Proponents of the Marr-Albus motor learning theory believe the climbing fiber contains either an error signal of directional information (16) or a binary teacher signal (17). Due to the periodic nature of climbing fiber input, Llinas et al. believe the climbing fiber is associated with the timing of movements (9). Keating and Thach did not find evidence for climbing fiber periodicity and believe that the climbing fiber activity is correlated with the onset of movement (18).

Several models have specifically addressed the role of the cerebellum in processing saccades (11, 19-22). These models of cerebellar motor control have not been detailed enough to: 1) account for burst patterns during saccades and 2) investigate the effects of Purkinje cell architecture at a compartment model level. Thanks to a novel approach to computational modeling, the Qualitative Reasoning Neuron (QRN) methodology (23), we have been able to construct a detailed, compartment model of the OV that accounts for bursting patterns in the cerebellum without sacrificing crucial neuro-molecular details (24). Qualitative reasoning began as an Artificial Intelligence technology to describe physical phenomena. The objective was to build systems that could reason about the physical world as do engineers and scientists (25). Our QRN model of the OV has the following features: 1) Bursting patterns during saccadic movement. 2) A signal of direction and magnitude from a population of broadly tuned Purkinje cells. 3) Adaptation without relying on a climbing fiber error signal. 4) Detailed processing within the Purkinje cell based on channel, receptor and second messenger interactions.

METHODS

The high-level architecture of the QRN OV model is based on the known neuro-anatomical architecture of the saccadic circuit (11, 26). Before movement, the higher cortex, visual system and pons (CTX) section of the model outputs a vector signifying the distance and direction from the current eye position to a target of interest. Both the superior colliculus (SC) and cerebellum (CBM) modules receive this signal. The modeled SC outputs a vector, with its direction from the CTX and a fixed magnitude, to the brainstem saccade generator (BSG). The CBM creates a motor command, based on the response of Purkinje cells to the CTX information. The CBM vector is passed to the BSG. The BSG adds the SC vector to the CBM vector and outputs a simulated motor command. The current muscle position is updated based on the BSG output. During movement, the SC calculates subsequent commands based on the motor efference copy from the BSG. Similarly, the CBM creates motor commands based on motor efference copy and muscle position information from the BSG. After movement is complete, the CTX outputs an error signal based on the distance between the current eye position and the desired eye position. This error signal is converted into a pattern of parallel fiber activity. The inferior olive module (IO) activates all climbing fibers regardless of error.

The main information processing of our model is performed through simulation of Purkinje cells (PC) within the CBM module. The following is a description of the qualitative reasoning algorithm that underlies our PC information processing. QRN simulates the behavior of a single PC by qualitatively describing the relationships among the various ionic currents and cell's membrane potential within different modeled compartments of the cell. Since the interaction between parameters is "reasoned", no quantitative values need to be calculated. In general, computation in the QRN algorithm is performed through addition, subtraction or table lookups.

Qualitative Reasoning Terminology. The terminology and algorithm of QRN is based on the Qualitative Simulation (QSIM) method developed by Kuipers (27). Qualitative reasoning uses *landmarks* to describe a critical value for a parameter at a given time. Time intervals are arbitrary and depict the occurrence of an interesting event such as reaching a landmark. A qualitative reasoning system has two types of parameters: *Continuous* and *Discrete*. The range of a continuous parameter is a continuous ordered set of landmark values called the *quantity space*. The *qualitative state* of a parameter consists of a *value*, a *direction*, and a *weight*. The value can either be at or between landmarks. The direction can be *increasing, decreasing* or *steady*. The weight is the relative magnitude of a parameter to the other parameters with which it interacts. A discrete parameter has an unordered discrete quantity space and has value but not direction. For example, a discrete parameter could describe a light switch with two landmark values; on and off.

Constraints ensure parameters stay within the range of the quantity space by containment rules that dictate changes between landmarks. Constraints can be mathematical functions or arbitrary functions. The equation, $a=b*c$, is described by "reasoning" rather than the traditional meaning: If b or c is zero, then a equals zero. If b is increasing and c is steady, a is increasing. Similarly, mathematical functions such as addition, subtraction, and division can be "reasoned" instead of calculated. Moreover, an arbitrary function, such as a monotonically increasing function, can be used to describe constraints between parameters. The constraints used by QRN are listed below:

1. M+(a,b): Monotonically increasing function. As a increases (decreases), b increases (decreases).
2. M-(a,b): Monotonically decreasing function. As a increases (decreases), b decreases (increases).
3. EXPD(a): Exponential decay function. a always decreases.
4. EXPI(a): Exponential increase function. a always increases.
5. TH(a,b): Threshold function. If $a.value > b.value$, $a = a$. Otherwise, the qualitative state of a has a value of 0 and a direction of steady.
6. $+ (a,b,c)$: Addition function. $c = a + b$.
7. $* (a,b,c)$: Multiplication function. $c = a * b$.

Qualitative Reasoning Simulation. Qualitative simulation of a system starts with a description of the known structure of the system, and an initial state. The simulation proceeds by resolving the constraints on each parameter to obtain new qualitative states for the system's parameters. Equations 1 through 3 describe the calculation of a new qualitative state for any parameter Q. The new qualitative state is based on the direction and relative weight of the parameters, q_1 through q_n, that constrain parameter Q.

$$TotalWeight = \sum_{i=1}^{n} q_i.qdir * q_i.qwgt \qquad (1)$$

$$Q.qdir = \begin{cases} decr.TotalWeight < 0, \\ incr.TotalWeight > 0, \\ stdy.Otherwise \end{cases} \qquad (2)$$

$$Q.qwgt = ABS(TotalWeight) \qquad (3)$$

Where: qdir is -1 for decreasing, 0 for steady, and +1 for increasing. ABS (n) represents the absolute value of n.

The new qualitative value of Q is dependent on the qualitative direction given by equation 2 and the quantity space. If the qualitative direction is increasing (decreasing), the qualitative value of Q is incremented (decremented). Otherwise, the qualitative value remains unchanged. The value of parameter Q is then checked to ensure it has not gone out of the bounds given by the quantity space landmarks.

Qualitative Reasoning Model of a Single Purkinje Cell. We constructed a qualitative reasoning model of the PC that has axon, soma, dendrite and dendritic spine compartments. In the oculomotor vermis network, PC's consisted of one soma compartment connected to one axon compartment and 128 dendrite compartments. Each dendrite compartment connected to 8 dendritic spine compartments (1024 dendritic spines per PC). A second qualitative reasoning PC model was developed to compare against a recent GENESIS model of the PC by DeSchutter and Bower (28, 29). This model has one soma that has 64 dendritic branches, each of which, have 24 dendritic spines (1536 total spines). The QRN model has approximately the same number of channels as the DeSchutter and Bower model with the following qualitative simplifications: 1) Collapse the qualitatively similar voltage-gated calcium channels into one dendrite parameter. 2) Collapse the qualitatively similar calcium-activated potassium channels into one dendrite parameter. 3) Build a uniform dendritic tree. 4) Change the soma into a summing junction of the membrane potential given by its child dendrites. There are two versions of this simplified model; one with passive spines (i.e. a monotonically increasing relationship between parallel fiber glutamate and membrane potential at the spine) and the other with active spines. The active spine version models sub-cellular processes that lead to calcium spikes and cerebellar LTD (based on Figure 2 of ref. 15). The constraint model for each compartment type is shown in Figure 1. A constraint model depicts the qualitative relationships between parameters. For example, Figure 1A illustrates a monotonically increasing function (M+) between parallel fiber (PF) activity and Phospholipase C (PLC) activity. The constraint models reveal the level of detail that the QRN model contains. All of the parameters have qualitative weights set to 1.

137

Qualitative Reasoning Model of the Oculomotor Vermis. The OV model executes a saccadic motor program as a state machine. That is, the determination of subsequent motor commands are based on previous motor commands and the current state of the motor system. Initially, the OV motor program receives eye position, the desired target and the state of the motor system as PF input. Subsequently, the OV motor program receives the previous motor command and the state of the motor system. At the end of the movement, the OV motor program receives eye position, the desired target, the state of the motor system, and climbing fiber (CF) input.

The OV model contains 256 QRN PCs. During the model's initialization phase, each PC is initialized, using cerebellar LTD as a learning rule, to fire: 1) maximally at a preferred direction, 2) either before and during a saccade (early burster) or towards the end of a saccade (late burster). The OV model has four sub-populations of 64 PCs: 1) Left early bursters. 2) Left late bursters. 3) Right early busters. 4) Right late bursters.

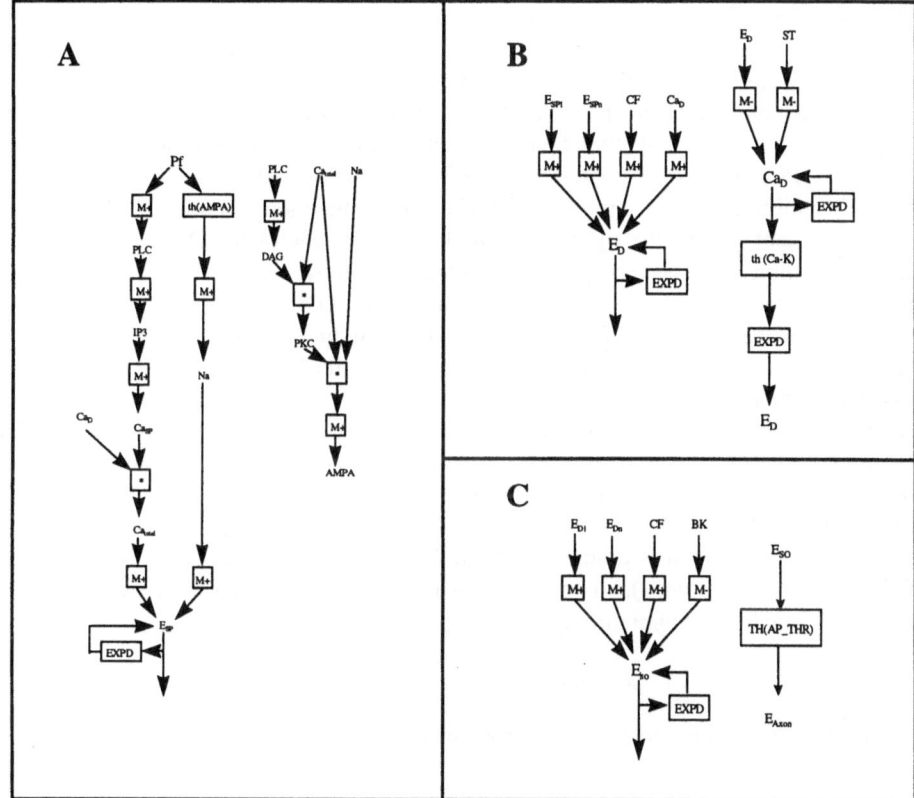

Figure 1. A. Constraints on the Dendritic Spine. Each dendritic spine in the model receives a parallel fiber (Pf) and calcium (Ca$_D$) from the neighboring dendritic branch as input. The membrane potential at the dendritic spine (E$_{SP}$), shown on the left, is based on the qualitative interaction between sodium ions (Na), internal stores of calcium at the spine (Ca$_{SP}$), and voltage attenuation (shown as the EXPD function). Phospholipase C (PLC) is activated in the presence of glutamate due to Pf activity. Ca$_{SP}$ levels increase due to activation of inositol-1,4,5-triphospate (IP3). Cerebellar long-term depression (LTD), shown on the right is due to a desensitization of the AMPA receptor to Pf input. LTD is based on the qualitative interaction of Na, Ca$_{SP}$, 1,2 diacyglycerol (DAG) and protein kinase C (PKC). **B. Constraints on the Dendrite.** Each dendritic branch in the model receives climbing fiber (CF) and a voltage signal from the

dendritic spines, E_{SP}, as input. The figure on the left illustrates the constraints involved in the calculation of dendritic membrane potential (E_D). E_D is based on the qualitative interaction between E_{SP}, CF, Ca_D and voltage attenuation of E_D (shown by EXPD function). The figure on the right illustrates the contstraints involved in the calculation of voltage-gated calcium at the dendrite. Ca_D is based on the qualitative interaction between E_D, stellate cell inhibition (st), and Ca_D current attenuation due to the opening of calcium activated potassium channels (Ca-K). **C. Constraints on the Soma and Axon.** The figure on the left illustrates the constraints involved in the calculation of somatic membrane potential (E_{SO}). The soma receives CF and a voltage signal from the dendrites, E_D, as input. E_{SO} is based on the qualitative interaction between E_D, CF, basket cell (BK) inhibition and voltage attenuation of E_{SO} (shown by EXPD function). The figure on the right illustrates the constraints involved in the calculation of the axon's membrane potential. E_{Axon} is active when E_{SO} is above a threshold value.

CF input is defined as a qualitative parameter that increases activity after movement is complete. All PCs receive the same CF input. PF input is defined as a qualitative parameter that represents a direction or a muscle position. Of the 1024 PF inputs to a PC, half are direction related and the other half are muscle position related. All PCs within a sub-population receive input from the same 1024 PFs.

At the start of an epoch, the OV model receives movement direction and eye muscle position information from the CTX and BSG respectively. An epoch lasts 20 time steps during which time no new input is received from the CTX. A given muscle position or direction is distributed among multiple PFs as a pattern of activity. During movement, PCs receive PF input from the SC and BSG, and a tonic level of stellate cell (ST) inhibition in order to prevent LTD during this motor phase. If PFs indicate early movement toward the right (left), the left (right) side of the OV receives basket cell (BK) inhibition. Conversely, if PFs indicate late movement toward the right (left), the right (left) side of the OV receives BK inhibition. During the first 10 time steps of an epoch, the active PFs, STs and BKs have a qualitative direction of *increasing*. During the last 10 time steps in the epoch, inputs to the OV model have a qualitative direction of *steady*.

The firing rate of a QRN PC is determined by its response to the PF, BK cell and ST cell input. The qualitative firing rate of a PC is converted into a quantitative vector via equations 4 and 5. Each vector has a directional component and a magnitude component. The directional component is given by the PC's preferred direction that is determined at initialization time. The magnitude component is based on the firing rate of the PC (see equation 4). The output of the OV model is a vector summation of every PC in each sub-population given by equation 6.

$$Mag_i = F_i * k \qquad (4)$$

Where: F_i is the firing rate (number of spikes during an epoch) of PC I k is a constant gain setting.

$$V_i.v1 = Mag_i * cos\ (Dir_i) \qquad (5)$$
$$V_i.v2 = Mag_i * sin\ (Dir_i)$$

Where: Dir_i is the preferred direction of PC i. V_i is the two dimensional vector for PC i.

$$PC_{pop} = \sum_i V_i \qquad (6)$$

Where: PC_{pop} is a two dimensional vector representing the output of the OV.

After movement is complete, the model enters an evaluation phase. All CFs are active. The amount of PF activity is based on the desired direction and the amount of error in movement. That is, the greater the error the more PF activity. If the error is a left (right) undershoot, only the left (right) early bursting PFs and the right (left) late bursting PFs are active. If the error is a right (left) overshoot, only the left (right) late bursting PFs and the left (right) early bursting PFs are active. ST inhibition is active to PCs with preferred directions away from the desired direction (greater than 45 degrees).

RESULTS

Single Purkinje Cell Experiments. The single QRN PC model was tested for its ability to simulate complex spikes and simple spikes. A simulated volley of CF input produced the burst of activity termed a complex spike in the QRN model. The inputs to the single cell models simulated asynchronous PF excitation from 1 to 100 Hz and asynchronous inhibitory ST inhibition from 1 to 30 Hz (30). Figure 2A illustrates the frequency responses to varying input activities. The model showed firing frequencies within normal simple spike firing ranges (31) and appeared to reach an asymptote at approximately 50-70 Hz. The firing frequency of these plateaus depended on the inhibition level. Figure 2B shows the interspike intervals for the QRN model in response to 30 Hz PF excitation and varying ST cell inhibition levels. The average interspike interval increased as the inhibition increases. However, the most common interspike interval for both models, independent of the inhibition level, is in the range of 15 to 20 ms. The results reported in Figure 2 were based on the passive spine model. The active spine model produced nearly identical firing characteristics.

The amount of computer time required by the different models were compared based on a 5 second simulation (250,000 time steps) with asynchronous PF input at 25 Hz and asynchronous ST cell input at 1 Hz. The QRN PC model with active dendritic spine processing was over 3 times faster than the GENESIS model (i.e. 138 minutes versus 464 minutes). The QRN model with passive spines was over 7 times faster than the GENESIS model (i.e. 65 minutes versus 464 minutes). All simulations were run on the same computer with the same number of time steps. The network model of the oculomotor vermis will use the QRN PC with active spines.

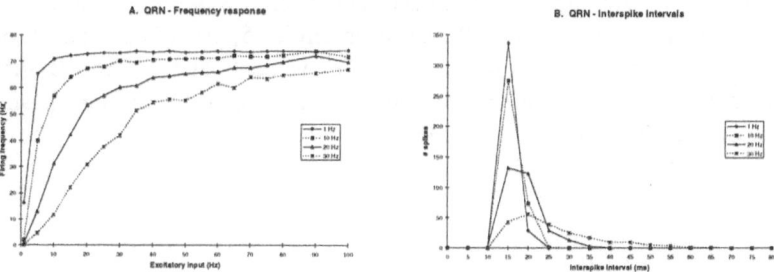

Figure 2. QRN Purkinje cell model response. A. Spike frequency response to varying excitation and inhibition. Parallel fibers provided the asynchronous excitatory input with firing rates ranging from 1 to 100 Hz. Stellate cells provided asynchronous inhibitory input with firing rates of 1, 10, 20 and 30 Hz.

B. Interspike intervals (ISI) in response to inhibition. The parallel fiber asynchronous firing rates were set at 30 Hz and asynchronous stellate inhibition was varied. Inhibition levels of 1, 10, 20, and 30 Hz have mean (± S.D.) ISI's of 13.7±0.9, 14.3±1.4, 16.6±3.8 and 23.8±10.3 ms respectively.

Oculomotor Vermis Experiments. The model of the oculomotor vermis was analyzed by simulating movements to test targets. Before running movement simulations, we investigated the population response to varying tuning widths. Population coding of direction implies that the broader the tuning width of a cell, the more fault tolerant the directional signal is to noise. We tested the OV model by making short movements from a center position to 16 equally spaced test targets (separated by 22.5° angles). The tuning half-width of cells was varied from narrowly tuned (5.625° half-width) to broadly tuned (90° half-width). Bilateral lesions of 5, 10 and 15 percent were simulated by randomly inactivating cells in the oculomotor vermis model. Each sub-population was tested in all directions. As expected, the broader tuned populations were more resistant to directional errors (see Figure 3). Subsequent experiments used the 45° half-width for the population codes.

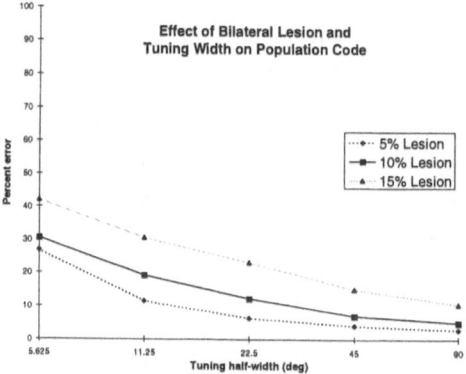

Figure 3. The effect of different tuning width on directional accuracy. Bilateral lesions were simulated by randomly inactivating PCs within the model. The model simulated movement to 16 equally spaced targets around a circle. The graph shows the amount of directional error with different tuning widths and lesion levels. A 100 percent error occurs when the model outputs a movement command with a direction that deviates 180 degrees from the desired direction.

In the next experiment, simulations of saccades from a center position to 8 equally spaced test targets (separated by 45°) placed on circles 5, 10, 15, 20 and 25 degrees were performed. Figure 4A shows 10 degree movements and Figure 4B shows 20 degree movements. All simulated movements to all targets of varying distance and direction were accurate within one degree.

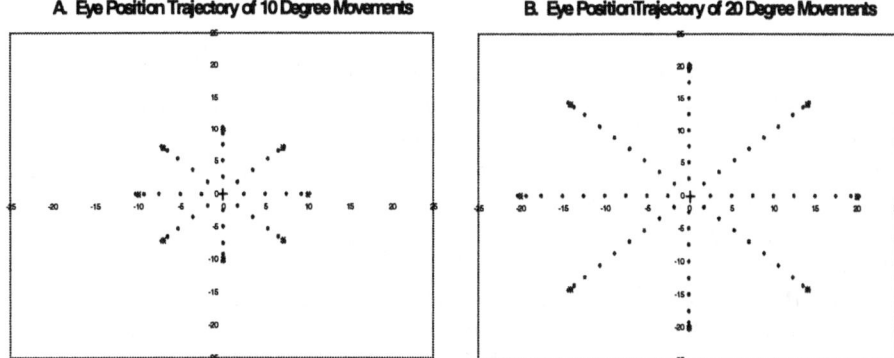

Figure 4. Simulation of saccades. Each chart illustrates 8 traces of movement over time to 8 different targets. Eye movement starts at a fixation point (the black cross in the middle). Eye movement can be to any one of the test targets (shown in green). The trajectory of eye movement are shown in red. **A.** 10 degree eye movements. **B.** 20 degree eye movements.

We simulated adaptation experiments with both the rapid and gradual adaptation paradigms (32). The rapid adaptation was simulated by presenting either a positive error signal (overshoot) or a negative error signal (undershoot) to the model after movement was complete. The OV model demonstrated the ability to both increase and decrease its gain in response to an error. The model showed the ability to decrease its gain 50% in response to a 75% overshoot. Similarly, the model increased its gain 45% in response to a 75% undershoot (see Figure 5A). Gradual adaptation was simulated by lowering the gain of one side of the modeled brainstem saccade generator. The OV model responded appropriately by maximally increasing its gain 28% (see Figure 5B). The time course of gradual adaptation was slower than rapid adaptation only when proprioceptive input from the brainstem saccade generator to the oculomotor vermis was cut proportionally to the lesion.

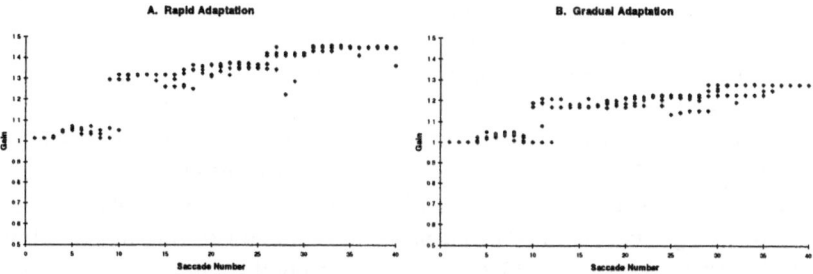

Figure 5. Rapid and gradual adaptation paradigms. Data from 4 trials combined to show change in gain over subsequent saccades. **A.** Response of the model to a 75% undershoot. The model increased its gain 45%. **B.** Response of the model to a 75% undershoot while simulating a 25% weakening of lateral eye muscles. The model increased its gain 28%. Rapid adaptation occurred 3 to 4 times faster than gradual adaptation.

DISCUSSION

Our main findings were: 1) The QRN PC model qualitatively matches the behavior demonstrated in PC recordings and is less computationally intensive than other current computational neuroscience models. 2) A population of QRN PCs, in which each individual cell is broadly tuned to a preferred direction, can accurately encode the direction and duration of saccadic movement. 3) State machine architecture is not only a feasible cerebellar strategy for the storage and execution of motor programs, but it is robust and preferable to feedback control. 4) Using an ensemble code and state machine control, the output of the cerebellar cortex model qualitatively matches actual recordings of bursting patterns in the OV during saccades. Simulations of saccades with the OV model agree with the saccadic metrics of test subjects. 5) Error information carried by PFs is sufficient for the cerebellar cortex to adapt to errors and store adjustments. In addition, and perhaps more importantly, these results demonstrate the power of the QRN algorithm. To our knowledge, the OV model, which contains approximately 300,000 compartments, is the largest, most detailed computational neuroscience model ever constructed. Our compartments specify parameters in a hierarchical way, from the macroscopic (i.e. membrane potential) down to the molecular level (i.e. activation of protein kinase C). The simulation time was short enough to make this type of modeling possible without the need for supercomputers.

The firing frequency of the QRN PC was comparable to simple spike frequencies reported *in vitro* (33) and *in vivo* (34) experiments and simple spike ISIs were comparable to ISIs reported *in vivo* (34) experiments and the GENESIS model (29). The efficiency of QRN allowed for the creation of a detailed network model of the cerebellar cortex.

The *in vivo* firing patterns of OV PCs appear to be broadly, directionally tuned (2, 12). The vector summation of a population of neurons in the motor cortex encoded an accurate direction of movement. In our model, the PCs were initialized with a preferred direction and a tuning width. As expected, the directional signal output by the model was more resistant to errors caused by simulated lesions as the PC tuning width increased (see Figure 3). Our results demonstrate the advantages of a broadly tuned ensemble coding of eye movement directions.

Most models of cerebellar saccadic control treat the saccadic burst as a single gain output from the cerebellum (19, 21, 22). Our model investigated the emergence of cerebellar bursting within a saccade. Although the visual input to the cerebellum is suppressed during saccadic movement, mossy fiber input carrying motor information is active during movement (3). One approach to using this information in a controller of eye movements is a state machine. In a state machine, the selection of a motor command is based on the current state of the motor system and a history of previous commands. By this notion, the cerebellum is receiving the current motor state in the form of muscle proprioceptors and a history in the form of a motor efference copy. Our results suggest that the state machine architecture was sufficient to accurately control movements of differing direction and amplitude (see Figure 4). The state machine architecture incorporated in this model is an alternative to the cerebellum as a feedback controller hypothesis (16, 35). The relationship between the cerebellum, superior colliculus and brainstem saccade generator could be designed as a negative feedback loop. As the feedback controller, the cerebellum would attempt to minimize the signal distance between current movement and desired movement received from the brainstem saccade generator. In a systems analysis of this design, we found the feedback approach, although stable, always overshot the desired target.

In experiments that tested normal saccade performance, our model was accurate, within one degree, in all directions and amplitudes tested (see Figure 4). We achieved these results by incorporating early bursting PCs in the ipsilateral OV and late bursting PCs in the contralateral OV (2) into our model. In our model, the late burst acted both to decelerate the saccade and signify the end of a saccade to the motor system.

In the present experiments, we treated the CF input as a timing signal and the PF as an error signal. This was a departure from recent models of cerebellar motor learning (16, 17, 19, 21, 22). However, considering the infrequent CF activity, which may be more concerned with timing of movement than error (9, 18), and the rich sensory information carried by the PFs (2, 3, 10), this approach to motor learning may not be so farfetched. Our OV model demonstrated that this approach was sufficient to adapt to errors in the rapid and gradual adaptation paradigms. In the rapid adaptation paradigm, subjects recover from inaccuracies in only 15 minutes; whereas in the gradual adaptation takes approximately a day (32). In our model, the correct timing of the rapid adaptation versus the gradual adaptation was only achieved by suppressing some of the PF error to the cerebellum. This would imply that weakening an eye muscle may also affect afferent information used by the cerebellum.

The present results, could only be possible by using a novel modeling technique such as QRN. The detail of each neuron was necessary to simulate firing patterns and a biologically plausible learning rule. A network model was necessary to emulate the bursting patterns created by an ensemble of PCs. The OV model presents a plausible solution to this complex oculomotor problem and suggests possible strategies for other animal or robotics motor control applications. In addition, the model of the cerebellar cortex is not restricted to simulating saccadic processing, but may be useful in the investigation of other cerebellar functions. However, what may be of more importance is that the qualitative reasoning methodology makes large scale network modeling possible without sacrificing crucial details.

References

1. Sparks, D.L., Barton, E.J. (1993) Neural control of saccadic eye movements. Curr. Opin. Neurobiol. **3**, 966-972.
2. Ohtsuka, K., Noda H. (1995) Discharge properties of Purkinje cells in the oculomotor vermis during visually guided saccades in the macaque monkey. J. Neurophysiol., **74**, No. 5,1828-1840.
3. Ohtsuka, K., Noda, H. (1992) Burst discharges of mossy fibers in the oculomotor vermis of macaque monkeys during saccadic eye movements. Neurosci. Res., **15**, 102-114.
4. Sato, H., Noda, H. (1992) Saccadic dysmetria induced by transient functional decortication of the cerebellar vermis Exp. Brain Res. **88**, 455-458.
5. Fuchs, A.F., Robinson, F.R., Straube, A. (1993) Role of the caudal fastigial nucleus in saccade generation. I. Neuronal discharge pattern. J. Neurophysiol. **70**, 1723-1740.

6. Robinson, F.R., Straube, A., Fuchs, A.F. (1993) Role of the caudal fastigial nucleus in saccade generation. II. Effects of muscimol inactivation. J. Neurophysiol. **70**, 1741-1758.
7. Ohtsuka, K., Noda, H. (1991) Saccadic burst neurons in the oculomotor region of the fastigial nucleus of macaque monkeys. J. Neurophysiol. **65**, 1422-1434.
8. Ito, M. (1989) Long-term depression. Ann. Rev. Neurosci. **12**,85-102.
9. Llinas, R., Welsh, J. P. (1993) On the cerebellum and motor learning. Curr. Opin. Neurobiol. **3**, 958-965.
10. Keifer, J., Houk, J. C. (1994) Motor function of the cerebellorubrospinal system. Physiol. Rev. **74**, 509-542.
11. Houk, J.C., Galiana, H.L., Guitton, D. (1992) in *Tutorials in Motor Behavior II*, eds. Stelmach, G. E., Requin, J. (Elsevier Science Publishers), pp. 443-474.
12. Helmchen, C., Buttner, U. (1995) Saccade-related Purkinje cell activity in the oculomotor vermis during spontaneous eye movements in light and darkness. Exp. Brain Res. **103**, 198-208.
13. Albus, J.S. (1971) A theory of cerebellar function. Math. Biosci. **10**, 25-61.
14. Marr, D. (1969) A theory of cerebellar cortex. J. Physiol. (London). **202**, 437-470.
15. Linden, D.J., Connor, J.A. (1995) Long-term synaptic depression. Ann. Rev. Neurosci. **18**, 318-357.
16. Kawato, M., Gomi, H. (1992) A computational model of four regions of the cerebellum based on feedback-error learning. Biol.Cybern. **68**, 95-103.
17. Houk, J.C., Barto, A.G. (1992) in *Tutorials in Motor Behavior II*, eds. Stelmach, G. E. & Requin, J. (Elsevier Science Publishers), pp. 71-100.
18. Keating, J.G., Thach, W.T. (1995) Nonclock behavior of inferior olive neurons: interspike interval of Purkinje cell complex spike discharge in the awake behaving monkey is random. J. Neurophysiol. **73**, 1329-1340.
19. Dean, P., Mayhew, J.E., Langdon, P. (1994) Learning and maintaining saccadic accuracy: A model of brainstem-cerebellar interactions J. Cog. Sci. **6**, 117-138.
20. Dean, P., (1995) Modelling the role of the cerebellar fastigial nuclei in producing accurate saccades: the importance of burst timing. Neuroscience **68**, 1059-1077.
21. Schweighofer, N., Arbib, M.A., Dominey, P.F. (1996) A model of the cerebellum in adaptive control of saccadic gain. I. The model and its biological substrate. Biol. Cybern. **75**, 19-28.
22. Schweighofer, N., Arbib, M.A., Dominey, P.F. (1996) A model of the cerebellum in adaptive control of saccadic gain. II. Simulation results. Biol. Cybern. **75**, 29-36.
23. Krichmar, J.L., Olds, J.L., Hunter, L. (1997) Qualitative reasoning as a modeling tool for computational neuroscience. Submitted to Biol. Cybern.
24. Krichmar, J.L., Hunter L., Olds J.L. (1996) A qualitative model of cerebral saccadic control. Soc. Neurosci. Abstr. **22**, 1093.
25. Weld, D. S. & De Kleer, J. (1990) *Readings in Qualitative Reasoning about Physical Systems.* (Morgan Kaufmann Publishers Inc., San Mateo, CA).
26. Pierrot-Deseilligny C, Rivaud S, Gaymard B, Muri R, Vermersch A.I. (1995) Cortical control of saccades. Annals of Neurology, **37**, 557-567.
27. Kuipers, B. (1986) Qualitative simulation. Artificial Intelligence **29**, 289-388.

28. De Schutter, E., Bower J.M. (1994) An active membrane model of the cerebellar Purkinje cell I. Simulation of current clamps in slice. J. Neurophysiol. **71**(1), 375-400.

29. De Schutter, E., Bower, J.M. (1994) An active membrane model of the cerebellar Purkinje cell II. Simulation of synaptic responses. J. Neurophysiol., **71**(1), 401-419.

30. Mitgaard, J. (1992) Membrane properties and synaptic responses of Golgi cells and stellate cells in the turtle cerebellum *in vitro*. J. Physiol. (London) **457**, 329-354

31. Llinas, R.R. (1981) Electrophysiology of the cerebellar networks. In: Handbook of physiology. The nervous system. Motor control. American Physiology Society, Bethesda, MD, Sect. 1, Vol. II, pp 831-876

32. Optican, L.M. (1985) in "Adaptive mechanisms in gaze control: Facts and theories.", eds. Berthoz & Melvill-Jones (Elsevier Science Publishers) pp. 71-79.

33. Llinas, R.R., Sugimori, M. (1980) Electrophysiological properties of *in vitro* Purkinje cell somata in mammalian slices. J. Physiol. (London) **305**, 171-195.

34. Sato, Y., Miura, M., Fushiki, H., Kawasaki, T. (1993) Barbiturate depresses simple spike activity in cerebellar Purkinje cells after climbing fiber input. J. Neurophysiol., **69**(4), 1082-1090.

35. Ito, M. (1984) *The cerebellum and neural control* (Raven Press, New York).

Balance Between Intercellular Coupling and Input Resistance as a Necessary Requirement for Oscillatory Electrical Activity in Pancreatic β-cells

E. Andreu*, R. Pomares, B. Soria and J.V. Sanchez-Andres

Dpt. Fisiología & Inst. Neurociencias. Universidad de Alicante. Aptdo Correos 374, Campus de San Juan. Alicante 03080. SPAIN

We studied the emergence of oscillatory electrical activity after addition of glucose to insulin secreting cells. In the physiological glucose range (7-20 mM), these cells show a typical square-wave bursting pattern when they are coupled in the islet. Islet of Langerhans consists of some thousands of beta cells, coupled through gap-junctions. When these cells are isolated they also become more excitable in presence of glucose, spiking continuously, but they fail to oscillate. We have hypothesized a role of cell coupling in the generation of oscillatory activity in this system. Now, we examine the phase of continuous activity that appears after a glucose challenge to check our hypothesis. Both experimental data and computer simulations of a small network of β cells, further supports our hypothesis on a role of intercellular coupling in the emergence of oscillatory patterns.

INTRODUCTION

We have already described that an optimal range of input resistance is necessary to keep the oscillatory behavior of the pancreatic beta cell (1,2). Inside this range, input resistance oscillates between two levels in a clear correlation with membrane potential. The absence of phase lag between the oscillations of the membrane potential and the input resistance of the cell points to the possibility of input resistance determining the appropriate balance of conductances allowing the cell to behave as an oscillator.

In a coupled network , input resistance reflects, not only the state of the cell conductances (non-junctional conductances) but also the contribution of the neighbor coupled cells through the coupling conductances (junctional conductance). A careful study of coupling changes (3) along the oscillatory phases, as well as in the silent periods, and continuous spiking periods, can help to understand the behavior of the input resistance for different glucose concentrations and thus to understand the electrical requirements of these cells to be able to show such a variety of electrical patterns. Besides, coupling seems to be crucial for this system to have a proper electrical response as far as isolated cells are unable to oscillate (4,5,6)

To whom correspondence should be addressed: ✉ *Departamento de Fisiología. Universidad de Alicante. Aptdo. de Correos 374. Campus de San Juan. 03080. Alicante. Spain.* ☎: *34-6-5903953*

Computational models of pancreatic beta cell have focused mainly on the modelization of ionic currents (7,8,9). Coupling between arrays of cells has also been implemented (6,10). However, computational models fail to explain some phases of the electrical activity of these cells. Determining the variability of input resistance and coupling conductances along all the possible physiological conditions can make possible to implement more realistic models.

METHODS.

Electrophysiological: Intracellular activity of β-cells from mouse islets of Langerhans was recorded with a bridge amplifier (Axoclamp 2A) as previously described (7). Recordings were made with thick wall borosilicate microelectrodes pulled with a PE2 Narishige microelectrode puller. Microelectrodes were filled with 3 M potassium citrate, resistance around 100 MΩ. The modified Krebs solution used had the following composition (mM): 120 NaCl, 25 NaHCO3, 5 KCl and 1 MgCl2, and was equilibrated with a gas mixture containing 95% O2 and 5% CO2 at 37°C. Islets of Langerhans were microdissected by hand. Data recorded were directly monitored on an oscilloscope and stored both in hard disk and tape for further analysis . Analysis of the data were performed using Origin 4.0, and Axoscope1.1. We recorded simultaneously neighbor cells of an islet, and passed different current protocols for different glucose concentrations (0-22 mM).

Computational: Single cell models were first simulated using XPPauto 2.5, a software for resolution of ordinary differential equations as well as for analysis of stability. Single cells are simulated with an 8-dimensional ordinary differential equation system (ODE) based on previous models by Smolen and Keizer (9), but with the modification of parameter values according to our experimental data. Network models were simulated using XSim, a neural network simulator developed by P. Varona and J.A. Siguenza of the I.I.C., UAM, Madrid. Islets of Langerhans are simulated like small (3x3) and medium (5x10) networks of beta cells, each one of these units is implemented using the same ODE developed with Xppauto. Each cell makes connection with all its neighbors.

RESULTS

1. Electrophysiology

At low glucose concentrations (0-5 mM), pancreatic beta cells remain silent and hyperpolarized (-80 mV). When exposed to stimulatory glucose concentrations (7-20 mM), cells depolarize until a threshold level where they begin to spike during 30-100 s. Afterwards cells repolarize to an intermediate level (-50 mV) and begin to oscillate between two states, active phase (depolarized, spiking) and silent phase (hiperpolarized, non spiking).

The first active phase after glucose application is longer than the following ones, and is called initial phase. A careful study of the time course (from 0 to 11mM glucose) of the input resistance and membrane potential is shown in figure 1.

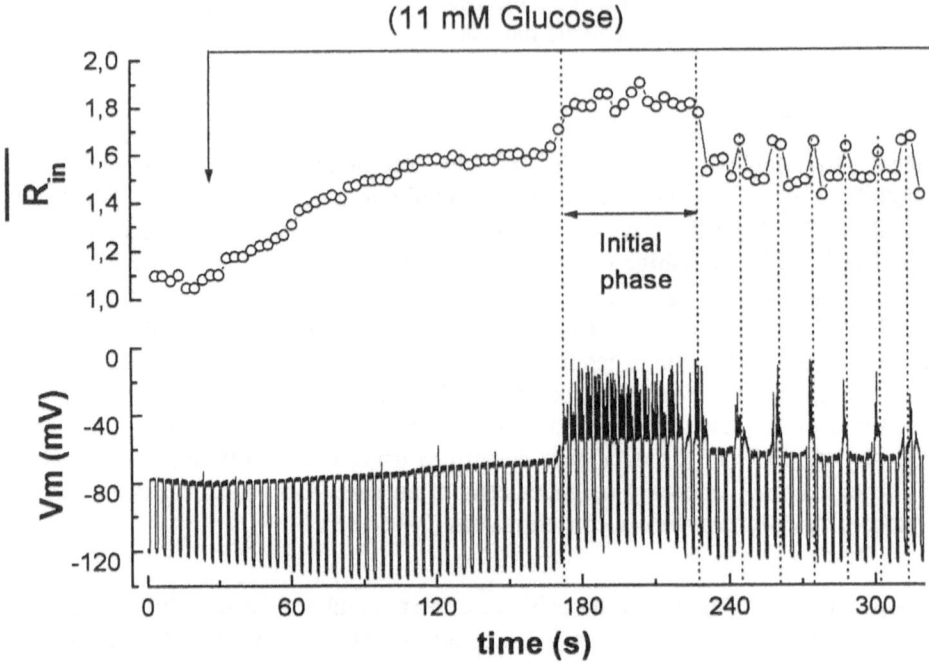

Figure 1 .-Input resistance measurement during glucose application. Top panel shows normalized input resistance values .Circles indicate the time where the input resistance measurements were done. Bottom : Corresponding membrane potential. Downward deflections correspond to current pulses (-0.3 nA amplitude, 600 ms duration) applied to obtain the measurements shown on top.

As it is shown in the figure, as the cell depolarizes, input resistance increases, arriving to its maximum level during the initial phase, then it decreases, and remains oscillating in an intermediate state during the oscillatory activity.

We studied the coupling characteristics, of the different electrical patterns. We recorded simultaneously neighbor cells from an islet, and passed different current protocols for different glucose concentrations. At low glucose levels, cells are uncoupled as can be proved by the application of current pulses on one of the cells. If there is coupling between them, an elicited voltage deflection should be noticeable in both cells.

Simultaneous recording of pairs of pancreatic β-cells shows that at low glucose concentrations cells are uncoupled *(fig2, left panel)*, as the application of current pulses in cell 1 is unable to elicit propagated voltage deflections in cell 2. *(fig2, left, bottom)*. Intermediate glucose concentrations induce oscillatory patterns *(fig2, central panel)* Under these conditions current injected into cell 1 *(fig2, central panel, top)* generates propagated deflections in cell 2. That current propagation also takes place for higher glucose concentrations.

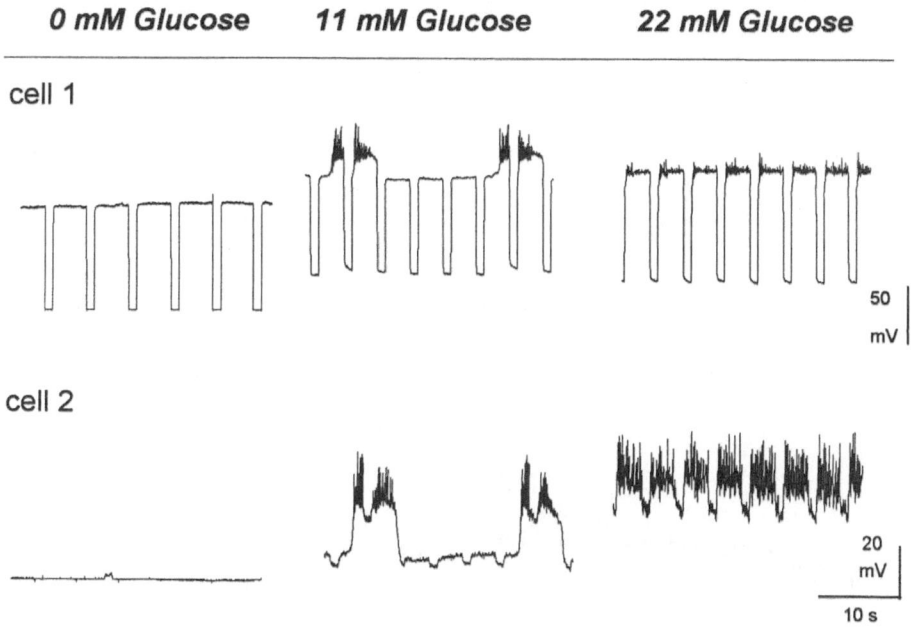

Figure 2.- Coupling measurements for different glucose concentrations. Simultaneous recording of two cells membrane potential during hyperpolarizing current pulses application to cell 1. Voltage deflections are recorded both in cell one and cell 2. Applied current : 0 mM glucose : -0.6 nA, 900 ms ; 11 mM Glucose :-0.3 nA, 900 ms ; 22 mM Glucose : -0.3 nA 900 ms.

The above shown data allows to propose that pancreatic beta cells in the electrical working range go from a non coupled state (silent periods, substimulatory glucose) to a coupled state (oscillatory periods, stimulatory glucose). The initial phase is a frontier period between non-oscillatory and oscillatory electrical patterns, as between non-coupled and coupled states, and also it is the period along which β cell reaches the highest input resistance. These properties make it a particularly interesting event to study the dependencies between the different electrical parameters.

In simultaneous intracellular recordings, neighbor cells (interelectrode distance < 100 mm) show a total synchronizity in their square-wave bursting activity. That synchronizity is not observed during the initial phase, but, as it is shown in figure 3 there is in the most part of the recorded pairs a total lack of synchronizity , existing a considerable lag between the onset of the spiking activity, as each cell needs different time to arrive to the spiking threshold.

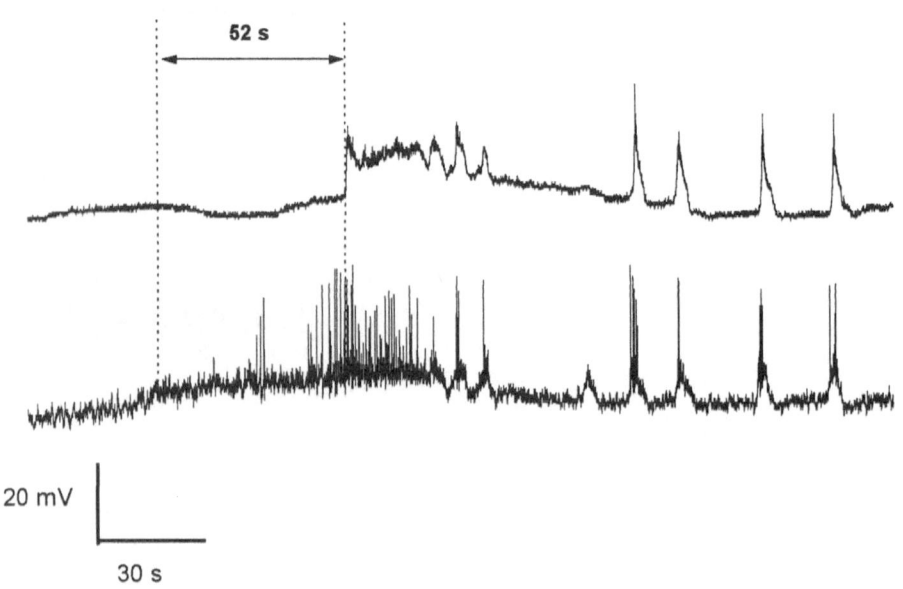

Figure 3 .- Asynchronizity of the initial phase between neighbor cells. Simultaneous voltage recording of two cells that were initially on 0 mM glucose and were exposed to 11 mM glucose. Dot lines indicate the time lag between onset of spiking.

2. Computational model
We simulated an small array of beta cells (3x3 cells) coupled through gap-junctions.

Single cell voltage is simulated by :

$$Cm \cdot dV/dt = (-I_{K\text{-}ATP} - I_{KDR} - I_{K(Ca)} - I_{Ca} - \sum_{j=1..n} I_{coupling\text{-}j})$$

Where $I_{K\text{-}ATP}$ is the ATP dependent potassium current. I_{KDR} is the delayed rectifier potassium channel. $I_{K(Ca)}$ is the calcium dependent potassium channel. I_{Ca} is the calcium current who has two components (FD fast-deactivating, and SD slow deactivating). Currents are simulated as previously described by Smolen and Keizer (9), but parametric values are those obtained from our experiments. Our single cell model does not oscillate by itself as far as there is not slow variable (ADP/ATP ratio) like in the previous models that makes the biestability inherent to the single cell model.

$I_{coupling-j}$ is the coupling current from the j-cell. Coupling can be abolished by setting the coupling conductance to zero. Glucose addition effect is simulated by blocking de ATP-dependent K^+ channel. In absence of glucose, model cells remain hyperpolarized. In presence of glucose, model cells spike continuously while not-coupled, but if we implement coupling between them, with a conductance in the range of our experimental data (3), cells begin to oscillate in a synchronous way.

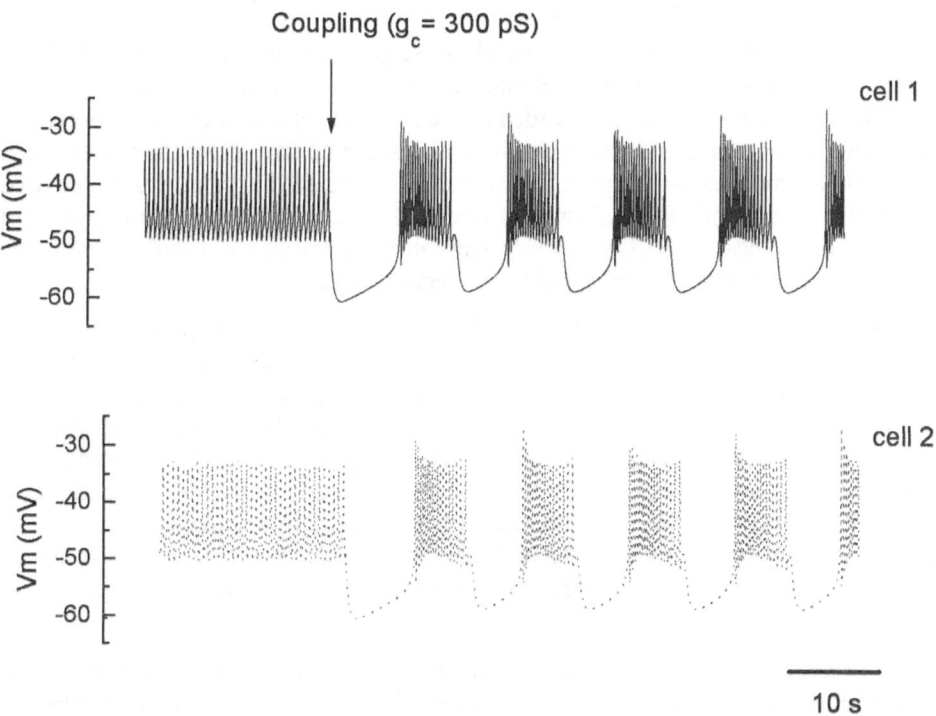

Fig 4. Computational simulation of an array of beta cells. Voltage plot of two cells of our (3x3) array. Cells are not coupled initially and are simulating the response to a stimulatory glucose concentration (12 mM). Arrow indicates the point where coupling is implemented.

DISCUSSION

We have shown that pancreatic β-cells do not exhibit neither oscillatory patterns nor coupling when are hyperpolarized (in the absence of glucose). Coupling appears in presence of intermediate glucose concentrations together with the onset of oscillatory activity. Higher glucose concentrations abolish the oscillatory pattern, while coupling is maintained. Consequently, oscillatory activity seems to be restricted to a rather limited range of conditions : oscillatory activity is disrupted out of an appropriate parameters balance. This is particularly well observed in the transitions from low to intermediate glucose levels : oscillatory activity is not immediately established, but an initial phase of continuous depolarization precedes its consolidation. Along the initial phase, synchronization between cells is lost and only happens after a period of minutes. These data support our hypothesis (2) on a role of input resistance limiting the oscillatory activity of pancreatic-β-cells. After this transition input resistance goes into permissive levels for the oscillatory pattern.

Lack of synchronizity along the initial phase suggests either a reduced intercellular feed-back or the inability of intercellular communication to induce synchronization. The former possibility can be discarded as we show in fig 2 that coupling is present for high glucose concentrations. Then it can be accepted that along the initial phase intracellular mechanisms produce a transitory increase in input resistance that depolarizes the membrane and impairs any role of intercellular communications in synchronizing these cells. Later, input resistance decreases and oscillates with the membrane potential as it does coupling conductance (3).

Oscillatory activity can not be based solely on the intrinsic biophysical properties of the cells, as isolated cells are unable to oscillate. On the other hand, the existence of high relative values of coupling conductance does not generate by itself the onset of oscillations as can be observed along the initial phase when intermediate glucose concentrations are applied or in the presence of higher glucose concentrations. We conclude that an appropriate interplay between coupling resistance and input resistance is necessary. We have checked this hypothesis in a computational realistic model showing that simulated isolated cells are unable to oscillate but the model reproduces the biological oscillatory pattern if coupling conductance is added.

Acknowledgments: We are indebted to Pablo Varona for helping with XSim installation, and for introducing us into XSim. We thank also A. Perez-Vergara and S.Moya for technical assistance. This study was partially supported by grants FIS94-0014, FIS96-2012, and European Union grant ERBSC1-CT9920833. E.A. was supported by a Formacion Personal Investigador (FPI) Doctoral Fellowship of the Dirección General de Investigación Cientifica y Técnica.

REFERENCES

(1) Sanchez-Andres, J.V. and Soria, B.(1993) The pancreatic B-cell as a voltage-controlled oscillator. *Lect. Notes Comp. Sci.* 686 : 37-42

(2) Andreu, E., Soria, B., Bolea, S., and Sanchez-Andres, J.V. (1995) Optimal range of Input Resistance in the Oscillatory Behavior of the pancreatic B-cell. *Lect. Notes Comp.Sci.* **930** *: 85-89*

(3) Andreu, E., Soria, B. and Sanchez-Andres, J.V. (1997) Oscillation of gap junction electrical coupling in the mouse pancreatic islets of Langerhans. *J.Physiol.(London) 498:753-761*

(4) Rorsman, P. and Trube, G.(1986) Calcium and delayed potassium currents in mouse pancreatic β-cells under voltage-clamp conditions. *J.Physiol. (London) 374 :531-550*

(5) Falke, L.C., Gillis, K.D., Pressel, D.M. and Misler, S. (1989). Perforated patch recording allows long-term monitoring of metabolite-induced electrical activity and voltage-dependent Ca2+ currents in pancreatic islet B cells. *FEBS Lett. 251 :167-172*

(6) Smolen, P., Rinzel, J. and Sherman, A. (1993). Why pancreatic islets burst but single β cells do not. *Biophys. J. 64 :1668-1680*

(7) Chay, T. and Keizer, J.(1983) Minimal model for membrane oscillations in the pancreatic β-cell. *Biophys.J. 42 :181-190*

(8) Chay,T. (1990). Effect of comparmentalized Ca ions on electrical bursting activity of pancreatic β-cells. Am.J.Physiol. **258** : C955-C965

(9) Smolen, P. and Keizer, J.(1992*)* Slow Voltage Inactivation of Ca^{2+} Currents and Bursting Mechanisms for the Mouse Pancreatic Beta-Cell. *J.Membrane Biol. 127 : 9-19*

(10) Sherman, A. and Rinzel, J.(1991). Model for synchronization of pancreatic β-cells by gap junction coupling. *Biophys. J. 59 :547-559*

Mechanisms of Synchronization in the Hippocampus and Its Role Along Development

L. Menendez de la Prida and Juan V. Sanchez-Andres.
Departamento de Fisiología, Instituto de Neurociencias, Universidad de Alicante, Campus de San Juan, aptdo. 374, Alicante 03080, Spain.

Abstract. Biological neuronal networks are subject to modifications along development. It is well known how specific mechanisms such as hebbian transformation of the synapsis contribute to the final configuration of the adult circuits. Synchronization and desynchronization of spontaneously firing cells could play a critical role in the reinforcement of the connectivity patterns. The purpose of this work is to investigate such synchronizing mechanism in the newborn hippocampus. We show that immature hippocampal CA3 and CA1 cells burst synchronously and discuss its implications in the operation of the implicated neural networks.

Introduction.

The hippocampus is an special part of the cerebral cortex which plays an important role in processes such as learning and memory [1]. Its neural networks are organized in a trisynaptic circuit (Fig.1) that involve neurons from three different regions as well as specific synaptic connections. The CA3 region shows the same structure than an autoassociative neural network [2] thus exhibiting properties such as noise resistance and learning based in the mechanisms of a content-addressable memory. The information in the trisynaptic circuit came from the enthorinal cortex via the perforant pathway; the first contact is made onto granule cells of the dentate gyrus which send their axons to the CA3 region. The CA1 region on the other hand, receives information from CA3 providing the output of this feedforward network.

Several electrophysiological studies have revealed that the hippocampus is involved in the processing of spatial and temporal patterns [3] but also is known as a structure specially prone to epilepsy [4]. Its high connectivity degree (a pyramidal CA3 cell receives around 80 synaptic contacts from granular neurons and 7700 recurrents inputs from other CA3 cell) has important effects in the synchronizing capability of this circuit. The role of such synchronizing mechanisms in the normal and abnormal hippocampal behaviour has been extensively investigated [5].

Recently, we have observed that the spontaneous activity of CA1 pyramidal cells is characterized by the presence of bursting, that dissappears by the end of the first month of postnatal life. Such a pattern is replaced by the adult one consisting in the presence of isolated action potentials [7]. This change seems to be a consequence of several modifications at the network level [8] and thus could be a manifestation of the plasticity mechanisms that are taking place. In these conditions the bursting recorded in CA1 is a strict feature of its immaturity. However, CA3 cells show also bursting activity in mature stages of developement. We wonder whether immature CA1 bursting is intrinsically generated in this region or is a consequence of CA3 activation. This possibility should be considered as far as CA3 region plays the role of input relay to CA1 area.

Two different approaches were addressed to assess this question: a. simultaneous recording of CA3-CA3 and CA3-CA1 cells, and b. extracellular stimulation of the Schaffer collaterals (the projection from CA3 to CA1). Simultaneous recording of cells from both areas should show if bursting appears correlated whatever synchronous or not. Extracellular stimulation of the CA3 projection (Schaffer collaterals) to CA1 should show if activity in CA3 is able to induce bursting in CA1, therefore if CA1 bursting can be attributed to a primary CA3 activation.

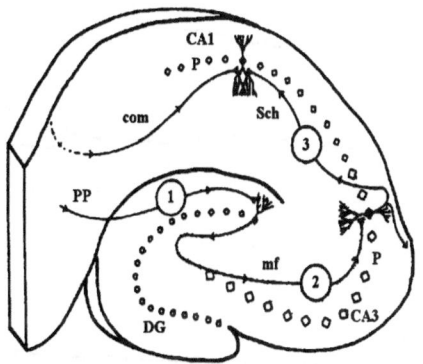

Figure 1.
The trisynaptic circuit in a hippocampal transverse slice. PP: perforant path ; DG: dentate gyrus; mf: mossy fibers; P: pyramidal cells in CA3 and CA1; Sch: Schaffer collaterals.

Materials and Methods.
Intracellular recordings were done in pyramidal (CA1 and CA3) cells from 500 μm thick hippocampal slices from New Zealand white rabbits (0-25 postnatal days, N=12 cells). The slices were prepared as previously described [7]. The impalements were made at the CA1 and CA3 stratum pyramidale. Extracellular stimulation was performed at the Schaffer collaterals (i.e. the CA3-CA1 connecting fibers). The criteria for a healthy impaled cell were: membrane resting potentials greater than -50 mV, spike train response to positive current injection, overshooting action potentials, and input resistance larger than 20 MΩ. Recordings were stored in magnetic tape for further analysis (Origin 3.73, Microcal Software).

Results.

Synchronicity in immature neurons from CA3 and CA1 hippocampal regions.

Simultaneous intracellular recordings from two CA3 pyramidal cells show a rhythmic spontaneous activity at around 0.3 Hz (Fig 2A, *top*). Extended computational models support the existence of partial synchronization in this circuit as a form of collective behavior that emerges from the structural organization of CA3 neural networks [9]. Tipically, we have found that these cells fire isolated spikes asynchronously, however we can not rule out the possibility of local mechanisms of synchronization since their firing frequency are quite similar. In any case the role of such mechanisms shoud be more to phase-lock the spiking frequency of the cells, rather than synchronizing their discharge. We have not detected any physical contact between these two neurons [10], thus the possible circuit conformed by them is as represented in Fig 2 *left*.

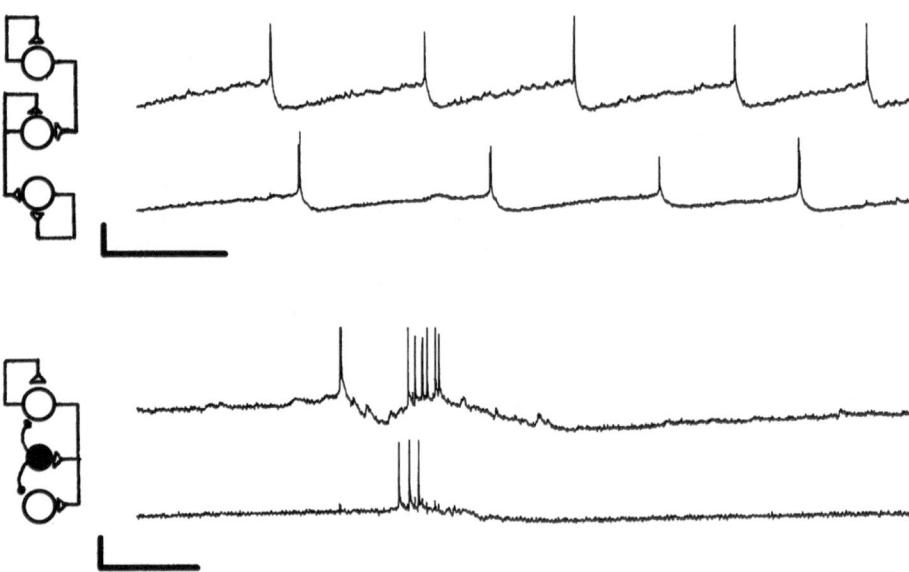

Figure 2. Simultaneous intracellular recordings in CA3-CA3 neurons (top diagram) and CA3-CA1 pyramidal cells (bottom). CA3 cells in both plots are represented in the upper trace. Calibration bars are 20 mV (vertical) and 2.5 s (horizontal). Schematic hypothetical circuits are represented at the left.

On the contrary, a different behavior is observed when the described immature bursts are fired. Figure 2 *bottom* shows a bursting episode in a CA3 cell recorded simultaneously with a CA1 one. This correlation of firing (lower part of fig. 2 *bottom*) demonstrates a clear synchronization between bursting activity in both areas.

Previous studies have shown that these bursts result from the synchronous activation of local inhibitory neurons since they are blocked by specific agents (bicuculline) and have similar electrical properties than the typical iPSP [7,8]. These explanation could seem paradoxical as involves an excitatory effect on the basis of the activation of an inhibitory neurotransmitter. Nevertheless, there are aboundant data supporting an excitatory role of the inhibitory neurotransmitter GABA in immature stages, although it is not still clear whether this is due to changes in the ions gradients or to more complex molecular mechanisms. All these support that the transient excitatory nature of these synapsis contribute to the onset of the bursts. However, this single mechanism does not explain by itself all the features of the bursts genesis [11].

The circuit shown on the left in Fig.2 *bottom*, represents the presumable processing that occurs in the real neural network. The activation of CA3 pyramidal cells produces the coupling of CA3 and CA1 inhibitory interneurons activity through a still unclear mechanism. This coupled activation of the immature interneuronal synapsis can be the responsible for producing a giant depolarization that constitute the burst.

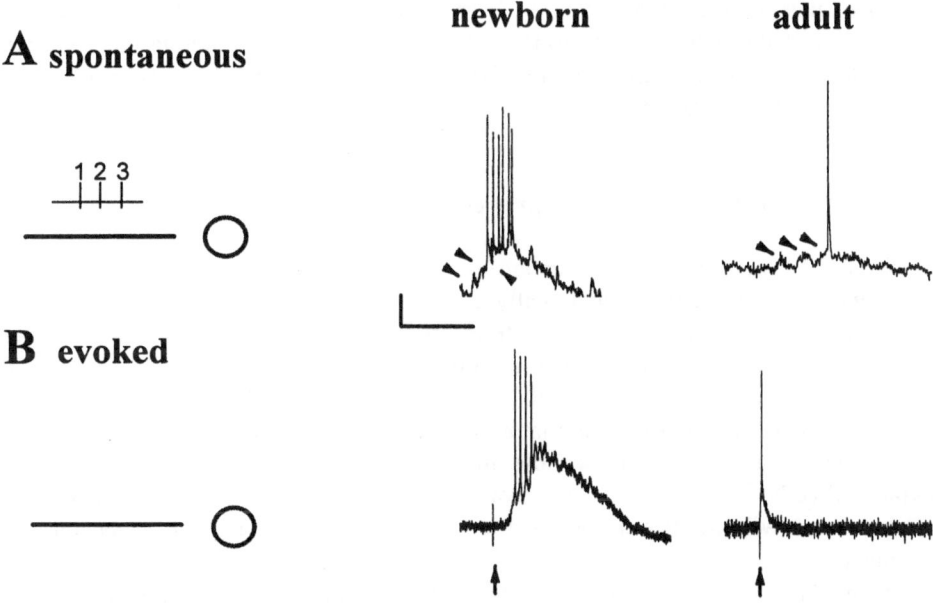

Figure 3. Changes in spontaneous and evoked responses of CA1 neurons along development. *A*. Three spikes reach the synaptic terminal. They evoke different responses in the postsynaptic cell according to its maturational degree. *B*. Similar output is obtained by the direct stimulation of Schaffer collaterals fibers. Calibration bars are 20 mV (vertical) and 500 ms (horizontal).

Evoked CA1 bursting through Schaffer collaterals extracellular stimulation.
The previous results provide evidences about the correlation among CA1 and CA3 immature cells when bursts are fired. Nevertheless, it is important to elucidate if such events have its origin at CA3 networks or emerge as a collective property of the entire undeveloped hippocampus.

Fig 3A (*central panel*) shows an spontaneous burst (*top*) resulting apparently from the summation of several ePSPs (see arrows). A very similar event can be experimentally induced by extracellular stimulation of the Schaffer collaterals (*bottom*). Like the spontaneous bursts, the evoked ones consist on a slow depolarization crowned by spikes and followed by a slow repolarization. Analysis of the biophysical properties of spontaneous and evoked bursts allows to state that both events are essentially equivalent (data not shown). This finding suggests that CA3 activation is capable by itself to induce bursting in CA1.

As it has been stated above bursting is not a permanent feature of CA1 activity. In fact, it disappears as development proceeds, and the pattern observed in adult cells consists on synaptic potentials that eventually are able to reach threshold and fire isolated action potentials (fig. 3, *right panel, top*). By the end of the first month of postnatal life CA1 bursting is rarely observed. Consistently, extracellular stimulation of the Schaffer colaterals is also unable to induce bursting (fig. 3, *right panel, bottom*), rather it produces regular ePSP that can trigger an action potential provided threshold is attained.

Modification of the CA1 input/output relationship along development.

In adult CA1 neurons, when bursting is not observed, simultaneous intracellular recording [12] shows that synaptically coupled neurons respond only with single action potentials. There are both electrophysiological and morphological evidences [13] demonstrating that individual inhibitory neurons can phase the activity of hippocampal pyramidal cells in adult. This mechanisms could be sufficient to explain the output characteristic of the immature network. An immature, and thus excitatory, GABA interneuron may be capable to synchronize the electrical firing of the pyramidal cells. The synchronous activation of these GABAergic neurons within the CA1 network will produce in the recorded cell a giant depolarization responsible for the burst.

As maturation proceeds, the excitatory nature of the interneuronal circuit is lost acquiring its normal inhibitory action. At this moment, the correlated activation of this circuit will not evoke depolarization as in the immature case. The output of the CA1 adult circuit is then different when compared with its immature response to the same stimulus (Fig.4).

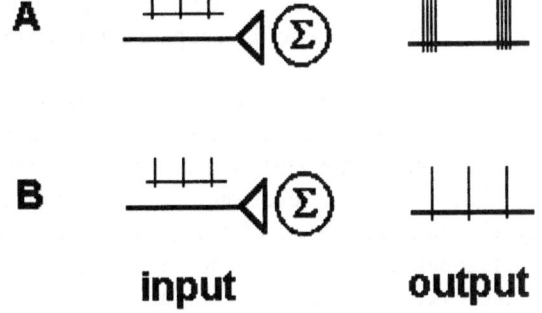

A

B

input   **output**

Figure 4. Transformation of the input/output relationship along development.

Discussion.

The data presented shows a high degree of synchronization in the developing hippocampus. Particularly, CA3 and CA1 bursting are synchronous, suggesting that the hippocampus works partially as a whole until advanced stages of postnatal development. Both areas show also the presence of no synchronous isolated spikes along this period (data not shown) Remains to be demonstrated the role of this high degree of synchronicity. It is well known the fact that intracellular calcium concentrations are higher in youngster than in adult animals [14]. Bursting implies a depolarization able to fire action potentials that can permit to keep relatively high calcium levels, playing essential roles in triggering plastic changes characteristic of this period of life.

A question arises: is the synchronicity intrinsic to the different hippocampal areas or is there an area playing the role of pacemaker?. We addressed this problem by studying if bursting in CA1 is intrinsically generated or dependent on CA3 activity. Our data strongly suggests that CA1 bursting requires CA3 activity as it can be induced by activation of the projection pathway from CA3 to CA1. Under this scope CA1 is not autonomous but dependent on CA3 which will assume the role of bursting pacemaker. Whether CA3 bursting results by the intrinsic activity of its cells or need a previous pacemaker is a question which answer would require further research.

On the other hand, we have observed a modification of the input/output relationship of the CA1 local circuitry along development (Fig. 4), consisting in the loss of the capability to burst as a response to synaptic stimulation. Stimulation of the Schaffer collaterals in newborns and adults produces in the newborn CA1 cells a burst, while upon the same stimulus, adult CA1 neurons fire a single action potential. This indicates that the input/output relationship of the CA1 local circuits suffers a modification as maturation proceeds. The question is which are the different mechanisms involved in the onset of such events and how they maturates to provide a quite different response in adult networks. This change can be partially explained in terms of the previously mentioned switch from the immature excitatory to inhibitory action of GABA. This alterations in the hippocampal circuits from newborn to adult seems to be a manifestation of the plasticity and more complex processes taking place during maturation. We wonder what role could have such process in the normal development of the neuronal networks i.e. as a response to the coupled action of the interneurons.

We propose that two physically connected immature hippocampal cells sustaining simultaneous activation in certain temporal intervals should cause the strengthening of its synaptic contacts in a hebbian manner. This electrical pairing also have important implications in the control of internal calcium levels and in the induction of long-term potentiation [15]. The synchronous activation of the bursts in all the hippocampal cells should thus have functional impact in at synaptic level. Not all the CA3-CA1 neurons are connected, but those cells which are, will fire in a simultaneous fashion thus reinforcing their synapsis. A similar role of spontaneous activity in the configuration of the adult circuitry with respect to other systems have been widely suggested [16]. Through this spontaneous synchronizing mechanisms the hippocampal formation would be able to build the connectivity patterns necessary for its normal adult function.

Acknowledgements: This work is supported by grant 96/2012 from Fondo de Investigacion Sanitaria. L.M.P was supported by Fellowships from GV.

References and Notes.

[1] See Shepherd GM, 1990, *The Synaptic Organization of the Brain*, Oxford Univ Press (3th Ed), for review.
[2] Kohonen, 1984. *Self-Organization and Associative Memory*, Springer-Verlag.
[3] Wilson MA and McNaughton BL, Science 261:1055. Dynamics of the hippocampal ensemble code for space.
[4] Taylor CP, 1988. TINS 11:375. How do seizures begin? Clues from hippocampal slices.
[5] Wong RKS, Traub RD and Miles R, 1984. In *Electrophysiology of Epilepsy* (Schwartzkroin PA and Wheal HV Eds) NYAcademic Press. Epileptogenesis mechanisms as revealed by the studies of the hippocampal slice.
[6] Brown TH, Chapman PF, Kairiss EW and Keenan CL, 1988. Science 242:724. Long-term synaptic potentiation.
[7] Menendez de la Prida L, Bolea S and Sanchez-Andres JV, 1996. Neurosc.Lett 218(3):185. Analytical characterization of spontaneous activity evolution during hippocampal development in the rabbit.
[8] Ben-Ari Y, Cherubini E, Corradetti R and Gaiarsa JL, 1989. JPhysiol 416:2870. Giant synaptic potentials in immature rat CA3 hippocampal neurones.
[9] Traub RD and Miles R, 1991 *Neuronal Networks of the Hippocampus*. Cambridge Univ. Press
[10] Current injections evoking a burst of spikes either in the first or the second neurons do not elicit any response in the other.

[11] Blockers of excitatory synapsis such as AP-5 also eliminated the bursts in the immature subjects (see reference 8) All this means that these events involve more complex networks mechanisms.

[12] Miles R, 1990, JPhysiol 428:61. Synaptic excitation of inhibitory cells by single CA3 hippocampal pyramidal cells of the guinea-pig *in vitro*.

[13] Cobb SR, Buhl EH, Halasy K, Paulsen O and Somogyi P, 1995, Science 378:75. Synchronization of neuronal activity in hippocampus by individual GABAergic interneurons.

[14] Spitzer NC, 1991. JNeurobiol. A developmental handshake: neuronal control of ionic currents and their control of neuronal differentiation.

[15] Magee JC and Johnston D, 1997. Science 275:209. A synaptically controled, associative signal for hebbian plasticity in hippocampal neurons.

[16] Shatz CJ, 1990. Neuron 5:745. Impulse activity and the pattering of connections during CNS development.

Analysis of Synfire Chains Above Saturation

R.M. Reyes and C.J. Pérez Vicente

Universidad de Barcelona, Departamento de Física Fundamental,
Diagonal 647, 08028 Barcelona, Spain

Abstract. We have studied a population of synfire chains above saturation. By considering a Hebbian-like learning strategy we show that to keep the propagation of activities controlled, the system reorganizes its structure leading to a new set of chains with shorter lengths distributed according to a non-trivial profile.

1 Introduction

After the experimental evidence [1, 2] of an accurate time structure in the activity of cortical neurons there is a great interest in providing mechanisms able to reproduce such phenomenon as well as to understand its possible relevance in the processing of information at a cortical level. Among the possible ideas which have appeared in the last years, perhaps the one which has attracted more attention is that based on the concept of synfire chain [3] since it not only provides an explanation for the spatio-temporal patterns of firing discovered from multi-cellular recordings but also because it might be a neural mechanism for dynamical binding [4, 5].

Basically, a synfire chain is a sequence of pools of neurons connected through feedforward links. Each pool is made of a small group of cells all of them receiving roughly the same type of inputs in a very short time window either from other pools or from an external source and sending outputs to another compact set of neurons. In this way, a volley of activity can propagate from pool to pool forming a chain of activity. The spatio-temporal pattern of activity is characterized by waves of synchronous spikes and therefore can be an explanation for the precise timing observed in experimental data. Several authors have proposed a synfire pattern of activity as the fundamental unit of computation [3, 6]. Under this assumption, since several waves can propagate simultaneously through the network, dynamical binding could be understood as the synchronization of such dynamical objects.

There has been several attempts to analyze theoretically the computational properties of a system formed by synfire chains [7, 6, 3]. It is a normal practise in almost all of these studies to assume that the physical network has been created a priori, so that a certain number of chains with a fixed length and well determined interconnections between pools, are distributed all over the network. This fact implies the existence of an underlying principle of self-organization plus a learning strategy which very often are not specified. Under such assumption the system is robust and efficient conveying information up to a certain saturation point. To be more precise, if N is the number of neurons of the system, s the size (width) of each pool, p the number of pools [1] and a the global activity in the network, then it has been shown [6] through a signal-to-noise ratio analysis that the quotient

$$\kappa = \frac{s}{\sqrt{pa\frac{s^2}{N^2}(1 + sa/N)}} \tag{1}$$

is a good measure to establish the capacity of the model. Simulations have shown that a value of $\kappa_c \approx 5$ denotes the limit above which synfire transmission is stable. Below κ_c the destructive interference between the stored patterns is so high that the propagation of information in terms of waves of activity is uncontrolled. In other words, if a pool of the chain, say j, is activated at a given time there is no guarantee that pool $j + 1$ will be activated at the next time step.

Let us consider a synfire-type system near saturation. Let us assume that each stored pattern has a fixed length and we want to add a new one, so that the new pattern will exceed the saturation limit. This situation is undesirable since the new item of information will make the whole network unstable in a very similar way as uses to happen in some models of associative memory. The goal of this paper is to show that a network of synfire chains has another mechanism to solve this problem based on a natural reorganization of its structure. As we will see this process will modify the length of the chains but keeping fixed the starting a end point points in such a way that the number of links or pools still fall below saturation. According to this procedure new patterns can be added to the network without observing a decrease in the computational properties of the system. The consequence of this internal reorganization is a tendency to create chains with lengths which follow a non-trivial distribution, being more likely short chains. This result is in agreement with recent results observed with integrate-and-fire neurons [8] when a dynamical learning procedure is used to set up the values of the synaptic connections.

The paper is structured as follows. Section 2 is devoted to describe the model. In section 3 we will present the most relevant results of our investigation. Finally, there is a short section for discussion and future work.

2 The model

First of all, let us to describe the basic features of the model studied in the paper. The system is formed by $N = 100$ neurons. We have chosen this size because few patterns are enough to reach the saturation limit and it is simpler to emphasize the relevant outcomes of our analysis.

When dealing with a system based on synfire chains one relevant point concerns the size s of the pools, since the saturation point given by (1) depends directly on such measure. It is well known that the level of activity in the cortex is rather small. Neurophysiological studies have shown that depending on the particular cortical area these levels oscillate between $\frac{a}{N} = 0.01$ and 0.06.

---

[1] Note that between two consecutive pools there is a set of links. For this reason $p - 1$ also denotes the number of links in the system.

Keeping in mind this fact and taking into acount that the network is small we have assumed that the number of neurons simultaneously active at a given time should be roughly the same of the width of a given pool [2]. Then, we have scaled s with the reduced size of the network and set $s = 3$. We have also considered larger values for the width of the pool but the main results does not depend of this particular choice.

One of the main roles of local inhibition is to control the level of activity of the whole network. It can be modeled in several manners, either by defining complex mechanisms based on an accurate distribution of time constants for the inhibitory synapses plus a precise topology for the network or by considering much simpler processes based on a suitable modulation of thresholds. In our case, we have assumed perhaps the simplest option by considering a winner-take-all strategy. Only those neurons which receive the largest postsinaptic potentials will be activated in the next time step. This hypothesis allows us to relax the necessity to introduce explicitly inhibitory synapses. In addition, we have assumed that the number of activated neurons coincides exactly with s.

Before learning the network does not have any definite structure. It is just the learning process which defines the feedforward architecture typical of a syn-fire chain. For this reason at the beginning of the process, the network is fully connected with synaptic values chosen randomly according to a non-negative uniform distribution that without loss of generality is defined in the interval $[0, 1]$. Learning is performed according to the following criterion: the first pool of neurons (the input) as well as the last one (the output) are randomly selected and then fixed. It mimics the fact that information is transmitted from sensitive regions to precise cortical areas. However, the flow activity which lead to the output region from the input region can follow a myriad of different pathways. These channels are learnt by non-linear local hebbian reinforcement. Each time that two connected neurons ζ_i^k, ζ_j^{k+1} in consecutive pools $k, k+1$ are simultaneously activated then their synaptic link is reinforced according to

$$\Delta J_{ij} = \alpha(1 - J_{ij})\zeta_i^k \zeta_j^{k+1} \tag{2}$$

where α is a (small) learning constant. In any other situation it remains invariant except if the total length of the chain exceeds a maximum value in which case it is depressed. In this way we punish the creation of large synfire chains, what makes sense since information is processed in a characteristic time scale, usually rather short.

Now, let us review the set of results that we have obtained from simulations performed on the model described above.

---

[2] For larger sizes it is clear that the global activity must be much larger than the size of a single pool. In this case it is plausible to consider several chains activated simultaneously or simply to assume a certain spontaneous activity. How does this situation affect learning is currently under study

3 Results

The results shown in this section are averages over 250 samples. In all the cases we have tried to store $l = 5$ patterns. According to the stability criterion described by eq. (1) the number of pools required to reach the saturation point assuming pools of width $s = 3$ is $p \approx 125$. This fact implies that the length of each chain to be stable should not exceed 25. To be sure that we are dealing with situations above saturation we have considered that the initial length of all the patterns is 30. Therefore, some of them are unstable which means that the wave of activity cannot propagate along the initial prescribed chain and other neurons not belonging to the pattern are activated, making the transimission of information unefficient. Once the patterns are defined we relearn them following the hebbian reinforcement strategy mentioned above with $\alpha = 0.2$. We present them sequentially [3] to the network, each one 30 epochs. If after the first sweep the patterns are not learnt we iterate the method till learning is over. This condition is achieved when all the patterns give the correct output and follow the same pathway at least for ten new trials. In the learning process some neurons that belonged to the old pattern are discarded while new ones form stable pools which lead to the desired output. As a consequence a new structure is associated to the chain. The distribution of lengths is shown in figure 1.

As we can see the synfire chains tend to be quite short with a peak located at $p = 3, 4$ followed by an exponential decayment for larger sizes. Further studies for larger N shows that this behavior is independent of the size of the system although the peak is moved to larger lengths of the chain. Note that there is a fraction of chains that still keep the same original length. These correspond to individual patterns that where stable even above saturation.

An interesting point concerns the percentage of neurons recruited by the patterns to form synfire chains. Before relearning almost all the units formed part of one of the 5 patterns stored by the network. However, after hebbian reinforcement figure 2 shows that such quantity follows a gaussian distribution centered below 50%. For larger sizes this distribution is shifted to smaller values. The main conclusion that can be extracted is that the system tends to create very robust structures with a very important connectivity between the neurons (pools) which form the chains and weak synapses (or simply absence) with the rest. The consequence is that information propagates very reliably since the new patterns stand far below saturation. Moreover, more patterns can be stored in the network without losing the computational properties of the system.

The results shown in the previous figure do not mean that just one neuron belongs to exactly one synfire chain. In fact, it is just the opposite since many neurons belong to several chains. As an example, we have shown in figure 3 the percentage of neurons that appear three times in different patterns. The distribution has also a gaussian shape centered around 10% of the active units. This result is in agreement with the majority of studies performed on this sort of systems which predict the existence of elements which form part of several path-

---

[3] The order is absolutely irrelevant

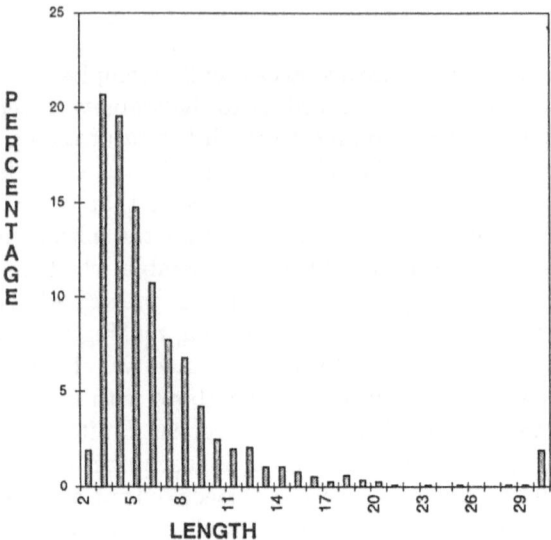

Fig. 1. Distribution of the lengths of the patterns after the reorganization of the system. The data are an average over 1250 synfire chains.

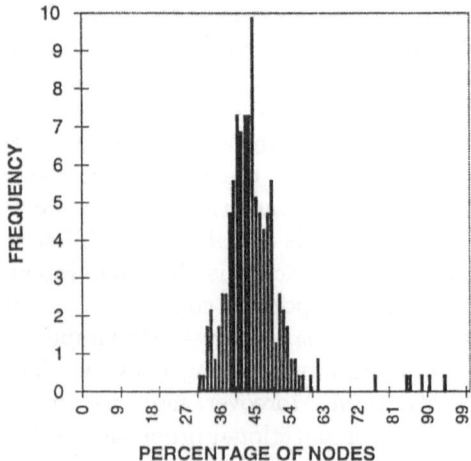

Fig. 2. Percentage of nodes involved to create $l = 5$ synfire chains. The figure describes the network after relearning. As in the previous figure are an average over 250 samples (1250 synfire chains).

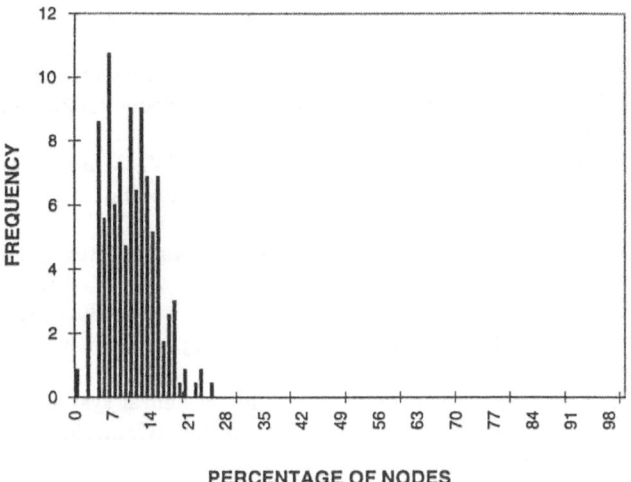

Fig. 3. Frequency of appearance of nodes belonging to three different synfire chains. The figure shows that with a non-vanishing frequency around 15% of the neurons belong simultaneously to 3 different patterns.

ways [3, 6, 4]. These units endow the system with a very rich dynamical behavior showing the existence of a complex architecture not completely feedforward that elicits reverberations within an assembly of neurons that excite themselves. In this way, the model displays features similar to those of realistic systems [9].

4   Discussion

We have seen that a network of synfire chains can modify its basic structure if the number of patterns stored in the network exceeds the saturation limit. Simply by considering a nonlinear hebbian-like learning procedure the system self-organizes with chains of variable length distributed in a non-trivial way being the shortest ones the more likely. The model is a first step to more realistic situations. In particular, some aspects which deserve special attention imply to consider networks with larger sizes, how the reorganization mechanism is affected when certain amount of spontaneous activity is superposed to the stored synfire patterns as well as further possible mechanisms of learning in a noisy environment. Furthermore, more realistic mechanisms of inhibition rather than a winner-take-all strategy as well as a fully non-supervised learning process are other ingredients that we will also analyze in the next future.

This work has been partially supported by DGYCIT under grant PB94-0897.

References

1. Gray C.M., Konig P., Engel A.K. and Singer W., Oscillatory responses in cat visual cortex exhibit intercolumnar synchronization which reflects global stimulus properties. Nature **338** (1989) 334-337.
2. Eckhorn R., Bauer R, Jordan W., Brosch M., Kruse W., Munk M and Reitboeck H.J., Coherent oscillations: a mechanism of feature linking in the visual cortex? Biol. Cybern. **60** (1988) 121-130.
3. Abeles M., Corticonics: Neuronal Circuits of the cerebral cortex. (Cambridge University Press) 1991.
4. Abeles M., Vaadia E., Bergman H, Prut Y., Haalman I. and Slovin H., Dynamics of neuronal interactions in the frontal cortex of behaving monkeys. Concepts Neurosci. **4** 131-158.
5. Bienenstock E. and Geman S., Compositionality in neural systems *The Handbook of Brain Theory and Neural Networks* ed MA Arbib (Cambridge, MA: Bradford books/MIT) Press 1995.
6. Bienenstock E., A model of neocortex. Network: Comput. Neur. Syst. **6** (1995) 179-224. See references therein.
7. Herrman M, Hertz J., and Prugel-Bennett A., Analysis of synfire chains. Network: Comput. Neur. Syst. **6** (1995) 403-414.
8. Hertz J. and Prugel-Bennett A., Learning short synfire chains by self-organization. Network: Comput. Neur. Syst. **7** (1996) 357-363.
9. Abeles M., Bergman H., Margalit E. and Vaadia E., Spatio temporal firing patterns in the frontal cortex of behaving monkeys. J. Neurophys. **70** (1993) 1629-1638.

Allometry in the Justo Gonzalo's Model of the Sensorial Cortex

Isabel Gonzalo

Departamento de Optica. Facultad de Ciencias Físicas
Universidad Complutense de Madrid. Ciudad Universitaria s/n. 28040-Madrid. Spain
E-mail: igonzalo@eucmax.sim.ucm.es

Abstract. We report on the interpretation of the "central" syndrome (described by J. Gonzalo and associated to a unilateral lesion in the parieto-occipital cortex, equidistant from the visual, tactile and auditory areas) as a reduction in the excitability of the cerebral system. The sensorial organization is maintained but, according to the principle of similitude, different sensory functions become altered differently (allometrically) depending on their excitability demands. This alteration reveals the functional behaviour of the sensorial structures. Taking J. Gonzalo's work as a starting point, we found allometric potential functions for some of the sensory visual functions.

1   Introduction

As exposed in previous work [1, 2, 3, 4], J. Gonzalo characterized what he termed the *central syndrome*. In this syndrome the organization of the cerebral system is maintained but on a smaller scale than in the normal case, and the exploration of the *functional* behaviour of the sensorial structures becomes possible.

The central syndrome arises from a unilateral lesion in the parieto-occipital cortex, equidistant from the visual, tactile and auditory projection areas (central zone). It is characterized by: (a) *repercussion* to all the sensory systems, in all their functions and with symmetric bilaterality; (b) *dissociation* of sensory qualities normally united in perception, and (c) *reinforcement* originated by the appearance of permeability to intersensorial summation, including muscular contraction, which partially compensates for the deficit of cerebral excitability due to the loss of neural mass in the lesion, improving the perception.

A complete gradation was found between the central syndrome, in which the neural mass lost has a rather unspecific physiological activity, and the syndrome originated by a lesion in the projection area, where there is a destruction of the nervous path. This led to the definition of sensory function *densities* with regional variation, called functional gradients [2], as shown in Fig. 1. In this model, the sensory functions themselves arise from the integration of their respective function densities throughout the cortex.

J. Gonzalo studied many patients with brain injuries from the Spanish Civil War (1936-39) and also analyzed the works of other authors of which we only refer to a few [5, 6, 7, 8, 9]. His contributions to brain theory are supported

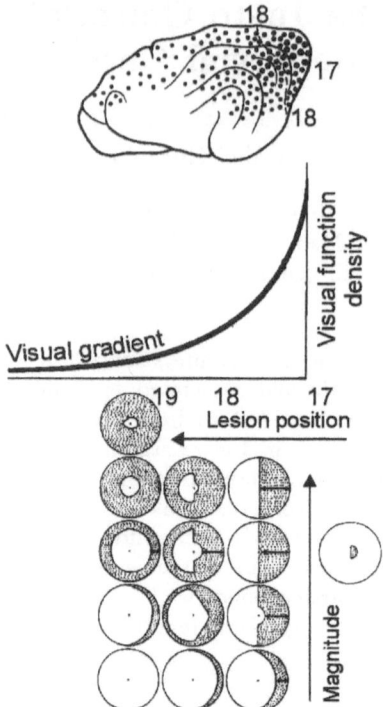

Fig. 1. Schematic representation of visual function density versus distance to the projection area 17. Lower part: visual field reductions according to the position and magnitude of the lesion. Concentric reductions correspond to central syndromes.

by physiological laws and data and have been considered of considerable scope and highly accurate [10, 11, 12, 13], lending themselves to a clear formalization [15]. His work is related to other approaches [14] and to recent treatments in which the distributed nature of cerebral processing, as well as the integrative and adaptative aspects of this processing, are involved [16, 17, 18, 19, 20].

Here we deal with that part of the model which, though presented by J. Gonzalo [21], has not been published except as a qualitative commentary [4]. In this paper we restrict ourselves to those aspects of the dissociation phenomenon which are necessary in order to find the allometric laws which govern the alteration of the sensory functions.

An example of dissociation in central syndrome is the following. If an individual is touched, only a general tactile sensation is perceived at first, but after some time, the stimulus can be localized. This is a functional dissociation or phasing out between tactile sensation and localization. Hence the loss of "central" cortical mass leads not only to a hypoexcitability of the nervous centres, as was shown in simple excitability experiences [1, 2], but also to a decomposition of the normal sensation. The individual first reaches the excitability thresholds of the most elementary sensorial levels (tactile sensation in the example) and when the stimulus becomes stronger (or as a result of temporal summation), reaches the next more complex or elaborated levels, which require more excitability (localization in the example).

One of the functions which appears on dissociation is the *direction* function which was systematically studied in humans for the first time by J. Gonzalo [1, 2]. As was explained in detail in the original bibliography and summarized in recent reports [3, 4], a vertical upright test arrow, for example, can be perceived to be inclined by individuals with central syndrome. The greater the central cortical mass lost, the more inclined the arrow is perceived to be, to the extent of being inverted if the arrow is poorly illuminated or far enough from the observer (low stimulation). From the detailed study of visual, tactile and auditory inversion, it was proposed that the perception which originates in the projection area, is inverted and constricted, but is then magnified and reinverted, i.e. elaborated or integrated, in the whole cortex and particularly in the central zone [2].

The concepts of similitude and allometry were applied to the central syndrome, as is shown in the following, where we restrict the exposition to the visual system.

2   Similitude and Allometry in the Central Syndrome

The essential mechanism in the central syndrome was considered to be a scale change in the excitability of the system. A new dynamic equilibrium appears, in which the general organization is maintained. It can be appreciated from the excitability and luminosity threshold curves [1, 2, 4], which are shifted with respect to the normal case, following the same law but with other values of the parameters, and from the concentric reduction of the visual fields and their sensibility profiles (Fig. 2) which both maintain the approximate shape of the normal case but on a reduced size.

J. Gonzalo applied the principle of similitude or dynamic similarity [24] to the central syndrome. According to this principle, the relative values between the parts of a system change differently under a scale change.

It is known that in the growth of a biological system the relation between the sizes of two parts, x and y, approximately follows an allometric law of the type [22, 23, 24],

$$y = ax^n \ . \tag{1}$$

This relation is obtained under the assumption of an exponential growth, as well as some other more general growth assumptions [25]. The above relation means that the rates of growth of the two parts (organs) compared are proportional, i.e. $(1/y)(dy/dt) = n(1/x)(dx/dt)$, this relation being obtained from (1) by taking the logarithm and differentiating. The constant n is called the allometric coefficient and characterizes the relation.

An analogy was established by J. Gonzalo between the biological growth and the "sensorial growth" by letting x and y now be sensorial functions of a given sensory system. The sensorial functions are parts of the system which vary

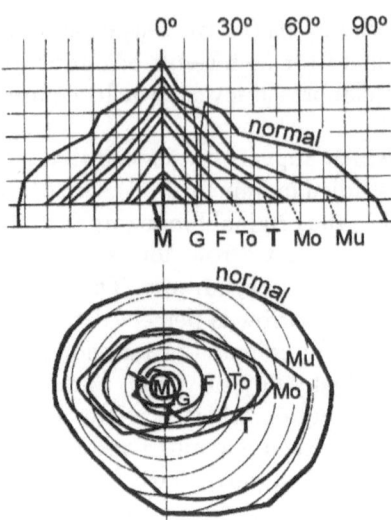

Fig. 2. Concentric reductions of visual fields and their sensibility profiles in cases of central syndrome. Cases M, T and that of the normal individual are shown, along with other cases studied.

under the reduction due to a central lesion, and could therefore be related by an allometric law. As a matter of fact, a correlation was found between the optic direction function and the width of the visual field of 24 patients [2].

Here we select, from among these 24 cases, only the 11 which can be considered *pure central* syndromes, to which the application of the similitude concept is more justifiable. The data of their visual field widths, optic direction and acuity in central vision of the observing eye, are given in table 1 [1, 2]. The extreme case of central syndrome manifested by the patient "M", the intermediate by "T", and the values for a normal individual (N) are indicated. The errors in the measurements are at least $2 - 3$ degrees in the visual field widths and in the inclinations perceived, and about 0.05 in the acuity. Absolute errors are greater in cases with intense central syndrome, such as case M, because the sensory degradation (in size, colours, form,...) is greater and also, because the high degree of permeability to temporal and intersensorial summation made the measurement difficult [1, 2].

From the data of table 1 we establish a correlation between the *loss* of a sensorial function, y, (optic direction or acuity) and the *loss* of the visual field, x. Since the visual field reduction is related to the magnitude of the central lesion, the variable x is then indicative of the scale reduction of the system. The variables x and y are expressed as percentages, e.g. 42 degrees of visual field (case T) corresponds to $x = (1 - \frac{42}{90}) \times 100$ per cent loss of visual field. The same patient perceived the test arrow inclined 18 degrees so that the percentage of loss of direction function is $\frac{18}{180} \times 100$ since a total loss of direction corresponds to 180 degrees (inversion). The acuity loss in this case is $(1 - 0.4) \times 100$ per cent.

Table 1. Visual field width, optic direction and acuity in cases of central syndrome and in a normal individual (N).

	Visual field (degrees)	Direction (degrees)	Acuity
M	6	160	0.04
	14	96	0.06
	30	45	0.33
T	42	18	0.4
	45	12	0.5
	49	14	
	52.5	12	0.5
	54.5	6	0.66
	60.5	5	0.66
	66.5	4	0.8
	74	2	0.9
N	90	0	1

If x and y are related by an allometric relation of type (1), its double logarithmic representation is a straight line whose slope is n. This representation is shown in Fig. 3 where x is the percentage of visual field lost while in one of the representations, y is the percentage of direction lost, and in the other, the percentage of acuity lost. Fitting the data to linear functions gives $n \simeq 2.66$ for the direction function and $n \simeq 1.33$ for the acuity.

At this point, first we must consider that the allometric law of type (1) may be a gross approximation in many biological examples. It is known that the double logarithmic representation presents in many cases concavities or inflexion points. The data are sometimes adjusted by linear relations of different slopes in different ranges so that brusque changes, even discontinuities, may occur. These critical

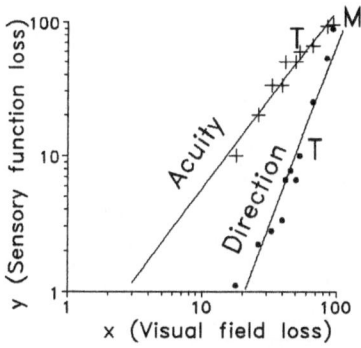

Fig. 3. Double logarithm representation of loss in acuity and direction versus loss in visual field width (expressed in percentages). The fitting to $y = ax^n$ gives $n \simeq 1.33$ for acuity and $n \simeq 2.66$ for direction.

points correspond to stages of the system where the rate of growth changes, i.e. an allometric law of type (1) changes to another one with different exponent [23, 24]. Second, some of the data in table 1 were difficult to determine with enough precision. Third, the cases selected as "pure" central syndromes, where the principle of similitude would apply, are perhaps not sufficiently similar, as can be appreciated in Fig. 2 where the visual field constrictions and profiles present differences in form which can be more or less pronounced.

From these remarks, the fitting of the data to the straight lines of Fig. 3 can be considered good enough to accept an allometric law of type (1) for the behaviour of the sensory functions considered, and justifies J. Gonzalo's proposal to extend this type of allometric behaviour to all sensorial functions of a given sensory system.

For the visual system, in Fig. 4 we represent the curves $y = ax^n$ obtained for the loss of acuity, $y = 0.27x^{1.33}$, and direction, $y = 3.13 \times 10^{-4} x^{2.66}$, together with a qualitative representation of the loss of two of the other sensory functions (there are many), according to the observed facts. The order of the split corresponds to the order of complexity (or excitability demand) of the sensory functions and to the order they are lost due to the shifts in their threshold excitabilities, as was observed in the cases studied [1, 2]. For example, when the excitability or intensity of the stimulus diminishes, an individual with central syndrome loses sensorial qualities in this order: aspects of the meaning of the test object ("scheme"), acuity, blue colour, yellow colour, red colour, direction, luminosity... When the test arrow is perceived to be very inclined, it is also perceived as very reduced in size, losing its form and colour, etc.

Very small differences of exictability between the different qualities occur already in the normal individual (e.g. there are slight differences in the excitability of colours) and such differences grow considerably on reduction of the size of the cerebral system due to a central lesion.

Now let us see the relation between these allometric changes and the functional gradients mentioned in the previous section (see Fig. 1). J. Gonzalo proposed that the visual function gradient or density in a normal individual, N,

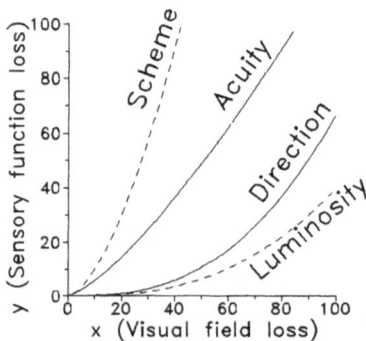

Fig. 4. Allometric curves $y = ax^n$ obtained for acuity and direction. Qualitative indications for the sensory functions "scheme" and luminosity as examples.

could be represented by an ensemble of different function densities or functional gradients $f_1^{(N)}, f_2^{(N)}, f_3^{(N)}, \ldots$ Their integrations throughout the cortex, say $F_i^{(N)} = \int_{cortex} f_i^{(N)}(s) \, ds$, by a phenomenon known cerebral recruitment, give normal sensory functions $F_1^{(N)}, F_2^{(N)}, F_3^{(N)}, \ldots$ (the origin in the curves of Fig. 4). The more complex the function F_i, the greater the excitability demand and the greater is the cerebral recruitment (the integration of f_i).

In the new equilibrium induced by a central lesion, the functional gradients or densities, f_i, take lower values than normal, so that, for two cases of different lesion magnitude, such as T and M, the integration throughout the cortex of the density functions $f_1^{(T)}, f_2^{(T)}, \ldots; f_1^{(M)}, f_2^{(M)}, \ldots$ gives smaller sensory normal values than normal, $F_1^{(T)}, F_2^{(T)}, \ldots; F_1^{(M)}, F_2^{(M)}, \ldots$ where $F_1^{(M)} < F_1^{(T)} < F_1^{(N)}$; $F_2^{(M)} < F_2^{(T)} < F_2^{(N)}$, and so on. In fact, M, T and N represent three different physiological levels.

In the notation employed above, the allometric relations established between acuity (say F_1), direction (say F_2) and visual field (say F_3) read $F_1^{(L)} = a_1 \left(F_3^{(L)} \right)^{n_1}$, $F_2^{(L)} = a_2 \left(F_3^{(L)} \right)^{n_2}$, where L denotes the different cases with central lesion and n_1, n_2 are the allometric coefficients obtained. It was shown that the acuity function (which has the greater excitability demand) suffers a bigger loss than the direction function (Fig. 4). In consequence, the acuity gradient or density, $f_1^{(L)}$, must suffer a greater decrease than the direction function density, $f_2^{(L)}$, for a given case L. The sensory function densities $f_i^{(L)}$ then become altered differently, i.e., allometrically.

Due to the special summation capabilities of the central syndrome, which are more pronounced the greater the deficit of cerebral excitability, an individual with this syndrome may experience a sensorial growth in three ways: by intensifying the stimulus, by iteration (temporal summation) and by reinforcement (intersensorial summation and strong muscular contraction). The best example is the extreme case presented by M who could approach the physiological level of the intermediate case T by the use of strong muscular contraction. The allometric laws also apply to this sensorial growth.

In the same way as in the visual system, J. Gonzalo analyzed other sensorial systems to which the same concepts and conclusions apply.

3 Conclusions and Discussion

The central syndrome, maintaining the same sensorial organization as in a normal case, reveals its structure through several dynamic phenomena, among them, the dissociation of a sensation into several components with different threshold excitabilities. Sensations usually taken as elementary can then be decomposed into several functions. One of the components appearing in this dissociation is the direction function.

J. Gonzalo considered this syndrome to be the result of a reduction of the system caused by a deficit of cerebral excitability, in turn due to the loss of central mass in the cortex as a consequence of the lesion, and proposed an allometric variation of the sensory functions.

In this article, we have shown that an allometric law of the the type $y = ax^n$ where x, y are sensory functions, holds for optic acuity, direction and visual field functions. The relations obtained are in agreement with the order in which the sensory functions are lost in the cases studied. The most complex functions, which demand more cerebral excitability, become reduced to a greater degree than simpler functions with a lower excitability demand. The dissociation of the system increases with the deficit of cerebral excitability.

An essential continuity, based on the quantity of excitability, can be established between the most simple functions and the most complex ones such as "scheme" or gnosia. The quantity of active cerebral central mass, this mass being rather unspecific and having a certain capacity of adaptation and learning, determines the excitability level and consequently the sensorial level.

The allometric alteration of sensory functions enables us to assert an allometric alteration of their corresponding sensory function *densities*.

The model explains the great variety of observed phenomena, relating them under a functional dynamic unity of the cerebral cortex.

Acknowledgements. The author expresses her gratitude to Miguel A. Porras and José Mira for helpful discussions and advice.

References

1. Gonzalo J.: *Investigaciones sobre la nueva dinámica cerebral. La actividad cerebral en función de las condiciones dinámicas de la excitabilidad nerviosa*, Publicaciones del Consejo Superior de Investigaciones Científicas, Instituto S. Ramón y Cajal. Vol. **I**: Optic functions, 342 pp., 81 Figs. Madrid 1945. Vol. **II**: Tactile functions, 435 pp., 79 Figs. Madrid 1950. (Copies can be requested from "Biblioteca del Instituto Cajal", C.S.I.C., Doctor Arce 37, 28002-Madrid, Spain. Phone: 34 1 5854747, Fax: 34 1 5854754).
2. Gonzalo J.: "Las funciones cerebrales humanas según nuevos datos y bases fisiológicas. Una introducción a los estudios de Dinámica Cerebral", Trabajos del Instituto Cajal de Investigaciones Biológicas, Consejo Superior de Investigaciones Científicas, Madrid. Vol. **XLIV**, 95-157 (1952).
3. Gonzalo I. and Gonzalo A.: "Functional gradients in cerebral dynamics: The J. Gonzalo theories of the sensorial cortex" in *Brain Processes, theories, and models. An international conference in honor of W.S. McCulloch 25 years after his death*, 78-87, Moreno-Díaz R. and Mira-Mira J. (Eds.), The MIT Press, Cambridge, Massachusetts 1996.
4. Gonzalo I. and Gonzalo A.: "Functional gradients, similitude and allometry in cerebral dynamics: Justo Gonzalo's findings and his theory on the sensorial cortex" in *Neurosciences and Computation in Brain Theory. In the Memory of W.S. McCulloch*, Moreno-Díaz R. and Mira-Mira J. (Eds.), The MIT Press (in press).
5. Gelb A. and Goldstein K.: *Psychologische Analysen hirnpathologischer Fälle*. Barth, Leipzig 1920. "Psychologische Analysen hirnpathologischer Fälle auf Grund Untersuchungen Hirnverletzter: VII Ueber Gesichtsfeldbefunde bei abnormer

Ermüdbarkeit des Auges (sog. Ringskotome)", Albrecht v. Graefes Arch. Ophthal., **109**, 387-403 (1922).

6. Köhler W.: *Gestalt Psychology*, Liveright, New York 1929. *Dynamics in Psychology*, Liveright, New York 1940.

7. Lashley K.S.: *Brain mechanisms and intelligence*, Univ. of Chicago Press, Chicago 1929. "Integrative functions of the cerebral cortex", Psychol. Rev., **13**, 1-42 (1933). "Studies of cerebral function in learning", Comp. Psychol. Monogr., **11** (2), 5-40 (1935). "Functional determinants of cerebral localization", Arch. Neurol. Psychiat. (Chicago), **30**, 371-387 (1937). "The problem of cerebral organization in vision", Biol. Symp., **7**, 301-322 (1942).

8. Piéron H.: *La connaissance sensorielle et les problèmes de la vision*, Hermann, Paris 1936. " Physiologie de la vision" in *Traité d'ophtalmologie*, Masson, Paris 1939.

9. Ramón y Cajal S.: *Histologia del sistema nervioso del hombre y de los vertebrados*, Vol. **II**, Madrid 1899.

10. Bender M.B. and Teuber H.L.:"Neuro-ophthalmology" in: Progress in Neurology and Psychiatry, Ed. Spiegel E.A., **III** , Chap. 8, 163-182 (1948)

11. Critchley Mc.D.: *The Parietal lobes*, Arnold, London 1953.

12. Delgado García A.G.: *Modelos Neurocibernéticos de Dinámica Cerebral*, Ph.D. Thesis. E.T.S. de Ingenieros de Telecomunicación. Univ. Politécnica de Madrid. Madrid 1978.

13. de Ajuriaguerra J. et Hécaen H.: *Le Cortex Cerebral. Etude Neuro-psychopathologique*, Masson, Paris 1949.

14. Luria A.R.: *Restoration of Function after Brain Injury*, Pergamon Press, Oxford 1963. *Traumatic Aphasia*, Mouton, Paris 1970.

15. Mira J., Delgado A.E. and Moreno-Diaz R.: "The fuzzy paradigm for knowledge representation in cerebral dynamics", Fuzzy Sets and Systems, **23**, 315-330 (1987).

16. Engel A.K. et al.: "Temporal coding in the visual cortex: new vistas on integration in the nervous system", TINS, **15**, 6, 218-226 (1992).

17. Llinás R.R.: "The intrinsic electrophysiological properties of mammalian neurons: Insights into central nervous system function", Science, **242**, 1654-1664 (1988).

18. Mira J., Delgado A.E., Manjarrés A., Ros S. and Alvarez J.R.: "Cooperative processes at the symbolic level in cerebral dynamics: reliability and fault tolerance" in *Brain Processes, theories, and models. An international conference in honor of W.S. McCulloch 25 years after his death*, 244-255, Moreno-Díaz R. and Mira-Mira J. (Eds.), The MIT Press, Cambridge, Massachusetts 1996.

19. Rakic D. and Singer W., (Editors): *Neurobiology of Neocortex*, J. Wiley and Sons, 1988.

20. Zeky S. and Shipp S.: "The functional logic of cortical connections", Nature, **335**, 311-317 (1988).

21. Gonzalo J.: Unpublished notes of doctoral courses on "Fisiopatología cerebral", Madrid 1956-1966.

22. Huxley J.S.: *Problems of relative growth*, The Johns Hopkins University Press, Baltimore, Md. 1993.

23. Teissier G.: "Les lois quantitatives de la croissance", Actualités scientifiques and industrielles, **455**, Hermann, Paris 1937.

24. Thompson D'A. W.: *Growth and form*, Vol. **I**, Cambridge University Press, 1952.

25. Perkkiö J. and Keskinen R.: "The relationship between growth and allometry", J. theor. Biol., **113**, 81-87 (1985).

Systems Models of Retinal Cells: A Classical Example

Roberto Moreno-Díaz

Centro Internacional de Investigación en Ciencias de la Computación.
Universidad de Las Palmas de Gran Canaria, Canary Islands, Spain, E-35017.

Motivation. Since the times of McCulloch, it became clear that computational models of neural networks are limited by the nature of the formal tools used to describe the correlations among experimental data. The nature of the conclusions from a model are implied by the nature of the operators in its formulation, operators that can be analytic, logical or inferential. For retinal cells, analytical non-linear operators are appropriate.

This was clear in the 60's. However in the great revival of neural computation, around 1985, these essential concept was forgotten, so that old models of the past were being copied but assuming that neurons were point neurons with irrelevant anatomy.

Fortunately, the horizon opens again when recognizing the nervous system complexity and the usefulness of analytics when certain methodological conditions are met. First, identification of the anatomical structure under consideration (input and output lines, processing layers, and other). Second, listing of physiological properties, separately from the anatomical details. Third, propose a set of hypothesis to correlate structure and function, which constitute the computational basis of the model and provides for the selection criteria of the proper mathematical operators. And fourth, simulation of the model and evaluation of the results, so that prediction of results and proposal of new experiments is possible. In this way, systems science, and physiology and anatomy interact fruitfully.

To illustrate this proposal you will find in this paper a reproduction, with minor changes, of a report by the author to the Massachusetts Institute of Technology, Instrumentation Laboratory, prepared in 1965, which is no longer available. It deals with the systems modeling of group 2 ganglion cell in the frog's retina and still nowadays it provides for a good illustration of how to proceed clearly from neurophysiological facts to hypothesis and to systems models which are non-trivial. The model does not rely for its verification on computer power, and therefore the basic approach, the formal developments and the performance of the model remain invariant. The model is very transparent to an engineer mind as well as to neurophysiologists, and it provides for a fresh illustration to contrast with some of the more opaque, simplistic and less effective neural models of today.

1. Introduction

Considerable efforts has been given to the simulation of the frog's retina in order to construct systems with similar properties. In the common leopard frog (Rana Pipiens), four main groups of ganglion cells have been distinguished by their anatomical and physiological characteristics (Lettvin et al., 1959; Maturana et al., 1960). These groups report to the tectum. They have been designated:

Group 1 - Edge Detectors.
Group 2 - Bug Detectors.
Group 3 - Event Detectors.
Group 4 - Dimming Detectors.

Work in the MIT Instrumentation Laboratory has indicated that one of the most intricate retinal structures to stimulate is the Group 2 (bug detector) ganglion cell. Simulation of this cell was, in part, achieved here, (Sutro et al., 1964) using logical methods. This paper describes an analytical approach to the problem based on the internal reports of the MIT Instrumentation Laboratory were the model is known as "Model 4" to distinguish it from three previous models.

2. Neurophysiological Basis

The model is based principally on the experimental results of Lettvin et al. and Maturana et al., referred above.

Figure 1 portrays vertical sections of two bipolar cells and one multi-level E-shaped ganglion cell, as drawn by Ramón y Cajal (1984) for the frog. Note that there are three ganglion cell dendritic levels.

Lettvin et al identified the multi-level E ganglion cell as a bug detector (Lettvin et al., 1961). Such a description could correspond to the ganglion cell "c" drawn in Fig. 1. Note that arborization exists between the outer dendritic levels.

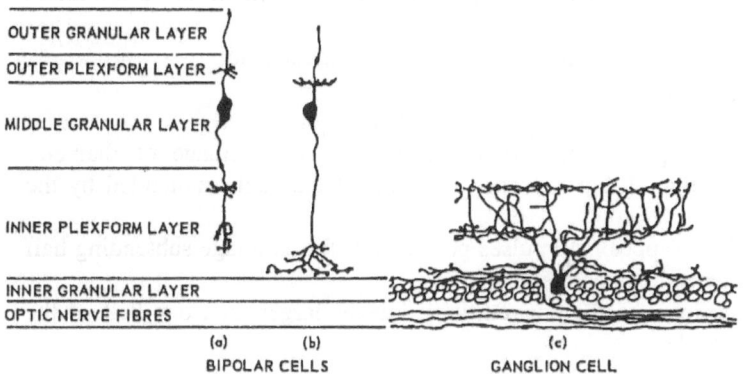

Fig. 1. Retinal sections of the frog (after Ramón-y-Cajal).

Two types of bipolar cells are represented in the model. Their designation "on" and "off", is that of Schypperhein (1965). The distinction between "on" and "off" cells was previously applied by Hartline to ganglion cells (Hartline, 1938).

A list of physiological properties of Group 2 ganglion cells has been abstracted (Sutro et al, 1964) and is summarized in Table 1in the next page.

3. General Structure of the Model

The main hypotheses used to describe the different cell types of the model are presented in Table 2. Most of the hypotheses correspond to characteristics of living nerve cells. Some are simplifications of neurophysiological processes. Information storage for biological temporal integration, for example, is given a rectangular time profile, being perfect for a constant time interval before instantaneous destruction.

Such simplifications may be biologically invalid, but the consequent discrepancies should not affect the model response appreciably.

Table-1:	Group 2 Ganglion Cell Properties	Refs..
1. Group 2 ganglion cells respond to an object:		
	a. that moves centripetally	1,2,4
	b. that is small (3° to 5°)	1,2,4
	c. that is darker than the background	1,2,4
	d. that has a sharp edge	1,2,4
2. There is no response to:		
	a. any object totally outside the responsive retinal field of diameter approximately 4°	2,3
	b. a straight edge of a dark band greater than 2° wide unless it is surrounded by a shield approx. 4° in diameter centered on the RRF	2,3
3. Response is independent of:		
	a. level of illumination	1,2,4
	b. rate of change as long as this is distinguishable	1,4
	c. speed of an edge if it travels between a maximum and a minimum speed.	1,2
	d. amount of contrast as long as this is distinguishable	1,2,4
4. Discharge rate is:		
	a. null for dark objects of width less than about 3'	1,2,4
	b. inversely proportional to the convexity of an image of diameter greater than 3' but less than one half of the angle subtended by the RRF	1
	c. maximal (approx. 40 pulses per second) for an image subtending half the angle subtended by the RRF	1,2,5
	d. proportional to the convexity on images larger than one half of the RRF angle	1
	e. feeble if the object is larger than RRF	1,2,4
	f. greater to movement broken into several steps than to continuous movement	1,5
	g. feeble in response to light spots	1,2,4
5. The response is maintained for several seconds if the object stops in the RRF		1,2,4

References: 1. Lettvin et al., 1959; 2. Maturana et al., 1960; 3. Gaze and Jacobson, 1963; 4. Lettvin et al, 1961; 5. Grüser et al., 1964

The structure of the model is diagrammed in Fig. 2. Each photoreceptor is assumed to have two outputs. That labeled 1 as function of the present illumination; that labeled 2 is a function of illumination at a previous instant. All photoreceptors are assumed to have the same properties.

Bipolar cells are assumed to be of two types. These are termed "on"-bipolar cells and "off"-bipolar cells and respond respectively to increases and decreases of the illumination. In Fig. 2 an on-bipolar cell is shown touching the left side of each photoreceptor, an off-bipolar cell the right.

Table-2: Fundamental Hypotheses Underlying the Model Design	Refs.
1. Photoreceptors	
a. are identical	
b. give two outputs which are respectively logarithmic functions of present and past illumination	
2. Bipolar Cells	1,2,4
a. are of two types ("on"- bipolar cells and "off"- bipolar cells)	
b. receive signals only from their corresponding photoreceptors	
c. receive both photoreceptor outputs	
d. show post-synaptic inhibition (subtractive inhibition)	
e. have thresholds	
f. have pulse outputs with a refractory periods inversely proportional to their total activities	
g. are randomly distributed	
3. Group 2 Ganglion Cells	
a. are identical	
b. have their dendrites distributed in three levels	1,2,6
c. have two kinds of presynaptic inhibition: lineal, divisional and nonlinear	
d. perform spatial and temporal synaptic integration	1,2
e. have threshold in their synaptic connections	
f. adapt	
g. multiply activities in sub-level 2 (b) (Fig. 6)	
h. exhibit post-synaptic inhibition	
i. have pulse outputs with refractory periods inversely proportional to their total activities	
j. respond maximally (40 pulses per second) to a dark disc when it subtends half the angle of RRF (Responsive Retinal Field)	1,2,4,6
k. respond maximally when such a disc has traversed a quarter of the RRF	
l. do not respond to a dark disc subtending the angle of less than 3'	1,2,6
m. do not respond to an image large than the RRF, a circular area of diameters 4°	1,2,6

References: 1. Lettvin et al., 1959; 2. Maturana et al., 1960; 3. Gaze and Jacobson, 1963; 4. Lettvin et al., 1961; 5. Grüser et al., 1964; 6. Ramón y Cajal, 1884

Each photoreceptor feeds only its associated on-bipolar cells and its associated off-bipolar cell. Therefore, each bipolar receptive field is restricted to a single photoreceptor. This is done for simplicity. Wider bipolar receptive fields could be supposed, but they would have to overlap in order to maintain the resolution.

The three dendritic levels are numbered as shown. In level 3, the dendrites spread over a circular area of radius R_1. All on-bipolar cells within this area are in synaptic contact with these dendrites. All off-bipolar cells of the anatomical type "b" in Fig. 1 would contact the ganglion cell "c" in level 3.

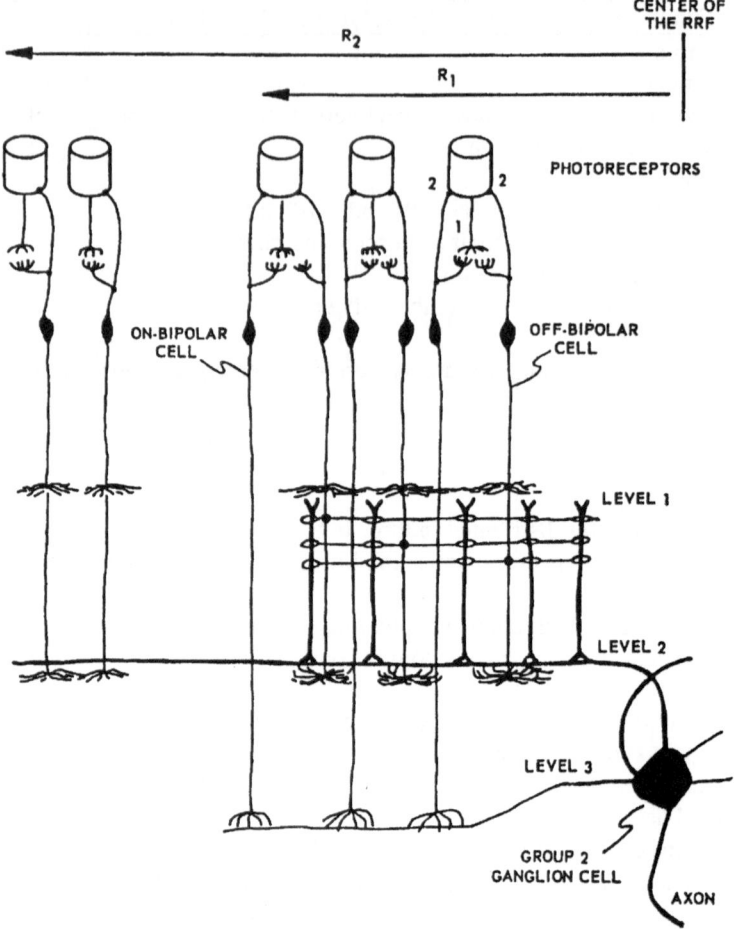

CENTER OF
THE RRF

R_2

R_1

PHOTORECEPTORS

ON-BIPOLAR
CELL

OFF-BIPOLAR
CELL

LEVEL 1

LEVEL 2

LEVEL 3

GROUP 2
GANGLION CELL

AXON

Fig.2 Section of the model

In level 2, the dendrites of the ganglion cells spread horizontally over the large area of radius R_2 and rise vertically in arborizations from level 2 to level 1 in the smaller area of radius R_1. All off-bipolar cells within the area of radius R_2 contact horizontal dendrites. All off-bipolar cells within area of radius R_1 have both excitatory (represented by bushy spreads) and inhibitory (represented by loops) synapses with the arborizations. Off-bipolar cells may have an anatomical configuration similar to "a" in Fig. 1. The cells illustrated in Fig. 2 are shown separately in Fig. 3.

The operation of the model may be explained geometrically. Figure 2 depicts the image of an object in three positions as it moves across the responsive retinal field of a Group 2 ganglion cell. The cell body lies at the center of the two areas of synaptic contact, the smaller the radius R_1 and the larger of radius R_2.

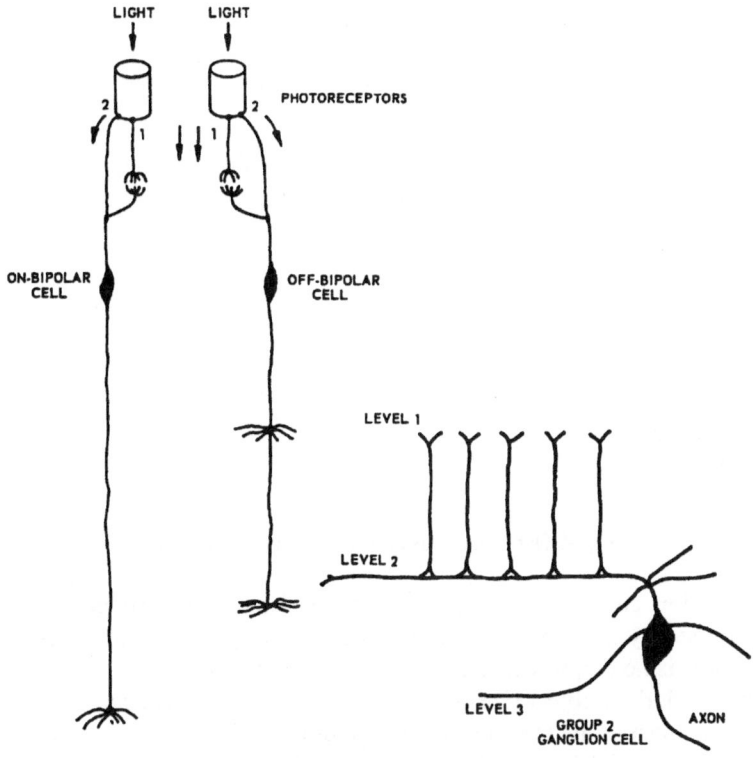

Fig.3 Details of Fig. 2

Consider an object that is darker than its background and is traveling in the direction marked with arrows. The moving image dims the retina along the heavily drawn portion of the object profile. d_2 is that part of the boundary length causing dimming within the circle of radius R_2. Similarly, d1 is a measure of length of dimming within the circle of radius R_1. As the dimming boundary proceeds, an area D_1 is mapped out within this smaller circle [Fig 4(b)]. The ratio $D_1/(d_1+\delta_1)$ where $\delta_1 \ll R_1$ is therefore a measure of the penetration of the object into the inner circle - a measure that can be computed by the model.

An "activity" A_{net} is defined as a function of d_2. the relationship is non-linear. A_{net} is never negative and is maximized for a particular value d_2. A^*_{net} is a similar function of the penetration $D_1/(d_1+\delta_1)$.

Proportional to the products of A_{net} and A^*_{net} is a quantity F_D. Thus F_D is zero if no part of the object lies within the circle of radius R_1. Therefore, this circle defines the responsive retinal filed (RRF) of the ganglion cell. The maximum F_D for a transit will depend on the image profile and trajectory.

Finally, image brightening, the complement of image dimming, must be considered. Brightening occurs along the lightly drawn edge of the object profile. The area generated within the responsive retinal field by this edge is denoted by B_1 [Fig 4(c)]. A function F_B is defined as proportional to B_1.

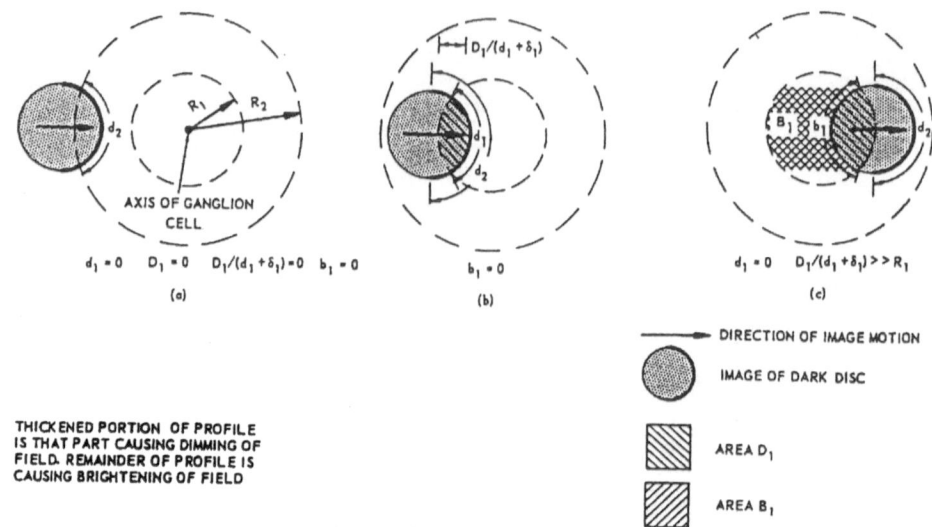

Fig.4 Dark disc crossing responsive retinal field

The ganglion cell output is assumed to consist of identical pulses. The pulse repetition frequency is proportional to the difference F_D-F_B, provided that this is not negative. There is no output when the difference is negative.

Consider a dark of object size comparable to RRF. B_1 is then considerably smaller than D_1 during most of the transit of the image of the objects across the RRF and F_B is correspondingly small. The model constants are such that F_B is much less than F_D. The cell output can be assumed to be governed by F_D alone.

However, for images brighter than their background B_1 is greater than D_1. In this case F_B inhibits F_D and the cell response is reduced.

The performance of the model may be summarized as follows. When a "bug" crosses the field of view, a ganglion cell yields no output until the object enters its RRF. The output pulse repetition frequency of the cell will then increase, reach a maximum when the "bug" is still moving centripetally and finally disappear. Objects darker than their background give much greater responses than bright objects. Indeed, the model has all the properties of living Group 2 ganglion cells that are listed in Table 1, with the exception of item 5. The latter may be explained by feedback from the tectum and may be simulated in a more complex model.

4. Mathematical Description

4.1 Photoreceptor Cell Layer

Lettvin has argued the probability of two degrees of freedom in the signals from the photoreceptors to bipolar cells (Lettvin, 1965). Correspondingly, each photoreceptor model has two outputs: P(t), which is a function of the illumination at time t, and H(t), which is a function of the previous illumination at time t-T (Hypotheses 1, 1 Table 2).

Thus:

$$P(t) = Kln\frac{E(t)}{E_0} \qquad (1)$$

$$H(t) = Kln\frac{E(t-T)}{E_0} \qquad (2)$$

where

$E(t)$ is the illumination at the time t.

$E(t-T)$ is the illumination at the time (t-T) and is therefore one measure of illumination history.

K, E_0 are constants defining the photoreceptor sensitivity.

Maximum and minimum speeds (Property num. 3c, Table 1) are determined by the parameters of the photoreceptors and bipolar cells. A large value of T permits detection of low speed objects.

4.2 Bipolar Cell Layer

4.2.1 General

Figure 5 is a functional diagram of the photoreceptor and bipolar cell systems. Each bipolar cell is fed by the outputs P(t) and H(t) from a single photoreceptor (Hypotheses 2b, c). Bipolar cells are assumed to act "all or none" in response to these signals (Hypothesis 2f). A bipolar cell that is producing an output is said to be "firing".

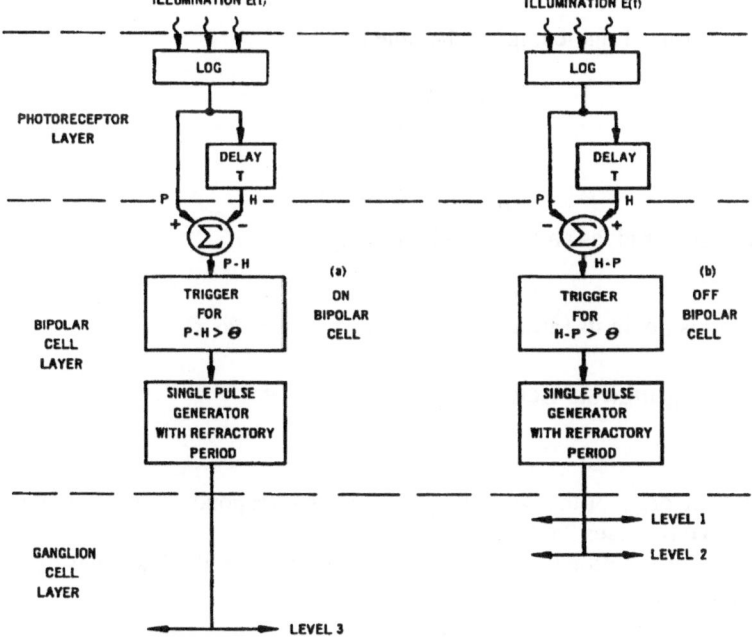

Fig.5 Block diagrams of photoreceptors and bipolar cells

4.2.2 Off-Bipolar Cells

Within each off-bipolar cell the signal H(t) is excitatory, but is post-synaptically inhibited by the signal P(t) (Hypothesis 2d). From the definition in the Appendix of subtractive (post-synaptic) inhibition.

$$A_{off}(t) = H(t) - P(t) = K ln \frac{E(t-T)}{E(t)} \tag{3}$$

We can define "dimming" at time t as

$$Dim(t) = ln \frac{E(t-T)}{E(t)} \tag{4}$$

For constant illumination, Dim(t) = 0 (zero). Hence,

$$A_{off}(t) = K[Dim(t)] \tag{5}$$

Thus, the activity of an off-bipolar is proportional to the dimming at the time t.

The off-bipolar cell yields an output in the form of a narrow pulse, of width δt and constant amplitude r, if it is not in a refractory period, and if

$$A_{off}(t) \geq \Theta \tag{6}$$

where Θ is a fixed threshold and $\Theta \geq 0$ (Hypotheses 2e, f).

4.2.3 On-Bipolar Cells

Conversely, the on-bipolar cell is excited by P(t) and subtractively inhibited by H(t) (Hypothesis 2d). Thus,

$$A_{on}(t) = P(t) - H(t) = K \ln \frac{E(t)}{E(t-T)} \tag{7}$$

We can define "brightening" at the time t as

$$Bri(t) = \ln \frac{E(t)}{E(t-T)} \tag{8}$$

Hence,

$$A_{on}(t) = K[Bri(t)] \tag{9}$$

The on-bipolar gives an output pulse of amplitude r if

$$A_{on}(t) \geq \Theta, \quad \Theta \geq 0 \tag{10}$$

and if the cell is not in its refractory period (Hypotheses 2e, f).

4.3 Group 2 ganglion Cells

4.3.1 Level 1

Figure 6 is a diagram of the dendritic processes that respond to the on and off-bipolar cell outputs (Hypotheses 3a, b).

Level 1 receives directly the off-bipolar cell axons in the RRF and has two Sub-levels 1(a) and 1(b).

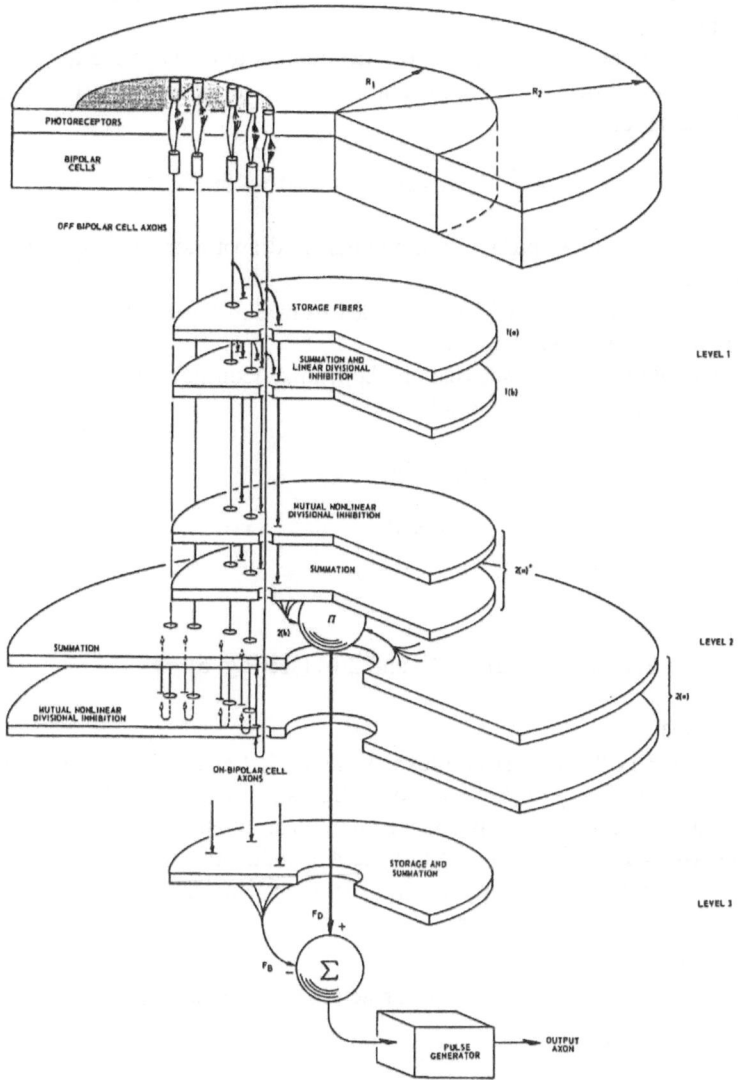

Fig.6 Diagram of models of photoreceptors, bipolar cells and a Group 2 ganglion cell

- *Sub-level* 1(a):

Each input to this level is stored in a fiber that is shown parallel to the off-bipolar cell in Fig. 6 and reached via the wide curved arrow shown there. Here each input is stored. Neural storage usually has the exponential decay characteristic displayed in Fig 7(a). However, for simplicity this storage is assumed to have a constant amplitude profile as shown in Fig. 7(b). After a constant time interval Δt, any stored signal is destroyed. Thus, whenever the jth bipolar cell fires, Sub-level 1(a) yields an output of constant amplitude a and duration Δt. Δt corresponds to the longest transit time for objects crossing the RRF that must be detected.

Let nd1 be the number of off-bipolar cells that are firing at the time t within the circle of radius R1.

Let ND1 be the number of Sub-level 1(a) outputs that are active at time t. Then ND1 is the number of off-bipolar cells that have fired in the interval (t - Δt, t) within the circle of radius R1.

$$N_{D1} = \frac{1}{\delta t} \int_{t-\Delta t}^{t} n_{d1} dt' \qquad \text{for } \Delta t >> \delta t \tag{11}$$

The lines carrying these ND1 stored outputs are drawn more heavily in Fig. 6.

- *Sub-level* 1(b):

At this level each of the ND1 active lines from Sub-level 1(a) suffers divisional inhibition by the nd1 active bipolar cells.

Divisional inhibition is discussed in Appendix. Consequently, the output from Sub-level 1(b) of each active line will be

$$\left. k \middle/ n_{d1} \right., \quad n_{d1} >> 1 \tag{12}$$

$$a, \quad n_{d1} = 0$$

where k and a are constants.

4.3.2 Level 2

Level 2 has three Sub-levels: 2(a), 2(a)*, and 2(b) (see Fig. 6).

- *Sub-level* 2(a):

This sub-level is the lowest sub-level in level 2. Let n_{d2} be the number of off-bipolar cells that are firing in the circle of radius R_2, at time t. In Sub-level 2(a), the outputs from these off-bipolar cells undergo nonlinear divisional inhibition of the kind described in Appendix. Equation (A.8) of the Appendix states that as a result of non-linear divisional inhibition, the output activity of the jth line is

$$A_j = c_2 r e^{-k_2 n_{d2}^r}$$

$$\text{for an active/inactive input} \tag{13}$$

$$A_j = 0$$

where c_2 and k_2 are constants.

The total activity, A_c, is the sum of the activities of all the fibbers that lie within a lateral distance R_2 of the ganglion cell.

$$A_c = \Sigma A_j = c_2 n_{d2} r e^{-k_2 n_{d2}^r} \tag{14}$$

If the threshold of total activity is θ (Hypothesis 3e), the net activity at Sub-level 2(a) is

$$A_{net} = A_c - \theta \qquad A_c \geq \theta \tag{15}$$

$$A_{net} = 0 \qquad A_c \leq \theta$$

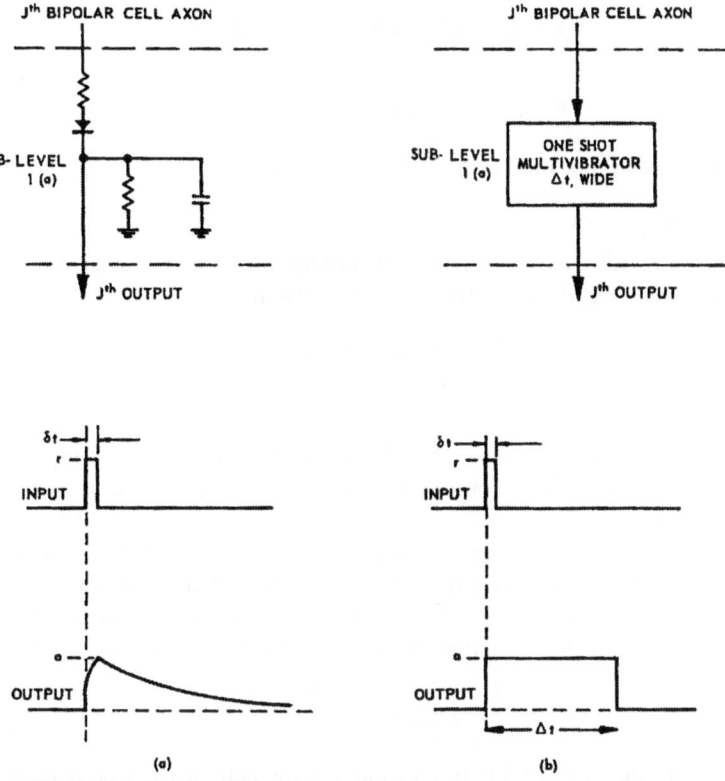

Fig.7 Electrical analogs of two forms of storage at level 1(a)

• *Sub-level 2(a)*:*

Similarly, the outputs of Sub-level 1(b) undergo nonlinear divisional inhibition in Sub-level 2(a)*. For $n_{d1} \gg 1$ there are N_{D1} active lines, each with a signal of magnitude k/n_{d1}. Thus the output activity in the jth line is

$$A_j^* = c_1^* (k/n_{d1}) e^{(-k_1^* N_{D1} \frac{k}{n_{d1}})} \qquad \text{for an active/inactive input} \qquad (16)$$

$$A_j^* = 0$$

The total activity is

$$A_c^* = c_1^* \left(kN_{D1}/n_{d1} \right) e^{\left[-k_1^* \left(kN_{D1}/n_{d1} \right) \right]} \qquad (17)$$

For small n_{d1}, $A_c^* \approx 0$, so that Eq. (17) is true for all n_{d1}.

Finally, the net activity at sub-level 2(a)* is

$$A^*_{net} = A^*_c - \theta^* \qquad A_c \geq \theta * \tag{18}$$

$$A^*_{net} = 0 \qquad A_c \leq \theta^*$$

θ^* being the threshold of total activity A_c^* (Hypothesis 3e).

- *Sub-level* 2(b):

The net activities A_{net} and A_{net}^* are multiplied to give the Level 2 output, F_D (Hypothesis 3g). If K_D is the proportionality constant,

$$F_D = K_D A_{net} A_{net}^* \tag{19}$$

4.3.3 Level 3

Let n_{b1} be the number of on-bipolar cells within the RRF that are firing at time t. The firing at time t. The firing pulses are of constant amplitude r and duration δt (Hypothesis 2f).

In level 3, these n_{b1} pulses are maintained for a time interval Δt, in the same manner that the off-bipolar cell outputs are stored in Sub-level 1(a). Thus N_{B1}, the number of active lines at time t after storage is equal to the number of on-bipolar cells that have fired in the circle of radius R_1 in the previous time interval (t - Δt, t)

$$N_{B1} = \frac{1}{\delta t} \int_{t-\Delta t}^{t} n_{b1} dt' \qquad \text{as } \Delta t >> \delta t \tag{20}$$

The output, F_B, of level 3 is proportional to the sum of these signals after storage. If K_B is the proportionality constant

$$F_B = K_N N_{B1} r \tag{21}$$

4.3.4 Ganglion Cell Output

The level 2 output is inhibited (post-synaptic or subtractive inhibition) by the Level 3 output (Hypothesis 3h).

Thus

$$\begin{array}{ll} F = F_D - F_B & \text{for } F_D \geq F_B \\ F = 0 & \text{for } F_D \leq F_B \end{array} \tag{22}$$

The pulse repetition frequency of the output along the ganglion cell axon is proportional to F (Hypothesis 3i).

4.4 Evaluation of the Constants and Thresholds

The documented results of physiological experiments on the frog retina are used here to determine the values of the various constants and thresholds of the model. First are described the spatio temporal properties of an object and its motion that corresponds to the numbers of active fibbers. Then successively it can be shown that k_2 depends on the size of the dark disc that maximizes the response; how an adaptation is necessary for the threshold of A_{net}; how R_2 may be evaluated from the width of the widest straight band that can excite the cell; how k_1^* is restricted by the result that the cell responds only to centripetally moving images and how the relative influence of Level 3 predicts the preferential response to dark rather than bright images.

We now consider the geometric significance of n_{d1}, N_{D1} and N_{B1}. The bipolar cells are randomly distributed (Hypothesis 2g) in such a manner that, given a line of constant length in the retina, the number of cells intersected is independent of the direction and position of the line. Consider the sharp edge of an object in contrast with its background, that is moving through the retinal field. The number n_{d1} of off-bipolar cell activated at any instant is proportional to the length of image dimming within the circle of radius R_1. And n_{d1} is also a function of image contrast and velocity, but is independent of them within limits determined by photoreceptors and bipolar cell parameters. Image contrast and velocity are assumed constant in this report.

$$n_{d1} \propto d_1 \tag{23}$$

where d1 is the length of the edge causing dimming - or "length of dimming" - within the RRF.

Similarly, n_{d2} is proportional to the "length of dimming" within the circle of radius R_2. Thus, we write

$$k_2 n_{d2} r = K_2 d_2 \tag{24}$$

where d_2 is the "length of dimming" within a circular area of radius R_2. In subsections 4.3 variables are the number of bipolar cells that fire under different conditions; lower case letters express the constants: c1*, k1*, c2 and k2. In subsection 4.4 variables are the length of the edge and the area scanned; upper case letters express the constants: C1*, K1*, C2, and K2.

N_{D1} is proportional to the dimmed area of the RRF.

$$N_{D1} \propto D_1 \tag{25}$$

Similarly, N_{B1} is proportional to the brightened area of the RRF. We put

$$N_{B1} r = B_1 \tag{26}$$

Variables d_1, d_2, b_1, D_1 and B_1 are assumed to be continuous. Although in fact, they are discontinuous because they are directly related to the number of firing bipolar cells. Since images with only the optimum velocity are considered here, these variables may be assumed dimensionless. The same comment is valid for R_1 and R_2, these being the numbers of bipolar cells along corresponding lengths.

By using the geometric equivalents above it is finally obtained that equation 22 becomes:

$$F = K_F \left[d_2 \left(e^{-2d_2/\pi R_1} \right) - 0.135 \right] \left(D_1/d_1 \right) e^{-D_1/0.26 R_1 d_1} \frac{K_B}{K_F} \tag{27}$$

for $F_D \geq F_B$, and also $F = 0$ for $F_D \leq F_B$.

The dimensionless constant (K_B/K_F) determines the difference between the responses to geometrically similar dark and light objects. The frequency output given by Eq. (27) is plotted for various circular dark objects in Fig. 8 (left), using $(K_B/K_F = 0.03)$ and $(K_F = 1.1 \cdot 10^3 R_1^{-2} \, \text{sec}^{-1})$ [i.e., the maximum frequency output is 40 pulses per second]. Figure 8 (right) shows the output frequency for light objects using the same values of K_F and (K_B/K_F).

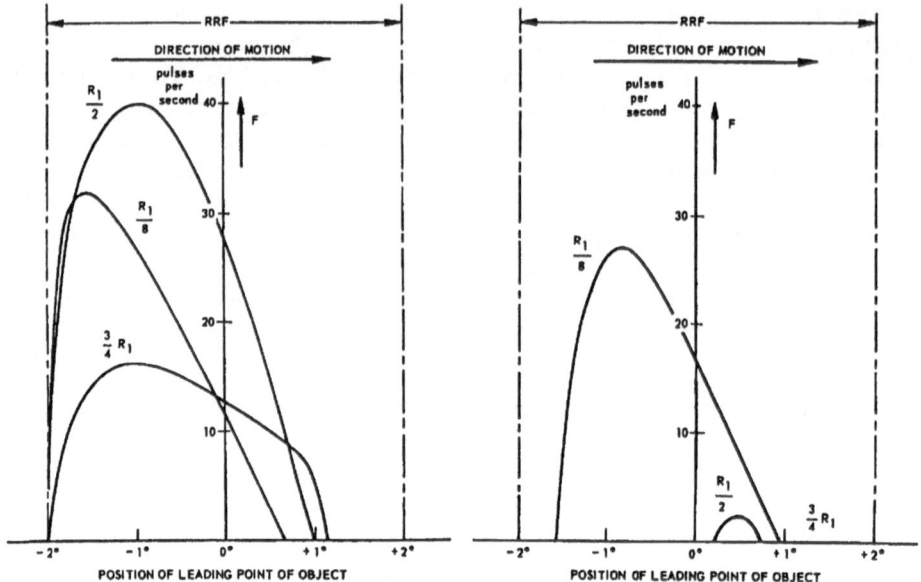

Fig.8 (left) Ganglion cell output for moving dark discs (various radii).
(right) Ganglion cell output for moving bright discs (various radii).

4.5 Mathematical-Equivalent Diagram of the Group 2 Ganglion Cell

Let us consider an array of photoreceptors followed by an array of bipolar cells, as described in subsections 4.1 and 4.2. Further, let us assume that these layers are followed by three new layers that sum the bipolar cell outputs to give n_{d1}, n_{d2}, n_{b1} respectively. In this way the magnitudes d_1, d_2 and b_1 are generated.

For an object moving across the field, D_1 and B_1 can be evaluated by time integration.

$$D_1 = \frac{1}{\tau} \int_{t-\Delta t}^{t} d_1 dt' \tag{28}$$

$$B_1 = \frac{1}{\tau} \int_{t-\Delta t}^{t} b_1 dt' \tag{29}$$

where τ is a constant.

Figure 9 shows that the behavior of the model may be derived in this way from the variables d_1, d_2 and b_1. This simpler system has all the properties and constants of the model, but does not correspond in the detailed operation of its components.

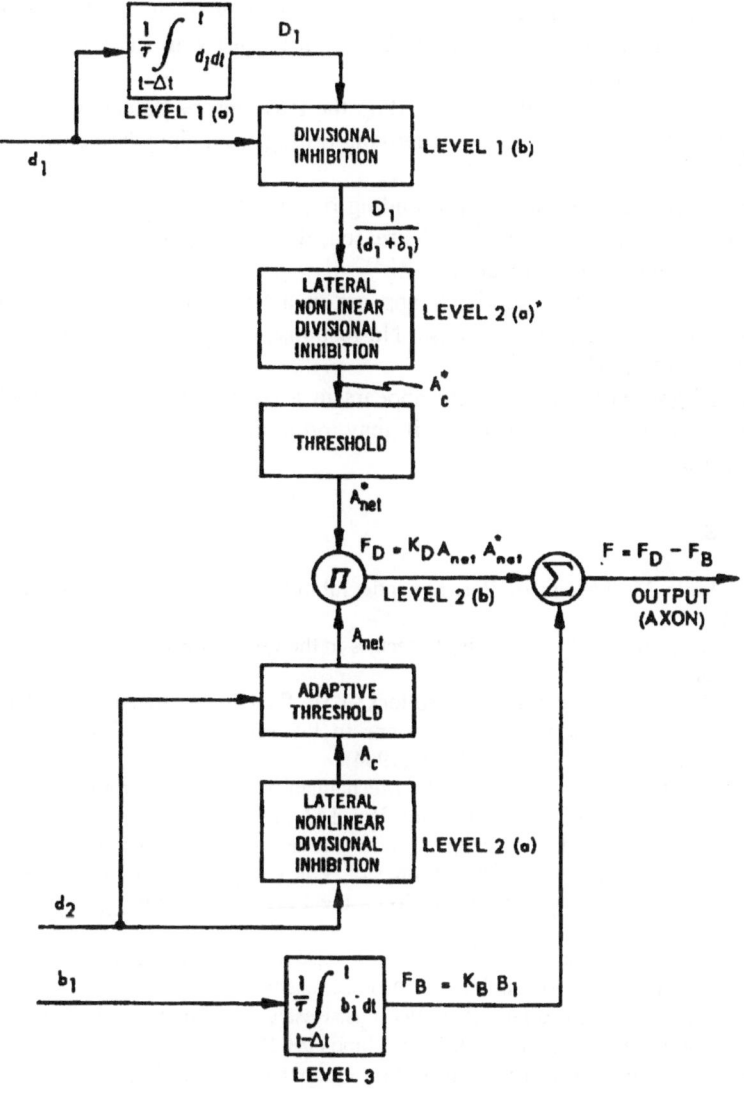

Fig.9 Equivalent block diagram of Group 2 ganglion cell

5. Conclusions

Responses of the model to transits across the RRF of discs of various radii are plotted in Fig. 8. The following are some of the consequences that can be derived from the model's characteristics.

    a. No response occurs to a general change in illumination. (In agreement with Refs. 1, 2, and of Table 1)

    b. A corner may produce a response. (In agreement with Ref. 2.)

c. Several small object images moving simultaneously in the RRF may produce either very small or null responses. This is in accord with Lettvin's observation quoted in Refs. 1 and 2.

d. No evidence of the annulus surrounding the RRF can be detected by fixed dark or light spots with a simultaneous moving testing spot. This is in agreement with Ref. 5

e. However, if the spots in the surrounding ring move, the response to the testing spot may be either increased or decreased, depending on the size of the spots. This is in agreement with Ref. 5

f. A small dark spot that suddenly appears in a fixed position in the RRF may produce an instantaneous response. No response occurs when it disappears. (As `remarked in Ref. 4)

g. A small light spot that suddenly appears in a fixed position in the RRF never produces a response. However, it may produce response when it suddenly disappears (Ref. 4)

References

Eccles, J.C., Ionic Mechanism of Post-Synaptic Inhibition, Science, Vol. 145, September, (1964).

Gaze, R.M. and Jacobson, M., Convexity Detectors in the Frog´s Visual System, Proc. Physiol. Soc., July, (1963).

Grusser, O.J., Grusser-Cornehls, U., and Bullock, T.H., Functional Organization of Receptive Fields of Movement Detecting Neurons in the Frog´s Retina, Pflugers Arch. Ges Physiol., 279 Band, 1 Helt, S. 88-93, (1964).

Hagiwara S., and Tasaky, I., A Study of the Mechanism of Impulse Transmission Across the Giant Synapse of the Squid, J. Physiol. Vol. 143, (1958).

Hartline, H.K., The Response of Single Optic Nerve Fibbers of the Vertebrate Eye to Illumination of the Retina, Amer. J. Physiol. Vol. 121, pp. 400-415, February, (1938).

Lettvin, J. Y. Et al, What the Frog's Eye Tells the Frog's Brain, Proc. I.R.E., November, (1959).

Lettvin, J.Y., et al., Two Remarks on the Visual Systems of the Frog, Sensory Communications, W. Rosenblith, de., Massachusetts Institute of Technology, (1961).

Lettvin, J.Y., Form-Function Relations in Neurons, Research Laboratory of Electronics, MIT Quarterly Progress Report, pp. 333-335, June (1962).

Lettvin, J.Y., Research Laboratory of Electronics Quarterly Progress Report No. 73, pp. 199-208, Massachusetts Institute of Tecnology, Cambridge, Massachusetts, 15 April, (1965).

Maturana, H.R., et al, Anatomy and Physiology of Vision in the Frog (Rana Pipiens), Journal of General Physiology, July , (1960).

Ramón y Cajal, S., Die Retina der Wirbelthiere, Tafel II, Fig. 6 Wiesbaden, Verlag von J.F. Bergmann , (1894).

Schypperheyn, J.J., Contrast Detection in Frog´s Retina, Acta Physiol. Pharmacol. Nerlandica 13, pp 231-277, (1965).

Sutro, L.L. et al., 1964 to September 1965 Advanced Sensor and Control Systems Studies, R-519, Instrumentation Laboratory, Massachusetts Institute of Technology, Cambridge, Massachusetts, January 1966.

A Generic Formulation of Neural Nets as a Model of Parallel and Self-Programming Computation

J. Mira, J.C. Herrero and A.E. Delgado

Dpto. de Inteligencia Artificial
Facultad de Ciencias. UNED
C/ Senda del Rey s/n. 28040 Madrid. SPAIN

Summary

In the same way the more conventional fields of computer science need some theory, including the mathematical foundations of the calculus and the establishment of formal models, neural computation also needs its own. The basic requirements of this model are modularity, "small grain", high connectivity, parametric local computation and some capacity of self-programming by means of the adjustment of these parameters.

We present here a proposal in this line that allows the integration in a single frame of all current models (analogic, logic and inferential) and makes clear the natural way to bridge the symbolic and connectionistic perspectives of AI extending the model of local computation to hierarchic graphs, building networks by joining graphs and studying the set of operators we need for modifying local computation parameters values.

1. The usual perspective in artificial neural nets

By neural computation is usually meant the calculation carried out by modular architectures organized in multi-layer nets with a great number of elementary processors with a high degree of interconnectedness and which locally carry out an analog, non-linear function (generally, a weighted sum followed by a threshold or sigmoid decision function). The distinguishing characteristic of this model of distributed computation are the fine grain parallelism and the partial elimination of the need for programming substituted by learning algorithms which modify the value of the parameters which define the computation by supervised or non-supervised procedures.

As a computation model, neural nets seek to complement the symbolic perspective of Artificial Intelligence (AI) for those problems having to do with changing and partially known environments, well-suited to parallel solutions in real time and which need to absorb the variability of this environment. These problem solving tasks include pattern recognition, perception, associative memory, non-linear functions identification, optimization, adaptive control, motion planning in robotics, and learning.

Fault tolerance and parameters adjustment are two basic advantages of neural nets. There are also disadvantages to neural computation as an alternative to symbolic AI and other algorithmic and von Neumann processes available to resolve these problems. The main one is the absence of a theory of computation and of a robust methodology with direct and inverse effective procedures which will allow us to:

1. *Given the description of a problem at the knowledge level, obtain the neural net which will resolve it.*

2. *Given a concrete neural net, find a description at the knowledge level of the computation it is carrying out.*

This inconvenient, which existed already in symbolic AI, becomes even more serious in neural nets which lack the intermediate symbolic level (formal languages and compilers) to facilitate the passage from the models of the task at the knowledge level to the primitives of the implementation level.

In this paper we shall see how the computational capacity of the neural nets can be extended without losing the essential aspects of the model of distributed and self-programmable computation they represent. Thus, we could speak not only of analog, logical or inferential neural nets but of local computation as hierarchic graphs, networks as parallel programs and learning as operators that modify the values of a set of parameters of the local programs. This generalization can be considered as an step forward in the stablisment of the mathematical foundations of neural computation, facilitating at the same time the connection with symbolic AI and opening, finally, the door to really hybrid systems [Mira & Delgado, 1995].

2. Formalization of Neural Computation in One Layer

In neural computation we begin by examining the global architecture of the net organized into functional groups of neurons of a same type (layers). Next we go down one layer to describe its frame of computation in terms of its input and output spaces as well as the local function.

Every neuron in a layer carries out a local computation of *parametric* nature, sampling data from one part of the input space (with a FIFO memory organization) called the receptive field (V_j), carrying out a local operation on these data and loading the result in its output space (also a FIFO memory). We thus say that this layer does not possess feedback, but feedforward. If we allow the field of data to also include the sampling of a region of the output space (V_j^*), we have a layer of *recurrent* neurons, with feedback. The shape and overlap of the receptive fields in the output space defines local connectivity between neighbouring neurons within a layer. The shape and overlap between receptive fields in the input space partially determine the global function of the net (figure 1).

The input and output spaces have a FIFO memory structure ("first in, fist out") with several time intervals delays. With this framework, the static and dynamic models are formally equivalent by simply moving from differential equations to finite difference formulations.

It is important to note here that these input and output spaces are representational spaces in which each input variable x_i is associated to its semantic T_i, both specifics of each application. The same thing is true with the outputs y_j and their meanings T_j. Hence, the neuron layer produces a set of outputs

$$y_j = f_j [x_i , y_j ; p_n(t)],$$

for data x_i belonging to its receptive field (V_j) and also for the feedback data y_j, coming from the sampled volume in its output space (V_j*). The meanings, T_i and T_j, always remain at the knowledge level (in the Newell sense), in the domain of the external observer (the net designer) where the layer's computation is interpreted.

The type of neuron and neural net (analog, logical or inferential) is related to the nature of the parametric operator f_j. If the operator is analog, we shall say that we have a layer of *"analog neurons"*. Non linear expansions (polinomic, logarithmic or application dependent) of the input and output spaces produce the so called "higher order" neurons. The most general case corresponds to spatio-temporal and recursive filters with knowledge based preprocessing (data analysis).

To go from analog to *"logical neural nets"* we simply have to modify the type of input variables and the set of operators used to fill the same skeletal model. Now the inputs, x_i, are logical variables which only can take on the values 0 or 1, the sums are replaced with logical sums and the products by logical products and the parameters are now also of boolean nature.

Finally, we can extend the computational power of neural nets with the same skeletal model again, substituting the logic operator f_j by a conditional multiple:

If <condition A₁> then <assignment B₁>

If <condition A₂> then <assignment B₂>

......

If <condition Aₙ> then <assignment Bₙ>

Where each branching condition, A_i, can be any arbitrary combination of logic and relational operators. The outputs, B_i, are labeled lines and assignment (A_i, B_i) is made by means of a look up table (LUT), for example. This allows us to include fuzzy formulations and probabilistic models as particular cases. Inferential neural nets (nets of conditionals) are then elementary parallel programs with learning capacity. The parameters that can be modified should be now in the branching condition and/or in the non-linear decision LUT.

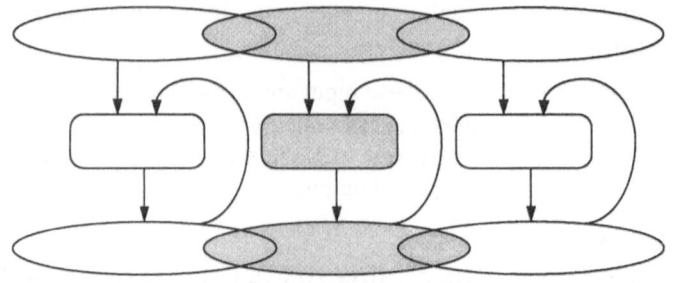

a)

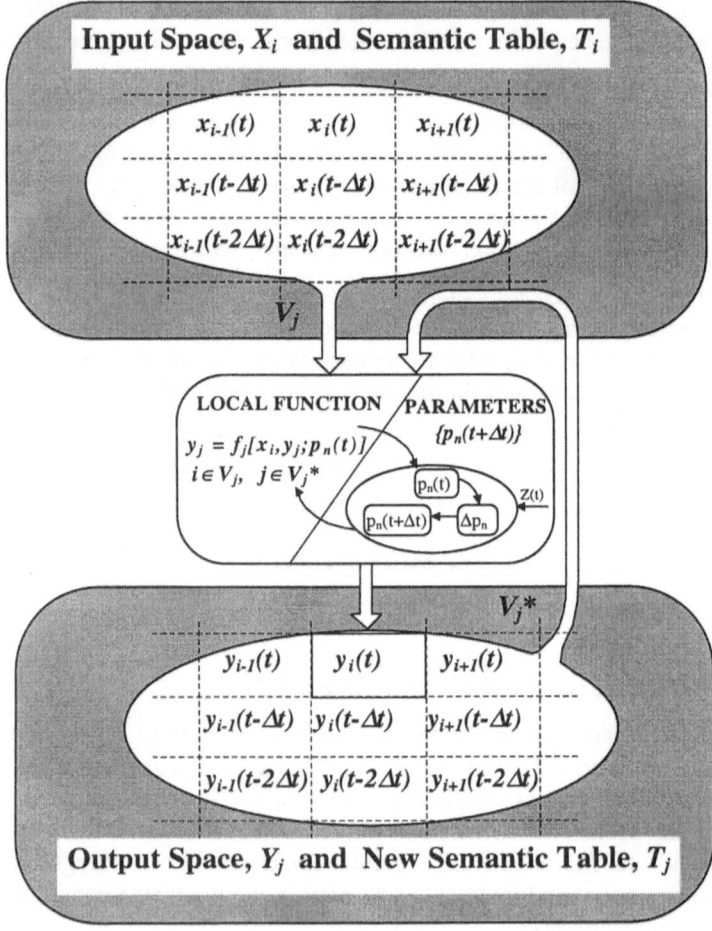

b)

Figure 1. a) Diagram of neural computation in one layer. b) Each neuron samples data in its receptive field, with a high degree of overlapping (connectivity), and calculating its response according to a function $\mathbf{f}$ (analog, logic or inferential) which is local, of parametric nature and adjusted by learning.

3.   A Parametric Description for a Local Computation Program

If we have a control structure like *"if ... then ... "*, *"if ... go to ... "*, etc., we can have any program. So the next step in the formalization of neural nets is to find a parametric description of generic programs that could be used as local computation functions, f_j, with capacity of learning.

Figure 2 is an instantiation of a *connected, directed, finite graph* [Nilsson, 1980][Winston, 1992][Mira, 1995a], one where *there is a node* (the starting node) *whose descent is the rest of the nodes* (and maybe itself); because of this node, there is a hierarchy between the nodes. Cycles are permitted. We label the graph itself by P_k, the nodes by S_i, the arcs by A_i and something associated to every node by M_i.

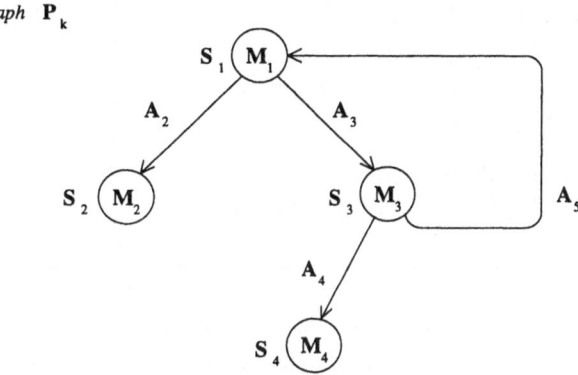

Figure 2. Connected, directed, finite, hierarchic graph, where cycles are permitted.

These symbols are intended to describe the graph, and we consider them not as mere labels, but as sets whose elements hold descriptions about the graph. So we can say that there is a set D of descriptions, such that we regard sets like $D_k \in \wp(D)$ and their elements $d_{ki} \in D_k$ are propositions that actually describe the graph. But, in order to achieve a complete description, we relate a node and its descent by means of a mapping like Φ, such that

$$\Phi(\, S_i \cup P_k\,, n\,) = S_s \cup P_k$$

that applies recursively. Analogously by means of mappings, we also relate a node and an arc that start at the node, and these to the successor of the node; we can also define the relationships "on the left" and "on the right". Therefore, we can describe an algorithm to go through the graph in a depth-first search way, starting from S_1 (the starting node), following first Φ, and then the relationship from the left to the right (that we impose). Due to there could be cycles, we must take precautions, in order to find out the descent of a node only once.

There is a relationship between this kind of graph and a program. If we interpret S_i as "step of a program"; M_i as "statements we have to execute on step S_i, except any IF"; and A_i as "expression we have to evaluate, belonging to an IF statement", we can interpret the graph in figure 2 as a program and describe it (as follows, on the left) and then we can write it (as follows, on the right):

Program "P_k"	$S_1:$ M_1
Starting step is "S_1"	IF (A_2) GOTO S_2
On step "S_1" applies "M_1"	IF (A_3) GOTO S_3
From step "S_1" we may go to "S_2", "S_3"	STOP
We go to step "S_2" if "A_2"	$S_2:$ M_2
On step "S_2" applies "M_2"	STOP
From step "S_2" we may go to "no more steps"	$S_3:$ M_3
We go to step "S_3" if "A_3"	IF (A_4) GOTO S_4
On step "S_3" applies "M_3"	IF (A_5) GOTO S_1
From step "S_3" we may go to "S_4", "S_1"	STOP
We go to step "S_4" if "A_4"	$S_4:$ M_4
On step "S_4" applies "M_4"	STOP
From step "S_4" we may go to "no more steps"	
We go to step "S_1" if "A_5"	

If we *a)* write $P_\varnothing$ instead of the expression "no more steps", *b)* take away all words, and *c)* write the result on a single line (as far as possible), we obtain a form we call $\otimes T_0 (\Pi^*_k)$ and so we have got to an unmistakably ordered parametric description for the program, in terms of the sets, as follows:

$$\otimes T_0 (\Pi^*_k) = (P_k , S_1 , S_1 , M_1 , S_1 , S_2 , S_3 , S_2 , A_2 , S_2 , M_2 , S_2 , P_\varnothing ,$$
$$S_3 , A_3 , S_3 , M_3 , S_3 , S_4 , S_1 , S_4 , A_4 , S_4 , M_4 , S_4 , P_\varnothing , S_1 , A_5)$$

where $\Pi^*_k = \{ \xi^*_{k1} , \dots , \xi^*_{kn} \}$ is an abstract ordered parametric description. We can also obtain a similar description in terms of more detailed program elements.

4. A Graph for a Network: Parallel Programming with Capability for Self-Programming

If the starting node in figure 2 were the only node in the graph, M_1 would describe all the statements of a program without any condition to evaluate. The same could be said of M_2 if we consider the second node alone. On the one hand, each node can be seen as a program without conditions to evaluate, that could be executed independently: we execute the program in the first node and then the program in the second node only if the evaluation of the condition is true. But, on the other hand, there could be a statement in each M_i that consist of a call to a subroutine or subprogram, i.e., another program including any number of conditions to evaluate. That means each node could be seen at different levels of detail: suppose we make zoom on a node and suddenly we can see a whole new graph inside.

But, what happens if we think of the nodes as programs that execute simultaneously instead of one after the other, i.e., if we think of parallel programs? In order to this parallel interpretation, we must not consider the arcs as conditions any more, but simply as permitted ways of communication between the programs of the nodes. Suppose that a program in a node u and a program in a node v share elements (think of program variables). From the viewpoint of S_v there were some elements M^u_v that changed in time between t and $t + \Delta T$, but from the viewpoint of S_u there were elements M^v_u that changed between $t - \Delta T$ and t. So, if we use quotes to distinguish the symbols in the parallel graph, we may write

$$\Phi\,(\,S''_u\,(t) \cup P''_k\,,\,n\,) =\ S''_v\,(t + \Delta T) \cup P''_k$$

$$M''_u\,(t) \cap M''_v\,(t + \Delta T) = M''^v_u\,(t) = M''^u_v\,(t + \Delta T) \neq \varnothing.$$

$$\otimes T_0\,(\,\Pi''^*_1\,) = (P''_1\,,\,S''_1\,,\,S''_1\,,\,M''_1\,,\,S''_1\,,\dots\,)$$

where the latter is the a parametric description for the parallel team of programs.

It is easy now to consider, according to figure 3, a set of these graphs that share something between any number of nodes, so we have a set

$$\Psi^*_n = \{\otimes_n T_0\,(\,\Pi''^*_i\,)\}_{i\,=\,1,\,m}$$

where every element of Ψ^*_n is a parametric description of a team of parallel programs and n identifies the way we choose to connect their nodes. Figure 3.b. is a graphical representation of this: the joining of the two graphs from figure 3.a.

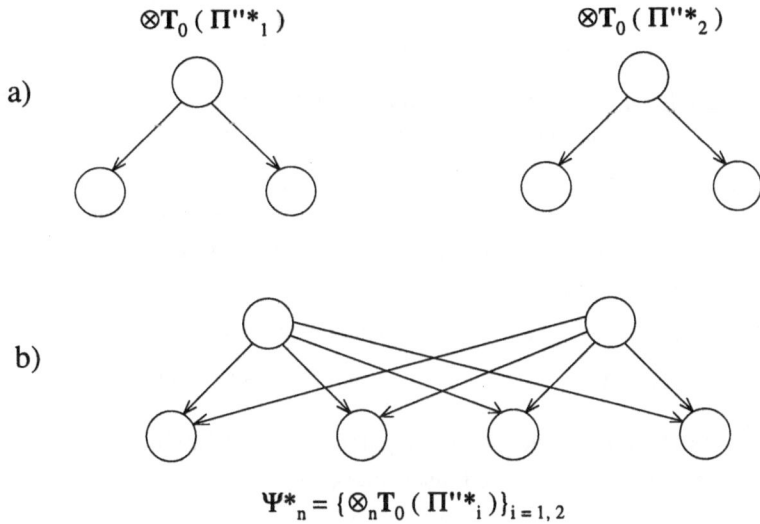

$$\Psi^*_n = \{\otimes_n T_0\,(\,\Pi''^*_i\,)\}_{i\,=\,1,\,2}$$

Figure 3. A network is built by joining graphs.

We can also describe the relationships between the elements of Ψ^*_n for any value of i; for each couple of elements P''_k and P''_q we have

$$\Phi\,(\,S''_u\,(t) \cup P''_k\,,\,n\,) = \;S''_v\,(t+\Delta T) \cup P''_q$$

$$M''_u\,(t) \cap M''_v\,(t+\Delta T) = M''^v_u\,(t) = M''^u_v\,(t+\Delta T) \neq \varnothing.$$

Thus, Ψ^*_n can represent a network that we call distributed computation without learning, where each S''_v represents a unit where the local computation is performed and all these units execute in parallel.

We may add *self-programming* capability if we include learning. Learning flow direction and computation flow direction are opposite, so we have

$$\Phi^\Lambda\,(\,S^\Lambda_v\,(t) \cup P^\Lambda_k\,,\,n\,) = \;S^\Lambda_u\,(t+\Delta T_\Lambda) \cup P^\Lambda_k$$

$$M^\Lambda_v\,(t) \cap M^\Lambda_u\,(t+\Delta T_\Lambda) = M^{\Lambda u}_v\,(t) = M^{\Lambda v}_u\,(t+\Delta T_\Lambda) \neq \varnothing.$$

So we use Φ^Λ to describe a graph like the one Φ describes, but with arcs pointing to the opposite direction with regard to Φ, and then we can build

$$\Psi^*_\Lambda = \{\otimes_\Lambda T_0\,(\,\Pi''^*_j\,)\}_{j\,=\,1,\,m}$$

where every step is a learning module that is executed in parallel.

Finally, we can obtain a set of pairs

$$\Psi^*_{\Lambda n} = \{\,(\,\otimes_\Lambda T_0\,(\,\Pi''^*_j\,)\,,\,\otimes_n T_0\,(\,\Pi''^*_i\,)\,)\,\}_{j,\,i\,=\,1,\,m}$$

to describe a network we call self-programmable distributed computation, whose complex nodes consist of a couple of single nodes, one for learning, one for local computation. The relationship

$$\Lambda\,(\,S^\Lambda_v\,(t)\,) = S''_v\,(t+\Delta t_\Lambda\,)$$

represents how local computation is modified by learning, where Δt_Λ is the time that the local learning program needs to modify the local computation program.

If every unit exchanges M''^u_v every ΔT with another unit and every unit performs a local computation, in parallel with other units, such computation must be finished in ΔT. But, if there is learning, there must be a share in the time between local learning and local computation. So, we have two situations:

$$\Delta T_\Lambda = \Delta t + \Delta t_\Lambda \qquad \text{when there is learning}$$

$$\Delta T = \Delta t \qquad \text{when there is not learning}$$

where Δt is the duration of the local computation and Δt_Λ is the duration of the local learning. If we want $\Delta T_\Lambda \approx \Delta T$, we must achieve $\Delta t_\Lambda \ll \Delta t$.

5. Operators for Modifying the Local Computation

It is possible to build a set of operators to be used by the learning module to introduce the suitable changes in the local computation program simply by handling the parametric description. We are going to show this for the case of XOR neural network. First, we are going to present a very trivial case, very useful to make clear how the operators modify the local computation. Then, we present the solution applied by Rumelhart et al. [Rumelhart, 1986], but achieved by using the concept of receptive field extension [Mira, 1995b]. To avoid confusion, we use ovals for the units in the graph representing the network, and circles to represent the steps in the graph of the program. We do not represent the graphs for learning.

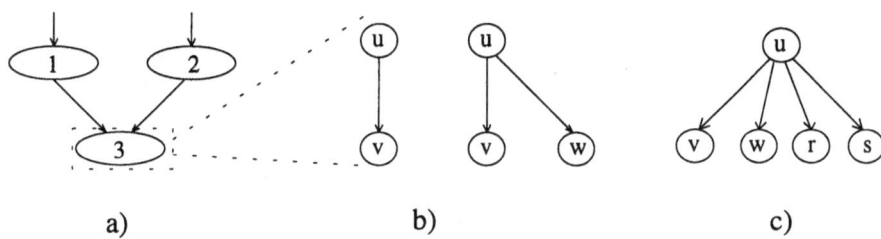

a)   b)   c)

Figure 4. The XOR gate network problems and solutions.

In figure 4.a. we can see two input units and one output unit. Computation for input units is trivial: output is equal to input. Output unit applies a learning algorithm, that considers the output error for every trainig pair; if an error occurs, local computation program must be corrected by adding a branch to the program. This is what we can see in figure 4.b., where at first there were only the nodes u and v, but now a new branch w is added. For units 1 and 2 we have

$$\otimes_\Lambda T_0 (\Pi''^*_i) = (P''_\varnothing)$$
$$\otimes_n T_0 (\Pi''^*_i) = (P''_i , S''_{i1} , S''_{i1} , M''_{i1} , S''_{i1} , P''_\varnothing) \qquad (i = 1, 2).$$

In both cases, M''_{i1} consists of only one element $m''_{i1,1} \in M''_{i1}$ that can be translated into the program code as $y_i = x_i$ $(i = 1, 2)$. For the output unit we have

$$\otimes_n T_0 (\Pi''^*_3) = (P''_3 , S''_{31} , S''_{31} , M''_{31} , S''_{31} , S''_{32}$$
$$, S''_{32} , A''_{32} , S''_{32} , M''_{32} , S''_{32} , P''_\varnothing).$$

In this unit, M''_{31} shares elements with M''_{i1}, i.e., $M''^i_{31} = \{ m''_{i1,1} \}$ $(i = 1,3)$. As for the learning module program, it contains the local computation parametric description as data. The algorithm is basically: if output is not suitable with regard to the input, add a branch to the program in order to correct the error for such input. This is somewhat similar to symbolic learning. Suppose that, in first, the local computation program in unit 3 looked like figure 5.a., where x_{31}, x_{32} are inputs to the unit 3 and y_{31} is the output. It is obvious that this program yields errors for three kinds of training pairs. So, if we present the training case $x_{31} = 0$, $x_{32} = 1$, the learning algorithm will detect the error and will add a branch (to a new node w, after u and on the left from v) by means of the operator

$$O*^S_N (\Pi''^*_3 , \Pi''^*_{3w} , \Pi''^*_{3u} , \varnothing , \Pi''^*_{3v} , 1) = \Pi^*_a .$$

a)	b)

$y_1 = x_1$
 $y_2 = x_2$
 $x_{31} = y_1$
 $x_{32} = y_2$
 if (x_{31} .eq. 0 .and. x_{32} .eq. 0) then
 $\qquad y_{31} = 0$
 end if
	$y_1 = x_1$
	 $y_2 = x_2$
	 $x_{31} = y_1$
	 $x_{32} = y_2$
	 if (x_{31} .eq. 0 .and. x_{32} .eq. 0) then
	 $\qquad y_{31} = 0$
	 else if (x_{31} .eq. 0 .and. x_{32} .eq. 1)
	 $\qquad y_{31} = 1$
	 end if

Figure 5. Local computation in unit 3, before and after learning.

As for the operator, a branch is a sequence like (S''_{3i} , A''_{3i} , S''_{3i} , M''_{3i}), where there is a condition to evaluate (x_{31} .eq. 0 .and. x_{32} .eq. 1) and some code to execute ($y_{31} = 1$) if the condition is true. Thus, the operator changes the local computation program and yields what is shown in figure 4.b. The new parametric description is

$$\otimes_n T_0 (\Pi''^*_a) = (P''_3 , S''_{31} , S''_{31} , M''_{31} , S''_{31} , S''_{32} , S''_{33}$$
$$, S''_{32} , A''_{32} , S''_{32} , M''_{32} , S''_{32} , P''_\varnothing$$
$$, S''_{33} , A''_{33} , S''_{33} , M''_{33} , S''_{33} , P''_\varnothing)$$

where

$$A''_{33} = \{ a''_{33,1} , a''_{33,2} \} \qquad\qquad a''_{33,1} \equiv [\, x_{31} \text{ .eq. } 0 \,]$$
$$M''_{33} = \{ m''_{33,1} \} \qquad\qquad a''_{33,2} \equiv [\, x_{32} \text{ .eq. } 1]$$
$$m''_{33,1} \equiv [\, y_{31} = 1]$$

This procedure poses the XOR gate as a local computation that regards all the possible cases (four cases, one for each possible input), as we see in figure 4.c. This can be applied to all logical gates and, not so trivially, to symbolic learning.

Figure 6 presents the XOR gate according to Rumelhart [Rumelhart, 1986]. Starting from 6.a., our learning algorithm consists of: first, a consensus algorithm [Mira, 1995b], where the hidden layer causes the extension of the receptive field when the output holds an error above ε during the training cycle; second, a backpropagation mechanism to calculate the weights.

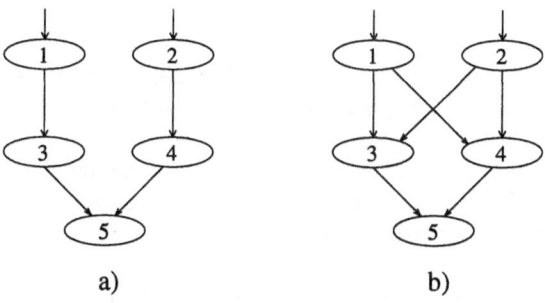

Figure 6. XOR gate according to Rumelhart.

On the other side, the local computation calculates the output by multiplying the input vector by the weight matrix. The parametric description for the local computation in the hidden layer is

$$\otimes_n T_0 \, (\, \Pi''^*_i \,) = (P''_i \, , S''_{i1} \, , S''_{i1} \, , M''_{i1} \, , S''_{i1} \, , P''_{\varnothing} \,) \qquad (i = 3, 4)$$

so, they are programs with only one step, without conditions to evaluate, where M''_{i1} consists of sums and multiplications, and such that

$$M''^j_{i1} = M''^i_{j1} \, , \text{ i.e., } M''_{j1} \, (t) = M''_{i1} \, (t + \Delta T) \qquad (i=3, \; j=1; \; i=4,$$
$$j=2)$$

i.e., some of the local computation results (the outputs) are transferred from a layer to the next (where they are inputs). If the learning algorithm decides to extend the receptive field, this is applied to units 3 and 4, so that the network becomes the one in figure 5.b. But this yields new elements in M''_{31} and M''_{41} (since the weight matrix increase its dimensions), by maybe several applications of operators like

$$O^C_A \, (M''_{j1} \, , m''_{j1,i} \,) = M''_{j1} \cup \{m''_{j1,i} \} \qquad (j=3,4)$$

and now we have that units 1 and 4 share data, as well as units 2 and 3:

$$M''^j_{i1} = M''^i_{j1} \, , \text{ i.e., } M''_{j1} \, (t) = M''_{i1} \, (t + \Delta T) \qquad (i=4, \; j=1; \; i=3,$$
$$j=2)$$

besides what we already had (units 1 and 3 share data, as well as units 2 and 4). We build these new elements (sums and multiplications) by means of operators like

$$O^F_p \, (m''_{j1,i} \,) = \Xi^m_{j1,i} \qquad (j=3,4)$$

and these elements are inserted or exchanged by means of operators like

$$O^{SM}_{IA} \, (\Pi_{31} \, , \xi_{kx} \, , \xi^v_n \,) = \Pi_{ai} \qquad O^{SM}_{IA} \, (\Pi_{41} \, , \xi_{ky} \, , \xi^u_m \,) = \Pi_{bi}$$

$$O^{SM}_{YS} \, (\Pi_3 \, , \Pi_{31} \,) = \Pi_c \qquad O^{SM}_{YS} \, (\Pi_4 \, , \Pi_{41} \,) = \Pi_d$$

$$O^{SM}_{YA} \, (\Pi_c \, , \xi_{kx} \, , \Pi_{ai} \,) = \Pi_a \qquad O^{SM}_{YA} \, (\Pi_d \, , \xi_{ky} \, , \Pi_{bi} \,) = \Pi_b$$

for the set of elements we wanted to add, so that the new operations are performed.

6. Conclusions

We just have seen how the computational capabilities of neural nets are modelled by a distributed and self-programmable computation, i.e., they consist of units where local computation and learning programs are executed in parallel. Due to the net as well as the programs can be represented by hierarchical graphs, as basic elements for building their structure, beside sets to describe their contents, we were able to achieve a parametric representation for them. From this viewpoint, there was possible to define operators used by the learning programs to modify the local computation, by modifying its parametric description. So the neural net concept becomes generalized.

7. Symbols Not Explained in the Text

Π''^*_{kj}	sequence	$\wp(\mathbf{D})$	the set of the parts of $\mathbf{D}$
$\Xi^m_{j1,i}$	program code	$\mathbf{p}_n$	parametric description
t	time coordinate	$\mathbf{P}_\varnothing, \varnothing$	the empty set

8. Acknowledgements

We acknowledge the financial support of the Spanish CICYT, under project TIC 94-95 and of the CAM research department under project I+D 11-94.

9. References

[Mira, 1995a] J. Mira, et al. *Aspectos básicos de la inteligencia artificial.* (Sanz y Torres, 1995).

[Mira, 1995b] J. Mira, et al. *Cooperative processes at the symbolic level in cerebral dynamics: reliability and fault tolerance.* In Brain Processes Theories and Models, R. Moreno-Díaz and J. Mira (Eds.) (The MIT Press. Cambridge, MA, 1995).

[Mira&Delgado, 1995] J. Mira and A.E. Delgado. *Computación neuronal,* in Aspectos básicos de la inteligencia artificial, Cap. 11, 485-575. J. Mira, et al. (Eds.) (Sanz y Torres, 1995).

[Nilsson, 1980] N. J. Nilsson. *Principles of Artificial Intelligence.* (Tioga Publishing, 1980).

[Rumelhart, 1986] D.E. Rumelhart, et al. *Learning internal representantions by error propagation,* in Parallel Distributed Processing: Explorations in the Microestructures of Cognition, Vol. 1, 318-362. D.E Rumelhart and J.L. McClelland (Eds.) (Cambridge, MA. The MIT Press , 1986)

[Winston, 1992] P. H. Winston. *Artificial Intelligence, Third Edition.* (Addison Wesley, 1992).

Using an Artificial Neural Network for Studying the Interneuronal Layer of a Leech Neuronal Circuit

Santos, Juan Miguel * and Szczupak, Lidia #
(*) Dpto. de Computación, F.C.E.y N., U.B.A. Ciudad Universitaria, 1428 Bs.As.
(#) Dpto. de Fisiología, Fac. de Medicina, U.B.A. Paraguay 2155, 1121 Bs.As., Argentina.

Abstract. We have modeled the neuronal circuit that conveys mechanosensory input onto a pair of serotonergic neurons in the nervous system of the leech. The objective of this work is to use an artificial neural networks (ANN) in the investigation of the interneuronal layer of this circuit, which represents an unknown population of cells. The ANN is a three layer perceptron, whose input units represent the mechanosensory neurons, the output unit represents the serotonergic neurons and the hidden units represent the unknown interneuronal layer. The connections between input and hidden units were represented as a saturation function rather than as a linear one. The weights and thresholds were adjusted using the backpropagation algorithm. The ANN parameters were correlated with specific biological parameters enabling to test and classify the resulting configurations according to physiological and experimental considerations. We obtained a finite and restricted number of solutions that can be experimentally tested.

1. INTRODUCTION

Artificial neuronal networks (ANN) that represent biological neuronal networks can be constructed on the basis of a detailed formulation of the properties of the neurons involved or through a simple but careful representation of their input/output functions (Segev, 1992). The first approach requires detailed morphological and biophysical knowledge of the neurons and of their synaptic interactions (which in certain cases is not readily accessible), and this approach involves high computational costs (Smith, 1992). In many cases, these considerations force the choice of simpler models, with algorithms that are simpler both in terms of their formulation and simulation (Cohen et al., 1992; Houk et al., 1993 and Chapeau-Blondeau & Chauvet, 1991). However, this kind of approach does not imply that the resulting models are dissociated from actual physiological considerations. On the contrary, specific correspondence between the parameters of the ANN and specific physiological variables of the biological network can be established.

The aim of this work is to formulate a model of the neuronal network linking sensory neurons with putative neuromodulatory neurons in the nervous system of the leech, to aid in the characterization of the cellular components of the interneuronal layer currently under investigation (Szczupak & Kristan, 1995, Szczupak & Kristan., 1996). The model is based on a multilayer perceptron whose parameters were optimized using the backpropagation algorithm (Rumelhart et al., 1986), an approach that has been used to model the local-bending and the shortening circuits in the leech (Lockery & Sejnowsky, 1993; Wittenberg & Kristan, 1992). Based on the biological interpretation of the variables of the model, we found that the networks obtained from the training sessions, which represent an infinite set of numerical solutions, can be reduced to a finite and restricted number of network configurations, testable through physiological experiments.

2. THE BIOLOGICAL QUESTION UNDER INVESTIGATION

The nervous system of the leech *H. medicinalis* (Fig 1A) is composed of 21 midbody ganglia (M1-M21), and rostral (R1-R4) and caudal (C1-C7) brains. Each ganglion controls the sensory and motor functions of the corresponding segment and is connected to the adjacent ganglia forming a chain that controls the behavior of the animal as a whole. In each midbody ganglion there is a pair of serotonergic neurons, the Retzius (Rz) neurons, which contain more than 60% of the serotonin content of the leech nervous system. Previous studies showed that stimulation of mechanosensory neurons, sensitive to pressure (P neurons) exerted on the skin, produces a strong excitatory signal onto the Rz neurons via a polysynaptic pathway (Wittenberg et al., 1990). This signal is not confined to the segment where the stimulus is applied but it spreads to Rz neurons in other ganglia all along the chain (Szczupak & Kristan, 1995).

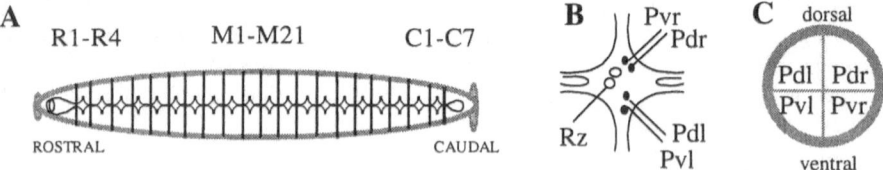

Figure 1. Diagram of the leech nervous system (**A**), localization of P and Rz neurons in the ganglion (**B**), and innervation pattern of the P cells (**C**).

Each ganglion contains one pair of Rz neurons and two pairs of P cells (Fig 1B): left and right dorsal and ventral P cells [P(dl), P(dr), P(vl) and P(vr), respectively]. Each P cell innervates the skin of the corresponding quadrant of the segment (Fig 1C). The electrical coupling between the pair of Rz neurons is high and they respond as a single unit. The P cells are not connected among themselves and each one of them was equipotential in its synaptic effect on the Rz neurons (Szczupak & Kristan, 1995). Stimulation of any P cell with a train of action potentials at 15 Hz produced excitatory postsynaptic potentials (e.p.p.s) that add in time producing a response with increasing amplitude up to a plateau that was usually reached at around the fourth action potential (Iscla and Szczupak, unpublished data). When the stimulation of one P cell, that produced a nearly maximal response in Rz neurons, was simultaneously applied with a subthreshold stimulation of the contralateral homologous P cell, the response virtually doubled. This observation suggests that contralateral homologous neurons share a common interneuron; and that the saturation of the response to the stimulus generated by a P cells is confined to the P-interneuron interaction.

As a first step in the study of the structure of the interneuronal layer we analyzed, at the single ganglion level, whether the four P cells use common or separate pathways in their connection to the Rz neurons. For that purpose, we studied the Rz responses to stimulation of different P cell pairs, following two different stimulation patterns:

Sub-Supra (Fig 2A). One P cell was stimulated eliciting the maximal number of action potentials that did not produce any measurable response in Rz (usually 1 action potential). We refer to this stimulation as "subthreshold". The second P cell was stimulated eliciting a number of action potentials that produced a small response in Rz (usually 3 action potentials). We refer to this stimulation as "suprathreshold". Note that "threshold" refers to Rz responsiveness to synaptic stimulation.

Supra-Supra (Fig 2B). Suprathreshold stimuli (as defined above) were applied to both P neurons.

At first, each P cell was stimulated individually (Fig 2.Aa & Ba) to measure the amplitude of the responses evoked in Rz. Then, both P cells were stimulated simultaneously (Fig 2.Ab & Bb) and the response to the combined action of both P cells was measured. We stimulated all possible pairs of P cells as the Rz neurons were held at a -70 mV throughout all the experiments. The unit value of the responses was computed as the average amplitude of the suprathreshold responses in the Rz neuron as each studied P cell was individually stimulated. The amplitude of the responses to the combined stimulation was expressed as the ratio to this value (Fig 2.Ac & Bc). It is noteworthy that this experimental design was aimed at uncovering the neuronal connectivity and not the dynamic interaction between these neurons. Stimulation patterns were applied at large enough intervals in order to minimize any plastic changes of the responses

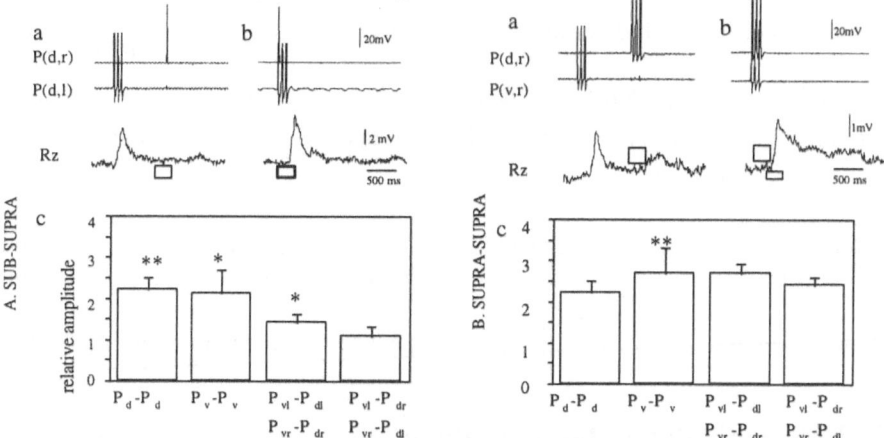

Figure 2. Rz responses to Sub-supra (A) and Supra-supra (B) patterns of P cell stimulation. **a.** response to individual P cell stimulation; **b.** response to combined stimulation; and **c.** average combined over average individual (suprathreshold) responses computed for each pair of P cells (n>4). Statistically different (t-test) than linear sum: * p <0.1 & ** p<0.05.

3. THE MODEL AND ITS BIOLOGICAL INTERPRETATION

3.1. The ANN architecture and definitions

The biological network was modeled on the basis of a multilayer perceptron with one hidden layer (Fig. 3). The input layer comprises 4 different units representing each one of the 4 P cells: $\{p_{dl}, p_{vl}, p_{vr}, p_{dr}\}$, for P(dl), P(vl), P(vr) and P(dr), respectively. It was important to assign an identity to each unit since the experimental data suggest specific interactions between specific P cells and the interneuronal layer. Input units do not interact among themselves as P neurons do not show synaptic interaction.

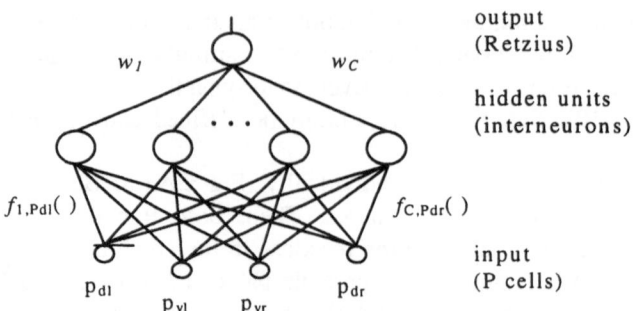

Figure 3. Scheme of the ANN, that models the P-Rz circuit

The synaptic interaction between one input unit i and one hidden unit j was modeled as a transference function, that is $f_{ji}(p_i) = (1 - e^{-\gamma_{ji} p_i}) \alpha_{ji}$; $i \in \{dl, vl, vr, dr\}$ [eq. 1] (see justification below in 3.2) where α_{ji} is the maximum transference of the connection and γ_{ji} is the slope parameter of the transference function $f_{ji}(\ )$ The activation state of the hidden unit j is given by $V_j = g(h_j) = \dfrac{1}{1 + e^{-\beta(h_j - \theta_j)}}$ [eq. 2], where β is the slope parameter of the activation function $g(\ )$, θ_j is the threshold of the hidden unit j, and h_j is the excitation state of the hidden unit j given by

$$h_j = \sum_{i \in \{dl, vl, vr, dr\}} f_{ji}(p_i) \quad \text{[eq. 3]}.$$

The interaction between the hidden unit j and the output unit follows a linear function with slope w_j. The output of the network is the excitation state of the output unit and is given by $s = \sum_{j=1}^{C} w_j * g(h_j)$ [eq. 4]; where C is the number of hidden units.

3.2. Biological interpretation of the model

The interpretation of the different parameters of the ANN in biological terms is, in principle, speculative. It has the value of a working hypothesis that will enable us to evaluate the physiological meaning of the results obtained.

We consider that the action potentials were directly correlated with a synaptic output from the P cells since, following the second action potential, there was a 1:1 relationship between the number of action potentials in P cell and the number of e.p.p.s in Rz cell. Besides, the e.p.p.s are time-locked to the action potentials. Thus, in our model the activation state of these units was determined according to the three levels of stimulation imposed onto the P cells. The connection between the input and the hidden units was modeled as an exponential function (eq. 1) representing the saturation level of the P-interneurons interaction. The sign and magnitude of the α_{ji} in the function $f_{ji}(\ )$, correspond to the polarity and the maximum amplitude of the synaptic responses of the interneurons. The sign depends on the type of ionic conductances associated with the postsynaptic receptors, that determines whether the postsynaptic neuron will be depolarized or hyperpolarized. The magnitude of α_{ji} is correlated with a series of factors that determine the synaptic efficacy, such as the maximal postsynaptic current density and the electronic distance between the synaptic site and the site of origin of regenerative responses (Barst & Egelhaaf, 1994). We assume that the interneurons that mediate the P to Retzius interaction produce regenerative responses on the basis of the regular latency between action potentials in P cell and the e.p.p.s in Rz cells, and on the robustness of the response.

The interneuron behavior was modeled using eq. 2, in agreement with the current conceptual view of the neuron. One unit may represent a population of neurons with the same behavior, or different units could represent separate physiological domains of a single neuron.

The weights of the connections between hidden and output units represents the efficacy of the synaptic interaction between the interneurons and the Rz cells. The sign and magnitude of this weight corresponds to the polarity and intensity of the synaptic connection involved, and it was set as a constant value., like in more traditional ANNs. This is based on the fact that, within the range of our experiments, the connection between the interneurons and the Rz neurons did not show signs of saturation. The pair of Rz neurons was modeled as a single unit (see Section 2).

4. TRAINING AND CLASSIFICATION OF THE NETWORKS

4.1. Training and validation sets

The input/output sets were obtained from the experiments described in Figure 2. The input patterns were determined according to the stimulation patterns applied to the P cells. Each pattern was defined by a vector with four elements, representing the activation state of each input unit. Each element was given a value of 0, 1 or 3, representing no-stimulation, subthreshold stimulation or suprathreshold stimulation respectively, following the definition given under Section 2. The desired output for each input pattern was determined based on the relative amplitude of the responses (Fig 2.Ac & Bc). The training set (Table 1) was obtained based on the sub-supra series of experiments (Fig. 2A).

Pdl	0	0	0	0	0	3	0	1	0	0	0	3	1	1	0	3	0	0	0	3	1
Pvl	0	0	3	1	0	0	0	0	0	3	1	0	0	3	0	1	0	3	1	0	0
Pvr	0	3	0	0	1	0	0	0	0	1	3	0	0	0	1	0	3	0	0	1	3
Pdr	0	0	0	0	0	0	3	0	1	0	0	1	3	0	3	0	1	1	3	0	0
Output	0	1	1	0	0	1	1	0	0	2.2	2.2	2.1	2.1	1.4	1.4	1.4	1.4	1.1	1.1	1.1	1.1

Table 1. Training set

With the aim of obtaining networks that generalize the supra-supra patterns we have used the corresponding data as the validation set rather than as part of the training set. A training set with the complete data could lead to networks with good memorization performance but without proper generalization (Hertz et al., 1990). Table 2 shows the complete set of input/output data used to validate the networks resulting from the training sessions. This set was obtained based on the supra-supra series of experiments (Fig. 2B).

Pdl	0	3	3	0	3	0
Pvl	3	0	0	3	3	0
Pvr	3	0	3	0	0	3
Pdr	0	3	0	3	0	3
Output	2.2	2.7	2.4	2.4	2.7	2.7

Table 2. Validation set

4.2. Number of hidden units and β parameter

Aiming at determining the number of units in the hidden layer necessary to reproduce the input/output pattern we have run two groups of 40,000 training sessions with ANNs containing 2 and 4 units in their hidden layer. Each group was subdivided in four subgroups (10,000 each) that were run using β values of 0.5, 1, 2 and 4. The training conditions were as follows: each training session had a maximum of 10^5 iterations (approximately 5000 epochs), momentum=0.9, learning rate η =0.1 and absolute error bound on each pattern ε =0.1. At the start of each training session the weight values were randomly assigned following a uniform distribution U (-1,1). The value of the γ parameter of all transference functions was constant and equal to 1. The equations used in the update algorithm are shown in the Appendix.

The results from these training sessions (Table 3) show that ANNs with 2 hidden units could not converge to any solution, while networks with 4 hidden units converge with all β values. Networks with 1 and 3 hidden units were considered as special cases in trainings with two and four hidden units (that is, some connection weights between one hidden unit and output unit is close to zero, or all connections between the inputs and some hidden units are close to zero).

Value of parameter β	0.5	1	2	4
# of hidden units				
2	0	0	0	0
4	24	8639	9225	9448

Table 3. Number of numeric solutions varying the number of hidden units and β parameter of its activation function.

To select the β value to be adopted throughout the training sessions we further tested the networks obtained with four hidden units, using the validation set and applying symmetry considerations. The obtained networks were tested with the validation set and those whose absolute error was greater than 0.5 were discarded. The bound of 0.5 was well within the variability range of the physiological experimental data. The symmetry considerations were:

Type A: a hidden unit is symmetric if its connections to the input units bear a symmetric pattern of connections around the rostro-caudal axis (p_{dl} *similar to* p_{dr} and p_{vl} *similar to* p_{vl}).

Type B: two hidden units form a symmetric pair if i) their connections to the output unit have *similar* weights (w_j); and ii) the pattern of connections from the input units of one hidden unit is a mirror image of the other, around the rostro-caudal axis.

The similarity between two scalar values was defined as follows: be a and b scalar values, *a* is *similar* to *b* if one of the following two conditions are met:

i) $|a| < \lambda$ and $|b| < \lambda$

ii) $(|a| > \lambda$ and $|b| > \lambda)$ and $((|a| > |b| \Rightarrow \frac{b}{a} \in [\delta,1])$ or $(|b| > |a| \Rightarrow \frac{a}{b} \in [\delta,1]))$

where λ and δ are constants settled according to a criterion compatible with biological or experimental considerations. We set $\lambda = 0.3$ and $\delta = 0.75$.

We defined a network as *symmetric*, if and only if, all its hidden units bear type A or type B symmetry, or a combination of both. We found that the number of networks that passed the validation and symmetry tests depended heavily on the β value (Table 4). Networks whose hidden units have a β value of 0.5 or 1 could not reproduce the validation input/output set.

Parameter β	Trained Networks	Validated Networks	Validated and symmetric network:
0.5	24	0	0
1	8639	15	0
2	9225	396	22
4	9448	53	6

Table 4. Number of numeric solutions before and after applying the validation and symmetry tests for 4 different β values using 4 units in the hidden layer. The values in the second column are the same as those in the second raw of Table 3.

4.3. Exploring the space of solutions

We trained 300,000 networks with 4 hidden units and $\beta = 2$, using the same training conditions described under Section 4.2.

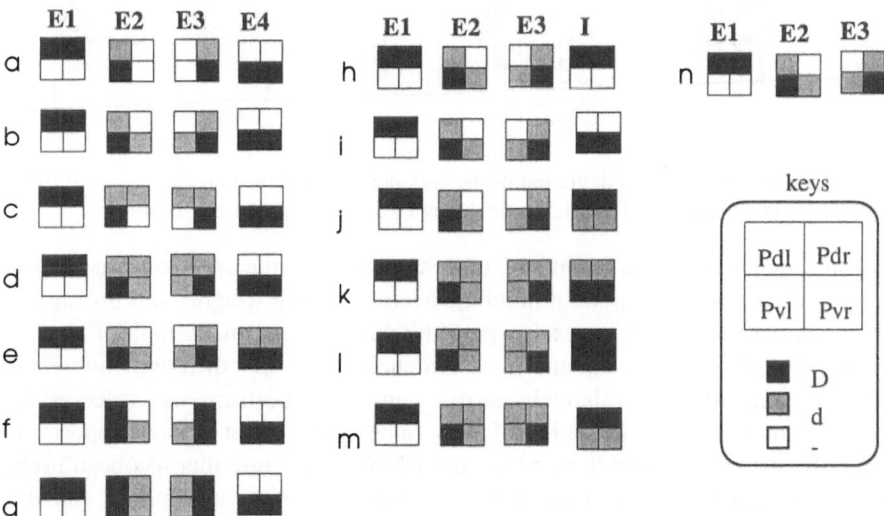

Figure 4. Configurations obtained through training, validation and classification. Each box represents one hidden unit. The sign of w, as excitatory (E) or inhibitory (I) is at the top of each column and the dedication to the input units is shown in the quadrants (see key inset).

The networks obtained were tested using the validation set and symmetry criteria, resulting in 10351 and 531 numeric solutions, respectively. These networks were classified according to the sign of the four weights of the connections between the hidden layer and the output (E for units with positive w_j and I for units with negative w_j) and according to the pattern of the maximum transference (α_{ji} of $f_{ji}(\ )$) of the connection of each input units to the hidden unit. Three degrees of *dedication* were defined, {D, d and -}, corresponding to major, minor and no-dedication of the hidden unit to each input unit; and they were determined as follows:

$$\text{Dedication}(\alpha)=\left\{- \ if \ \left|\alpha/\alpha_{max}\right|\leq 0.3 \ ; \ d \ if \ 0.3<\left|\alpha/\alpha_{max}\right|<0.75 \ ; \ D \ if \ \left|\alpha/\alpha_{max}\right|>0.75\right\}$$ were α_{max} is the maximal absolute value of α_{ji}, $\forall i \in \{dl, vl, vr, dr\}$ for each hidden unit j. Figure 4 shows the 14 different types of qualitative solutions or *configurations* obtained from this classification.

5. DISCUSSION

5.1. General

The set of numerical solutions fitting the mapping described in Table 1 could be infinite. Therefore, we propose to partition it by defining class-representatives, and thus, obtain a finite number of classes.

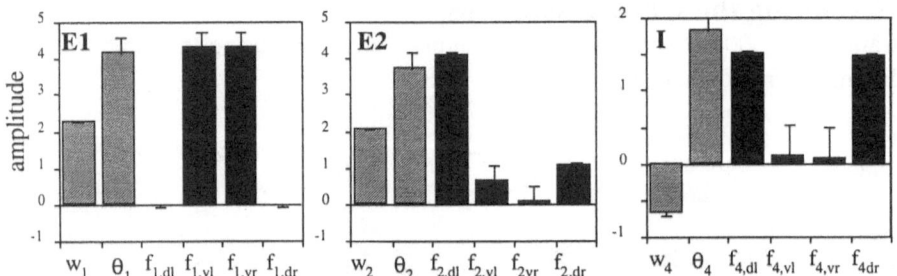

Figure 5. Average and standard deviation of the parameters (displayed on the x axis) of the configuration shown in Fig 4.i, that includes 15 different numerical solutions.

To study the quantitative variability of the parameters of the individual networks belonging to the same subclass we plotted the averages of the weights and thresholds with their standard deviation for each type of hidden unit. As an example, Figure 5 shows an example of such an analysis, evidencing that the different numerical solutions are confined into a relatively narrow numerical spectrum, as evidenced by the relatively small standard deviation of each value. These observations support our classification criteria, showing that, within the ANN architecture that has been used, the dedication definition using local network information was appropriate and the training and validation set can be reproduced by a finite number of configurations.

Using combinatorial analysis, we found 7700 and 780 possible symmetric configurations with four and three hidden units, respectively. However, the classification of the trained and validated networks rendered only 14 configurations.

To analyze whether the configurations found within the 300,000 runs represent the whole population of configurations, we computed the ratio of cumulative number of new configurations over the number of training session as a function of the number of training sessions (Fig. 6), dividing the runs in three groups. In order to avoid any possible skew derived from the choice of a particular sequence in which these groups were considered in the graph, we interchanged this sequence three times. In all cases the plot shows that after 200,000 training sessions no new class

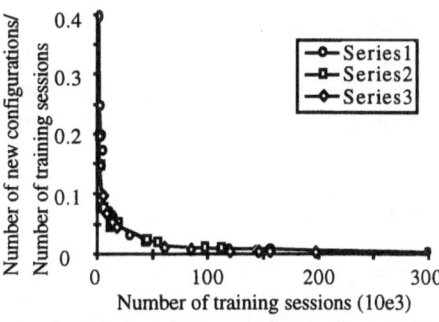

Figure 6. Results of 300,000 training session. Series 1, 2 and 3 result from changing the sequence in which three groups of 100,000 runs were considered.

appeared. Although we cannot claim that no new configuration could appear with a yet larger number of trainings, we do not expect a considerable number of new configurations in view of the maximum-likelihood estimate.

5. 2. Biological interpretation of the results

The results suggest that the interneuronal layer of the P-Rz network in the leech is composed of two types of interneurons: a) neurons with a distributed sensory input (units E2 and E3 in all the configurations, and unit I), and b) neurons dedicated to the homologous contralateral dorsal or ventral P cells (units E1 in all the configurations, and E4 in all the EEEE group, respectively). Lockery and collaborators identified 9 interneurons involved in the leech local bending reflex and studied their sensory input profile (Lockery and Kristan, 1990). In general, the connectivity pattern of the hidden units E2 and E3 (Fig 4) resembles that of the local bending interneurons. Specifically, the hidden units E2 and E3 in configuration f show a strong resemblance to the connectivity pattern of neuron 212, and units E2 and E3 in configuration g resembles neuron 125. No interneurons with exclusive dedication to the contralateral homologues have been described so far. However, it was found that the annulus erector motor neuron receives input from both dorsal P cells (Gu et al., 1991), suggesting that the leech nervous system contains both types of arrangements. The modeling of the P-Rz network shows that its input/output function can be solved by means of a single hidden layer with a minimal number of three units (Fig 4n). In current experiments we are searching for the neurons that constitute the interneuronal layer. The results from the present computational investigation will be used as a guide to design physiological experiments and interpret their results. For example, the involvement of inhibitory interneurons can be tested in a general way by modifying the extracellular concentration of the two ions usually involved in inhibitory responses (namely chloride and potassium), shifting their equilibrium potential to more depolarized values than in control conditions. If this experimental procedure was devoid of any effect on the P-Rz interaction the EEEI configurations could be discarded. As neurons, excited by P cells and generating a response in Rz neurons, will be identified as interneurons of the P-Rz network, their physiological and connectivity properties will be included in the model. This, in turn, will allow to clarify the role of the identifyed neurons, decide whether they are sufficient or else what are the properties of the neurons to be yet identified.

APPENDIX

Be $E = \frac{1}{2}(\varsigma^\mu - s^\mu)^2$ the cost function for any pattern ξ^μ, where ς^μ is the desired output and s^μ is the output of the network. Using eq. 4, the update for the weights w_j is given by $\Delta w_j = -\eta \frac{\partial E}{\partial w_j} = \eta(\varsigma^\mu - s^\mu)V_j$, where V_j is given by eq. 2.

The update for the maximum transference α_{ji} in the connection $f_{ji}(\ )$ is

$$\Delta\alpha_{ji} = -\eta\frac{\partial E}{\partial\alpha_{ji}} = -\eta\frac{\partial E}{\partial V_j}\frac{\partial V_j}{\partial\alpha_{ji}} = \eta(\varsigma^\mu - s^\mu)w_j g'(h_j)f(\xi_i^\mu),$$ where ξ_i^μ is the scalar

value of the ith element in the input pattern ξ^μ with $i \in \{dl, vl, vr, dr\}$.

The update for γ_{ji} in weight connection $f_{ji}(\ )$ is

$$\Delta\gamma_{ji} = -\eta\frac{\partial E}{\partial\gamma_{ji}} = \frac{\partial E}{\partial V_j}\frac{\partial V_j}{\partial f_{ji}}\frac{\partial f_{ji}}{\partial\gamma_{ji}} = \eta(\varsigma^\mu - s^\mu)w_j g'(h_j)\alpha_{ji}f'(\xi_i^\mu)\xi_i^\mu,$$ where

$g'_\beta(x) = \beta g(x)(1 - g(x))$ and $f'(x) = \alpha_{ji}\gamma_{ji}e^{-\gamma_{ji}x}$.

REFERENCES
Borst, A. and Egelhaaf, M. (1994) TINS, **17**:257-263
Chapeau-Blondeau, F. and Chauvet G. (1991) Biol. Cyb., **65**: 267-279
Hertz J. A.; Krogh, A. and Palmer R. G. (1990) **Introduction to the theory of neural computation**, Addison-Wesley Publishing
Houk, J.; Singh, S.; Fisher, C. and Barto, A. G. (1992) In: **Neural Network for control**, Chapter 13, pgs. 301-348.
Gu, X.; Muller, K.J. and Young, S.R. (1991) J. Physiol. **441**: 733-754.
Lockery, S.R. and Sejnowsky T.J. (1993) TINS **16**: 283-290
Lockery, S.R. and Kristan, W.B. Jr. (1990) J. Neurosci. **10**:1816-1829.
Rumelhart D. and McClelland J. (1986) **Parallel Distributed Processing**, V.1, MIT Press.
Segev I. (1992) TINS **15**: 414-421.
Smith R.G., (1993) In: **User's Manual Version 3.4**, , Univ. of PA Medical School, Phila.,
Szczupak, L. and Kristan W.B. Jr. (1995) J. Neurophysiol. **74**:2614-2623.
Szczupak, L. and Kristan, W.B. Jr. (1996) Soc. Neurosci. Abs.811.4.
Wittenberg, G. and Kristan, W.B. Jr. (1992) J. Neurophysiol. **68**:1693-1707.

Capacity and Parasitic Fixed Points Control in a Recursive Neural Network

V. Giménez[1], M.Pérez-Castellanos[2], J. Rios Carrion[3] and F. de Mingo[3]

[1] Departamento de Matemática Aplicada
[2] Departamento de Arquitectura y Tecnología de Sistemas Informáticos
[3] Departamento de Inteligencia Artificial
Facultad de Informática, Universidad Politécnica de Madrid,
Campus de Montegancedo s/n, Boadilla del Monte, 28660, Madrid, SPAIN
Phone: + 34.1.336.74.29, Fax: + 34.1.336.74.12
E-mail:vgimenez@fi.upm.es

Abstract - This paper describes a new method for controlling the capacity and for diminishing the number of parasitic fixed points in a Recursive Neural Network RNN. Based on preliminary researches [1] a Recursive Neural Network may be seen as a graph. The matrix of weights W presents certain properties for which it may be called a *tetrahedral matrix* [2]. The geometrical properties of these kind of matrices may be used for classifying the n-dimensional *state-vector* space in n classes[2]. In the recall stage, a parameter vector σ may be introduced, which is related with the capacity of the network [3]. It may be shown that the bigger is the value of the *i-th* component the vector σ the higher became the capacity of the i class of the *state-vector* space[2]. Once the capacity has been controlled with the parameter σ, we introduce a new parameter that use the *statistical deviation* of the prototypes to compare them with those that appears as fixed points, eliminating in this way a great number of parasitic fixed points.

1. TRAINING

A RNN may be expressed as a *Complete Graph G* [4], with n vertices $\{v_1, ..., v_n\}$, and *one* bi-directional *edge* a_{ij} connecting every possible pair of different vertices. At the *training* stage; initially, the weight matrix W is initialized with null values; afterwards, when a pattern ξ^μ belonging to $\{0,1\}^n$, is presented to the net, this value is modified as follows:

$$
\Delta w_{ij}^\mu = \begin{cases} +1 & \text{if } \xi_i^\mu = \xi_j^\mu = 1,\ i \neq j, \\ -1 & \text{if } \xi_i^\mu = \xi_j^\mu = 0,\ i \neq j, \\ 0 & \text{otherwise.} \end{cases}
\tag{1}
$$

This procedure is also based in Hebb's law but must be interpreted as a mapping or a *coloring* of the edges in G [4]: the *coloring* associated with the pattern ξ^μ is defined in such a way that the edges connecting vertices in correspondence with every component $\xi_i^\mu = 1$ are, for example, *red* colored; reversibly, those edges connecting vertices in correspondence with every component $\xi_i^\mu = 0$, are *blue* colored. Then all the edges in the *red* subgraph are positively reinforced and all the edges in the *blue* subgraph are negatively reinforced; those edges connecting both subgraphs remain unchanged. Once acquired the pattern ξ^μ the colors are erased and we repeat the same color assignation with the next pattern to be acquired by the net; when every vector in the *training set* has been integrated in the net, the training stage is finished [1], the *Resulting Graph G* has become edge-valued and its adjacency matrix W is the *weight matrix* of the net. We have used this method taking the 37 eight-bit patterns corresponding with a certain codification of the most relevant phonetic sounds for the *Spanish* language given in [5], as the training pattern set. After applying the *training algorithm* [1] we obtained as adjacency matrix W of the *Resulting Graph G* the one in figure 1. As it may be observed by our example, the matrix W of weights has a very interesting property: if the four corners of any one of the possible rectangles in this matrix (except those with a corner in the diagonal), are chosen,

$$
W = \begin{pmatrix}
0 & [21] & 7 & 16 & [12] & 16 & 4 & 1 \\
21 & 0 & -1 & 8 & 4 & 8 & -4 & -7 \\
7 & [-1] & 0 & -6 & [-10] & -6 & -18 & -21 \\
16 & 8 & -6 & 0 & -1 & 3 & -9 & -12 \\
12 & 4 & -10 & -1 & 0 & -1 & -13 & -16 \\
16 & 8 & -6 & 3 & -1 & 0 & -9 & -12 \\
4 & -4 & -18 & -9 & -13 & -9 & 0 & -24 \\
1 & -7 & -21 & -12 & -16 & -24 & -24 & 0
\end{pmatrix}
$$

Figure 1. *Tetrahedral property (matrix).*

then the sum of the two values on any one of the two pairs of opposite corners are equivalent, for example:

$$w_{12} + w_{35} = w_{15} + w_{25} = 11. \tag{2}$$

The reader may observe that this is true for every possible rectangle which we would chose. The formal proof of this *Tetrahedral property* in the matrices resulting from the *graph based training* explained before, may be seen in [2], the matrices with that property will be named *Tetrahedral Matrices*. The *Tetrahedral property* is used for obtaining a new set with only n parameters, these new parameters will be the ones that will control the dynamic of the system. These parameter are obtained defining the next set of basic tetrahedral matrices

$$U^k = (u_{ij}^k) = \begin{cases} 1 & \text{if } (i = k \text{ or } j = k) \text{ and } (i \neq j) \\ 0 & \text{if } (i \neq k) \text{ and } (j \neq k) \end{cases} \qquad k = 1, ..., n. \tag{3}$$

where the matrix U^k is the zero diagonal matrix with all its elements equal to zero except the elements in the *k-th* row and column that are equal to the unity. It is very easy to prove that these matrices U^k are tetrahedral matrices too. Using now the properties derived from the n square matrix *vector space*, we may say that every possible weight matrix W or adjacency matrix of the *Resulting Graph* after training, may be generated by the *Basic tetrahedral matrices* $\{U^1, U^2, ..., U^n\}$, the *new parameters* of the net are then the coordinates of the vector W, or in other words, W may be expressed as

$$W = p_1 U^1 + p_2 U^2 + ... + p_n U^n \quad \Rightarrow \quad W = (p_1, p_2, ..., p_n) \tag{4}$$

In our application, it may be proved that

$$
\frac{29}{2}\begin{pmatrix}
0&1&1&1&1&1&1&1\\
1&0&0&0&0&0&0&0\\
1&0&0&0&0&0&0&0\\
1&0&0&0&0&0&0&0\\
1&0&0&0&0&0&0&0\\
1&0&0&0&0&0&0&0\\
1&0&0&0&0&0&0&0\\
1&0&0&0&0&0&0&0
\end{pmatrix}
+\frac{13}{2}\begin{pmatrix}
0&1&0&0&0&0&0&0\\
1&0&1&1&1&1&1&1\\
0&1&0&0&0&0&0&0\\
0&1&0&0&0&0&0&0\\
0&1&0&0&0&0&0&0\\
0&1&0&0&0&0&0&0\\
0&1&0&0&0&0&0&0\\
0&1&0&0&0&0&0&0
\end{pmatrix}
+\frac{-15}{2}\begin{pmatrix}
0&0&1&0&0&0&0&0\\
0&0&1&0&0&0&0&0\\
1&1&0&1&1&1&1&1\\
0&0&1&0&0&0&0&0\\
0&0&1&0&0&0&0&0\\
0&0&1&0&0&0&0&0\\
0&0&1&0&0&0&0&0\\
0&0&1&0&0&0&0&0
\end{pmatrix}
+\frac{3}{2}\begin{pmatrix}
0&0&0&1&0&0&0&0\\
0&0&0&1&0&0&0&0\\
0&0&0&1&0&0&0&0\\
1&1&1&0&1&1&1&1\\
0&0&0&1&0&0&0&0\\
0&0&0&1&0&0&0&0\\
0&0&0&1&0&0&0&0\\
0&0&0&1&0&0&0&0
\end{pmatrix}+
$$

$$
\frac{-5}{2}\begin{pmatrix}
0&0&0&0&1&0&0&0\\
0&0&0&0&1&0&0&0\\
0&0&0&0&1&0&0&0\\
0&0&0&0&1&0&0&0\\
1&1&1&1&0&1&1&1\\
0&0&0&0&1&0&0&0\\
0&0&0&0&1&0&0&0\\
0&0&0&0&1&0&0&0
\end{pmatrix}
+\frac{3}{2}\begin{pmatrix}
0&0&0&0&0&1&0&0\\
0&0&0&0&0&1&0&0\\
0&0&0&0&0&1&0&0\\
0&0&0&0&0&1&0&0\\
0&0&0&0&0&1&0&0\\
1&1&1&1&1&0&1&1\\
0&0&0&0&0&1&0&0\\
0&0&0&0&0&1&0&0
\end{pmatrix}
+\frac{-2}{2}\begin{pmatrix}
0&0&0&0&0&0&1&0\\
0&0&0&0&0&0&1&0\\
0&0&0&0&0&0&1&0\\
0&0&0&0&0&0&1&0\\
0&0&0&0&0&0&1&0\\
0&0&0&0&0&0&1&0\\
1&1&1&1&1&1&0&1\\
0&0&0&0&0&0&1&0
\end{pmatrix}
\frac{-27}{2}\begin{pmatrix}
0&0&0&0&0&0&0&1\\
0&0&0&0&0&0&0&1\\
0&0&0&0&0&0&0&1\\
0&0&0&0&0&0&0&1\\
0&0&0&0&0&0&0&1\\
0&0&0&0&0&0&0&1\\
0&0&0&0&0&0&0&1\\
1&1&1&1&1&1&0&0
\end{pmatrix}
$$

$$
=\begin{pmatrix}
0&21&7&16&12&16&4&1\\
21&0&-1&8&4&8&-4&-7\\
7&-1&0&-6&-10&-6&-18&-21\\
16&8&-6&0&-1&3&-9&-12\\
12&4&-10&-1&0&-1&-13&-16\\
16&8&-6&3&-1&0&-9&-12\\
4&-4&-18&-9&-13&-9&0&-24\\
1&-7&-21&-12&-16&-24&-24&0
\end{pmatrix}
$$

In other words, in our case, the matrix of weights may be written as:

$$W = \frac{29}{2}U^1 + \frac{13}{2}U^2 - \frac{15}{2}U^3 + \frac{3}{2}U^4 - \frac{5}{2}U^5 + \frac{3}{2}U^6 - \frac{2}{2}U^7 - \frac{27}{2}U^8$$

and the matrix W may then be replaced by the vector $1/2(29, 13, -15, 3, 5, 3, -2, 27)$

2. DYNAMICS OF THE SYSTEM

The state vector x at time t could also be interpreted as a *coloring* of the edges in G, but now this coloring is going to be used for retrieving the data. The summation of all the edges in the subgraph colored with the same color may be interpreted as the *gain* of that subgraph. If the graph G is colored with the coloring associated with the pattern $x(t)$, thinking in how the training algorithm were designed,, it is easy to understand that the bigger is the summation of all the edges in the *red* subgraph and the lower is the summation of all the edges in the *blue subgraph* the more correlated must be the pattern *x(t)* with those that were used in the training stage. With these considerations in mind, a different *dynamic equation* (taking into account the relation between both of these *gains*) may be defined. In [2] we explain how this system is designed. At time *t* when the net is in state *x(t)*, we define the *Energy Pair Number* EPN, associated to the net as a pair of numbers *{I(t), O(t)}*, where *I(t)*, represents the summation of all the values on the edges of the *red* subgraph and the second pair element *O(t)*, represents the summation of all the values on the edges of the *blue*. In a mathematical way, the pair of numbers *{I(t), O(t)}* are defined as the quadratic forms

$$I(t) = \frac{1}{2} x(t).W.x(t)^t$$

and   (5)

$$O(t) = \frac{1}{2} \overline{x}(t).W.\overline{x}(t)^t$$

But if we replace the matrix W by its equivalent $W = p_1 U^1 + p_2 U^2 + \ ..... \ + p_n U^n$, we obtain that

$$I(t) = \frac{1}{2}.[p_1(x_1(t)U^1(x_1(t))^t) + p_2(x_2(t)U^2(x_2(t))^t) + ...+ p_n(x_n(t)U^n(x_n(t))^t)]$$

$$O(t) = \frac{1}{2}.[p_1\overline{(x_1(t)}U^1\overline{(x_1(t))^t}) + p_2\overline{(x_2(t)}U^2\overline{(x_2(t))^t}) + ...+ p_n\overline{(x_n(t)}U^n\overline{(x_n(t))^t})]$$

in other words, if n_1 is the number of unitary components of *x(t)* and n_0 is the number of the null ones, the EPN may be expressed as a function of the new parameter p_i as

$$I(x) = (n_1 - 1)\, p.x(t) \qquad \text{and} \qquad O(x) = (n_0 - 1)\, p.\overline{x}(t) \qquad (6)$$

Taking as *Energy Field* the plain of axis *(I, O)*, every state *x(t)* of the net has a point *{I(t), O(t)}* associated with it, this point may be considered as the projection of the *vector state x(t)* over the *Energy Field*. We may then establish the *dynamic* of the *state vectors* trough theirs projections over the *Energy Field*. In (2) we proved that if n_1 is the number of unitary components of *x(t)* and n_0 is the number of null ones, then *{I(t), O(t)}* is a point of a line (energy line) with equation

$$\frac{n-1}{n_1-1}x + \frac{n-1}{n_0-1}y = K \qquad (K = p_1 + p_2 + \ ..... \ + p_n) \qquad (7)$$

In other words, all the EPN's associated with vector-states sharing the same number of unitary components are placed in the same line of the *Energy Field*, the equation of this line is the one represented in (7). In this way the *State Vector Space* so as the *Energy Field* are classified in so many classes as the dimension n of the space. In our application the distribution of the energy of the state vector space over the energy field is the one that may be seen in [2]. For defining the *Dynamic Equation* we must consider three important basic point

i) The energy associated with every possible state must take into account the correlation degree of that state with the training prototypes.

ii) x_i *(t)* must change its state if the new state $x/t+1)$ has a higher degree of correlation with the training prototypes which must mean that $x(t)$ has a lower energy.

iii) The dynamic of the states must avoid the cycles, and this must be proved trough the projection of the state dynamic over the *Energy Field*.

The way for reaching this goal will be easier to trace if the *energy lines* could be passing through the plane origin. Doing a constant translation, of value minus the arithmetic mean of its components, to the vector parameter p, we obtain

$$q = p + \frac{1}{2}(m_2.I), \quad \text{where} \quad m_2 = -\frac{1}{n}2\sum_{i=1}^{n}p_i = -\frac{1}{n}\cdot\frac{2K}{(n-1)} \tag{8}$$

In our example $m_2 = -\frac{1}{n}\cdot\frac{2K}{(n-1)} = \frac{5}{2}$, so

$$q = p + \frac{1}{2}(m_2.I) \Rightarrow q = p + \frac{5}{4}.I \Rightarrow (q_1, q_2, ..., q_n) = (p_1, p_2, ..., p_n) + \frac{5}{4}.I =$$

$$\left(\frac{29}{2}, \frac{13}{2}, \frac{-15}{2}, \frac{3}{2}, \frac{-5}{2}, \frac{3}{2}, \frac{-2}{2}, \frac{-27}{2}\right) + \frac{5}{4}.I = \left(\frac{63}{4}, \frac{31}{4}, \frac{-25}{4}, \frac{11}{4}, \frac{-5}{4}, \frac{11}{4}, \frac{-37}{4}, \frac{-49}{4}\right)$$

The energy lines r_i of the space where the parameters controlling the net are those represented in the vector q will be those having as equations the homogeneous equations associated to the equations obtained before (7). The representations of the energy will be then a set of lines, as may be seen for our example in the figure 2, where the big points are the EPN's of the training prototypes. If the net is in the $x(t)$ state (with its associated coloring), the contribution of the neuron x_i *(t)* to its "color" may be defined in different ways depending on the degree of *threshold color* compromise that we want to establish among the cells when the net is in $x(t)$, so if x_i *(t)* is, for example in the *red* subgraph. We define the *relative weight* of x_i $t)$ as the relation between all the *red* edges connecting with x_i *(t)* and all the *red* edges of the net when its state is $x(t)$.

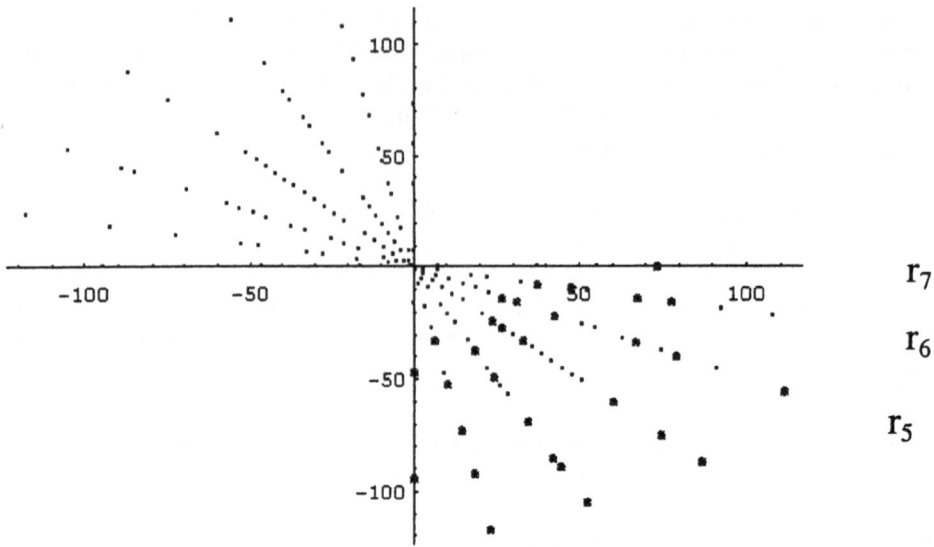

Figure 2. *Energy Lines.*

Those *relative weights* may be also expressed as a function of the vector parameters $p = \{p_{1,,} ..,p_n \}$. We have proved that the *relative weight* of x_i *(t)* which may be written as w_i *(t)* is

$$w_i(t) = \frac{x_i(t)}{n_1 - 1} + \frac{n_1 - 2}{n_1 - 1} \cdot \frac{p_i \cdot x_i(t)}{p \cdot x(t)} \quad , \text{ if } x_i = 1, \text{ (if } x_i = 0, \text{ } \bar{x}(t) \text{ must be taken)} \quad (9)$$

The *Relative Weight Vector* associated to $x(t)$ is defined as $w(t) = (w_1(t), .., w_n(t))$

Once defining the relative weight $w_i(t)$. The dynamic equation is defined as:

$$\text{If } x(t) \in [j], \text{ then } x_i(t+1) = f_h\left[f_b(x_i(t))\left(w_i(t) - \theta_j\right)\right] \quad (10)$$

where *[j]* the class of state vector whose EPS´s are in the r_j line, and

$$f_b(x) = \begin{cases} 1 \text{ si } x = 1 \\ -1 \text{ si } x = 0 \end{cases} \quad \text{and} \quad f_h(x) = \begin{cases} 1 \text{ si } x \geq 0 \\ 0 \text{ si } x < 0 \end{cases}$$

with the convention that x_i *(t+1)=$x_i(t)$* if w_i *(t)=θ_j.* This equation must be interpreted in this way: if $x(t) \in [j]$ and the *relative weight* of x_i *(t)* is less than the parameter θ_j, (which may be consider as the *capacity parameter* that is controlling the capacity of the *[j]* class) then x_i *(t)* must change its value, in other case this value remains unchanged. Defining the energy associated to $x(t)$ as $|I(t).O(t)|$ we have proved that if x_i (t) change its state, the new state $x(t+1)$ has its EPN placed in a utter hyperbolic surface, look to figure 3.

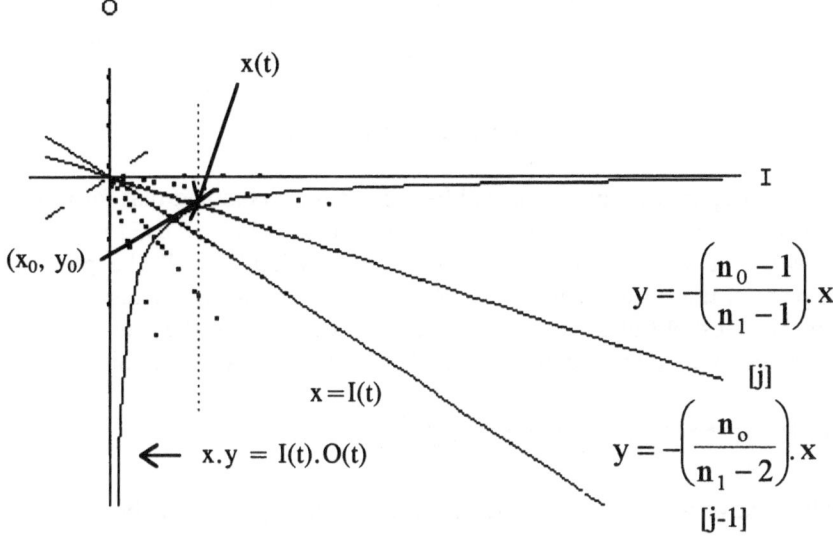

Figure 3. Capacity Parameter θ_i.

Solving the equation system (12), where the first equation represents the *level line*, where $\{I(t), O(t)\}$ is placed, and the second equation represents the adjacent *energy line* where the point $\{I(t), O(t)\}$ will be placed in the case that the net would change its state

$$\begin{cases} x.y = I(t).O(t) \\ y = -\left(\dfrac{n_0}{n_1 - 2}\right)x \end{cases} \tag{11}$$

If (x_0, y_0) is the solution of the system. We have evaluate the maximum value of w_i (t) which ensure that $I(t+1)$ or $O(t+1)$ will be placed in a point of a hyperbolic surface verifying that $|I(t).O(t)| > |I(t+1).O(t+1)|$. This value depends on the line r_j (α_j is the scope of r_j) over which the point $\{I(t), O(t)\}$ is placed

$$w_i(t) < 1 - \sqrt{\frac{\alpha_j}{\alpha_{j-1}}} = 1 - \sqrt{\frac{(n-j-1)(j-2)}{(j-1)(n-j)}} = 1 - m_j \tag{12}$$

We have applied this procedure for recognizing the Arabian digits, in figure 4

Figure 4. *Arabian Digits*

After applying the training explained before, we have obtained that for the *capacity parameter vector {-0.0462338, 0.0741306, 0.0741306, 0.053286, 0.0741306, 0.0660969, 0.0106075, 0.0250541, 0.0660969, 0.0755759}* all the prototypes are fixed point of the system. The next problem is to eliminate the parasitic fixed point. When the threshold of a class is very low (as for example the one that control de digit number one) a lot of points of this class will appear as fixed points.

3. PARASITIC FIXED POINTS

The relative weight vector distribution of the noisily digit number two, is the one that you can see in figure 5

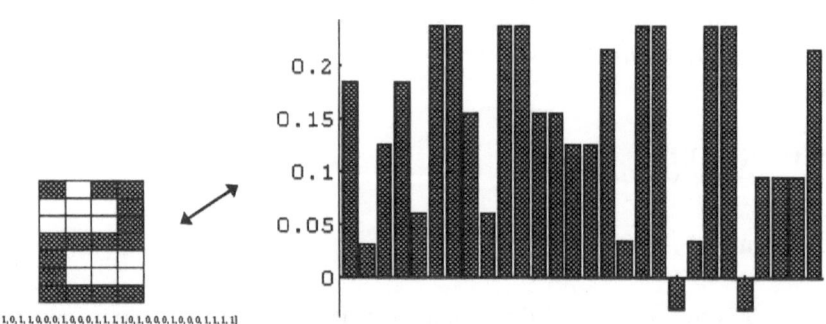

Figure 5 *Relative Weight Vector distribution of the noisily digit number two*

The noisy digit number *two* is in the class *[15]*, but the digits numbers *six* an the number *nine* are also in the same class. So if we want these digits as fixed points, the capacity parameter for the class *[15]* must be very low assigned; that is the reason for which the noisy digit number *two* appears as a parasitic fixed point. Now is when we are going to use the *statistical* properties of the distribution of the *Relative Weight Vector* distribution of the *state vectors*. Taking into account that for every relative weight vector *w(t)* associated to the state vector *x(t)* of the space is (2)

$$\sum_{i=1}^{n} \left[w_i(t) \right] = 4 \tag{13}$$

We define the *deviation* of $x(t)$ to the mean vector $w(\overline{x}) = (4/n,\ 4/n,\ 4/n)$ as

$$\text{Desv}[x(t)] = \sqrt{\sum_{i=1}^{n} \left[w_i(t) - w_i(\overline{x}) \right]^2} \tag{14}$$

if there had exist any state vector whose *Relative Weight Vector* components were the same as those of mean vector $w(\overline{x}) = (4/n,\ 4/n,\ 4/n)$, that vector would be the maximum equilibrium point of the space. The deviation of the prototype vectors may be taken as new parameter for eliminating the parasitic points: if the deviation of a fixed point is different of every prototype belonging to the same class as that fixed point, that prototype is eliminated as fixed point: in other words, the fixed points of the system will be those that having all its *Relative Weight Vector* components greater that its corresponding *threshold parameter capacity* have also the same *statistical deviation* to the mean distribution as the prototypes of its class. We have used this method and we have eliminating a big numbers of parasitic fixed points.

5. CONCLUSIONS

The conclusions which could be derived from the present research can be seen under the point of view of the re-interpretation of classical *Neural Network Theory* as *Graph Formalisms*. The *state-vector* space may be then classified in n classes according to the n different possible distances from any of the state-vectors to the *zero* vector The $(n \times n)$ matrix of weights may also be reduced to a n-vector of weights, in this way the computational time and the memory space required for obtaining the weights is optimized and simplified. In the recall stage, a parameter vector θ is introduced, this parameter is used for controlling the capacity of the net: it may be proved that the bigger is the θ_i component of θ, the lower is the number of fixed points located in the r_i energy line. Once the capacity of the net has been controlled by the θ parameter, we introduced other parameter, obtained as the *relative weight vector deviation* parameter, in order to reduce the number of spurious states.

Acknowledgments: This research is carried out under Grants Ns. TIC95 0122

6. REFERENCES

[1] V. Giménez, P. Gómez-Vilda, M. Pérez-Castellanos and V. Rodellar, *A New Approach for Finding the Weights in a Neural Network using Graphs*, Proc. of the 36th Midwest Symposium on Circuits and Systems, Detroit, August 16-18, 1993, pp. 113-116.

[2] V. Giménez, E. Torrano, P. Gómez-Vilda and M. Pérez-Castellanos, *A Class of Recursive Neural Networks Based on Analytic Geometry*, Proc. of the International Conference on Brain Processes, Theories and Models. Canary Islands, Spain, November 12-17, 1995. pp.330.339.

[3] V. Giménez, P. Gómez-Vilda, M. Pérez-Castellanos and E. Torrano, *A New Approach for improving the capacity limit on a Recursive Neural Network*, Proc. of the AMS'94. IASTED, Lugano, Switzerland, June 20-22, 1994, pp. 90-93.

[4] V. Giménez, P. Gómez-Vilda, E. Torrano and M. Pérez-Castellanos, *A New Algorithm for Implementing a Recursive Neural Network*, Proc. of the IWANN'95 Málaga-Torremolinos, Spain, June 1995, pp. 252-259.

[5] V. Rodellar, P. Gómez, M. Hermida and R. W. Newcomb, *An Auditory Neural System for Speech Processing and Recognition*, Proceedings of the ICARCV'92, Singapore, September 16-18, 1992, pp. INV-6.2.1-5.

[6] Yves Kamp and Martin Hasler, *Recursive Neural Networks for Associative Memory*, Wiley-Interscience Series in Systems and Optimization, England, 1990, pp. 10-34.

The Use of Prior Knowledge in Neural Network Configuration and Training

Melanie Hilario and Ahmed Rida

University of Geneva (CUI),
CH-1211 Geneva 4, Switzerland,
E-mail: hilario|rida@cui.unige.ch

Abstract. This paper describes an attempt to improve neural network design through the use of prior knowledge. The idea is to give preference to design techniques which exploit available domain knowledge (either approximate domain theories or partial knowledge about the target function), using search-based techniques only as a last resort. Many of these techniques have been integrated into SCANDAL, whose aim is to provide a unified framework for knowledge-based neural network design. Experimental results on the impact of knowledge on neural network accuracy and training time are presented.

1 Introduction

Neural network (NN) design can be roughly decomposed into three stages: pre-processing, configuration or the choice of topology, and training. NN design techniques can be situated along an axis representing the amount of domain knowledge available—going from one endpoint where no knowledge is available to the other extreme where domain knowledge forms an approximately perfect (i.e., almost complete and correct) theory. Between these two extremes are a number of intermediate points representing the availability of specific and partial domain knowledge, also known as hints (see Fig. 1.)

Domain knowledge	NN Design Techniques
Almost perfect domain knowledge	*Knowledge-intensive or translational techniques* e.g. Rule compilation
Partial domain knowledge	*Knowledge-primed or hint-based techniques* e.g. Weight sharing
No domain knowledge	*Knowledge-free or search-based techniques* e.g. Constructive algorithms

Fig. 1. Prior Knowledge and NN Design Techniques

This paper focuses on the configuration and training phases of the NN design process. In the following sections, we investigate a number of techniques from the three main subregions of the (domain) knowledge spectrum—knowledge-intensive (Section 2), knowledge-primed (Section 3), and knowledge-free techniques (Section 4). These three approaches are integrated within the SCANDAL architecture, which is described in Section 5. Section 6 reports on experimental results and Section 7 concludes.

2 Knowledge-intensive or translational techniques

When prior knowledge is rich enough to form an approximately complete and correct domain theory, it can be mapped directly onto a neural network using translational techniques. Most often, the source symbolic formalisms are propositional rules, whether categorical as in KBANN [24] and TopGen [16], probabilistic as in BANNER [19], or rules with certainty factors as in RAPTURE [13] and EN [10]. Attempts have also been made to build neural networks from decision trees [4, 21] and deterministic finite state automata [12].

Translational methods can be summarized in two steps: map the domain theory onto a neural net, then train the configured network. All rule formalisms, for instance, are translated in basically the same way. Final conclusions of the rulebase are mapped onto output units of the network, intermediate conclusions onto hidden units, and initial facts onto input units. Dependencies among assertions map onto weighted connections between the units. The difference lies in the semantics of the weights of links between rule antecedents and conclusions: they indicate positive or negative dependencies in categorical rulebases, certainty factors in certainty-factor rulebases, measures of probability in probabilistic rulebases, or degrees of membership in fuzzy rulebases. During the training phase, these techniques are distinguished by the learning algorithms used: while KBANN uses a slightly modified backpropagation procedure, most of the others like TopGen and RAPTURE use dynamic learning algorithms which add or prune nodes and links. Translational methods are complete NN design techniques in the sense that prior domain knowledge, together with the technique-specific training regime, totally determines a neural network's final topology.

3 Knowledge-primed or hint-based techniques

Almost perfect domain theories are hard to come by in real-world applications. The typical situation in neural network design is one in which only partial—often piecemeal—domain knowledge is available. Though insufficient to completely determine the final neural network structure, such partial knowledge (aka hints) can be used to prime NN design and training. Hints come in many flavors: some concern the structure of the application task, others involve characteristics of the application domain, and still other express global constraints on the function to be learned, or partial information about its implementation. Approaches based on such hints can be divided into two broad classes—hint-based configuration and hint-based training—depending on the design phase into which prior knowledge is incorporated.

In **hint-based configuration**, any information about the problem domain can be used to determine some aspect of the initial structure of the neural system. First of all, when a task is too complex to be solved by a single network, knowledge of the task decomposition can be used to *generate a neural macrostructure*: the task is broken down into subtasks which can be solved sequentially or in parallel by a set of neural networks. This gives rise to modular neural networks, which have been the object of a long research history that goes beyond the scope of this paper.

For a single network, the configuration task involves the following subtasks: choose the input and output units (and their encoding), determine the number of hidden layers and units, and the connectivity pattern. Hints can be used to support any of these subtasks. For instance, prior knowledge can be injected as *extra outputs of a neural network*. Practically, this means constraining the network to learn both the original task and a second task that is known to be related to the first. For instance, it has been shown that non monotonic functions like the fuzzy predicate 'medium height' are more difficult to learn than monotonic functions such as 'tallness'. A network that was constrained to learn both concepts simultaneously succeeded in half the time needed to learn the first concept alone [23]. Prior knowledge can be formulated as *permanent constraints on connections and weights*. In pattern recognition, for instance, one can encode invariance knowledge by structuring the network in such a way that transformed versions of the same input produce the same output. The best-known technique for doing this is weight sharing—i.e., constraining different connections to have equal weights [11].

In **hint-based training** prior knowledge is incorporated via the training process rather than being wired directly into the network structure. One way to do this is to augment the training set by *generating a sufficiently large number of examples that illustrate the hint*. Abu-Mostafa [1] proposes two ways of doing this. One is by means of duplicate examples: available training examples are copied and modified to express a particular hint. An alternative way is the use of virtual examples; for instance, to reflect the knowledge that the target function is even, a virtual example takes the form $y(x) = y(-x)$ for a particular input x. Since virtual examples do not provide target values, the objective function used during the training process should be modified. Other variants of hint-based generation of examples have been used to preset the rolling force in a steel rollmill [20], to diagnose coronary heart disease [17], and to generate error-correcting codes [2]. An alternative approach to hint-based training is to *incorporate hints into the learning algorithm* itself. In TangentProp, knowledge of invariance to transformations in pattern recognition was used to modify the standard back-propagation algorithm [22]. The main advantage of this technique is that training time is not increased due to artificial expansion of the dataset. Hints can also be used to *determine initial connection weights* in a more task-specific fashion than the usual random method. In domains where prior knowledge can be cast as finite state automata, transition rules have been used to initialize connection weights in both first-order [6] and second-order [7] recurrent networks.

4 Knowledge-free or search-based techniques

At the other extreme, where domain knowledge is scarce or unusable, NN design techniques rely mainly on guided search. These knowledge-free methods can be subdivided into dynamic and static methods. In static configuration the network topology is chosen *before* the training process, whereas in dynamic configuration it is dynamically modified *during* training.

An example of static configuration is Vysniauskas et al.'s method, which estimates the number of hidden units h needed by a single-hidden-layer network to approximate a function to a desired accuracy [26]. For classification tasks, [3] propose a design algorithm based on the construction of Voronoi diagrams over the set of data points in the training set. Contrary to the preceding technique, which estimates only the number of hidden units, this is a complete configuration procedure that determines the number of hidden layers and hidden units per layer, as well as the connections and corresponding weights of the nework before training.

Dynamic configuration methods can be divided into constructive (or growing) and destructive (or pruning) algorithms. Constructive methods start out with a minimal network structure (e.g., the input and output layers) and progressively add hidden layers and/or units until a stopping criterion is met [5]. Destructive methods adopt the reverse approach: they start from an initially complex model and prune units and/or links until a the desired out-of-same accuracy is attained. A comprehensive review of pruning methods can be found in [9].

Knowledge-free methods do not produce knowledge-based neural networks, but they are indispensable as a last resort in knowledge-poor domains. To compensate for the lack of domain knowldge, knowledge-free methods exploit an considerable amount of metaknowledge concerning neural networks in order to reduce the space of search for a reasonable network architecture.

5 A Workbench for Building Prior Knowledge into Neural Networks

SCANDAL (Symbolic-Connectionist Architecture for Neural Network Design and Learning) is an experimental workbench for exploring the use of prior knowledge—both domain and metalevel—in the design of neural networks. Its focus is on feedforward neural networks. As shown in the preceding sections, there is now a varied repertoire of techniques have been proposed for building knowledge into neural networks; the goal of SCANDAL is to select well-founded and useful methods in view of integrating them into a coherent system for knowledge-based NN design. However, it is by now a matter of consensus that there are no universally superior methods; a given technique is better than another under very precise conditions concerning the nature of the application task, the available domain knowledge, the distribution of training and test data, and other factors. An important research problem is thus to determine the conditions in which one method or technique outperforms others in doing a given task. To be of any use, such conditions should be represented explicitly whenever possible; for this reason, SCANDAL provides for the declarative representation of NN design techniques, their application conditions, and the resulting networks.

The architecture consists of a connectionist baselevel and a symbolic metalevel (Figure 2). At both levels, the main components are implemented as agents interacting via a distributed platform [8]. At the baselevel can be found a number of NN simulators whose task is to train the configured neural networks. At the metalevel, a supervisor receives the application task definition and, on the basis of its knowledge of the domain and of neural networks in general, selects one or several design agents. Each design agent embodies a class of techniques for accomplishing one or several design subtasks. For instance, in knowledge-rich domains where an approximate domain theory is available in the form of propositional rules, a rule compilation agent translates the rulebase into a feed-forward network. In knowledge-lean domains, other agents are activated, such as the constructive configuration agent and the pruning agent shown in Fig. 2.

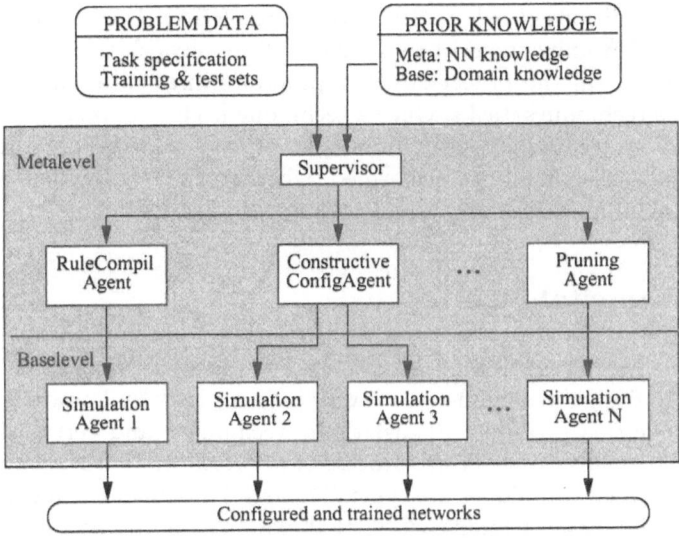

Fig. 2. The SCANDAL Architecture

A comprehensive ontology provides a unifying framework for domain as well as metalevel knowledge. Prior domain knowledge is represented in a specific subtree of this ontology and can be inspected by the different design agents, which are themselves represented in another region of the hierarchy. In addition, entities which typically remain inaccessible in the baselevel simulator, such as training algorithms, neural networks, and their components, have associated meta-objects which can be used for explicit reasoning and decision making by the supervisor, the design agents and the user.

To facilitate the incremental integration of novel techniques, each NN design agent is based on a carefully chosen set of related techniques; these are subjected to a thorough comparative analysis in order to unveil their common task structure as well as their individual particularities. Analysis results are then incorporated into the three main components of the metalevel knowledge base. First, the *ontological framework* is extended to represent *what* characterizes the different techniques. For instance, objects representing knowledge compilation techniques must specify their source symbolic formalism, while those representing incremental learning techniques must define a set of criteria for various decisions (e.g., when to stop training candidate hidden nodes, which candidate nodes to tenure, when to stop network expansion, etc). Second, *production rules* specify *when* to use each technique or its component tasks. The application constraints uncovered during the analysis phase (e.g., that Cascade Correlation can be used only for classification while OIL [25] can also be used for regression), are expressed as production rules. In addition, technique-specific limitations uncovered during use of a given technique can give rise to choice heuristics. For instance, when a categorical rulebase is not nearly complete, it is preferable to use TopGen, whose incremental network construction algorithm allows for the dynamic additions of rule nodes, rather than KBANN, where the number of additional hidden units should be specified before network training. Third, specialized *methods* (in the object-oriented programming sense) prescribe *how* to execute these techniques/tasks. Once a technique is chosen, its component tasks are executed in accordance with an associated task schedule. Methods specify how these subtasks should be performed, whether directly by the design agent itself or by sending service requests to simulator agents.

6   Experimental results

The implementation agenda of SCANDAL has taken account of the fact that complete NN design techniques are available only at the knowledge-intensive and knowledge-free extremes. Translational techniques and certain knowledge-free methods can ensure the entire process of network configuration and training, whereas hint-based techniques are typically partial in scope. For instance, weight sharing and hint-based training via virtual examples both rely on the existence of a preconfigured network. For this reason, priority has been given to the integration of translational and search-based techniques. The experiments described below explore the impact of prior knowledge by comparing the performance of these two technique classes. Knowledge-intensive techniques are represented by KB-RP, a variant of KBANN [24] which uses RPROP [18] instead of the original KBANN's backpropagation method. A "lesioned" version of KB-RP, called LKB-RP below, retains the network structure generated from the rulebase but randomly reinitializes network weights. Thus only a part of the prior domain theory is effectively reflected in the initial network to represent an intermediate zone where only partial knowledge is used. At the knowledge-free extreme, we chose OIL [25] to represent the set of incremental network configuration algorithms used in SCANDAL.

All experiments were conducted using 10-fold stratified cross-validation. With this method, each dataset is divided into 10 partitions or folds which reproduce overall class distributions; on each run, one fold is reserved exclusively for testing. Of the 9 remaining folds, 8 are used for training and 1 for validation every 10 epochs in view of early stopping. Training is stopped when the mean squared error falls below 0.001, or when response accuracy stagnates or degrades over 5 validation points, or when the limit of 3000 training epochs is attained. Each experiment was repeated 10 times, each time with a different network initialization; performance means reported for the different classification tasks below are thus based on 100 network training/test runs each. Several benchmarks from the UCI Machine Learning Repository [14] were used.

The Japanese credit screening dataset is one of two credit screening benchmarks available in the UCI repository. Prior domain knowledge is impossible to use on the Australian dataset, since attribute names and values have been changed to meaningless symbols for confidentiality reasons. The Japanese benchmark retains the real-world semantics of the credit application; it has the added advantage of being accompanied by a domain theory developed in cooperation with domain experts from a Japanese credit company. The dataset contains 125 records representing applicants, of which 85 were granted and 40 were refused credit. Thus the baseline accuracy, attained by simply choosing the majority class, is 68%. The original first-order rules were rewritten in propositional logic and the original dataset was preprocessed accordingly. The derived ruleset contains 17 propositional Horn-clause rules; the preprocessed dataset contains 2 output classes (credit granted or refused) and 22 input variables (8 attributes in 1-of-c encoding). Table 1 presents experimental results. The two performance measures used are classification accuracy on the test set and the number of training epochs taken to achieve this accuracy. For each performance measure, the mean and standard deviation are given together with their minimal and maximal values. Both the off-sample accuracy and the number of training epochs confirmed our expectations: as knowledge decreases (i.e., as we go down the table), mean accuracy decreases and training time increases.

	Accuracy				Training Epochs			
	Mean	StDev	Min	Max	Mean	StDev	Min	Max
KB-RP	78.54	9.38	61.54	100.00	11.00	0.0	11.0	11.0
LKB-RP	75.22	12.18	50.00	100.00	26.20	15.5	1.0	111.0
OIL	65.23	16.81	30.77	92.31	35.80	17.1	1.0	131.0

Table 1. Results of tests on the Japanese credit screening dataset.

The student loan benchmark is also based on a relational dataset and rulebase which had to be rewritten in propositional logic. The domain task is to decide whether a student is exempted from reimbursing a student loan. The propositional version has 18 rules for 2 output classes and 21 input variables (7 attributes in 1-of-c encoding). Results of experiments are shown in Table 2. Here again, mean test accuracy is proportional to the amount of knowledge built into the neural network. Note that the difference is accuracy between the knowledge-intensive and the search-based method is lower than in the Japanese Credit dataset. This is explained by the larger size of the Student loan dataset; its 1000 (643 positive and 357 negative) instances gives OIL greater chances of learning the target concept without any boost from prior knowledge. However, it takes 264 cycles to learn, around 26 times the training time required by KB-RP.

	Accuracy				Training Epochs			
	Mean	StDev	Min	Max	Mean	StDev	Min	Max
KB-RP	95.05	2.28	91.92	98.99	11.00	0.00	11.0	11.0
LKB-RP	94.85	2.22	89.90	98.99	34.00	5.20	31.0	51.0
OIL	88.80	6.26	72.00	96.00	263.50	160.57	31.0	491.0

Table 2. Results of tests on the Student Loan dataset.

In the tic-tac-toe endgame benchmark, the task is to determine in which cases player X will win, assuming she played first. The benchmark does not include a domain theory, but it was quite straightforward to write propositional rules which define winning endgame configurations. The interesting feature of this benchmark is its complete dataset which encodes all possible board configurations at the end of a game where X played first. Since prior knowledge is expected to add information value when a dataset is deficient, in this particular case there should be little to add. This is confirmed by the results shown in Table 3; the difference in the final accuracy attained by knowledge-based KB-RP and search-based OIL is less than 1%. However, the decisive difference lies in the complexity value added by the prior knowledge: whereas KB-RP attains near-perfect accuracy in around 10 training epochs, OIL takes 6 times longer to approach a comparable performance level.

Various results have been reported on this dataset in the past: accuracy results range from 76.7% to 99.1% for seven symbolic learning algorithms. Some of these learn from both theory and data, but their reported accuracies are significantly lower than the 99.97% scored by SCANDAL using KB-RP (see summary of previous results in [14]).

	Accuracy				Training Epochs			
	Mean	StDev	Min	Max	Mean	StDev	Min	Max
KB-RP	99.97	0.23	99.89	100.00	11.00	0.00	11.0	11.0
LKB-RP	98.08	4.54	85.26	100.00	45.90	6.40	31.0	61.0
OIL	99.05	0.65	91.91	100.00	64.70	10.80	41.0	91.0

Table 3. Results of tests on the Tic-tac-toe Endgame database.

7 Conclusion

From the experiments reported in the preceding section, we can draw the following conclusions. First, generalization power (as measured by off-sample accuracy) increases with the amount of knowledge built into a neural network. Second, training time (in terms of number of training cycles) drecreases as the amount of prior knowledge increases. Finally, as the size of the dataset increases, the impact of background knowledge on final accuracy diminishes in the sense that a knowledge-free technique will eventually attain accuracy levels comparable to that of a knowledge-based neural network. However, it will do so in a considerably larger number of training epochs.

These conclusions confirm the utility of the SCANDAL strategy which consists in giving priority to techniques which exploit as much of the available knowledge as is possible. However, in cases where domain knowledge is not only incomplete but also incorrect, prior knowledge can hinder learning. In such cases, search-based techniques have proved to be more effective in finding a reasonable solution than knowledge-based approaches. Experiments bearing this out will be reported in a forthcoming paper.

References

1. Y. S. Abu-Mostafa. Hints. *Neural Computation*, 7:639–671, 1995.
2. K. A. Al-Mashouq and I. S. Reed. Including hints in training neural networks. *Neural Computation*, 3:418–427, 1991.
3. N. K. Bose and A. K. Garga. Neural network design using Voronoi diagrams. *IEEE Transactions on Neural Networks*, 4(5):778–787, 1993.
4. R. P. Brent. Fast training algorithms for multilayer neural nets. *IEEE Transactions on Neural Networks*, 2(3):346–353, 1991.
5. S. E. Fahlman and C. Lebiere. The cascade-correlation learning architecture. Technical Report CMU-CS-90-100, Carnegie Mellon University, 1990.
6. P. Frasconi, M. Gori, M. Maggini, and G. Soda. Unified integration of explicit knowledge and learning by example in recurrent networks. *IEEE Transactions on Knowledge and Data Engineering*, 7(2):340–346, 1995.
7. C. L. Giles and C. W. Omlin. Extraction, insertion and refinement of symbolic rules in dynamicallly driven recurrent neural networks. *Connection Science*, 5(3-4):307–337, 1993.
8. J. C. Gonzalez, J. R. Velasco, C. A. Iglesias, J. Alvarez, and A. Escobero. A multiagent architecture for symbolic-connectionist integration. Esprit Project MIX Deliverable D1, March 1995.
9. C. Jutten and O. Fambon. Pruning methods: A review. In M. Verleysen, editor, *European Symposium on Artificial Neural Networks*, pages 129–140, Brussels, April 1995. D facto.
10. R. C. Lacher, S. I. Hruska, and D. C. Kuncicky. Backpropagatin learning in expert networks. *IEEE Transactions on Neural Networks*, 3:63–72, 1992.
11. Y. LeCun, B. Boser, J. S. Denker, et al. Handwritten digit recognition with a back-propagation network. In D.S. Touretzky, editor, *Advances in Neural Information Processing, 2*, pages 396–404. Morgan-Kaufmann, 1990.

12. R. Maclin and J. W. Shavlik. Using knowledge-based neural networks to improve algorithms: refining the Chou-Fasman algorithm for protein folding. In R.S. Michalski, editor, *Machine Learning, Special Issue on Multistrategy Learning*, volume 11, pages 195–215. Kluwer, 1993.

13. J. J. Mahoney and R. J. Mooney. Combining connectionist and symbolic learning to refine certainty factor rule bases. *Connection Science*, 5(3 & 4):339–393, 1993.

14. C. J. Merz and P. M. Murphy. Uci repository of machine learning databases. http://www.ics.uci.edu/ mlearn/MLRepository.html, 1996. lrvine, CA: University of California, Dept. of Information and Computer Science.

15. J. E. Moody, S. Hanson, and R. P. Lippman, editors. *Advances in Neural Information Processing, 4*. Morgan-Kaufmann, 1992.

16. D. W. Opitz and J. W. Shavlik. Heuristically expanding knowledge-based neural networks. In *Proc. of the 13th IJCAI*, pages 1360–1365, Chambéry, France, 1993. Morgan Kaufmann.

17. E. Prem, M. Mackinger, G. Dorffner, G. Porenta, and H. Sochor. Concept support as a method for programming neural networks with symbolic knowledge. In *GAI-92: Advances in Artificial Intelligence*. Springer, 1993.

18. M. Riedmiller and H. Braun. A direct adaptive method for faster backpropagation learning: the RPROP algorithm. In *Proc. IEEE International Conference on Neural Networks*, San Francisco, CA, 1993.

19. S. Romachandran. Revising bayesian network parameters using backpropagation. In *Proc. IEEE International Conference on Neural Networks*, pages 82–87, Washington, DC, 1996.

20. M. Röscheisen, R. Hofmann, and V. Tresp. Neural control for rolling mills: incorporating domain theories to overcome data deficiency. In Moody et al. [15], pages 659–666.

21. I. K. Sethi and M. Otten. Comparison between entropy net and decision tree classifiers. In *International Joint Conference on Neural Networks*, volume III, pages 63–68, San Diego, California, 1990.

22. P. Simard, B. Victorri, Y. LeCun, and J. Denker. TangentProp—a formalism for specifying selected invariances in an adaptive network. In Moody et al. [15], pages 895–903.

23. S. C. Suddarth and A. D. C. Holden. Symbolic-neural systems and the use of hints for developing complex systems. *International Journal of Man-Machine Studies*, 35(291-311), 1991.

24. G. G. Towell and J. W. Shavlik. Knowledge-based artificial neural networks. *Artificial Intelligence*, 70:119–165, 1994.

25. V. Vysniauskas, F. C. A. Groen, and B. J. A. Kröse. Orthogonal incremental learning of a feedforward network. In *International Conference on Artificial Neural Networks*, Paris, 1995.

26. V. Vysniauskas, F. C. A. Groen, and B.J.A. Kröse. The optimal number of learning samples and hidden units in function approximation with a feedforward network. Technical Report CS-93-15, CSD, University of Amsterdam, 1993.

A Model for Heterogeneous Neurons and Its Use in Configuring Neural Networks for Classification Problems

Julio J. Valdés[1] and Ricardo García

Dept. of Languages and Informatic Systems. Polytechnical University of Catalonia,
Jordi Girona Salgado, 1-3, 08034 Barcelona, Spain.

Abstract. In the classical neuron model inputs are continuous real-valued quantities. However in many important domains from the real world, objects are described by a mixture of continuous and discrete variables and usually containg missing information. A general class of neuron models accepting heterogeneous inputs in the form mixtures of continous and discrete quantities admiting missing data is presented. From these, several particular models can be derived as instances and also different neural architectures can be constructed with them. In particular, hybrid feedforward neural networks composed by layers of heterogeneous and classical neurons are studied here, and a training procedure for them is constructed using genetic algoritmhs. Their possibilities in solving classi-fication and diagnostic problems are illustrated by experiments with data sets from known repositories. The experiments shows that these networks are robust and that they can both learn and classify complex data very effectively and without preprocessing or variable transformations, also in the presence of missing information.

1 Introduction

Neural networks have been widely used in classification and prediction tasks. They provide a general model-free framework for the representation of non-linear mappings between two sets of variables, performing the transformation of one vector space into another, possibly with different dimensions. The basic operation is the one taking place inside the artificial neuron, which according to the classical feedforward perceptron model is a two-step process. First, the scalar product of the input vector with the one containing neuron weights is obtained, and then a non-linear (activation) function is applied, having such dot product as argument and an image given by a subset of the reals (possibly the whole). Usual activation functions are the sigmoid, the hyperbolic tangent, etc. Clearly, both the input vector and that of neuron weights are composed by real-valued quantities, such that they represent vectors from an n-dimensional space given by a cartesian product of n-copies of the reals. Neurons of this type are organized in networks according to different architectures, and in particular, for the feed-forward networks, training algorithms have been derived (eg. backpropagation and the like). The response of a neuron is basically controlled by the *transfer function* given by the scalar product, since the sigmoidal function is a nonlinear, but monotonically increasing function. Since the scalar product is a measure of the degree of colinearity between two vectors, the response is higher the more coincident both vectors are with respect to their direction and orientation. It is interesting to observe that since long ago the scalar product has been used as a meassure of similarity between objects in multivariate data analysis [5].

The classical neuron model and its associated network architectures have proved to be useful and efective. However, when applied to complex domains with data described by mixtures of numeric and qualitative variables, also with *missing data*, problems arises. In particular qualitative variables might have different nature. Some are *ordinal* in the usual statistical sense (i.e. with a discrete domain composed by k-categories, but totally ordered w.r.t a given relation) or *nominal* (discrete but without an ordering relation). A special kind of qualitative variable is the *binary* or *presence/absence* one. In problems described by a mixture of variables the above mentioned neural network models are applied in an *a fortiori* manner by transforming input variables according to different approaches. All variables are treated as real quantities, in some cases also reescaled, and in the case of nominal variables usually splitted into k new binary variables per original nominal descriptor. Concerning missing information, extended practices are: computation of statistical estimates in order to fill *the holes* (normally based on assumed probability distributions models of questionable validity in real domains), exclusion of the objects having incomplete descriptions (sacrifying the information given by the rest of available variables, and lossing the effort made in their registration).

A general consequence is that the original data are distorted in order to fit into the neural network model. Many times this "preprocessing" involves heuristic or tricky manipulations, difficulting the interpretation of final results specially when they are poor or unsatisfactory.

The purpose of this paper is to present a class of general models for neurons accepting heterogeneous inputs (possible missing too), from which different instance or particular models or *heterogeneous neurons* can be derived. Such derived or particular neuron models can be used for conforming a variety or network architectures, and in particular, a *hybrid feed-forward* network composed by heterogeneous and classical neurons is studied. A training procedure is constructed for them, and shown appropiate for solving classification problems, with remarcable robustness as illustrated by numerical experiments with known data sets.

The paper is organized as follows: section 2 presents the general model for a class of heterogeneous neurons, discuss some of its possible instatiations, proposes their use in constructing networks with different composition and layout, and finally introduces a hybrid feed-forward network; section 3 discusses this network further and presents training procedures, among which the one based on genetic algorithms is developed; finally, section 4 presents some comparative results of the performance of the network using heterogeneous data sets from known repositories and also using other classification procedures like the k-nn non-parametric pattern recognition method [7].

2 The Heterogeneous Neuron Model

2.1 The general class and some of its instances.

Let $\mathcal{R}$ denote the set of real numbers, $\mathcal{N}$ the set of natural numbers, and X a special symbol having the property of behaving as an incomparable element w.r.t any ordering relation in any set to which it belongs. Now let $\mathcal{N}^+ = \mathcal{N} - \{0\}$, $\hat{\mathcal{N}} = \mathcal{N}^+ \cup$ X, and $\hat{\mathcal{R}} = \mathcal{R} \cup$ X. Let us introduce three quantities $n_r, n_o, n_m \in \mathcal{N}$.

Now let $\mathcal{O}_i$, $1 \leq i \leq n_o$ be a family of finite sets with cardinalities $k_i^o \in \mathcal{N}^+$ respectively, composed by arbitrary elements, such that each set has a fully ordering relation $\leq_{\mathcal{O}_i}$. Construct the sets $\hat{\mathcal{O}}_i = \mathcal{O}_i \cup$ X, and for each of them define a partial ordering $\hat{\leq}_{\mathcal{O}_i}$ by extending $\leq_{\mathcal{O}_i}$ according to the definition of X.

Also let $\mathcal{M}_i$, $1 \leq i \leq n_m$ be a family of finite sets with cardinalities $k_i^m \in \mathcal{N}^+$ composed by arbitrary elements but such that no ordering relation is defined on any of the $\mathcal{M}_i$ sets, and construct the sets $\hat{\mathcal{M}}_i = \mathcal{M}_i \cup$ X. Let $\hat{\mathcal{R}}^{n_r} = \underbrace{\hat{\mathcal{R}}_1 \times \hat{\mathcal{R}}_2 \times \ldots \times \hat{\mathcal{R}}_{n_r}}_{n_r \text{ times}}$ be the cartesian product of n_r copies of the set $\hat{\mathcal{R}}$,

and analogously construct the sets $\hat{\mathcal{O}}^{n_o} = \underbrace{\hat{\mathcal{O}}_1 \times \hat{\mathcal{O}}_2 \times \ldots \times \hat{\mathcal{O}}_{n_o}}_{n_o \text{ times}}$ as the cartesian

product of n_o copies of the set $\hat{\mathcal{O}}_i$, and $\hat{\mathcal{M}}^{n_m} = \underbrace{\hat{\mathcal{M}}_1 \times \hat{\mathcal{M}}_2 \times \ldots \times \hat{\mathcal{M}}_{n_m}}_{n_m \text{ times}}$ as n_m

products of the set $\hat{\mathcal{M}}_i$. Define $\hat{\mathcal{R}}^0 = \hat{\mathcal{O}}^0 = \hat{\mathcal{M}}^0 = \hat{\mathcal{H}}^0 = \phi$ (the empty set), and consider the triplet $\hat{\mathcal{H}}^n = < \hat{\mathcal{R}}^{n_r}, \hat{\mathcal{O}}^{n_o}, \hat{\mathcal{M}}^{n_m} >$, where $n = n_r + n_o + n_m$.

An heterogeneous neuron is defined as a mapping $h : \hat{\mathcal{H}}^n \to \mathcal{R}_{out} \subseteq \mathcal{R}$, satisfying $h(\phi) = 0$. According to this more general definition, inputs are arbitrary tuples (possibly empty), composed by n elements among which there might be reals, ordinals, nominals and X, with the latter denoting *missing information*.

Naturally, even more general neuron models can be constructed in this way by considering mappings with a wider class of sets as domains and images, but as will be seen bellow, the one defined above has some interesting properties: generalizes sufficiently the classical neuron definition, and leads to straightforward implementations with useful performance. This does not mean, however, that they should not be considered in future studies.

Particular classes of submodels can be constructed by imposing further restrictions on the nature of the mapping h representing the desired neuron. For instance, consider a structural representation for h given by the composition of two mappings, that is, $h = f \circ s$, such that $s : \hat{\mathcal{H}}^n \to \mathcal{R}' \subseteq \mathcal{R}$ and $f : \mathcal{R}' \to \mathcal{R}_{out} \subseteq \mathcal{R}$. The mapping h can be considered as a $n - ary$ function parametrized by a $n - ary$ tuple $\hat{w} \in \hat{\mathcal{H}}^n$ representing neuron's weights. i.e $h(\hat{x}, \hat{w}) = f(s(\hat{x}, \hat{w}))$. From this interpretation the classical neuron model can be easily derived: takeing $n_r \neq 0$, $n_o = n_m = 0$, and assuming that there is no incomplete information. Then $\hat{\mathcal{H}}^n = \mathcal{H}^n = \mathcal{R}^n$, and therefore $\hat{\mathbf{x}}, \hat{\mathbf{w}} \; \epsilon \; \mathcal{R}^n$. Now define $s(\hat{\mathbf{x}}, \hat{\mathbf{w}}) = \hat{\mathbf{x}} \cdot \hat{\mathbf{w}} = \sum_{i=1}^{n} x_i \times w_i$ and $f(r) = 1/(1 + e^{-r})$, where $r \; \epsilon \; \mathcal{R}$. Thus, $h = f(s)$ is the neuron response, where s is the scalar product of $\mathbf{x}$ and $\mathbf{w}$ and f is the activation function, which in this case is the usual sigmoid. A general *two-mappings* submodel is shown in Fig 1.

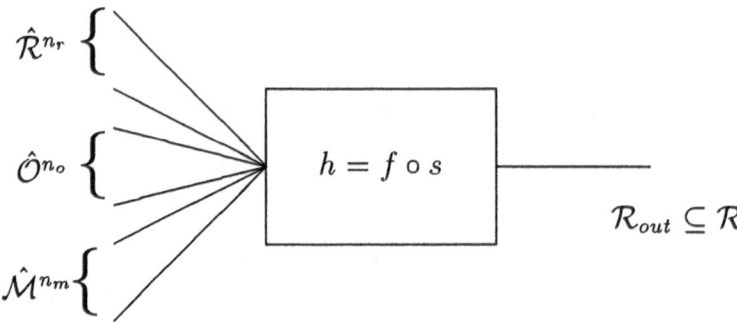

Fig. 1. The *two-mappings* submodel with its similarity and activation components.

In these *two-mappings* compositional submodels, a conceptual parallelism with the classical neuron definition can be preserved if the s mapping always express a *similarity* component for the neuron (i.e. the similarity of the input pattern with respect to neuron weights), and if the f mapping represents the *squashing* part, in the form of a suitable nonlinear behaviour. Keeping this interpretation when returning to the more interesting general case of $n_r, n_o, n_m \neq 0$, particular models can be constructed using

Observe that even more general models can be constructed by allowing the neuron output to be a magnitude belonging to more abstract domains like the complex numbers or other kind of sets. The same can be extended when considering neuron's inputs.

2.2 A concrete instance of a two-mappings heterogeneous neuron.

Now, restricting the analysis to the case of two-mappings neuron models, a neuron very close to the classical one, but more general than it, is one accepting heterogeneous inputs (i.e. elements from the set $\hat{\mathcal{H}}^n$, while producing the same output (i.e. a real in the interval $[0, 1]$). In order to obtain such case it will be enough to take as mapping s a similarity coefficient accepting as arguments tuples of the desired type, and a possible choice is Gower's general similarity coefficient [9], well known in the literature on multivariate data analysis.

For any two objects i, j described in terms of k variables, this coefficient has its values in the real interval $[0, 1]$ and is given by the expression.

$$G_{ij} = \frac{\sum_{k=1}^{n} g_{ijk} \, \delta_{ijk}}{\sum_{k=1}^{n} \delta_{ijk}}$$

where:

g_{ijk} : is a similarity *score* for objects i, j according to the value of variable k. Their values are in the interval $[0,1]$ and are computed according to different schemes for numeric and qualitative variables (see [9] for details).

δ_{ijk}: is a binary function expressing whether both objects are comparable or not according to their values w.r.t. variable k. It is 1 if and only if both objects have values different from X for variable k, and 0 otherwise.

Clearly, other similarity coefficients of this type might be used. As activation function, a modified version of the classical sigmoid should be used since now its domain will be that of the similarity coefficient ($[0,1]$), and not the whole reals as in the classical case. Again, there are many possible choices for non-linear monotone increasing and symmetric one-one mappings of $[0,1]$ onto $[0,1]$. A suitable family might be:

$$f(x,k) = \begin{cases} \left(\frac{-k}{(x-0.5)-a(k)} - a(k)\right) & \text{if } x \leq 0.5 \\ \left(\frac{-k}{(x-0.5)+a(k)} + a(k) + 1\right) & \text{otherwise} \end{cases}$$

where $a(k)$ is an auxiliary function given by

$$a(k) = \frac{-0.5 + \sqrt{0.5^2 + 4*k}}{2}$$

and k is a real-valued parameter controlling the curvature.

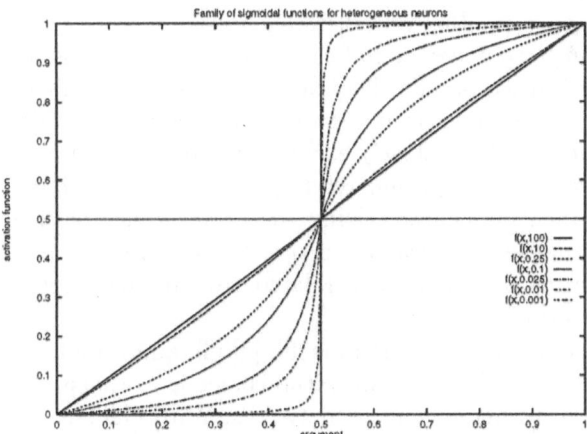

Fig. 2. Family of sigmoidal functions.

The behaviour of this family of functions is shown in Fig 2.

2.3 Network architectures and the hybrid feed-forward layout.

There are many possible ways in which particular heterogeneous neurons could be used when constructing definite purpose networks. The whole network can be composed by the same type of heterogeneous neurons, or different kind of them may be present. Also, they might be arranged in layers, or scattered in an unstructured pattern. Moreover, they might be allowed to have *feed-back* connections or not so complex interactions might be prefered. All these gives a wide variety of possible choices, but obviously, their properties must be subjected to both theoretical and experimental study, and their relative merits and drawbacks clarified.

Possibly the simplest of all these designs is that of a *hybrid feed-forward network*. This architecture results from selecting one hidden layer and one output layer, being the hidden layer composed by *heterogeneous neurons* of the same kind, and the output layer composed by *classical neurons*. In this case there is no problem from the point of view of internal network compatibility, since heterogeneous neurons's outputs are always real quantities. Therefore, they can be easily coupled with the classical neurons.

What remains is to devise a learning procedure for such kind of networks, in order to study their behaviour and classificatory capabilities. This will be described in the sequel.

3 Training procedures

The use of such hybrid feed-forward networks does not differ from the one of classical: A pattern (in this case heterogeneous) is presented as input to the network, and an output vector indicates the resulting class according to what the network has learned previously about the problem. The fact that the network inputs are tuples composed from a mixture of continous and discrete quantities, even with unknown values, clearly excludes approaches of constructing learning algorithms *a la backpropagation*, relying on the notion of partial derivatives. Since an error function characterizing the performance of the network can be defined exactly in the same way as in the classical case, an inmediate approach is the minimization of a functional related with it, like for instance, the mean squared classification error.

In this respect Genetic Algorithms [8], [3] [6], and Simulated Annealing [12], [10], are suitable candidates for an initial choice as cornerstones for learning algorithms oriented to heterogeneous networks. With Genetic Algorithms in particular, some of their known features are that of been free from restrictive assumptions concerning the search space, as continuity requirements, existence of derivatives, etc. Also, they have known drawbacks like no guarantee of finding a real optimum, the problem of epistasis (unwanted gen interactions), premature convergence, high computing time demands, and others. However, they are also robust, and usually able to find reasonable good solutions for most problems, although maybe not optimal. In the present case a coding scheme was chosen which recovers from the chromosomes all real and discrete weights associated with both heterogeneous and classical neurons present in the network. Concerning the parameters of the genetic algorithm, the following were chosen: Population size of 100 entities, linear rank scaling with amplification 1.5 [1], selection by stochastic universal [2], replacement procedure by excluding the worse entities, one-point crossover with probability 0.6, 0.01 mutation probability (Iris data) and (0.006 for the rest). Stopping criteria were: 20000 maximum iterations, no new optimum found after 2000 iterations, and complete classification of all entities (never happens).

4   Experimental results

Experiments were made in order to study the properties of the hybrid feed-forward heterogeneous network model. In this respect, several data sets from known repositories like the PROBEN1 group [11] were chosen. They are relatively well documented and in particular most of them are from real domains. Thus, they are complex enough from the point of view of their composition and illness, and therefore, suitable for the present study. They were the following: Iris (a classical data in pattern recognition in the domain of flowers. There are 150 objects with 4 variables, all numeric and no missing values); Horse (364 horses and 20 variables of mixed type with 30% of missing values. The decision variable is the prediction on whether the horse will survive, died or euthanized.); and Card (assignment of credit cards to bank clients. There are 690 clients described by 15 variables of mixed type with 0.65% of missing information).

Experimental data sets were divided into three subgroups corresponding to training, test and validation, with respective sizes of 50%, 25% and 25%. However, Iris data were divided in two halves for simplicity. The network configuration was: one hidden layer (with 16 heterogeneous neurons), another with 8 classical neurons, and an output layer with as many classical neurons as the number of classes. In the case of Iris data there were 15 h-neurons in the hidden layer, 8 in the second and 3 in the output.

For each data set the network was first trained and later tested with the corresponding data sets, introducing a progressive number of missing values uniformly distributed within *both* the training and the data sets (Observe that this network model can be *trained* in the presence of missing information). For comparisson, also the *k-nn* method with Gower's distance [9] was used. In all figures the different curves corresponds to the percentage of missing values in the training set. The horizontal axis is the percentage of missing data in the test set, and the vertical axis is the classification accuracy obtained.

Fig 3 shows the results using the k-nn rule and the heterogeneous network for the Card data. In this figure and in the sequel, ordinate values at the −10 abscise indicate pure training set performance (in the absence of test data). Also, as reference, the horizontal dashed line indicates the frequency of the mostly represented class in the data set.

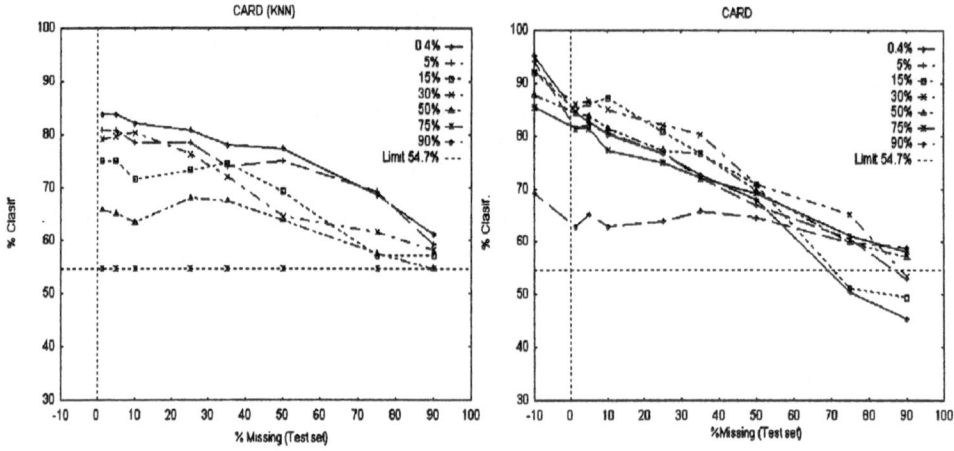

Fig. 3. Card data: k-nn rule (left) and the heterogeneous network (right).

As can be appreciated in Fig 3, the neural network is about 10% more accurate than the k-nn rule, and also suffers considerably less as the training set degrades. This can be observed by noting that in the k-nn case the different curves (corresponding to various proportions of missing data in the training set), are much more spread down, whilst in the neural network they tend to be clustered. This indicate more robustness. However, in some cases the network performs slightly worse than the k-nn rule, but the higher learning rates in the absence of test set could indicate an overfitting. The behavior of the network in the extreme case of 90% of missing data in the training set seems to be relatively insensitive to the increase of missing data in the test set. In all cases the effectivity is clearly above the a-priori frequency of the most represented class (54.7%). For the k-nn rule all cases with more than 50% of missing data in the test set fuse together al the 54.7% horizon.

Fig 4 shows the results using the k-nn rule (left) and the heterogeneous network (right) for the Horse data. The k-nn rule is not much informative, since for all cases it produces a prediction almost equal to the a-priori probability of the most frequent class, whilst the network performs worse only when the amount of missing data goes above 75% for the training set and 50% for the test set. Two other efects might have also influenced this case: in PROBEN1, a binary variable was excluded from the data set, and some amount of overfitting. Observe, for example, that with half information in both sets, the neural network is still gives 65% of classification accuracy.

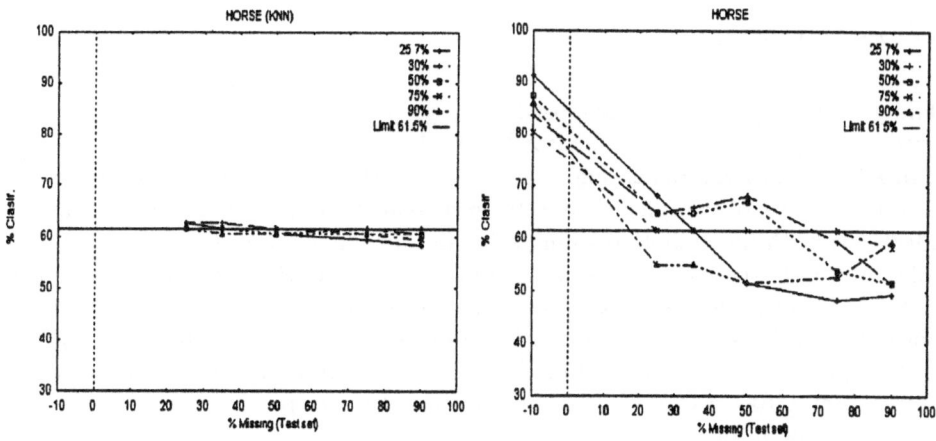

Fig. 4. Horse data: k-nn rule (left) and the heterogeneous network (right).

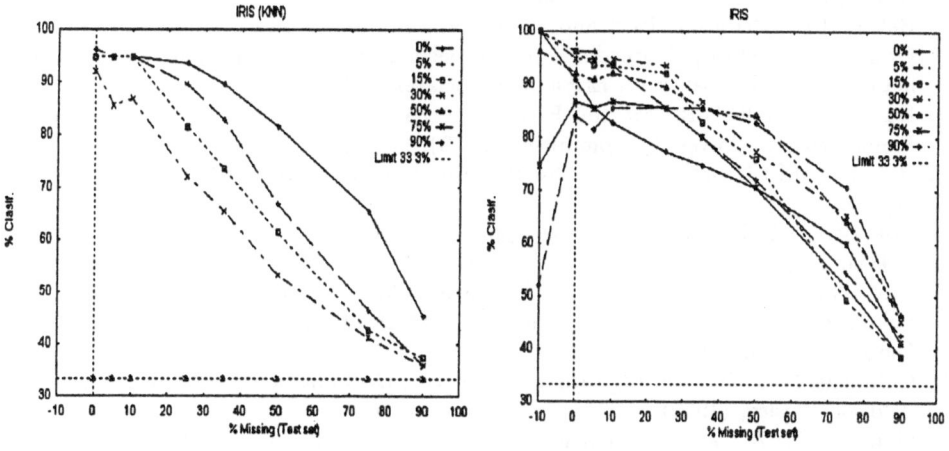

Fig. 5. Iris data: k-nn rule (left) and the heterogeneous network (right).

Finally, Fig 5 shows the results for the classical Iris data. The k-nn rule is rapidly affected by the increase of missing information, specially for the test set, whereas the neural network suffers much less. The effect of the dillution of the test set on the network is higher after the 50% barrier. Also note that, for example, network predictions are over 80% with 50% of missing values in both sets.

5 Conclusions

The general neuron model proposed leads to particular submodels of potential practical importance in classification tasks. Hybrid neural networks composed by a mixture of heterogeneous and classical networks are able to learn effectively in the presence of complex data sets from real domains. They can be trained effectively with incomplete training sets, and also have proved to be robust and tolerant to the increase of missing information in both the training and the test set. They performs comparably with traditional classification methods in normal conditions and better when the information degrades, while keeping the advantages of been model free function estimators. Moreover, they need no data preprocessing like transformation of variables, thus respecting more the nature of the original information.

References

1. Baker, J.E.: Adaptative selection methods for genetic algorithms. In J.J. Grefenstette, editor, Proceedings of the First International Conference on Genetic Algorithms, Lawrence Erlbaum Associates, , (1985), 101-111.
2. Baker, J.E.: Reducing Bias and Inefficiency in the Selection Algorithm. In J.J. Grefenstette, editor, Proceedings of the Second International Conference on Genetic Algorithms, Lawrence Erlbaum Associates, , (1987), 14-21.
3. Beasey, D.; Bull, D.R.; Martin, R.R.: An Overview of Genetic Algorithms: Part1, Fundamentals. University Computing, 15(2), (1993) 58-69.
4. Bishop, C.M.: Neural Networks for Pattern Recognition. Oxford University Press, (1995).
5. Chandon, J.L, Pinson, S: Analyse Typologique. Théorie et Applications. Masson, (1981), 254.
6. Davis, L.D.: Handbook of Genetic Algorithms. Van Nostrand Reinhold, (1991).
7. Fukunaga, K.: Introduction to Statistical Pattern Recognition (Second Ed.) San Diego : Academic Press, (1990).
8. Goldberg, D.E.: Genetic Algorithms for Search, Optimization & Machine Learning. Addison-Wesley, (1989) .
9. Gower, J.C.: A General Coefficient of Similarity and some of its Properties. Biometrics 27, (1971), 857-871.
10. Ingber, A.L. : Very Fast Simulated Annealing. Jour. Math. Comp. Modeling. 12, (1989) 967-973.
11. Prechelt, L.;: Proben1: A Set of Neural Network Benchmark Problems and Benchmarking Rules. Fakultät für Informatik. Universität Karlsruhe, Germany. Technical Report 21/94, ftp://ftp.ira.uka:/pub/neuron, (1994).
12. Rutenbar, R. A. : Simulated Annealing Algorithms: An overview. IEEE Circuits and Devices Mag. Jan. (1989) 19-26.

A Computation Theory for Orientation-Selective Simple Cells Based on the MAP Estimation Principle and Markov Random Fields*

Mehdi N. Shirazi and Yoshikazu Nishikawa

Osaka Institute of Technology
Faculty of Information Science
Kitayama, Hirakata-shi, Osaka, Japan
Email: shirazi@ij.oit.ac.jp

Abstract

A computation theory is proposed for the orientation-selective simple cells of the striate cortex. The theory consists of three parts: (a) a probabilistic computation theory based on MRFs and the MAP estimation principle which is assumed to be carried out by the visual pathway lying between the lateral geniculate nuclei and the simple cells; (b) a deterministic parallel algorithm which compute the MAP estimation approximately; and (c) a neural implementation of the algorithm.

1   Introduction

The vertebrate retina consists of several layers of nerve cells and these contain four kinds of cells: rods and cones (photoreceptors), horizontal, bipolar, amacrine and ganglion cells. Photoreceptors synapse with both horizontal and bipolar cells, and bipolar cells synapse with both amacrine and ganglion cells. The axons of the retinal ganglion cells (RG-cells) gather into a bundle at the optic disc and leave the eye to synapse with the lateral geniculate nuclei cells (LGN-cells). The axons of the LGN-cells in turn form the optic radiations and project to the striate cortex.

The first recordings from the cells in the striate cortex of monkeys were made by Hubel/Wiesel (Hubel and Wiesel, 1968). They found that some cells in the striate cortex have concentric receptive fields similar to RG- and LGN-cells, but others have a quite different kind of receptive fields, in which elongated excitatory and inhibitory areas lie adjacent and parallel to each other. They classified these

---

*This research was supported by the Japan SCAT (Support Center for Advanced Telecommunications Technology Research) Foundation.

cells as simple cells. Because of the elongated shape of their receptive fields, they respond most to a bar or edge with a particular orientation.

How is the wiring between the cells from the photoreceptors to the simple cells and what are the information-processing tasks carried out by these cells which endow the simple cells with this very orientation-selective property?

In this article, we try to answer this question partially by providing a computation theory for the visual pathway lying between the LGN and the striate cortex where the simple cells reside. The theory consists of three parts: (a) a probabilistic computation theory based on Markov random field modeling and the Maximum-A-Posteriori (MAP) estimation principle which is envisaged as the computation carried out in this part of visual pathway; (b) a deterministic parallel algorithm which compute the MAP estimate approximately; and (c) a neural implementation of the algorithm which gives possible cell-connectivities among and between the LGN-cells, the cortical cells lying between the LGN-cells and the cortical simple cells.

The proposed computation theory can be considered as a theory in the Marr's paradigm (Marr, 1982) and the cell-connectivities which it suggests for this part of the visual pathway is compatible with the neurophysiological results (Colonnier, 1968) which suggest that the first cortical cells have an excitatory input from the LGN-cells combined with and modified by lateral intracortical inhibition.

2 A Probabilistic Computation Theory

Let us consider the following explanatory scenario. The input to an orientation-selective cell (OS-cell) is provided by an assembly of intermediate (I-cells) lying between the OS-cell and its receptive field cells (RF-Cells) which are LGN-cells in this article. The I-cells are driven by the RF-cells. The RF-cells and I-cells are locally interconnected and interact in such a way that (a) all or most of the I-cells fire when the activity pattern of the RF-cells happens to be elongated in a particular orientation to which the OS-cell has been tuned and (b) all or most of the I-cells are silent otherwise.

In the following, we will introduce our probabilistic computation theory by giving a mathematical formulation of this scenario by using the Markov random field (MRF) modeling and the Maximum-A-Posteriori (MAP) estimation principle. To do so, we assume that the RF- and I-cells are located on two identical lattices denoted by $\mathcal{L}$ whose activity patterns are realizations of two interactive stochastic spatial processes called the RF-process and I-process, respectively. We denote the interactive RF-process and I-process by $Y_{\mathcal{L}}$ and $X_{\mathcal{L}}$, respectively and give thier joint statistical description in the following.

2.1 A Markov Random Field

Let us consider a rectangular lattice of units. Let $\mathcal{L} = \{(i,j)\}$; $i = 1, 2, \ldots, N$ and $j = 1, 2, \ldots, M$ denote a finite set of sites on the lattice.

As it is the case in Markov random field (MRF) theory, the units interact and the interactions are maintained through the so-called neighborhood system $\mathcal{N} = \{\mathcal{N}_{ij}\}$ and clique system $\mathcal{C} = \{\mathcal{C}_{ij}\}$, where $\mathcal{N}_{ij}$ is a neighborhood associated with the (i,j)-unit and $\mathcal{C}_{ij}$ is a set of cliques defined on $\mathcal{N}_{ij}$. The neighborhood $\mathcal{N}_{ij}$ can be any subset of $\mathcal{L}$ which contains the (i,j)-unit, i.e., $(i,j) \in \mathcal{N}_{ij} \subset \mathcal{L}$ and a clique defined on $\mathcal{N}_{ij}$ can be any subset of $\mathcal{N}_{ij}$ which contains the (i,j)-unit, i.e., $(i,j) \in C \subset \mathcal{N}_{ij}$. For example, for the first-order neighborhood system, we have $\mathcal{N}_{ij} = \{(i,j),(i,j-1),(i-1,j),(i,j+1),(i+1,j)\}$ which consists of five sites and $\mathcal{C}_{ij} = \{\{(i,j)\},\{(i,j),(i,j-1)\},\{(i,j),(i-1,j)\},\{(i,j),(i,j+1)\},\{(i,j),(i+1,j)\}\}$ which consists of one singleton and four doubleton cliques.

Let $X_{\mathcal{L}} = \{X_{ij}\}; (i,j) \in \mathcal{L}$ [1] be a Markov random field defined on $\mathcal{L}$ where X_{ij}s take a value from a common local state space Q. MRFs can be described by Gibbs distributions, having the following form

$$P(x_{\mathcal{L}}) = Z^{-1}e^{-\mathcal{E}(x_{\mathcal{L}})}. \tag{1}$$

Here, $x_{\mathcal{L}}$ denotes a realization of $X_{\mathcal{L}}$ from the configuration space $\Omega = Q^{|\mathcal{L}|}$; where $|\mathcal{L}|$ is the number of units on the lattice, $\mathcal{E}(x_{\mathcal{L}})$ is the global energy function and

$$Z = \sum_{x_{\mathcal{L}} \in \Omega} e^{-\mathcal{E}(x_{\mathcal{L}})} \tag{2}$$

is the partition function. For details on MRFs and related concepts such as neighborhoods, cliques and boundary conditions, see (Besag, 1974; Geman and Geman, 1984).

The global energy function can be defined as

$$\mathcal{E}(x_{\mathcal{L}}) = \sum_{(i,j) \in \mathcal{L}} \sum_{C \in \mathcal{C}_{ij}} \mathcal{E}(x_C) \tag{3}$$

where $\mathcal{E}(x_C)$ is the clique energy function associated with the clique C. For example, for a doubleton clique $C = \{(i,j),(i,j-1)\}$, $\mathcal{E}(x_C)$ denotes $\mathcal{E}(x_{ij-1},x_{ij})$.

2.2 A Two-layer Hierarchical Markov Random Field with one Hidden Process

A two-layer hierarchical Markov random field (HMRF) with one hidden process consists of two layers:(a) a hidden layer which incorporates an unobservable homogeneous MRF label process (I-process in this article); and (b) an observation layer which incorporates an observable nonhomogeneous MRF observation process (RF-process in this article) being composed of a finite number of homogeneous MRF processes. The spatial locations and extensions of the constituent observation processes are controlled by the label process. The realizations of such

---

[1]For a given $\mathcal{A} = \{a_1,\dots,a_l\}$, $x_{\mathcal{A}}$ and $f(x_{\mathcal{A}})$ denote $\{x_{a1},\dots,x_{al}\}$ and $f(x_{a1},\dots,x_{al})$, respectively.

a two-layer HMRF can be envisaged as spatial patterns comprising of patches of distinct textures.

Assuming that the lattices on the two layer have the same configurations, the composite random field $(X_\mathcal{L}, Y_\mathcal{L})$ comprising of the label process $X_\mathcal{L}$ and the observation process $Y_\mathcal{L}$ can be statistically characterized by a joint probability function $P(x_\mathcal{L}, y_\mathcal{L})$ or equivalently, according to Bayes' rule, by $P(x_\mathcal{L})$ and $P(y_\mathcal{L} \mid x_\mathcal{L})$.

Assuming that the unobservable (hidden) label process $X_\mathcal{L}$ is an MRF, it can be statistically characterized by $P(x_\mathcal{L})$ defined by (1)-(2). Assuming the same neighborhood structures and local state spaces for the MRF-constituents of the observation process $Y_\mathcal{L}$, $P(y_\mathcal{L} \mid x_\mathcal{L})$ is given as

$$P(y_\mathcal{L} \mid x_\mathcal{L}) = (Z^{Y|X})^{-1} e^{-\mathcal{E}(y_\mathcal{L} \mid x_\mathcal{L})} \tag{4}$$

where the global conditional energy function is

$$\mathcal{E}(y_\mathcal{L} \mid x_\mathcal{L}) = \sum_{(i,j) \in \mathcal{L}} \sum_{C \in \mathcal{C}^Y_{ij}} \mathcal{E}(y_C \mid x_{ij}) \tag{5}$$

and the conditional partition function is

$$Z^{Y|X} = \sum_{y_\mathcal{L} \in \Omega_Y} e^{-\mathcal{E}(y_\mathcal{L} \mid x_\mathcal{L})}. \tag{6}$$

In the above, $\Omega_Y = \prod_{(i,j) \in \mathcal{L}} Q_Y$, where Q_Y is a set from which Y_{ij}s take a value. For detail, see (Shirazi, 1993).

2.3 The Maximum-A-Posteriori Estimation Principle

The computation goal of the network is formulated as the Maximum-A-Posteriori estimation of the I-cells' activity pattern generating process $X_\mathcal{L}$ from a given RF-cells' pattern $y_\mathcal{L}$.

The MAP estimation of $X_\mathcal{L}$ is defined as

$$\hat{x}_\mathcal{L} = \arg \max_{x_\mathcal{L} \in \Omega_X} P(x_\mathcal{L} \mid y_\mathcal{L}) \tag{7}$$

where $\Omega_X = Q_X^{|\mathcal{L}|}$ and $|\mathcal{L}|$ denoting the number of sites on the lattice.

3 A Parallel Algorithm

To solve the MAP estimation problem (7) in a general case where the covering patterns are spatially correlated, i.e., when $p(y_\mathcal{L} \mid x_\mathcal{L}) \neq \prod_{(i,j) \in \mathcal{L}} p(y_{ij} \mid x_{ij})$, a deterministic parallel algorithm have been proposed (Shirazi, 1993). The proposed parallel algorithm called the GICM algorithm can be considered as a generalization of the ICM algorithm (Besag, 1986) which is applicable to the restricted cases where there are no spatial correlations within the constituent observation processes, i.e., where $p(y_{\mathcal{L}} \mid x_{\mathcal{L}}) = \prod_{(i,j)\in\mathcal{L}} p(y_{ij} \mid x_{ij})$.

The GICM algorithm is defined by the following updating rule

$$\hat{x}_{ij}^{(p+1)} = \arg \max_{x_{ij}\in Q_X} P(y_{ij} \mid y_{\mathcal{N}'_{ij}Y}, x_{ij}) P(x_{ij} \mid \hat{x}_{\mathcal{N}'_{ij}X}^{(p)}) \tag{8}$$

where

$$P(x_{ij} \mid x_{\mathcal{N}'_{ij}X}) = (Z_{ij}^X)^{-1} e^{-\mathcal{E}(x_{ij}\mid x_{\mathcal{N}'_{ij}X})} \tag{9}$$

$$Z_{ij}^X = \sum_{q\in Q_X} e^{-\mathcal{E}(q\mid x_{\mathcal{N}'_{ij}X})} \tag{10}$$

$$\mathcal{E}(x_{ij} \mid x_{\mathcal{N}'_{ij}X}) = \sum_{C\in\mathcal{C}_{ij}^X} \mathcal{E}(x_{ij} \mid x_{C'}) \tag{11}$$

and

$$P(y_{ij} \mid y_{\mathcal{N}'_{ij}Y}, x_{ij}) = (Z_{ij}^{Y|X})^{-1} e^{-\mathcal{E}(y_{ij}\mid y_{\mathcal{N}'_{ij}Y}, x_{ij})} \tag{12}$$

$$Z_{ij}^{Y|X} = \sum_{p\in Q_Y} e^{-\mathcal{E}(p\mid y_{\mathcal{N}'_{ij}Y}, x_{ij})} \tag{13}$$

$$\mathcal{E}(y_{ij} \mid y_{\mathcal{N}'_{ij}Y}, x_{ij}) = \sum_{C\in\mathcal{C}_{ij}^Y} \mathcal{E}(y_{ij} \mid y_{C'}, x_{ij}). \tag{14}$$

In the above, $\mathcal{N}_{ij}'^X$, $\mathcal{N}_{ij}'^Y$ and C' denote $\mathcal{N}_{ij}^X$, $\mathcal{N}_{ij}^Y$ and C, respectively, with the (i, j)-element deleted.

3.1 Initialization

To apply the GICM algorithm (8), the initial activity pattern of the I-cells $\hat{x}_{\mathcal{L}}^{(0)}$ should be provided. We can initialize the algorithm by a binary random pattern or by a checkerboard pattern. However, the MAP estimation can be obtained with considerably less iteration by initializing the algorithm with the Maximum Likelihood (ML) estimate, i.e., with $\hat{x}_{\mathcal{L}}^{(0)}$ given by

$$\hat{x}_{\mathcal{L}}^{(0)} = \arg \max_{x_{\mathcal{L}}\in\Omega_X} P(y_{\mathcal{L}} \mid x_{\mathcal{L}}). \tag{15}$$

The ML estimate can be calculated approximately using the following locally updating rule

$$\hat{x}_{ij}^{(0)} = \arg \max_{x_{ij}\in Q_X} P(y_{ij} \mid y_{\mathcal{N}'_{ij}Y}, x_{ij}). \tag{16}$$

4 Horizontally Tuned OS-cells' Clique Energies

To complete the description of the computation theory defined by (1)-(7) and the computing parallel algorithm characterized by (8)-(16) for horizontally-tuned OS-cells, we should specify:(a) the neighborhood topologies $\mathcal{N}_{ij}^X$ and $\mathcal{N}_{ij}^Y$; (b) the local state spaces Q_X and Q_Y; and (c) the clique energy functions $\mathcal{E}(x_{ij} \mid x_{C'})$ and $\mathcal{E}(y_{ij} \mid y_{C'}, x_{ij})$ of (11) and (14), respectively.

We assume 2nd-order neighborhood topologies. Two distinct states are assumed for the RF- and I-cells. we denote the states of the RF-cells by p_{ON} and p_{OFF} and of the I-cells by q_{ON} and q_{OFF}, i.e., $Q_Y = \{p_{ON}, p_{OFF}\}$ and $Q_X = \{q_{ON}, q_{OFF}\}$.

Furthermore, we assume that all clique energies are zero except the doubleton clique energies, i.e., it is assumed that $\mathcal{C}_{ij}^X$ and $\mathcal{C}_{ij}^Y$ consist of doubleton cliques only.

- **Doubleton Clique Energies for Horizontally Oriented Patterns**

  For $C \in \mathcal{C}_{ij}^{Y,H}$

  $$\mathcal{E}(y_{ij} \mid y_{C'}, q_{ON}) = -\beta\delta(y_{ij}, y_{C'}) \qquad (17)$$

  and for $C \in \mathcal{C}_{ij}^Y - \mathcal{C}_{ij}^{Y,H}$

  $$\mathcal{E}(y_{ij} \mid y_{C'}, q_{ON}) = \beta\delta(y_{ij}, y_{C'}). \qquad (18)$$

- **Doubleton Clique Energies for Non-horizontally Oriented Patterns**

  For $C \in \mathcal{C}_{ij}^{Y,H}$

  $$\mathcal{E}(y_{ij} \mid y_{C'}, q_{OFF}) = \beta\delta(y_{ij}, y_{C'}) \qquad (19)$$

  and for $C \in \mathcal{C}_{ij}^Y - \mathcal{C}_{ij}^{Y,H}$

  $$\mathcal{E}(y_{ij} \mid y_{C'}, q_{OFF}) = -\beta\delta(y_{ij}, y_{C'}). \qquad (20)$$

- **Doubleton Clique Energies for Non-directional (Blob-like) Patterns**

  For $C \in \mathcal{C}_{ij}^X$

  $$\mathcal{E}(x_{ij} \mid x_{C'}) = -\gamma\delta(x_{ij}, x_{C'}). \qquad (21)$$

In the above, $\beta, \gamma > 0$ and $\mathcal{C}_{ij}^{Y,H}$ denotes the set of horizontal doubleton cliques, i.e., $\mathcal{C}_{ij}^{Y,H} = \{\{(i,j), (i,j-1)\}, \{(i,j), (i,j+1)\}\}$ and

$$\delta(x,y) = \begin{cases} 1 & \text{if } x = y \\ 0 & \text{otherwise.} \end{cases} \qquad (22)$$

5   Neural Implementation

Equation (8) can be written equivalently as

$$\hat{x}_{ij}^{(p+1)} = \arg\max_{x_{ij}\in Q_X}\left\{\ln P(y_{ij}\mid y_{\mathcal{N}_{ij}'Y}, x_{ij}) + \ln P(x_{ij}\mid \hat{x}_{\mathcal{N}_{ij}'X}^{(p)})\right\}. \tag{23}$$

Replacing $P(x_{ij}\mid \hat{x}_{\mathcal{N}_{ij}'X}^{(p)})$ and $P(y_{ij}\mid y_{\mathcal{N}_{ij}'Y}, x_{ij})$ by using (9)-(14), it can be shown that (23) is equivalent to

$$\hat{x}_{ij}^{(p+1)} = \arg\max_{x_{ij}\in Q_X}\left\{\; -\;\mathcal{E}(y_{ij}\mid y_{\mathcal{N}_{ij}'Y}, x_{ij}) - \mathcal{E}(x_{ij}\mid \hat{x}_{\mathcal{N}_{ij}'X}^{(p)})\right.$$
$$\left. -\;\ln\sum_{p\in Q_Y}e^{-\mathcal{E}(p\mid y_{\mathcal{N}_{ij}'Y}, x_{ij})}\right\}. \tag{24}$$

Using the following approximation

$$\ln\sum_{p\in Q_Y}e^{-\mathcal{E}(p\mid y_{\mathcal{N}_{ij}'Y}, x_{ij})} \simeq \ln e^{\max_{p\in Q_Y}\{-\mathcal{E}(p\mid y_{\mathcal{N}_{ij}'Y}, x_{ij})\}}$$
$$= \max_{p\in Q_Y}\{-\mathcal{E}(p\mid y_{\mathcal{N}_{ij}'Y}, x_{ij})\} \tag{25}$$

(24) can be written as

$$\hat{x}_{ij}^{(p+1)} = \arg\max_{x_{ij}\in Q_X}\left\{\; -\;\mathcal{E}(y_{ij}\mid y_{\mathcal{N}_{ij}'Y}, x_{ij}) - \mathcal{E}(x_{ij}\mid \hat{x}_{\mathcal{N}_{ij}'X}^{(p)})\right.$$
$$\left. -\;\max_{p\in Q_Y}(-\mathcal{E}(p\mid y_{\mathcal{N}_{ij}'Y}, x_{ij}))\right\}. \tag{26}$$

Using (11) and (14), (26) can be written as

$$\hat{x}_{ij}^{(p+1)} = \arg\max_{x_{ij}\in Q_X}\left\{\; -\;\sum_{C\in\mathcal{C}_{ij}^Y}\mathcal{E}(y_{ij}\mid y_{C'}, x_{ij}) - \sum_{C\in\mathcal{C}_{ij}^X}\mathcal{E}(x_{ij}\mid x_{C'})\right.$$
$$\left. -\;\max_{p\in Q_Y}\{-\sum_{C\in\mathcal{C}_{ij}^Y}\mathcal{E}(y_{ij}\mid y_{C'}, x_{ij})\}\right\}. \tag{27}$$

Provided that the doubleton clique energies are given by (17)-(21), the local updating rule (27) contains three binary operators, δ, max, and arg max. The δ-operator by its definition given in (22) is the complement of the XOR-operator which can be implemented simply by a two-layer feedforward or a recurrent neural network. They lie between the LGN-cells and the S_1-cells of Fig. 1, however for the sake of simplicity they have not been shown in the figure. The max- and (arg max)-operators can be implemented by recurrent neural networks as shown in Fig.1. The networks consisting of S_1-, S_2- and S_3-cells implement the max-operator and the network consisting of T_1- and T_2-cells implements the (arg max)-operator. The S-cells are linear cells which their outputs are the sums of their inputs and the T-cells are threshold cells.

6 Simulation Results

The neural-network implementable version of the parallel algorithm defined by the local updating rule (27) was applied to a rectangular receptive field consisting of 7×5 LGN-cells.

Regarding to the initialization problem, a close look at Eqs. (8) and (16) reveals that (8) degenerates to (16), when the doubleton clique parameter associated with $X_{\mathcal{L}}$ is equal to zero. This led us to get around the initialization problem by starting with $\gamma = 0$ and then increase it as the algorithm proceeds until it saturates to a pre-assigned value. We used the following rule

$$\gamma^{(p)} = \gamma_{\infty} \tanh p. \tag{28}$$

The free boundary scheme (Geman and Geman, 1984) was used and the parameters β and γ_{∞} were set to 1.0 and 2.0, respectively. The simulation results has been shown in Fig. 1.

The first column of Fig. 2 displays the LGN-cells' activity patterns $y_{\mathcal{L}}$ where the firing cells has been painted black. The rest of the columns from left to right display the activity patterns of the I-cells $\hat{x}_{\mathcal{L}}^{(0)}, \hat{x}_{\mathcal{L}}^{(1)}, \ldots, \hat{x}_{\mathcal{L}}^{(4)}$, respectively. The I-cells' activity patterns have been shown up to the point beyond that there were no changes.

References

Besag, J.E. (1974). Spatial interaction and the statistical analysis of lattice systems. J. Roy. Stat. Soc. ser. B, 36, 192-226.

Besag, J.E. (1986). On the statistical analysis of dirty ictures. J. Roy. Stat. Soc. ser. B, 48, 259-302.

Colonnier, M. (1968). Synaptic patterns on different cell types in the different laminae of the cat visual cortex. An electron microscope study. Brain Res., 9, 268-187.

Hubel, D.H., and Wiesel, T.N. (1968). Receptive fields and functional architecture of monkey striate cortex. Journal of Physiology, 195, 215-243.

Geman, S. and Geman, D. (1984). Stochastic relaxation, Gibbs distributions, and the Bayesian restoration of images. IEEE Trans. on Pattern Anal. Mach. Intell., 6, 721-741.

Marr , D. (1982). Vision:A computational investigation into the human representation and processing of visual information. San Francisco, CA: Freeman.

Shirazi, M.N. and H. Noda, H. (1993). A deterministic iterative algorithm for HMRF-textured image segmentation. Proceedings of International Joint Conference on Neural Networks, Nagoya Japan, 3, 2189-2194.

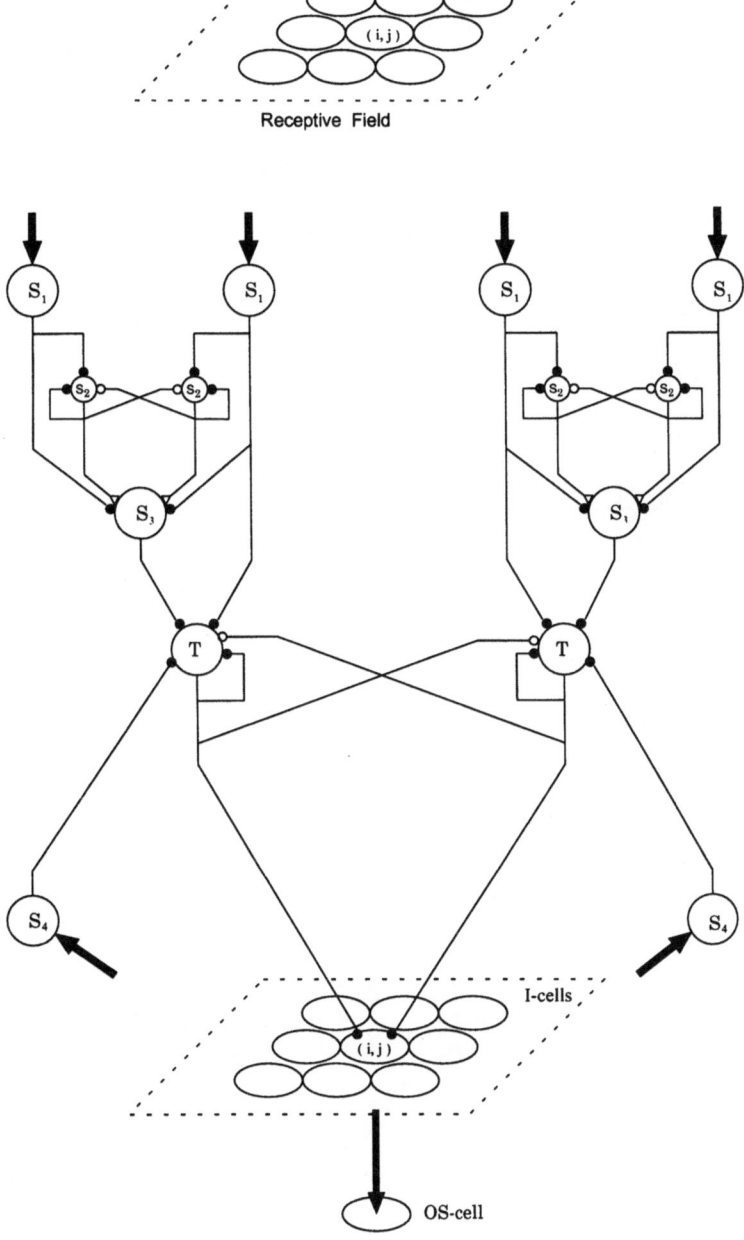

Figure 1: A Neural implementation.

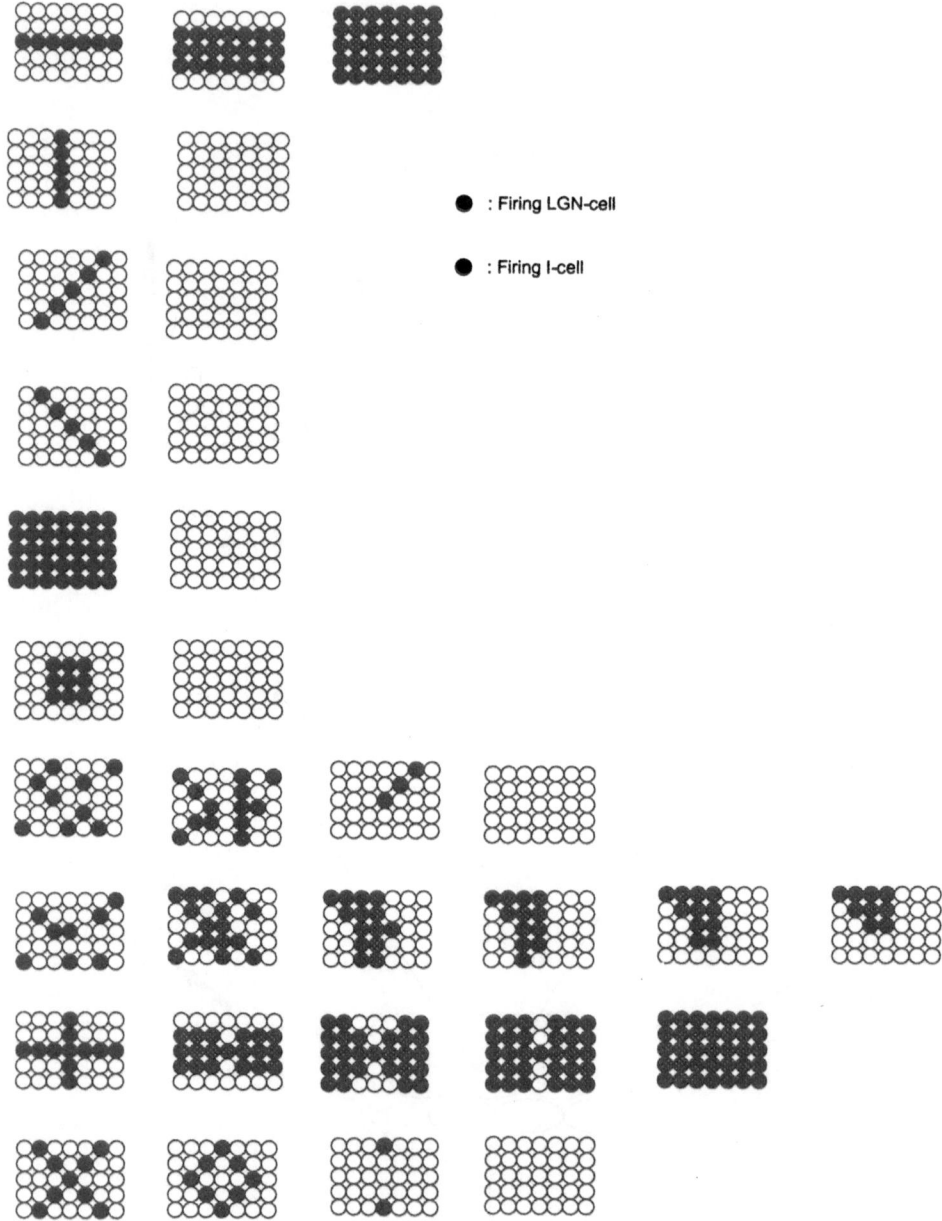

Figure 2: The simulation results. The first column displays the LGN-cells' activity patterns, $y_{\mathcal{L}}$ whereas the rest from left to right display the activity patterns of the I-cells $\hat{x}_{\mathcal{L}}^{(0)}, \hat{x}_{\mathcal{L}}^{(1)}, \ldots, \hat{x}_{\mathcal{L}}^{(4)}$, respectively.

Competition Between Feed-Forward and Lateral Information Processing in Layered Neural Networks

A.C.C. Coolen[1] and L. Viana[2]

[1] Department of Mathematics, King's College London
Strand, London WC2R 2LS, U.K.
[2] Laboratorio de Ensenada, Instituto de Física, UNAM
A. Postal 2681, Ensenada B.C., Mexico

Abstract. We use replica theory to analyse layered networks of binary neurons, with lateral Hebbian-type synapses within individual layers, in combination with strictly feed-forward Hebbian-type synapses between successive layers. The competition between two qualitatively different modes of operation, feed-forward versus lateral, induces interesting transitions, and allows for the identification of an optimal value for the balance between the two synapse types.

1 Introduction

The operation of networks with purely feed-foward interactions differs fundamentally from that of networks with recurrent interactions. In feed-forward networks the final state depends only on stimuli from previous layers, not on the initial state. The final state of recurrent networks, on the other hand, often depends strongly on the initial state. Both have specific advantages and disadvantages, and it seems sensible to try to have the best of both worlds by building networks with feed-forward and lateral interactions. Feed-forward interactions provide information transport, but do not enforce temporal continuity; inputs are responded to strictly, even when wildly changing over time. Lateral interactions introduce a dependence on the system's history, and thereby enforce temporal continuity. The optimal balance between feed-forward and lateral processing appears to be the onset of ergodicity breaking, where temporal continuity is maximised under the contraint that no attractors have yet been created. Here we present and solve a model in which this competition between feed-forward and lateral processing can be studied analytically.

2 Model Definitions

We study feedforward chains of L recurrent layers, with N binary neurons each. The variable $\sigma_i^\ell \in \{-1, 1\}$ denotes the state (non-firing/firing) of neuron i in layer ℓ, and the collective state of the N neurons in layer ℓ is written as $\sigma^\ell = (\sigma_1^\ell, \ldots, \sigma_N^\ell)$. The dynamics is a sequential stochastic alignment of the neurons

σ_i^ℓ to post-synaptic potentials h_i^ℓ, to which only neurons from within layer ℓ (via lateral interactions) and from the previous layer $\ell - 1$ (via feed-forward interactions) can contribute:

$$\text{Prob}[\sigma_i^\ell \to -\sigma_i^\ell] = \frac{1}{2}[1 - \tanh(\beta \sigma_i^\ell h_i)] \qquad h_i^\ell = \sum_{j \neq i} J_{ij}^\ell \sigma_i^\ell + \sum_j W_{ij}^\ell \sigma_i^{\ell-1} \quad (1)$$

The parameter $\beta = T^{-1}$ controls the amount of noise.

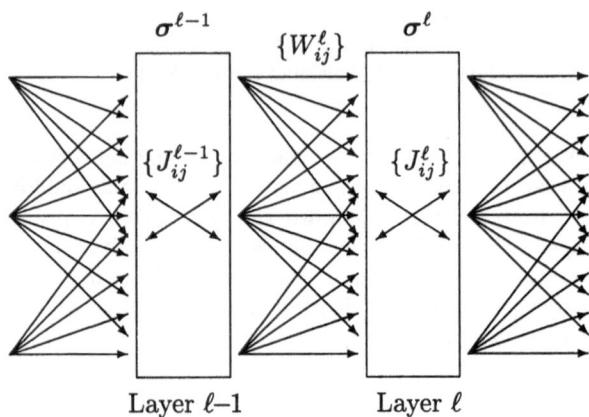

Layer ℓ–1   Layer ℓ

The $2L \times N^2$ synaptic interactions are defined as the result of the system having learned $p = \alpha N$ specific neural configurations (patterns) with a Hebbian-type rule:

$$J_{ij}^\ell = \frac{J_0}{N} \sum_{\mu=1}^{\alpha N} \xi_i^{\mu,\ell} \xi_j^{\mu,\ell} \qquad W_{ij}^\ell = \frac{J}{N} \sum_{\mu=1}^{\alpha N} \xi_i^{\mu,\ell} \xi_j^{\mu,\ell-1} \quad (2)$$

with $\xi_i^{\mu,\ell} \in \{-1, 1\}$ denoting bit i in layer ℓ of pattern μ. For simplicity all patterns are drawn at random, and N is assumed to be large (eventually $N \to \infty$). The parameters (J_0, J) control the relative strength of the two interaction types.

For $J = 0$ we find decoupled symmetric attractor networks, which can be solved in equilibrium (Amit, Gutfreund and Sompolinsky, 1985), showing (for small T) a breakdown at $\alpha_c \sim 0.138$, and non-trivial ergodicity breaking. For $J_0 = 0$ we have feedforward synapses only, and one can solve for the L layers iteratively (Domany, Meir and Kinzel, 1986), showing (for small T) a breakdown at $\alpha_c \sim 0.269$, and only trivial ergodicity breaking. When *both* interaction types are present, neither of the two simplifying properties (interaction symmetry and lack of recurrence, respectively) can be exploited. However, at $T = 0$ all layers will, in order of layer label, eventually go to a stationary state, so for each layer ℓ there will be a stage after which the input from layer $\ell-1$ is stationary, enabling a replica-theoretic analysis in the $T \to 0$ limit. This analysis will be slightly more complicated than usual for attractor networks with external inputs, due to the fact that each pattern bit to be averaged out appears both in the lateral and in the feed-forward interactions. Finally, for technical reasons we will take the limit $T \to 0$ after the limit $N \to \infty$.

3 Solution with Replica Theory

Once for layer ℓ the external terms in the potentials h_i^ℓ (1) are stationary, the dynamics obeys detailed balance and we know the state probabilities in equilibrium:

$$p_\infty(\sigma^\ell) \sim e^{-\beta H(\sigma^\ell)} \qquad H(\sigma^\ell) = -\frac{1}{2} \sum_{ij} J_{ij}^\ell \sigma_i^\ell \sigma_j^\ell - \sum_{ij} \sigma_i^\ell W_{ij}^\ell \sigma_i^{\ell-1}$$

From the free energy $f = -(\beta N)^{-1} \log \sum_{\sigma \in \{-1,1\}^N} e^{-\beta H(\sigma)}$ all relevant observables are generated by differentiation. For large N it is self-averaging with respect to pattern details, i.e. $\lim_{N\to\infty} f = \lim_{N\to\infty} \langle f \rangle_\xi$, so its calculation can be simplified by performing the pattern average, with the replica identity $\log z = \lim_{n\to 0} \frac{1}{n}[z^n - 1]$. After several manipulations, assuming only a finite number k of patterns to have an $\mathcal{O}(1)$ overlap with the system state σ^ℓ, and taking the limit $N \to \infty$, we obtain:

$$f = \frac{1}{2}\alpha J_0 - T \log 2 - \lim_{n\to 0} \frac{1}{\beta n} \, \text{extr} \left\{ i \sum_{\alpha\beta} \hat{q}_{\alpha\beta} q_{\alpha\beta} - \frac{1}{2}\beta J_0 \sum_\alpha \sum_{\mu=1}^k [m_\alpha^\mu]^2 \right.$$

$$-\frac{1}{2}\alpha \log \det \left[\mathbf{I} - \beta J_0 q \right] + \frac{1}{2}\beta^2 J^2 [\sum_{\mu > k} \tilde{m}_\mu^2] \sum_{\alpha\beta\gamma} q_{\alpha\beta} \left[\mathbf{I} - \beta J_0 q \right]_{\beta\gamma}^{-1}$$

$$+ \sum_{\xi \in \{-1,1\}^k} \log \sum_{\sigma \in \{-1,1\}^n} e^{\beta \sum_\alpha \sigma^\alpha \sum_{\mu=1}^k [J_0 m_\alpha^\mu + J\tilde{m}_\mu]\xi_\mu - i \sum_{\alpha\beta} \sigma_\alpha \hat{q}_{\alpha\beta} \sigma_\beta} \left. \right\} \tag{3}$$

with $\tilde{m}_\mu = \frac{1}{N}\sum_i \xi_i^{\mu,\ell-1} \sigma_i^{\ell-1}$. The extremum in (3) refers to the saddle point (by variation of the parameters $\{\hat{q}_{\alpha\beta}, q_{\alpha\beta}, m_\alpha^\mu\}$), which for $n \geq 1$ minimises f. Equation (3) simplifies if we make the replica-symmetric ansatz, i.e. if we assume that non-ergodicity will take the form of large ergodic subspaces, giving $q_{\alpha\beta} = \delta_{\alpha\beta} + q[1 - \delta_{\alpha\beta}]$ and $m_\alpha^\mu = m_\mu$ ($\mu \leq k$). We can now work out the saddle-point equations and establish a recurrent relation to deal with $\sum_{\mu > k} \tilde{m}_\mu^2$, giving recurrent relations $(m, q, r) \to (m', q', r')$ for the order parameters in subsequent layers $\ell - 1 \to \ell$:

$$m' = \langle \xi \int Dz \, \tanh \beta \left[\xi \cdot (J_0 m' + Jm) + z\sqrt{\alpha r'} \right] \rangle_\xi \tag{4}$$

$$q' = \langle \int Dz \, \tanh^2 \beta \left[\xi \cdot (J_0 m' + Jm) + z\sqrt{\alpha r'} \right] \rangle_\xi \tag{5}$$

$$r'\left[1-\beta J_0(1-q')\right]^2 - J_0^2 q' = \beta^2 J^2 r(1-q)^2 - J^2 q + \frac{J^2(1+q)}{1-\beta J_0(1-q)} \tag{6}$$

with $\langle G(\boldsymbol{\xi})\rangle_{\boldsymbol{\xi}} = 2^{-k}\sum_{\boldsymbol{\xi}\in\{-1,1\}^k} G(\boldsymbol{\xi})$ and $Dz = (2\pi)^{-\frac{1}{2}}e^{-\frac{1}{2}z^2}dz$. The overlaps $m_\mu = \frac{1}{N}\sum_i \xi_i^{\mu,\ell}\sigma_i^\ell$ measure the resemblance between system state and pattern μ in layer ℓ, and $\boldsymbol{m} = (m_1,\ldots,m_k)$. The only exception is the case where $\ell = 2$ and the first layer is clamped into a configuration $\boldsymbol{\sigma}^1$, in which case (6) is to be replaced by

$$r'\left[1-\beta J_0(1-q')\right]^2 - J_0^2 q' = J^2 \tag{7}$$

Equations (4-7) constitute the solution of our model. We will now analyse their consequences for $T \to 0$, and validate their predictions with simulation experiments. We eliminate the parameter redundancy by putting $J_0 = \frac{1}{2}[1+\omega]$ and $J = \frac{1}{2}[1-\omega]$.

4 Transitions

We first calculate the information storage capacity α_c for long chains. A stable state is reached along the chain when $(\boldsymbol{m},q,r) = (\boldsymbol{m}',q',r')$ in (4-6), giving

$$\boldsymbol{m} = \langle\boldsymbol{\xi}\int Dz\ \tanh\beta\left[\boldsymbol{\xi}\cdot\boldsymbol{m}+z\sqrt{\alpha r}\right]\rangle_{\boldsymbol{\xi}} \qquad q = \langle\int Dz\ \tanh^2\beta\left[\boldsymbol{\xi}\cdot\boldsymbol{m}+z\sqrt{\alpha r}\right]\rangle_{\boldsymbol{\xi}}$$

$$r = \frac{(1-\omega)^2+q(1+\omega)^2-2\omega\beta q(1+\omega)(1-q)}{4[1-\frac{1}{2}\beta(1+\omega)(1-q)]\left[1-\beta(1+\omega)(1-q)+\omega\beta^2(1-q)^2\right]}$$

Upon taking the limit $T \to 0$ and concentrating on pure states, where $m_\mu = m\delta_{\mu\lambda}$, we can perform the averages and integrations, substitute $x = m/\sqrt{2\alpha r}$ and reduce the above set of equations to a single transcendental equation (with $m = \mathrm{erf}(x)$):

$$x\sqrt{2\alpha} = \frac{\mathrm{erf}(x) - \frac{2x}{\sqrt{\pi}}e^{-x^2}}{\sqrt{\frac{1}{2}(1+\omega^2)}}\left\{\frac{\left[\mathrm{erf}(x)-\frac{2\omega x}{\sqrt{\pi}}e^{-x^2}\right]\left[\mathrm{erf}(x)-\frac{(1+\omega)x}{\sqrt{\pi}}e^{-x^2}\right]}{\left[\mathrm{erf}(x)-\frac{2x}{\sqrt{\pi}}e^{-x^2}\right]\left[\mathrm{erf}(x)-\frac{\omega^2+\omega}{\omega^2+1}\frac{2x}{\sqrt{\pi}}e^{-x^2}\right]}\right\}^{\frac{1}{2}} \tag{8}$$

The storage capacity α_c is the value for α for which the nonzero solutions of (8) vanish, resulting in figure 1 (left picture). For $\omega = -1$ and $\omega = 1$, equation (8) reduces to the results of Domany et al (1986) and Amit et al (1985), as it should. The optimal balance turns out to be $\omega \sim -0.12$, giving $\alpha_c \sim 0.317$. We support our analytical results with numerical simulations on systems with $L = 60$ layers of $N = 900$ neurons, for $T = 0$ and $\omega = 0$. Figure 1 (right picture) shows the equilibrium values of the overlap m_1 as a function of the layer number. For $\alpha \in \{0.26, 0.28\}$ the desired state $m_1 \sim 1$ appears stable, for $\alpha \in \{0.33, 0.35\}$ the state $m_1 \sim 1$ is unstable, whereas for $\alpha \in \{0.29, 0.30\}$ we appear to be close to the critical value, with $m_1 \sim 1$ appearing stable initially, but eventually destabilising further down along the chain. This is in reasonable

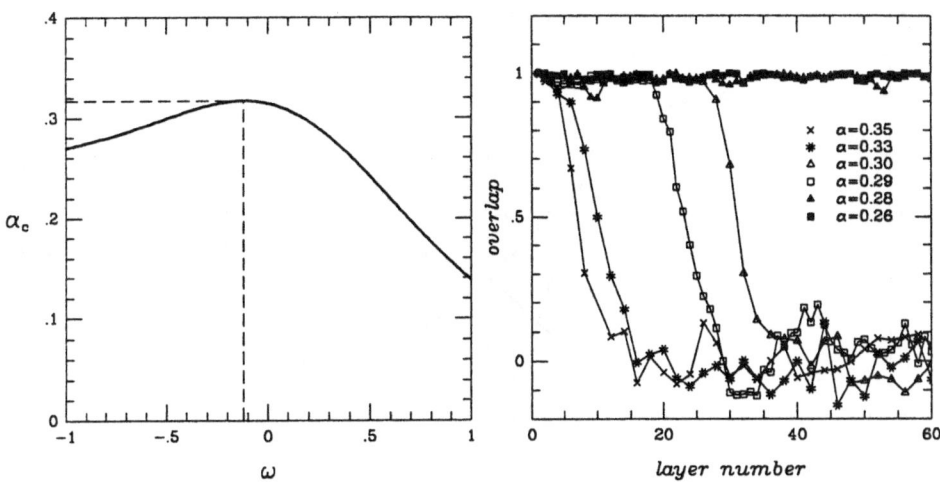

Fig. 1. Left picture - theory: storage capacity α_c for long chains at $T = 0$ (dashed: location of the optimum). Right picture - numerical simulations ($N = 900$): equilibrium overlap m_1 as a function of layer number ℓ, for $T = 0$ and $\omega = 0$.

agreement with the theory, which predicts $\alpha_c \sim 0.314$ for $\omega = 0$, if we take finite size effects into account. Next we turn to transitions marking the appearance of multiple coexistent stable states. We restrict ourselves to pure states, and to the junction between layer $\ell = 1$ and layer $\ell = 2$, where the first such transitions are expected. Here we present results only for the case where the first (input) layer is clamped into a state with overlap m_{in}. Insertion of pure state solutions, and taking the limit $T \to 0$ in (4-6) again allows us to perform the averages and integrations. Substitution of $y = [J_0 m' + J m_{\text{in}}]/\sqrt{2\alpha r'}$ now leads to the transcendental equation (with $m' = \text{erf}(y)$):

$$y\sqrt{2\alpha}\left[1 + \left(\frac{1-\omega}{1+\omega}\right)^2\right]^{\frac{1}{2}} - m_{\text{in}}\left[\frac{1-\omega}{1+\omega}\right] = \text{erf}(y) - \frac{2y}{\sqrt{\pi}}e^{-y^2} \qquad (9)$$

Derivation with respect to y subsequently allows us to identify the condition for new solutions of (9) to bifurcate as

$$\alpha = \frac{2}{\pi}y^4 e^{-2y^2}\frac{(1+\omega)^2}{1+\omega^2} \qquad m_{\text{in}}\left[\frac{1-\omega}{1+\omega}\right] = \frac{2y}{\sqrt{\pi}}(1+2y^2)e^{-y^2} - \text{erf}(y) \qquad (10)$$

These equations (10), to be solved simultaneously, give the value(s) $\alpha_{\text{bif}}(m_{\text{in}}, \omega)$ for which multiple states come into existence. The result is shown in figure 2. The left picture shows the transition lines for input overlap $m_{\text{in}} = 1.0$ (as an example): left and above the solid lines is the region with a unique stable state, at the other side of the solid lines (close to the $\omega = 1$ axis) multiple stable states exist. The right picture shows the result of combining all regions with multiple stable states, for $m_{\text{in}} \in [0, 1]$. Left of the solid line there is a single stable state

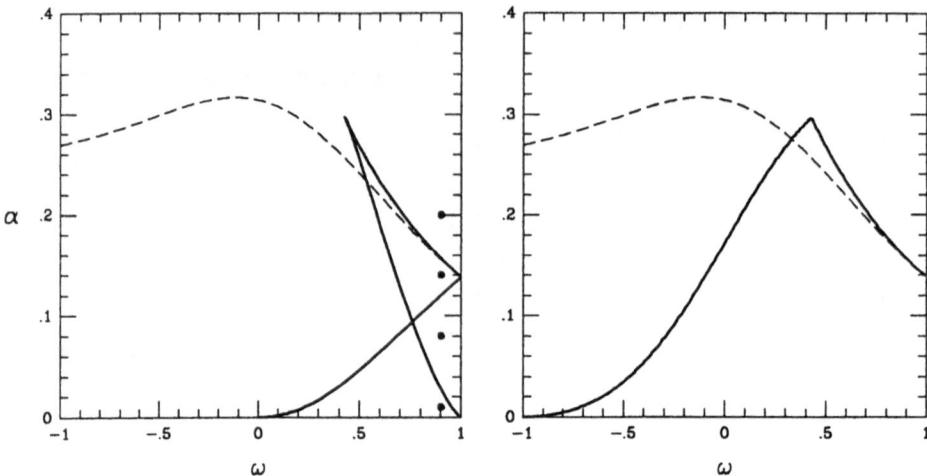

Fig. 2. Left picture: bifurcation lines for $m_{\text{in}} = 1.0$. Left of the solid lines: single stable state in layer 2; close to the $\omega = 1$ line: multiple stable states. Right picture: boundary of the cumulative region where multiple stable states exist (obtained by combining the results for input overlaps $m_{\text{in}} \in [0,1]$). Dashed: large-chain storage capacity. Dots in left picture: parameters for numerical simulations described below.

in the second layer, irrespective of m_{in}, right of the solid line input overlaps $m_{\text{in}} \in [0,1]$ can be found for which multiple stable states exist. These results are again supported by numerical simulations: figure 3 shows the relation between initial and (almost) final overlaps in layer 2, for $N = 12,000$ and the four (ω, α) combinations indicated with dots in figure 2. The number of stable states predicted (figure 2) are $(2,3,2,1)$, for the four cases $\alpha = (0.01, 0.08, 0.14, 0.20)$, respectively. This is in perfect agreement with the simulation results of figure 3, where for each graph the number of stable states is the number of discontinuities plus one. For a system in equilibrium one expects horizontal line segments between the discontinuities; apparently the state at $t = 200$ is not an equilibrium state yet (due to the large relaxation times involved).

The third and final transition to be analysed is the de so-called AT-line, where the relevant saddle-point of (3) ceases to be replica-symmetric, marked by an instability with respect to so-called replicon fluctuations $q_{\alpha\beta} \to \delta_{\alpha\beta} + q[1-\delta_{\alpha\beta}] + \eta_{\alpha\beta}$ (with $\sum_\alpha \eta_{\alpha\beta} = 0$). Working out the mathematics for the present model leads to the following equation, to be solved in combination with (4-7):

$$1 = \frac{\alpha(\beta J_0)^2}{[1 - \beta J_0(1-q)]^2} \langle \int Dz \ \cosh^{-4} \beta \left[\boldsymbol{\xi} \cdot (J_0 \boldsymbol{m}' + J\boldsymbol{m}) + z\sqrt{\alpha r'} \right] \rangle_{\boldsymbol{\xi}} \qquad (11)$$

The solution is shown in figure 4 for $\omega \in \{-0.5, 0, 0.5, 1.0\}$ (for $\omega = -1$ the AT line collapses onto the line $T = 0$). As we move away from the fully recurrent case $\omega = 1$, the noise level T at which replica symmetry breaks decreases monotonically, so that we can be confident that for the present model, as for the fully

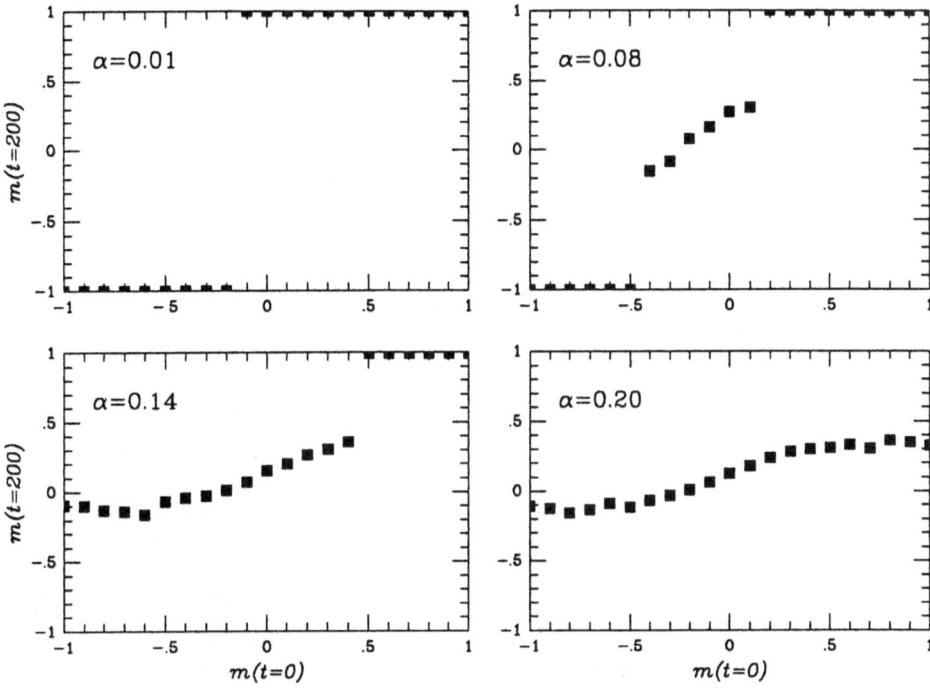

Fig. 3. Simulation results ($N = 12,000$), indicating the number of stable states in layer 2. Each picture shows the overlap at time $t = 200$ versus the overlap at $t = 0$, for $\omega = 0.9$ and $m_{in} = 1$. The α values correspond to the dots in figure 2.

recurrent case, replica symmetry breaking effects will be of a minor quantitative nature only.

5   Summary

We have studied the competition between lateral and feed-forward information processing in a feedforward chain of recurrent neural networks with Hebbian-type synapses. This model can be solved analytically, in spite of the absence of the simplifying properties like synaptic symmetry or strictly feed-forward projection on which analysis of this type of model usually relies (see e.g. Engel et al, 1989, Fassnacht and Zippelius, 1991, Kurchan et al, 1994, and also the more recent studies by Kurchan et al, 1994, English et al, 1995, and Wong et al, 1995). In two extreme limits, fully recurrent and fully feed-forward operation, respectively, the existing solutions of Amit et al (1985) and Domany et al (1986) are recovered correctly. In the intermediate regime, we have calculated the relevant transitions in the parameter plane: saturation breakdown, ergodicity-breaking and replica-symmetry-breaking. Our results, which are supported by numerical

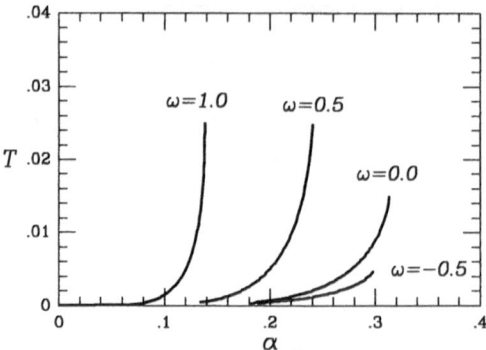

Fig. 4. Location of the replica symmetry breaking transition in the (α, T) plane.

simulations, allow for the identification of an optimal balance between lateral and feed-forward synapses (see e.g. figure 2). Recent experimental evidence suggests that in certain brain regions the competition between feed-forward and lateral processing is actively controlled by neuromodulators[3]; our study might also play a role in describing such processes (upon suitable adaptation of model ingredients). For simplicity we have restricted our analysis to binary neurons, random patterns and uniform layer sizes, extensions to other neuron types, pattern types or different layer sizes are straightforward.

Acknowledgements We wish to thank the McDonnel-Pew Foundation (Visiting Fellowship, LV) and the National University of Mexico (UNAM) (DGAPA Project IN100294) for support

References

1. Amit D.J., Gutfreund H. and Sompolinsky H. (1985) *Phys. Rev. Lett.* **55** 1530
2. Domany E., Meir R. and Kinzel W. (1986) *Europhys. Lett.* **2** 175
3. Fassnacht C. and Zippelius A. (1991) *Network* **2** 63
4. Kurchan J., Peliti L. and Saber M. (1994) *J. Phys I (France)* **4** 1627
5. Engel A., English H. and Schütte A. (1989) *Europhys. Lett.* **8** 393
6. English H., Mastropietro V. and Tirozzi B. (1995) *J. Phys. I France* **5**, 85
7. Wong, K.Y.M., Campbell, C. and Sherrington D. (1995) *J. Phys. A* **28**, 1603

---

[3] P.F.M.J. Verschure (private communication)

Computing Functions with Spiking Neurons in Temporal Coding

Berthold Ruf

Institute for Theoretical Computer Science, Technische Universität Graz
Klosterwiesgasse 32/2, A-8010 Graz, Austria
E-mail: bruf@igi.tu-graz.ac.at
Homepage: http://www.cis.tu-graz.ac.at/igi/bruf

Abstract. For fast neural computations within the brain it is very likely that the timing of single firing events is relevant. Recently Maass has shown that under certain weak assumptions functions can be computed in temporal coding by leaky integrate-and-fire neurons. Here we demonstrate with the help of computer simulations using GENESIS that biologically more realistic neurons can compute linear functions in a natural and straightforward way based on the basic principles of the construction given by Maass. One only has to assume that a neuron receives all its inputs in a time intervall of approximately the length of the rising segment of its excitatory postsynaptic potentials. We also show that under certain assumptions there exists within this construction some type of activation function being computed by such neurons, which allows the fast computation of arbitrary continuous bounded functions.

1 Introduction

There exist several models for explaining how the brain can perform computations and how the underlying values are encoded. Especially for fast processing tasks it is very likely that the timing of single spikes is relevant [8, 9, 10] and can be very precise [3]. For example Thorpe and Imbert have shown that humans can classify reliably patterns within 100 msec. Because of the architecture of the visual cortex and the firing rates of the neurons involved it is clear that there is not sufficient time to sample firing rates [9]. Instead of encoding values in firing rates, Thorpe suggests considering the relative order of firing times of different neurons within a certain time window [10]. Recent results by Markram and Tsodyks showed that a synapse is much more sensitive to the first stimulus in a train of stimuli, depressing its response very fast to subsequent stimuli and that this effect is increased by long-term potentiation [6]. This might indicate that for a neuron, which receives stimuli through several of its synapses, the onset of those stimuli encodes the relevant information.

In a theoretical work, Maass has recently shown how one can compute functions with leaky integrate-and-fire neurons in temporal coding [4] assuming a linear initial segment of the excitatory postsynaptic potentials (EPSP's). This computation is based on the idea that one can make a neuron fire such that its firing time represents the weighted sum of its analogue-valued inputs, given by the firing times of its input neurons.

In this paper we focus on the question in how far the basic principles of this construction still can be used for biologically more realistic models, i.e. where the EPSP's are described by some α-function and where one also considers the nonlinear effects in the dendrites and Hodgkin-Huxley kinetics at the soma. With the help of the "general neural simulation system" (GENESIS) by Bower et Beeman [1] we demonstrate in section 2 that such a neuron can compute in a very natural way linear functions in temporal coding, only assuming that the neuron receives all its inputs within a certain time window.

In section 3 we investigate the question how the firing time of such a neuron relates to the weighted sum of its temporally encoded inputs. We demonstrate that there is a functional dependence in a way which gives rise to an easy possibility of computing arbitrary bounded continuous functions within temporal coding.

2 Computing linear functions

Assuming that the initial segments of the EPSP's are linear, leaky integrate-and-fire neurons can compute linear functions in the following way, as shown in [4]: We consider a neuron v, which receives excitatory input from neurons $u_1, \ldots, u_n$. The corresponding weights w_i and input values $s_i \in [0, T_{in}], 1 \leq i \leq n$ are given, with T_{in} being a sufficiently small constant. The goal is to make v fire at a time which is determined by $\sum w_i s_i$. We assume that the delays, i.e. the difference between the firing time of the presynaptic neuron and the time the resulting EPSP starts to increase, are all equal. Each neuron $u_i, 1 \leq i \leq n$ fires exactly once at time $t_i = T_{in} - s_i$. Furthermore v receives excitatory input from an additional neuron u_0, which fires at time $t_0 = T_{in}$ and thus produces the input $s_0 = 0$. For proper choices of T_{in} and w_0 such that v fires at a time t_v, when all its EPSP's are still in their linear segments, it is easy to see that for t_v the following holds:

$$\Theta = \sum_{i=0}^{n} w_i(t_v - t_i)$$

where Θ is the threshold of v. Hence

$$
\begin{aligned}
t_v &= \frac{\Theta + \sum_{i=0}^{n} w_i t_i}{\sum_{i=0}^{n} w_i} \\
&= \frac{\Theta + \sum_{i=0}^{n} w_i(T_{in} - s_i)}{\sum_{i=0}^{n} w_i}
\end{aligned}
\tag{1}
$$

and t_v is (up to constants) determined by $\sum_{i=0}^{n} w_i s_i$. Thus this weighted sum is computed by v in a very natural way in this construction, we only have to assume that the input spikes arrive sufficiently close together and that w_0 is chosen properly.

The question arises in how far biologically more realistic models of neurons, which can only approximate the abovementioned assumptions, can perform this computation. GENESIS was used in order to find out if the time t_v still depends in a linear way on the inputs s_i where nonlinear effects could be caused by

- the interaction of different EPSP's in the dendritic tree
- the Hodgkin-Huxley kinetics at the soma of neuron v
- various types of noise.

Neuron v was built out of $n + 3$ compartments: the soma, the root of the dendritic tree and for each of the $n + 1$ inputs a separate dendritic branch. Each input was made fire exactly once at a random time point within a certain time intervall of size Δt. All weights including w_0 were set to random values within a certain intervall and then multiplied with a scaling factor, in order to guarantee that the incoming EPSP's suffice to make v fire. k trials with one fixed weight setting but varying input firing times t_i were performed in order to find out if the resulting output firing times depend linearly on these t_i. We used linear regression for this and measured the quality of the linear regression with the coefficient of determination. It is defined as the ratio between the variance of the output values estimated by the regression and the variance of the observed output values. The closer this coefficient is to one, the closer the linear regression explains the observed values.

The quality of the results depended strongly on Δt and the size of the weights. Figure 1 shows, how the coefficient of determination depends on the scaling of the weights with $\Delta t = 1.5 msec$. For this experiment, neuron v received 10 synaptic inputs, for each synapse the maximum sodium conductivity (i.e. the weight) was set initially to a random value between 0 and $0.31 nS$. 100 experiments with varying input firing times were performed and then the resulting regression function and its coefficient of determination computed. These experiments were repeated with different weight scalings, i.e. all weights were multiplied with a scaling factor, which was varied between 1 and 100. For too small factors, v did not fire at all. Hodgin-Huxley kinetics at the soma generally deteriorate the linear behavior. However, as can be seen in figure 1 even in this case due to the very small Δt the behavior of v can be described as linear.

It is dubious if biological neurons can fire with such a high precision. If one relaxes the timing constraints on the input neurons to $\Delta t = 3 msec$ (approximately the length of the rising phase of the EPSP's in these experiments) the range of weights where the computation of the neuron can be described as linear becomes smaller because the firing time of v is more strongly influenced by the non-linear segments of the EPSP's and by the fact that for strong weights the peak of the resulting EPSP depends sublinearly on the weights. For small factors which suffice to make v fire, no linear dependency could be observed, because the potential of v increases in this case rather slowly. Thus many EPSP's will have reached their non-linear saturation segment when v fires. Figure 2 shows that nevertheless linear computations are possible within a wide range of weight values.

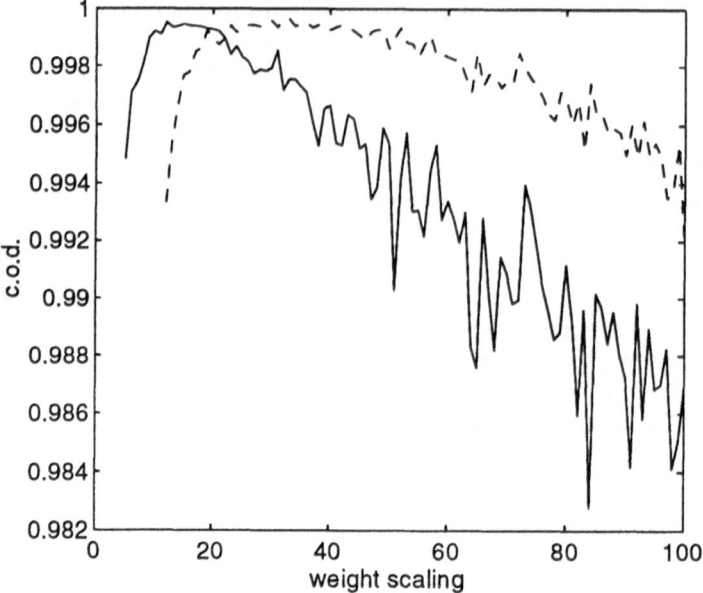

Fig. 1. Coefficient of determination in dependence of the weight scaling with $\Delta t = 1.5msec$. The solid (dashed) line shows the results for the case where the soma contained (no) Hodgkin Huxley channels.

A Δt which is larger than the length of the rising segment of the EPSP can also be achieved by assuming for example that a presynaptic neuron forms several synapses to the postsynaptic neuron with different delays and weights, such that if the presynaptic neuron fires the changes of the postsynaptic potential result in a long linear increasing segment (see [5] for details).

For the construction presented in this section a proper weight scaling is needed in order to guarantee that v actually fires. However, this assumption is not really necessary if one makes use of sufficiently many auxiliary EPSP's which increase the resting potential of v by a constant amount, such that a smaller input to v now suffices to exceed its threshold.

The results in this section indicate that a neuron can compute in a very simple and straightforward way a linear function in temporal coding. In the next section we investigate how the weighted sum of the temporally encoded inputs is related to the firing time of the output neuron and in how far this can be used to approximate arbitrary functions.

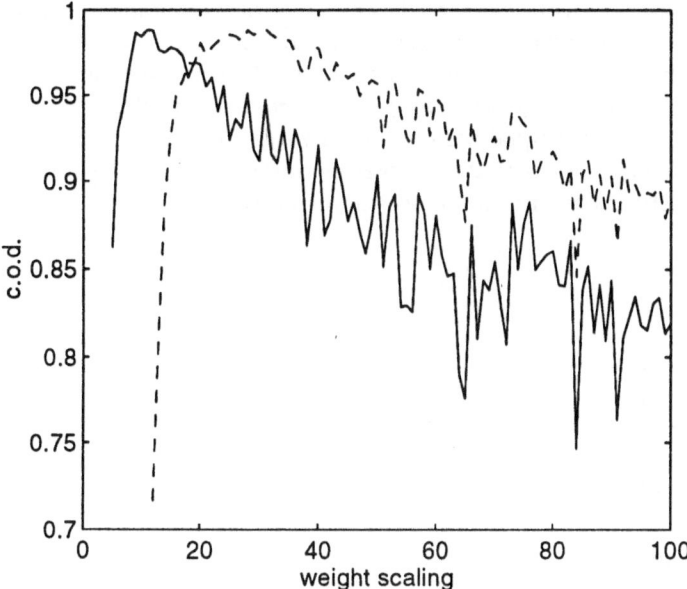

Fig. 2. As in figure 1 but with $\Delta t = 3 msec$. Observe the different scaling of the y-axis.

3 Activation functions

For artificial neural networks one usually considers neurons having a certain activation function σ. Such a neuron computes first the weighted sum $\sum w_i s_i$ of its inputs s_i and delivers as output $\sigma(\sum w_i s_i)$. The most common choice for activation functions are sigmoidal ones, i.e. which are non-decreasing, (almost everywhere) differentiable and where $\lim_{x \to -\infty} \sigma(x)$ and $\lim_{x \to \infty} \sigma(x)$ have finite values. It has been shown that sigmoidal feedforward networks with one layer of hidden units are universal approximators (i.e. they can approximate to any degree of accuracy any given bounded continuous function), if they use activation functions which are not a polynomial [2]. In [4] Maass has shown how the construction described in section 2 can be extended such that non-polynomial sigmoidal activation functions can be computed. Here the question arises whether biologically more realistic neurons have an activation function, i.e. whether there is a functional dependence between the weighted sum $\sum w_i s_i$ and the firing time of v and of what type it might be.

First we observe that given a neuron v with n inputs, fixed weights $w_i, 1 \leq i \leq n$ and fixed inputs $s_i, 2 \leq i \leq n$ the influence of varying s_1 on the firing time t_v of v is basically determined by the shape of the corresponding EPSP. Figure 3 shows the result of a GENESIS simulation, where $n = 10$ with each input being again on a seperate branch of the dendritic tree and s_1 was allowed to vary within a time interval of size $\Delta t = 9msec$. 200 trials were executed, where the output value was computed each time by subtracting the current firing time t_v of v from the largest t_v occuring in this experiment. Note that the output values shown in figure 3 equal zero for small weighted sums, since in this case s_1 is very small, which means that the corresponding input neurons fires very late, such that its EPSP does not influence the firing time of v.

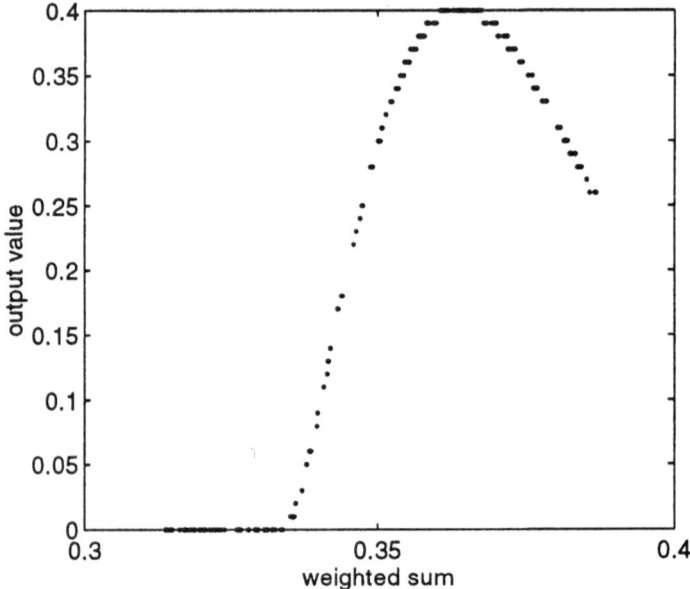

Fig. 3. Dependence of the output value of neuron v from the weighted sum $\sum w_i s_i$, where only one input neuron varied its firing time.

When all the weights are fixed and all s_i may vary within Δt, where Δt is of approximately the length of the rising segment of the EPSP's, the firing time of v should depend on $\sum w_i s_i$ approximately in a linear fashion. As in section 2 this can be only true, if the weights are not too large (large weights would cause non-linear effects) or too small. An example is shown in figure 4.

If one varies both the weights and the inputs, the choices of the intervals where the inputs respectively the weights are chosen from are critical. For Δt being approximately of the length of the rising phase of the EPSP and weights chosen from an interval being within the range where the slope of the EPSP is influenced linearly by the corresponding weight, one can no longer say without

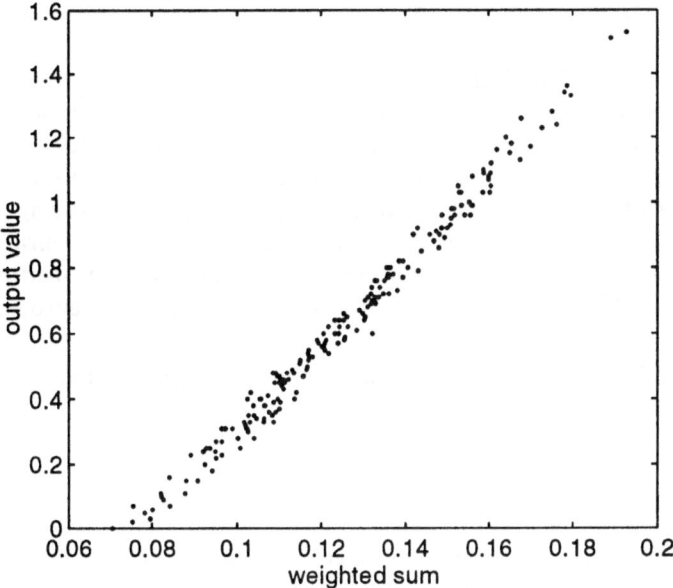

Fig. 4. As figure 3, but where all input neuron could fire at random times within $\Delta t = 3msec$.

any further assumptions that neuron v computes some function in the sense described above: It is possible that for two different sets of weights and inputs s_i, w_i and s_i', w_i' with $\sum w_i s_i = \sum w_i' s_i'$ the resulting firing times of v differ considerably. However, if one increases the lower bound of the weight interval (we have seen in section 2 that small weights cause non-linear effects) and uses a Δt being smaller than 4 msec, one gets a dependence which is approximately linear. We used for our experiments the weight interval $[3.2nS, 4.6nS]$ and $\Delta t = 3msec$, which resulted in a plot very simular to the one shown in figure 4.

All experiments in this section were performed without Hodkin-Huxley channels. As in section 2 including these channels does simply reduce the range of possible parameters and adds some noise to the results, but the type of computations described above is still possible.

We finally note that a linear activation function can be easily extended to a sigmoidal one in the above construction, if one uses additional auxiliary EPSP's and IPSP's which guarantee that v definitely fires within a certain time-intervall (for details see [4]). This gives rise to the universal approximation property mentioned at the beginning of this section.

4 Conclusions

We have shown that biologically realistic models of neurons receiving their inputs within a small time window can compute naturally linear functions very fast and that this idea gives rise to approximating arbitrary functions. These results motivate further research about possible computational models for biological systems, especially within the context of fast information processing. They also make it possible to implement various concepts from artificial neural networks. E.g. in [7] it has been shown, how the construction described in section 2 can be used to realize self-organizing maps with networks of spiking neurons.

Acknowledgements: I would like to thank Wolfgang Maass and Thomas Natschläger for several helpful discussions.

References

1. Bower., J. and Beeman, D.: The book of genesis. Springer-Verlag, New York (1994).
2. Leshno, M., Lin, V.Y., Pinkus, A., and Schocken, S.: Multilayer feedforward networks with a nonpolynomial activation function can approximate any function. Neural Networks **6** 861–867.
3. Mainen, Z.F. and Sejnowski,T.J.: Reliability of spike timing in neocortical neurons. Science **268** (1995) 1503–1506.
4. Maass, W.: Fast sigmoidal networks via spiking neurons. To appear in Neural Computation.
5. Maass, W. and Natschläger, T.: Networks of spiking neurons can emulate arbitrary hopfield nets in temporal coding. Submitted for publication.
6. Markram, H. and Tsodyks, M.: Redistribution of synaptic efficacy between neocortical pyramidal neurons. Nature **382** (1996) 807–810.
7. Ruf, B. and Schmitt, M.: Self-organizing maps with networks of spiking neurons using temporal coding. Submitted for publication.
8. Sejnowski, T. J.: Time for a new neural code? Nature **376** (1995) 21–22.
9. Thorpe, S. J. and Imbert, M.: Biological constraints on connectionist modelling. In: Connectionism in Perspective, Pfeifer, R., Schreter, Z., Fogelman-Soulié, F., and Steels, L., eds., Elsevier, North-Holland (1989).
10. Thorpe, S. J. and Gautrais, J.: Rapid visual processing using spike asynchrony. To appear in: Advances in neural information processing systems 10, Morgan Kaufmann, San Mateo (1997).

An Introduction to Fuzzy State Automata

L.M. Reyneri

Dipartimento di Elettronica - Politecnico di Torino
C.so Duca Abruzzi, 24 - 10129 Torino - ITALY
e.mail reyneri@polito.it; phone ++39 11 568 4038; fax ++39 11 568 4099

Abstract. This paper introduces Fuzzy State Automata for control applications. They are derived from the integration of traditional finite state automata and neuro-fuzzy systems, where a finite state automaton tracks the state of the plant under control and modifies the characteristic of the neuro-fuzzy system accordingly. The main difference with respect to existing systems is that the states of the automata are identified by fuzzy, instead of crisp variables, therefore state transitions and the corresponding controller characteristic are smoother and easier to train or tune.

1 Introduction

Fuzzy logic [2] and neural networks [1] are currently used in several control applications [3] due to their interesting performance. They allow either to include "human knowledge" into a controller design, or to design learning controllers which can be trained "by examples", or both of them (neuro-fuzzy integration). Neuro-fuzzy systems are also applied in several other fields, such as pattern classification, function approximation, and many others.

Unfortunately, all fuzzy systems and most neural networks are feed-forward systems, with no feedback, and therefore with no memory. This limits the applicability of neuro-fuzzy systems, especially in the field of control, where most controllers must have memory.

In particular, recurrent neural networks [1] do have memory, but they are difficult to train and they cannot use the human knowledge directly (this is valid for any neural network), unlike fuzzy system which can learn from human experience but which are seldom used in a recurrent architecture, due to the difficulty to deal with feedback in fuzzy logic. Furthermore, feedback in recurrent networks is often limited to simple variables (for instance, speed, acceleration, etc.).

This paper introduces Fuzzy State Automata, a method to design recurrent neuro-fuzzy systems and to deal in a neuro-fuzzy fashion with sequences of states and events. Section 2 first describes how neuro-fuzzy systems can be interfaced with traditional Finite State Automata, and then introduces the theory of Fuzzy State Automata. Section 3 describes their operating principles, while section 4 describes few applications.

Throughout the paper, FFSA are described bearing in mind their application to control systems, but the proposed theory can also be applied to other domains.

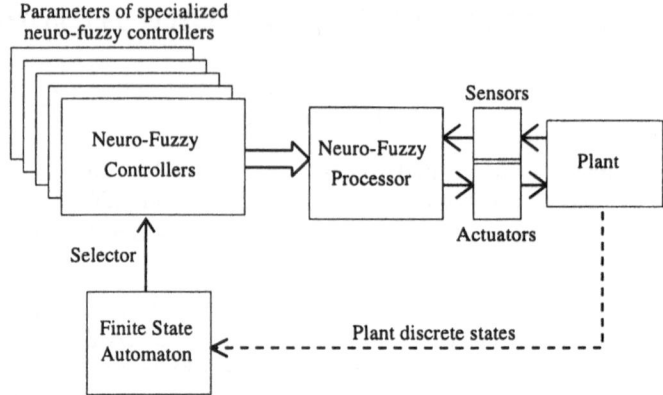

Fig. 1. Interaction of neuro-fuzzy controllers with FSA in control applications.

2 State of the Art

Many plants may have many different states, according to operating conditions. Examples are: the different strokes of strongly non-linear trajectories, control of robot arms with different load conditions; forward and backward steps of walking robots; optimal operation of tooling machines with different tools; etc.

Neuro-fuzzy control methods can be used in conjunction with traditional Finite State Automata (**FSA**) to deal with the different plant states, for the following reasons:

1. fuzzy logic allows to include the available "human knowledge" into the solution of a problem [2], but is not well suited to deal with sequences of events;
2. neural networks may deal with time-dependent signals (with an appropriate training), but they cannot easily acquire the human knowledge;
3. FSA are ideally suited to deal with sequences of events, but they are not good at handling continuous signal and human knowledge;
4. fuzzy systems are mostly feed-forward systems, without feedback, as fuzzy logic is mostly associated with memoryless systems. This is because no method has been found so far to describe efficiently recurrent fuzzy systems;
5. mixing neuro-fuzzy systems with FSA allows to include human knowledge and training capabilities into recurrent systems (therefore, with memory) which can therefore deal with sequence of events, different plant states, etc.

Neuro-fuzzy controllers interacting with FSA have already been used in control applications [5, 6]. The basic idea is sketched in fig. 1: a neuro-fuzzy system controls a plant through a set of sensors and actuators, in a quite well known fashion. A traditional FSA tracks the discrete plant states and modifies the controller characteristic accordingly, by selecting one out of many weight matrices (or knowledge bases, for fuzzy controllers). A different weight matrix (or knowledge base) is associated with each state of the FSA.

Plants operating with a set of well defined states may either be controlled by a single controller, or by a set of simpler controllers interacting with a FSA. The advantages of the second solution are the following:

1. the overall controller is subdivided into a set of simpler controllers; each of these may often be as simple as a linear controller;
2. each controller will be trained only over a limited subset of the state space;
3. each controller has a reduced size and can be implemented in an optimal way, also by using different paradigms for each of them;
4. controllers can be trained independently of each other, therefore training one of them does not affect any of the others.
5.

Unfortunately, traditional FSA interacting with neuro-fuzzy controllers have few major drawbacks:

1. The overall controller characteristic may change abruptly when the FSA has a state transition. This may often cause very high accelerations and jerks which may increase mechanical stresses, vibrations, and often reduce comfort and possibly also the lifetime of the plant.
2. To reduce discontinuities, there is the need for either an additional "smoothing" subsystem (but this often worsens the overall system performance), or the individual controllers should be designed to take care of state changes (but this increases the complexity and the duration of training)

The following sections describe *Fuzzy State Automata* (**FFSA**), which are an alternative solution to the problems listed above. They are derived from the tight integration of neuro-fuzzy systems and more traditional FSA, but they have better performance and are more suited to design smooth characteristics and to deal with sequences of events.

2.1 Fuzzy State Automata

The main difference of FFSA with respect to traditional FSA is that transitions in the automata are not triggered by crisp events but by fuzzy variables, and state transitions are fuzzy as well. It immediately results that, at any time, the whole system is not necessarily in one and only one well-defined state, but it may well be in more states at the same time, each one associated with its own *membership value*.

State transitions are therefore smoother and slower, even if the controllers activated by each state are not designed to smooth the transitions. As a consequence, all the individual controllers may become as simple as a traditional PID, each one designed for a different target specific of that state. The FFSA then takes care of smoothing discontinuities between partial local characteristics, but the overall system is much simpler to design and tune than a traditional controller with similar characteristics.

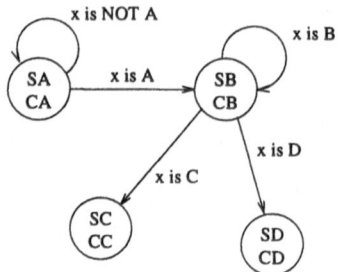

Fig. 2. An example of an FFSA.

Unfortunately FFSA have a drawback, that is, as the automata are often in more than one state, they have to process more than one controller at any time. This increase of computing time often counterbalances the higher speed that can be achieved through the use of simpler controllers. In practice, FFSA achieve approximately the same speed of traditional FSA controlling neuro-fuzzy controllers, but they are easier to design (especially for complex systems) and produces smoother trajectories.

So far, the theory of FFSA has been tested on some applications in the field of control, for instance to design the Motion and Leg Coordination Controls of a walking hexapod [6], where FFSA have been used as a way to specify the cooperation between individual legs. This example is shortly described in section 4.

3 Operation of Fuzzy State Automata

This section describes theory and operation of FFSA, and in particular how FFSA can be converted to fuzzy descriptions and the corresponding mathematical models. The theory will be described by means of the example shown in fig. 2. This is nothing but a simple example with only one input variable, but the diagram could well be part of a more complex FFSA. Furthermore, the theory applies also to FFSA with more than one input.

An FFSA looks very similar to a traditional FSA, in the sense that it can be represented as a collection of *fuzzy states* S_j (namely, the circles), connected by *fuzzy transitions* (namely, the arrows). Each fuzzy transition is labeled by a fuzzy expression which can be as complex as desired (A, B, C, D are traditional *fuzzy sets*).

As in a traditional FSA, a state represents univocally the operating conditions of an FFSA, at any given instant. But, unlike FSA, a system need not be in only one state at a time. Each state S_j is therefore associated with a *fuzzy state activity* $\mu_{S_j} \in [0, 1]$, which represents "how much" the system is in that particular state. The state activity is somehow equivalent to the degree of membership of a fuzzy variable, yet note that the state in an FFSA is not a fuzzy variable.

If we associated a state activity to states in a traditional FSA, it would be:

$$\mu_{S_j} = \begin{cases} 1, & \text{for the active state} \\ 0, & \text{for all other states} \end{cases} \tag{1}$$

with the constraint that the *total activity*:

$$\sum_j \mu_{S_j} = 1, \tag{2}$$

meaning that the system is always in one and only one well defined state S_j (the one with $\mu_{S_j} = 1$) and in no other state.

Constraint (2) applies also to FFSA, therefore activity can be distributed among several states and can partially move from one state to another (possibly more than one), but the total activity shall always be constant and equal to 1.

FFSA can be of different types: *time-independent* and *time-dependent*; the former can either be *synchronous* or *asynchronous*, which somewhat reflect synchronous and asynchronous FSA, respectively. They differ only in the way state activity moves from one state to another, according to the degrees of membership of the fuzzy transitions. Roughly we could say that:

1. in asynchronous, time-independent FFSA, state transitions may take place as soon as inputs vary (yet transitions are not so abrupt as in traditional FSA);
2. in synchronous, time-independent FFSA, state transitions are computed in a way similar to asynchronous FFSA, but they are applied only at the next clock cycle;
3. in time-dependent FFSA, there is always an intrinsic delay (usually larger than the clock period, if any) between input variations and the corresponding state transitions.

3.1 Time-independent automata

In time-independent FFSA, the activity moves from each state to one or more other states, as a function of the present state activity and the degrees of membership of the state transitions, but independently of time (hence the name). State activity may "move" only along state transitions.

In more details, a time-independent FFSA (either synchronous or asynchronous) can be translated into a more traditional fuzzy description made of one rule per each transition, where each rule is in the form:

$$\text{IF (STATE is oldstate) AND (fuzzyexpression) THEN (STATE is newstate)} \tag{3}$$

where the term (STATE is S_j) has, by definition, a degree of membership equal to the state activity μ_{S_j}, whereas the AND, OR and NOT implicators have the traditional meaning of any fuzzy system [2]. Two rules with the same consequent are supposed to be connected by an OR implicator.

Transitions starting from and ending into the same state must also be taken into account. For instance, for the example shown in fig. 2, the equivalent time-independent fuzzy description is made of five rules:

$$
\begin{aligned}
&\text{IF (STATE is SA) AND (x is NOT A) THEN (STATE is SA)}\\
&\quad\text{IF (STATE is SA) AND (x is A) THEN (STATE is SB)}\\
&\quad\text{IF (STATE is SB) AND (x is B) THEN (STATE is SB)}\\
&\quad\text{IF (STATE is SB) AND (x is C) THEN (STATE is SC)}\\
&\quad\text{IF (STATE is SB) AND (x is D) THEN (STATE is SD)} \quad (4)
\end{aligned}
$$

A constraint has to be placed on the membership functions $\mu_i(x)$ associated with all the transitions exiting from a state, to let the FFSA operate properly (not proven here):

$$
\sum_{i \in T_j} \mu_i(x) \approx 1 \quad \forall x \in \mathcal{U}_x, \quad \forall j \tag{5}
$$

where T_j is the set of transitions exiting from state S_j and $\mathcal{U}_x$ is the *universe of discourse* of the input variable x. For the example of fig. 2, the above identity holds if and only if $B = \text{NOT } (C \text{ OR } D)$.

In traditional FSA, the transitions exiting from a state must be both *exhaustive* (meaning that the logic sum of all transition labels must always be true) and *mutually exclusive* (meaning that no two transitions must be active at the same time). Similarly, in FFSA, constraint (5) is equivalent to exhaustivity, but there is no equivalent of mutual exclusion.

As any other fuzzy system, the fuzzy equivalent of an FFSA also has a mathematical model. Supposing to use the min and max operators as AND and OR implicators, respectively, the fuzzy system (4) results into the following mathematical expression, valid only for synchronous FFSA (one equation per state):

$$
\begin{aligned}
\mu'_{\text{SA}}(t + \Delta t) &= \max\{\min\{\mu_{\text{SA}}(t), 1 - \mu_{\text{A}}(x)\}, \ldots\}\\
\mu'_{\text{SB}}(t + \Delta t) &= \max\{\min\{\mu_{\text{SA}}(t), \mu_{\text{A}}(x)\}, \min\{\mu_{\text{SB}}(t), \mu_{\text{B}}(x)\} \ldots\}\\
\mu'_{\text{SC}}(t + \Delta t) &= \max\{\min\{\mu_{\text{SB}}(t), \mu_{\text{C}}(x)\}, \ldots\}\\
\mu'_{\text{SD}}(t + \Delta t) &= \max\{\min\{\mu_{\text{SB}}(t), \mu_{\text{D}}(x)\}, \ldots\} \tag{6}
\end{aligned}
$$

where the dots stand for any other transition reaching the corresponding states (not shown in fig. 2). The values $\mu'_{S_j}(t + \Delta t)$ represent the *new state activities* at time $(t + \Delta t)$, where Δt is the sampling period.

This might seem a straightforward implementation of an FFSA, but it does not guarantee that total activity remains constant. In practice, it suffers from the problem depicted in fig. 3.a: suppose that, at $t = 0$, the FFSA is mostly in state SA (namely, $\mu_{\text{SA}} \approx 1$, $\mu_{\text{SB}}, \mu_{\text{SC}}, \mu_{\text{SD}} \approx 0$), $\mu_{\text{A}}(x), \mu_{\text{C}}(x), \mu_{\text{D}}(x) \approx 0$, therefore $\mu_{\text{B}}(x) \approx 1$ (from (5)). In this case equations (6) simplify and, as soon as $\mu_{\text{A}}(x)$ begins to monotonically increase:

$$
\begin{aligned}
\mu'_{\text{SA}}(t + \Delta t) &= \min\{\mu_{\text{SA}}(t), (1 - \mu_{\text{A}}(x))\} = (1 - \mu_{\text{A}}(x))\\
\mu'_{\text{SB}}(t + \Delta t) &= \max\{\min\{\mu_{\text{SA}}(t), \mu_{\text{A}}(x)\}, \mu_{\text{SB}}(t)\} =\\
&= \max\{\min\{(1 - \mu_{\text{A}}(x)), \mu_{\text{A}}(x)\}, \mu_{\text{SB}}(t)\} \tag{7}
\end{aligned}
$$

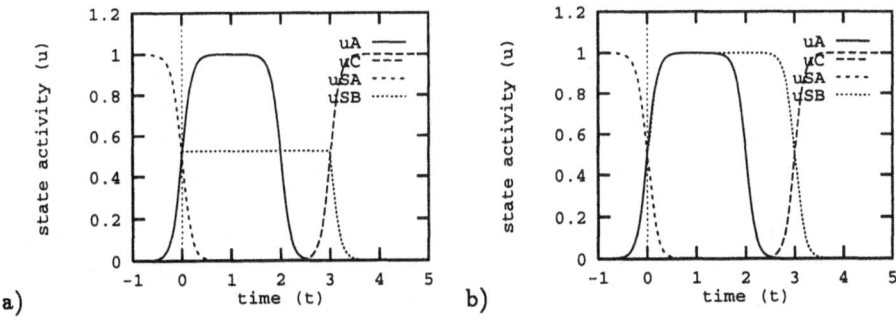

Fig. 3. a) a deadlock in an FFSA; b) correct behavior of an asynchronous time-independent FFSA. Simulation results with $\mu_D(x) \approx 0$ and $\mu_B(x) \approx (1 - \mu_C(x))$

Therefore, as long as $\mu_A(x) \leq 0.5$, the more $\mu_A(x)$ increases, the more the activity moves from SA to SB; in particular, the activity of SB increases and tracks $\mu_A(x)$, and that of SA decreases accordingly, thus total activity remains constant.

The problems arises when $\mu_A(x) > 0.5$. From that point, activity of SA keeps decreasing but that of SB stops and gets stuck at a value of 0.5 (from (7)), as shown in fig. 3.a.

This problem is solved by using constraint (2) to normalize the state activities given by equation (6), therefore:

$$\mu_{S_j}(t + \Delta t) = \frac{\mu'_{S_j}(t + \Delta t)}{\sum_k \mu'_{S_k}(t + \Delta t)} \tag{8}$$

By doing so, the total activity is guaranteed to remains equal to 1.

Instead, asynchronous time-independent FFSA use an alternative criterion to update μ_{S_j}; the mathematical model of asynchronous FFSA can be obtained by modifying formulae (6) and (8) as, respectively:

$$\begin{cases} \mu'_{S_j}(t) = f_j(\ldots \mu_i(x) \ldots; \ldots \mu_{S_k}(t) \ldots) \\ \mu_{S_j}(t) = \frac{\mu'_{S_j}(t)}{\sum_k \mu'_{S_k}(t)} \end{cases} \tag{9}$$

where $f_j(\cdot)$ are appropriate functions obtained by analyzing the state diagram (for instance, formulae (6)).

The difference with respect to the synchronous FFSA is that formulae (9) give the value of μ_{S_j} at time t (and not $(t + \Delta t)$), therefore they are a set of non-linear equations in the variables μ_{S_j} (which appear in both sides of (9)) and must be solved appropriately (for instance, by means of relaxation methods).

Also in this case, total activity remains equal to 1; see fig. 3.b, which shows how the activity of SB keeps tracking $\mu_A(x)$ also when larger than 0.5.

The difference between synchronous and asynchronous FFSA is that the the latter is more complex to compute, but the activity of SB tracks perfectly $\mu_A(x)$ for any value of it, while the former is faster to compute, but the activity of SB tracks $\mu_A(x) > 0.5$ with a slight time delay.

3.2 Time-dependent automata

Like in time-independent FFSA, the activity moves from each state to one or more other states, as a function of the state activity and the degrees of membership of state transitions. But, unlikely time-independent FFSA, variations are slower and depend on time (hence the name). State activity "moves" only along transitions.

In more details, a time-dependent FFSA can be translated into a more traditional fuzzy description made of one rule per each state transition, where each rule is in the form:

$$\text{IF (STATE is oldstate) AND (fuzzyexpression) THEN}$$
$$\text{(STATE flows to newstate)} \tag{10}$$

where all terms have the same meaning described in section 3.1, except for the new operator (STATE flows to newstate), whose mathematical model is expressed by two differential equations:

$$\frac{d}{dt}\,\mu_{\text{newstate}}(t) = \frac{1}{\tau}\,\mu_{\text{fuzzyexpression}}$$
$$\frac{d}{dt}\,\mu_{\text{oldstate}}(t) = -\frac{d}{dt}\,\mu_{\text{newstate}}(t) \tag{11}$$

For instance, for the example shown in fig. 2, the equivalent time-dependent fuzzy description is:

$$\text{IF (STATE is SA) AND (x is NOT A) THEN (STATE flows to SA)}$$
$$\text{IF (STATE is SA) AND (x is A) THEN (STATE flows to SB)}$$
$$\text{IF (STATE is SB) AND (x is B) THEN (STATE flows to SB)}$$
$$\text{IF (STATE is SB) AND (x is C) THEN (STATE flows to SC)}$$
$$\text{IF (STATE is SB) AND (x is D) THEN (STATE flows to SD)} \tag{12}$$

which produces to the following mathematical model (two equations per rule):

$$\frac{d}{dt}\,\mu_{\text{SA}}(t) = \frac{1}{\tau}\min\{\mu_{\text{SA}}(t), 1 - \mu_{\text{A}}(x)\}\ ; \quad \frac{d}{dt}\,\mu_{\text{SA}}(t) = -\frac{d}{dt}\,\mu_{\text{SA}}(t) \tag{13}$$

$$\frac{d}{dt}\,\mu_{\text{SB}}(t) = \frac{1}{\tau}\min\{\mu_{\text{SA}}(t), \mu_{\text{A}}(x)\}\ ; \quad \frac{d}{dt}\,\mu_{\text{SA}}(t) = -\frac{d}{dt}\,\mu_{\text{SB}}(t) \tag{14}$$

$$\frac{d}{dt}\,\mu_{\text{SB}}(t) = \frac{1}{\tau}\min\{\mu_{\text{SB}}(t), \mu_{\text{B}}(x)\}\ ; \quad \frac{d}{dt}\,\mu_{\text{SB}}(t) = -\frac{d}{dt}\,\mu_{\text{SB}}(t) \tag{15}$$

$$\frac{d}{dt}\,\mu_{\text{SC}}(t) = \frac{1}{\tau}\min\{\mu_{\text{SB}}(t), \mu_{\text{C}}(x)\}\ ; \quad \frac{d}{dt}\,\mu_{\text{SB}}(t) = -\frac{d}{dt}\,\mu_{\text{SC}}(t) \tag{16}$$

$$\frac{d}{dt}\,\mu_{\text{SD}}(t) = \frac{1}{\tau}\min\{\mu_{\text{SB}}(t), \mu_{\text{D}}(x)\}\ ; \quad \frac{d}{dt}\,\mu_{\text{SB}}(t) = -\frac{d}{dt}\,\mu_{\text{SD}}(t) \tag{17}$$

$$\tag{18}$$

where all terms $\frac{d}{dt}\,\mu_{\text{S}_j}$ with the same j, at the left sides of the above set of equations must be summed up together. It is clear from (13) and (15) that,

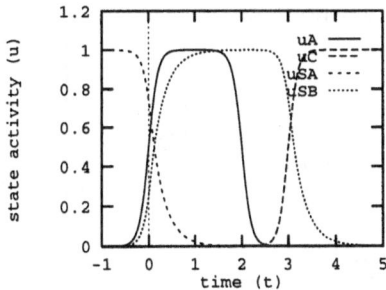

Fig. 4. Behavior of a time-dependent FFSA.

unlikely time-independent FFSA, transitions starting from and ending into the same state have no effect at all, as they produce two opposite terms which cancel each other out. Therefore such transitions need not be taken into account.

The main difference with respect to time-independent FFSA is that in time-dependent FFSA, state activity moves from one state to another with a time constant τ (which may also be different for each transition), therefore there is a time-lag between variations of inputs and of state activities, as shown in fig. 4.

3.3 Integration with neuro-fuzzy systems

So far, FFSA alone have described in details; this section explains how they can be integrated with neuro-fuzzy systems to produce operational controllers to be used in practical applications. As for FFSA, the technique proposed below can also be applied to other domains.

To integrate FFSA with neuro-fuzzy systems(or traditional controllers, as well), it is necessary to associate each state S_j with one controller C_j, as shown in the example of fig. 2.

Controllers can be either linear, non-linear, neural, fuzzy, neuro-fuzzy, look-up tables, or any combination thereof. Furthermore, each controller can be implemented using the technique which best suits the requirements of that particular state, therefore the individual controllers need not be of the same type.

The only constraint is that all controllers must be functions of the same input variables and must produce the same output variables. In vector form:

$$\mathbf{Y}^j = C_j\left(\mathbf{X};t\right) \tag{19}$$

where $\mathbf{X}$ and $\mathbf{Y}$ are the input and output vector, respectively. The overall characteristic, resulting from the integration of the controllers with the FFSA becomes:

$$\mathbf{Y} = \sum_j \mu_{S_j}(t) \cdot C_j\left(\mathbf{X};t\right) \tag{20}$$

As state activity $\mu_{S_j}(t)$ moves smoothly from one state to another, when input conditions vary, the overall controller characteristic varies smoothly as well, removing all abrupt changes and variations. This has a lot of benefits in control applications, as described in section 1.

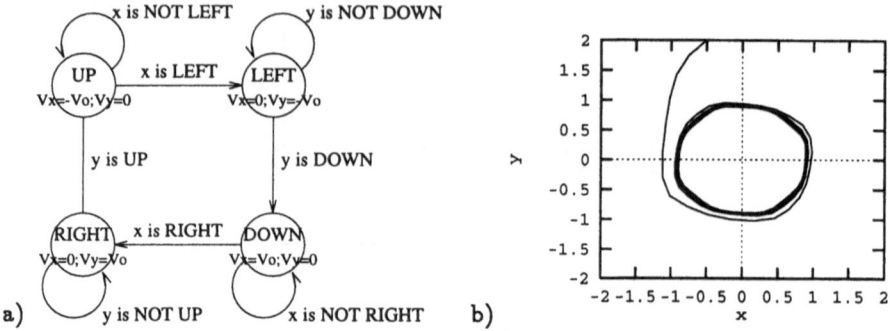

Fig. 5. Circular trajectory controlled by a simple FFSA. The trajectory shown in fig. 6.b looks more octagonal than circular; this is due to sampling time in simulations.

4 Application Examples

This section describes shortly two application examples of FFSA in the field of control, and in particular to control the trajectory of two robots.

The first one is a simple method to define circular (or elliptical) trajectories, by means of an FFSA. Fig. 5.a shows an FFSA with four states and two inputs x and y, which are the coordinates of a moving spot (for instance, the robot's tip). Each fuzzy state activates one of four simple controllers, which move the spot towards either left, right, up, or down, at a constant speed. The four fuzzy sets LEFT, RIGHT, UP and DOWN are self-explanatory.

Fig. 5.b shows the resulting trajectory, starting from a point external to the desired trajectory. The speed of the moving spot is almost constant and equal to V_0 (the target speed of the four controllers), while the diameter depends on the membership functions LEFT, RIGHT, UP and DOWN. When the four fuzzy sets are symmetrical with respect to axis origin, the trajectory is circular, but it may easily become elliptical, or its center can be moved, by shifting the membership functions. Yet, spot speed remains constant.

Another application is to define the Motion and Leg Coordination Controls of a six-legged robot [6]. This is a quite tough task, which requires a tight interaction between different controllers. A much simpler approach can be obtained using the FFSA shown in fig. 6, which tracks the state of each leg of the robot (for instance, forward step, backward step, leg lift and contact phases). Each state is associated with a different controller, each one tailored to the specific system state. For instance, controllers C1, C2, C3 and C4 are simple PID controllers tuned to, respectively: hold body weight and move the leg backwards; rise the leg at maximum speed; return the leg forward at maximum speed; lower the leg at reduced speed.

The FFSA of fig 6 has provisions for the normal leg step (namely, walking on smooth surfaces without obstacles) and to recover from obstacles encountered during leg movements. The actual FFSA is more complex, as it includes provi-

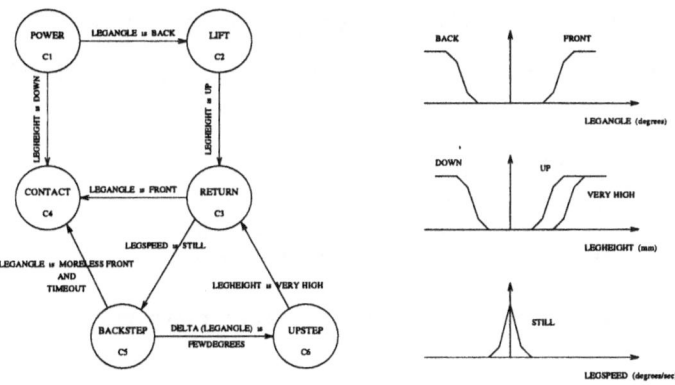

Fig. 6. FFSA of the Leg Control.

sions for: i) modification of hexapod attitude in presence of complex obstacles (e.g. to walk inside low tunnels), ii) modification of leg trajectory when walking sideways, iii) control of alternative paces for dynamic gaits (e.g. galloping, or climbing staircases); iv) other special cases a leg might have to face.

5 Conclusion

The paper has introduced Fuzzy State Automata, as a method which allows a tight integration between finite state automata and neuro-fuzzy systems. Fuzzy State Automata can be converted to more traditional fuzzy systems, therefore they can run on either traditional fuzzy processors, or on systems with fuzzy computing capabilities. Fuzzy State Automata are also a method to describe recurrent fuzzy systems. Two examples have been described in the field of control.

References

1. S. Haykin, "Neural Networks: A Comprehensive Foundation", *Mc Millan College Publishing Company*, New York, 1994.
2. L. Wang, "Adaptive Fuzzy Systems and Control", *Prentice Hall*, Englewood Cliffs, New Jersey, 1994.
3. D.A. White and D.A. Sofge, "Handbook of Intelligent Control", Van Nostrand Reinhold, 1992.
4. L.M. Reyneri, "Weighted Radial Basis Functions for Improved Pattern Recognition and Signal Processing", submitted to *Neural Letters*.
5. L.M. Reyneri, M. Chiaberge, L. Zocca, "CINTIA: A Neuro-Fuzzy Real Time Controller for Low Power Embedded Systems", *IEEE MICRO, special issue on Hardware for Artificial Neural Networks*, June 1995, pp. 40-47.
6. F. Berardi, M. Chiaberge, E. Miranda, L.M. Reyneri, "A Walking Hexapod Controlled by a Neuro-Fuzzy System", Proc. of MCPA'97, Pisa (I), February 1997 (to be published).

Statistical Analysis of Regularization Constant — From Bayes, MDL and NIC Points of View

Shun-ichi Amari and Noboru Murata

RIKEN Frontier Research Program
Wako-shi, Hirosawa 2-1, Saitama 351-01, JAPAN
fax: +81-48-462-9881
amari@zoo.riken.go.jp; mura@zoo.riken.go.jp

Abstract. In order to avoid overfitting in neural learning, a regularization term is added to the loss function to be minimized. It is naturally derived from the Bayesian standpoint. The present paper studies how to determine the regularization constant from the points of view of the empirical Bayes approach, the maximum description length (MDL) approach, and the network information criterion (NIC) approach. The asymptotic statistical analysis is given to elucidate their differences. These approaches are tightly connected with the method of model selection. The superiority of the NIC is shown from this analysis.

1   Introduction

Overfitting is a serious problem in neural learning. Model selection, an important method to avoid overfitting, chooses a model of an adequate size depending on the number of examples. Therefore, a number of criteria for fitness of models are proposed. One criterion is the MDL (Rissanen, 1989), which minimizes the coding or description length for a given set of examples. It is closely related to the empirical Bayes criterion of model selection (McKay, 1992, Rissanen, 1989).

Another criterion for model selection is the NIC (Murata et al., 1994), which is a generalization of the AIC (Akaike, 1974). It is another formulation of the method of effective degrees of freedom (Moody, 1992). This criterion minimizes an unbiased estimate of the generalization error.

Brake et al. (1995) compared MDL and NIC and reported that no general superiority was found between these criteria. In other words, one is better in some cases and the other one is better in other cases, see also Ripley (1995).

The introduction of a regularization term is another commonly used technique to avoid overfitting (Poggio and Girosi, 1990). In this case, it is necessary to determine the regularization constant adequately. The empirical Bayes method (McKay, 1992) or the cross validation method are widely used for this purpose.

The present paper uses the above-mentioned criteria of fitness of models to determine the regularization constant. We give an asymptotic statistical analysis of the optimal choice of the regularization constant under various criteria. This analysis elucidates the relation between the model selection and the regularization theory. It is shown that the NIC is useful for this purpose (Murata et al., 1994). The NIC is expected to give asymptotically better results than the Bayesian method.

2 Stochastic Framework of Neural Learning

Let us consider a parametric model of stochastic input-output systems including neural networks. Let x and y be an input vector and an output scalar, respectively, where the conditional distribution of y conditioned on x is given by $p(y|x; w)$ specified by w. Here, $w = (w_1, \ldots, w_d)$ is a d-dimensional vector to parameterize this model. The system is given by the set of conditional probability distributions,

$$M = \{p(y|x; w), \ w \in \boldsymbol{R}^d\}. \tag{1}$$

A typical example is a multilayer perceptron with an additive Gaussian noise n, where output y is given by

$$y = f(x; w) + n. \tag{2}$$

Here, $f(x; w)$ is the deterministic (noiseless) output from the multilayer perceptron specified by the vector w consisting of all the modifiable parameters (synaptic weights and thresholds). Then,

$$p(y|x; w) = \frac{1}{\sqrt{2\pi}\sigma} \exp\left[-\frac{1}{2\sigma^2}|y - f(x; w)|^2\right] \tag{3}$$

where σ^2 is the variance of the noise, that is, $n \sim N(0, \sigma^2)$.

Let $D_T = \{(x_1, y_1), \ldots, (x_T, y_T)\}$ be a set of T input-output training examples generated independently from an unknown but fixed probability distribution $q(x)q(y|x)$, where $q(x)$ is the probability distribution of input x and $q(y|x)$ represents the stochastic mechanism of the teacher that generates the output y from the input x.

Let us consider a loss function given by

$$l(x, y; w) = -\log q(x)p(y|x; w). \tag{4}$$

The empirical loss evaluated by the examples D_T is called the training error, and is given by

$$L_{\text{train}}(w) = \sum_{t=1}^{T} l(x_t, y_t; w). \tag{5}$$

The maximum likelihood estimator $\hat{w}$ is the minimizer of $L_{\text{train}}(w)$,

$$\hat{w} = \underset{w}{\operatorname{argmin}} L_{\text{train}}(w). \tag{6}$$

In order to avoid overfitting, we add a regularization term $r(w)$ which penalizes large values w_i of the components of w. A typical example is

$$r(w) = \frac{1}{2} \sum_{i=1}^{d} (w_i)^2. \tag{7}$$

The loss function to be minimized is then written as

$$l(x, y; w, \lambda') = l(x, y; w) + \lambda' r(w), \qquad (8)$$

where λ' is called the regularization constant.

Given λ, the minimizer of

$$L_{\text{train}}(w, \lambda) = L_{\text{train}}(w) + \lambda r(w) \qquad (9)$$

is denoted by

$$\hat{w}_\lambda = \text{argmin} \, L_{\text{train}}(w, \lambda). \qquad (10)$$

The loss and the estimator without the regularization term correspond to the case with $\lambda = 0$.

The estimator $\hat{w}_\lambda$ should be evaluated by the generalization error without the regularization term. It is given by

$$L_{\text{gen}}(w) = E[l(x, y; w)], \qquad (11)$$

which is the expectation of the loss with respect to future input-output pairs (x, y). This is equivalent to the Kullback-Leibler divergence given by

$$L_{\text{gen}}^{KL}(w) = D[q(x)q(y|x) : q(x)p(y|x; w)], \qquad (12)$$

except for constant difference. Let w^* be the optimal parameter such that

$$w^* = \underset{w}{\text{argmin}} \, L_{\text{gen}}^{KL}(w). \qquad (13)$$

3 Bayesian, MDL and NIC Criteria for Model Selection

We first explain the Bayesian standpoint (McKay, 1992). The Bayesian statistics assumes that the unknown parameter w is subject to the prior probability $\pi(w)$. However, we do not know $\pi(w)$ in many cases. In such a case, there is a method of determining $\pi(w)$ from the observed data. This is called the empirical Bayes method. It assumes a parametric family $\pi(w; \lambda)$ of the prior distributions specified by an extra parameter λ called the hyper parameter.

In the case of neural learning, it is typically written as

$$\pi(w; \lambda) = \exp\{-\lambda r(w) + c(\lambda)\}, \qquad (14)$$

where

$$c(\lambda) = -\log \int \exp\{-\lambda r(w)\} dw. \qquad (15)$$

When $r(w)$ is given by the quadratic form

$$r(w) = \frac{1}{2} \sum_{i=1}^{d} (w_i)^2,$$

$$c(\lambda) = \frac{1}{2} \log \lambda - \frac{1}{2} \log(2\pi), \qquad (16)$$

it is the Gaussian prior. The present paper assumes that $\pi(w; \lambda)$ is differentiable with respect to w. The non-differentiable case is also important, and will be studied in a forthcoming paper.

Given λ, the joint probability of D_T and w is given by

$$p(D_T, w, \lambda) = \pi(w; \lambda) \prod_{t=1}^{T} q(x_t) p(y_t | x_t; w). \tag{17}$$

The parameter $\hat{w}_\lambda = \hat{w}_\lambda(D_T)$ that maximizes $p(D_T, w, \lambda)$, or equivalently that maximizes the posterior distribution of w under the condition that D_T is observed,

$$p(w | D_T) = \frac{p(D_T, w, \lambda)}{p(D_T, \lambda)}, \tag{18}$$

is called the maximum posterior estimator, where $p(D_T, \lambda)$ is the marginal distribution of D_T,

$$p(D_T, \lambda) = \int p(D_T, w, \lambda) dw. \tag{19}$$

The maximum likelihood estimator is given by $\hat{w}_0$ with $\lambda = 0$ for the Gaussian prior (16). We denote the negative logarithm of the joint probability by

$$L_{\text{train}}(D_T, w, \lambda) = -\log \pi(w; \lambda) + \sum_{t=1}^{T} l(y_t | x_t; w) + c \tag{20}$$

where

$$l(y | x; w) = -\log p(y | x; w) \tag{21}$$

is the negative logarithm of the conditional probability and

$$c = -\sum_{t=1}^{T} \log q(x_t)$$

does not depend on w and λ. The term c is neglected hereafter. In the case of the multilayer perceptron,

$$L_{\text{train}}(D_T, w, \lambda) = \frac{1}{2\sigma^2} \sum_{t=1}^{T} |y_t - f(x_t; w)|^2 + \frac{1}{2} \log 2\pi\sigma^2 + \lambda r(w) - c(\lambda). \tag{22}$$

Hence, the maximum posterior estimator $\hat{w}_\lambda$ is the parameter that minimizes the sum of the squared error and the regularization term $\lambda r(w)$.

In order to evaluate the likelihood of the observed data D_T, we consider a number of statistical models $M_1, M_2, M_3, \ldots$, where each M_i specifies a conditional probability $p_{M_i}(y | x; w^{(i)})$. The log likelihood of data D_T under the Bayesian model M_i with prior $\pi(w; \lambda)$ is given by

$$L_{M_i}(D_T, \lambda) = -\log p_{M_i}(D_T, \lambda). \tag{23}$$

From the Bayesian standpoint, the model M_{i_0} that maximizes $L_{M_i}(D_T, \lambda)$ is regarded as the best model that explains the observed data D_T, where λ is a given constant. Hence, the Bayesian criterion of model selection is given by $L_{M_i}(D_T, \lambda)$. When T is large, we have the following asymptotic expansion.

Theorem 1.

$$L_M(D_T, \lambda) = L_{M,\text{train}}(D_T, \hat{w}_\lambda, \lambda) + d \log T + O_p(1), \tag{24}$$

where the index M denotes the specific model.

Proof is given in Appendix.

This shows the Bayesian criterion of model selection.

The $L_M(D_T, \lambda)$ depends on the hyper parameter λ. In order to explain data D_T, it is wise to choose such λ that minimizes the negative of the likelihood. So we define

$$L_M(D_T) = \min_\lambda L_M(D_T, \lambda). \tag{25}$$

This gives us the criterion of model selection which does not depend on λ.

We now describe the three criteria of the model selection. For various models M and given data D_T, the model M^* that minimizes one of the following criteria should be selected.

I. Bayesian criterion of model selection: $L_{\text{train}}^{\text{Bayes}}(D_T, \lambda)$ defined by (23).
II. MDL criterion :

$$L_M^{\text{MDL}}(D_T) = L_{M,\text{train}}(\hat{w}) + d \log T.$$

III. NIC criterion :

$$L_M^{\text{NIC}}(D_T) = L_{M,\text{train}}(\hat{w}) + \text{tr}(K^{-1}G),$$

where K and G are matrices defined by

$$K = E[\nabla_w \nabla_w l(x, y; \hat{w})],$$
$$G = E[\nabla_w l(x, y; \hat{w}) \nabla_w l(x, y; \hat{w})],$$

∇_w denoting the gradient with respect to w.

When the optimal model includes the true distribution, $G = K$ and is the Fisher information matrix. In this case, we have

$$L_M^{\text{NIC}}(D_T) = L_{M,\text{train}}(D_T) + d.$$

The MDL and NIC criteria do not directly include the regularization term, whereas the Bayesian criterion depends on λ. When $\lambda = 0$, it coincides with the MDL criterion that minimizes the description length. It is possible to generalize the NIC such that it depends on λ, because the NIC is defined for a general loss function including the regularization term other than the log loss.

We analyze in the following the effect of the regularization term. Here the value of λ is determined such that it minimizes the model selection criteria.

4 Bayesian Method of Determining λ

The optimal λ is determined form the Bayesian point of view to minimize $L_M(D_T, \lambda)$. We have the following theorem.

Theorem 2. The Bayesian optimal λ is given by the solution of

$$\frac{d}{d\lambda} \log \pi(\hat{w}_\lambda, \lambda) = 0. \tag{26}$$

In particular, for the multilayer perceptron with the Gaussian prior (16),

$$\lambda_{\text{opt}} = \frac{d}{2r(\hat{w}_0)} + O_p\left(\frac{1}{T}\right) = \frac{d}{\sum (\hat{w}_i^0)^2} + O_p\left(\frac{1}{T}\right). \tag{27}$$

Proof. The Bayesian optimal λ maximizes $L_M(D_T, \hat{w}_\lambda, \lambda)$. Therefore, we have

$$\frac{d}{d\lambda} L_M(D_T, \hat{w}_\lambda, \lambda) = \nabla_w L_M(D_T, \hat{w}_\lambda, \lambda) \cdot \frac{d}{d\lambda} \hat{w}_\lambda + \frac{\partial}{\partial \lambda} L_M(D_T, \hat{w}_\lambda, \lambda) = 0. \tag{28}$$

Since $\nabla_w L_M(D_T, \hat{w}_\lambda, \lambda) = 0$, from (20) we have (26). In the case of the multilayer perceptron, we have (27) from (26) and (16). $\square$

Now we give an asymptotic relation which shows how the regularization term modifies the maximum likelihood estimator $\hat{w}_0$ into $\hat{w}_\lambda$. From (20), by putting

$$\hat{w}_\lambda = \hat{w}_0 + \varepsilon, \tag{29}$$

we have

$$-\nabla_w \log \pi(\hat{w}_0 + \varepsilon, \lambda) + \sum_{t=1}^{T} \nabla_w l(y_t | x_t; \hat{w}_0 + \varepsilon) = 0.$$

By expansion,

$$\sum \nabla_w l(y_t | x_t; \hat{w}_0 + \varepsilon) = \sum \nabla_w \nabla_w l(y_t | x_t; \hat{w}_0) \varepsilon.$$

Let us put

$$A = \frac{1}{T} \sum \nabla_w \nabla_w l(y_t | x_t; \hat{w}_0),$$

and we have

$$\varepsilon = \frac{A^{-1}}{T} \nabla_w \log \pi(\hat{w}_0).$$

It should be noted that A converges to $K = E[\nabla_w \nabla_w l(y | x; \hat{w}_0)]$. In the case of the multilayer perceptron with the Gaussian prior, we have

$$\hat{w}_\lambda = \hat{w}_0 - \frac{\lambda}{T} K^{-1} \nabla_w r(\hat{w}_0), \tag{30}$$

which is a shrinkage estimator of the form

$$\hat{w}_\lambda = \left(I - \frac{\lambda}{T} K^{-1}\right) \hat{w}_0. \tag{31}$$

It should be remarked that K is not the identity matrix in general. This is the Fisher information matrix in the realizable case of

$$q(y|x) = p(y|x; w^*).$$

5 Evaluation of the optimal λ by NIC

We have obtained the model selection criterion and adaptive selection of λ based on data from the Bayesian standpoint at the same time. However, it is still not certain how good they are. So we search for the method of minimizing the generalization error directly, and compare the results.

The generalization error $L_{\text{gen}}(w)$ of the network belonging to M with parameter w is given by

$$L_{M,\text{gen}}(w) = E[l(y|x;w)]. \tag{32}$$

This does not include the regularization term. This can be written as

$$L_{M,\text{gen}}(w) = \frac{1}{2\sigma^2} E[|y - f(x;w)|^2] + \frac{1}{2}\log 2\pi\sigma^2 \tag{33}$$

for the multilayer perceptron (3). The purpose of introducing the regularization term is to decrease the generalization error preventing from overfitting. In order to see its effect, we calculate the generalization error of the trained network with parameter λ.

Let w^* be the minimizer of $L_{M,\text{gen}}(w)$,

$$w^* = \underset{w}{\text{argmin}}\, L_{M,\text{gen}}(w). \tag{34}$$

We then have

$$L_{M,\text{gen}}(\hat{w}_\lambda) = L_{M,\text{gen}}(w^*) + \frac{1}{2}(\hat{w}_\lambda - w^*)^T K(\hat{w}_\lambda - w^*) \tag{35}$$

where higher order terms are neglected. Let e be the deviation error of the maximum likelihood estimator $\hat{w}_0$ from the optimal one,

$$\hat{w}_0 = w^* + e. \tag{36}$$

The asymptotic behavior of the maximum likelihood estimator is well known in statistics. It is asymptotically subject to the normal distribution. It has a bias of order $1/T$, and its variance is given by matrix $(1/T)K^{-1}GK^{-1}$, where

$$G = E[\nabla_w l(y|x;w^*)\nabla_w l(y|x;w^*)].$$

The bias term b is given by

$$E[\hat{w}_0] = w^* + \frac{b}{T} + O\left(\frac{1}{T^2}\right), \tag{37}$$

where the bias vector $b = (b^i)$ is calculated as

$$b^i = \sum_{l,m,n} k^{il} k^{mn} s_{lmn} - \frac{1}{2} \sum_{l,m,n,r,s} k^{il} k^{mn} k^{rs} t_{lmr} g_{ns}, \tag{38}$$

where (k^{ik}) is the inverse of matrix K, $G = (g_{ms})$, and

$$s_{lmn} = E\left[\frac{\partial^2}{\partial w^l \partial w^m} l(y|x;w) \frac{\partial}{\partial w^n} l(y|x;w)\right] \tag{39}$$

$$t_{lmn} = E\left[\frac{\partial^3}{\partial w^l \partial w^m \partial w^n} l(y|x;w)\right] \tag{40}$$

Taking account of (31), we have

$$L_{M,\text{gen}}(\hat{w}_\lambda)$$
$$= L_{M,\text{gen}}(w^*) + \frac{1}{2}\left\{e - \frac{\lambda}{T}K^{-1}\nabla_w r(\hat{w}_\lambda)\right\}^T K \left\{e - \frac{\lambda}{T}K^{-1}\nabla_w r(\hat{w}_\lambda)\right\}$$

Now we expand

$$\nabla_w r(\hat{w}_\lambda) = \nabla_w r(w^*) + Re - \frac{\lambda}{T}RK^{-1}\nabla_w r(w^*) \tag{41}$$

where

$$R = \nabla_w \nabla_w r(w^*). \tag{42}$$

We also decompose e as

$$e = \tilde{e} + \frac{b}{T} \tag{43}$$

where $\tilde{e}$ is non-biased. Then, the term

$$\hat{w}_\lambda - w^* = e - \frac{\lambda}{T}K^{-1}\nabla_w r(\hat{w}_\lambda) \tag{44}$$

is decomposed as the sum of the fluctuating term and bias term,

$$\hat{w}_\lambda - w^* = \left(I - \frac{\lambda}{T}K^{-1}R\right)\tilde{e} - \frac{\lambda}{T}K^{-1}\nabla_w r + \frac{b}{T}, \tag{45}$$

where higher order terms are neglected. Neglecting higher order terms again, the expectation of $L_{M,\text{gen}}(\hat{w}_\lambda)$ is written as

$$E[L_{M,\text{gen}}(\hat{w}_\lambda)]$$
$$= L_{M,\text{gen}}(w^*) + \frac{1}{2}E\left[\tilde{e}^T\left(I - \frac{\lambda}{T}K^{-1}R\right)^T K \left(I - \frac{\lambda}{T}K^{-1}R\right)\tilde{e}\right]$$
$$+ \frac{\lambda^2}{2T^2}(\nabla_w r)^T K^{-1}\nabla_w r + \frac{\lambda}{T^2}b^T \nabla_w r.$$

Taking account of

$$E[\tilde{e}\tilde{e}^T] = \frac{1}{T}G,$$

we have the following theorem.

Theorem 3. The optimal λ that minimizes the expected generalization error (NIC) is given by

$$\lambda_{\text{opt}} = \frac{\text{tr}(RK^{-1}GK^{-1}) + b^T \nabla_w r}{(\nabla_w r)^T K^{-1} \nabla_w r}. \tag{46}$$

When r is the quadratic form (7), which corresponds to the Gaussian prior,

$$\lambda_{\text{opt}} \simeq \frac{\text{tr}(K^{-1}GK^{-1}) + b^T w^*}{w^{*T} K^{-1} w^*} \simeq \frac{\text{tr}(K^{-1}GK^{-1}) + \hat{b}^T \hat{w}_0}{\hat{w}_0^T K^{-1} \hat{w}_0}. \tag{47}$$

Except for the case of $K = G =$ identity matrix, this is different from the Bayesian principle. The Bayesian λ_{opt} does not depend on the structure of the underlying neural network, whereas λ_{opt} derived from the least generalization error is sensitive to the underlying structures of K and G. This is the λ that directly minimizes the generalization error. So it is plausible that this λ gives a better performance of minimizing the generalization error.

6 Appendix A

Proof of Theorem 1.

We expand (20) at $\hat{w}_\lambda$, giving

$$L_{\text{train}}(D_T, w, \lambda)$$
$$= L_{\text{train}}(D_T, \hat{w}_\lambda, \lambda)$$
$$\quad + \frac{1}{2}(w - \hat{w}_\lambda)^T \left\{ \sum \nabla_w \nabla_w l(y_t | x_t; \hat{w}_\lambda) + \nabla_w \nabla_w \log \pi(w; \lambda) \right\} (w - \hat{w}_\lambda)$$
$$\quad + \text{higher order terms,}$$

because

$$\nabla_w L_{train}(D_T, \hat{w}_\lambda, \lambda) = 0.$$

Therefore,

$$p(D_T, \lambda) = \exp\left\{-L_{\text{train}}(D_T, \hat{w}_\lambda, \lambda)\right\} \int \exp\left[-\frac{T}{2}(w - \hat{w}_\lambda)^T A(w - \hat{w}_\lambda)\right] dw,$$

where

$$A = \frac{1}{T} \sum \nabla_w \nabla_w l(y_t | x_t; \hat{w}_\lambda) + \frac{1}{T} \nabla_w \nabla_w \log \pi.$$

The matrix A converges to

$$K = E[\nabla_w \nabla_w l(y | x; w^*)].$$

Therefore, by integration, we have (24) by neglecting higher order terms.

7 Conclusions

The present paper studies the effect of the regularization constant in neural learning by asymptotic statistical analysis. The Bayesian method is used to determine the optimal constant, and its relation to MDL is elucidated. The generalization error is analyzed when learning takes place from the Bayesian standpoint. It is then shown that the NIC criterion is useful for obtaining the optimal regularization constant, given a better generalization error.

References

1. H. Akaike (1974) A new look at statistical model identification, *IEEE Transactions on Automatic Control*, **19**, 716–723.
2. G. Brake, J.N. Kok and P.M.B. Vitányi (1995) Model Selection for Neural Networks: Comparing MDL and NIC, NeuroCOLT Technical Report NC-TR-95-021.
3. D.J.C. McKay (1992) Bayesian interpolation, *Neural Computation*, **4**, 415–447.
4. J.E. Moody (1992) The effective number of parameters: an analysis of generalization and regularization in nonlinear learning systems, in *NIPS4*, pp.847–854.
5. N. Murata, S. Yoshizawa and S. Amari (1994) Network information criterion — determining the number of hidden units for artificial neural network models, *IEEE Transactions on Neural Networks*, **5**, 865–872.
6. T. Poggio and F. Girosi (1990) Regularization algorithms for learning that are equivalent to multilayer networks, *Science*, **247**, 978–982.
7. B.D. Ripley (1996) *Pattern Recognition and Neural Networks*, Cambridge University Press.
8. J.Rissanen (1989) *Stochastic Complexity in Statistical Inquiry*, Singapore: World Scientific Publishing Co.

Building Digital Libraries from Paper Documents, Using ART Based Neuro-fuzzy Systems

R. Sanz Guadarrama[1]   Y. A. Dimitriadis[1]   G.I. Sainz Palmero[2]

J. M. Cano Izquierdo[2]   J. Lopez Coronado[2]

[1] Department of Signal Theory, Communications and Telematics Engineering
School of Telecommunications Engineering - University of Valladolid
Real de Burgos s/n, 47011 Valladolid, Spain
[2] Department of Systems Engineering and Control
School of Industrial Engineering - University of Valladolid
Paseo del Cauce s/n, 47011 Valladolid, Spain
e-mail: rausan@tel.uva.es, yannis@tel.uva.es

Abstract. In this paper a new neuro-fuzzy system is proposed for both tasks of document analysis and Optical Character Recognition. FasART (Fuzzy adaptive system ART based) inherits the stability, flexibility and modularity properties of ART supervised models, but with a formal description as a Fuzzy Logic System, and increased functionality. On the other hand Recursive FasART permits us to exploit context information, crucial aspect in document understanding. Satisfactory experimental results are presented for the global application of building a digital library of scientific papers, giving special emphasis on the creation of links between items in table of contents and paper first pages.

Keywords: Neuro-fuzzy systems, Adaptive Resonance Theory, Digital Library, Document Understanding, OCR, HTML.

1 Introduction

Document concept has been changing due to advances in computer and communications technology. While previously it was considered as simple information contained on a piece of paper, nowadays a document is increasingly assumed to be electronic or digital. Such a digital document can be defined as structured information containing numerous types of information (text, graphics, video, voice ...), coming from multiple sources that can be stored or transmitted .

Digital media have many advantages over paper media: flexibility with regard to editing and annotating, quick transport, fast and more convenient search mechanisms [13]; they also resolve six traditional problems associated with paper: paper is expensive, heavy, slow in being transported, cumbersome, restricted in format and static with regard to updates.

Digital library systems have been emerging, supported by the massive use of digital documents and by technical innovations in networks and personal computer areas. A digital library is a distributed environment which reduces barriers to the creation, dissemination, manipulation, storage, integration and reuse of information by individual and groups [8].

The service provided by a digital library must exploit the capacity of computation, storage, search mechanisms and dissemination of digital media, besides extending the functions provided by conventional libraries based on paper, that have been described as four-fold: collection of information; organization and representation; access and retrieval; and analysis, synthesis and dissemination of information. In addition to being able to process documents created in a computer, digital libraries have to manage conventional information existing in paper documents. Then, we could be able to retrieve complete documents as in conventional libraries, and not only document references like most present information systems [14].

Paper documents have to go through several processes, globally known as Document Image Understanding (DIU) [19] [20], in order to be usable in a digital library. The first task refers to image analysis, where we extract physical information from the blocks that compose the document. This paper treats the second major part of DIU, document understanding, that involves logical labeling and Optical Character Recognition (OCR). While by logical labeling we obtain information about the type and structure of the document, instead of simply physical blocks [1] [7], OCR achieves a conversion of scanned images of characters into a computer processable format (such as ASCII) [10].

In literature, several approaches has been proposed to document understanding based on: *Knowledge or rules*, knowledge must be extracted from the experts, and unexpected or distorted documents can't be dealt with robustness [15]; *Formal Grammars*, that have similar problems as above [2]; *Fuzzy Logic*, that permits to work with imprecise knowledge about the documents, but still uses rules [9]; *Artificial Neural Networks*, that learn from examples of documents [12] [16] [17]. On the other hand, we dispose of a great variety of solutions for OCR, from simple matrix matching to feature-based intelligent techniques [10], whose success for machine-printed characters mainly depends on the document quality and number of fonts.

The main goal of the proposed system is to process existing paper documents, such as table of contents and articles of a scientific journal, and the posterior construction of a digital library. A unified approach is proposed for both logical labeling and OCR problems, based on a neuro-fuzzy system that exploits in a synergistic form the symbolic and imprecise representation of a fuzzy system, as well as the capacity of neural networks for learning by examples. The model used in this system is called FasART (Fuzzy adaptive system ART based), originally proposed by us for a nonlinear system identification problem [4]. This neuro-fuzzy architecture has significant advantages over its predecessor Fuzzy ARTMAP [5], a supervised-learning architecture of the Adaptive Resonance Theory family [6], especially in its theoretical formulation. RFasART, a recurrent version of FasArt is also used, that takes advantage of the context information to analyze the document structure, as in expert systems but in a self-organizing way. This network has also been tested successfully in a MIME mailing system for business letters [18].

In section 2, we present the main characteristics of the proposed neuro-fuzzy architectures FasART and RFasART. In the next section we initially describe the experimental environment, and explain the specific objectives and modules of the working prototype for digital library construction. The experimental results for both logical labeling problems and OCR are later presented, together with the establishment of links between table of contents and first pages of the articles in a scientific journal. We finally conclude with a discussion of the main conclusions and ongoing research work.

2 FasArt and RFasArt: the neuro-fuzzy systems

The main underlying neural architecture of the proposed neuro-fuzzy systems is Fuzzy ARTMAP, fuzzy version of ARTMAP [5], where ART-1 modules have been replaced by Fuzzy ART modules [6]. Mainly used for pattern classification problems and rule extraction from a data collection, Fuzzy ARTMAP presents the following relevant characteristics:

- The architecture performs its task without need to know the nature of the underlying system. This characteristic makes it applicable to all types of systems where numeric data are available.
- Capacity for incremental learning. The real-time property of all ART based architectures permits the system to keep learning during performance phase, thus providing a continuous adaptation of the system to the real one.
- Respect to the stability-plasticity dilemma, since the system memory keeps increasing according to the learning needs. The system does not forget the already accumulated knowledge, when new learning is performed, avoiding that way the phenomenon of catastrophic forgetting, common characteristic of back-propagation based networks.
- Generation of fuzzy sets in form of hyperboxes, that makes them easily interpretable as ranges of variation of the associated variables.
- Availability of various parameters for vigilance, learning etc. that the designer can use in order to adapt the architecture to the particular specifications of a system.

Nevertheless, it was considered by us necessary to enhance several aspects of Fuzzy ARTMAP in order to use it in real pattern recognition and formalize its relationship with fuzzy systems:

- Operations with fuzzy sets are used in Fuzzy ART modules, considering that input values are membership function values in certain fuzzy sets. On the contrary, Fuzzy AND operations would have no sense. Then fuzzy logic should be incorporated in better terms to ART models.
- The way that Inter-ART map works has no clear connection to fuzzy set theory, thus complicating its interpretation in terms of fuzzy rules.

- The output provided by the system is an interval of values for the variables and not a concrete point.
- No defuzzification algorithm can be applied in the output of Fuzzy ART-b.

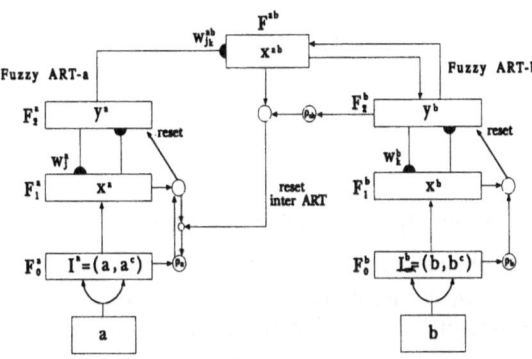

Fig. 1. Diagram of the Fuzzy ARTMAP neural fuzzy architecture

These inconveniences are avoided with FasArt (Fuzzy Adaptive System ART-based) [4], that adds the following features, while keeping the same general structure of Fuzzy ARTMAP, shown in figure 1:

- The new activation function of the neurons can be considered equivalent to the membership function of a fuzzy set.
- The degree of fuzziness can be regulated through values of factor γ. Values of $\gamma \to 0$ produce crispier fuzzy sets, while values of $\gamma \to \infty$ increase the fuzzy nature of sets.
- Inter-ART map can now be considered as an association of the output fuzzy sets in Fuzzy ART-a and Fuzzy ART-b modules. The equivalent symbolic form in terms of rules is: IF i IS R_j THEN y IS R_x
  that can be decomposed in turn into:
  IF$((i_1$ IS $r_{j1})$ AND $(i_2$ IS $r_{j2})$ AND... AND $(i_n$ IS $r_{jn}))$
  THEN $(y_1$ IS $r_{j1})$ AND $(y_2$ IS $r_{j2})$ AND... AND $(y_n$ IS $r_{jn})$
- A confidence degree is attached to any FasArt prediction. This value is specially significant if the system performance space has not been entirely sampled in the learning phase.

RFasArt [18], a recurrent version of FasArt, that uses contextual information in a integrated way. In this case, a document of a certain type can be considered as a temporal sequence, in the wide sense, where the block labels depend on the decisions taken for previous sequence items.

The performance of RFasArt whose block diagram is Figure 2, can be described as follows:

- Each part or block as a document logical component: author, page, title, etc...

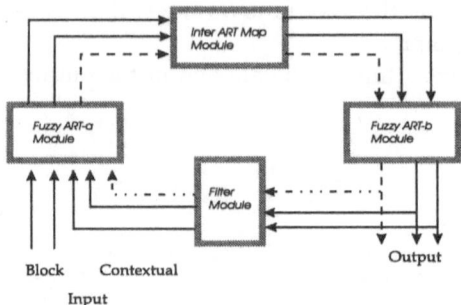

Fig. 2. Block diagram of the proposed neuro-fuzzy RFasArt.

- Each document block is assigned at the output to the logical object class, that presents the highest membership function value.
- Neural networks outputs are feed to the input contextual units, as feedback signals in the filter module. Then, the output for a certain block reflects also its relation with the label of preceding blocks.

3 Experimental system description and results

3.1 Design and structure of the system

The hardware platform of the experimental environment consists of a commercial HP scanner attached to a front-end PC, a Unix Sun Sparcstation that serves as high-level processing and storage unit, as well as a WWW server. Both machines are connected in an Ethernet-based TCP/IP network.

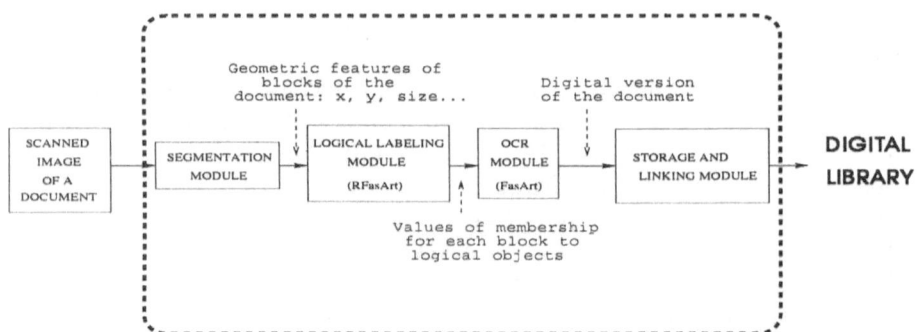

Fig. 3. Overview of the working system

The input of the system consists of scanned images of first pages of scientific papers and tables of contents of the issue where they are contained. The final goal is the establishment of hypertext links between both in a digital library. Then the following system modules, shown in figure 3, were designed and implemented:

- *Segmentation module*, that extracts the hierarchical structure of physical blocks of the document. Through vertical and horizontal segmentations using techniques such as Hough transform, histograms and connected components, we obtain the geometric information of each physical block (position, size, number of lines and number of words).

- *Logical labeling module*, an RFasArt architecture, that uses all features extracted for each block from the segmentation module. This neural network classifies each document block by means of a value that expresses its degree of membership to the fuzzy set that represents a logical object: author, title, abstract, page number, etc... using contextual information about the precedent document block. As a consequence, the entire document can be classified to a certain document class, e.g. table of contents of a specific journal.
- *OCR module*, composed by the FasArt system, that uses features extracted from the images of each character. This set of features includes quantitative information of the character (ratio of black vs total pixels, number and position of vertical and horizontal strokes), and geometric information of the character (position of the character inside the line of text). The results obtained in the previous module are exploited here, since a logical label of e.g. a page number implies the candidate character classes are reduced to those of digits.
- *Storage and linking module*, that stores structured information of the processed document in a digital library, so that it can be possible to be retrieved efficiently. This is carried in two phases:
  1. Information related with the journal structure is extracted from the results of the logical labeling module. Then the tables of the relational database can be defined.
  2. Hypertext links are created, based on a matching procedure between contents of table of contents and articles.

3.2 Experimental results

The working prototype was experimentally tested on a data set of 14 issues from different volumes of the *Neural Networks* journal, i.e. using images scanned in black and white at 300 dpi, thus obtaining 14 journal tables of contents and 31 first pages of articles.

In the initial preprocessing steps, we obtained images with null skew, applying the well known Hough transform to connected image components [11]. Then, applying the image analysis techniques, we extracted physical blocks and the associated features. Such a physical information together with context information (sequence of blocks) is utilized by the RFasArt module to assign each physical block to a logical class.

In order to perform logical labeling of tables of contents, we employed 198 blocks and 702 blocks in the learning and test phase, respectively. The results, shown in Table 1 confirm the efficiency of the employed neuro-fuzzy system. The errors originated in the new logical blocks added by the editors in the most recent journal issues, that were not present in the issues used in the learning set.

We should point out that just one training cycle was required, and the system complexity was rather low, since the dimension of the input pattern was 16 (6 elements with physical information and 10 elements with context information) and the number of categories in Fuzzy ART_a was only 25. Then the computational resources needed for this process are low.

Logical Blocks of Tables of Contents	Results
Title	99 %
Authors	98.8 %
Volume	100 %
Year	100 %
Page number	99 %
Section	97.3 %
TOTAL	99 %

Table 1. Results in Table of contents. RFasArt parameters: $\gamma = 3.5$, $\rho = 0.3$

For the logical labeling process of article first pages, we employed 68 blocks and 387 blocks in the learning and test phase, respectively. The results, shown in Table 2 indicate that more errors are committed in this process, due to the lower discriminating power of the pattern features related with the geometrical aspects of the blocks. Then, sometimes no decision with sufficient confidence degree can be taken, and therefore the context information is incomplete.

Logical Blocks of First Pages	Results
Title	96.8%
Authors	96.8 %
Abstract	80.6 %
Keywords	80.6 %
Page number	96.7 %
Block of text	94.5 %
TOTAL	94.3 %

Table 2. Results in article first pages. RFasArt parameters: $\gamma = 2$, $\rho = 0.3$

The next step in the whole system refers to optical character character recognition, using FasART with an input feature vector of 20 geometrical elements about each character matrix. Note that we did not try the best feature set, since we do not require perfect character recognition for the subsequent task of establishing hypertext links. Thus we could demonstrate that the primary functionality of the build digital libraries can be obtained, even in the case of imperfect OCR.

In a first experiment, we tried to show how previous logical labeling can enhance OCR efficiency. Then, we tried to recognize characters within blocks labeled as *page number*. Then, type of font is uniform and only digits are considered as candidate classes. For tables of contents we employed 50 characters and 556 characters in the learning and test phase respectively, while for first pages we employed 47 characters and 90 characters in the learning and test phase respectively. The results, shown in Table 3 indicate that very few errors are produced, mainly because of digitization noise.

In a second experiment, we processed all blocks and considered all possible classes (94 in an ASCII table). Learning set consisted of 668 (table of contents) and 332 (first pages) characters, while for test we used 17266 and 2952 characters respectively. The results, shown in Table 3, confirm that the increased number of fonts and character classes produced more errors, and therefore the fact that logical labeling contributes in a more efficient performance.

The final task was to establish links between first pages of articles and tables

	In First Pages	In Tables of Contents
Page number	96.6 %	99.1 %
Total characters	89.3 %	87 %

Table 3. OCR results in first pages and table of contents. FasArt parameters: $\gamma = 3.5$, $\rho = 0.3$

of contents, through matching page numbers, and in case of failure, by matching based on article titles. Such a problem is an typical case of string matching [14]. In order to compensate for OCR errors, canonical characters were obtained from the confusion matrix and the Levenshtein distance was measured [21]. For the first type of matching we required exact matching, while for article titles, that are long enough we allowed for a 20 % for errors. For this experiment we disposed of 31 first pages of articles, and we obtained the results, shown in Table 4. The two cases of error in establishing links were due to an error of the image processing module, because the skew of the digitized image was excessive. We included this case, since we think that this is an example of a problem in a real-world application.

Total Number of Links	By Page Number	By Title	Errors
31	21 (67.7 %)	8 (26.3 %)	2 (6 %)

Table 4. Establishment of link through string matching

In Figure 4 we finally show a view of the digital library obtained from the system.

4 Conclusions

Two new neuro-fuzzy architectures were proposed and validated in a real application in this paper.

These models extend the Fuzzy ARTMAP model and formalize the relationship with fuzzy systems, thus being able to consider them as Fuzzy Logic Systems and extract efficiently fuzzy rules form a data collection. Besides FasArt, originally used in a system identification problem, its recurrent version RFasArt permits us to process sequences and thus exploit context information in better terms than in a symbolic system.

These networks served as a unified framework for the solution of the problem of building digital libraries form paper documents. Both tasks of document understanding, logical labeling and OCR, were handled with the proposed systems in an efficient way in terms of recognition accuracy and consumption of computational resources.

Extensive experimental work permitted us to handle tables of contents and article first pages, in order to establish hypertext links, essential for information recovery in digital libraries.

The system was shown to be able to cope with OCR errors and take advantage of logical labeling results for efficient OCR.

Current research work deal with the development of high level architectures that integrate in a better form context and geometrical information, and enhance the information retrieval procedures.

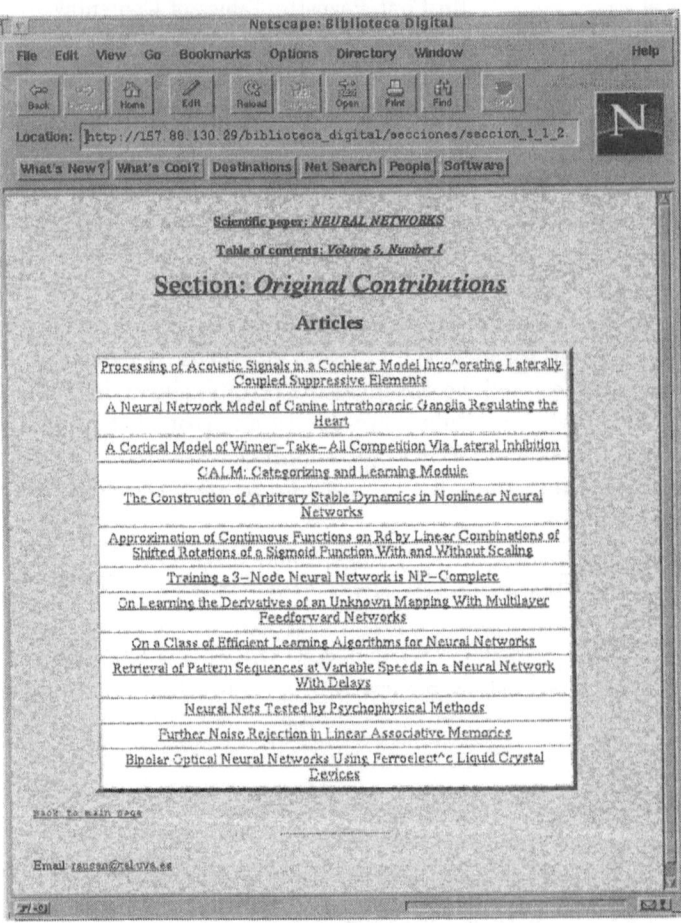

Fig. 4. View of Digital Library

Acknowledgment:

We would like to thank the members of the Neural Networks and of the Document Processing Groups for their contributions to the preparation of this paper.

References

1. Bayer, T. A. Understanding structured text documents by a model based document analysis system. Proc. of the 2nd International Conference on Document Analysis and Recognition, Sukuba Science City, Japan, October 20-22 1993.
2. Bayer, T. A. , Walischewski H. Experiments on extracting structural information from paper documents using syntactic pattern analysis. Proc. of the 3rd International Conference on Document Analysis and Recognition, Montreal, Canada, August 14-16, 1995, 476–479.

3. Cano, J. M. , Dimitriadis, Y.A. Lopez, J. A fuzzy neural arquitecture for supervised learning and classification of temporal sequences. Proc. of International Congress on Artificial Neural Networks , Amsterdam, The Netherlands, 1993.

4. Cano, J. M. , Dimitriadis, Y. A., Lopez, J. FasArt: A New Neuro-Fuzzy Architecture for Incremental Learning in System Identification. Proc. of Congress of International Federation on Automatic Control '96, Vol. F, 133–138, San Francisco, USA, June 30-July 5 1996.

5. Carpenter, G., Grossberg, S., Markuzon, N., Reynolds, J. Fuzzy ARTMAP: A Neural Network Architecture for Incremental Supervised Learning of Analog Multidimensional Maps. IEEE Transactions on Neural Networks, 3 698–713 1992.

6. Carpenter, G. A., Grossberg, S. A massively Parallel architecture for self-organizing neural pattern recognition machine. Computer Vision, Graphics, and Image Processing, 37 54–115 1987.

7. Dengel, A., Bleisinger, R., Hoch, R. From paper to office document standard representation. Computer, 25 63–67 1992.

8. Gladney, H. M., Ahmed, Z., Ashany, R., Belkin, N. J., Fox, E. A., Zemankova, M. Digital Library: Gross structure and Requirements. IBM Research Report RJ 9840 (May 1994).

9. Fujihara, H., Babiker, E., Simmons, D. B. Fuzzy Approach to Document Recognition. Proc. of the 2nd IEEE International Conference on Fuzzy Systems, 980–985, San Francisco, USA, March-April 1993.

10. Govindan, V. K., Shivasprased, A. P. Character recognition: A review. Pattern Recognition, 23 671–683 1990.

11. O' Gorman, L., Kasturi, R. Document Image Analysis. Chapter 4, IEEE Computer Society Press. 161–181, 1994

12. Le, D. X., Thomas, G. R., Wechsler, H. Document image analysis using integrated image and neural processing. Proc. of the 3rd International Conference on Document Analysis and Recognition, 327–330, Montreal, Canada, August 14-16 1995.

13. Myka, A. Putting Paper Documents in the World-Wide-Web. Proc. of the 2nd International WWW conference, October 17-20, 1994, Chicago, 199–208.

14. Myka, A., Guntzer, U. Full-text searches in OCR databses. In Adam, N.; Bhargava, B.; Halem, N.; Yesha, Y. (editors): Advances in Digital Libraries. Springer Verlag, 1996.

15. Nagy, G., Seth, S. C., Stoddard, S. D. Document analysis with an expert system. Proc. of Recognition in Practice II, Amsterdam, June 19-21 1985.

16. Sainz, G. I., Cano, J. M., Dimitriadis, Y. A., Lopez, J. Document Understanding Based on a Neuro-Fuzzy Approach. Proc. of International Conference on Engineering Applications of Neural Networks, Part II, London, UK, June 17-19 1996.

17. Sainz , G. I., Cano, J. M. , Dimitriadis, Y. A., Lopez, J. A New Neuro-Fuzzy System for Logical Labeling of Documents. Proc. of The International Conference on Pattern Recognition, volume Track D: Parallel and Connectionist Systems, pp 431-436, Vienna, Austria, August 25-29 1996.

18. Sainz , G. I., Cano, J. M. , Dimitriadis, Y. A., Lopez, J. Structured document labeling and rule extraction using a new recurrent fuzzy-neural system. Submited to ICDAR'97.

19. Satoh, S., Takasu, A., Katsura, E. An Automated Generation of Electronic Library based on Document Image Understanding. Proc. of International Conference on Document Analysis and Recognition, Vol. I, 63–166, Montreal, August 1995.

20. Srihari, S. N., Lam, S. W., Govindaraju, V., Srihari, R. K., Hull, J. J. Document Image Understanding. CEDAR-TR-92-1. (May 1992). Available at http://www.cedar.buffalo.edu/Publications/TechReps/Survey/survey.html

21. Wu, S., Manber, U. Fast text searching allowing errors. Communications of ACM, 35 83–91 1992.

Parallelization of Connectionist Models Based on a Symbolic Formalism*

J. Santos,[1], M. Cabarcos,[2], R.P. Otero[1] and J. Mira[3]

[1]Dept. of Computer Science
University of A Coruña
E-15071 A Coruña, Spain
e-mail: {santos, otero}@dc.fi.udc.es
[2] Dept. of Computer Languages and Systems
University of Vigo
Ourense, Spain
email: cabarcos@uvigo.es
[3] Dept. of Artificial Intelligence
UNED
E-28040, Avd. Senda del Rey s/n, Madrid, Spain

Abstract. In this paper we study the parallelization of the inference process for connectionist models. We use a symbolic formalism for the representation of the connectionist models. With this translation, the training mechanism is local in the elements of the network, the computing power is improved in the network nodes and a local hybridization with symbolic parts is achieved. The inference in the final knowledge network can be parallelized, whether the knowledge corresponds to a symbolic module, a connectionist model or a hybrid connectionist-symbolic module. Besides, the concurrency for knowledge networks corresponding to connectionist models is presented for the phases of processing and training. The parallelization is studied for a multiprocessor architecture with shared memory.

1 Introduction

The solution usually applied for the implementation of connectionist models is the simulation by software. The training and processing mechanisms of the network are implemented on a hosts, that drives the simulation, activates the network nodes and updates the connection weights. Another approach, more related to the biological model, uses autonomous processors that locally implement the network mechanisms. These two extreme solutions have been called [7]: a) "all in the host, nothing in the network", b) "all in the network, nothing in the host". Our approach is situated between these two alternatives as the processors have a high autonomy, but they are activated by a host.

---

* This research was supported in part by the Government of Galicia (Spain), grant XUGA10503B/96.

We implement the connectionist models using a symbolic formalism. This translation introduces new capabilities for the nodes of the connectionist model. All the knowledge needed for the processing and training phases of the network elements are locally placed into the representation of each element. This distribution of knowledge makes our connectionist models to be massively parallel, as in the biological approach. This characteristic has been exploited for improving the parallel execution of the network.

Besides, the symbolic representation of the connectionist models increases the processing capability of the network nodes by adding the possibility of conditional calculus. In this way, we generalize the classical solution based on analogical and logical calculus in the network elements. This concept, that has been defined as *inferential networks* [8], represents a bridge between connectionist and symbolic representations.

Finally, the translation allows a local hybridization between the connectionist and the symbolic parts of a system. The hybridization is introduced at low level, where the network elements (information transfer elements and learning elements) can interact with the symbolic ones. This idea is on the foundations of Nettool [10], a connectionist-symbolic hybridization environment.

We could think of a hardware implementation of the network processing, supported by hardware processors with enough processing capability for the symbolic formalism. Instead of this, our approach to the problem is the evaluation using a multiprocessor system. The extraction of concurrency is automatic, and we can dynamically schedule the evaluation of the network elements according to different criteria.

An example of a parallel solution with hybrid processing is DESCARTES [4]. It is a development tool that simulates heterogeneous connectionist networks, and combines characteristics of distributed, localist and symbolic marker-passing networks. The nodes and the links of the network may have different processing functions. All the nodes are activated synchronously, therefore the propagation of a vector from input to output in a feedforward neural network require as much cycles as layers, although only are evaluated the nodes that need to update their value. The parallelization is implemented on a SIMD machine, where a processor is allocated for each network node. Each active processor in a SIMD machine must execute the same instruction, so each specific processing characteristic of the active nodes or links must be simulated sequentially, while processors not using that function must remain idle.

In our approach, the extraction of parallelism works in the two main phases of a neural network: the processing phase and the learning phase. This is a consequence of the representation of the learning mechanisms using our model. On the contrary, other hybrid solutions that can be parallelized do not implement the learning phase. The object oriented model presented in [3] allows the processing of heterogeneous networks, with symbolic and numerical nodes. The connections also can be numerical between numerical nodes, symbolic between symbolic nodes and "hybrid" after a cast from symbolic concepts to numerical entities. The final system can be executed in a sequential machine as well as in a multiprocessor transputer system.

This paper is organized as follows. First, we comment on the implementation of connectionist models using our symbolic formalism. In section 3, we study the automatic parallelization of the inference for those models. Finally, we present the results of a simulation study for the parallelization, and the conclusions.

2 Implementation of connectionist models

The symbolic formalism we use is based on "microframes" named Generalized Magnitudes (GMs). The facts and events of the system are grouped in GMs. The GMs have the property that only one of the facts inside it can be true in each situation, while the remaining facts of the GM are false [6].

We distinguish between *primitive GMs* whose facts are input facts, and *derived GMs* whose facts are conclusion facts. Each derived GMs incorporates a slot named *Knowledge Module (KM)*, that contains an expression used for obtaining the true fact of the GM. The syntax of a KM allows multiple conditional levels or "clauses" that are evaluated in sequential order. The KM can include references to other facts, expressing dependencies between GMs. The language of representation incorporates a set of logical, mathematical and relational operators, including the conditional, for the definition of relations between numerical or non numerical GMs, as well as operators for temporal relations.

As for the implementation aspects of a neural network, we have distinguished between *transfer nodes* and *learning nodes*. The first ones correspond to the nodes of a neural network, and the second ones correspond, among others, to the connections between the transfer nodes. The learning nodes incorporate the connectionist learning mechanisms locally, in the knowledge module of the corresponding GM. The GMs that define a connection learning node are of the following form:

GM: Connection (C)
KM: current value of C if MODE = recall;
modification of C according to learning algorithm if MODE = training;
initial value if MODE = initiation;

where we use a "control GM" named MODE that especifies which clause in the connection GM must be evaluated, so the KM expression is valid for the two phases of the network. The conditional relation with this MODE GM, allows the definition of the possible modes of activation of the network.

The temporal reasoning mechanisms of the underlying formalism are basic for the network representation. The possibility of reference to the information of the previous situation avoids an extra process of actualization in many nodes, that otherwise would be done by an external host when changes occurs in the activation mode of the network. This is done by associating each one of the consecutives activations of the network (in recall or learning mode), with situations or consecutive states in the model represented over the GM temporal formalism.

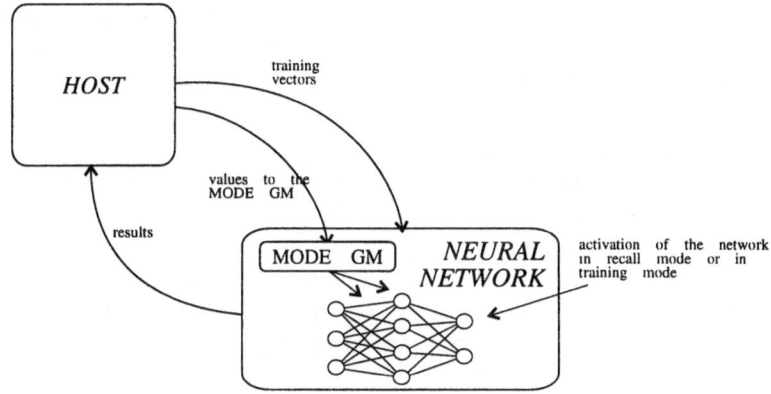

Fig. 1. External control of the evaluation mode of a network

Our solution is represented in figure 1. The transfer and learning mechanisms are local into each corresponding GM, while the host is only responsible of the activation of the knowledge, by means of the true values assigned to the MODE GM. Examples of implementation of several learning algorithms can be found in [10].

Finally, the global operation is as follows. The transfer and learning GMs, together with the MODE GM, define a knowledge base (KB) wich will be used by the application of knowledge mechanism. In the recall phase we assign the values of an input vector to the primitive input GMs of the network, and the "recall" value to the MODE GM. Then, a new value is inferred for the transfer nodes or GMs. In the training phase, the "training" value is assigned to the MODE GM, target values are assigned if necessary to target GMs, and new values are inferred in learning nodes. Thus, the evaluation in recall mode and in training mode (or in any other phase) needs a different application of knowledge.

The presented methodology is general enough to be extended to any model that incorporates this two (or more) forms of functioning. Furthermore, the learning methods are not restricted to the classical connectionist learning algorithms, and new methods can be included for learning at inferential level, in an intermediate position with symbolic learning.

3 Parallelization of the inference in connectionist models

Most of the parallel implementations of connectionist models are for particular machines. These solutions can exploit efficiently the parallelism of the neural network, and can benefit from the characteristics of the machine, but they lack portability.

Our solution includes the automatic parallelization of the neural net inference, and it can adapt to any machine with variable number of processors. It is based on the parallelization of representations in the GM formalism, that allows an automatic extraction of concurrency in the models represented, whether the knowledge corresponds to a symbolic module, a connectionist module or a hybrid connectionist-symbolic module.

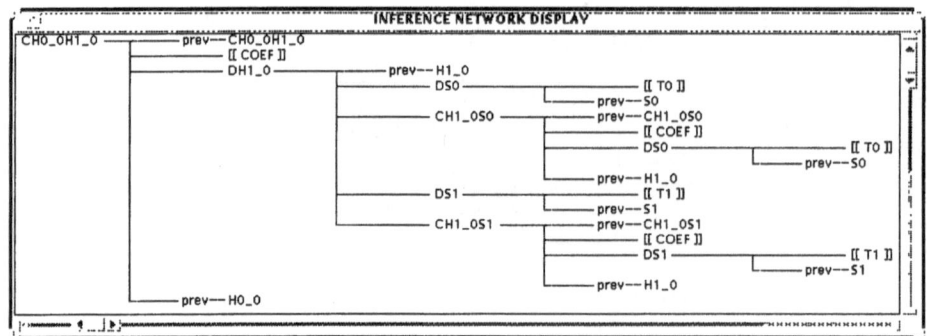

Fig. 2. Data flow between GMs in training mode

As the knowledge is local to the GMs, and only part of it will be used in the recall phase or in the learning phase, we can distinguish two different inference networks: one that define the static relationship between the GMs in the recall phase, and another for the relationship in the learning phase. The inference network in recall mode expresses the dependencies between the transfer nodes. In learning mode it represents the dependencies between the learning nodes, and shows how the learning algorithm is distributed among the nodes. Figure 2 shows the dependencies between some of the learning GMs for a network with the backpropagation training algorithm. Four levels of dependencies are represented and it can be seen that the GMs of one level can be evaluated in parallel. The "prev" label represents a dependence modified by the *previous* operator, that returns the previous value of a GM. It is necessary for updating the values of many GMs when the mode of activation of the network changes. The COEF GM is the learning coefficient.

Our purpose is to automate the analysis of the topology of the inference network, scheduling the evaluation of the different GMs over a multiprocessor architecture with shared-memory. In an architecture without shared-memory, the problem will be transformed in finding an adequate partition of the knowledge network with similar evaluation times and with minimized communications among processors. This partition will not be possible for a great variety of KBs.

In [9] several levels of concurrency in connectionist models are commented. We consider the parallelization at the GM or node network level. A finest grain, as for example the different clauses of the knowledge expression, will introduce important overheads due to the required control compared with the usually small size of the clauses. On the contrary, a coarser grain size will require a previous topology analysis.

A bottleneck appears in the parallelization of connectionist models when the activation mode changes. There is many required applications of knowledge, althought it is possible in some models, like backpropagation, to process the training phase with the total accumulated error. This possibility allows the temporal parallelism by means of pipelined hardware structures [1], that process different training vectors in different segmented phases, in recall mode and in learning mode.

The temporal parallelism is studied in [5] for expert systems using the GM formalism, introducing the concept of *swell processing*. This technique allows starting the change propagation (wave) for the next situation when the current one has not finished yet. In the evaluation of a KB only a subset of GMs are updated. Those GMs are said to be "pertinent", and their value must be updated because they depend directly or indirectly on the new input values. This is not the case in connectionist models where it is usual to enter a whole input vector with new values for all the input GMs, and this way all the transfer nodes become pertinent. The same reasoning can be made for the learning phase. In general, in expert systems and hybrid connectionist systems, the applications of knowledge are not necessarily started with new input values for all the primitive GMs.

However, we must consider separately the parallelization of the inference in the recall phase and in the learning phase. We generate two knowledge networks, one related to the evaluation of the transfer nodes, and another when infer new values the learning nodes.

4   Simulation results

The applicability of this parallelization techniques in connectionist models has been studied by simulation. We have defined the *effective current layer* as the set of GMs ready for evaluation, that is, the GMs with all their dependencies solved. This layer may contain GMs belonging to different static dependence layers as a consequence of the dynamics of the KB evaluation. The evaluation of a GM solves dependencies of another GMs in the next layers, and, in turn, they can be included in the effective current layer.

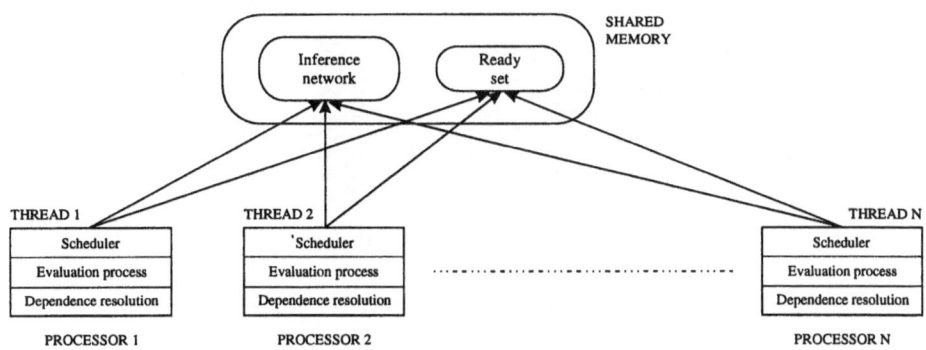

Fig. 3. Solution based in evaluation with threads.

The transition from simulation to real execution is starightforward. The basic idea is to benefit from multiprocessor operating systems facilities, as in [2], in which each of the processes created in concurrent C++ corresponds with a process directly managed by the operating system. In our case a good solution is the generation of a thread for each processor, and a global ready set for storing the GMs belonging to the effective current layer. Each thread will execute a loop with three basic operations, as represented in Figure 3: 1) retrieve a ready GM by applying the scheduling criterion; 2) evaluate the GM; and 3) mark as solved all the dependencies from the GM.

We have simulated dynamic algorithms with minimum overheads at runtime. They are based on static information obtained from a priori analysis of the KB. This information considers several GMs properties, as for example the connectivity between layers, the number of dependencies of a GM, the number of output connections to other GMs, the number of dependence layer of a GM, the processing time of a GM, and the accumulated time to obtain the final conclusions from the evaluation of the GM. Previous studies with symbolic KBs have showed that the better criteria are those with higher priority to the GMs with maximum accumulated time (that favours GMs far away in time respect to the final conclusions), and next, the algorithms that have the layer number of a GM as a priority criterion (that favours GMs far away in intermediate dependencies to the final GMs) [5].

For connectionist models with total regularity in the network, in the recall phase almost all the scheduling algorithms have the same behaviour. All the GMs have the same processing time, the same accumulated time to the final conclusions, the same number of dependencies with the nodes of the next layer, etc..., and the current effective layer can not include GMs of different levels of dependence. These characteristics imply that all the algorithms would behave like the FIFO algorithm, but it is important to remember that the extraction of concurrency is automatic, and when the total connectionism between layers does not hold, the scheduling criteria will perform different.

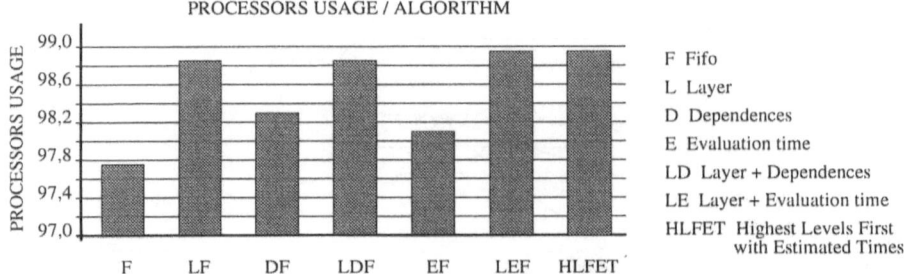

Fig. 4. Processors usage in the learning phase

However, there is not total connectionism between the levels of dependence of the learning GMs in the learning phase. In this stage, the connections infer a new value with local information from the previous dependence layer, and the dependence of the learning GMs in that level is not revealed. Figure 4 shows the time percentage of processors usage (eight in this case) achieved by the different scheduling algorithms, for a backpropagation network with 30 input nodes, 10 output nodes, and two hidden layers of 15 and 10 nodes respectively. The criteria with better results have been the HLFET algorithm (Highest Levels First with Estimated Times) and LEF (number of Layer, Evaluation time and Fifo).

The speedup has been also measured by simulation, and the results have been very positive. When the number of processors is not greater than the average number of GMs in each dependence level, the speedup increases almost linearly with the number of processors. The parallel efficiency is basically established by the relationship between the number of nodes of a dependence layer and the number of processors.

5 Conclusions

We have studied the parallelization of connectionist models represented using a symbolic formalism. This translation of the processing and learning mechanisms of connectionist models has several properties, as the localization of the algorithms in the network elements. It allows an automatic extraction of concurrency with any knowledge base – connectionist, symbolic or hybrid. In this sense we schedule the evaluation of the knowledge elements in a shared-memory multiprocessor. In connectionist models, the extraction of parallelism also works in the two phases of the network, the recall phase and the learning phase.

Our approach also allows the inclusion of priority criteria for the use of the KBs in run-time systems, where it is important to advance the evaluation of knowledge elements that contribute to critical decisions.

References

1. H. Chung, J-H. Yoon and S.R. Maeng. A sistolic array exploiting the inherent parallelism of artificial neural networks. *Microprocessing and Microprogramming*, 33(3):145–159, 1992.
2. M. Heileman, G.L. Georgiopoulos and Roome W.D. A general framework for concurrent simulation of neural network models. *IEEE Transactions on Software Engineering*, 18(7):551–562, 1992.
3. T. Lallement, Y. Cornu and Vialle S. An abstract machine for implementing connectionist and hybrid systems on multi-processor architectures. *Parallel Processing for Artificial Intelligence, Machine Intelligence and Pattern Recognition Series, H. Kitano, V. Kumer and C.B. Sutter (Eds.), Elsevier Science*, (2), 1994.
4. T.E. Lange. Simulation of heterogeneous neural networks on serial and parallel machines. *Parallel Computing*, (4):287–303, 1990.
5. Cabarcos M., Otero M., Cabalar P., and Otero R.P. Efficient concurrent execution of medtool expert systems. *EXPERSYS-96: Expert systems applications and artificial intelligence, Technology Transfer series, A. Niku-Lari (Ed)*, pages 389–394, 1996.

6. The medtool project and related activities are described in web documents http://www.dc.fi.udc.es/ai/medtool.html, 1997.
7. J. Mira. Computación neuronal. *Curso de la UNED*, 1992.
8. J. Mira and A.E. Delgado. Linear and algorithmic formulation of cooperative computation in neural nets. In F. Pichler and Moreno-Diaz, editors, *Computer Aided Systems Theory*, pages 2–20. Springer-Verlag, 1991. Lecture Notes in Computer Science, Vol. 585.
9. T. Nordstrom and B. Svensson. Using and designing massively parallel computers for artificial neural networks. *J. Parallel and Distributed Computing*, 14(3):260–285, 1992.
10. J. Santos, R.P. Otero, and J. Mira. Nettool: A hydrid connectionist-symbolic development environment. In J. Mira and F. Sandoval, editors, *From Natural to Artificial Neural Computation*, pages 658–665. Springer-Verlag, 1995. Lecture Notes in Computer Science, Vol. 930.

Generic Neural Network Model and Simulation Toolkit

Montserrat García del Valle, Carlos García-Orellana, Francisco J. López-Aligué and Isabel Acevedo-Sotoca.

Departamento de Electrónica e Ingeniería Electromecánica
Universidad de Extremadura
Avda. de Elvas s/n. 06071 - Badajoz - SPAIN
E.mail: montse@nernet.unex.es
Phone: +34-24-289542
Fax: +34-24-289543

Abstract: This work presents a generic higher-order model of neuron behaviour, connection scheme and learning rule, suited for high-speed parallel processing. In contrast to the construction of a real application, it would be more operational to expend effort in parameterizing the problem-solving architecture, offering a testbed as a useful simulation tool for experimenting with a variety of network designs within the said generic model. We include some initial simulation results applied to image processing and pattern recognition tasks.

1. INTRODUCTION

Artificial neural network models have been studied for many years with the aim of achieving human-like performance in fields such as searching, representation and learning, closely related to the associative property and self-organizing capability of the brain.

Models of artificial neural networks are specified by three basic factors: the neuron's behaviour itself (transfer function and neuron characteristics), the network topology (neuron interconnection structure) and the learning rules (mechanisms for adjusting the weights). Differences in the theoretical bases of these aspects support the variety of models developed to date.

Within this framework, we propose a generic multilayer higher-order neural model. In the following sections of this paper, we therefore present a new methodology for describing the functioning of artificial neural networks, including a new mathematical approach to the neuron's behaviour as well as a general connection pattern.

Because it is important to researchers to be able to implement and test ideas, we present a toolkit that supports the construction and simulation of a wide variety of networks as far as the flexibility of the generality allows. With this simulation toolkit, it will no longer be necessary to make any changes in the network design philosophy, but just to make a choice from a proposed library of connection patterns, activation functions and training algorithms, so that the generic model can be molded to whatever practical realization one likes.

The last section of the paper presents some early simulation results of two models resulting from particularizations of the generic case, applied to pattern recognition and image processing tasks.

2. A GENERIC NEURAL MODEL

In the course of their evolution, neural networks have often been used as associative memories. One way to construct these associative memories has been to develope a generalized higher-order equation for defining the neuron's behaviour [1].

From the conventional formulation for higher-order neurons, we propose a general and flexible non-restrictive model that leaves room for a wide variety of applications to be designed by specifying the values corresponding to a series of coefficients:

$$F_{ordp} = C_{ordp}\left[\sum_{J_1-J_p=1}^{n} m^i_{J_1-J_p} \varphi^{orap}_{J_1}\left(x_{J_1}, \varphi^{ordp}_{J_2}\left(x_{J_2}, \ldots \left(x_{J_{p-1}}, x_{J_p}\right)\right)\right)\right) - \theta_p\right]$$

Then we can take as neuron function

$$y_i = \psi\left[\varphi_{ord1}\left(F_{ord1}, \varphi_{ord2}\left(F_{ord2}, \ldots\right)\right)\right]$$

where the elements φ_{ordk} and $\varphi^{ordp}_{J_k}$ can be chosen from among the functions listed in Table I. The function ψ is chosen to be either linear and monotonically increasing, such as the sigmoid function, or with a sharp change, such as the threshold function, depending on the type of inputs x_j being dealt with (continuous or binary). The weight m^i_j represents the strength of the connection between the neuron i and the neuron j, and C_{ordp} is chosen to normalize the function.

Once the activation function of the neuron has been defined, the behaviour of the network is determined by the pattern of connections and the weights on the links. Neurons in our network form layers, each one receiving at the same time information that computes synchronously, suiting the network to high-speed parallel processing. We propose a complete connectivity scheme for a multilayer network comprising feed-forward links where the signal flow from one layer to the next is unidirectional, feedback links where backward connections from one layer to an "earlier" layer exist, and lateral interactions that allow feedback paths within the same layer.

Unit:	$\forall\, x_i, x_j \in [0,1]$	$\varphi_{UNIT}[x_i, x_j] = x_i$
Binary AND:	$\forall\, x_i, x_j \in \{0,1\}$	$\varphi_{BAND}[x_i, x_j] = x_i \bullet x_j$
Binary OR:	$\forall\, x_i, x_j \in \{0,1\}$	$\varphi_{BOR}[x_i, x_j] = x_i + x_j$
Binary EXOR:	$\forall\, x_i, x_j \in \{0,1\}$	$\varphi_{BXOR}[x_i, x_j] = x_i \oplus x_j$
Analog Multiplication:	$\forall\, x_i, x_j \in [0,1]$	$\varphi_{AMUL}[x_i, x_j] = x_i * x_j$
Analog Addition:	$\forall\, x_i, x_j \in [0,1]$	$\varphi_{AADD}[x_i, x_j] = x_i + x_j$
Fuzzy AND:	$\forall\, x_i, x_j \in [0,1]$	$\varphi_{FAND}[x_i, x_j] = \min(x_i, x_j)$
Fuzzy OR:	$\forall\, x_i, x_j \in [0,1]$	$\varphi_{FOR}[x_i, x_j] = \max(x_i, x_j)$
Fuzzy EXOR:	$\forall\, x_i, x_j \in [0,1]$	$\varphi_{FXOR}[x_i, x_j] = |x_i - x_j|$

Table I. The set of different functions φ

The connection scheme proposed is based on the concept of neighbourhood found in cellular logic [2][3][4]. Each neuron in a layer is directly connected to its neighbour cells (Figure 1). The symmetric neurons in the remaining layers of the network together with their respective neighbourhoods can also interact with the neuron under consideration. The size of these neighbourhoods, measured by their radius, depends on the distance from the input layer (Figures 2 and 3).

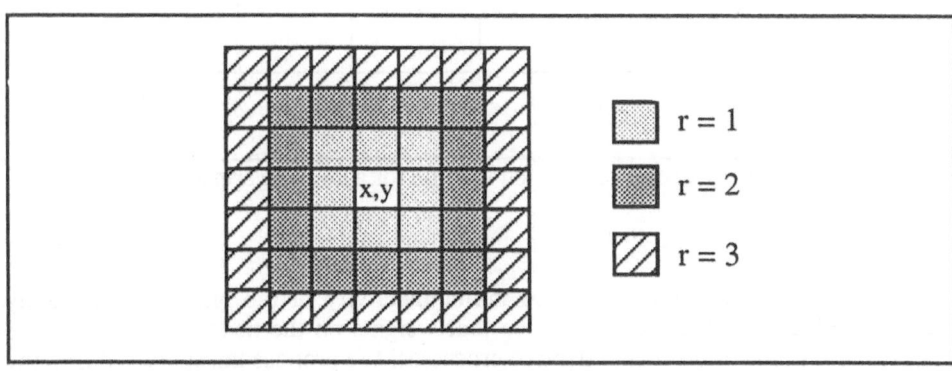

Figure 1. The neighbourhood of the neuron (x, y)

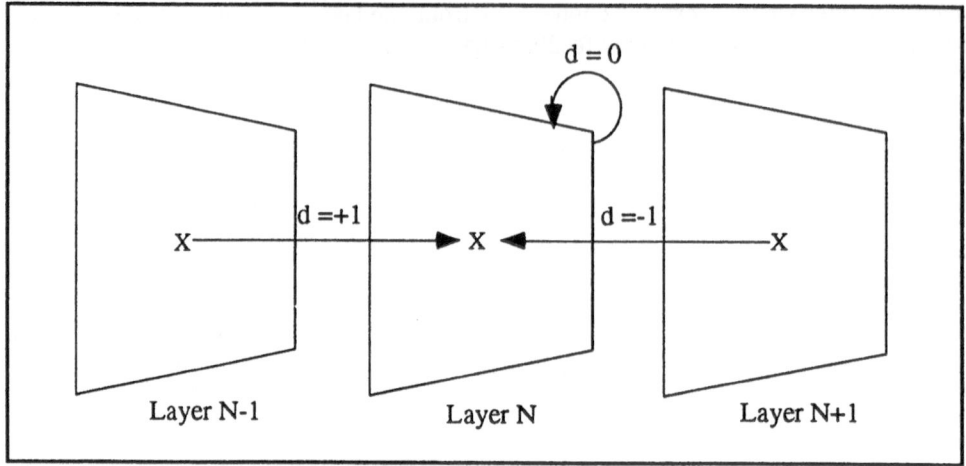

Figure 2. Distance between layers

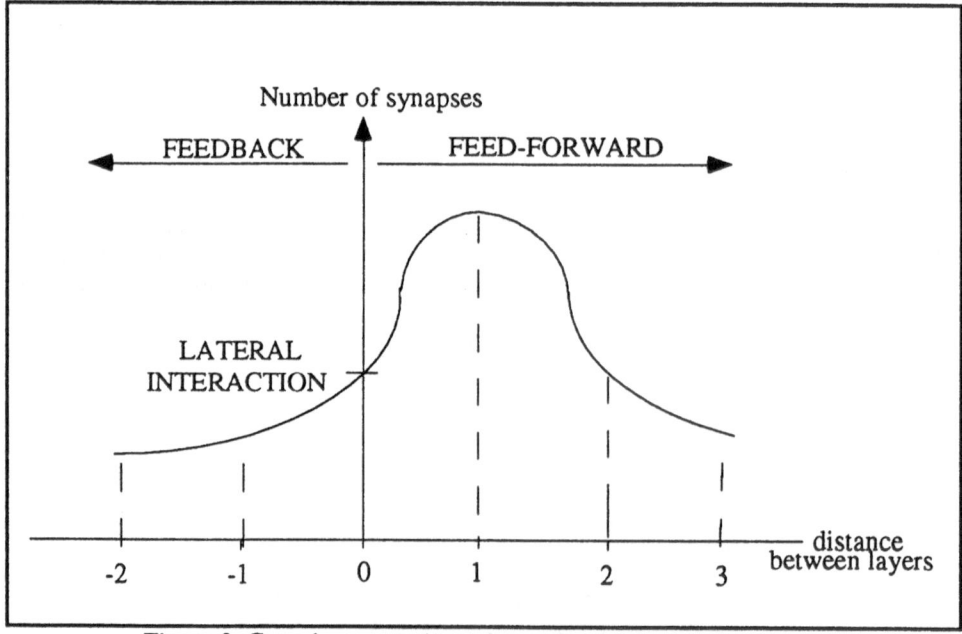

Figure 3. Generic connection scheme for a three layer network

In 1949 Hebb proposed that the connectivity of the brain is continually changing as an organism learns [5], so that weights on the neuron links are to be adjusted by means of different adaptive processes. We propose unsupervised training procedures which construct internal models that capture regularities in input signals. The learning rules state that changes in the synaptic weight connecting two neurons are directly proportional to their activities (measured in terms of the magnitudes of the inputs or outputs). At the same time, as that synaptic weight comes close to its bounds, sensitivity to change decreases.

The above learning mechanism corresponds to the following equations:

first order

$$m_j^i(t+1) = sign\left(m_j^i(t)\right) * \sigma\left[\overline{x_j y_i}(t+1)\right]$$

$$\overline{x_j y_i}(t+1) = \overline{x_j y_i}(t) + \frac{\varphi\left[x_j(t+1), y_i(t+1)\right] - \overline{x_j y_i}(t)}{\Delta} \quad ,, \quad \overline{x_j y_i}(0) = 0$$

higher orders

$$m_{j_1 \cdots j_r}^i(t+1) = sign\left(m_{j_1 \cdots j_r}^i(t)\right) * \sigma\left[\overline{x_{j_1} .. x_{j_r} y_i}(t+1)\right]$$

$$\overline{x_{j_1} .. x_{j_r} y_i}(t+1) = \overline{x_{j_1} .. x_{j_r} y_i}(t) +$$

$$+ \frac{\varphi\left\{\varphi_{j_1}^{ord\varphi}\left[x_{j_1}, \varphi_{j_2}^{ord\varphi}\left[\ldots \varphi_{j_{r-1}}^{ord\varphi}\left(x_{j_{r-1}}, x_{j_r}\right)\right]\right], y_i\right\}(t+1) - \overline{x_{j_1} .. x_{j_r} y_i}(t)}{\Delta}$$

with $\overline{x_{j_1} .. x_{j_r} y_i}(0) = 0$ and where Δ is a learning coefficient and the elements φ can be chosen from among the functions listed in Table I.

Because of the high cost in computing time that a large value of the order p may cause, we propose an alternative simplified formulation for higher orders:

$$m_{j_1 \cdots j_r}^i(t+1) = \varphi_{j_1}^{ord\varphi}\left[m_{j_1}^i, \varphi_{j_2}^{ord\varphi}\left[\ldots \varphi_{j_{r-1}}^{ord\varphi}\left(m_{j_{r-1}}^i, m_{j_r}^i\right)\right]\right](t+1)$$

where m_j^i has previously been adapted.

3. SIMULATION

Once a generic model has been designed, the problem of implementing and testing different types of networks is one of experimental method. In order to avoid a complete rework for every new application, a toolkit for defining and configuring the networks must be made available.

The proposed neural network development environment incorporates the following basic functional modules:

Network designer

This module facilitates the construction task by supplying menu-driven specification of coefficients and parameters for neurons, links, and learning strategies in order to initialize the testbed network. These data may be either written by the user or supplied from a library. Researchers may add their own structuring functions to the library, so that, when displaying, saving or reloading the network, user-supplied functions are called by the simulator to handle these user-defined structures.

Network simulator

This module provides support for executing an application and facilitates experimental runs. The main goal of a simulator is to run a system under varying environmental scenarios, in this case, different topologies, different activation functions and different learning rules, to test the networks before their being implemented.

Network analyzer

In order to help in the analysis of the results of an experiment, this module offers optional facilities such as performance monitoring, that provides capabilities for displaying weights and outputs, or fuzzy classification, through which one may know the degree to which an output of the network belongs to a class [6].

Network implementer

This module handles some mechanism to start up and run the simulated network on a real parallel environment. The access to a multi-CPU system with several processors increases the speed and capacity of network simulations taking advantage of the high degree of parallelism that characterizes connective architectures.

Besides these main well-defined modules, the toolkit incorporates an external connection module to interface with the data acquisition peripherals as such as cameras, audiorecorders, sensors, etc.

4. RESULTS

In this section, we present early results of the opening applications simulated with the proposed toolkit. These initial simulations were applied to pattern recognition and image processing tasks.

The first simulation we carried out was based on a very simple neural model design. The network consists only of two layers of $100x100$ neurons each, with a feedforward connection scheme, and no learning. Choosing a second order equation for the neuron behaviour, we obtained exciting results closely related to cellular logic theory (Figure 4).

Let us complement the above network design by adding an adaptive process and a complete connection scheme, including feedforward and feedback links as well as lateral interactions. As in the former example, we take the following selection for the neuron function:

$$y_i(t+1) = \psi \left\{ C_{ord1} \left[\sum_{j_1=1}^{n} m_{j_1}^i x_{j_1}(t) - \theta_1 \right] + C_{ord2} \left[\sum_{j_1 j_2=1}^{n} m_{j_1 j_2}^i \varphi_{j_1}^{ord2} \left[x_{j_1}, x_{j_2} \right] - \theta_2 \right] \right\}$$

At the same time, we subject the synaptic weights to an adaptive mechanism:

$$m_{j_1 j_2}^i(t+1) = \varphi_{j_1}^{ord2} \left[m_{j_1}^i(t+1), m_{j_2}^i(t+1) \right]$$

where

$$m_j^i(t+1) = sign\left(m_j^i(t) \right) * \sigma\left[\overline{x_j}(t+1) \right] \quad \wedge \quad j = j_1, j_2$$

For the simulation of this model design, we used two different networks. First, we considered a three layer network of 1 600 neurons each, where a neuron was physically connected to 121 neurons per layer at most (equivalent to a radius 5 neighbourhood). Thus we obtained a network with 6 400 neurons and 2 323 200 real links. Only one neuron in a layer receives 644 inputs; 484, coming from the neurons which it is physically connected to, contribute to the first-order term of the neuron behaviour; the remaining 160 are virtual inputs (result of applying a function to two real activities) and define the second-order expression of the activity function. Therefore, we are working with a network with a maximum of 4 121 600 inputs. With the proper values given to the simulation parameters and the second-order term emphasized, the said network learns to detect edges on a picture (Figure 5).

Based on the same model design, a one layer network of 10 000 neurons with 178 inputs each was used for pattern recognition. Choosing the functions suitably, we obtained high performance for the identification of noisy images (Figure 6). Connecting the network response to the fuzzy classifier that the simulation toolkit provides, we were able to observe a really good classification for all the noisy pictures presented (Table II).

categorization %	Noisy input P1	Noisy input P2
Pattern1 class	169.26	31.88
Pattern2 class	0.00	100.78
Pattern3 class	0.00	8.40
Pattern4 class	0.00	63.47

Table II. Fuzzy classifier results

320

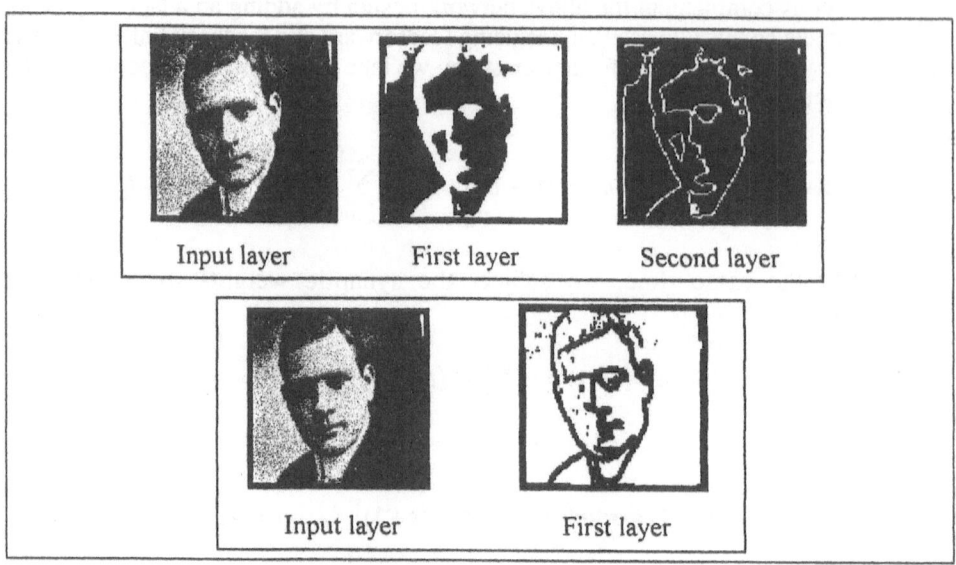

Figure 4. Image Processing

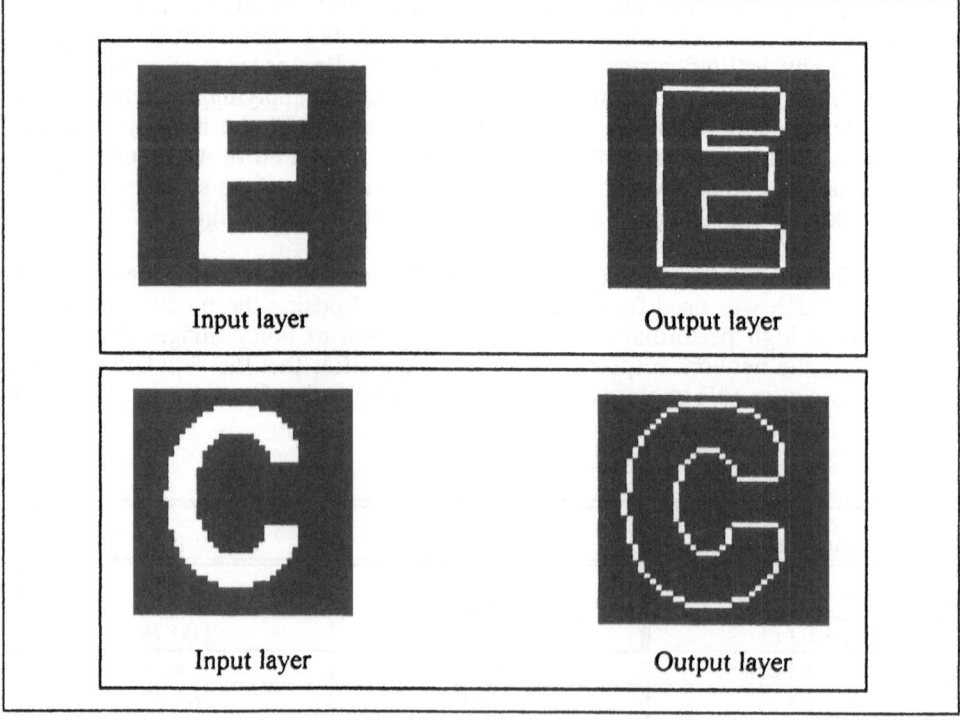

Figure 5. Simulation for edge detection

Figure 6. Pattern recognition

Network response to patterns 1 and 2

Pattern 1

Input layer

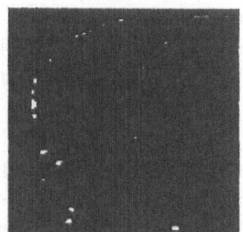

Output layer

Pattern 2

Input layer

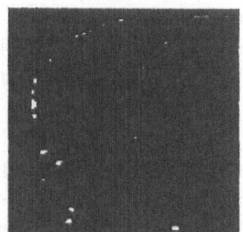

Output layer

Network response to noisy inputs

Noisy input P1

Input layer

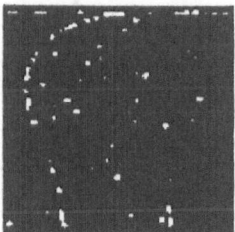

Output layer

Noisy input P2

Input layer

Output layer

5. CONCLUSIONS

This work provides a complete and powerful working environment, since the model designed allows the simulation of a great number of neural networks of different categories and functionalities.

We have presented an approach to a higher-order model with a cellular connection scheme, remembering that the higher-order models have already allowed image treatment processes when they had previously seemed reserved for other methodologies such as cellular logic [2]. Cellular logic is currently the object of attention because of its possible synthesis by means of neural networks [4][7] in which there has also been a powerful incidence of the use of higher-order neural models [8][9], given the evident similarity between the two structures.

Therefore, image processing and pattern recognition tasks have been the purpose of our initial simulations. This does not mean it would not be possible to find other possibilities in new fields such as in process control. As we design and simulate more and more neural network models, our system will grow and extend in capabilities.

Moreover, the implementation of neural networks in multi-CPU systems for high performance parallel processing acquires a special importance since it is possible to handle ambitious networks of considerable scope.

REFERENCES

[1] X. Xu and T. Tsai, "Constructing associative memories using neural networks", *Neural Networks*, vol.3, no.3, pp. 301-309, 1990.

[2] A. Rosenfeld and A.K. Kak, *Digital Picture Processing*, New York: Academic Press, 1982.

[3] K. Preston Jr., "Cellular logic computers for Pattern Recognition", *Computer*, 1, 36-47, 1983.

[4] L.O. Chua and L. Yang, "Cellular Neural Networks: Theory", *IEEE Transactions on CAS*, vol.35, no.10, pp. 1257-1272, 1988.

[5] V. Vemuri, "Artificial Neural Networks: An Introduction", *Artificial Neural Networks: Theoretical Concepts*, IEEE Computer Society Press, 1988.

[6] F.J. López, M.I. Acevedo and M.A. Jaramillo, "The fuzziness of fuzzy partitions", *Pattern Rcognition Letters*, 12, 265-271, North-Holland, 1991.

[7] L.O. Chua and L. Yang, "Cellular Neural Networks: Applications", *IEEE Transactions on CAS*, vol.35, no.10, pp. 1273-1290, 1988.

A Neural-Fuzzy Technique for Interpolating Spatial Data via the Use of Learning Curve

P.M. Wong[1], K.W. Wong[2], C.C. Fung[2] and T.D. Gedeon[3]

[1]Centre for Petroleum Engineering, The University of New South Wales, Sydney 2052, Australia

[2]School of Electrical and Computer Engineering, Curtin University of Technology, GPO Box U1987, Perth 6001, Australia

[3]School of Computer Science and Engineering, The University of New South Wales, Sydney 2052, Australia

ABSTRACT

In this paper, we present a new and simple method for function interpolation based on the use of neural networks and fuzzy logic. We particularly discuss the application of the technique in spatial data analysis. In this application domain, conventional early-stopping criteria to avoid over-training in neural networks based on the use of minimum error on validation set, may not be suitable. In the proposed method, we use "interpolated error" to stop training. The trained networks are used as fuzzy rules, and these rules are interpolated to the location of interest. We demonstrate the methodology in petroleum reservoir modelling in which properties are estimated between two oil wells. Data from a third well, which are withheld from the training process, is used to evaluate different prediction models. We also compare our method with the recent data-splitting approach using self-organising map (SOM) with the use of early-stopping in neural training. The results of this study show that the SOM approach is only applicable to wells in which their locations are half-way between the two given wells. The proposed methodology, however, provides the best results in the test well and is also suitable for any location of interest. The end result is a simple and computationally-cheap method in engineering studies.

Keywords: Neural, Fuzzy, Interpolation, Spatial Data, Well Logs.

INTRODUCTION

Function approximation and interpolation are important in most engineering disciplines. The former involves the fitting a curve or a surface through existing data points. These data points are usually clustered. One example of function approximation method is linear regression in which the total error of predictions is minimised. The later is similar to function approximation, however, the available data points are more scattered and separated in space (e.g. distance). One example is distance-weighted interpolation method in which the weighting factors of samples are calculated based on the location of the point-of-interest.

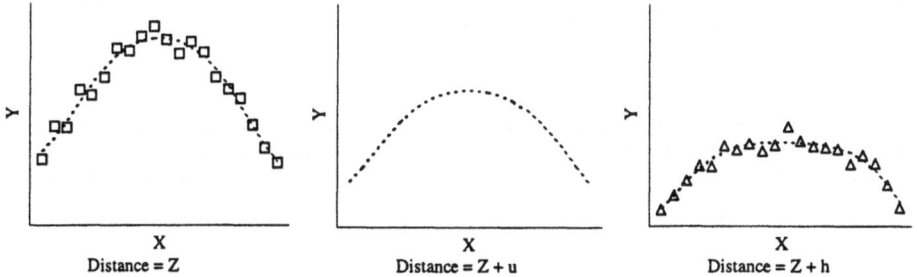

Figure 1. Example for function approximation and interpolation.

Figure 1 shows an example for the need to approximate and interpolate functions. In this figure, we have two sets of noisy data measured at two sample locations, Z and $Z+h$, where Z is the distance between the first sample location and a reference location, and h is the separation distance between the two sample locations. The notion of function approximation is to fit two noise-free curves through the data points. This is done to ensure that we can obtain function values, Y (the dependent variable), for any X values (the independent variable).

In some application domains, there is also a need to predict function values at location $Z+u$ where $0<u<h$ if the measurable X values are available at that location. This is the purpose of function interpolation. If, however, $u<0$ or $u>h$, this becomes an extrapolation problem. There is usually a larger uncertainty in extrapolating functions than interpolating functions, especially when $u<<0$ or $u>>h$.

In this paper, we are proposing a methodology to achieve these tasks by the use of neural-fuzzy technique. The next section will revisit pattern selection criteria in neural networks, followed by the introduction of our new methodology. In the later sections, we will demonstrate the use of this technique in petroleum reservoir modelling.

PATTERN SELECTION

Choosing proper training and validation patterns is crucial in neural learning. Wong et al. [1] has recently presented a systematic way to select training and validation patterns using self-organising map (SOM). SOM is first applied to classify the available data. Then the training and validation data are selected from each class. This is to ensure that the training and validation sets are generalised to cover the entire data range. The results showed that the SOM method in data-splitting consistently provided good generalised networks and the amount of training time was dramatically reduced.

While the above technique is applicable in most situations, it may not be appropriate in spatial function interpolation, especially when there is a continuity between the functions approximated at locations Z and $Z+h$. Under this situation, if u is close to 0, one may expect that its function relation should be similar to that at location Z. One solution to this problem is to construct a training set based on the relative proportions of patterns at locations Z and $Z+h$. The disadvantage of this method is that it is a computationally-expensive process as we need to run these analyses for a large number of times in order to avoid bias introduced to the estimates.

NEURAL-FUZZY TECHNIQUE

Some related works have been done using neural nets to interpolate spatial data in both mining and petroleum industries [2,3], however only areal data maps are produced and there is no indications on how to extend the methodology to vertical data interpolation. The proposed methodology is appropriate for interpolating vertical measurements. It is developed based on the use of learning curve in neural networks and fuzzy rules interpolation [4,5]. It consists of three steps: 1) pattern selection for neural learning; 2) neural learning to extract fuzzy rules; 3) interpolating the results of fuzzy rules to obtain final estimate.

1) Pattern Selection

In the proposed technique, the selection of training and validation sets is straightforward. All the patterns at location Z are used for the training set, and all the patterns at location $Z+h$ are used for the validation set. The statistics of these two sets may not necessarily the same, but they need to be correlated.

2) Neural Learning

The purpose of this step is to approximate functions at the sample locations. Standard BPNN procedure is utilised. Most neural nets avoid over-training by stopping iteration at minimum error on the validation set (see Figure 2a). The proposed technique uses minimum "interpolated error" or IErr to stop training. This error is defined as:

$$IErr = \frac{\sum\limits_{j}^{n} W_j E_j}{\sum\limits_{j}^{n} W_j} \qquad \qquad \ldots(1)$$

and,

$$W_j = \frac{1}{D_j} \qquad \qquad \ldots(2)$$

where W_j is the weighting factor of the jth data set, D_j is the separation distance between the point-of-interest and j^{th} sample location. In the case shown in Figure 1 (n=2), D_1 and D_2 are simply u and $(h-u)$ respectively. E_j represents the error on the learning curve using j^{th} data set for training (and the rest for validation). Thus, when u equals 0, this is equivalent to over-learning of the training set, and when u equals h, this is equivalent to the standard early-stopping method in BPNN (i.e. stop training when minimum validation error is reached). In cases where $0<u<h$, the interpolated error curve is between the validation and training errors as displayed in Figure 2b. When the minimum interpolated error is reached, training is stopped.

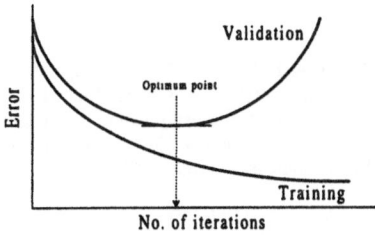
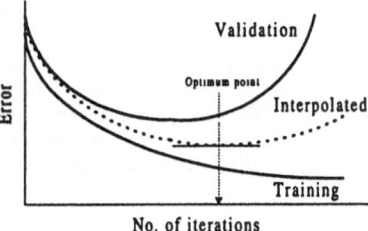

a. Optimum point based on validation error.   b. Optimum point based on interpolated error.

Figure 2. Stopping criteria for BPNN learning.

3) Rules Interpolation

Once each data set has been used as the training set, we will have n trained neural nets. These networks are used as fuzzy rules. These are disjoint rules in a large sparse rule-base. Each of these rules can be expressed as:

$$\text{Rule}_j: \text{ IF } X_j = (x_1,\ldots, x_m) \text{ is } A_j \text{ THEN } Y_j = NN_j(x_1,\ldots,x_m) \qquad \ldots(3)$$

where m is the dimension of the input vector X, A_j is the fuzzy set of the jth partitioned rule space, and Y_j is the output from j^{th} trained neural net or NN_j.

The conceptualisation of the use of neural nets as fuzzy rules allows us to use the developed formal machinery of fuzzy set theory such as fuzzy interpolation and defuzzification. We can then interpolate the relevant rules to location u, and only recombine the effects of the interpolated rules subsequently. The interpolated estimates can be expressed as:

$$Y_u = \frac{\sum\limits_{j}^{n} W_j Y_j}{\sum\limits_{j}^{n} W_j} \qquad \qquad \ldots(4)$$

where Y_u is the final estimate at location u.

In the case displayed in Figure 1, if **u** equals 0, Y_u is the result of the fuzzy rule using data at location **Z** as the training patterns. Similarly, if **u** equals **h**, Y_u is the result of the fuzzy rule using data at location **Z+h** as the training patterns. Since data set is trained independently at each location and if more data become available at new sample locations, it is straightforward to incorporate the new information in the existing system and there is no need to re-learn all the data.

CASE EXAMPLE

1) Background of petroleum reservoir modelling

The proposed technique was applied to petroleum reservoir modelling. In reservoir modelling, we often drill only a few holes at different locations. Reservoir properties are measured at the laboratory from the rock samples (or cores) retrieved at different depths. One of these properties is permeability, which is a measure of fluid conductivity of the rock sample. Unlike "well-logs", a series of digital measurements at different depths, which are available in every hole, retrieving rock samples (or "coring") is not a routine process because it is an expensive activity, and hence permeability values will be available at every hole. Thus, estimation of permeability has to be relied on well-logs alone [6].

2) Data description

In this example, we had data from two oil wells, namely Well 1 and Well 2. They are separated by approximately 3 km. The data was obtained from the North West Shelf, offshore Australia. Wells 1 and 2 recorded 222 and 166 data respectively at various depths, together with the corresponding permeability measurements. Each data point consisted 5 well-log measurements, namely gamma ray (GR), neutron porosity (NPHI), bulk density (RHOB), photoelectric adsorption index (PEF) and sonic travel times (DT).

3) Objective

The objective of this example is to predict the permeability values of another well, namely Well 3, using only well-logs. This well is located between Well 1 and Well 2. It is 2 km away from Well 1 and 1 km away from Well 2. It had 174 measurements of the same five well-logs as in Well 1 and Well 2. We will first split the data according to SOM method described in Wong et al. [1]. Then we will use the standard BPNN to predict the permeability values at Well 3 from its well-log data. We will compare the predictions with those obtained from the proposed neural-fuzzy technique. Actual permeability values are available at this well for performance evaluation.

4) Network setup

Data from Well 1 and 2 are first combined together. These data are then classified by the SOM approach. The SOM matrix is a two-dimensional 14-by-14 network that gives a total of 196 classes. Splitting of data is made in such a way that the two sets of data are selected from each class. The number of data in each sets, namely Set A and Set B, were 226 and 162, respectively. We will first use the data from Set A for training and the data from Set B for validation. We will then swap the use of the data (i.e. Set B for training, Set A for validation). Final estimates are obtained by averaging the results from the two trained networks.

For the proposed technique, we will use the data from Well 1 for training and the data from Well 2 for validation. We then swapped the use of the data as before. We will predict the permeability values by first assuming that the location of Well 3 is half-way between Well 1 and Well 2 (1.5 km away from each well). Then we will re-run the analyses using the actual separation distances.

All the well-logs (the independent variables) were all normalised in the range of (0,1), while the permeability values (the dependent variable) were scaled in the range of (0.1,0.9). All the BPNN models were based on the same configuration. It had 1 hidden layer with 5 hidden neurons, the learning rate and the momentum term were both 0.01. Sigmoid functions were used as the transfer functions. The maximum allowable epoch was 10,000.

5) Results

Figure 3 shows the BPNN learning profiles for the SOM data-splitting approach. We stopped training based on the minimum validation errors. Note that the shapes of the learning curves are similar in both training schemes. This is because the statistics of training and validation sets are similar. The final estimates were averaged. Figure 4 shows the learning profiles for the proposed neural-fuzzy method. We stopped training based on minimum interpolated errors. The interpolated error curves for different well locations are also shown in Figure 4. The shapes of the curves are dissimilar because the statistics of the training and validation sets are different. The final estimates were calculated based on the separation distances as in Equation (4). Figure 5 displays the predicted permeability profiles at Well 3 together with the actual data. This figure shows only range of the predicted values (0.35-0.65). Total sum-of-squares were used to evaluate the performance of different models, and the results of the comparisons are summarised in Figure 6.

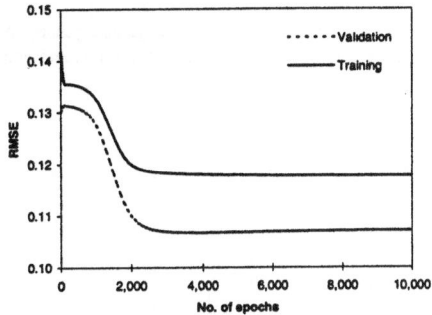 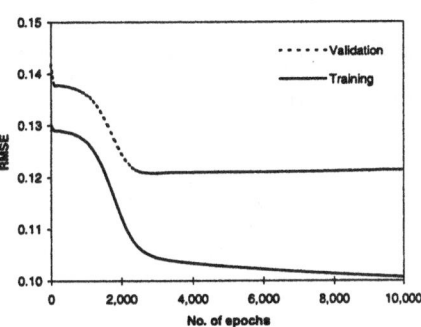

a) Learning profile using Set A for training, Set B for validation.   b) Learning profile using Set B for training, Set A for validation.

Figure 3. BPNN learning profiles for the SOM data-splitting approach with root-mean-squares error (RMSE) versus number of epochs.

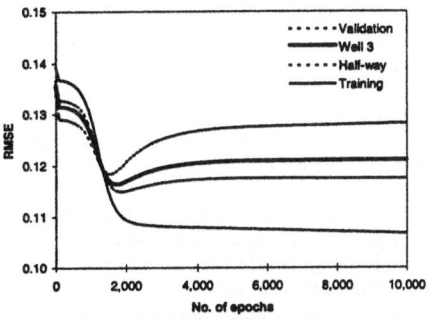 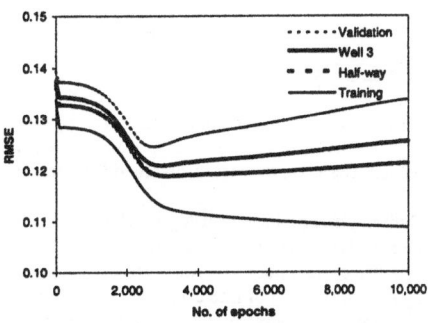

a) Learning profile using Well 1 for training, Well 2 for validation.   b) Learning profile using Well 2 for training, Well 1 for validation.

Figure 4. BPNN learning profiles for the neural-fuzzy approach with root-mean-squares error (RMSE) versus number of epochs.

The results showed that the SOM method (based on minimum validation errors) gave approximately the same performance as the neural-fuzzy method (based on minimum interpolated errors) when we assumed the location of Well 3 is half-way between Well 1 and Well 2. This shows that the SOM method is only applicable if the well location is "averaged" among the locations of the well-log databases. When the actual location of Well 3 is used, the neural-fuzzy method provided the best results (lowest error). This concludes that the neural-fuzzy method is applicable in all well locations.

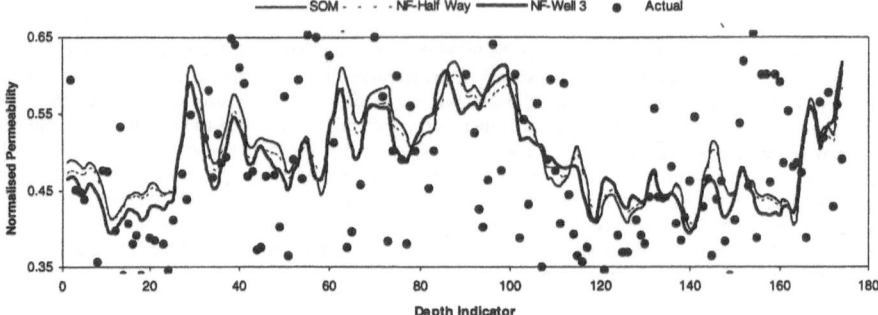

Figure 5. Permeability profiles at Well 3. "SOM" denotes results by data splitting by self-organising map, "NF-Half Way" and "NF-Well 3" mean neural-fuzzy estimates at half way between Well 1 and Well 2, and actual Well 3 location, respectively.

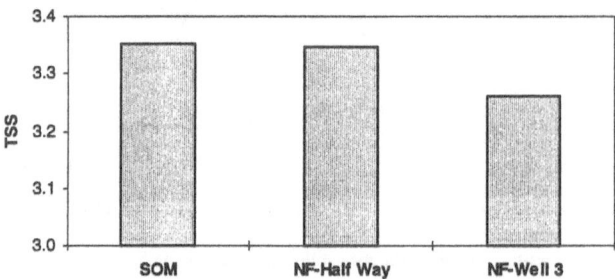

Figure 6. Summary of analyses.

CONCLUSIONS

The paper uses a new technique in function approximation and interpolation based on neural networks and fuzzy rules interpolation. The technique was applied to petroleum well-log analysis in which data is spatially correlated. Based on the results obtained from this paper, the major findings are:

1) Splitting a combined well-log database into training and validation patterns is not inappropriate in spatial data analysis, unless the location of interest is averaged among all the locations of the database.
2) The use of minimum validation error to stop neural training is not applicable in spatial data analysis, unless, again, the location of interest is averaged among all the locations of the database.
3) The proposed neural-fuzzy interpolation method, based on the use of minimum interpolated error, provides reliable estimates at any location of interest.
4) The current methodology assumes that the spatial functions are fully correlated which is not always true in geological modelling (e.g. in the presence of faulting). Hence, further work is required to extend the methodology for interpolating partial correlated functions.

REFERENCES

1. Wong, K.W., Fung, C.C. and Eren, H., 1996. "A Study of the Use of Self-Organising Map for Splitting Training and Validation Sets for Backpropagation Neural Network", *IEEE Region Ten Conference (TENCON) on Digital Signal Processing Applications*, November 27-29, Perth, Australia, vol. 1, pp. 157-162.

2. Shibli, S.A.R., Wong, P.M. and Maignan, M., 1996. "Interpolation Neural Networks: A New Methods for Interpolation", *5th International Workshop on New Computing Techniques in Physics Research*, September 2-6, Lausanne, Switzerland, 4 pages.

3. Wu, X. and Zhou, Y., 1993. "Reserve Estimation using Neural Network Techniques", *Computer and Geosciences*, vol. 19, no. 4, pp. 567-575.

4. Baranyi, P., Gedeon, T.D. and Kóczy, L.T., 1995. "A General Method for Fuzzy Interpolation: Specialised for Crisp Triangular and Trapezoidal Rules", *Proceedings of the 3rd European Congress on Intelligent Techniques and Soft Computing*, Aachen, 5 pages.

5. Kóczy, L.T. and Hirota, K., 1993. "Approximate Reasoning by Linear Rule Interpolation and General Approximation", *International Journal of Approximate Reasoning*, no. 9, pp. 197-225.

6. Wong, P.M., Taggart, I.J. and Gedeon, T.D., 1995. "Use of Neural Network Methods to Predict Porosity and Permeability of a Petroleum Reservoir", *AI Applications*, vol. 9, no. 2, pp. 27-37.

Task Decomposition Based on Class Relations: A Modular Neural Network Architecture for Pattern Classification

Bao-Liang Lu and Masami Ito

Bio-Mimetic Control Research Center,
the Institute of Physical and Chemical Research (RIKEN)
3-8-31 Rokuban, Atsuta-ku, Nagoya 456, Japan
lbl@nagoya.bmc.riken.go.jp; itom@nagoya.bmc.riken.go.jp

Abstract. In this paper, we propose a new methodology for decomposing pattern classification problems based on the class relations among training data. We also propose two combination principles for integrating individual modules to solve the original problem. By using the decomposition methodology, we can divide a K-class classification problem into $\binom{K}{2}$ relatively smaller two-class classification problems. If the two-class problems are still hard to be learned, we can further break down them into a set of smaller and simpler two-class problems. Each of the two-class problem can be learned by a modular network independently. After learning, we can easily integrate all of the modules according to the combination principles to get the solution of the original problem. Consequently, a K-class classification problem can be solved effortlessly by learning a set of smaller and simpler two-class classification problems in parallel.

1 Introduction

One of the most important difficulties in using artificial neural networks for solving large-scale, real-world problems is how to divide a problem into smaller and simpler subproblems; how to assign a modular network to learn each of the subproblems independently; and how to combine the individual modules to get the solution of the original problem. In the last several years, many researchers have studied modular neural network systems for dealing with this problem, for example see [8, 3, 2, 1, 7]. Up to now, various problem decomposition methods have been developed based on the divide-and-conquer strategy. These methods can be roughly classified into three classes as follows.

Explicit decomposition: Before learning, a problem is divided into a set of subproblems by a designer who should have some domain knowledge and deep prior knowledge concerning the decomposition of the problem. Several modular systems have been developed based on this decomposition method, see for instance [10, 4]. The limitation of this method is that sufficient prior knowledge concerning the problem is necessary.

Class decomposition: Before learning, a problem is broken down into a set of subproblems according to the inherent relations among training data. Anand et al. [1] first introduced this method for decomposing a K-class classification problem into K two-class problems by using the class relations among the training data. In contrast to the explicit decomposition, this method only needs some common knowledge concerning the training data,

Automatic decomposition: A problem is decomposed into a set of subproblems with the progressing of the learning. Most of the existing decomposition methods fall into this category, see for instance [2, 7]. From computational complexity's point of view, the former two methods are more efficient than this one because the problems have been decomposed into subproblems before learning, and therefore, they are suitable for solving large-scale and complex problems. The advantage of this method is that it is more general than the former ones because it can work when prior knowledge concerning the problem is absent.

In this paper, we propose a new methodology for decomposing classification problems. The basic idea behind this methodology is to use the class relations among the training data, similar to the method developed by Anand *et al.* [1]. In comparison with Anand's method, our methodology has two main advantages as follows. (a) The two-class problem obtained by our method is to discriminate between every pair classes, i.e., class C_i and class C_j for $i = 1, \cdots, K$ and $j = i+1$. The existence of the training data of the other $K-2$ classes is ignored. Therefore, the number of training data for each of the two-class problems is $2N$. However, the two-class problem obtained by Anand's method has to discriminate between one class and the remaining classes. Therefore, the number of training data for each of the two-class problems is $K \cdot N$. When K is large, learning of the two-class problems obtained by Anand's method may be still problematic. Here, for simplicity of description, the assumption we made is that each of the classes has the same number of training data N. (b) By using our method, the two-class problem can be further divided into $N_i \cdot N_j$ smaller and simpler two-class problems, where N_i and N_j are the numbers of training subsets belonging to C_i and C_j, respectively. However, Anand's method can not be applied to decomposing two-class problems. Since the two-class problems obtained by our method can be much smaller and simpler than those obtained by Anand's method, it is easier to assign a smaller modular network to learn each of the two-class problems. We also propose two combination principles for integrating individual modules to solve the original problem. After training each of the two-class problem with a modular network, we can easily integrate all of the modules according to the combination principles to create a solution to the original problem. Consequently, a K-class classification problem can be solved effortlessly by learning a set of smaller and simpler two-class problems in parallel.

The remainder of the article is organized as follows. In Section 2, we present a new decomposition methodology. In Section 3, we introduce three integrating units for constructing modular networks and describe two combination principles. Section 4 gives several examples and simulation results. Finally, conclusions are given in Section 5.

2 The Task Decomposition Methodology

The decomposition of a task is the first step to implement a modular neural network system. In this section, we present a new methodology for decomposing a K-class classification problem into a set of smaller and simpler two-class classification problems.

2.1 Decomposition of K-class problems

We address K-class ($K > 1$) classification problems. Suppose that grandmother cells are used as output representation. Let $\mathcal{T}$ be the training set for a K-class classification problem:

$$\mathcal{T} = \{(X_l, Y_l)\}_{l=1}^L, \tag{1}$$

where $X_l \in \mathbf{R}^d$ is the input vector, and $Y_l \in \mathbf{R}^K$ is the desired output.

A K-class problem can be divided into K two-class problems [1]. The training set for each of the two-class problems is defined as follows:

$$\mathcal{T}_i = \{(X_l, y_l^{(i)})\}_{l=1}^L \quad \text{for } i = 1, \cdots, K \tag{2}$$

where $X_l \in \mathbf{R}^d$ and $y_l^{(i)} \in \mathbf{R}^1$. The desired output $y_l^{(i)}$ is defined as:

$$y_l^{(i)} = \begin{cases} 1 - \epsilon & \text{if } X_l \text{ belongs to class } \mathcal{C}_i \\ \epsilon & \text{if } X_l \text{ belongs to } \bar{\mathcal{C}}_i \end{cases} \tag{3}$$

where ϵ is a small positive real number, $\bar{\mathcal{C}}_i$ denotes all the classes except $\mathcal{C}_i$. That is, $\bar{\mathcal{C}}_i$ is $\mathcal{C}_i$'s complement.

If the original K-class problem is large and complex, learning of the two-class problems as defined in Eq. (2) may be still problematic. One may ask: whether can the two-class classification problems be further decomposed into simpler two-class problems ? We will give an answer to this question in the remainder of the article.

2.2 Decomposition of two-class problems

From Eq. (1), the input vectors can be easily partitioned into K sets:

$$\mathcal{X}_i = \{X_l^{(i)}\}_{l=1}^{L_i} \quad \text{for } i = 1, 2, \cdots, K, \tag{4}$$

where $X_l^{(i)} \in \mathbf{R}^d$ is the input vector, all of the $X_l^{(i)} \in \mathcal{X}_i$ have the same desired outputs, and $\sum_{i=1}^K L_i = L$. Note that this partition is unique.

We suggest that the two-class problems as defined in Eq. (2) can be further divided into $K - 1$ smaller two-class problems. The training set for each of the smaller two-class problems is defined as follows:

$$\mathcal{T}_{ij} = \{(X_l^{(i)}, 1 - \epsilon)\}_{l=1}^{L_i} \cup \{(X_l^{(j)}, \epsilon)\}_{l=1}^{L_j} \text{ for } j = 1, \cdots, K \text{ and } j \neq i \tag{5}$$

where $X_l^{(i)} \in \mathcal{X}_i$ and $X_l^{(j)} \in \mathcal{X}_j$ are the input vectors belonging to class $\mathcal{C}_i$ and class $\mathcal{C}_j$, respectively. For task T_{ij}, the existence of the training data belonging to the other $K-2$ classes is ignored.

From Eq. (5), we see that partitioning of the two-class problem as defined in Eq. (2) into $K-1$ smaller two-class problem is simple and straightforward. No domain specialists or prior knowledge concerning the decomposition of the learning problems are required. Consequently, any designer can perform this decomposition easily if he or she knows the number of training patterns belonging to each of the classes.

From Eq. (5), we see that a K-class problem can be broken down into $K \cdot (K-1)$ two-class problems, which are represented as a $K \times K$-matrix as follows:

$$
\begin{bmatrix}
\emptyset & T_{12} & T_{13} & \dots & T_{iK} \\
T_{21} & \emptyset & T_{23} & \dots & T_{2K} \\
\multicolumn{5}{c}{\dotfill} \\
T_{K1} & T_{K2} & T_{K3} & \dots & \emptyset
\end{bmatrix}
\tag{6}
$$

where $\emptyset$ represents empty set.

In fact, among the the above problems, only $\binom{K}{2}$ two-class problems in the upper triangular are different, and other $\binom{K}{2}$ ones in the lower triangular can be solved by inverting the former ones by using the INV units (see Section 3). Therefore, the number of two-class problems that need to be learned can be reduced to $\binom{K}{2}$. Comparing Eq. (5) with Eq. (2), we see that the two-class problem defined in Eq. (5) is much smaller than that defined in Eq. (2) if the K is large and the number of patterns for each of the K-classes is roughly equal.

2.3 Fine decomposition of two-class problems

Even though a K-class problem can be broken down into $\binom{K}{2}$ relatively smaller two-class problems, some of them may be still hard to be learned: for instance, the "two-spirals" problem [5]. In order to deal with this problem, we propose a method for further decomposing the two-class problem T_{ij} as defined in Eq. (5) into a set of smaller and simpler two-class problems.

Assume that the input set $\mathcal{X}_i$ is further partitioned into N_i ($N_i \geq 1$) subsets:

$$
\mathcal{X}_{ij} = \{X_l^{(ij)}\}_{l=1}^{L_i^{(j)}} \quad \text{for } j = 1, \cdots, N_i,
\tag{7}
$$

where $X_l^{(ij)} \in \mathbf{R}^d$ is the input vector and $\sum_{j=1}^{N_j} L_i^{(j)} = L_i$. This partition is not unique in general. One can give a partition randomly or by using prior knowledge concerning the decomposition of the learning problems.

The training set for each of the smaller and simpler two-class problems is defined as follows:

$$
T_{ij}^{(uv)} = \{(X_l^{(iu)}, 1 - \epsilon)\}_{l=1}^{L_i^{(u)}} \cup \{(X_l^{(jv)}, \epsilon)\}_{l=1}^{L_j^{(v)}}
\tag{8}
$$
$$
\text{for } u = 1, \cdots, N_i, \ v = 1, \cdots, N_j, \text{ and } j \neq i
$$

where $X_l^{(iu)} \in \mathcal{X}_{iu}$ and $X_l^{(jv)} \in \mathcal{X}_{jv}$ are the input vectors belonging to class $\mathcal{C}_i$ and class $\mathcal{C}_j$, respectively.

3   The Modular Network Architecture

After solving each of the smaller two-class problems as defined in Eq. (5) or Eq. (8) by a modular network, we need to organize the individual modules and construct a modular system to get the solution of the original problem. In this section, we will first introduce three integrating units for constructing the modular networks, and then we will give two combination principles for integrating the individual modules.

3.1   Three Integrating Units

Before describing our modular neural network architecture, we introduce three integrating units, namely MIN, MAX, and INV respectively.

The basic function of a MIN unit is to find a minimum value from its multiple inputs. The transfer function of a MIN unit is given by

$$q = \text{Minimize}\,\{p_1,\, \cdots,\, p_n\} \tag{9}$$

where $p_1,\, \cdots,\, p_n$ and q are the inputs and output, respectively, $p_i \in \mathbf{R}^1$ for $i = 1,\, \cdots,\, n$, and $q \in \mathbf{R}^1$.

The basic function of a MAX unit is to find a maximum value from its multiple inputs. The transfer function of a MAX unit is given by

$$q = \text{Maximize}\,\{p_1,\, \cdots,\, p_n\} \tag{10}$$

where $p_1,\, \cdots,\, p_n$ and q are the inputs and output, respectively.

The basic function of an INV unit is to invert its single input. The transfer function of an INV unit is given by

$$q = b - p \tag{11}$$

where b, p and q are the upper limit of its input, input, and output, respectively.

3.2   The Combination Principles

Suppose that each of the two-class problems has been learned by a modular network completely. One may ask a question: how to combine the outputs of the individual modules to get the solution of the whole problem ? In this subsection, we will present two combination principles which give the designer a systematic method for organizing the modules.

Minimization Principle: *The modules, which were trained on the same train-ing inputs corresponding to the desired outputs $1 - \epsilon$, should be integrated by the MIN unit.*

Consider the two-class problems $\mathcal{T}_{i1}, \mathcal{T}_{i2}, \cdots, \mathcal{T}_{iK}$ as defined in Eq. (5). These problems have the same training inputs corresponding to the desired outputs $1 - \epsilon$. Suppose that the $K - 1$ modules, which are represented as $\mathcal{M}_{i1}, \mathcal{M}_{i2}, \cdots, \mathcal{M}_{iK}$, were trained, respectively, on $\mathcal{T}_{i1}, \mathcal{T}_{i2}, \cdots, \mathcal{T}_{iK}$. According to the minimization principle, we can organize the $K \cdot (K - 1)$ modules into a modu-lar network as illustrated in Fig. 1(a), where, for simplicity of illustration, the assumption we made is that all of the $K \cdot (K - 1)$ two-class problems as defined in Eq. (5) are learned and no INV unit is used.

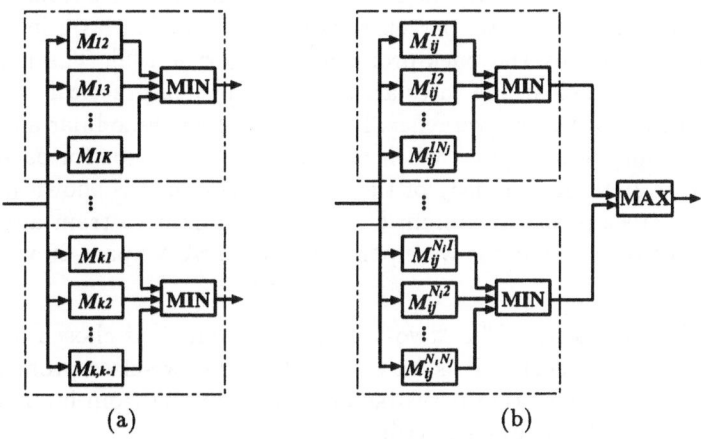

(a)   (b)

Fig. 1. The organization of the $K \cdot (K - 1)$ modules by using the MIN units (a) and the organization of the $N_i \cdot N_j$ modules by using the MIN and MAX units (b).

Maximization Principle: *The modules, which were trained on the same train-ing inputs corresponding to the desired outputs ϵ, should be integrated by the MAX unit.*

Consider the combination of the modules which were trained on the following $N_i \cdot N_j$ two-class problems as defined in Eq. (8).

$$
\begin{array}{llll}
\mathcal{T}_{ij}^{(11)} & \mathcal{T}_{ij}^{(12)} & \ldots & \mathcal{T}_{ij}^{(1, N_j)} \\
\\
\mathcal{T}_{ij}^{(21)} & \mathcal{T}_{ij}^{(22)} & \ldots & \mathcal{T}_{ij}^{(2, N_j)} \\
\multicolumn{4}{c}{\dotfill} \\
\mathcal{T}_{ij}^{(N_i, 1)} & \mathcal{T}_{ij}^{(N_i, 2)} & \ldots & \mathcal{T}_{ij}^{(N_i, N_j)}
\end{array}
\tag{12}
$$

According to the decomposition method defined in Eq. (8), the N_j training sets in each of row of Eq. (12) have the same training inputs corresponding to the desired outputs $1 - \epsilon$. In contrast, the N_i training sets in each column of Eq. (12)

have the same training inputs corresponding to the desired outputs ϵ. Following the minimization and maximization principles, the $N_i \cdot N_j$ modules that were trained on the $N_i \cdot N_j$ two-class problems can be organized as illustrated in Fig. 1(b).

4 Examples and Simulations

To evaluate the effectiveness of the proposed decomposition methodology, the two combination principles, and the modular network architecture, several benchmark learning problems have been simulated in this section. In the following simulations, the structure of all the nonmodular and modular networks are chosen to be the three-layer quadratic perceptrons with one hidden layer [6]. All of the networks are trained by the back-propagation algorithm [9]. The momentums are set all to 0.9. The learning rates are selected through practical experiments. They are optimal for fast convergence. For each of the nonmodular and modular networks, training was stopped when the mean square error for each network was reduced to 0.01. A summary of the simulation results is shown in Table 1, where "Max." means the maximum CPU time required to train any modular network. All of the simulations were performed on a SUN Sparc-20 workstation.

Two-Spirals Problem: The "two-spirals" problem [5] is chosen as a benchmark for this study because it is an extremely hard two-class problem for the conventional backpropagation networks and the mapping from input to output formed by each of the modules is visible.

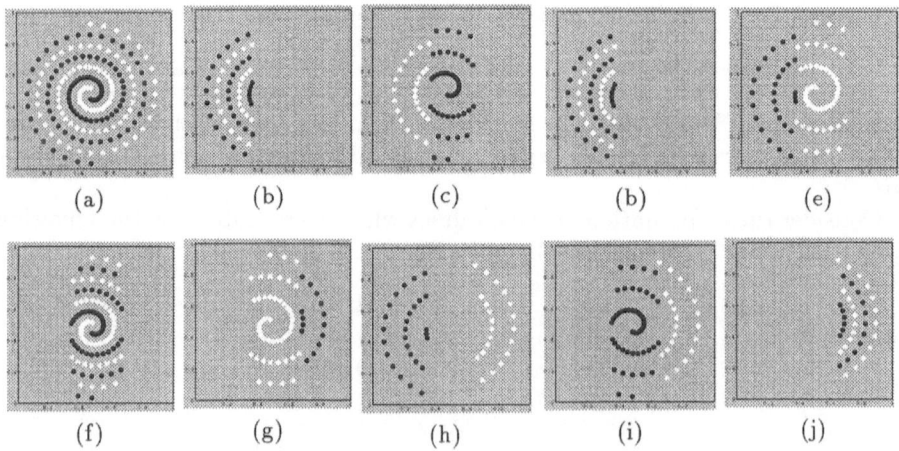

Fig. 2. The training inputs for the original two-spirals problem (a). The training inputs for the nine subproblems (b) through (j), respectively. The black and white points represent the desired outputs of "0" and "1", respectively.

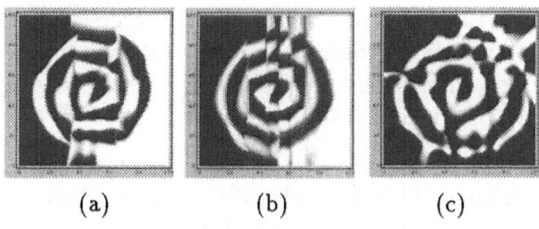

(a)   (b)   (c)

Fig. 3. The responses of the modular network with the 9 modules (a), the modular network with the 36 modules (b), and the single network with 40 hidden units (c). Black and white represent the outputs of "0" and "1", respectively.

The 194 training inputs for the original two-spirals problem are shown in Fig. 2(a). We performed three comparative simulations on this problem. In the first simulation, the original problem was divided into nine subproblems by partitioning the input variable through the axis of abscissas into three overlapping intervals. The training inputs for the nine subproblems are shown in Figs. 2(b) through 2(j), respectively. All of the nine modular networks were selected to be five hidden units except that the fifth module was selected to be twenty-five hidden units because the fifth task (see Fig. 2(f)) is the hardest problem to be learned in the nine problems. The combination of the outputs of the nine trained modules is shown in Fig. 3(a). In the second simulation, the original problem was divided into 36 subproblems by partitioning the input variable through the axis of abscissas into 6 overlapping intervals. The numbers of hidden units of the 1st, the 8th, the 15th, the 22nd and the 29th modules were chosen to be 10, and the others were chosen to be 1. The response of the modular network which consists of 36 trained modules is shown in Fig. 3(b). For comparing with the above results, this problem was also learned by a single network with 40 hidden units. After 200,000 iterations, the mean square error was still about 0.57. The response of the single network is shown in Fig. 3(c). All of the CPU times required to train the single and modular networks are shown in Table 1.

Table 1. Performance comparison of nonmodular and the proposed modular networks

Task	Network	♯ Modules	CPU time		Success rate (%)	
			Max.	Total	Training data	Test data
Two-spirals	Nonmodular	1	105447	105447	99.48%	-
	Modular	9	5513	5983	100.00%	-
	Modular	36	648	1439	100.00%	-
Image	Nonmodular	1	50828	50828	99.95%	91.19%
	Modular	21	350	1121	100.00%	90.76%
Vehicle	Nonmodular	1	134971	134971	99.76%	72.34%
	Modular	6	3456	4567	100.00%	73.05%

Image Segmentation: The image segmentation problem was obtained from the University of California at Irvine (UCI) repository of machine learning databases. This real problem consists of 210 training data and 2100 test data. The number of attributes is 19 and the number of classes is 7. The original problem is decomposed into $\binom{7}{2}$ two-class problems according to the decomposition method defined in Eq. (5). Each of the two-class problems consists of 60 training data. Each of the 21 two-class problems was learned by a modular network with 3 hidden units. The original problem was also learned by a single network with 10 hidden units. The simulation results are shown in Table 1.

Vehicle Classification: This real classification problem was also obtained from UCI repository of machine learning databases. The problem is to classify a given silhouette as one of four types of vehicle by using a set of features extracted from the silhouette. We divided the original data set into training and test sets. Each of the two sets consists of 423 data. The number of attributes is 18 and the number of classes is 4. The original problem was decomposed into $\binom{4}{2}$ two-class problems. All of the 6 modules were selected to be 4 hidden units, except that the module used to train on T_{23} was selected to be 8 hidden units. The 6 trained modules are organized as illustrated in Fig. 4. This original problem was also learned by a single network with 24 hidden units. The simulation results are shown in Table 1.

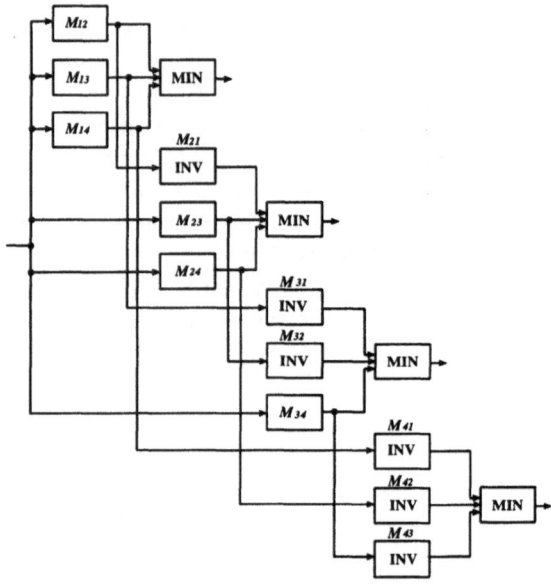

Fig. 4. The modular network architecture for learning the Vehicle classification problem. Corss lines do not represent connections unless there is a dot on the intersection.

5 Conclusions

In this paper, we have proposed a new decomposition methodology, two combination principles for integrating modules, and a new modular neural network architecture. The basic idea of the methodology is based on the class relations among the training data. Given a K-class classification problem, by using the proposed decomposition methodology, we can divide the problem into a set of smaller and simpler two-class problems. Several attractive features of this methodology can be summarized as follows: (a) we can break down a problem into a set of smaller subproblems even though we are not domain specialists or we have no any prior knowledge concerning the decomposition of the problem; (b) training of each of the two-class problems can be greatly simplified and achieved independently; and (c) different network structures or different learning algorithms can be used to learn each of the problems. The two combination principles gives us a systematic method for organizing the individual modules. By using three integrating units, we can combine the outputs of all the individual modules to create a solution to the original problem. The simulation results (see Table 1) indicate that (a) the speedups of up to one order of magnitude can be obtained with our modular network architecture and (b) the generalization performance of trained single and modular networks are about the same. The importance of the proposed decomposition methodology lies in the fact that it provides us a promising approach to solving large-scale, real-world pattern classification problems.

References

1. Anand, R., Mehrotra, K. G., Mohan, C. K., and Ranka, S.: Efficient classification for multiclass problems using modular neural networks, *IEEE Transaction on Neural Networks*, 1995, **6**(1), 117-124.
2. Jacobs, R. A., Jordan, M. I., Nowlan, M. I., and Hinton, G. E.: Adaptive mixtures of local experts, *Neural Computation*, 1991, **3**, 79-87.
3. Hrycej, T.: *Modular Learning in Neural Networks*, 1992, John-Wiley & Sons, Inc.
4. Jenkins, R., and Yuhas, B.: A simplified neural network solution through problem decomposition: The case of the truck backer-upper, *IEEE Transaction on Neural Networks*,1993, **4**(4), 718-722.
5. Lang, K., and Witbrock, M.: Learning to tell two spirals apart, *Proceedings of 1988 Connectionist Models Summer School*, 1988, 52-59. Morgan Kaufmann.
6. Lu, B. L., Bai, Y., Kita, H., and Nishikawa, Y.: An efficient multilayer quadratic perceptron for pattern classification and function approximation, *Proceedings. of International Joint Conference on Neural Networks*, Nagoya, 1993, 1385-1388.
7. Lu, B.-L., Kita, H., and Nishikawa, Y.: A multi-sieving neural network architecture that decomposes learning tasks automatically, *Proceedings of IEEE Conference on Neural Networks*, 1994, 1319-1324.
8. Murre, J. M. J.:*Learning and Categorization in Modular Neural Networks*, 1992, Harvester Wheatsheaf.
9. Rumelhart, D. E., Hinton, G. E., and Williams, R. J.: Learning internal representations by error propagation, in *Parallel Distributed Processing: Explorations in the Microstructure of Cognition*, 1996, 1, D. E. Rumelhart, J. L. McClelland, and PDP Research Group eds, MIT Press.
10. Thiria, S., Mejia, C., Badran, F., and Crepon, M.: Multimodular architecture for remote sensing operations, *Advances in Neural Information processing Systems 4*, 1992, 675-682.

Lower Bounds of Computational Power of a Synaptic Calculus[*]

João Pedro Neto, J. Félix Costa and Helder Coelho
{jpn, fgc, hcoelho}@di.fc.ul.pt

Dept. Informática, Faculdade de Ciências da Universidade de Lisboa, Bloco C5 - Piso 1,
Campo Grande 1700 Lisboa - Portugal. Tel. 7573141. Fax. 7570084

Abstract

The majority of neural net models presented in the literature focus mainly in the neural structure of nets, leaving aside many details about synapses and dendrites. This can be very reductionist if we want to approach our model to real neural nets. These structures tend to be very elaborate, and are able to process information in very complex ways (see [MEL 94] for details).

We will introduce a new model, the S-Net (Synaptic-Net), in order to represent neural nets with special emphasis on synaptic and dendritic transmission. First, we present the supporting mathematical structure of S-Nets, initially inspired on Petri-Net formalism, adding a transition to transition connection type. There are two main components of S-Nets, neurones and synaptic/dendritic units (s/d units). All activation values are integers. Neurones are similar to McCulloch-Pitts neurones, and s/d units will process information within certain class of functions.

S-Nets are able to represent spatial nets representations in a very natural way. We can easily modulate the length of an axon, the connection or branching of two dendrites or synaptic connections. Some examples are shown.

Next, the focus will be on what kind of functions are suited to s/d units. We will present three function types: sum, maximization and simple negation (changing an excitatory impulse to an inhibitory one, or vice-versa). With these functions for s/d units and with simple neurones, we will prove that all recursive functions can be computed by at least one specific S-Net. In order to achieve this, we will use the Register Machine, and show a way to build for each symbolic program, a S-Net capable of computing the function defined for that specific program. It is shown a simple application example. This computational power will be achieved without any use of synaptic weights (i.e., all weights are one as in McCulloch's model) or neurones activation values (i.e., all values are set to zero).

Finally, some aspects for future investigation are presented, namely, the possibility of synaptic-synaptic connections, how can noise be handled, and some other features intended to approach this mathematical model to our reality.

Keywords: Neural Networks, Synapses, Dentritic Trees, Neural Computation, Computational Theory, Spatial Representation.

Introduction

Many neural network models proposed in the literature have their major focus in neurones. Although they were inspired initially in the way brain processes information (see [McCulloch 43]), almost all systems leave aside the complex structure involving neurones. Namely, dentritic trees that process spatial integration of signals sent from presynaptic cells; the terminal arborization at the end of each axon; the subtle structure of synaptic transmissions, that can make direct contact with dendrites, with axons, with the soma, or even with other synapses (a good overview can be found in [Shepherd 94]).

Usually, the ultimate reference are synaptic weights between neurones. If it is true that some of these models had succeeded on classification and learning tasks, it is also a fact that they put themselves far away in modelling real neural networks.

Several neurobiological studies had consistently shown how complex and intricate are dentritic trees and synaptic connections. These structures possess a great computational symbolic potential. This capability is not only an effect of how the single units process information, but it is also a consequence of its spatial configuration, of its different

---

[*] This work was supported by JNICT PBIC/TIT/2527/95

external influences or even of its capacity for multiple pseudoindependent processing subunits inside the same dentritic tree (see [Mel 94]).

Synaptic-Nets

Our aim with this paper, is to propose a mathematical system able to model some of those complex connection structures, in a direct and natural way. The present result, still in development, is termed *Synaptic-Nets* (S-Nets).

The earliest idea came from the common way we see an Artificial Recorrent Neural Net (ARNN): as a direct graph. Usually, an ARNN uses graph nodes to represent neurones, and arcs to represent connections between neurones, also called synapses.

Here arises the first difference. In this new structure, we are able to perform calculations in neurones as well as in dentritic and synaptic units. So, one type of node is no longer sufficient. The first question is, how many different nodes should we allow? We think that two types are enough, one for neurones, and one for all others. It is not relevant to introduce more node types. What seems important is the connection point. They can be dentrite to dentrite connections, axon to dentrite, axon to axon, and so on... What is common to all, is the linking synapses. We also decide to include with this type, the concept dendritic convergence and divergence.

With two types of nodes, the next question is whether S-Nets are Petri-Nets? In fact, there are some similarities. Several concepts and ideas from Petri-Nets formalism can be imported to our advantage. However, there is a fundamental difference between both. In Petri-Nets, we must have an alternate sequence event to transition to event, but in S-Nets, the existence of arbitrary complex synapse to synapse connections is necessary. The binary relation of Petri-Nets must be generalized.

Definition: A Synaptic-Net (**S-Net**) is a tuple $< G, S, R, F >$, where G is a finite non-empty set whose elements are from now on called *neurones*; S is a finite set whose elements are from now on called *synapses*; R is a binary relation, $R \subseteq (G \times S) \cup (S \times G) \cup (S \times S)$; and F is a function, $F:(G \cup S) \times \omega \rightarrow Z$.

Remark: R determines the corresponding S-Net structure; F associates to each neurone or synapse an *activation value* in Z, for each time $t \in \omega$. To compute this value, each element $x \in G \cup S$ has a specific *information processing function*, $\phi_x:Z^n \rightarrow Z$, being n the number of inputs of x, i.e., $n = \#\{y: (y,x) \in R\}$.

In graphical terms, we will use circles to identify neurones, and squares to identify synaptic connections and dendritic trees (s/d units). These graphs are admissable S-Nets:

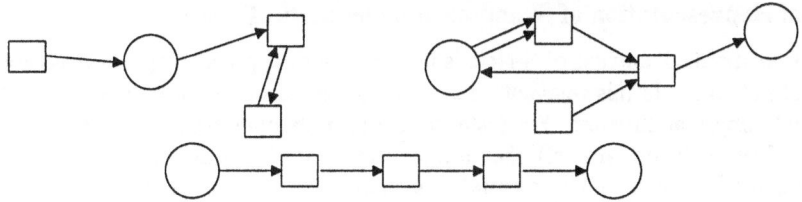

Figure 1 - S-Nets examples.

For each $x \in G \cup S$,

$F(x,0) \quad = 0,$

$F(x,t+1) \quad = \phi_x(F(x_1,t), ..., F(x_n,t)) \quad \forall_{i=1..n}: (x_i,x) \in R$

All activation values are set to zero at time t=0, and the activation value at time t+1 is computed by the values of its predecessors at time t.

We can associate to each element $x \in G \cup S$, a specific function ϕ_x. We will focus on a few set of simple functions, in order to get as close as possible of what might happen in real neural nets. One of the main concerns is to obtain functions able to support n inputs, and to consider all inputs in the same way (what could or could not be correct...). If there is more than one output, the signal is the same to all.

For neurones, there is only one function. Each neuron adds its inputs, both excitatory and inhibitory (it does not exist absolute inhibition), and checks if the sum is greater or equal to zero. If it is, the neuron becomes active and sends an excitatory signal through its outputs (value 1 for each output). If it is not, the neuron is inactive (value 0).

For all $g \in G$,

$$\phi_g(\vec{y}) = \phi_G(\vec{y}) = \begin{cases} 1, & \sum_{i=1}^{n} y_i \geq 0 \\ 0, & \sum_{i=1}^{n} y_i < 0 \end{cases}$$

All internal activities are set to zero, so all neurones have an identical internal structure. There are 3 functions for s/d units: sum, maximum and negation. If $s \in S$ then $\phi_s \in \{\phi_\Sigma, \phi_M, \phi_\neg\}$. For each $s \in S$, we have one of the following functions,

$$\phi_s(\vec{y}) = \phi_\Sigma(\vec{y}) = \sum_{i=1}^{n} y_i$$

$$\phi_s(\vec{y}) = \phi_M(\vec{y}) = \begin{cases} max(y_1,...,y_n), & \forall_{y_i \in G \cup S} : y_i \geq 0 \\ 0 & , \exists_{y_i \in G \cup S} : y_i < 0 \end{cases}$$

$$\phi_s(\vec{y}) = \phi_\neg(\vec{y}) = -\phi_\Sigma(\vec{y})$$

Therefore, $\phi_x \in \{\phi_G, \phi_\Sigma, \phi_M, \phi_\neg\}$, for all $x \in G \cup S$.

Spatial Representation of Synapses and Dendritic Trees

One of the main advantages of S-Nets is its capacity to apprehend spatial structures of real neural networks. Is this relevant? Yes, because some neural proprieties depend on the neuron's physical structure. For instance, action potentials travel between neurones with some finite velocity, after all, they are just flows of electrical current along nerves. So, the length of those nerves is essential. We can model axons with different sizes very directly, adding more s/d units in proportion to their sizes. This also has an effect on transmission time, as required (long axons take more time to transmit the action potential to their postsynaptic cells).

Hence, these different lengths determine *when* the action potential arrives at the next neurones, which is fundamental to determine if a neuron is activated or not. Moreover, each neuron has an absolute refractory period (the period, after the neuron activation, during which it is impossible to stimulate the nerve a second time) and a relative refractory period (the period during which it can be stimulated, but only by using a larger current than usual), both functions of time.

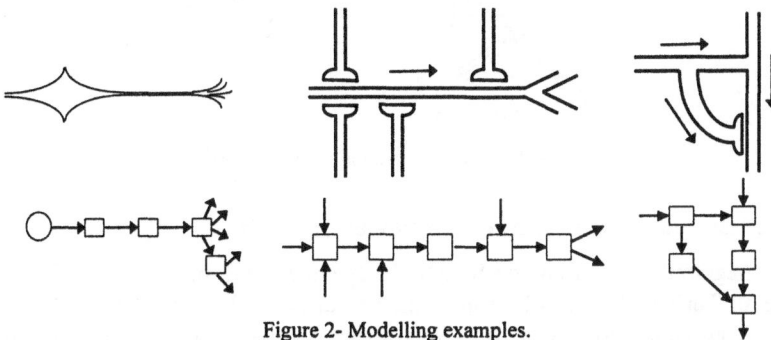

Figure 2- Modelling examples.

There are important spatiotemporal interactions inside neural nets as we will see in the following examples. The geometry of dendritic trees or the complex organization of synaptic contacts are more than just a useful way to connect neurones. They are special mechanisms that delay, attenuate and synchronize action potentials.

A well-known example, first introduced by Wilfred Rall in 1964, and very common in the literature (see, e.g., [Mel 94, Arbib 94, Anderson 95]), confirms that a passive dendritic branch, because of its spatial extension, can act as a filter that selects specific sequences of synaptic inputs. In Fig. 3, if input I_1 is activated at time t, I_2 at time t+1, and I_3 at time t+2, the neuron is activated. If the activation sequence is $I_3 \rightarrow I_2 \rightarrow I_1$, the neuron is not activated.

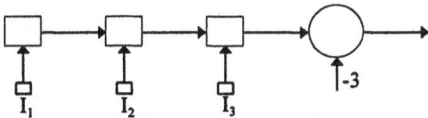

Figure 3- Input sequence selection.

Another example is taken from [Arbib 89], about *lateral inhibition*. This happens when a network structure is made so that neurones inhibit all but their close neighbours. Inhibition can be recurrent or nonrecurrent. In *nonrecurrent inhibition*, the inhibitory signal is a combination of the current excitations, $O_1 = I_1 - k.I_2$ and $O_2 = I_2 - k.I_1$, in which the local excitation is reduced by k times the neighbours excitation.

In *recurrent inhibition*, the signal sent to the neighbours is itself subject to their own inhibitory effect. In fig. 4b, $O_1(t+2) = I_1 - k.O_2(t)$ and $O_2(t+2) = I_2 - k.O_1(t)$. This type of inhibition is observed in the lateral eye of *Limulus*, the horseshoe crab. These two equations can be seen as a dynamic system, and the final outputs O_1 and O_2 will held an eventual equilibrium pair for this system.

The events of nonrecurrent inhibition are strictly localized, since the outputs depend exclusively on the inputs. In recurrent inhibition, the effects can spread to distant neighbours, since the actual changes effect their immediate neighbours, and these changes produce further changes in the next neighbours, and so on. Recurrent inhibition has an important feature, *disinhibition*, the inhibitions can themselves be inhibited.

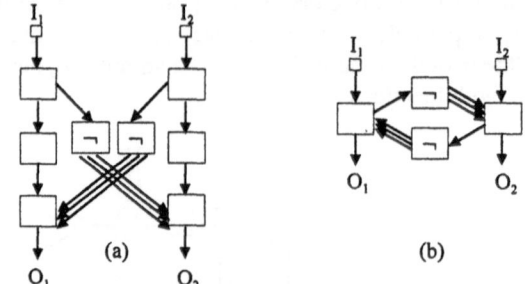

(a)   (b)

Figure 4- (a) Nonrecurrent inhibition (b) Recurrent inhibition.

Lower Bounds of Computational Power of S-Nets

Hava Siegelmann and Eduardo Sontag, [Siegelmann 95], proved that it is possible to simulate all Turing Machines by finite size neural networks. As in their paper, we will show the computational power of S-Nets, using another mathematical formalism (but equivalent to Turing Machines), the Universal Register Machine (URM).

Having $\{\phi_G, \phi_\Sigma, \phi_M, \phi_\neg\}$, we state the following.

Proposition: All partial recursive functions can be computed by at least one specific S-Net, using only $\phi \in \{\phi_G, \phi_\Sigma, \phi_M, \phi_\neg\}$.

Proof:

We will show that each URM instruction (see appendix) can be simulated by a specific S-Net. First, we see how to make a register with only 4 s/d units. So, for each program P, we need $4*\rho(P)$ s/d units to simulate all necessary memory ($\rho(P)$ is the greatest register index used in P).

Graphically, each s/d function is represented by one of the following diagrams:

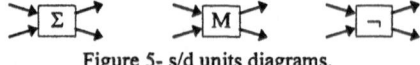

Figure 5- s/d units diagrams.

We can increment, reset and access a register value, that is done by the following sub-net,

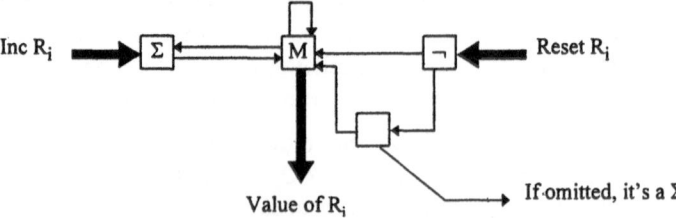

Value of R_i

If omitted, it's a Σ

Figure 6- Memory Register.

In the following diagrams (where we build each URM instruction), the signals mean:

345

I_{n-1} - activate this instruction; I_0 - activate first instruction;
I_{n+1} - activate next instruction; I_f - end of the program.

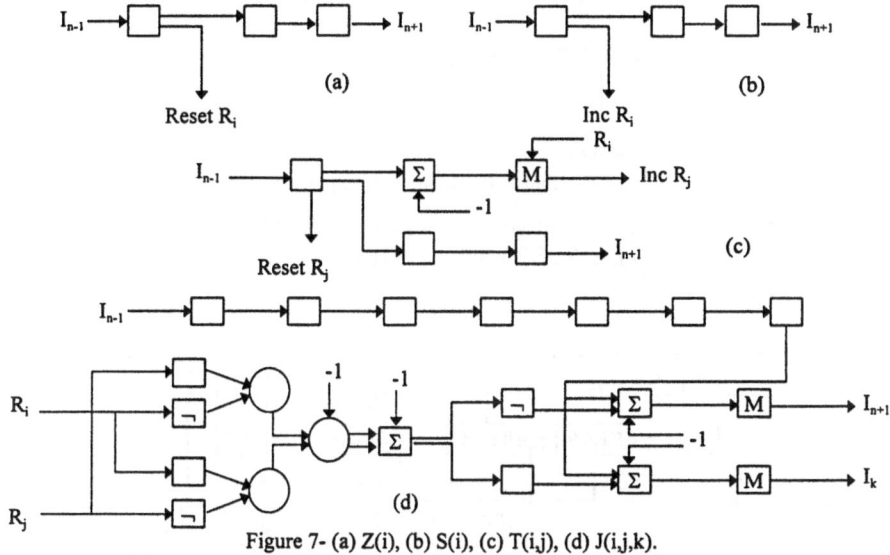

Figure 7- (a) Z(i), (b) S(i), (c) T(i,j), (d) J(i,j,k).

At this point, we already know how to build all instruction types and also to make memory registers. The next two diagrams show how to receive inputs, and how to return outputs. There are two special activation channels indicating when the input and the output are available. They are activated when they have value one.

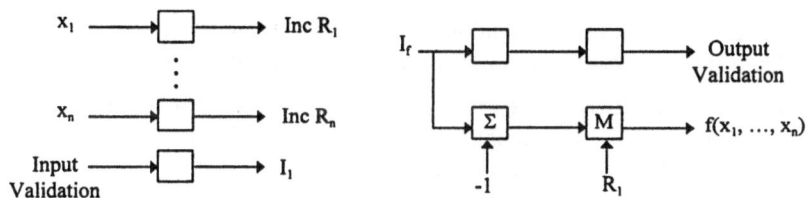

Figure 8 - Input and Output.

Finally, to create constant -1, we use the following sub-net,

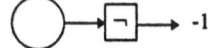

Figure 9 - Constant definition.

Using this method we see that each computable function $f:\omega^n \to \omega^m$, is also computed by a S-Net with 3.J+1 neurones and 1+(n+1)+4.R+3.(Z+S)+18.J+5.T+(2.m+2) s/d units, where R is the number of registers used in P, and, Z, S, J and T, the number of Z, S, J and T instructions of P. We have a linear complexity in time and in space, with respect to the Register Machine. ∎

How this is done? Find P, and then for each sequential instruction of P, use the respective sub-net, and link them in the same sequential order.

To present an example, we will enclose each instruction in a box, labelling each one with its specific instruction (e.g., fig. 10). We only explicit the inputs and outputs. Also, we do not represent any register sub-net, only some of their reset, increment and value connections.

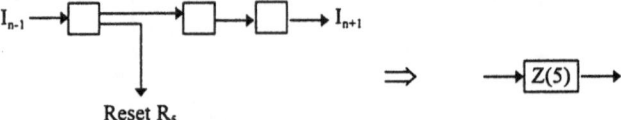

Figure 10 - Input and Output.

Let's compute binary multiplication, x*y. A possible URM program is, P = <J(1,4,9), J(2,5,6), S(5), S(3), J(1,1,2), Z(5), S(4), J(1,1,1), T(3,1)>. We will need ρ(P)=5 memory registers. The S-Net will be,

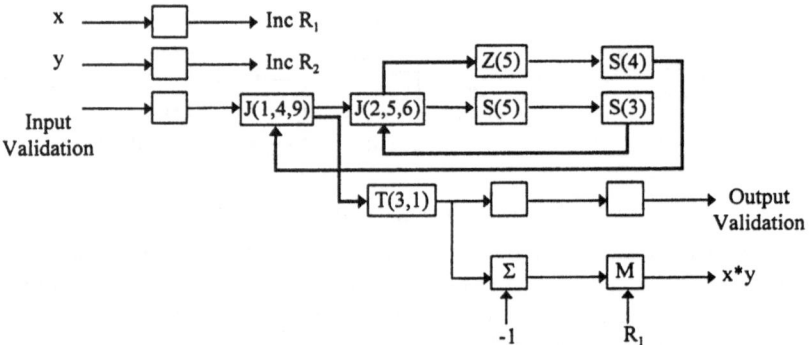

Figure 11 - Computing the product.

The URM program has two inconditional jumps, J(1,1,2) and J(1,1,1). For these, we do not use a Jump sub-net, but only a straight arrow to the next instruction.

Obviously, if there is a countable infinity of URM programs that compute each partial recursive function (and in fact, there is), it is also possible to construct a countable infinity of S-Nets capable of handle those same computations.

A S-Net is intrinsically a massive parallel machine. However, the method used to compute a function is based on the URM, which is a sequential machine. It is important to understand that our main goal in this section is to check the lower computational bounds of S-Nets, not to seek the fastest or simplest way to compute those same functions.

Since S-Nets are made of local units of information processing, it is not difficult to generalize this method to perform parallel computations. There are only the usual problems, like synchronising shared resources access. We believe that S-Nets do not introduce new fundamental problems in this area.

Future Developments

There are several paths open to exploration in S-Nets. We shall point here some of the most promising.

S/d units refer to synaptic connections with axons, dendrites or with the soma. But they cannot model synaptic-synaptic connections. In graphical terms, we need something like this:

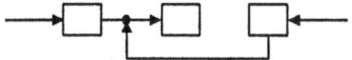

Figure 12- Synaptic-synaptic connection.

What kind of functions can be used in this new connection type? Some proposals fall into partial or total inhibition of that specific connection (or perhaps a probabilistic inhibition); simple addition of both currents, or even multiplication (introducing synaptic weights).

In the Definition of S-Nets, R is defined as a sub-set of $(G \times S) \cup (S \times G) \cup (S \times S)$. It remains one possibility, $(G \times G)$ connections. This means, neurones linked directly, soma with soma. Some approaches are possible:

a) The introduction of Slow Potential Theory (see, e.g., [Anderson 95]). This theory suggests that the important feature of cell activity is the value of its slow potential, not the presence/absence of action potentials. Action potentials are used only as a way to transmit information through, an otherwise long and poor conductor, the axon. If neurons are close together, perhaps action potentials are not needed.

b) The event of neuronal death or neuronal merging (interesting as a simplification method in a future S-Net construction algorithm).

Another interest point is noise. Several components of the neuron are intrinsically noise sources, like ion flows through ion channels, or the rate of neurotransmitter release stored on synaptic vesicles. Complex dendritic trees and intricate synaptic connections can increase noise and create distortion, which can affect information transmission. These can be approached by S-Nets, if we change the information processing functions in order to handle noise. Noise probably inserts the need of rational numbers in activation values.

Conclusion

We have presented a model that tries to grasp the internal complexity of real neural networks. With only four simple types of information processing units, we have shown that S-Nets can compute all partial recursive functions.

We think that S-Nets have potential to represent many subtle structures existing in central nervous systems, and perhaps eventually, they can help us to understand a little more of what is going on. There is a lot to do, but fortunately, there are many new directions to improve this model, as shown in the previous section.

Bibliography

ANDERSON, J. (1995). *An Introduction to Neural Networks*. Massachusetts Institute of Technology.

ARBIB, M (1989). *The Metaphorical Brain 2. Neural Networks and Beyond*. John Wiley & Sons.

CUTLAND, N. (1988). *Computability. An Introduction to Recursive Function Theory*. Cambridge University Press.

McCULLOCH, W.; PITTS, W.(1943). *A logical calculus of the ideas immanent in nervous activity. Bulletin of Mathematical Biophysics*, 5, pp.115-33.

MEL, B. (1994). "Information Processing in Dendritic Trees". *Neural Computation*, 6, pp. 1031-85. Massachusetts Institute of Technology.

SHEPHERD, G. (1994). *Neurobiology*. Oxford University Press.

SIEGELMANN, H.; SONTAG, E. (1995). "On the Computational Power of Neural Nets", in *Journal of Computer and System Sciences*, Vol. 50, No.1. Academic Press.

Appendix: The Unlimited Register Machine (URM)

The URM has an infinite number of *registers* labelled R_1, R_2, ... each one containing a natural number. The value contained in R_i, is denoted by r_i. These values can be altered by the URM in reply of some very simple *instructions* that the machine do recognise. A finite list of instructions establishes a *program*. There are four types of instructions,

- Zero instructions. Syntax: Z(n); $n \in \omega^+$. Change value of R_n to 0 ($r_n := 0$).
- Successor instructions. Syntax: S(n); $n \in \omega^+$. Increase value of R_n by 1 ($r_n := r_n+1$).
- Transfer instructions. Syntax: T(n,m); $n,m \in \omega^+$. Replace content of R_n, by r_m ($r_n := r_m$).
- Jump instructions. Syntax: J(n,m,k); $n,m,k \in \omega^+$. If the values of R_n and R_m are equal, jump to the k*th* instruction. If not, proceed to the next instruction (if $r_n=r_m$ then goto k else nil). Jump instructions do not change any registers, only the program *execution*.

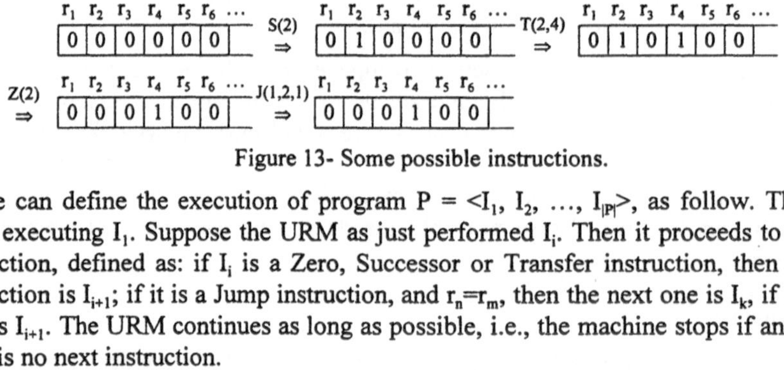

Figure 13- Some possible instructions.

We can define the execution of program P = $<I_1, I_2, ..., I_{|P|}>$, as follow. The URM starts executing I_1. Suppose the URM as just performed I_i. Then it proceeds to the next instruction, defined as: if I_i is a Zero, Successor or Transfer instruction, then the next instruction is I_{i+1}; if it is a Jump instruction, and $r_n=r_m$, then the next one is I_k, if $r_n \neq r_m$ the next is I_{i+1}. The URM continues as long as possible, i.e., the machine stops if and only if there is no next instruction.

A URM-program P computes a function $f:\omega^n \to \omega$, iff, $\forall(x_1,..., x_n) \in Dom(f)$, P with input $(x_1,..., x_n)$, converges to $f(x_1,..., x_n)$, i.e., P ends and $f(x_1,..., x_n)$ is stored in some register (usually in R_1). A function is URM-computable if there is a program that computes f. All partial recursive functions are URM-computable (for a good introduction see [Cutland 88]).

Feed Forward Neural Network Entities

Andreas Hadjiprocopis, Peter Smith

The City University,
Department of Computer Science,
Northampton Square, London EC1V 0HB,
Britain

Abstract. Feed Forward Neural Networks (FFNNs) are computational techniques inspired by the physiology of the brain and used in the approximation of general mappings from one finite dimensional space to another. They present a practical application of the theoretical resolution of *Hilbert's* 13th *problem* by Kolmogorov and Lorenz, and have been used with success in a variety of applications. However, as the training data grows in both dimension and size, larger network implementations are required. As a consequence, in most cases *scaling problems* arise; the existing training algorithms can not handle the *vast search space*, *saturation* occurs at the output of the hidden layer nodes and, in general, the network becomes inflexible, slow and inefficient. Considering all of the above, we are proposing a methodology for breaking down the traditionally single and rigid FFNN into an *entity* of simpler units, in line with the *connectionist view* of the distributive representation of knowledge and using Kolmogorov's paradigm of approximating functions of many variables by compositions of one-variable functions. Although the entities' concept is still developing, some preliminary results indicate superiority over the single FFNN model when applicable to problems involving high-dimensional data (e.g. financial/meteorological data analysis, etc.).

1 Introduction

In the past several years research in exploring the ability of Feed Forward Neural Networks to approximate general mappings from one finite dimensional space to another has yielded many successes in a wide variety of applications (see for example [Fre91], [Was89], [Bis95], [WHR91], [Wei93]).

From a theoretical point of view, Kolmogorov ([Kol57] and also in [HN87]) showed that all continuous functions of n variables can be represented by superpositions and compositions of a number of *different* nonlinear functions. Later, it was shown, [Cyb89], [HN89], [Fun89], [HSW92] and [Hor91], that, in general, a FFNN with a single hidden layer, *employing as many hidden units as required*, and *sigmoidal* activation functions, can approximate any continuous function arbitrarily well.

However, despite the aforementioned research activity on, mostly, theoretical aspects of FFNNs, the practical realisation of such a system often relies on rules of thumb (e.g. architecture, training set selection) and, for problems of increased complexity, is sometimes inefficient and non-optimal. Most of the *gradient-descent* based training algorithms for FFNNs (e.g. back-propagation as in [RHW86], [Wer74], [Wer88], [Hin89]), suffer from a basic limitation; there is no inherent, simple and effective mechanism that can avoid local minima, [MG90], [PFT], [MLBS89],[MP88].

In addition, one has to be aware of the fact that as the number of input variables increases, both the training set and the neural network have to expand in order to be able to capture and convey the increased complexity of the approximated mapping which, in many cases, is *multi-modal* and *non-differentiable*. It has been suggested that the network complexity increases at a rate approximately proportional to the size of the training data, [Cyb89], [Whi89].

Some other weight optimisation techniques, e.g. using *genetic algorithms* [Yao93], although very successful in avoiding local minima, can become very inefficient and computationally demanding when dealing with the big search space produced by the large number of weights.

Finally, it has been shown that in order to produce a good solution the majority of the hidden layer units must operate predominantly in the linear region[1], with small excursions into nonlinear regions but with very little *saturation*, [BN93]. However, with a large number of input nodes, saturation is hard to be avoided without setting any constraints to weight values and network connectivity.

Summarising, the FFNN is a very strong mathematical tool in approximating multi-dimensional mappings by supervised training, but it has practical limitations, especially when the size of the network has to be increased in order to accommodate the increased complexity of the input data.

In this paper, we are proposing four methods for breaking down the traditionally single, rigid neural network into an *entity* of simpler FFNNs, each of which is trained with only some of the input variables of the original training data set.

2 Related Work

Weight elimination, [Bis95], [WRH90], [HP89], is a commonly used method for partitioning a single network into smaller units, creating several centres of neuronal activity. However, despite the fact that this method eliminates redundant weight connections and, in most cases, makes the network more efficient, the basic limitations of scaling still apply. It may well be that after weight elimination the network size has been reduced significantly but the number of weights is still sufficient to cause the problems mentioned in section 1.

A method of actually connecting several neural networks together has been suggested in [Per94]. It consists of L networks which are trained with the same output for the same or different subsets of the input data and at different local minima of the error function (using different initial weight values). The total output of this *committee of networks* is the (weighted) average of the individual outputs. Averaging is a generally acceptable method of error reduction and, here, in the best case, reduces the individual error by a factor of L, if we make the assumption that the individual errors are *uncorrelated* and have *zero mean*, [Bis95].

Cross-validation is another method to increase generalisation ability by involving several neural networks. The *mixture of experts*, introduced in [RAJH91], uses a *gating network* to select the most appropriate neural network from a pool of trained candidates according to their previous performance.

Another implementation of cross-validation strategies was suggested by Wolpert in [Wol92]. *Stacked generalisation* is a scheme for minimising generalisation error

---

[1] In reference to their sigmoidal activation.

by introducing a second space of generalisers whose inputs come from the guesses of a first space of units after trained with part of the learning set and generalising on the rest.

3 Feed Forward Neural Network Entities

3.1 Definitions

Definition 1. Let the family of all *affine* functions in $\mathbb{R}^n$, $\mathcal{A}^n$, be:

$$\mathcal{A}^n = \{A : \mathbb{R}^n \to \mathbb{R} \mid A(\mathbf{x}) = \mathbf{w} \cdot \mathbf{x}^T + b, b \in \mathbb{R}, \mathbf{x}, \mathbf{w} \in \mathbb{R}^n\}$$

Definition 2. Let $\mathcal{S}$ be the family of all (sigmoidal) functions which are *monotonic*, *non-constant*, *bounded* (e.g. $\mathbb{R} \to [0,1]$) and *continuous* in $\mathbb{R}$. We further introduce the additional constraint that $\mathcal{S}$ be closed under *multiplication*[2].

Definition 3. Let a measure of the difference between two reals, x_1 and x_2 be given by,

$$h : \mathbb{R}^2 \to \mathbb{R} \mid h(x_1, x_2) = (x_1 - x_2)^2$$

Definition 4. Let $\mathcal{G}^n$ be the family of all the transfer functions a FFNN with n inputs, a single hidden layer of q units and a single output can implement:

$$\mathcal{G}^n = \{g_n : \mathbb{R}^n \to \mathbb{R} \mid g_n(\mathbf{x}) = \sum_{i=1}^{q} V_i \sigma(A_i(\mathbf{x})), q \in \mathbb{N}, V_i \in \mathbb{R}, A_i \in \mathcal{A}^n, \sigma \in \mathcal{S}\}$$

3.2 Class 1 FFNN Entity

Class 1 (C_1) entities consist of a number of *arbitrarily*[3] interconnected FFNNs. Some of them are trained to map subsets of the input space $\mathbf{X}_i$, $\mathbf{X}_i \subset \mathbf{X}, \mathbf{X} \in \mathbb{R}^n$, to the output, $\mathbf{Y} \in \mathbb{R}$. Some others are trained to map combinations of elements of the input space and outputs of other FFNNs. Figure 3 shows one possible implementation of a C_1 entity. This scheme is structurally equivalent to a single, partially interconnected (after, say, weight elimination) FFNN. However, instead of having a unique training process involving all its nodes and weights, we have several *interdependent*, yet *localised* training processes. Here, we aim to implement the mapping, $\mathbb{R}^n \to \mathbb{R}$ by an arbitrary composition of m mappings, $g_{r_i} : \mathbb{R}^{r_i} \to \mathbb{R}, g_{r_i} \in \mathcal{G}^{r_i}, r_i < n, i = 1, 2, \cdots, m$. The family of all possible C_1 entity transfer functions is,

---

[2] For example, the product of $\frac{1}{1+e^{-x}} \frac{1}{1+e^{-y}} = \frac{e^{x+y}}{e^{x+y}+e^x+e^y+1}$ belongs to $\mathcal{S}$ as it fulfills all the requirements mentioned above.

[3] The only constraint that is imposed on the connectivity scheme of the entity is that before attempting training, any FFNN should have all its inputs defined and available. Circular links will result in a deadlock.

Definition 5.

$$C_1^n = \{f : \mathbb{R}^n \to \mathbb{R} \mid$$
$$f(\mathbf{a}, \mathbf{b}) = g_r(f(\mathbf{a}), \mathbf{b}), \mathbf{a}, \mathbf{b} \subset \{x_1, x_2, \cdots, x_n\},$$
$$f_0(\emptyset) = \emptyset, f_1(x) = x, g_r \in \mathcal{G}^r\}$$

Lemma 6. *The family of functions C_1^n is closed under* **scalar multiplication**:
Proof.

$$c \cdot f(\cdot) = n \cdot \sum_i V_i^a \sigma(\cdot) = \sum_i V_i^b \sigma(\cdot), \; c \in \mathbb{R}$$

Lemma 7. C_1^n *is closed under* **addition**:

Proof.

$$f^a(\cdot) + f^b(\cdot) = \sum_i V_i^a \sigma(\cdot) + \sum_j V_j^b \sigma(\cdot) = \sum_k V_k^c \sigma(\cdot) \in \mathcal{G}.$$

Lemma 8. C_1^n *is closed under* **multiplication**:

Proof.

$$f^a(\cdot) f^b(\cdot) = \sum_i V_i^a \sigma^a(\cdot) \sum_j V_j^b \sigma^b(\cdot) = \sum_k V_k^c \sigma^c(\cdot) \in \mathcal{G} \quad \text{iff} \quad \sigma^c \in \mathcal{S}.$$

Theorem 9. *The family C_1^n is an algebra of functions as it is closed under scalar multiplication, addition and multiplication.*

Lemma 10. *The family C_1^n separates points on $\mathbb{R}^n$, that is:*

$$\forall \mathbf{x_1}, \mathbf{x_2} \in \mathbb{R}^n; \mathbf{x_1} \neq \mathbf{x_2} \quad \exists f \in C_1^n \text{ such that } f(\mathbf{x_1}) \neq f(\mathbf{x_2})$$

Proof.

Let $\quad \mathbf{x_1} = \{f(\mathbf{a_1}), \mathbf{b_1}\}$ and $\mathbf{x_2} = \{f(\mathbf{a_2}), \mathbf{b_2}\}$, then, $\forall \mathbf{x_1}, \mathbf{x_2} \in \mathbb{R}^n$,

$$f(\mathbf{x_1}) = f(\mathbf{x_2}) \Leftrightarrow g_n(\mathbf{x_1}) = g_n(\mathbf{x_2}) \Leftrightarrow \sigma(A(\mathbf{x_1})) = \sigma(A(\mathbf{x_2})) \Leftrightarrow A(\mathbf{x_1}) = A(\mathbf{x_2}) \Leftrightarrow$$
$$\Leftrightarrow \mathbf{w}\mathbf{x_1}^T + b_0 = \mathbf{w}\mathbf{x_2}^T + b_0 \Leftrightarrow \mathbf{x_1} = \mathbf{x_2}, \text{provided that } \mathbf{w} \neq \emptyset, b_0 \neq 0$$
$$\text{therefore,} \quad \mathbf{x_1} \neq \mathbf{x_2} \Rightarrow f(\mathbf{x_1}) \neq f(\mathbf{x_2})$$

Lemma 11. *The family C_1^n vanishes at no point on $\mathbb{R}^n$, that is:*

$$\forall \mathbf{x} \in \mathbb{R}^n \quad \exists f \in C_1^n \text{ such that } f(\mathbf{x}) = c \in \mathbb{R} \neq 0$$

Proof.

Let $\quad \mathbf{x} = \{f(\mathbf{a}), \mathbf{b}\}$ then, $\forall \mathbf{x} \in \mathbb{R}^n$ and c constant,

$$f(\mathbf{x}) = c \Leftrightarrow g_n(\mathbf{x}) = c \Leftrightarrow \sigma(A(\mathbf{x})) = c \Leftrightarrow A(\mathbf{x}) = c_1 \Leftrightarrow \mathbf{w}\mathbf{x}^T + b_0 = c_2 \Leftrightarrow \mathbf{w} = \emptyset$$

3.3 Class Two FFNN Entity

A Class 2 (C_2) entity is a special case of the C_1 entity, where interconnections between FFNNs are not arbitrary, but follow a pattern as indicated in figure 4. Firstly, a FFNN ($\mathcal{N}_1$) is trained to implement $g_k(x_1, \cdots, x_k) = y$. In most cases this will not be sufficient as y may depend on some other inputs too. Therefore, a second FFNN ($\mathcal{N}_2$) is trained to implement $g_{l-k}(x_{k+1}, \cdots, x_l) = y$ and so on, until $\mathcal{N}_i$ is trained to implement $g_{n-q}(x_{q+1}, \cdots, x_n) = y$. The next FFNN ($\mathcal{N}_{i+1}$) will attempt to implement $g_2(g_{q-p}(x_{p+1}, \cdots, x_q), g_{n-q}(x_{q+1}, \cdots, x_n))$. The rest of the FFNNs are trained in a similar fashion. The family of functions implemented by C_2 entities, C_2^n, is,

Definition 12.

$$C_2^n = \{f_n : \mathbb{R}^n \to \mathbb{R} \mid$$
$$f_n(x_1, x_2, \cdots, x_n) = g_2(g_k(x_1, \cdots, x_k), f_{n-k}(x_{k+1}, \cdots, x_n)),$$
$$f_0(\emptyset) = \emptyset, f_1(x) = x, x_i \in \mathbb{R}, g_i \in \mathcal{G}^i\}$$

3.4 Class Three FFNN Entity

Class 3 (C_3) entities have the same interconnection scheme as C_2's (see figure 4). They differ, however, in that the training target (output) of every FFNN is not y. Instead, it is a measure of the discrepancy between *desired* and *actual* outputs of the previous FFNN. Training commences with $\mathcal{N}_1$. It is trained to implement $g_k(x_1, \cdots, x_k) = y$. In most cases, its actual output, o_1, after training, will not be exactly y. A measure of this discrepancy is given by $e_1 = h(y, o_1)$, as in Def. 3, and will be used as the training target for the next FFNN, $\mathcal{N}_2$. Table 1 summarises the training procedure for C_3 entities. C_3 entities implement the same transfer functions as those implemented by C_2 and belong to the C_2^n family of functions, as in Def. 1

FFNN	Input (training)	Output (training)	Discrepancy
N_1	$x_1 \cdots x_k$	y	$e_1 = h(y, o_1)$
N_2	$x_{k+1} \cdots x_l$	e_1	$e_2 = h(e_1, o_2)$
N_3	$x_{l+1} \cdots x_m$	e_2	$e_3 = h(e_2, o_3)$
$\vdots$	$\vdots$	$\vdots$	$\vdots$
N_{i-2}	$x_o \cdots x_p$	e_{i-3}	$e_{i-2} = h(e_{i-3}, o_{i-2})$
N_{i-1}	$x_{p+1} \cdots x_q$	e_{i-2}	$e_{i-1} = h(e_{i-2}, o_{i-1})$
N_i	$x_{q+1} \cdots x_n$	e_{i-1}	$e_i = h(e_{i-1}, o_i)$
N_{i+1}	o_{i-1}, o_i	e_{i-2}	$e_{i+1} = h(e_{i-2}, o_{i+1})$
$\vdots$	$\vdots$	$\vdots$	$\vdots$
N_{j-1}	o_{j-2}, o_2	e_1	not required
N_j	o_{j-1}, o_1	y	not required

Table 1. Training procedure for C_3 entities

3.5 Class Four FFNN Entity

C_4 entities also make use of the discrepancies in the approximations of previous FFNNs in order to train the rest of the FFNNs. It is a slight variation of C_3 entity in the sense that this discrepancy is not used as a target output in the training process but, rather, as an additional input. The first FFNN ($\mathcal{N}_1$), in figure 5, is trained to implement $g_{n-q}(x_{q+1}, \cdots, x_n) = y$. The discrepancy of the attempted approximation is calculated as $e_1 = h(y, g_{n-q}(x_{q+1}, \cdots, x_n))$, as in Def. 3. The second FFNN ($\mathcal{N}_2$) is trained to implement $g_{q-p+1}(e_1, x_{p+1}, \cdots, x_q) = y$, and so on. The family of functions implemented by C_4 entities, C_4^n, is,

Definition 13.

$$C_4^n = \{f_n : \mathbb{R}^n \to \mathbb{R} \mid$$
$$f_n(x_1, x_2, \cdots, x_n) = g_k(x_1, \cdots, x_k, h(f_{n-k}(x_{k+1}, \cdots, x_n))),$$
$$f_0(\emptyset) = \emptyset, f_1(x) = x, g_k \in \mathcal{G}^k, x_i \in \mathbb{R}\}$$

3.6 FFNN Entities as Universal Function Approximators

Theorem 14. *It follows from theorem 9 and lemmas 10 and 11 and by application of the* **Stone–Weierstrass** *theorem, [Rud64], that the family of functions C_1^n is capable of arbitrarily accurate approximation to any real continuous function over a compact set.*

The same method of proving theorem 14 may be used to prove that the family of functions C_2^n, C_3^n and C_4^n defined on $\mathbb{R}^n$ are *algebras* and that they *separate points* on $\mathbb{R}^n$ and *vanish at no point* of $\mathbb{R}^n$. It follows that C_2^n, C_3^n and C_4^n are also capable of arbitrarily accurate approximation to any real continuous function over a compact set and $\mathbb{R}^n$ in general.

4 Experimental results

Experiments comparing the performance of a *single FFNN* and a C_1, C_3 entities were carried out with data of different dimensions. Data for the evaluation of C_1 entities was created by a randomly generated function of the form,

$$\mathcal{F}_1(x_1, \cdots, x_r) = \pm \sum_{i=1}^{r} \alpha_i \sin(2\pi x_i) \cos(2\pi x_i)$$
$$\pm \sum_{i,j=1}^{r} \beta_{i,j} \cos(2\pi x_i) \cos(2\pi x_j) \tag{1}$$
$$\pm \sum_{i,j=1}^{r} \gamma_{i,j} \sin(2\pi x_i) \sin(2\pi x_j),$$
$$0 < x_{i,j} < 1, \alpha, \beta, \gamma \in \mathbb{R}$$

and data for the evaluation of C_3 entities was created by,

$$\mathcal{F}_2(x_1, \cdots, x_r) = \frac{1}{10}\sin(3\pi x_1)^2$$

$$+ \frac{1}{1.2\sqrt{r}}\sum_{i=1}^{r-1}(x_i - 1)^2(1 + 10\sin(3\pi x_{i+1})^2) \qquad (2)$$

$$+ \frac{1}{10}(x_r - 1)(1 + \sin(2\pi x_n)),$$

$$0 < x_i < 1$$

modified from [LM85][4].

We have used 2500 input vectors[5] for testing each network, which is, in our opinion, sufficient in order to measure the generalisation ability of a neural network.

For various r in the range of 50 to 1000, different sets of about 250 training vectors and 2500 testing vectors each, were created. The measure of performance of each network was given by $\frac{1}{2500}\sum_{i=1}^{2500}|y_i - o_i|$, o_i was the actual output when y_i was expected.

The first test was done on C_1 entities (figure 1(a)). Overall, the entity had a 20% better performance than that of the single FFNN.

The second and third tests (figures 1(b) and 2) used C_3 entities. The performance of the entities was better than the single FFNN's by 35% and 45% respectively.

In addition, the time taken for the various tests to be completed is shown in table 2[6].

5 Discussion

The single FFNN model exhibits functional limitations when is used for the analysis of multi-dimensional data (e.g. more than 200 input variables). *Scaling problems, increased computational requirements* due to *inefficiency, saturation* and *failure to generalise* are some of the problems which were identified and discussed in section 1. Although there are methods to minimise the effect of some of the problems above, for example *weight elimination* and *node pruning* for the saturation problem, it is our opinion that they are setting too many constraints to the learning process apart from introducing their own side effects. The concept of connecting FFNNs together has been used in the past (section 2) in order to increase the *confidence of prediction* and *generalisation ability* when dealing with unknown data. In this paper, we are proposing four FFNN entity schemes aiming to resolve the problems encountered when dealing with multi-dimensional data. The first class of FFNN entities is an

---

[4] The use of functions for the generation of data is practical as it implies a virtually unlimited test set.

[5] For the case a 1000-input function this amounts to a 15 M bytes file!

[6] This is hardly indicative of performance as it is not CPU time. Also, not all the comparisons took place on the same CPU, although comparisons with same data were always run on the same machine. It can only be used as a relative measure of training times for C_3 entities and single FFNN with the same number of inputs, i.e. must be in the same row.

arbitrarily interconnected assembly of FFNNs. Training is done locally, but because of the several and unconstrained interconnections, the decision process is global. On the other hand, C_2, C_3 and C_4 entities are more organised and solid in their structure with a fixed configuration.

The results of the somewhat *restricted* tests outlined in section 4, indicate that the performance of the entities is better than that of the single FFNN by about 30% on average. In addition, the entities have *faster training times* than single FFNNs.

Using the proposed entity schemes, one is able to analyse data of extremely large dimensions without the need for gigantic networks. Possible applications might include the analysis of financial or meteorological data. We believe that with the possibility of investigating as many input variables as required, without the limitations set so far by the network size, more complex and accurate models may be built.

Number of Inputs	Entity (hours:secs)	Single (hours:secs)
100	0:17	0:25
200	11:25	13:09
350	8:56	22:31
400	20:01	28:33
500	4:15	8:54
600	16:30	19:56
700	14:57	39:33
800	12:19	32:40
1000	27:06	96:51

Table 2. Time taken to complete training for C_3 entity and single FFNN with $\mathcal{F}_2$

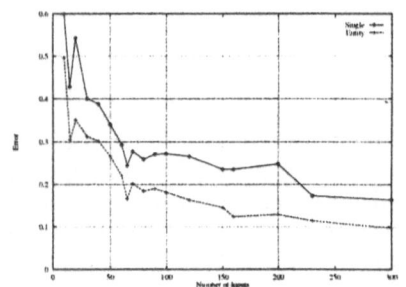 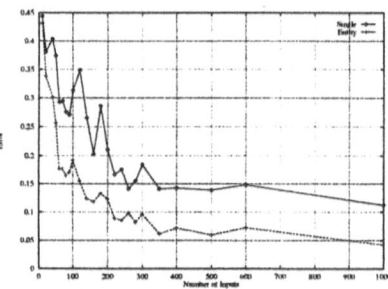

Fig. 1. C_1(a) and C_3(b) entities versus single FFNN in terms of $\mathcal{F}_1$

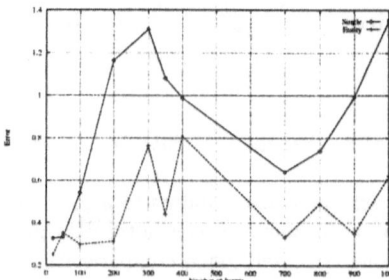

Fig. 2. C_3 entity versus single FFNN in terms of $\mathcal{F}_2$

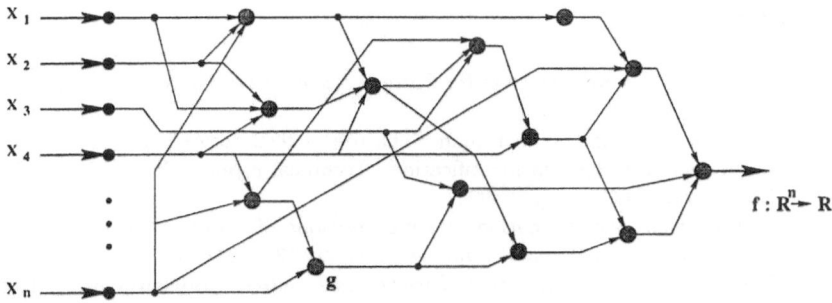

Fig. 3. Schematic of a possible implementation of a C_1 entity

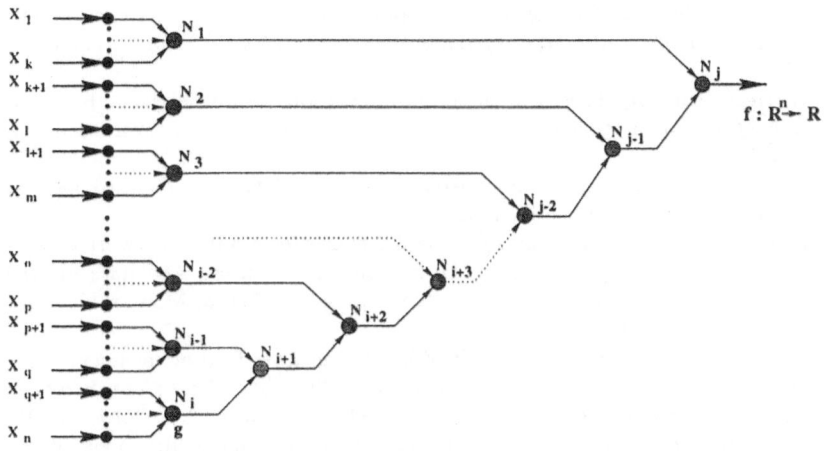

Fig. 4. Schematic of a C_2 & C_3 entity

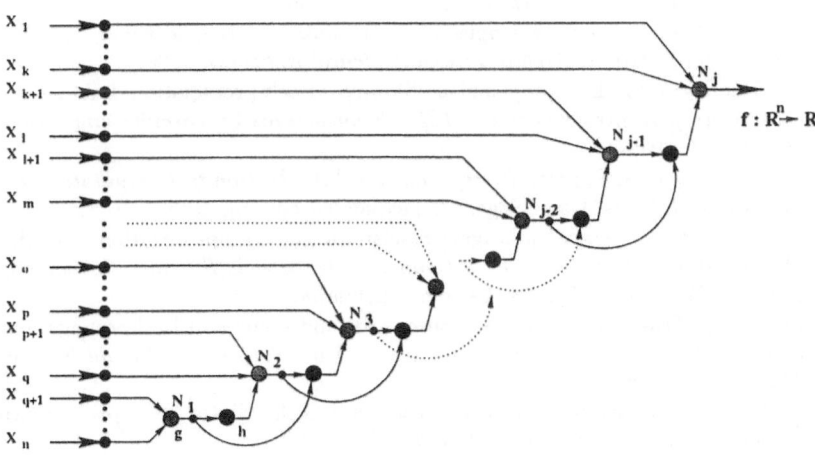

Fig. 5. Schematic of a C_4 entity

References

[Bis95]   C. Bishop. *Neural Networks for Pattern Recognition*. Clarendon Press, Oxford, 1995.

[BN93]   T. L. Burrows and M. Niranjan. The use of feed–forward and recurrent neural networks for system identification. Technical report, Cambridge University Engineering Department, 1993.

[Cyb89]   G. Cybenko. Approximation by superpositions of a sigmoidal function. *Mathematics of Control, Signals, and Systems*, 2(4):303–314, 1989.

[Fre91]   J. Freeman. *Neural Networks: Theory and Practice*. Addison–Wesley, 1991.

[Fun89]   K. Funahashi. On the approximate realization of continuous mappings by neural networks. *Neural Networks*, 2:183–192, 1989.

[Hin89]   G. E. Hinton. Connectionist learning procedures. *Artificial Intelligence*, 40:185–234, 1989.

[HN87]   R. Hecht-Nielsen. Kolmogorov's mapping neural network existence theorem. *IEEE First International Conference on Neural Networks, San Diego*, 3:11–14, 1987.

[HN89]   R. Hecht-Nielsen. Theory of the backpropagation neural network. In *Proceedings of the International Joint Conference on Neural Networks*, volume 1, pages 593–606, 1989.

[Hor91]   K. Hornik. Approximation capabilities of multilayer feedforward networks. *Neural Networks*, 4:251–257, 1991.

[HP89]   S. J. Hanson and L. Y. Pratt. Comparing biases for minimal network construction with back-propagation. In D. S. Touretzky, editor, *Advances in Neural Information Processing Systems*, volume 1, pages 177–185. Morgan Kaufmann, San Mateo, CA, 1989.

[HSW92]   K. Hornik, M. Stinchcombe, and H. White. Multilayer feedforward networks are universal approximators. In Halber White, editor, *Artificial Neural Networks: Approximation and Learning Theory*, pages 12–28. Blackwell, Oxford, UK, 1992.

[Kol57]   A. N. Kolmogorov. On the representation of continuous functions of many variables by superposition of continuous functions of one variable and addition. *Doklady Akademii Nauk SSR*, 114:953–956, 1957.

[LM85]   A. V. Levy and A. Montalvo. The tunnelling algorithm for the global minimization of functions. *SIAM J. Sci. Stat. Comput.*, 6:15–29, 1985.

[MG90]   A. Tesi M. Gori. Some examples of local minima during learning with back-propagation. *Parallel Architectures and Neural Networks*, 1990.

[MLBS89] R. Raghavan M. L. Brady and J. Slawny. Back-propagation fails to separate where perceptrons succeed. *IEEE Transactions on Circuits and Systems*, 36:665–674, 1989.

[MP88]   M. Minsky and S. Papert. *Perceptrons: An Introduction to Computational Geometry*. MIT Press, Cambridge, MA, expanded edition, 1988.

[Per94]   M. P. Perrone. General averaging results for convex optimization. In M. C. Mozer, editor, *Proceedings 1993 Connectionist Models Summer School*, pages 364–371, Hillsdale, NJ, 1994. Lawrence Erlbaum.

[PFT]   M. Gori P. Frasconi and A. Tesi. Susccesses and failures of backpropagation: a theoretical investigation. In O. Omidvar, editor, *Progress in Neural Networks*. Ablex Publishing.

[RAJH91] S. J. Nowlan R. A. Jacobs, M. I. Jordan and G. E. Hinton. Adaptive mixture of local experts. *Neural Computation*, 3:79–87, 1991.

[RHW86] D. E. Rumelhart, G. E. Hinton, and R. J. Williams. Learning internal representations by back-propagating errors. In D. E. Rumelhart and J. L. McClelland,
editors, *Parallel Distributed Processing: Explorations in the Microstructure of
Cognition.* MIT Press, Cambridge, MA, 1986.

[Rud64]   W. Rudin. *Principles of Mathematical Analysis.* McGraw-Hill, New York, 1964.

[Was89]   P. D. Wasserman. *Neural Computing: Theory and practice.* Van Nostrand
Reinhold, 1989.

[Wei93]   A. S. Weigend. *Time Series Prediction : Forecasting the Future and Understanding the Past.* Addison–Wesley, 1993.

[Wer74]   P. J. Werbos. *Beyond Regression: New Tools for Prediction and Analysis in
the Behavioral Sciences.* Doctoral Dissertation, Applied Mathematics, Harvard
University, Boston, MA, November 1974.

[Wer88]   P. J Werbos. Back-propacation: Past and future. In *Proceedings of IEEE International Conference on Neural Networks,* volume 1, pages 343–353. IEEE Press,
New York, 1988.

[Whi89]   H. White. Learning in artificial neural networks: A statistical perspective. *Neural Computation,* 1(4):425–464, 1989.

[WHR91] A. S. Weigend, B. A. Huberman, and D. E. Rumelhart. Predicting sunspots
and exchange rates with connectionist networks. In M. Casdagli and S. Eubank, editors, *Nonlinear Modeling and Forcasting, SFI Studies in the Sciences
of Complexity,* volume 12. Addison–Wesley, 1991.

[Wol92]   D. H. Wolpert. Stacked generalisation. *Neural Networks,* 5:241–259, 1992.

[WRH90] A. S. Weigend, D. E. Rumelhart, and B. A. Huberman. Back–propagation,
weight elimination and time series prediction. In *Proceedings of the 1990 Connectionist Models Summer School,* pages 65–80. Morgan Kaufmann, 1990.

[Yao93]   X. Yao. A review of evolutionary artificial neural networks. *International Journal of Intelligent Systems,* 8:539–567, 1993.

Astrocytes and Slow Learning in the Formation of Distal Cortical Associations

J. G. Wallace and K. Bluff

Swinburne University of Technology
PO Box 218, Hawthorn, Victoria Australia 3122

E-mail: jgwallace@swin.edu.au

ABSTRACT

An extension is proposed of the range of biologically inspired models for local computation architectures and learning in ANNs. The specific focus is on learning processes underlying the formation of connections providing distal access between local neural areas engaging in subsymbolic processing. A neural network architecture underlying the construction of distal cortical associations representing shared, specific contextual relevance between local regions is described. The model involves the evolutionary co-option of hippocampal neural structures to new functions and the adoption of a slow competitive learning process involving interaction between astrocytic and neural processes to enable identification of appropriate associations between cortical and motivational/emotional activity which proceed on different time scales.

INTRODUCTION

There is a need for an extension of the range of biologically inspired models available as a basis for local computation architectures and learning in ANNs. In the search for new models our work focuses on learning processes underlying the formation of connections providing distal access between local neural areas engaging in subsymbolic processing.

Cortical rhythmic activity provides a means of defining local neural processing. Local resonances are relatively high frequency rhythmic processes in the 30 - 80 Hz range with a spatial dimension in the millimetric range. A major distinction between local and lower frequency global/regional resonant modes is that the local resonant modes are not coherent over distances of more than a few millimeters. Human regional and global resonant modes, in contrast, are always coherent over a distance of centimeters. With Silberstein (1994) we assert that cognitive behavioral or perceptual states requiring a high degree of spatio-temporal specificity in activation are only consistent with local states.

Synchronization of responses of spatially separated, local cortical areas with zero - phase lag is adopted as the definition of association, (Singer, 1994). We adopt synchronous modulation of the excitability of local areas as a means of influencing the probability of them engaging in synchronous firing. The source of the modulation is a proposed mechanism for resolving the problems posed by cortical separation of local areas and the determination of their shared, specific contextual relevance.

The mechanism requires prior recruitment of hippocampal neural structures to new functions through evolution. Extracellular recordings of pyramidal cells in freely moving rats show that most cells in hippocampal regions CA1 and CA3 are 'place' cells which only fire when the rat is in a particular portion of its environment. Burgess at al. (1994) consider place cells in terms of their receptive fields and suggest how place cell firing fields or 'place fields' are constructed. Region CA1 provides many place fields each restricted to a portion of the environment. Extracellular recording data indicate that less than 5% of place cells have multiple place fields and that place fields are single, peaked, smooth functions. Place cells are controlled by sensory cues from different modalities. It is believed that the sensory inputs come from the entorhinal cortex which has access to multimodal sensory information. Our mechanism involves the assumption that, in human cognition, local cortical areas can be substituted for sensory input from portions of the environment and via entorhinal cortex become represented as place fields in the hippocampus. It is indisputable that co-option of pre-existing features into new features has occurred during the course of evolution, (Raff, 1996). Our proposal conforms to the definition of 'co-option' in that pre-existing structures which perform a particular function are presumed to be enlisted for a new use that is selectively advantageous in evolutionary terms.

LEARNING DISTAL CORTICAL ASSOCIATIONS

The neural network architecture underlying the construction of distal cortical associations representing shared, specific contextual relevance between local regions is illustrated in Figure 1. The nature of the representations of cortical structure available to the hippocampal system is problematic. Consensus favours a form of attenuated representation. In our model, sparse connections from cortical local regions, via layers 2 and 3, connect to entorhinal cortex in a functionally similar manner to environmental, sensory inputs. These connections are reciprocal and impermanent since the flexibility required for the development of new environment representations is, also, necessary for construction of a sequence of distal cortical associations.

Each cortical local region (CLR) currently represented in the hippocampal system is connected to a cell (EC) in the entorhinal layer. The layer of place cells (PCs) in hippocampal area CA1 is arranged in place fields (PFs). Each EC has a reciprocal connection with PCs from 0.5 PFs. Excitation of CLRs produces competitive learning in PFs. Acceptable resolution of competitive learning is defined as emergence of a minimum of two active PCs.

Our approach to establishing shared, specific contextual relevance involves further connections to PCs in PFs. These represent input to the hippocampal system from motivational/emotional activity in the brainstem. The structure of the hippocampal system as a whole features a functional subdivision from input to output in a large number of lamellae. This is exemplified in the mossy fibre connections between the dentate gyrus and CA3 pyramidal cells and the Schaffer collaterals connecting the pyramidal cells of CA3 with those of CA1. In our model individual lamellae between the dentate gyrus and CA3 represent motivational/emotional elements (MEs). The architecture and operation of connectivity between the ME layer and CA3 place cells organised in ME fields (MEFs) is identical to the EC-PF arrangements already described. The same competitive learning process operates in MEFs.

Construction of Distal Cortical Associations

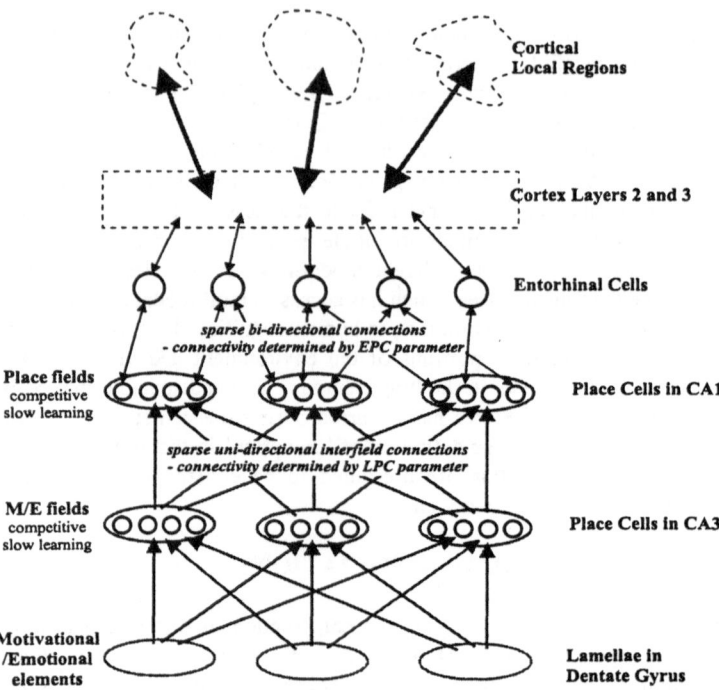

Figure 1

Feedforward connections from the CA3 layer to CA1 place cells represent resolved MEFs. Each MEF has connections with PCs from 0.5 PFs. Critical to establishing shared contextual relevance is the requirement that acceptable resolution of competitive learning in any PF involves inclusion of at least one MEF connected PC in the active group.

Cortical activity and motivational/emotional activity proceed on different time scales. Detection of appropriate associations in our model is achieved by adoption of a slow competitive learning process in PFs, (Wallace and Bluff, 1995 a,b). This involves local aggregation of the effect of EC-PF input to bring this aspect of PF competitive learning resolution into temporal line with the neuromodulation rate of ME processes.

SLOW LEARNING

The significance of slow learning processes has emerged in a number of neural network research areas. Drawing on the successes and failures of connectionist models of learning and memory McClelland et al. (1995) propose complementary learning systems in the hippocampus and neocortex. Their connectionist model of neocortex discovers the structure in ensembles of items if learning of each item is gradual and interleaved with learning about other items. The hippocampal system is responsible for satisfying these conditions. Hippocampal synaptic changes permit rapid learning of new items without disturbing the current cortical structure. Over time these changes support reinstatement of recent memories in the neocortex and produce slow learning as neocortical synapses change a little on each reinstatement. The reinstatement process interleaves new memories with others and facilitates their integration into structured neocortical memory systems.

A learning mechanism operating on a time base consistent with the rate of neuromodulation is a key feature of Wickens (1993) natural neural network theory of the striatum. In presenting an account of reinforcement learning he highlights the critical significance of maintaining raised postsynaptic calcium levels until the arrival of dopamine produces differential reinforcement.

The occurrence of LTP as a result of synaptic plasticity continues to be the major focus of the search for a neural mechanism underlying learning and associative memory. LTP begins to develop within 10 seconds of the application of stimulation and its effects are fully present within 20 - 30 seconds, (Gustafsson and Wigstrom, 1990). The survival value of a learning mechanism linking long-term synaptic facilitation with the results of experience of very short duration is highly questionable.

Direct experimental evidence supports the involvement of much slower processes in synaptic facilitation. Williams et al. (1989), in addition to producing LTP in hippocampal tissue, explored the effects of a combination of weak stimulation and added arachidonic acid. This produced an increase in potentiation comparable with maximal LTP but, in contrast to the rapid onset of LTP, the time course of potentiation revealed a slow onset and a gradual climb to a plateau during a 60 - 120 minute period.

A comprehensive review of experimental data, (Bliss et al., 1993), reveals a wide range of relationships between biochemical states and temporal characteristics of LTP. In general, LTP of longer duration is associated with candidate mechanisms operating at the neuromodulation rate or slower. These mechanisms include protein synthesis from existing mRNAs and gene transcription.

A POSSIBLE SLOW LEARNING MECHANISM

As already indicated, our model requires a competitive learning mechanism which reconciles neural network operations proceeding at neurotransmitter rate with motivational and emotional processes varying at a rate comparable with neuromodulation. The proposed mechanism represents an extension of the range of biologically inspired models since it deviates from the current norm by assigning a critical information processing role to glia and, more specifically, astrocytes and their interactions with neurons. Although one of the two major cell types composing the brain, glia have been assigned an entirely auxiliary role in which they provide physical, trophic and metabolic support to neurons. Occasional proposals that glia may, in addition, have a more direct and active part in determining the results of neural computation have had little impact (Pomerat, 1952; Galambos, 1961; Hertz, 1965; Laming, 1989). Recently, however, it has been established that, although astrocytes show no evidence of the type of electrical excitability involved in signaling between neurons, they exhibit intracellular calcium dynamics that provides an alternative basis for signaling between them. Finkbeiner (1992, 1993), Smith (1992, 1994) and others have, accordingly, proposed that astrocyte networks might mediate slow modulations of neural function, like those underlying arousal, selective attention, motivational state, mood change, learning and memory. In support of this view Smith (1994) points out that many neuronal actions, including most of those mediated by neuropeptides and biogenic amines, like norepinephrine, dopamine and serotonin, are just as slow or even slower than signaling in astrocyte networks.

To enable inclusion of an astrocyte based mechanism in our model we have reviewed a wide range of experimental evidence, the vast majority of it derived from studies of the hippocampus. The results will be considered in relation to three fundamental issues: the existence in astrocytes of processes consistent with the functional requirements of learning and associative memory; the presence of input to astrocytes as a result of adjacent neural functioning thus enabling aggregation of information on neural performance over time; the mechanisms available for feedback from astrocytes to neurons to produce modulation of neural functioning.

ASTROCYTES AND SLOW LEARNING

Astrocytes exhibit a degree of sensitivity, plasticity and associativity consistent with the occurrence of learning and associative memory. There is, however, no empirical data directly relevant to the issue. Establishing clear evidence that behaviour is dependent on particular neurochemical events poses difficult problems. In the case of LTP, for example, its relevance to learning is based on an increased concentration of glutamate separately associated with LTP and training in a variety of learning tasks in rats (Richter-Levin et al., 1994, 1995). We will, also, focus on the complex changes in astrocytes in response to the excitatory neurotransmitter glutamate.

The sensitivity of astrocytes is revealed in the high degree of differentiation of their temporal and spatial responses to exposure to increasing concentrations of glutamate (Kim et al. 1994). They respond to glutamate within seconds with a single, initial calcium, Ca^{2+}, spike of an amplitude typically an order of magnitude greater than the baseline Ca^{2+} level. The spike is not spatially uniform. The Ca^{2+} level rises in one part of the cell and propagates through the rest of the cytoplasm. It, also, propagates rapidly and irregularly between different cells. The occurrence and amplitude of the initial spike is largely independent of extracellular Ca^{2+}. Subsequently, cells exhibit a sustained elevation in Ca^{2+} levels and/or oscillatory responses. The sustained elevation involves an influx of extracellular Ca^{2+} and lasts as long as glutamate and Ca^{2+} are present. Ca^{2+} oscillations initially depend on the release of Ca^{2+} from intracellular stores but also require extracellular Ca^{2+} for sustained cycles. They comprise intracellular waves that propagate through individual cells from one region in the cytoplasm to another with lower amplitude and at slower speeds than the initial spike.

In the continued presence of glutamate for minutes, astrocytes exhibit delayed intercellular Ca^{2+} waves of high amplitude and broad shape that propagate smoothly from cell to cell for distances as long as 100 cell lengths. The rise in Ca^{2+} peaks early and remains elevated for 10 to 20 seconds. Extracellular Ca^{2+} is required. Intercellular waves pass from cell to cell through gap junctions. Their spatial and temporal characteristics markedly differ from those of the initial spike and oscillatory intracellular waves. Accordingly, it is generally agreed that two related but distinct cascades of intracellular physiological effects produce the variations.

A wide range of characteristics of astrocytes provide plasticity capable of representing differential effects of experience. The most striking evidence is the extent of variation between individual cells in the latency, frequency, amplitude and overall pattern of response to a given stimulation, (Rooney, 1989). In addition, individual cells consistently reproduce their own patterns of Ca^{2+} response when repeatedly stimulated by the same agonist. The factors regulating these cell specific patterns are unknown although they appear to be experience based since genetically identical cells produce different patterns of Ca^{2+} response.

Two specific sources of plasticity are relevant to our model. The first involves variations in the intracellular level of free Ca^{2+} as determined by processes governing the entry of extracellular Ca^{2+} and the quantity of Ca^{2+} held in the intracellular Ca^{2+} stores or pools. These variations interact with the second source of plasticity. This is the gap junctions forming direct pathways for intercellular communication between astrocytes. The degree of coupling is not a static feature but subject to high plasticity regulated by developmental or functional factors.

These two forms of plasticity provide a basis for considering associativity between astrocytes. Gap junctions offer both structural and operational features consistent with varying degrees of association. They are composed of connexin proteins that vary from area to area in the CNS. In adult brains the predominant connexin is connexin 43 which is abundant in astrocytes. Connexin 32 has been demonstrated among neuronal sub-populations in different cerebral cortical layers. Establishing whether connexin 43 figures significantly in neurons in situ is complicated by its high level of presence in surrounding glia. Gap junction formation appears to be a symmetrical process between cells expressing the same connexin. Where cells with different connexins are involved an asymmetric movement of molecules is established providing a basis for a hierarchy of command between interconnected cells, (Robinson et al., 1993).

Astrocytes from different CNS regions vary in the frequency and strength of gap junction coupling. Hippocampal astrocytes exhibit the strongest level of gap junction coupling and the greatest resistance to inhibition by the uncoupling agent octanol, (Lee et al., 1994). These findings are consistent with their proposed role in slow learning and associative memory. Further support is derived from experimental results on cell division. Unlike neurons, astrocytes maintain the ability to proliferate postnatally. Increasing frequency of gap junctions in an area is associated with a marked decrease in cell proliferation. This is consistent with the preservation of associations represented in gap junction coupling once learning has produced a stable relational structure in an area.

As with many other membrane channels, gap junctions are modulated by the action of second messengers. This produces both increases and decreases in junctional conductance. The role of Ca^{2+} as a modulator suggests a mechanism for the strengthening of gap junction connections between astrocytes as a result of the experience of individual cells. As indicated above, a sufficient rise in the level of intracellular free Ca^{2+} increases gap junctional conductance and produces intercellular waves. A critical determinant of the rate of increase of free Ca^{2+} is the level of Ca^{2+} held in the intracellular stores or pools. Our model involves the assumption that repeated intercellular waves result in an increase in the base level of Ca^{2+} held in the stores. This strengthens the connection between cells by increasing the probability of attaining the Ca^{2+} level required to trigger intercellular waves.

The focus on Ca^{2+} provides a complete mechanism since large increases in free Ca^{2+} have been shown to uncouple gap junctions, (Pappas et al., 1996; Bennett and Verselis, 1992). There is, thus, a time window of operational intercellular connectivity defined by the level of stored Ca^{2+} and the trigger level for uncoupling. The suggested increase in the stored Ca^{2+} base level narrows the time window and produces a relatively faster, more precise interaction between the astrocytes. On this basis, gap junction connectivity begins to suggest a phylogenetic precursor of neuronal connectivity.

We have presented evidence in support of the existence of astrocytic processes capable of sustaining learning and associative memory. It remains to establish the feasibility of astrocytes aggregating information on the performance of adjacent neurons over time and, as a result, generating feedback that modulates neural operations. Since they represent a complete interactive cycle between astrocytes and neurons the two processes will be discussed together.

Evidence has already be presented of a sufficient mechanism enabling the internal state of astrocytes to reflect the level of adjacent neural activity over time. The level of interstitial glutamate produced by synaptic activity is reflected in the quantity of Ca^{2+} in the intracellular stores and free within the cell. Since astrocytes possess receptors for many neurotransmitters and neuropeptides there are many additional possible sources of information on neuronal functioning (Martin, 1992). Neurotransmitter sources and information on neural states are not restricted to the synaptic regions of neurons. A mechanism has been detected by which glutamate release from axons produces intracellular spiking in glial cells (Chiu and Kriegler, 1994).

There is a wide range of possible mechanisms enabling feedback from astrocytes to modulate neural performance. Astrocytes contain, synthesize or release at least 20 potentially neuroactive compounds (Martin, 1992). More specific proposed mechanisms for astrocyte-neuron communication include: astrocyte regulation of interstitial K^+ or Ca^{2+}, altering of neuronal function by control of interstitial neurotransmitter levels, for example, by releasing arachidonic acid which increases free glutamate in the synaptic area; modulating neural function by modulating the dimensions of extracellular spaces via Ca^{2+} dependent changes in astrocyte structure; modulation of neural activity by astrocytic regulation of cerebrovascular function via production of nitric oxide, NO (Finkbeiner, 1993; Smith, 1992; Wallace and Bluff, 1995b).

Importantly, this range of possibilities has recently been joined by experimental evidence supporting direct signalling from astrocytes to neurons (Nedergaard, 1994; Parpura et al., 1994). In both cases it was demonstrated that astrocytic Ca^{2+} waves, produced by methods not directly affecting neurons, result in large increases in intracellular Ca^{2+} in adjacent neurons. Parpura et al. attribute this result to the calcium-dependent release of glutamate by astrocytes while Nedergaard proposes unidirectional signalling between astrocytes and neurons via gap junctions as a mechanism. This is consistent with the asymmetric movement of molecules between interconnected cells reported by Robinson et al. (1993) when, as in the case of astrocytes and neurons, the cells have different connexins.

As indicated in Figure 1, prior to the resolution of neuronal competitive learning our model requires the detection of appropriate associations between cortical and motivational/emotional activity which proceed on different time scales. It is the development of associations between adjoining astrocytes as a result of the aggregated detection of neural activity in their immediate spatial proximity and the subsequent intervention of the astrocytes in the learning of the proximal neurons to which we attribute the means of achieving this. The direction we are taking in computational modelling of this interaction between fast (neural) and slow (astrocytic) learning is briefly described in the final section.

INTERACTION BETWEEN SLOW AND FAST LEARNING

A standard depiction of neural learning is the Hebbian relationship which can be expressed in terms of the weight changes resulting from pre- and postsynaptic activity as,

$$\frac{dw_{ij}}{dt} = \lambda \, F[a_i(t), a_j(t)] - \delta \tag{1}$$

where $w_{ij}(t)$ is the weighted connection strength at time t from the jth (presynaptic) element to the ith (postsynaptic) element and $a_j(t)$ and $a_i(t)$ are measures of presynaptic and postsynaptic activity respectively, λ is the learning constant and δ is a passive decay.

In the current model postsynaptic activity at a dendritic site is influenced by presynaptic activity at the site, by postsynaptic activity at other nearby sites and a correlate of their distance (Brown et al., 1992).

Slow learning is again a Hebbian interaction which is a response to signal transmission between astrocytes, across gap junctions. The transmission occurs when the spatio-temporal aggregation of the products of synaptic activity by the astrocytes on both sides of the gap junction reaches appropriate levels (of calcium) and affects the level in the recipient astrocyte. If $S_K(t)$ is the level of slow learning aggregation (calcium level) in astrocyte K then the function $ST_{KI}[S_K,S_I]$ which determines the effect on the slow learning level in recipient K of the signal transmission across the junction from I to K could be expressed as

$$ST_{KI}[S_K,\ S_I] = \begin{cases} H[\ S_K\ ]: \varphi_{init} < S_I\ \sigma\,[W_{KI}] < \varphi_{close}\ \text{and}\ \varphi_{rec} < S_K\ \sigma\,[W_{KI}] < \varphi_{close} \\ 0: \text{otherwise} \end{cases} \tag{2}$$

where $\varphi_{close} > \varphi_{init} > \varphi_{rec}$ and they are respectively thresholds for gap junction uncoupling, signal initiation and signal reception. W_{KI} is a connection weight representing the plasticity of the junction, which is modified by a co-occurrence rule. The common sigmoid squashing function (σ) is used to map the weight values to the range 0 to 1.

Currently the aggregation of synaptic activity is modelled in a global fashion for each astrocyte. $A_K(t)$, the aggregated synaptic activity measure is

$$A_K(t) = \sum_{i}^{n_K} a_i(t)\ /\ n_K \tag{3}$$

where n_K is the number of dendritic sites adjacent to astrocyte K.

If $\chi_K(t)$ is a moving average of the synaptic activity,

$$\chi_K(t) = \sum_{m=0}^{T} A_K(t - m)\ /\ T \tag{4}$$

where T is the number of fast learning time steps that are considered to constitute the time window for response by the slow learning mechanism, then S_K, the aggregated slow learning value in astrocyte K is expressed as follows

if $\qquad P_K(t) = S_K(t - \Delta t) - a\,\Gamma[t] \tag{5}$

and $\qquad Q_K(t) = P_K(t) + \Gamma[t]\,G[P_K(t), \chi_K(t)] \tag{6}$

then $\qquad S_K(t) = Q_K(t) + \sum_{I}^{N} \Gamma[t] ST_{KI}[Q_K(t),\ Q_I(t)] \tag{7}$

where α is a decay, the second term in equation (6) represents the contribution of fast learning activity to the slow learning and the second term in equation (7) represents the effect of signal transmission from the N neighbouring slow learning sites (astrocytes) with which gap junctions exist. $\Gamma[t]$ is currently simplistically defined as zero unless t is a multiple of T (the slow learning time window as above).

If the value of $S(t)$ in an element reaches a critical threshold then interaction with the current fast learning relationships occurs. This is modelled by adjusting the learning and decay constants of equation (1).

REFERENCES

Bennett, M.V.L. and Verselis, V.K., (1992). Biophysics of gap junctions. *Cell Biology.* 3, 29-47.

Bliss, T.V.P., Collingridge, G.L., (1993). A synaptic model of memory: long-term potentiation in the hippocampus. *Nature.* 361, 31-39.

Brown, T.H., Zador, A.M., Mainen, Z.F., and Claiborne, B.J. (1992). Hebbian computations in hippocampal dendrites and spines. In T. McKenna, J. Davis, S.F. Zornetzer (Eds.), *Single Neuron Computation.* Academic Press, 81-116.

Burgess, N., Reece, M. and O'Keefe, J., (1994). A model of hippocampal function. *Neural Networks.* 7, 1065-1082.

Chiu, S.Y. and Kriegler, S., (1994). Neurotransmitter-mediated signaling between axons and glial cells. *Glia.* 11, 191-200.

Finkbeiner, S., (1992). Calcium waves in astrocytes-filling in the gaps. *Neuron.* 8, 1101-1108.

Finkbeiner, S.M., (1993). Glial Calcium. *Glia.* 9 83-104.

Galambos, R., (1961). A glial-neural theory of brain function. *Proc. natn. Acad. Sci.* 47, 129-136.

Gustafsson, B., Wigstrom, H., (1990). Long-term potentiation in the CA1 region: its induction and early temporal development. *Progress in Brain Research.* 83, 223-232.

Hertz, L. (1965). Possible role of neuroglia: a potassium-mediated neuronal-neuroglia-neuronal impulse transmission system. *Nature.* 4889, 1091-1094.

Kim, W.T., Rioult, M.G. and Cornell-Bell, A.H., (1994). Glutamate-induced calcium signaling in astrocytes. *Glia.* 11, 173-184.

Laming, P.R., (1989). Do glia contribute to behaviour? A neuromodulatory review. *Comp. Biochem. Physiol.* 94A, No.4, 555-568.

Lee, S.H., Kim, W.T., Cornell-Rell, A.H. and Sontheimer, H. (1994). Astrocytes exhibit regional specificity in gap-junction coupling. *Glia.* 11, 315-325.

Martin, D.L. (1992). Synthesis and release of neuroactive substances by glial cells. *Glia.* 5, 81-94.

McClelland, J.L., McNaughton, B.L., and O'Reilly, R.C. (1995). Why there are complemenary learning systems in the hippocampus and neocortex: insights from the success and failures of connectionist models of learning and memory. *Psychological Review.* 102(3), 419-457.

Nedergaard, M. (1994). Direct signaling from astrocytes to neurons in cultures of mammalian brain cells. *Science.* 263, 1768-1771.

Pomerat, C.M. (1952) Dynamic neurogliology. *Texas Rep. Biol. Med.* 10, 883-913.

Pappas, C.A., Rioult, M.G., and Ransom, B.R. (1996). Octanol, a gap junction uncoupling agent, changes intracellular [H+] in rat astrocytes. *Glia.* 16, 7-15.

Parpura, V., Basarsky, T.A., Liu, F., Jeftinija, K., Jeftinija, S., and Haydon, P.G. (1994). Glutamate-mediated astrocyte-neuron signalling. *Nature.* 369, 744-747.

Raff, R.A. (1996). *The Shape of Life.* University of Chicago Press, Chicago.

Richter-Levin, G., Errington, M.L., Maegawa, H., and Bliss, T.V.P., (1994), Activation of metabotropic glutamate receptors is necessary for long-term potentiation in the dentate gyrus and for spatial learning. *Neuropharmacolog.,* 33, No. 7, 853-857.

Richter-Levin, G., Canevari, L., Bliss, T.V.P. (1995). Long-term potentiation and glutamate release in the dentate gyrus: links to spatial learning. *Behavioural Brain Research.* 66, 37-40.

Robinson, S.R., Hampson, E.C.G.M., Munro, M.N., Vaney, D.I. (1993). Unidirectional coupling of gap junctions between neuroglia. *Science.* 262, 1072-1074.

Rooney, T.A., Sass, E.J. and Thomas, A.P. (1989). Characterization of cytosolic calcium oscillations induced by phenylephrine and vasopressin in single fura-2-loaded hepatocytes. *The Journal of Biological Chemistry.* 264, No. 29, Issue of October 15, 17131-17141.

Silberstein, R.B. (1994). Neuromodulation of Neocortical Dynamics. In P.L. Nunez (Ed.), *Neocortical Dynamics and Human EEG Rhythms.* Oxford University Press, 591-627.

Singer, W. (1994). The Role of Synchrony and Neocortical Processing and Synaptic Plasticity. In E. Domany, J.L. van Hemmen, K. Schulten (Eds.), *Models of Neural Networks II*. Springer-Verlag, New York.

Smith, S.J., (1992). Do astrocytes process neural information? *Progress in Brain Research*. **94**, 119-136.

Smith, S.J., (1994). Neuromodulatory astrocytes. *Current Biology*. **4**, 807-810.

Wickens, J., (1993). *A Theory of the Striatum*. Pergamon Press, Oxford.

Wallace, J.G., and Bluff, K. (1995a). Neurons, glia and the borderline between subsymbolic and symbolic processing. In C. Pinto-Ferreira and N.J. Mamede (Eds.). *Progress in Artificial Intelligence*. Berlin, Springer, 201-211.

Wallace, J.G., and Bluff, K. (1995b). Should ANN be ANGN? In J. Mira and F. Sandoval (Eds.). From Natural to Artificial Neural Computation. Berlin, Springer, 53-60.

Williams, J.H., Errington, M.L., Lynch, M.A., & Bliss, T.V.P. (1989). Arachidonic acid induces a long-term activity-dependent enhancement of synaptic transmission in the hippocampus. Nature, 341, 739-742.

Adaptation and Other Dynamic Effects on Neural Signal Transfer

László Orzó, Elemér Lábos

Laboratory of Neurobiology, Department of Anatomy, Semmelweis Medical School, Tűzoltó u. 58, H-1450, Budapest, HUNGARY

ABSTRACT

In the light of the latest results concerning the dynamics of synaptic transmission (Markram et al., 1996) and spike frequency adaptation, the question how the signals are transferred between neurons and what are the meaningful signals in the neural information processing, has to be reconsidered. We constructed simple models of these phenomena and computed qualitatively the neural transfer properties. In these models we examined only the transmission of two parameters of the neurons' membrane current: the mean and the standard deviation. In the cerebral cortex there is usually significant convergence between neurons and so, if the inputs follow the Central Limit Theorem, these two parameters can fully describe the summed input of the cell. From these transfer properties we can conclude, that the big observed coefficient of variation (Softky and Koch, 1993) can be produced, at least partly, by these mechanisms. The low firing rate and spike frequency adaptation do not allow us to use simple firing rate code for longer period simulation of real neurons, as it is assumed in the majority of regular artificial neural network models (ANN) (Gerstner et al., 1992). The measured effects of long term potentiation on the dynamics of synaptic transmission show, that the transfer of membrane currents mean does not change appropriately. Namely, its modification can not be considered as synaptic weight change. Our result shows, that there could be an additional parameter, the standard deviation of the synaptic currents, which can provide the same sort of transfer properties as the average firing rate in typical ANN models. The formation and operation of this type of "code" is discussed.

INTRODUCTION

It is a crucial and sensitive question in the neuroscience and in the neural network theory, what is the principal neural code (Perkel & Bullock, 1968; Mitchison & Miall, 1990). Presumably there is no universal code. That is, different parts of nervous systems can use quite special type of codes (Bialek et al., 1991). It is still an open question, however, what is the neural code within central nervous system (CNS), e.g. within the cortex (Ferster & Spruston, 1995). There are different theories, which emphasize different kind of neural response properties. The majority of ANN models imply, that firing rate transmits information between neurons (Gerstner & van Hemmen, 1992). Some of the experimental results demonstrate, that this variable really changes for adequate stimulation in the CNS (Richmond et al., 1990; Tovee et al., 1993) and can operate as a code. Other models and experiments claim, that fine temporal structure of neural responses can integrate information in the CNS (Singer, 1993). This type of code could be more efficient than the 'robust' rate code (Softky, 1994), but in this case a more precise processing is necessary and the sensitivity for noise of this code is high. The temporal precision of the neurons signal processing and the biological relevance of these models

(König et al., 1996) are open questions. There are considerable doubts about that neurons in the CNS are capable of processing at this level accuracy (Shadlen & Newsome, 1994). There are problems, nonetheless, with the use of firing rate code as well. In the CNS the background activity of neurons is very low (Abeles et al., 1990), the coefficient of variation of neuronal firing is high (Dean, 1981; Softky & Koch, 1993) and the principal cells show considerable rate of spike-frequency adaptation. These factors reduce the firing rate code efficiency. It takes longer time to be able to determine the output activity of a neuron's firing at very low rate with high variability, and during this longer time spike-frequency adaptation occurs. No doubt, that in the peripheral nervous system the code is usually the firing rate. Hence the experiments could not reach a final decision in the question of the existing neural code. Therefore an indirect method which try to use the recent experimental results concerning the dynamic changes of synaptic transmission and synaptic modification rules (Sejnowski, Markram & Tsodicks, 1996) were chosen. The authors demanded, that the experience dependent synaptic plasticity do not results in simple change of gain, but redistribution of synaptic efficacy. In this paper we investigate the neural transfer properties in the light of these data. These transfer properties could be misleading, if spike frequency adaptation is ignored from our investigation. We show simple models of the above phenomena and analyze simulated neuron's transfer properties. The aim is to determine a reasonable code candidate, using these data.

The synaptic release of neurotransmitters is a random process (e.g.: Burnod & Korn, 1989). At an appropriately low rate of stimulation the presynaptic neuron the evoked postsynaptic potentials have identical amplitudes (Volgushev et al., 1995). Astonishingly the experience dependent long term synaptic modifications, like long term potentiation (LTP) or long term depression (LTD), which have been claimed to be the fundamental mechanisms of learning in neural networks, do not alter the amplitude of excitatory postsynaptic potentials (EPSP) or the number of active synaptic contacts, but the probability of synaptic transmission (Stevens & Wang, 1994). Other kind of experience dependent long term synaptic modifications should exist within the CNS. but limited duration experiments indicate only the previous type of changes. There is an other striking property of synaptic transmission: Using a higher rate of stimulation on the presynaptic neuron, which still evokes a 'relatively' low rate firing, then the consecutive amplitude of EPSPs in average decrease considerably, until it reaches some static level (Markram & Tsodicks, 1996). If postsynaptic potentials are close to each other in time, then they are integrated. So called dynamic redistribution of synaptic efficacy during prolonged stimulation and synaptic modifications are established by these processes. Behind these phenomena "simple" mechanisms could be assumed. For example, the random release might be caused by a certain increase of calcium level, which is accumulated in the presynaptic terminal after an incoming action potential and may lead, through some cascade mechanism, to the all or none emission of a quantum of transmitter. Various exhaustion processes also can be the cause of EPSPs' amplitude decrease for repetitive stimulation.

Although a there is a lot of available details about spike-frequency adaptation (Sah, 1996), we used a simple model. One mechanism could be as follows: Due to the emitted action potentials there is appropriate depolarization and it leads to the rise of intracellular calcium level. This enhanced level causes the opening of slow calcium dependent potassium channels and it can maintain the long term change of the neuron's responsibility.

MODEL

Phenomenological models of spike frequency-adaptation and dynamic redistribution of synaptic efficacy have been formulated. Spike details are neglected because the goal was the investigation of synaptic transfer of spike trains. Neither accurate shape and time course of evoked action potentials nor the description of exact dendritic propagation of EPSPs were considered in this work. We could not incorporate such important phenomena in our examination like burst-firing, pacemaker activity or even the effects of inhibitory neurons on the neural response. It is believed that a wide rang of neural events can be explained without consideration of these factors.

DYNAMIC REDISTRIBUTION OF SYNAPTIC EFFICACY

The dynamic redistribution of the synaptic efficacy was modeled as follows. A simple linear differential equation describes the available resources. As an approximation of it we used:

$$Res_{n+1} = Tint*Amax + (1-Tint)*Res_n - Amp \qquad (1)$$
(The used parameter values are: Tint=0.002; Amax=10; Amp=Res_n*0.3*Prob;)

Where Res variable corresponds to the available resources and so to instantaneous efficacy of the synapse. Tint and Amax determined by the integration time constant and the maximal strength of the synapse appropriately. Amp is the decrease of resource level, supposed that action potential leads to transmitter release. It is proportional to the instantaneous resources. The parameter values were chosen for a good representation and to fit experimental data. Integration time constant was adjusted to that low value because of the measured slow recovery of synaptic efficacy. This simple iteration can assure that, for low rate stimulation resources reach Tmax and so evoked postsynaptic current's amplitude tends to maximum. A further parameter determines the probability of synaptic transfer. Effects of LTP and LTD are considered as increase and decrease of this probability.

The model reproduces several findings. The modeled EPSPs have no fix amplitudes, depending on the previous cell activity. If a train of action potentials evoke postsynaptic potentials, amplitudes of consecutive EPSPs decrease until they reach a stable value, which depends on stimulus frequency. The probability of synaptic transmission, however, determines the average size of consecutive postsynaptic potentials in a spike train. These phenomena have to have extreme influence on the signal transmission properties of neurons. For its characterization we computed the average and standard deviation of evoked synaptic current as functions of interspike interval's average and deviation (Fig. 1). The input interspike interval distribution was taken to be normal.

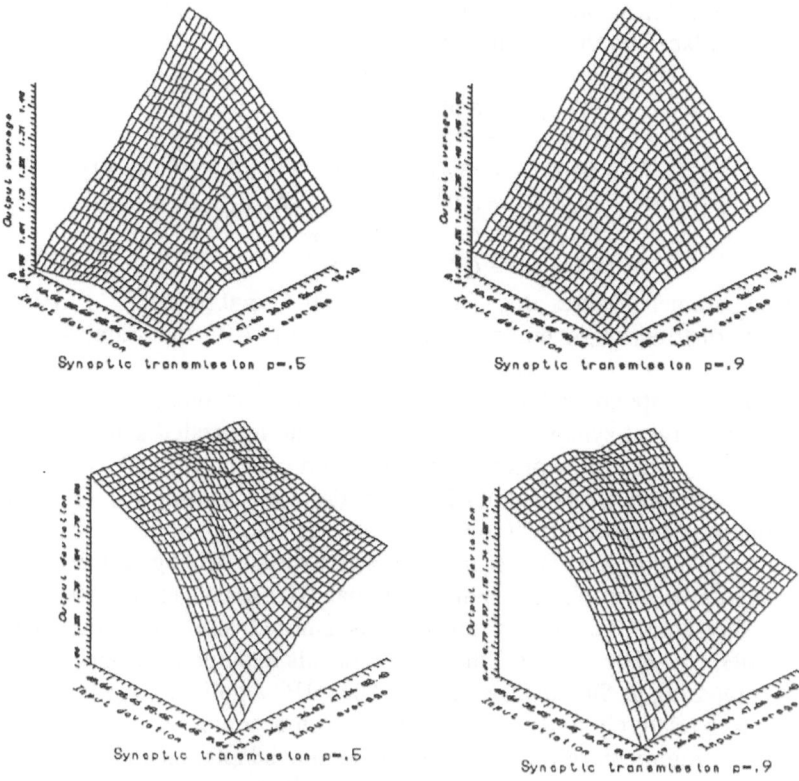

Figure 1 - The averaged output postsynaptic currents and its deviation as functions of interspike interval average and deviation at two different levels of transmission probability.

The results do not show startling changes in the shape of functions, unlike the range of the output, for modification of transmission probability. However, the interspike interval distribution also can play essential role in the signal transfer. To obtain a useful transfer function we need comparable input and output variables and a more realistic interspike interval distribution. Therefore we simulated neuronal presynaptic and postsynaptic currents using model sequence of action potentials as input.

MODELING SPIKE FREQUENCY ADAPTATION

An integrate and fire type neuron model with spike frequency adaptation was constructed. Its input was taken to be a normally distributed random variable. It corresponds to the sum of presynaptic action potentials weighted by synaptic strengths. Normal distribution is supported by the Central Limit Theorem, since the number of converging synapses is large. Uncorrelated firing of pre-presynaptic neurons was supposed. Instead of the widely accepted alpha function for synaptic current simulation, linear differential equations were applied. These equations can be effectively replaced by the next iteration:

$$
\begin{aligned}
ITr_{n+1} &= a*ITr_n +Inp \\
ICu_{n+1} &= b*ICu_n +ITr_n \\
AHC_{n+1} &= c*AHC_n +AIC \\
U_{n+1} &= d*U_n +ICu_n -AHC_n \\
OTr_{n+1} &= OTr_n +Samp \\
OCu_{n+1} &= OCu_n +OTr_n
\end{aligned}
\qquad (2)
$$

(The used parameter values are: $a=0.9$; $b=0.9$; $c=0.9995$; $d=0.4$; Inp=Input distribution (average,standard deviation); $AIC=0.3$; $Samp=0.5$;)

(1) ITr variable corresponds to the summed quantity of transmitters reaching the cell produced by convergent synaptic inputs; (2) ICu is the integrated synaptic currents induced by ITr; (3) AHC is the afterhyperpolarization current; (4) OTr is the amount of transmitter released postsynaptically from the test neuron; (5) OCu is the corresponding postsynaptic current to OTr.

The incoming synaptic currents (ICu) and afterhyperpolarizing currents (AHC) determine simulated neuron's membrane potential (U). If this membrane potential reaches the threshold, then action potential is emitted and afterhyperpolarizing current is generated subsequently. Time course of this hyperpolarization is adjusted to be slow comparing to the other potentials. The AHC amplitude was chosen to be small, so it takes relatively long time until the neuron becomes adapted. This model can reproduce a rudimentary form of spike-frequency adaptation. Mechanisms can be diverse in different neural systems but comparisons are beyond the goal of this study.

To be able to compare input and output properties of modeled neuron, we computed the postsynaptic currents evoked by simulated train of action potentials. The parameters of these currents were identical to those of presynaptic ones.

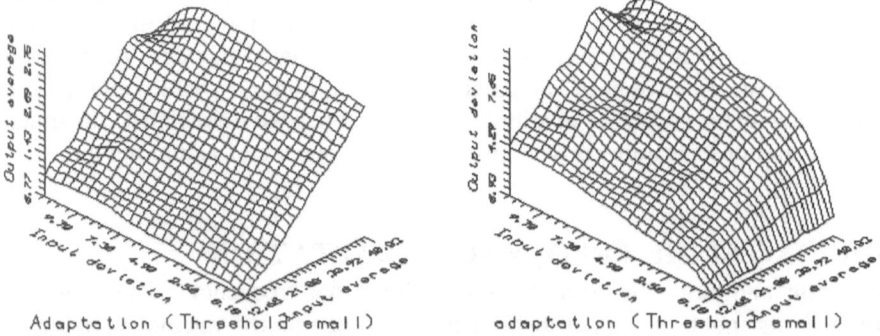

Figure 2 - Transfer properties of the model neuron. - Average and standard deviation of postsynaptic (output) current is plotted against average and standard deviation of input current.

Only the input and output synaptic currents were compared, because the membrane potential includes, for example, afterhyperpolarization and other currents as well. In this way the transfer properties of the simulated neuron can be characterized as it is seen on Fig. 2.

The applied threshold ensures, that within the whole tested range of input the neuron is active. Furthermore, it can be seen, that average output current is almost proportional to that of the input current. Its dependence on the deviation of input current is not significant. However, the standard deviation of the output current depends on both average input currents and its deviation. Since the output current's coefficient of variation is high, therefore it is hard to measure this variable.

At a much higher threshold the transfer properties do not change significantly, but bellow a minimum of the input current no transmission takes place.

Threshold was set to low value to ensure high sensitivity of neuron. High coefficient of variation of neural firing has been explained by Shadlen (Shadlen & Newsome, 1994) with a particular balance of input inhibition and excitation, leading to the wandering membrane potential near the threshold. Our result shows, that this balance also can be achieved by adaptation. Softky (Softky & Koch, 1993) could reach the previous high variation, only using very low threshold value. It is supposed, that adaptive currents can influence neuronal behavior at very low level of activity too. This can expand the dynamic range of neuronal responsiveness. Other mechanisms, like large correlation between synaptic inputs, also can explain big variation of firing rate (Bair & Koch, 1996).

COMBINATION OF THE MODELS

Having a simple model of spiking activity and that of the dynamic changes of synaptic transmission, a composition of them can be constructed. The simulated neural activities are displayed on Fig. 3.

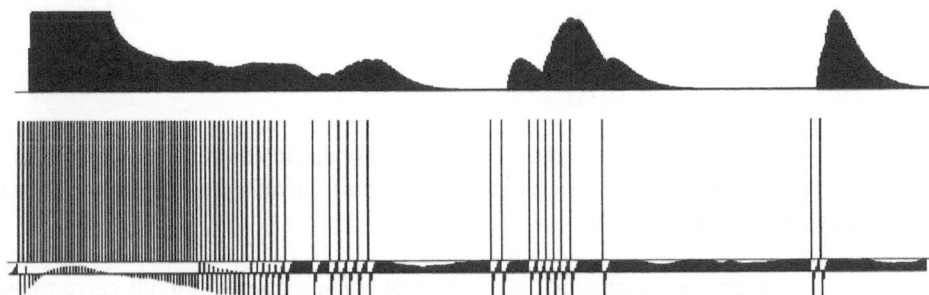

Figure 3 - Effects of adaptation and redistribution of synaptic efficacy on postsynaptic currents. Lower line indicates presynaptic neuronal membrane potential. Upper line shows postsynaptic currents elicited by presynaptic activity.

We can see nice spike-frequency adaptation and consequences of random transmitter release. Additionally, gradual decrease of synaptic efficacy as a result of stimulus train, together with integration of consecutive EPSCs can be observed.

Transfer properties of composite model neuron are displayed on Fig. 5, at two different levels of the transmission probability (corresponding to LTP).

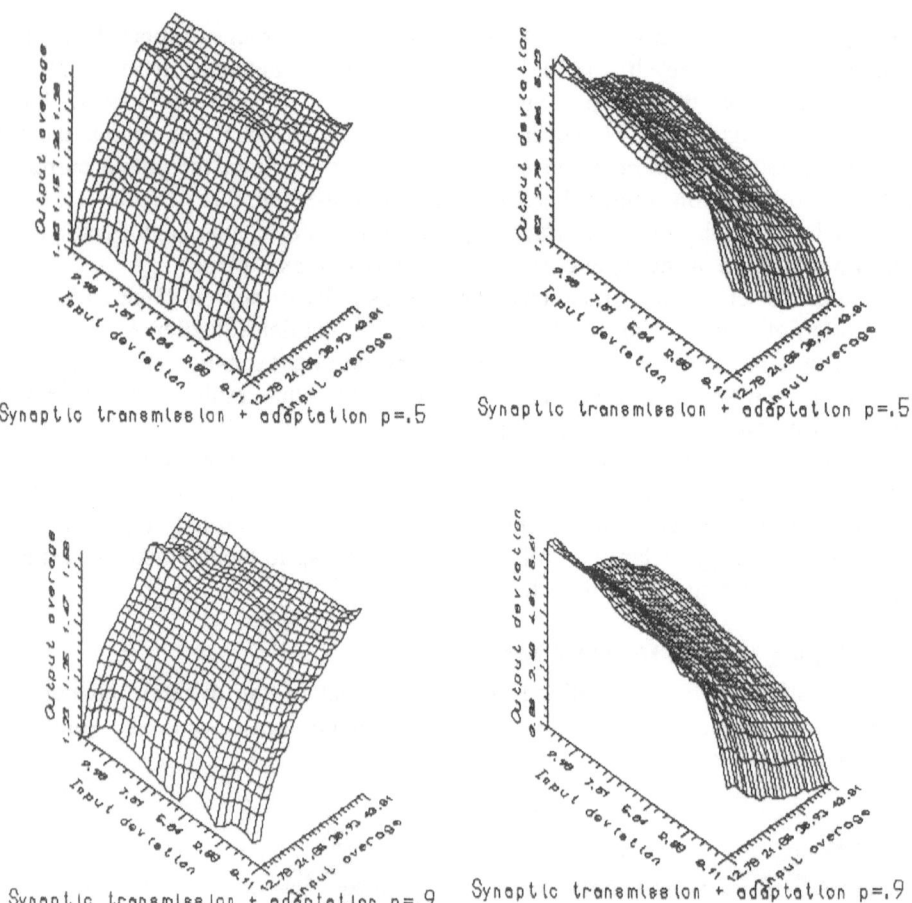

Figure 4 - Model neuron's average evoked postsynaptic (output) current and its deviation as functions of average presynaptic (input) currents and deviation at two different transmission probability.

Figure 4 shows almost linear relation between average output and input currents. In addition, the mean output current increases together with transmission probability, as it was expected. The striking behavior is in this case is, that there is no enhancement in the slope of transfer function, only a shift of average at any input current level. It means, that changes of synaptic transmission is not manifested in increase of synaptic weights, but only in a shift of threshold! This result is further examined on Fig 5.

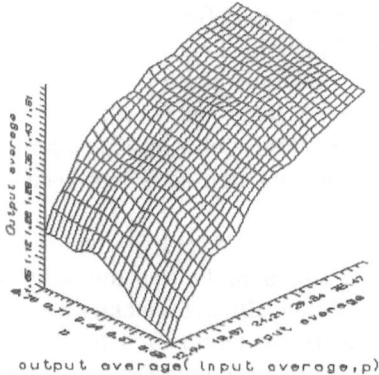

Figure 5 - Average evoked postsynaptic current as a function of average input current and transmission probability.

It can be concluded from Fig. 5, that at any level of transmission probability, there is no considerable change in the gradient. Small observable changes points to direction contrary to expectation. The other measured parameter on Fig. 4, behaves quite a different way. On Fig. 6 transfer properties of deviation at various levels of transmission probability and a fixed value of average input current is demonstrated.

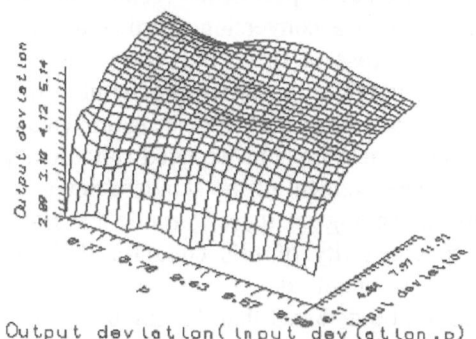

Figure 6 - Dependence of deviation of evoked postsynaptyic current standard deviation of input current and transmission probability.

It can be seen, that dynamic range of deviation transfer function is increasing as transmission probability is growing (corresponding to LTP). This finding can indicates, that transmission of this variable behaves like synaptic weights in usual ANN models. Change in transmission probability leads to simultaneous change in the slope of deviation transfer function (DTF). Metaphorically slope of DTF can be regarded as a "synaptic weight".

DISCUSSION
There are considerable limits of firing rate code. It can not be efficient for a prolonged period of time (see in the Introduction). Our results indicate further

doubts, that is, long term synaptic modifications do not lead to change of synaptic weights, but only that of the threshold. Two additional consequences can arise:

(1) - The LTP primarily does not modify the synaptic weight. Some indirect mechanism should exist, which is able to adjust it later.

(2) - The firing rate is a code only for a restricted period of time in the CNS, or it works quite a different way, not like as it has been assumed in many neural network models.

Above conclusions are not directly valid for hypothetical codes, when fine temporal structure of spike trains are supposed to transmit information. The probabilistic properties of synaptic transmission, although, make them less probable.

An other variable, the standard deviation of synaptic currents, can also convey information within CNS and can be regarded as a code. This variable is as straightforwardly determined as the average synaptic current. As it has been mentioned above, a parameter of DTF works like a "pseudo synaptic weight" and appropriately reacts to LTP and LTD. For the use of this type of code (Orzó, 1993), it is necessary to solve the transformation of frequency code to this code. Our results show, that the model neuron transfer properties make this transformation possible. There is almost a linear dependence of the output deviation on input average at any level of LTP (corresponding to transmission probability). Namely, the higher is the presynaptic currents averaged amplitude the lower will be the evoked postsynaptic current deviation (Fig 4.). Certain experimental result (Softky and Koch, 1993) can indicate the possibility of such a conversion. Further neurocomputation based on this code may occur like in case of firing rate code - as in usual ANN models - because the corresponding variable transfers follow similar rules. At the output side of neural processing firing rate code may appear again. This transformation can be achieved using same type of neurons, but we need suitable adjustment of threshold parameter. The accurate measurement of the standard deviation can be much easier than that of the mean, depending on these variable's variabilities, and so on the whole distribution. The sensitivity of this code for different type of noise and propagation uncertainties is small, due to its specific determination. From our results, however, it can not be deduced that deviation of synaptic current transmits information between neurons in reality. It has only some plausibility.

To identify codes which are in use within CNS, experimental measurements are obviously required. A more or less final decision concerning the relevance any type of neural code can be made only on empirical grounds of physiological experiments. Nevertheless, emerging noise is usually suppressed in physiological experiments by computing the average of measured data. If there is some, not only stimulus dependent temporal variation of neural response, which can transmit information, then this method eliminates it. However, cautious new experimental paradigms help to avoid these obstacles.

ACKNOWLEDGMENTS

The support of the grant No. T013014 (OTKA), and the comments and help of professor József Hámori is gratefully acknowledged.

REFERENCES

Abeles, M., Vaadia, E., Bergman, H. (1990) Firing patterns of single units in the prefrontal cortex and neural network models. Network. Vol. 1, pp. 13-25.

Bair, W., Koch, C. (1996) Temporal Precision of Spike Trains in Extrastriate Cortex of Behaving Macaque Monkey. Neural Computation. Vol. 8, pp. 1185-1202.

Bialek, W., Rieke, F., de Ruyter van Steveninck, R.R., Warland, D. (1991) Reading a Neural Code. Science Vol. 252, pp. 1854-1857.

Burnod, Y., Korn, H. (1989) Consequences of stochastic release of neurotransmitters for network computation in the central nervous system. Pro. Natl. Acad. Sci. USA Vol. 86, pp. 352-356.

Dean, A.F. (1981) The Variability of Discharge of Simple Cells in the Cat Striate Cortex. Exp. Brain Res. Vol. 44, pp. 437-440.

Ferster, D., Spruston, N. (1995) Cracking the Neural Code. Science Vol. 270. pp. 756-757.

Gerstner, W., van Hemmen, J.L. (1992) Universality in neural networks: the importance of the 'mean firing rate' . Biol. Cybern. Vol. 67, pp. 195-205.

König, P., Engel, A., Singer, W. (1996) Integrator or coincidence detector? The role of the cortical neuron revisited. TINS. Vol. 19, No. 4, pp. 130-137.

Markram, H., Tsodyks, M. (1996) Redistribution of synaptic efficacy between neocortical pyramidal neurons. Nature Vol. 382, pp. 807-810.

Mitchison, G., Miall, C. (1990) The Enigma of cortical code. TINS. Vol. 13, No. 2, pp. 41-43.

Orzó, L. (1993) Deviation code is a prospective candidate of the communication between adapting neurons. Neurobiology. Vol. 1 (3), pp. 223-234.

Perkel, H., Bullock, T.H. (1968) Neural Coding. Neuroscience Research Program Bulletin. Vol. 6, pp. 221-344.

Richmond, B.J., Optican, L.M. (1990) Temporal Encoding of Two-Dimensional Patterns by Single Units in Primate Primary Visual Cortex II. Information Transmission. J. of Neurobiol. Vol. 64, No. 2, pp. 370-380.

Sah, P. (1996) Ca^{++}-activated K^+ currents in neurons: types, physiological roles and modulation. Trends in neuroscience Vol. 19 (4), pp. 150-154.

Sejnowski, T.J. (1996) Synapses get smarter. Nature Vol. 382, pp. 759-760.

Singer, W. (1993) Synchronization of cortical activity and its putative role in information processing and learning. Annu. Rev. Physiol. Vol. 55. pp. 349-74.

Shadlen, M.N., Newsome, W.T. (1994) Noise, neural codes and cortical organization. Current Opinion in Neurobiology. Vol. 4, pp. 569-579.

Shadlen, M.N., Newsome, W.T. (1995) Is there a signal in the noise? Current Opinion in Neurobiology. Vol. 5, pp. 248-250.

Softky, W., Koch, C. (1993) The Highly Irregular Firing of Cortical Cells Is Inconsistent with Temporal Integration of Random EPSPs. J. of Neurosci. Vol. 13(1), pp. 334-350.

Softky, W.R. (1995) Simple codes versus efficient codes. Current Opinion in Neurobiology. Vol. 5, pp. 239-247.

Stevens, C.F., Wang, Y. (1994) Changes in reliability of synaptic function as a mechanism for plasticity. Nature Vol. 371, pp. 704-707.

Tovee, M.J., Rolls, E.T., Treves, A., Bellis, R.P. (1993) Information Encoding and the Responses of Single Neurons in the Primate Temporal Visual Cortex. J. of Neurophysiol. Vol. 70, No. 2, pp. 640-654.

Volgushev, M.,Voronin, L.L., Chistiakova, M., Artola, A., Singer, W. (1995) All-or-none Excitatory Postsynaptic Potentials in the Rat Visual Cortex. European Journal of Neuroscience. Vol. 7, pp. 1751-1760.

Hebbian Learning in Networks of Spiking Neurons Using Temporal Coding

Berthold Ruf and Michael Schmitt

Institute for Theoretical Computer Science, Technische Universität Graz
Klosterwiesgasse 32/2, A-8010 Graz, Austria
E-mail: {bruf, mschmitt}@igi.tu-graz.ac.at

Abstract. Computational tasks in biological systems that require short response times can be implemented in a straightforward way by networks of spiking neurons that encode analogue values in temporal coding. We investigate the question how spiking neurons can learn on the basis of differences between firing times. In particular, we provide learning rules of the Hebbian type in terms of single spiking events of the pre- and post-synaptic neuron and show that the weights approach some value given by the difference between pre- and postsynaptic firing times with arbitrary high precision. Our learning rules give rise to a straightforward possibility for realizing very fast pattern analysis tasks with spiking neurons.

1 Introduction

Analogue variables within computations in artificial neural networks are usually represented by the average firing rate of some neuron. However, biological neural systems can perform certain computations within a time window being too small for sampling firing rates. It seems more likely that these computations are based on single firing events (see e.g. [1, 10, 13, 12]). For example, Thorpe and Imbert [13] were able to show that humans can analyze and classify visual patterns in 100 msec, although at least 10 synaptic stages are involved. Neurons participating in such computations usually have a firing rate of less than 100 Hz, hence 10 msec are not sufficient to estimate the current firing rate of some neuron. Results like these motivated the interest in investigating networks of spiking neurons, which can base their computation on single firing events.

In this article we consider spiking neuron networks (SNN's), as introduced by Maass [6], where each neuron is basically a *leaky integrate-and-fire* neuron and can be considered as a noise free version of the spike response model by Gerstner and van Hemmen [4]. These SNN's are besides their biological realism also because of their computational power of great interest. In [7, 5] it has been shown that SNN's are computationally more powerful than McCulloch-Pitts neurons (i.e. threshold gates) and also than sigmoidal gates. It has turned out that especially neurons receiving their input as the time difference between firing times can be used to compute in a biologically plausible way the product between its weight vector and its input vector. In [5] it has been shown how this approach can be used for simulating arbitrary feedforward sigmoidal neural nets in a way which is much faster and more consistent with experimental results about fast information processing in biological neural systems than the usual method using average firing rates. Hence these SNN's can approximate any continuous function of several variables in temporal coding. This approach gives also rise to a very simple and straightforward way of realizing pattern analysis tasks with SNN's [5, 9].

Within this model one considers neurons with given fixed weights that compute functions where the inputs and output are temporally encoded. For example in order to realize pattern analysis one may assume that the learned patterns are stored in weights, whereas new patterns to be analyzed are presented as temporally encoded input. Generally for realizing neural computations based on single firing events it seems to be important to understand the relation between weights and temporally encoded inputs. Thorpe and Gautrais have recently shown that exactly this question could be crucial for fast information processing of neurons in the visual system [12]. They demonstrated that feature extraction might be achieved by the visual system using the order of firing and a proper tuning of synaptic strenghts. Hence the question arises, how a synapse can modify its weight simply on the basis of time differences between firing times and not on firing rates.

Weight changes for some learning process are usually realized according to the Hebb rule, where a synapse is strengthened if both the presynaptic and postsynaptic neuron have at the same time a high firing rate. In this article we want to focus on single firing events, thus we consider some kind of Hebb rule, where the weight change depends on the time difference of single presynaptic and postsynaptic firing times. This type of learning is motivated by recent neurobiological results, where it was shown that the action potential in neocortical pyramidal cells is propagated backwards from the soma to the dendritic tree, such that information about the pre- and postsynaptic firing times is available at the synapse [11] and that synaptic efficacy in those neurons can be actually changed due to this time difference [8].

In this article we focus on supervised learning, where the weight is supposed to assume a certain value and where the weight changes during the learning process are only based on differences between pre- and postsynaptic firing times. This approach is of importance e.g. for the question how a neural system can store information about certain stimuli which are presented repeatedly in the same temporal coding.

We examine how Hebbian learning can be implemented in this context. After introducing the precise model in Section 2 we will show in Section 3 how a single synapse can learn a given weight value simply on the time difference between the presynaptic and the postsynaptic spike. Furthermore we show in Section 4 how several synapses of some neuron can learn given weight values in parallel.

2 Basic Assumptions and Definitions

We consider the common model of *leaky integrate-and-fire neurons*, without noise. As in [5, 6, 7] we make only one additional assumption for realizing computations in such a model: The postsynaptic potential is assumed to grow linearly during an initial segment of fixed but arbitrarily small length.

More precisely, an SNN consists of a finite set V of neurons, a set $E \subseteq V \times V$ of synapses and for each synapse $\langle u, v \rangle \in E$ a response-function $\varepsilon_{u,v}(t) : \mathbf{R}^+ \to \mathbf{R}$ where $\mathbf{R}^+$ denotes the set of all positive reals. One usually models for each neuron v in addition the refractory period of v by a time dependent threshold. However in our constructions, each neuron fires only once within one learning cycle, hence it suffices if each neuron has a constant threshold value Θ. For each synape $\langle u, v \rangle \in E$, where u is the presynaptic and v the postsynaptic neuron, we assume that either $\varepsilon_{u,v}(t) \geq 0$ for all $t \in \mathbf{R}^+$ (excitatory postsynaptic potential, EPSP) or that $\varepsilon_{u,v}(t) \leq 0$ for all $t \in \mathbf{R}^+$ (inhibitory postsynaptic potential, IPSP). The potential $P_v(t)$ of some neuron v at time t is given by

$$P_v(t) = \sum_{u:\langle u,v \rangle \in E} \sum_{s \in F_u : s < t} w_{u,v} \cdot \varepsilon_{u,v}(t - s) \tag{1}$$

where $F_u \subseteq \mathbf{R}^+$ is the set of firing times of neuron u and $w_{u,v} \geq 0$ is the weight of the synapse from neuron u to neuron v. A neuron v fires at time t if $P_v(t)$ reaches Θ from below.

By P_v^{rest} we denote the resting potential of neuron v, i.e. the potential of v when it receives no input. According to (1), $P_v^{\text{rest}} = 0$. However, by continuously receiving sufficiently many EPSP's in addition, P_v^{rest} can be brought arbitrarily close to Θ. (Obviously P_v^{rest} is then no longer time-independent, which corresponds to a noisy membrane potential, however with sufficiently many EPSP's one can reduce this noise arbitrarily.)

Furthermore we assume that each excitatory synapse $\langle u, v \rangle$ has a response function $\varepsilon_{u,v}(t)$ such that

$$\varepsilon_{u,v}(t) = \begin{cases} 0 & \text{for all } t \in [0, d_{u,v}] \\ t - d_{u,v} & \text{for all } t \in [d_{u,v}, d_{u,v} + \delta], \end{cases}$$

where $\delta \geq 0$ is the length of the initial linear segment of the EPSP. The parameter $d_{u,v} \geq 0$ describes the delay between the generation of the presynaptic action potential of neuron u and the time when this action potential reaches the synapse $\langle u, v \rangle$. Throughout this work we consider only passive dendrites, hence we assume that the propagation time in the dendritic tree can be neglected. However even without this assumption one can show similar results as the ones shown in this paper.

3 Monosynaptic Learning

In this section we introduce a Hebbian-type learning rule for monosynaptic learning which is based solely on single pre- and postsynaptic spikes. Let neurons u

and v be connected by an excitatory synapse $\langle u, v \rangle$ with properties described in the previous section. If u fires then a spike arrives at $\langle u, v \rangle$ after a delay of $d_{u,v}$. The synapse does not perceive the actual firing but the arrival of the spike instead. Therefore we will describe firing times of u in terms of arrival times at $\langle u, v \rangle$ in the following.

The learning rule is applied once during a certain period of time which we call learning cycle. The spike trains that are emitted from u and v during such a cycle are shown in Figure 1. The firing of u at t_u is supposed to make v fire at t_v. A second firing of u at t_0 close to t_v is essential for learning in the monosynaptic model in two ways. Firstly, together with t_u it defines a time interval $[t_u, t_0]$ having length $t_0 - t_u$ which can be considered as the input value for v at $\langle u, v \rangle$. Secondly, together with t_v it will be used in the learning rule that depends only on these two firing times during each learning cycle.

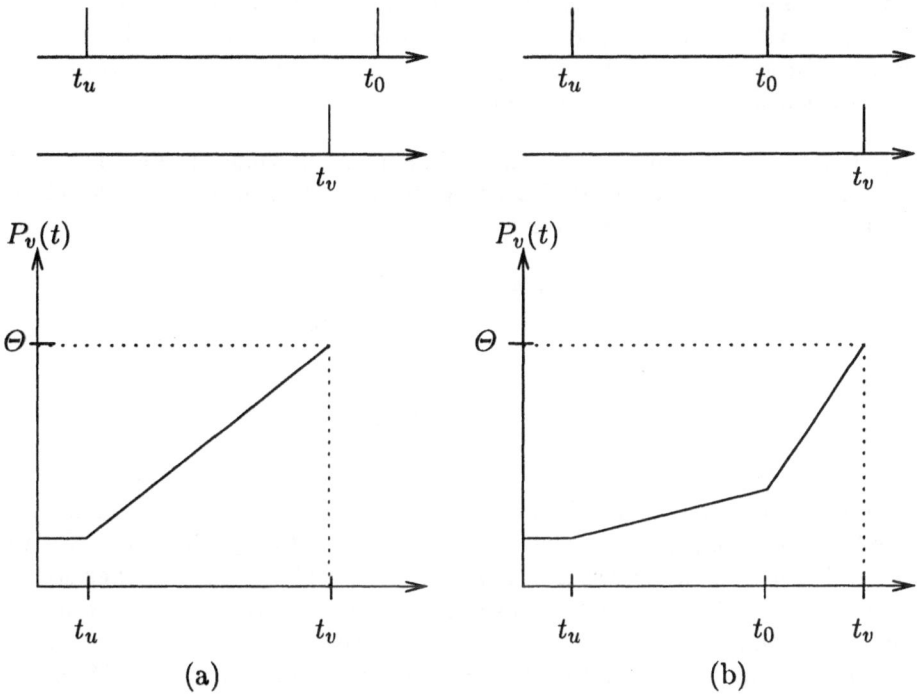

Fig. 1. Spikes at synapse $\langle u, v \rangle$ and potential of neuron v during a learning cycle. In (a) the first spike of u is sufficient to make v fire before the second spike of u; in (b) the firing of v is determined by two spikes of u.

As monosynaptic, Hebbian-type learning rule we propose a change of $w_{u,v}$ by

$$\Delta w_{u,v} = \eta(t_v - t_0) \tag{2}$$

where $\eta > 0$ is the learning rate, a parameter that has to be chosen before learning starts but is fixed during all learning cycles. The time difference $t_v - t_0$ between the presynaptic spike of u at t_0 and the postsynaptic spike of v at t_v can be considered as an error that has to be reduced during learning.

The main result of this section is a proof that by repeating the learning cycles $w_{u,v}$ approaches a value $\widetilde{w}$ representing the input; in fact, $\widetilde{w}$ is proportional to $1/(t_0 - t_u)$.

Theorem 1. *Given two neurons u, v with excitatory synapse $\langle u, v \rangle$ let the interval $[w_{\min}, w_{\max}] \subseteq \mathbf{R}^+$ be the range of possible values for weight $w_{u,v}$. Let the learning rate η be any arbitrary real number satisfying $0 < \eta \leq w_{\min}^2/(\Theta - P_v^{\mathrm{rest}})$. If neuron u fires twice during each learning cycle with fixed time difference $t_0 - t_u$ and $w_{u,v}$ is updated according to rule (2) with learning rate η then, for any arbitrary initial weight from $[w_{\min}, w_{\max}]$, the sequence of weight values converges to $\widetilde{w}$ such that $\widetilde{w} = (\Theta - P_v^{\mathrm{rest}})/(t_0 - t_u)$.*

Proof. Let $\widetilde{w}$ be the value of $w_{u,v}$ when it remains unchanged. This happens if and only if the second firing of u at t_0 and the firing of v at t_v take place simultaneously, i.e. $t_v = t_0$. Then $\widetilde{w}$ satisfies

$$P_v^{\mathrm{rest}} + \widetilde{w}(t_v - t_u) = \Theta.$$

Substituting t_0 for t_v we obtain

$$t_0 - t_u = \frac{\Theta - P_v^{\mathrm{rest}}}{\widetilde{w}}. \tag{3}$$

Thus, the fixed point $\widetilde{w}$ of the learning rule is proportional to $1/(t_0 - t_u)$.

It remains to show that $\widetilde{w}$ is also an attracting fixed point for all weight values from $[w_{\min}, w_{\max}]$. Let w be the current value of $w_{u,v}$ where $w \neq \widetilde{w}$. We distinguish between two cases, $w > \widetilde{w}$ and $w < \widetilde{w}$. This is necessary because if $w > \widetilde{w}$ then the firing of v at t_v takes place before the second firing of u at t_0. Hence, t_v depends only on P_v^{rest}, Θ, and t_u. In the case $w < \widetilde{w}$ neuron u fires twice, at t_u and t_0, before $P_v(t)$ reaches Θ. Hence t_v depends also on t_0. We show that in both cases the value of $w_{u,v}$ stays on the same side of $\widetilde{w}$ and approaches $\widetilde{w}$ successively. This will be done using a fixed point theorem from numerical analysis [2].

Case 1: $w > \widetilde{w}$. In this case, which is shown in Figure 1(a), the firing of v at t_v satisfies

$$P_v^{\mathrm{rest}} + w(t_v - t_u) = \Theta$$

from which we get

$$t_v = \frac{\Theta - P_v^{\mathrm{rest}}}{w} + t_u.$$

Using this together with (3) we obtain an expression for the change of w by a single application of rule (2) by

$$\Delta w = \eta(t_v - t_0) = \eta \left(\frac{\Theta - P_v^{\text{rest}}}{w} - \frac{\Theta - P_v^{\text{rest}}}{\tilde{w}} \right).$$

Thus we can describe the progression of w by an iterative map $f : \mathbf{R} \to \mathbf{R}$ defined as

$$f(w) = w + \kappa \left(\frac{1}{w} - \frac{1}{\tilde{w}} \right) \quad \text{with } \kappa = \eta(\Theta - P_v^{\text{rest}}).$$

Obviously, $\tilde{w}$ is a fixed point of f. We will show that f maps $[\tilde{w}, w_{\text{max}}]$ into itself and is contractive on $[\tilde{w}, w_{\text{max}}]$. Then it follows by a well known fixed point theorem (see, e.g., [2, Theorem 5.1.1]) that f has a unique fixed point in $[\tilde{w}, w_{\text{max}}]$, which is then $\tilde{w}$, and that the sequence $w^{(n+1)} = f(w^{(n)})$ converges to that fixed point for any choice of $w^{(0)} \in [\tilde{w}, w_{\text{max}}]$.

First we show $f(w) \geq \tilde{w}$ for all $w \in [\tilde{w}, w_{\text{max}}]$. This is equivalent to

$$w + \kappa/w \geq \tilde{w} + \kappa/\tilde{w} \quad \text{for all } w \in [\tilde{w}, w_{\text{max}}]. \tag{4}$$

The condition for the learning rate $\eta \leq w_{\text{min}}^2/(\Theta - P_v^{\text{rest}})$ can be rewritten as $w_{\text{min}}^2 \geq \kappa$ from which

$$1 - \frac{\kappa}{w^2} \geq 0 \quad \text{for all } w \geq w_{\text{min}} \tag{5}$$

follows. The expression $1 - (\kappa/w^2)$ is the derivative of $g(w) = w + (\kappa/w)$ which is therefore increasing on $[w_{\text{min}}, w_{\text{max}}]$. Hence (4) is proven and $f(w) \geq \tilde{w}$ holds for all $w \in [\tilde{w}, w_{\text{max}}]$.

From $w > \tilde{w}$ we get

$$\kappa \left(\frac{1}{w} - \frac{1}{\tilde{w}} \right) < 0.$$

Thus $f(w) < w$ holds which implies $f(w) < w_{\text{max}}$ for all $w \in [\tilde{w}, w_{\text{max}}]$. It follows from this, together with $f(w) \geq \tilde{w}$, that f maps $[\tilde{w}, w_{\text{max}}]$ into itself.

To see that f is contractive on $[\tilde{w}, w_{\text{max}}]$ we consider f', the derivative of f, and show that $|f'(w)| \leq L$ for all $[\tilde{w}, w_{\text{max}}]$ with Lipschitz constant $L < 1$. The derivative of f is

$$f'(w) = 1 - \frac{\kappa}{w^2}.$$

In (5) this expression was already seen to be non-negative. A Lipschitz constant $L < 1$ such that $f'(w) \leq L$ for all $w \in [\tilde{w}, w_{\text{max}}]$ is then provided by $L = 1 - (\kappa/w_{\text{max}}^2)$.

Case 2: $w < \tilde{w}$. Here the potential of neuron v is determined by two spikes of neuron u, at t_u and at t_0 (see Figure 1(b)). The firing time t_v of v then satisfies

$$P_v^{\text{rest}} + w(t_v - t_u) + w(t_v - t_0) = \Theta$$

from which we derive

$$t_v = \frac{1}{2} \left(\frac{\Theta - P_v^{\text{rest}}}{w} + t_0 + t_u \right).$$

Solving (3) for t_0 and substituting we can rewrite this as

$$t_v = \frac{1}{2}\left(\frac{\Theta - P_v^{\text{rest}}}{w} + \frac{\Theta - P_v^{\text{rest}}}{\widetilde{w}}\right) + t_u.$$

Combined with (3) this yields a synaptic change of

$$\Delta w = \eta(t_v - t_0) = \frac{\eta}{2}\left(\frac{\Theta - P_v^{\text{rest}}}{w} - \frac{\Theta - P_v^{\text{rest}}}{\widetilde{w}}\right).$$

The sequence of weights is then determined by iterating the function $h : \mathbf{R} \to \mathbf{R}$ defined as

$$h(w) = w + \lambda\left(\frac{1}{w} - \frac{1}{\widetilde{w}}\right) \quad \text{with } \lambda = \eta(\Theta - P_v^{\text{rest}})/2.$$

Obviously, $\widetilde{w}$ is a fixed point of h. It remains to show that h maps $[w_{\min}, \widetilde{w}]$ into itself and is contractive on $[w_{\min}, \widetilde{w}]$. From $\lambda = \kappa/2$ and (5) we get

$$w + \lambda/w \leq \widetilde{w} + \lambda/\widetilde{w} \quad \text{for all } w \in [w_{\min}, \widetilde{w}]$$

which is equivalent to $h(w) \leq \widetilde{w}$ for all $w \in [w_{\min}, \widetilde{w}]$. From $w < \widetilde{w}$ we get $\lambda(1/w - 1/\widetilde{w}) > 0$ and hence $h(w) > w_{\min}$ for all $w \in [w_{\min}, \widetilde{w}]$. Thus h maps $[w_{\min}, \widetilde{w}]$ into itself. Finally, along the same lines of reasoning as in case 1, it is now easy to conclude that h is contractive on $[w_{\min}, \widetilde{w}]$ with Lipschitz constant $L = 1 - (\lambda/w_{\max}^2) < 1$.

This completes the proof of the theorem. $\square$

With regard to the convergence rate we can make the following observation: It is well known that for contractive iterative mappings the rate of convergence is governed by

$$|w^{(n)} - \widetilde{w}| \leq L^n |w^{(0)} - \widetilde{w}|$$

where $L < 1$ is a Lipschitz constant [2]. Thus the sequence of weights generated according to the learning rule (2) satisfies

$$|w^{(n)} - \widetilde{w}| \leq \left(1 - \frac{\mu}{w_{\max}^2}\right)^n |w^{(0)} - \widetilde{w}|$$

where $\mu = \eta(\Theta - P_v^{\text{rest}})$ if $w^{(0)} > \widetilde{w}$ and $\mu = \eta(\Theta - P_v^{\text{rest}})/2$ if $w^{(0)} < \widetilde{w}$.

In the theorem we put no restrictions on values $w_{\min}$ and $w_{\max}$, except that they have to be positive, in order to keep the statement and the proof as general as possible. However, for learning to be biologically plausible, restrictions are certainly necessary. For instance, it seems adequate that the pre- and postsynaptic firing times t_0 and t_v, on which the weight change is based, are not too far from each other. On the other hand, our construction relies on the fact that the pair of pre- and postsynaptic spikes at t_u and t_v does not result in a change. Hence they should not be too close to each other. Furthermore, it must be guaranteed that the firing of v caused by u takes place while the EPSP is still in its initial linear segment. By raising the resting potential P_v^{rest} as described in Section 2 and by restricting the range of possible weight values these requirements can be met easily.

The type of learning we have considered in this section is based on three spikes during each learning cycle, two presynaptic and one postsynaptic. The pair of presynaptic spikes is used to describe the input value, whereas one presynaptic and the postsynaptic spike define the synaptic modification. In the context of supervised learning this seems to be the minimal number of spikes required for biologically plausible, Hebbian-type learning.

4 Parallel Learning

The method described in the previous section makes it possible to learn single weights. However, usually a neuron v receives its input from several neurons $u_1, \ldots, u_n$. The corresponding weights $w_{u_1,v}, \ldots, w_{u_n,v}$ can still be learned in the above described way, if the weights are learned sequentially, such that v receives only inputs through one synapse during one learning cycle. Hence the question arises, how several weights can be learned in parallel without too strong additional assumptions. If one would try to learn weights in parallel in a similar fashion as described in the previous section, the problem would arise of how to distribute the error encoded by $t_0 - t_v$ among the weights. Additional error signals would make Hebbian learning, where the weight change is only influenced by the time difference between some pre- and postsynaptic firing time, impossible.

Since the traditional formulation of the Hebb-rule allows only an increase of the weights, one frequently assumes in addition that after each weight change all weights of the neuron are normalized (see e.g. [3]). Hence if one normalizes the weight vector $\underline{w} = (w_{u_1,v}, \ldots, w_{u_n,v})$ such that $||\underline{w}|| = 1$ then a simple Hebbian learning rule can be formulated: The goal is that $\underline{w} = \underline{\widetilde{w}}$ for some desired weight vector $\underline{\widetilde{w}}$ with $||\underline{\widetilde{w}}|| = 1$. In contrast to Section 3 v receives here through some additional synapse with a sufficiently strong weight an input spike such that v fires at time t_0. The i-th synapse $\langle u_i, v \rangle$ receives a spike from neuron u_i at time t_{u_i} such that $t_0 - t_{u_i} = \widetilde{w}_{u_i,v}$. We ensure that v does not fire before t_0 by providing sufficiently many IPSP's through additional inhibitory synapses. One gets the learning rule

$$\Delta w_{u_i,v} = \eta(t_0 - t_{u_i}) / ||\underline{w}||, \qquad 1 \leq i \leq n, \tag{6}$$

where $\eta > 0$ is the learning rate, that has the following convergence property:

Theorem 2. Let $\underline{\widetilde{w}}$ be a weight vector satisfying $||\underline{\widetilde{w}}|| = 1$ and let $\eta > 0$. Then the sequence of weight vectors generated by the parallel learning rule (6) converges to $\underline{\widetilde{w}}$ for any arbitrary initial weight vector.

Proof. We immediately have $\underline{w}^{(n)} \cdot \underline{\widetilde{w}} \leq 1$ for $n \geq 1$ by the Cauchy-Schwarz inequality. It is not hard to see then that $\underline{w}^{(n)} \cdot \underline{\widetilde{w}} < \underline{w}^{(n+1)} \cdot \underline{\widetilde{w}}$ holds for $\underline{w}^{(n)} \neq \underline{\widetilde{w}}$, which implies the convergence. $\square$

388

It is easy to verify that by some proper transformation of the normalization not only weights from $[0, 1]$ but from some arbitrary interval $[w_{min}, w_{max}] \subseteq \mathbf{R}^+$ can be learned.

We finally observe that the type of learning in this section is not based on an error signal. Furthermore, we do not require here that EPSP's have a linear segment.

5 Conclusions

We have introduced some type of Hebbian learning for spiking neurons by providing rules for synaptic modification that are based on single firing events of the pre- and postsynaptic neuron. To the best of our knowledge, this is the first approach for this model where it is shown that the weights provably converge to the desired values. Thus, a spiking neuron can learn weights from its firing times, which appears to be essential for fast information processing in biological neural systems. Furthermore, our constructions give rise to simple methods for learning weights within pattern analysis tasks where the inputs are temporally encoded.

In the model of a spiking neuron we have explicitly excluded the presence of noise. However, our constructions in Sections 3 and 4 should also work in the case of a certain amount of noise, but they are harder to analyze theoretically. We expect that similar results can be shown using standard methods dealing with noise.

References

1. Abeles, M., Bergman, H., Margalit, E., and Vaadia, E.: Spatiotemporal firing patterns in the frontal cortex of behaving monkeys. J. of Neurophysiology **70** (1993) 1629–1638.
2. Blum, E.K.: Numerical Analysis and Computation: Theory and Practice. Addison-Wesley, Reading, Mass. (1972).
3. Brown, T.H. and Chattarji, S.: Hebbian synaptic plasticity. In: Handbook of brain theory and neural networks, Arbib, M., ed., MIT Press, Cambridge, (1995) 454–459.
4. Gerstner, W. and van Hemmen, L.H.: How to describe neuronal activity: spikes, rates, or assemblies? Advances in Neural Information Processing Systems 6, Morgan Kaufmann, San Mateo (1994) 463–470.
5. Maass, W.: Fast sigmoidal networks via spiking neurons. To appear in Neural Computation.
6. Maass, W.: Lower bounds for the computational power of networks of spiking neurons. Neural Computation 8 (1996) 1–40.
7. Maass, W.: Networks of spiking neurons: the third generation of neural network models. Proc. of the 7th Australian Conference on Neural Networks, Canberra, Australia (1996) 1–10.
8. Markram, H.: Neocortical pyramidal neurons scale the efficacy of synaptic input according to arrival time: a proposed selection principle of the most appropriate synaptic information. Proc. of Cortical Dynamics in Jerusalem (1995) 10–11.

9. Ruf, B.: Pattern analysis with networks of spiking neurons. Submitted for publication.
10. Sejnowski, T. J.: Time for a new neural code? Nature **376** (1995) 21–22.
11. Stuart, G. J. and Sakmann, B.: Active propagation of somatic action potentials into neocortical pyramidal cell dendrites. Nature **367** (1994) 69–72.
12. Thorpe, S. J. and Gautrais, J.: Rapid visual processing using spike asynchrony. To appear in Advances in Neural Information Processing Systems 9, MIT Press, Cambridge, 1997.
13. Thorpe, S. J. and Imbert, M.: Biological constraints on connectionist modelling. In: Connectionism in Perspective, Pfeifer, R., Schreter, Z., Fogelman-Soulié, F., and Steels, L., eds., Elsevier, North-Holland (1989).

An Associative Learning Model for Coupled Neural Oscillators

Jun Nishii

Laboratory for Neural Modeling
The institute of physical and chemical research (RIKEN)
2-1 Hirosawa, Wako, Saitama 351-01, Japan
nishii@postman.riken.go.jp

Abstract. Neurophysiological experiments have shown that many motor commands in living systems are generated by coupled oscillatory components, such as neural oscillators, which show periodic activities. In this paper a learning model for coupled neural oscillators is proposed. The learning rule is given in a simple associative form and makes the storage of an instructed phase pattern in coupled neural oscillators possible.

1 Introduction

Results of theoretical studies have demonstrated that neural networks have the high ability of nonlinear transformation for static signals, and many learning rules for neural networks have been proposed and discussed from both an engineering and a neurobiological point of view. Although it is known that neural networks with recurrent connections can generate temporal signals, few rules for learning temporal signals have been discussed because the analysis of the behavior of the recurrent network is difficult, and most proposed learning rules are for artificial use and seem unacceptable as biological models. One reason for the difficulty of the analysis is the complex behavior of the recurrent network, which consists of nonlinear components. Other reasons involve the choice of the model of the component in the network.

From the viewpoint of neurophysiology, many motor commands, including those for heartbeats[2], mastication[6], basic locomotor patterns[4, 5, 13], and movement of gastric mills[3], are generated by coupled oscillatory components, such as neural cells and neural circuits. That is, these dynamic components are associated with dynamic behavior in living systems. In contrast, the basic components of conventional neural network model are static components, such as nonlinear maps or first-order low-pass filters. Considering these neurophysiological facts, a discussion of the behavior of the coupled oscillatory components would benefit our understanding of information processing in living systems. The authors have investigated learning models for coupled oscillators and have shown that temporal patterns and simple locomotor patterns can be learned by simple associative rules for phase oscillators[7, 9, 11]. Furthermore, it was recently reported that the proposed learning rule can be applied for learning of a locomotor pattern by a neural oscillator [8]. In this report, from the learning rule for coupled phase oscillators we derive a learning rule to enable coupled two-dimensional neural oscillators to learn an instructed phase pattern.

In section 2, the proposed learning rule to enable coupled phase oscillators to learn an instructed phase pattern is briefly summarized. In section 3, a learning rule for coupled neural oscillators is derived, and the results of computer simulations are shown in section 4.

2 A learning model for coupled phase oscillators

In this section we summarize the learning model for an instructed phase pattern in coupled phase oscillators proposed by Nishii [9, 11]. Assume that the coupled oscillators display phase dynamics in the form,

$$\begin{cases} \dot{\theta}_i = \omega_i + R_i + \epsilon_f F_i, \\ \dot{\tilde{\theta}}_i = \Omega, \quad (i = 1, \dots, N), \end{cases} \tag{1}$$

$$R_i \equiv \begin{cases} 0 \qquad\qquad (i = 1) \\ \displaystyle\sum_{k \in J_i} \sum_{l=1}^{L_{ik}} w_{ik}^l R_{ik}^l \ (i = 2, \dots, N), \end{cases} \quad R_{ik}^l \equiv R(\theta_i, \theta_k - \psi_{ik}^l), \quad F_i \equiv F(\theta_i, \tilde{\theta}_i),$$

where $\theta_i, \tilde{\theta}_i \in \mathbf{S}^{(1}$, $(i = 1, \dots, N)$ are the phases of component oscillators and the teacher signal, respectively, N is the number of component oscillators, J_i indicates the ensemble of indices of oscillators sending signals to the i-th oscillator, and ω_i is the intrinsic frequency of the i-th oscillator. The frequencies of the teacher signals are assumed to be the same as the value of Ω. Supposing various phase delayed couplings between oscillators w_{ik}^l and ψ_{ik}^l $(l = 1, \dots, L_{ik})$ show the coupling strength and the phase delay of the l-th coupling from the k-th oscillator to the i-th one, assume that $\sum_{k \in J_i} \sum_{l=1}^{L_{ik}} w_{ik}^l \ll 1$, i.e., the total effect of the signals from the coupled oscillator is less than the effect of the intrinsic dynamics, and L_{ik} indicates the number of different phase delayed couplings from the k-th oscillator to the i-th one. R and F are functions showing the effects of signals from the coupled oscillator and the teacher signal, respectively. The small constant value $\epsilon_f \ll 1$ indicates the strength of the teacher signal.

The proposed learning rule for the coupling weight w_{ij} and the intrinsic frequency ω_i to obtain the same phase pattern as the teacher signal takes the form

$$\begin{cases} \dot{\omega}_i = \varepsilon < \epsilon_f F_i + R_i > \\ \dot{w}_{kj}^l = \varepsilon\gamma < \epsilon_f F_k > \cdot < R_{kj}^l >, \end{cases} \tag{2}$$

where $\varepsilon, \gamma \ll 1$ are constants determining the learning velocity, and $< * >$ shows the time averaged term. This learning rule implies that the intrinsic frequency changes according to the total effect of the input signals, and the coupling strength changes according to the correlation between the effects of the signal

---

[1] Here, we define $\mathbf{S} = \mathbf{R} \pmod 1$.

from the coupled oscillator and the teacher signal. With this learning rule, the same phase pattern and the same frequency as the teacher signal are obtained in the coupled oscillators under certain conditions (see detail in [9]).

3 A learning model for coupled neural oscillators

In this section we derive a learning model for neural oscillators from learning rule (2). When each component oscillator is composed of neural cells, we must know the relation between the state of the cells and the function R in eq. (2). The basic idea for the derivation is shown in [8].

Consider a simple case: two-dimensional dynamics with a circular limit cycle around the origin which receives a small input signal $\epsilon = \epsilon(1,0)^t$ ($\epsilon \ll 1$) takes the form

$$\dot{x} = f(x) + \epsilon, \tag{3}$$

where $x = (x, y)^t \in \mathbf{R}^2$ is a state vector. If the attraction to the limit cycle is large and the above dynamics can be transformed to a normal form for the Hopf-bifurcation, the effect of the input signal on the phase dynamics of eq. (3) $R(x, \epsilon)$ (R_i in eq.(1)) is given by the following form in the lowest approximation [12],

$$R(x, \epsilon) = c\epsilon P(x), \tag{4}$$

where $c > 0$ is a positive constant and $P(x)$ (ψ: constant) is a function which shows the phase dependence of the effect of the input signal and is almost equivalent to the phase response curve (PRC) [14]. The relation between the function P and the state variable x is roughly estimated by the following idea (the analytical study is shown in [10]).

Suppose that the state changes anti-clockwise along the limit cycle (Fig. 1). If the input signal is given when $y = 0$, the phase is not affected. If the input signal is given when $x = 0$ and y takes its maximum value, the phase is delayed considerably. On the other hand, the phase is advanced considerably if the input signal is given when $x = 0$ and y assumes its minimum value. From such considerations, we obtain $P(x) \sim -y$ and $P(x) \sim \dot{x}$. When the state changes clockwise, we obtain $P(x) \sim y$ and $P(x) \sim \dot{x}$ by the analogy. Because $P(x) \sim \dot{x}$ preserves its sign in these two situations, we adopt this relation to the learning rule.

Second, we derive a learning rule for coupled two-dimensional neural oscillators by applying these results. Suppose a neural oscillator is composed of two cells, and one cell receives external input signals from coupled oscillators and a teacher signal $(\tilde{y}_i, 0)^t$. The dynamics of the oscillator is given in the following form instead of eq. (1).

$$\dot{u}_i = f(u_i) + (\sum_{j \in J,} \sum_{l=1}^{2} w_{ij}^l y_j^l, 0)^t + (\tilde{y}_i, 0)^t, \tag{5}$$

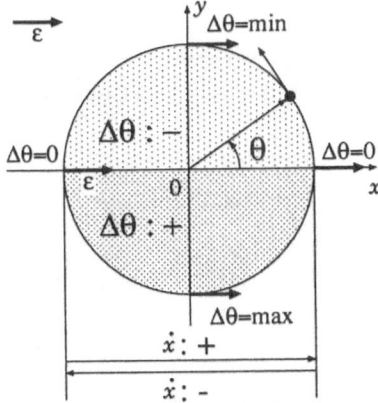

Fig. 1. Effect of input signal on phase dynamics. When the limit cycle is circular on the xy-plane, the PRC for the input $\epsilon = (\epsilon, 0)$ is given by $\Delta\theta = c\dot{x}$ or $\Delta\theta = \pm cx$.

where $\boldsymbol{f}(\boldsymbol{u}) = (f^1(\boldsymbol{u}), f^2(\boldsymbol{u}))^t$, each variable of the state vector $\boldsymbol{u}_i = (u_i^1, u_i^2)^t$ shows the state of the component cell, $y_j^l = y(\boldsymbol{u}_j^l)$ denotes the output signal from the l-th cell ($l = 1, 2$) in the j-th oscillator, and w_{ij}^l denotes the coupling weight.

Putting the function showing the effect of the external input signal on the phase dynamics as $p(\boldsymbol{u}_i)$, learning rule (2) can be expressed in the form,

$$
\begin{cases}
\dot{\omega}_i = \epsilon < p(\boldsymbol{u}_i) \displaystyle\sum_{j \in J} \sum_{l=1}^{2} w_{ij}^l y_j^l > \\
\dot{w}_{ij}^l = \epsilon\gamma < p(\boldsymbol{u}_i)\tilde{y}_i >< p(\boldsymbol{x}_i)y_j^l >,
\end{cases}
\tag{6}
$$

where

$$
p(\boldsymbol{u}_i) = f^1(\boldsymbol{u}_i)
\tag{7}
$$

In the above equation f^1 is used instead of $\dot{u}_i^1$ because p should be given analytically by the zeroth-order term of the input signal. The learning rule obtained in this manner means that the coupling weight changes according to the correlation between the time averaged effects of the input signals, which are given by the product of the input signal and the temporal change of the state of the postsynaptic cell which receives the signal.

4 Results of simulations

In this section, results of simulations of learning an instructed phase relation between neural oscillators are shown. As a neural oscillator, a Wilson-Cowan oscillator was used, the dynamics of which takes the form,

$$
\begin{cases}
\tau\dot{u}_k^e = -u_k^e + g^{ee}y_k^e - g^{ie}y_k^i + s_k^e \equiv f^e(\boldsymbol{u}_k) + s_k^e \\
\tau\dot{u}_k^i = -u_k^e + g^{ei}y_k^e - g^{ii}y_k^i + s_k^i \equiv f^i(\boldsymbol{u}_k) + s_k^i
\end{cases}
\tag{8}
$$

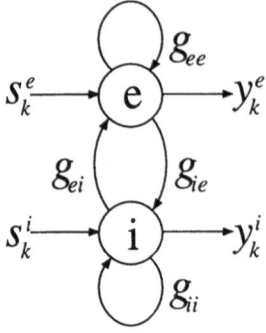

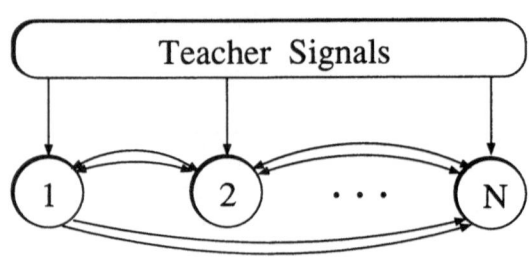

Fig. 2. A Wilson-Cowan oscillator. The symbols "e" and "i" indicate the excitatory cell and inhibitory cell, respectively.

Fig. 3. Wilson-Cowan oscillators with all-to-all coupling. Each circle indicates one Wilson-Cowan oscillator. Each cell in the oscillators sends signals to the other oscillators, but only the excitatory cell receives the input signals in the simulation.

$$y_k^e = h(u_k^e), \quad y_k^i = h(u_k^i), \qquad (9)$$

where the subscript k indicates the index of the oscillator, the superscripts (e, i) indicate the excitatory cell and the inhibitory cell, respectively, $\boldsymbol{u}_k = (u_k^e, u_k^i)$ is the state vector, g^{ee}, g^{ei}, g^{ie}, and g^{ii} are the coupling weights between cells, y_k^e and y_k^i are the output signals from the cells, h is a function giving the output, and s_k^e and s_k^i are the external input signals to the cells (see fig.2). The dynamics (8) shows periodic activity for some parameter sets when h is a sigmoidal function[1]. In this simulation we set $h(x) = 2/(1 + \exp(-x)) - 1$, $g^{ee} = 6.0$, $g^{ei} = 5.0$, $g^{ie} = 5.0$, and $g^{ii} = 0.0$.

Wilson-Cowan oscillators with all-to-all coupling were examined in this simulation (Fig.3). Each excitatory cell receives signals from all other cells belonging to other oscillators. The input signals to the k-th oscillator $s_k^e, s_k^i, (k = 1, \ldots, N)$ in the learning mode are given by

$$s_k^e = \sum_{j \neq k}^{N} (w_{kj}^e y_j^e + w_{kj}^i y_j^i) + \epsilon_f F_k(t), \quad s_k^i = 0, \quad (k = 1, \ldots, N), \qquad (10)$$

where w_{kj}^e and w_{kj}^i are the coupling weights for the input signals from excitatory and inhibitory cells in the j-th oscillator to the k-th oscillator, respectively, and $F_k(t) = \cos 2\pi(t - k/N)$ is the teacher signal for the k-th oscillator. Because the intrinsic frequency is not explicitly given in eq.(8), we apply the learning rule for the frequency for the inverse of the time constant, $1/\tau_k$, instead. The coupling weight is expressed by the sigmoidal function of the parameter α so as to restrict the value, i.e., $w_{kj}^l = W(\alpha_{kj}^l)$, $W(\alpha) = 2w_{max}/(1 + \exp(-\alpha/T)) - 1$, ($w_{max}$, T:

constant). The learning rule for the parameter is given in the following form.

$$
\begin{cases}
\dot{\tau}_k &= -\varepsilon(\tau_k)^2 < p(\boldsymbol{x}_k) \sum_{j \neq k}^{N} (w_{kj}^e y_j^e + w_{kj}^i y_j^i) > \\
\dot{\alpha}_{kj}^l &= \varepsilon \gamma < p(\boldsymbol{u}_k) \tilde{y}_k >< p(\boldsymbol{u}_k) y_j^l >,
\end{cases}
\qquad (k, j = 1, \dots, n, \ l = e, i),
$$

$$(11)$$

where $p(\boldsymbol{u}_k) = f^e(\boldsymbol{u}_k)/\tau_k$ represents the effect of the input signal on the phase dynamics of the Wilson-Cowan oscillator. The time averaged term in the above equation is given by the first-order low-pass filter: $\tau_0 < \dot{*} > = - < * > + *$, where $\tau_0 = 3$ s and each initial value of the time averaged value was set to zero. The time constants of the oscillators τ_i were set randomly in the range from 0.2 s to 0.133 s. This corresponds to a frequency range of about from 1 Hz to 1.5 Hz.

Figure 4 shows the phase pattern of oscillators during and after learning, where we plot the time when the state of the excitatory cells in the oscillators changes from a negative value to a positive one. Convergence to the pattern instructed in the learning mode and rapid recall of the pattern learned were achieved.

5 Discussion

A learning rule for coupled neural oscillators was derived from the learning model for phase oscillators. Although the derivation was obtained in an intuitive way in the simple case and the analytical derivation must still be done [10], the learning performance seems good in the computer simulations.

In this learning model, the teacher signal works not only as a modulatory system which communicates the correct activity to the oscillator but also as a drive system which makes the activity of the oscillatory networks approach the correct one by forcing oscillation. The proposed learning rule for coupling weights depends on the correlation between the time averaged "effect" of input signals, and these effects are given by the product of the input signal and the time derivative of the state in the postsynaptic cell. Although the time derivatives of the states of cells are not discussed much in relation to learning models in neural systems, they can contribute to the learning of temporal signals.

Acknowledgment

This study was supported by the special postdoctoral researchers program at RIKEN.

References

1. S. Amari. Characteristics of random nets of analog neuron-like elements. *IEEE Transactions on Systems, Man, and Cybernetics*, smc-2:643–657, 1972.

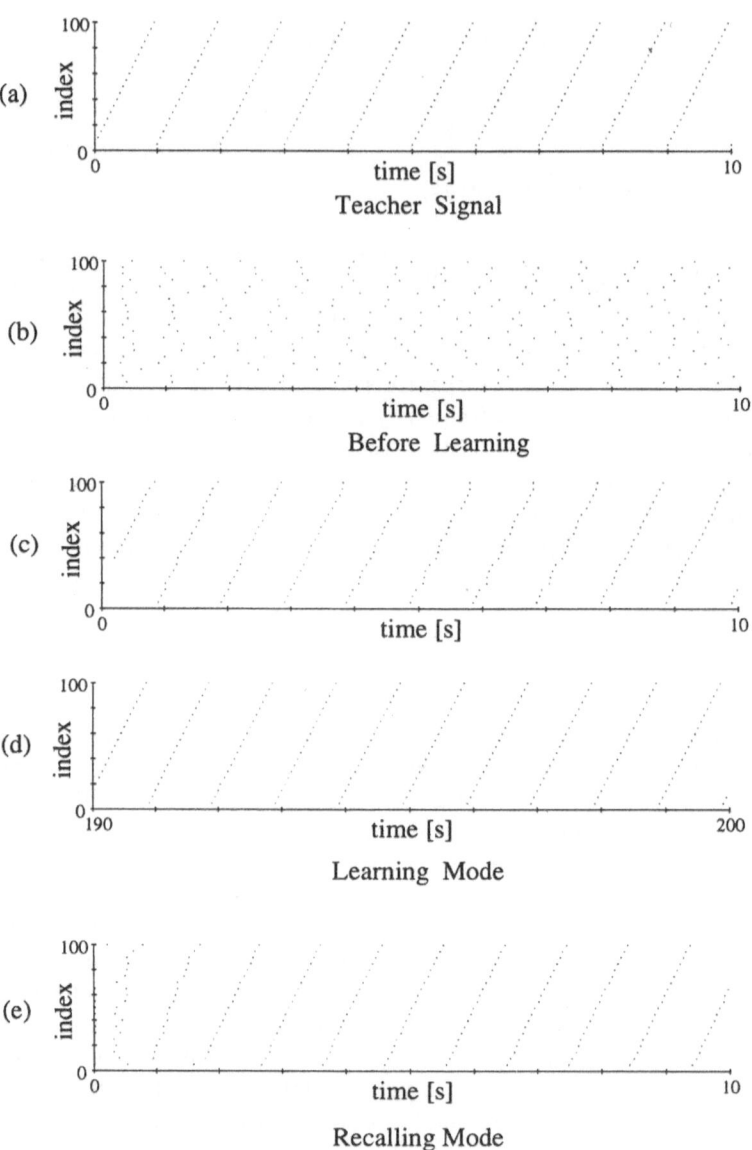

Fig. 4. The firing pattern of the coupled Wilson-Cowan oscillators during learning and after learning; (a) teacher signal, (b) before learning (c)(d)during learning, (e) after learning (recalling mode). The ordinate shows the index of oscillators and the dots show the firing time of each oscillator. During the learning the firing pattern approaches the teacher signal. At t=200s, learning was stopped and random phases were set to each oscillator. The memorized pattern is regenerated within 10s. Parameters: $\Omega = 1$ Hz, $w_{max} = 1.0$, $T = 0.2$ $\epsilon_f = 4.0$, $\varepsilon = 0.005$, $\gamma = 0.1$, $\tau = 3.0$s.

2. R. Eckert, D. Randall, and G. Augustine, editors. *Animal physiology: mechanisms and adaptations*. W.H. Freeman and Company, New York, third edition, 1988.

3. R. E. Flamm and R. M. Harris-warrick. Aminergic modulation in lobster stomatogastric ganglion. I. effects on motor pattern and activity of neurons within the pyloric circuit. *J Neurophysiol*, 55(5):847–865, 1986.

4. P. A. Getting. Mechanisms of pattern generation underlying swimming in *tritonia* I. neuronal network formed by monosynaptic connections. *J Neurophysiol*, 46:65–79, 1981.

5. S. Grillner, P. Wallen, and L. Brodin. Neuronal network generating locomotor behavior in lamprey: circuitry, transmitters, membrane, properties and simulation. *Annu Rev Neurosci*, 14:169–199, 1991.

6. J. P. Lund and S. Enomoto. The generation of mastication by the mammalian central nervous system. In A. H. Cohen, S. Rossignol, and S. Grillner, editors, *Neural control of rhythmic movements in vertebrates*, pages 41–72. John Wiley and Sons, 1988.

7. Jun Nishii. An adaptive control model of a locomotion by the central pattern generator. In *From Natural to Artificial Neural Computation, Lecture Notes in Computer Science, 930*, pages 151–157. Springer, 1995.

8. Jun Nishii. A learning model for a neural oscillator to generate a locomotor pattern. *Proc. of 2nd International Congress of Computational Inteligence and Neuroscience*, in press, 1997.

9. Jun Nishii. A learning model for the oscillatory network, submitted.

10. Jun Nishii and Kaoru Nakano. An adaptive control model of a locomotor pattern by a neural oscillator (in Japanese). *Tech Rep of IEICE*, NLP94-111:269–276, 1995.

11. Jun Nishii and Ryoji Suzuki. Oscillatory network model which learns a rhythmic pattern of an external signal. *Proc. of IFAC Symposium*, pages 501–502, 1994.

12. Jun Nishii, Yoji Uno, and Ryoji Suzuki. Mathematical models for the swimming pattern of a lamprey. I: analysis of collective oscillators with time delayed interaction and multiple coupling. *Biological Cybernetics*, 72:1–9, 1994.

13. K. Pearson. The control of walking. *Scientific American*, 235(6):72–86, 1976.

14. A. T. Winfree. *Geometry of biological time*. Springer, New York, 1980.

Random Perturbations to Hebbian Synapses of Associative Memory Using a Genetic Algorithm

Akira Imada[1] and Keijiro Araki[2]

[1] Graduate School of Information Science
Nara Institute of Science and Technology
8916-5 Takayama, Ikoma, Nara, 630-01 Japan
akira-i@is.aist-nara.ac.jp

[2] Department of Computer Science and Computer Engineering
Graduate School of Information Science and Electrical Engineering
Kyusyu University
6-1 Kasuga-koen, Kasuga, Fukuoka, 816 Japan
araki@c.csce.kyusyu-u.ac.jp

Abstract. We apply evolutionary algorithms to Hopfield model of associative memory. Previously we reported that a genetic algorithm using *ternary* chromosomes evolves the Hebb-rule associative memory to enhance its storage capacity by pruning some connections. This paper describes a genetic algorithm using *real-encoded* chromosomes which successfully evolves over-loaded Hebbian synaptic weights to function as an associative memory. The goal of this study is to shed new light on the analysis of the Hopfield model, which also enables us to use the model as more challenging test suite for evolutionary computations.

1 Introduction

In the field of evolutionary computations, a lot of complicated functions have been proposed as test suites for comparing and evaluating the effectiveness of various evolutionary computations [1, 2, 3, 4, 5, 6, 7, 8]. However as Mülenbein [9] pointed out:

> But are such problems typical applications? We have not encountered such a problem.

these functions are far from real problems. We know the number of their local optima, values, locations etc. This makes us think of using the Hopfield model of associative memory as a test function of evolutionary computations. The model per se is very simple, and have favorable properties as a test function of real-valued parameter optimization problems. Firstly, as we will describe later in detail, for a fixed set of patterns to be stored, there exist almost infinite number of combinations of synaptic weights which give the network a function of associative memory (multi modality). Secondly, the model has a global optimum in the sense of Gardner [10]. Finally, there is a trade off between the storage capacity and error correcting capability (multi-objectivity). However, we have no knowledge such as the exact distribution of these local optima. In fact, many problems concerning the model are still open, despite a lot of discussions have been made since Hopfield proposed the model. For example, Sompolinsky [11] studied an effect of small perturbations of Hebbian synaptic weights. Namely, he modified the weights by

$$w_{ij} = J_{ij} + \eta_{ij},$$

where J_{ij}'s are Hebbian synaptic weights and η_{ij}'s are small normal random variables. His analysis was only under the condition that the number of stored pattern was less than the storage capacity, and modification were made *randomly*. What Sompolinsky showed is the robustness of the Hopfield model against small synaptic noises. But what will happen if we make the experiment for more patterns than the capacity? In this paper, we experiment it with patterns more than the saturation level, and we modify synaptic weights *adaptively* using a genetic algorithm. The genetic algorithm uses a continuous representation instead of traditional binary representation. We believe the study shed new light to the analysis of the model, as well as give a more challenging test suite to evolutionary computations.

The remainder of this paper is organized as follows. The other evolutionary approaches to the Hopfield model will be described in the following section. Since we only modify synaptic weights strictly maintaining the framework of Hopfield model, the model is described more fully in the third section. Our genetic algorithm is not the standard one and these are addressed in the fourth section. After presenting experimental results and discussion, the final section describe the concluding remarks.

2 Related Works

Since the late 1980's, genetic algorithms have been used extensively to search the optimal set of synaptic weights of neural networks and/or their optimal architectures (see [12], [13], and references quoted therein). However all of these researches were for layered neural networks, and applications of genetic algorithms to Hopfield networks remain few so far.

Previously, we successfully evolved both random weight matrices and over-loaded Hebbian weight matrices to function as associative memories, using ternary chromosomes [14, 15]. The chromosomes of those implementations were made up mostly 1 but a few components were 0 or -1. Optimization was made both by pruning some synaptic connections and by balancing the number of excitatory/inhibitory synapses with these 0 and -1, respectively. That is, the modified synaptic weights were obtained as follows

$$w_{ij} = c_{ij} \times J_{ij},$$

or

$$w_{ij} = c_{ij} \times R_{ij}$$

where c_{ij}'s are components of the chromosome and R_{ij}'s are Random synaptic weights.

The simulations were made with 49 neurons as in this paper. The random matrix evolved to store eventually 7 patterns as fixed points. And starting with over-loaded Hebbian matrix which learned 16 patterns it was evolved to stabilize all these 16 patterns as fixed points. Note that this size of Hebbian matrix would store at most 8 patterns.

3 Hoppfield Model of Associative Memory

Associative memory is a dynamical system which has a number of stable states with a domain of attraction around them [16]. Applying a vector x_0 as the initial state of the dynamical system results in the state trajectory $\{x_t\}_{t\geq0}$. If we start the system with $x_0 = u + \delta u$ in the vicinity of some fixed point u, and the state trajectory converges to u, i.e., $\lim_{t\to\infty} x_t = u$, then we can regard the system as an associative memory.

In 1982, Hopfield proposed his fully connected neural network model of associa-tive memory [17]. The model consists of N neurons and N^2 synapses. Each neu-ron can be in one of two states ±1, and p bipolar patterns $\xi^\nu = (\xi_1^\nu, \cdots, \xi_N^\nu)$ $(\nu = 1, \cdots, p)$ are memorized as equilibrium states. He employed a discrete-time, asyn-chronous update scheme. Namely, at most one neuron updates its state at a time, according to

$$s_i(t+1) = f\left(\sum_{j\neq i}^N w_{ij}s_j(t)\right),$$

where $s_i(t)$ is a state of i-th neuron at time t, w_{ij} is a synaptic weight from neuron j to neuron i, and $f(z) = 1$ if $z \geq 0$ and -1 otherwise.

The behaviors of the collective states of the individual neurons are characterized by the synaptic weights. When these synaptic weights are determined appropri-ately, networks store some number of patterns as fixed points. Hopfield specified w_{ij}'s by the Hebbian rule [18], i.e.,

$$w_{ij} = \frac{1}{N}\sum_{\nu=1}^p \xi_i^\nu \xi_j^\nu \ (i \neq j), \ w_{ii} = 0.$$

Then giving one of the memorized patterns, possibly including a few errors, to the network as an initial state will result in the stable state after certain time steps. Hopfield suggested by using computer simulation that the maximum num-ber p of the patterns to be stored in a network with N neurons is $0.15N$, if a small error in recalling is allowed. Later, this was theoretically calculated by Amit et al. [19] by using *replica* method. They showed that the storage capacity is $p = 0.14N$. In 1987, McEliece et al. [20] proved that when $p < N/4\ln N$ holds, the Hopfield's model will recall the memory without error.

The storage capacity depends strongly on how synaptic weights are specified. The specification of the synaptic weights is conventionally referred to as *learning*. Other learning schemes instead of Hebbian rule have been proposed for their increased storage capacity. The *pseud-inverse matrix method* by Kohonen et al. [21] and *spectral algorithm* by Pancha et al. [22] are the examples, and the capacity is enlarged to $p = 0.5N$ and $p = N$, respectively. Both are extensions of the Hebbian learning.

Then, what is the ultimate capacity when only the learning scheme is explored within the Hopfield's framework? Generally the storage capacity is traded off against basin size. Gardner[10] studied the optimal storage capacity as a function of size of basin of attraction. She showed that as basin size tends to zero, the ultimate capacity will be $p = 2N$, proposing an algorithm to obtain the weight matrix.

4 GA Implementation

In this simulation, a set of random bipolar patterns ξ^ν ($\nu = 1, 2, \cdots, p$) are produced before each run. The goal of the evolutionary programming is to find a weight matrix which stores all these patterns as fixed points. A weight matrix $\{W_{ij}\}$ is also pre-determined as an ancestor. This matrix remains fixed during evolution. Then a population of chromosomes are initialized at the beginning of a run. Each chromosome is a N^2-dimensional real-valued vector. Its component is called as allele. We denote the chromosome as

$$(\eta_{11}, \eta_{12}, \cdots, \eta_{1N}, \eta_{21}, \cdots\cdots \eta_{NN})$$

In each generation, a population of chromosomes make ancestor's copies by adding each allele to its corresponding component of the ancestor matrix. Namely,

$$w_{ij}^k = W_{ij} + \eta_{ij}^k,$$

where w_{ij}^k is ij-component of the copied matrix, η_{ij}^k is $(iN + j)$-th allele of the chromosome, and the superscript k denotes the individual number in the population. The chromosomes are modified through crossover and mutation operation. The fitness values of the corresponding phenotype are evaluated. According to the fitness values, individuals of the next generation are selected, using a $(\mu+\lambda)$-strategy in ES terminology. The cycle of reconstructing the new population with better individuals and restarting the search is repeated until a perfect solution is found or a maximum number of generation has been reached. Specific details of the operations are as follows.

Initialization: In the first generation, all chromosomes are initialized so that each allele value is Gaussian random variables with mean 0 and standard deviation σ.

Fitness evaluation: When one of memorized patterns ξ^ν is given to the network as an initial state, the state of neurons varies from time to time afterwards (unless ξ^ν is a fixed point). In order for the network to function as associative memory, the instantaneous network state must be similar to the input pattern. The similarity is defined by,

$$m^\nu(t) = \frac{1}{N} \sum_{i=1}^{N} \xi_i^\nu s_i^\nu(t),$$

where $s_i^\nu(t)$ is the state of neuron i at time t. This $m^\nu(t)$ is referred to as *overlap*. The quality of retrieval of memorized patterns are represented by $\langle m^\mu \rangle$, a temporal average of $m^\nu(t)$ over certain time interval t_0. We evaluate the fitness value of each network by further averaging the $\langle m^\mu \rangle$ over all memorized patterns. Namely, the fitness f is

$$f = \frac{1}{t_0 \cdot p} \sum_{t=1}^{t_0} \sum_{\nu=1}^{p} m^\nu(t).$$

In this paper, t_0 is set to $2N$, twice the number of neurons. Note that the fitness 1 implies all the p patterns are stored as fixed points, while fitness less than 1 includes many other cases.

Selection: We use the truncation-selection [9], i.e., two parent chromosomes are chosen *randomly* from the best $T\%$ of the population.

Recombination: Discrete Recombination [9] which is similar to *uniform crossover* [23] in binary chromosomes, operates on alleles of the selected parents' chromosomes, i.e., two parents $(u_1, \cdots, u_n)$ and $(v_1, \cdots, v_n)$ produce an offspring $(w_1, \cdots, w_n)$ such that w_i is either u_i or v_i with equal probability.

Mutation: Mutation is made by randomly picking up an allele with probability p_m, and replacing it with uniformly random number selected from $[-\delta, \delta]$.

5 Experiment

The parameters we used in the genetic algorithm are as follows. The population number is 256. This is just because of our computer resources. The standard deviation σ in initializing allele values in the first generation is set to 0.01. The mutation probability p_m and the mutation range δ are set to 0.05 and 0.1, respectively. In truncation selection T is chosen to be 40%. Searching procedure is iterated until 12000 generation unless perfect solution is not found. These were determined mainly on the basis of trial and error.

6 Results and Discussion

In this experiment, we start the genetic algorithm with a Hebbian weight matrix which learned a set of patterns more than capacity. The networks of

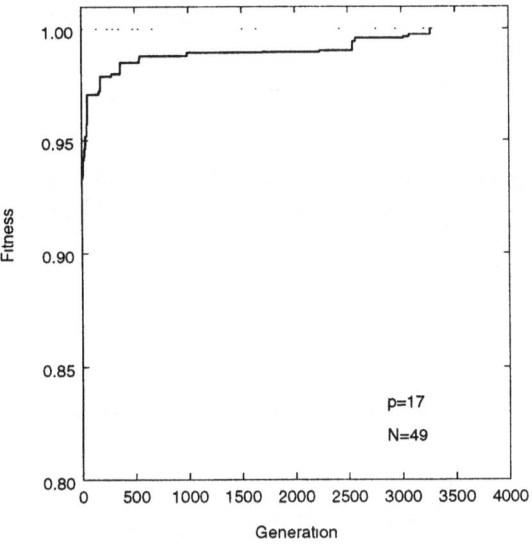

Fig. 1. Fitness vs. generation

this type were extensively investigated by Amit et al. [19, 24]. As the number of patterns increases, the correlation of these patterns can not be neglected. when the number exceeds the capacity, some memory states are shifted to their neighborhood, and spurious attractors around memory states merge into spin-glas attractors.

Let us take an example. We made an experiment for 17 patterns to be memorized on a network with 49 neurons. Since the storage capacity of the Hebbian synaptic weights of this size would be around 8 patterns, these 17 patterns are a little more than twice of the capacity. Each input of memorized patterns results in the convergence to the stable attractors though, most of them are not memorized states. These attractors can be studied by the temporal evolution of Hamming distance of instantaneous network state from the corresponding memorized pattern. We observed that the Hamming distances of the 17 stable states from their corresponding input were 0, 0, 0, 0, 1, 1, 1, 6, 7, 7, 8, 10, 10, 14, 15, 18 and 19. The four are fixed-point attractors and the rest are distributed from near-fixed-point attractors to spin-glass-attractors.

We use this weight matrix as an ancestor matrix for our genetic algorithm as described above. Surprisingly we obtained a evolved matrix which store all these 17 patterns as fixed points perfectly after 3273 generations. We show the fitness evolution in Figure 1.

While we obtained the perfect solutions more easily for less than 16 patterns (not in the Figure), we have not observed the same success for 18 patterns among 30 runs with different random number seed.

As Hertz et al. [25] and Parisi [26] suggested in their analysis using Ising model, we conjectured that the above success is due to the destabilization of the spin-glass attractors by the asymmetry of synaptic weights introduced by the genetic algorithm.

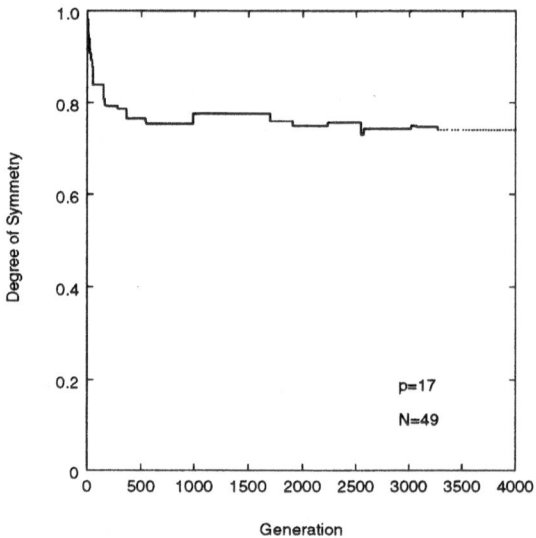

Fig. 2. Degree of symmetry of the matrix which evolved to perfect solution.

To see this, we investigated the time evolution of degree-of-symmetry of weight matrices. The degree-of-symmetry is defined by

$$\sum_{i=1}^{N}\sum_{j=1}^{N} w_{ij}w_{ji} \cdot \left(\sum_{i=1}^{N}\sum_{j=1}^{N} w_{ij}^2 \right)^{-1} ,$$

following after Krauth et al. [27]. We show the results of both experiments in Figure 2. At the beginning the Hebbian matrix, however over-loaded it may be, is totally symmetric. As evolution proceeds the degree of symmetry decreases and asymptotically approaches to the value around 0.8 and retrieval states emerge. Our conjecture is that the asymmetry destabilized the many spin-glass attractors which existed at the beginning.

However the role of asymmetry of the synaptic weights is still an open question. To say a network an associative memory system, the tolerance for certain amount of noises should be shown. That is, each memorized pattern must have a certain domain of basin of attraction.

We observed the basin size of the network of the above experiment as follows. One of the memorized patterns are chosen and flipped d bits at random. The pattern is given to the network, and temporal average of *overlaps* of the network state visited by the dynamics is calculated. This is repeated 800 times for each incremented d. In Figure 3, the averaged overlap over repetition are plotted against noisy-bit d, together with the same experiment with Gardner's weight matrix for the same set of memorized patterns. Although the size of basin of attraction is much smaller than that of Gardner's optimized matrix, it still shows some tolerance for noises.

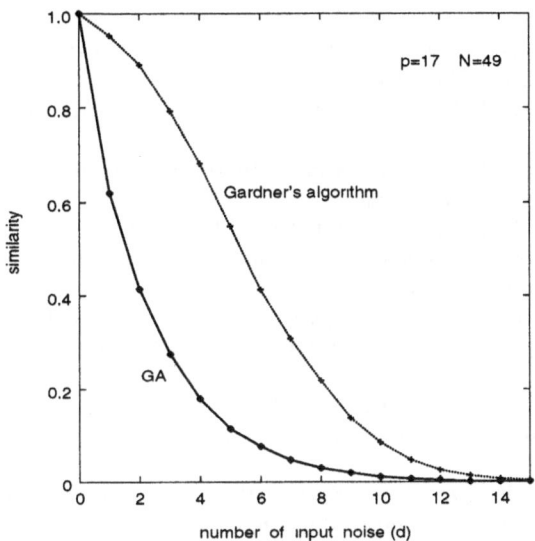

Fig. 3. Similarity of updated states with noisy input

7 Conclusion

We have described an application of genetic algorithms to Hopfield model of associative memory. We give small perturbations to Hebbian synaptic weights. This is essentially the same as Sompolinsky's model, with difference being that we use an over-saturated network and choose synapses *adaptively*, while Sompolinsky used an unsaturated network and chose synapses *randomly.* The simulations in this paper were made with 49 neurons. This size of Hebb-rule associative memory would store at most 8 patterns as fixed points. We succeeded in evolving a network which learned 17 patterns by Hebbian rule to store eventually all these patterns as fixed points.

Although our investigations have not been conclusive so far, we believe this study shed new light on the analysis of Hopfield model from new aspect, and give new test suite for evolutionary computations.

Acknowledgments

We thank Peter Davis at Advanced Telecommunication Research Institute (ATR) for providing us great insight into the dynamics of Hopfield neural networks.

References

1. K. DeJong (1975) *An Analysis of the Behavior of a Class of Genetic Adaptive Systems.* Ph.D thesis, University of Michigan.
2. D. Ackley (1987) *A Connectionist Machine for Genetic Hillclimbing.* Boston: Kluwer.
3. K. Deb (1989) *Genetic Algorithms in Multimodal Function Optimization.* Master's thesis. The University of Alabama.
4. H. Mühlenbein, M. Schomish, and J. Born (1991) *The Parallel Genetic Algorithm as Function Optimizer.* Parallel Computing **17** pp619–632.
5. J. D. Schaffer, B. A. Caruana, L. J. Eshlman, and R. Das (1991) *A Study of Control Parameters Affecting Online Performance of Genetic Algorithms for Function Optimization.* Proceedings of the 3rd International Conference on Genetic Algorithms. pp51–60.
6. D. Whitley, K. Mathias, S. Rana, and J. Dzubera (1995) *Building Better Test Functions.* Proceedings of the 6th International Conference on Genetic Algorithms. pp239-247.
7. S. M. Mahfoud (1995) *A. Comparison of Parallel and Sequential Niching Methods.* Proceedings of the 6th International Conference on Genetic Algorithms. pp239-247.
8. D. Schlierkamp-Voosen, and H. Mühlenbein (1996) *Adaptation of Population Sizes by Competing Subpopulation.* Proceedings of IEEE International Conference on Evolutionary Computation. pp330-335.
9. H. Mühlenbein, and D. Schlierkamp-Voosen (1995) *Predictive Models for the Breeder Genetic Algorithm I. Continuous Parameter Optimization.* Evolutionary Computation **1** (1996) pp25–49.
10. E. Gardner (1988) *The Phase Space of Interactions in Neural Network Models.* J. Phys A: Math Gen. **21**. pp257–270.
11. H. Sompolinsky (1986) *Neural Network with Non-linear Synapses and Static Noise.* Phys. Rev. **A34** pp2571-2574.
12. J. D. Schaffer, D. Whitley, and L. J. Eshelman (1992) *Combinations of Genetic Algorithms and Neural Networks: A Survey of the State of the Art.* Proceedings of the Workshop on Combinations of Genetic Algorithms and Neural Networks. pp1–37.

13. X. Yao (1993) *A Review of Evolutionary Artificial Neural Networks.* International Journal of Intelligent Systems **8**. pp539–567.

14. A. Imada, and K. Araki (1997) *Evolved Asymmetry and Dilution of Random Synaptic Weights in Hopfield Network Turn a Spin-glass Phase into Associative Memory.* The 2nd International Conference on Computational Intelligence and Neuroscience (to appear)

15. A. Imada, and K. Araki (1995) *Genetic Algorithm Enlarges the Capacity of Associative Memory.* Proceedings of the 6th International Conference on Genetic Algorithms. pp413–420.

16. J. Komlós, and R. Paturi (1988) *Convergence Results in an Associative Memory Model.* Neural Networks **1**. pp239–250.

17. J. J. Hopfield (1982) *Neural Networks and Physical Systems with Emergent Collective Computational Abilities.* Proceedings of the National Academy of Sciences, USA. **79**. pp2554–2558.

18. D. O. Hebb (1949) *The Organization of Behavior.* Wiley.

19. D. J. Amit, H. Gutfreund, and H. Sompolinsky (1985) *Statistical Mechanics of Neural Networks near Saturation* Annals. of Physics 173, pp30-67.

20. R. J. McEliced, E. C. Posner, E. R. Rodemick, and S. S. Venkatesh (1987) *The Capacity of the Hopfield Associative Memory.* IEEE Trans. Information Theory **IT-33**. pp461-482.

21. T. Kohonen, and M. Ruohonen (1973) *Representation of Associated Data by Matrix Operators.* IEEE Trans. Computers **C-22(7)**. pp701–702.

22. G. Pancha, S. S. VenKatesh (1993) *Feature and Memory-Selective Error correction in Neural Associative Memory.* in M. H. Hassoun eds. Associative Neural Memories: Theory and Implementation. Oxford University Press. pp225-238.

23. G. Syswerda (1989) *Uniform Crossover in Genetic Algorithms.* Proceedings of the 3rd International Conference on Genetic Algorithms, 2.

24. D. J. Amit, H. Gutfreund, and H. Sompolinsky (1985) *Storing Infinite Number of Patterns in a Spin-glass Model of Neural Networks.* Physical Review Letters **55**. pp1530-1533.

25. J. A. Hertz, G. Grinstein, and S. A. Solla (1987) *Irreversible Spin Glasses an Neural Networks.* in L. N. van Hemmen and I. Morgenstern eds. Heidelberg Colloquium on Glassy Dynamics. Lecture Notes in Physics No.275 Springer-Verlag, pp538–546.

26. G. Parisi (1986) *Asymmetric Neural Networks and the Process of Learning.* J. Phys. **A19** ppL675–L680.

27. W. Krauth, J.-P. Nadal, and M.Mezard (1988) *The Roles of Stability and Symmetry in the Dynamics of Neural Networks.* J. Phys. A: Math. Gen. **21** pp2995–3011.

Phase Memory in Oscillatory Networks

Kuzmina M.G.,
Keldysh Institute of Applied Mathematics, Russian Academy of Sciences,
e-mail: kuzmina@spp.keldysh.ru; kuzmina@applmat.msk.su

Surina I.I.
RRC *Kurchatov Institute*, e-mail: edmany@nlodep.kiae.su

As it was shown earlier, oscillatory networks consisting of limit-cycle oscillators interacting via complex-valued connections can be used for associative memory design. Phase memory as a special type of associative memory in oscillatory networks has been invented and studied. Detailed analysis of phase memory features of phasor networks related to oscillatory networks has been performed. It has been found that under special choice of parameters the oscillatory networks possess high memory storage capacity and low extraneous memory.

The designed networks can be interpreted as networks consisting of complex-valued neurons. They could be useful in problems of invariant pattern recognition and in recognition of colored patterns.

1. Introduction

Dynamical system governing the dynamics of nonlinear coupled limit cycle oscillators [1-3] is studied here from the viewpoint of associative memory modelling.

Similar dynamical systems and their limit cases ("phase" systems) were studied in different aspects for a long time, refs in [1-3] . The main interest was concentrated around phase transition into the state of synchronization.

As it is known, in the problem of associative memory network design the following subproblems arise:

- development of an algorithm for calculation of the whole set of stable equilibria of the dynamical system at given W (network memory);

- development of an algorithm of imposing a prescribed set of stable equilibria with natural, large enough basins of attraction;

- calculation of maximum number M of equilibria at a given finite number N of processing elements in a network;

- estimation of "loading ratio" $\alpha = M/N$ in the limit $N \to \infty$, $M \to \infty$, $M/N < \infty$ (memory storage capacity);

- study of "extraneous" memory (additional equilibria arising in a network together with the desired).

In addition, a proper learning algorithm should be developed.

As it was shown [1-3], the oscillatory networks of associative memory with Hebbian matrix of connections (W is taken in the form of proper outer product on memory vectors) can be designed. A number of questions from the listed subproblems have been elucidated.

The designed oscillatory networks can be implemented in semiconductor laser systems [4], since the dynamics in these systems is governed by the same equations as in oscillatory networks. At the present time such implementation is under development.

2. Oscillatory Networks and Phase Memories

Let us recall the dynamical system of N coupled limit cycle oscillators [1-3]:

$$\dot{z}_j = (1 + i\omega_j - |z_j|^2)z_j + \kappa \sum_{k=1}^{N} W_{jk}(z_k - z_j), \quad j = 1, ..., N. \quad (1)$$

Here $z(t)$ is a complex-valued N-dimensional vector representing the states of oscillators as functions of independent variable t, $z_j(t) = r_j exp(i\theta_j) = x_j + iy_j$. Each oscillator has a natural frequency ω_j. Complex-valued $N \times N$ Hermitian matrix $W = [W_{jk}]$ specifies the weights of connections. Non-negative parameter κ defines the absolute value of interaction strength in oscillatory system. Matrix W satisfies the following natural restrictions:

$$W = W^+, \quad |W_{jk}| \leq 1, \quad \sum_{k=1}^{N} |W_{jk}| = 1. \quad (2)$$

Matrix W is constant in the phase space C^N.

The system (1) can be rewritten in matrix form:

$$\dot{z} = (A - D_z)z, \quad (3)$$

$A = D_0 + \kappa W$. The diagonal matrix $D_0 = diag(D_{01}, \ldots, D_{0N})$,

$$D_{0j} = 1 + i \cdot \omega_j - \kappa \eta_j, \quad \eta_j = \sum_{k=1}^{N} W_{jk},$$

is constant. In contrast, the diagonal matrix $D_z = diag(|z_1|^2, \ldots, |z_N|^2)$ depends on absolute values of z_j.

In Cartesian coordinates the matrix A can be rewritten as follows:

$$A = \begin{pmatrix} g_1 + ih_1 & b_{12} + ic_{11} & \cdots & b_{1N} + ic_{1N} \\ b_{12} - ic_{11} & g_2 + ih_2 & \cdots & b_{2N} + ic_{2N} \\ \vdots & \vdots & \ddots & \vdots \\ b_{1N} - ic_{1N} & b_{2N} - ic_{2N} & \cdots & g_N + ih_N \end{pmatrix} \quad (4)$$

Here $b_{jk} + ic_{jk}$ denote the weights κW_{jk}, $g_j = 1 - \kappa Re(\eta_j)$, $h_j = \omega_j - \kappa Im(\eta_j)$.

Dynamical system (1-3) demonstrates a great variety of complicated dynamical regimes at different values of its parameters ω_j, κ, W. Among them the regime of mutual synchronization at some threshold value κ^* of interaction strength exists. Parametric domain for synchronization regime can be roughly specified as $\kappa > \Omega$, where $\Omega = max_j |\omega_j|$. The dynamical regime is simple in the domain of synchronization: if $\sum_j \omega_j = 0$, this is relaxation to stable equilibria. This condition for frequencies can be always satisfied by proper rescaling of (1-3), and below we assume it satisfied.

As it was found out ([1-3] and refs.), the system (1), being imposed into parametric domain of synchronization, can possess sufficiently rich set of stable equilibria. So, the problem of recurrent associative memory design was posed. Let us remind the problem.

Given M arbitrary points $V^1, \ldots, V^M$ in the phase space C^N of the dynamical system, it is necessary to point out the parameters (ω_j, κ, and matrix W) to provide the following properties:

1) the points $V^1, \ldots, V^M$ are stable fixed points of dynamics;
2) basin of attraction for each V^k is as large as possible;
3) total number of extraneous stable fixed points (other than $V^1, \ldots, V^M$) is as small as possible.

The set $V^1, \ldots, V^M$ is just the network memory.

Obviously, if we fix amplitudes r_j of oscillators, then the corresponding equilibrium points of eq.(3) are defined by the linear system with constant coefficients:

$$(A - D_z)z = 0.$$

Hence, the determinant of the matrix $(A - D_z)$ must be zero to provide at least one non-zero equilibrium point.

Jacobian of the system (3) in a point z can be written as follows:

$$J = \begin{pmatrix} G_1 & BC_{12} & \cdots & BC_{1N} \\ BC_{12}^T & G_2 & \cdots & BC_{2N} \\ \vdots & \vdots & \ddots & \vdots \\ BC_{1N}^T & BC_{2N}^T & \cdots & G_N \end{pmatrix} \tag{5}$$

Here G_j is the matrix of order 2:

$$G_j = \begin{pmatrix} g_j - 3x_j^2 - y_j^2 & -h_j - 2x_jy_j \\ h_j - 2x_jy_j & g_j - x_j^2 - 3y_j^2 \end{pmatrix},$$

and BC_{jk} is also the matrix of order 2:

$$BC_{jk} = \begin{pmatrix} b_j & -c_j \\ c_j & b_j \end{pmatrix},$$

BC_{jk}^T denotes the transposed matrix.

An equilibrium point is stable iff one eigenvalue of J is zero (due to invariance of solutions relative to multiplication with $e^{i\varphi}$, where φ is any constant angle), and all others are negative.

We call the stable points that have constant amplitudes $r_j \equiv const.$ *phase memories*. By convention, we call the stable points with non-constant amplitudes *amplitude memories*. In general, both phase and amplitude stable points exist in the system (3).

Phase memories have peculiar properties as associative memory and can be studied exhaustively. First of all, let us note that if some phase vector is an eigenvector of the matrix A with a positive eigenvalue, then, taking this vector with an appropriate normalizing factor, we obtain an equilibrium point of the system (3). Conversely, a phase equilibrium point is an eigenvector of the matrix A. Thus, the one-to-one correspondence exists between the eigenvectors and equilibrium points of the system (3). Consequently, we can design a matrix A with a prescribed set of eigenvectors so that all extraneous memories will be amplitude ones. This property looks very promising, because this is probably that a hardware method might discriminate amplitude from phase memories.

Now, we describe the method permitting to load up to $N - 1$ phase memories. Aiming at this goal, let us introduce the "phase basis".

If we take N phases $(0, \beta_2^1, \ldots, \beta_N^1)$ with arbitrary $\beta_2^1, \ldots, \beta_N^1$ and calculate recurrently:

$$
\begin{bmatrix}
0 \\
\beta_2^m \\
\beta_3^m \\
\cdot \\
\cdot \\
\cdot \\
\beta_N^m
\end{bmatrix}
\longrightarrow
\begin{bmatrix}
0 \\
\beta_2^m + \varphi \\
\beta_3^m + 2\varphi \\
\cdot \\
\cdot \\
\cdot \\
\beta_N^m + (N-1)\varphi
\end{bmatrix}
\tag{6}
$$

$\varphi = 2\pi/N$, then we obtain N linearly independent orthogonal phase vectors $V_j^m, m = 1, \ldots, N, j = 1, \ldots, N$ in C^N, if N is a prime number.

Any subset of these vectors can be loaded as phase memories.

Noteworthy is that extraneous memories (that can be easily revealed as memories with different amplitudes) are more or less abundant for one or another combinations of M vectors from the phase basis for the same values of M. Examples with $N \leq 20$ can be shown where M is close to $M/2$ and extraneous memories do not appear in computer simulation of the retrieval process (this means that if extraneous memories exist in these cases, then they have very small basins of attraction).

3. Phase Memories in Oscillatory Networks of Low Dimensions

1. $N = 2$. In this case the strict analytical analysis of dynamical system (3) with arbitrary parameters has been fulfilled and exact solution of the system has been found.

Only one stable point can exist in the network of two coupled oscillators. Its polar coordinates: $(r_1, r_2 e^{\theta_2})$, $r_1 = r_2 = \sqrt{g + \sqrt{d - h^2}}$, $\theta_2 = -i ln\left(\frac{b-ic}{ih+\sqrt{d-h^2}}\right)$. Here $g = 1 - b_{12}$, $h = \omega - c_{12}$. So, in this case, only phase memory exists. This point exists and is stable iff

$$d > h^2, \quad AND \quad |g - 1| \leq \sqrt{d} \quad AND \quad (g > 0 \quad OR \quad g^2 + h^2 < d).$$

Conversely, for an arbitrary point $(r_1, r_2 e^{\theta_2})$ the parameters providing its stability can be presented.

2. $N = 3$. In this case amplitude stable points exist. One arbitrary phase vector can be loaded and two orthogonal phase vectors can be loaded as well. The values g_j in (4) have to be real and equal, if phase vectors are loaded. Consequently, b_{jk} are equal as well. This is not

true for $N \geq 4$, but in any case the matrices W providing storage of a prescribed set of the vectors from the phase basis (8) are quite special.

4. Phasor Networks Related to Oscillatory Networks

Analysis of dynamical system (1) with arbitrary frequencies ω_j and arbitrary matrix W represents a complicated mathematical problem. Only a few number of rigorous results was obtained for the system (1) and for its limit case — so-called phase model. Those results concern mainly the case of the special architecture of connections — homogeneous all-to-all connections $(W_{jk} = N^{-1}(1 - \delta_{jk}))$.

Eq. (1) with $\omega_j \equiv 0$ represents an important special case of oscillatory system which can be regarded as phasor networks. These phasor networks can be viewed as natural generalization of the known "clock" neural networks.

The equilibria of oscillatory networks and corresponding phasor networks proved to be closely related. Namely, the following proposition is valid.

Let $\mathcal{N}(\{\omega_j\}, \kappa, W)$ be an oscillatory network with arbitrary frequencies ω_j satisfying the condition $\sum_j \omega_j = 0$.

Let $\mathcal{N}(\{0\}, \kappa, W)$ be the corresponding phasor network possessing the collection of M phase memory vectors $\{U^1, \ldots U^M\}$.

Define $\tilde{\kappa} > \kappa$ satisfying the condition: $\gamma \equiv \Omega/\tilde{\kappa} \ll 1$, where $\Omega = max_j|\omega_j|$.

Then oscillatory network $\mathcal{N}(\{\omega_j\}, \tilde{\kappa}, W)$ has phase memory vectors $\tilde{U}^1, \ldots, \tilde{U}^M$, which represent slight perturbations of the corresponding $U^1, \ldots, U^M$.

The proof of this proposition has been obtained using the perturbation method on small parameter γ. This proposition is also confirmed by computer studies of phase portraits of the dynamical system (1) for small N.

5. The Class of Phasor Networks with Guaranteed Memory Characteristics.

Phasor networks governed by dynamical system (1) at $\omega_j = 0$ can be considered as basic ones among all oscillatory networks of given architecture (defined by the same matrix W). Their memory has the most symmetrical structure preserving at the same time all the features inherent to oscillatory network memory.

As it was shown [2], special class of phasor networks with Hebbian matrix of connections W^H, possessing the guaranteed memory of high storage capacity, can be designed. The construction of W^H is based on the important property of interaction term in eq. (1). As one can see, the interaction between two oscillators of the network has the form $W_{jk}(z_j - z_k)$. This means that the matrices with nonzero diagonal are admissible for specification of network connections (unlike to the case of neural networks). This permits to use the matrices of projection operators in construction of Hebbian-like matrices of connection.

The most essential step in architecture design is introduction of a special set of orthogonal vectors in N-dimensional complex space C^N — "phase" basis (6):

$$\mathcal{B}_N = \{ \ V^m \ | \ (V^s)^+ V^m = N\delta_{sm} \ \ m, s = 1, \ldots, N.\}$$

The phase basis is defined by single generating vector $V^0 = (1, \ldots 1)^\top$ and the single parameter $\varphi = 2\pi/N$. All other vectors are of $\mathcal{B}_N$ can be calculated with the help of recurrent transformation or, the same, by multiple action on vector V^0 of irreducible group representation operator

$$T_g = diag(1, exp(i\varphi), \ldots, exp(i(N-1)\varphi).$$

The basis $\mathcal{B}_N$ is an eigenbasis of any weight Hermitian $N \times N$ matrix W satisfying the conditions (2). Therefore, any W can be represented in the form

$$W = N^{-1} \sum_{m=1}^{N} \lambda^m V^m (V^m)^+,$$

where $\lambda^m, m = 1, \ldots N$, are real numbers, V^m is column-vector $(V_1^m, \ldots V_N^m)^\top$ and $(V^m)^+$ is the corresponding conjugated row-vector: $(V^m)^+ = (\bar{V}_1^m, \ldots, \bar{V}_N^m)$. For zero-diagonal W, obviously, $\sum_{m=1}^{N} \lambda^m = 0$.

The matrix W^H of rank M,

$$W^H = \sum_{m=1}^{M} V^m (V^m)^+, \qquad M = rankW, \qquad (7)$$

is the matrix of the projection operator into M-dimensional subspace of C^N spanned on $V^1, \ldots, V^M$.

Note, that both the basis $\mathcal{B}_N$ and the matrices W^H are cyclical.

The following results are valid for phasor networks with matrices of connections W^H.

1. Let N to be a prime number.

Define basis $\mathcal{B}_N$ and choose any subset of $M \leq N$ vectors from this basis $\{V^1, \ldots V^M\}$. Construct W^H in accordance with (7).
Then phasor network has memory vectors

$$U^1, \ldots U^M, \quad U^m = cV^m,$$

where $c = 1$ if $V^0 \in \{V^1, \ldots, V^M\}$ and $c = (1 + \kappa)^{1/2}$ if $V^0 \notin \{V^1, \ldots, V^M\}$.
All memory vectors $U_1, \ldots U^M$ have equal basins of attraction.

The sizes of the basins can be controlled if weighted Hebbian matrix

$$\tilde{W}^H = \sum_{m=1}^{M} \lambda^m V^m (V^m)^+$$

is used. The values of λ^m should be slightly different and all close to unit.

It should be noted that all matrices W^H are irreducible if N is prime.

2. Let the number of oscillators N to be not prime.

The main feature of the network memory in this case is that the memory is not completely controllable unlike to the previous case. Namely, only special odd numbers M of vectors from the basis $\mathcal{B}_N$ can be imposed into network memory. If M is different from mentioned special numbers, recalling process is impossible at all: the dynamical system (1) demonstrates continual set of degenerated equilibria.
The matrices W^H are reducible in this case. Under the interaction specified by these matrices the phasor system is decomposed into non-interacting subsystems.

Conclusions
The special type of oscillatory associative memory have been designed. The class of oscillatory and corresponding phasor networks of high performance can be pointed out. It is characterized by fully controllable memory of high storage capacity: up to $N - 1$ memory vectors from some specific set ("phase" basis) can be loaded into the memory of the net-

work consisting of N processing units. The weight matrix is designed in Hebbian form generalized to complex-valued connections.

Extraneous memory exists, but it can be easily discriminated due to its non-phase character.

Oscillatory networks are promising from many viewpoints, in particular, in view of possibility of physical (optical) implementations.

Acknowledgments

This work is partially funded by Russian Foundation for Fundamental Researches, grant n. 96-01-00084.

References.
1. Kuzmina M. G., Manykin E. A., Surina I. I., Oscillatory networks with Hebbian matrix of connections - *Lecture Notes in Computer Science, v.930: Proc. of IWANN'95*, 1995, pp.246-251.
2. Kuzmina M. G., Manykin E. A., Surina I. I., Oscillatory networks with guaranteed memory characteristics - *Proc. of EUFIT'96, v.1*, 1996, pp.320-324.
3. Kuzmina M. G., Manykin E. A., Surina I. I., Oscillatory networks of associative memory - *Optical memory and neural networks, v.5, n.2*, 1996, pp.91-103.
4. Kurchatov S.Yu., Lihanskii V.V., Napartovich A.P., Teoriya sinkhronizatsii laserov pri opticheskoi svyazi "kazhdyi s kazhdym" - JETP, v. 107, n.5, 1995, pp. 1491-1502 (in Russian).

Strategies for Autonomous Adaptation and Learning in Dynamical Networks

Nabil H. Farhat, Emilio Del Moral Hernandez, and Gee-Hyuk Lee
University of Pennsylvania
The Moore School of Electrical Engineering
Philadelphia, PA 19104, USA

Abstract

The complexity of dynamical networks* , which compute with diverse attractors, renders them inaccessible, at present anyway, to entirely analytical treatment. Therefore, exploration and development of a learning algorithm for such nets, would need to rely mostly on numerical simulations. Here, we discuss strategies for the development of autonomous adaptation and learning algorithms for dynamical networks that are driven by entropy related information theoretic measures. A net of parametrically coupled logistic processing elements, an instance of a dynamical network, is used to illustrate the rationale, detail, and features of the strategies developed.

1. INTRODUCTION

There is considerable evidence, stemming from anatomical, physiological, and modeling work (see for example [1]-[5] and references therein) that the basic functional unit in the cortex is the *neuronal assembly* or *netlet*. It is suggested that cortical networks can be viewed as interacting populations of netlets. A netlet consists of randomly interconnected probabilistic neurons, and netlet behavior is simplest described in terms of the activity $A(n)$ $\varepsilon[0,1]$ which represents the fraction of neurons firing at any discrete integer time n [1]-[3]. The temporal and spatial fine structure in neuronal activity within a netlet are considered to be of secondary significance and are subsumed by netlet dynamics.

We have noted that plots of $A(n+1)$ vs. $A(n)$, prepared from netlet evolution in [1]-[3], resemble the logistic (quadratic) map [6]. This leads us to suggest that networks of parametrically coupled logistic maps, resembling those studied also in [7]-[10], be used to model and investigate cortical dynamics and therefore higher-level brain functions. The orbits or state space trajectories $X(n)$ produced by the logistic map, like other nonlinear mapping on the unit interval, are known to exhibit periodic (period-m) orbits, chaotic orbits, and bifurcation between them depending on the value of the control or bifurcation parameter of the map, which can be viewed, as in [7], as the input or driving signal of the map. One can expect netlets to behave in a like manner where the activity $A(n)$, analogous to $X(n)$ of the logistic map, would exhibit similar complex orbits. Indeed, such behavior has been observed in some of the netlet models studied in [1]-[5]. What can be significant, is that periodicity, bifurcation, and chaos can emerge on the driven netlet level despite the well known imprecision of neuronal firing. This picture goes a long way towards elucidating the *neuronal code* and the way cortical nets, encode and process information. It provides a potential answer

---

* Dynamical nets exhibit in their state-space not only fixed-point (static) attractors but also dynamic (periodic and chaotic) attractors.

for an important question in neuroscience, namely how coherence, synchronicity, phase-locking, and chaos can operate on the netlet and cognitive levels despite the imprecise nature of neuronal firing.

This novel intriguing proposal on the nature of the neuronal code has prompted us to investigate the dynamics of parametrically coupled nets of logistic processing elements (PEs) [7],[11]. We refer to these nets as dynamical nets because they are capable of exhibiting, in their state space, not only static (fixed-point) attractors, but also dynamic (periodic and chaotic) attractors. In [7] and [11] we have studied the rich dynamics of externally driven dynamical nets of logistic PE with nonlinear global coupling between the elements.

This paper focuses on the feasibility of developing autonomous adaptation and learning algorithms for dynamical networks.

2. STRATEGIES FOR AUTONOMOUS ADAPTATION AND LEARNING

A first step in the development of such an autonomous adaptation process is to identify a meaningful measure of the activities of any two PEs or neurons[*] in a dynamical network that could logically serve as a local mechanism for driving the adaptation of the coupling weights between them during the learning stage of the network. To this end we have identified the mutual information (MI) I_{ij} between the activities (orbits) of any two logistic PEs of a network (the i-th and j-th PEs) as such a measure. In particular we identified the normalized mutual information $I_{ij} \varepsilon [0,1]$ as a measure of the dependence between the two orbits or state variables $X_i(n)$ and $X_j(n)$. (For discussion of MI and information measures see for example [12] and [13].)

The justification for selecting the mutual information is appreciated from the following reasoning. The MI I_{ij} can be interpreted as the information found about one variable $X_j(n)$ in the other variable $X_i(n)$. Thus if we regard $X_i(n)$ as the orbit of the target PE (equivalent to a postsynaptic neuron) and $X_j(n)$ as the orbit of one of the PEs feeding into the target PE (equivalent to a presynaptic neuron) by participating in the formation and modulation of its control parameter $\mu_i(n)$, then I_{ij} can be interpreted as a measure of the degree with which a presynaptic PE (neuron) influences a postsynaptic neuron PE (neuron). It makes sense therefore to make the adaptation of the coupling weight W_{ij} between the j-th and i-th neurons PEs (neurons) be a function of the mutual information. One can recognize this strategy to be basically a kind of Hebbian adaptation process except it applies now to a dynamical network in which the nature of the orbits of the neurons[*] observed over a prescribed time window of sufficient length to allow computing the needed probabilities and entropies, determines the adaptation of the weights through MI or a function of MI, as δ_{ij} shown in table 1, computed over that window. (The reasons for choosing δ_{ij} will be given below). This is the basic idea underlying the notion of MI driven adaptation of the coupling weights in dynamical networks we are beginning to explore.

---

[*] Whenever reference is made to neuron, it is meant to be a bifurcation neuron.
[*] In [7] and [11] we show that the state variable of a spiking bifurcation neuron is the normalized phase $\theta(n)$ of its periodic activation potential at the instants of the n-th firing and that $\theta(n)$ describes a sequence or orbit analogous to the state variable $X(n)$ of the logistic PE.

$X_i(n)$	$X_j(n)$	$H_i/log(N_b)$	$H_j/log(N_b)$	coordination X_i / X_j	$H_{ij}/2.log(N_b)$	I_{ij}	$\delta_{ij} = 1 - 4.H_{ij}/(H_i + H_j)$
fixed	fixed	0	0	always none	0	0	$1 - 0/0 \Rightarrow 0$ *
fixed	period m	0	$log(m)/log(N_b)$	always none	$(log(m)/log(N_b))/2$	0	$1 - 2 = -1$
fixed	chaotic	0	1	always none	.5	0	$1 - 2 = -1$
period m	period m	$log(m)/log(N_b)$	$log(m)/log(N_b)$	always complete	$(log(m)/log(N_b))/2$	$log(m)/log(N_b)$	$1 - 1 = 0$
period m	chaotic	$log(m)/log(N_b)$	1	always none	$(log(m)/log(N_b) + 1)/2$	0	$1 - 2 = -1$
chaotic	chaotic	1	1	none	1	0	$1 - 2 = -1$
chaotic	chaotic	1	1	complete	.5	1	$1 - 1 = 0$

*Here we take $0/0 \to 1$

Legend:

$X_i(n)$ — orbit of i-th PE
$X_j(n)$ — orbit of j-th PE

$H_i/log\ N_b$ — normalized entropy H_i' of $X_i(n)$ computed using N_b bins over the range of X_i.

$H_j/log\ N_b$ — normalized entropy of H_j' $X_j(n)$ computed using N_b bins over the range of X_j.

$H_{ij}/2log\ N_b$ — normalized entropy H_{ij}' of the joint variable (X_i, X_j) computed using $N_b \times N_b$ bins.

$I_{ij} = H_i' + H_j' - 2H_{ij}'$ is the normalized mutual information of the orbits of the i-th and j-th PEs.

Table1:   Qualitative evaluation of the value of the normalized mutual information I_{ij} and associated quantities, for specific classes of activities X_i and X_j and joint activity (X_i, X_j). Notes: a) $log\ (m)/log\ (N_b)$ ranges in (0 to 1) as m ranges in (1 to N_b); b) $(log\ (m)/log(N_b))/2$ ranges in (0 to .5) as m ranges in (1 to N_b); c) in this table, chaotic means fully chaotic. i.e., the activity visits all N_b bins with equal probability; d) fixed means fixed-point orbit.

Factors that could potentially influence and guide such exploration are summarized in table form (table 1). Table 1 shows the normalized self entropies (H_i and H_j), the normalized joint entropy (H_{ij}), and two entropy functionals (I_{ij} and δ_{ij}) that one can associate with specific pair-wise orbits of PEs in the network. We expect the entries in table 1 to help in formulating, and elucidating further the idea of autonomous adaptation and learning in dynamical neural networks, or at least help us gain an idea of the type of self-adapting dynamical net a given adaptation strategy could lead to.

The entries in table 1 reflect the fact that the mutual information between the activities (orbits) of PEs (or neurons) in a dynamical network is governed by two factors:

1. The degree of spread or variance of the activity: High values of mutual information occur only if the activities (orbit values) of both PEs (neurons) are spread widely over the range [0,1] of the state variables. If that is not the case and one of the two PEs has narrowly spread activity, meaning that the PE is a poor source of information, then no matter how rich an information source is the other PE, the mutual information will be low.

2. The degree of coordination of the two activities: Having spread activity in both PEs (neurons), i.e. both of them are rich sources of information, does not guarantee that the mutual information is high, since the MI value depends also on the degree of coordination between the two activities or orbits.

Table 2 illustrates this idea of dependency of mutual information on both aspects: the individual activity and the joint activity.

X_i	X_j	Coordination X_i / X_j	I_{ij}
narrow	narrow	low or high	low
narrow	spread	low or high	low
spread	spread	low	low to medium
spread	spread	high	high

Table 2.   Qualitative evaluation of the value of mutual information for specific classes of joint activity for a given pair of neurons.

3. A CANDIDATE LEARNING STRATEGY

Next we will briefly illustrate how the information in table 1 might be used to develop the rudiments for a strategy or algorithm for autonomous adaptation and learning in dynamical networks. First however, it is instructive to list some desirable features we would like to see in such an algorithm:

a) Autonomous unsupervised adaptation of the coupling weights between PEs because unsupervised learning is the norm in biological networks.

b) Uniform input patterns are ignored by the adaptation process because these are structureless and do not convey any meaningful information. The adaptation process is activated however by all other structured input patterns.

c) Adaptation automatically halts when the network collapses into a periodic (period-m) attractor that uniquely characterizes the input pattern.

d) Episodes of chaotic orbits of the PEs, when they occur, do not induce adaptation but act only to move or shift the network trajectory in state-space so as to help it find a *loss-region* compatible with the input where the trajectory falls suddenly (collapses) into a periodic attractor of low dimension[*]

e) The adaptation process leads to sparse (distributed) memory with minimum or no cross-talk between the memories formed for the different inputs. Intolerable cross-talk defines *storage capacity*. After learning is completed there is a list of patterns stored in memory and considered to be familiar.

f) As long as the number of input patterns learned by the net is below the storage capacity limit, the application of an already learned input pattern causes the net to rapidly collapse into the associated periodic (period-m) attractor so that no adaptation could occur. A novel input on the other hand, would produce complex orbits for which adaptation will take place and continue according to the adaptations algorithm outlined below, until a distinct periodic attractor develops for that particular input thus adding it to the repertoire of the network. Adaptation in the network is "frozen" when a familiar input pattern is presented and is "fluid" for a novel one. The algorithm recognizes novelty just as for example antibodies are able to recognize a foreign or strange object.

g) To encourage convergence to attractors, the evolution of the network during adaptation should proceed at two time rates: a fast rate for the state variables, and a slower rate for the adaptation of coupling weights and the control parameters of the PEs.

h) Local rather than global interconnections between the PEs is preferable in order to minimize computational requirement and encourage the network to partition itself into distinct clusters with PEs within a cluster being synchronized i.e. having identical period-m orbits and all clusters being phase-locked into distinct period-m attractors. An instant of such behavior is described in [7] for a globally connected net of parametrically coupled logistic PEs where the coupling between PEs was determined by a nonlinear function of state variables.

A possible adaptation and learning procedure with potential to satisfy the above attributes is described next. This procedure may not necessarily be the exact one that leads to successful adaptation and convergence to a periodic attractor, but is meant to illustrate how information like that in table 1 might be applied to developing such an algorithm.

Consider a locally connected network of NxN parametrically coupled bipolar logistic (quadratic) PEs arranged on a rectangular lattice with the state variable $X_{ij}(n)$ of the (i,j)th parametrically coupled PE obeying the iterative mapping,

---

[*] Such low-dimensional periodic attractor accompanies clustering wherein the PEs in the net partition themselves into groups or clusters, with PEs within each cluster having identical (synchronized) period-m orbits and the period-m orbits of the different clusters being distinct but phase-locked. This type of behavior was observed in the dynamical net studied in [7].

$$X_{ij}(n+1) = 1 - \mu_{ij}\, X^2_{ij}(n) \qquad\qquad i,j = 1,2,...N \qquad\qquad (1)$$

where $X_{ij}(n)\varepsilon\,[-1,1]$, $\mu_{ij}\varepsilon\,[0,2]$, and $n=1,2,...$ is discrete integer time. μ_{ij} in eq. (1) is not fixed but undergoes punctuated change at selected values of n in a manner determined by the combined effect of a normalized external input pattern $I_{ij}^{\,v}\,\varepsilon$ [0,2] applied to the network and by nearest neighbor parametric coupling i.e., by the weighted inputs from the nearest N_n neighbors of the (i,j) site in accordance to,

$$\mu_{ij}\,(p) = I_{ij}^{v} + \frac{\eta}{N_n} \sum_{k\ell=1}^{N_n} \sum_{m=1}^{p} \delta_{ijk\ell}(m)\, X_{k\ell}\,[n = T + (p-1)q] \qquad (2)$$

where $q > 1$ is a constant, $p=1,2,...$ indexes the adaptation times and $\delta_{ijk\ell}$. $p=1$ in eq. (1) is interpreted as giving the first adaptation at n=T (see Fig. 1) namely,

$$\mu_{ij}(1) = I_{ij}^{v} + \frac{\eta}{N_n} \sum_{k\ell=1}^{N_n} \delta_{ijk\ell}(1)\, X_{k\ell}\,(n = T) \qquad (3)$$

and similarly p=2 designates the second adaptation at n = T+q and p=3 giving the third adaptation at n=T+2q and so on. In these equations $\upsilon = 1,2,...M$ where M is the number of patterns to be autonomously learned or classified by the net by associating each pattern with its own distinguishing period-m attractor with prescribed basin of attraction. In eq. (1), $q > 1$ is a constant determining the number of elapsed network iteration before the coupling weights are adapted again by the entropy functional

$$\delta_{ijk\ell}(p) = 1 - \frac{H'_{ijk\ell}(p)}{H'_{ij(p)} + H'_{k\ell(p)}} \qquad (4)$$

where $p=1,2,...$ is the adaptation index. $\delta_{ijk\ell}$ is equivalent to δ_{ij} in table 1 but expressed now in a 2-D lattice format. Like δ_{ij} in table 1, $\delta_{ijk\ell}$ is a function of the normalized self entropies H'_{ij} and $H'_{k\ell}$ of the orbits $X_{ij}(n)$ and $X_{k\ell}(n)$ of the (ij)th and $(k\ell)$th PEs respectively that precede the p-th adaptation, and similarly $H'_{ijk\ell}$ is the normalized entropy of the joint process or orbit $\left(X_{ij}, X_{k\ell}\right)$ preceding the p-th iteration. The entropies and $\delta_{ijk\ell}(p)$ are always determined from orbits T iterations long, as depicted schematically in Fig. 1, where T is sufficiently long to ensure meaningful computation of the probability distributions needed for computing the entropies and effecting adaptation. Like δ_{ij} in table 1, $\delta_{ijk\ell}(p)$ will range in [-1,0]. Then, because $X_{k\ell}\varepsilon\,[-1,1]$, $\mu_{ij}(p)$ in eq. (1) and eq. (2) need to be taken as modulus 2 with negative values of $\mu_{ij}(p)$ when they occur being discarded in order to confine it in the range [0,2] required by eq. (1).

The coupling weights at each adaptation instant p are altered according to

$$W_{ijk\ell}(p) = \sum_{m=1}^{p} \Delta W_{ijk\ell}(m) \qquad p = 1,2,.... \qquad (5)$$

where

$$\Delta W_{ijk\ell}(m) = \eta\, \delta_{ijk\ell}(m) \qquad m = 1,2,...p \qquad (6)$$

with $p = 1$ in eq. (1) being interpreted again as giving the first adaptation of coupling weights at $n = T$ namely $W_{ijk\ell}(1) = \eta\, \delta_{ijk\ell}(1)$. The size of the

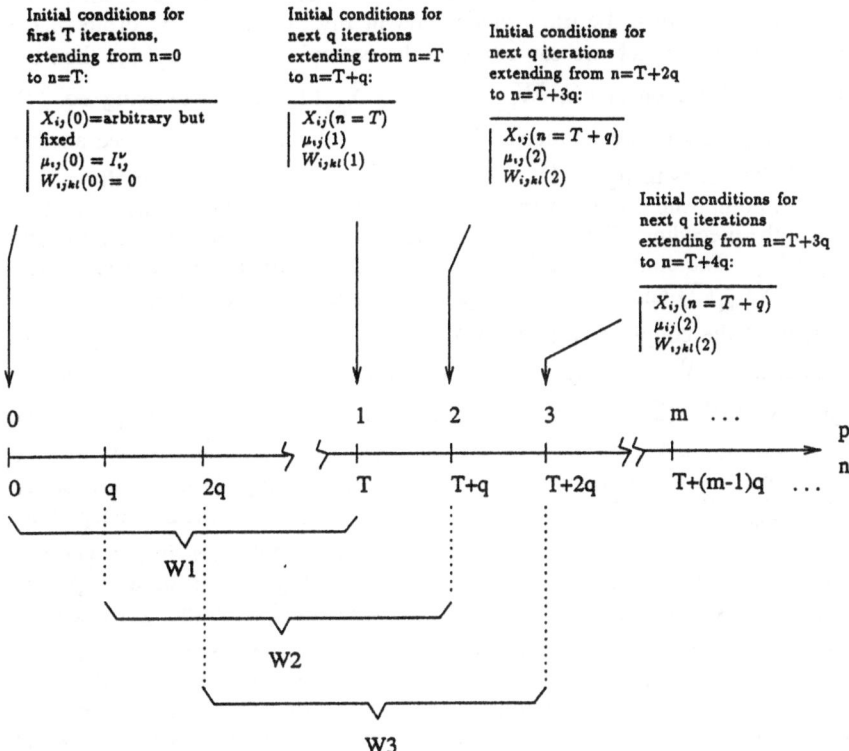

Figure 1: Time scales for the for network adaptation: $p=1,2,3$ ... designating the instants of computing weights adaptation, and n designating discrete integer time. Orbits over the time window W_1 are used to compute $\delta_{ijkl}(1)$ and to execute the first ($p=1$) weights adaptation at $n=T$ (see text). Orbits over W_2 are used to compute $\delta_{ijkl}(2)$ for the second ($p=2$) adaptation at $n=T+q$, and so on. Notice that W_1, W_2, W_3, etc... form a set of sliding windows all of which are T iterations long, and that W_2 overlaps W_1 in that it is formed from (T-q) iterations from W_1 plus q new iterations (from $n=T$ to T+q), and similarly, W_3 overlaps W_1 and W_2. The degree of overlap between windows, determined by choosing the value for q, makes the increments in δ_{ijkl}, W_{ijkl} and μ_{ij} from adaptation to adaptation under our control. The adaptation process schematically depicted here continues until presumably the network converges to a periodic (period-m) attractor, at which time $\delta_{ijkl} \to 0$ by design (see table 1) and the adaptation halts.

neighborhood N_n in computing all $\mu_{ij}(p)$ and $\delta_{ijk\ell}(p)$ is taken to be small e.g. N_n = 4 or 8 nearest neighbors to minimize computation requirements.

The evolution of the network proceeds as follows. Referring to Fig. 1, the network is iterated T times from initial conditions $\mu_{ij}(o) = I^v{}_{ij}$, and $X_{ij}(o) =$ arbitrary but fixed, say .5. Note that the initial weights matrix $W_{ijk\ell}(o)$ is immaterial at this stage and therefore can be taken to be zero i.e. the network self-adapts from a state of *tabula rasa* i.e. with no initial coupling between its PEs. The coupling weights would grow however as the network evolves in time. At the end of T iterations, the time window W_1 in Fig. 1 provides N^2 orbits $X_{ij}(n)$ i,j = 1,2,...N from which we can compute the entropies, then $\delta_{ijk\ell}(1)$, then the first weight adaptation matrix $W_{ijk\ell}(1) = \eta\, \delta_{ijk\ell}(1)$ and $\mu_{ij}(1)$ using eq. (4). The coupling weights are set then to $W_{ijk\ell}(1)$ and the network is iterated another $1 < q < T$ iterations using $\mu_{ij}(1)$ and $X_{ij}(1) = X_{ij}(n = T)$ as initial conditions. A new set of composite orbits is formed at this point by adding the q new iteration values to the preceding (T-q) iterations from W_1 to form a new composite orbits window W_2. The composite orbits in W_2 are next used to compute entropies, $\delta_{ijk\ell}(2)$, $W_{ijk\ell}(2)$ and $\mu_{ij}(2)$, using eqs. (2)-(6), and these used to effect a second adaptation of the coupling weights at $\eta = T+ q$. The coupling weights are altered next to $W_{ijk\ell}(2)$ (equivalent to incrementing the existing coupling weights by $\Delta W_{ijk\ell}(2)$) and the network is iterated again for another q iterations extending from n = T + q to n = T + 29. A composite orbits window W_3 is now formed consisting of iterations n = 2q to n = T + 2q (see Fig. 1) and a new set of matrixes $\delta_{ijk\ell}(3)$, $\mu_{ij}(3)$ and $W_{ijk\ell}(3)$ is computed to effect the p=3 adaptation. This adaptation process is continued until the network collapses into a periodic (period-m) attractor compatible with the input pattern $I_{ij}{}^v$. At this time adaptation halts by virtue of the fact that $\delta_{ijk\ell}(p) \Rightarrow 0$ when the PEs in the network form period-m orbits. (See entry in table 1 for period-m/period-m.) Such collapse into a period-m attractor, as result of the outlined adaptation process, would satisfy requirements (a) and (c) of Section 3. Evidence of the strong tendency of dynamical networks to collapse into period (period-m) attractors with *clustering* has been encountered in [7]. The clustering, described in [7] is synonymous with the appearance of periodic (period-m) attractors with low dimension.

We note next that when the applied input pattern $I_{ij}{}^v \,\varepsilon\, [0,2]$ is constant the resulting orbits in W_1 produce according to table 1 $\delta_{ijk\ell}(1) = 0$, because for $I_{ij}{}^v$ = constant, the first T iterations give identical orbits for all PEs and no matter whether the orbits are all fixed-point, period-m, or chaotic (depending on the constant value of $I_{ij}{}^v$), the associated entries in table 1 show $\delta_{ijk\ell} = 0$. This would satisfy requirement (b) in Section 3.

Depending on the nature of $I_{ij}{}^v$, i.e., the intensity distribution and shape of input pattern, all PEs at sites (i,j) for which $I_{ij}{}^v = 0$, will be in fixed point orbits $X_{ij}(n) = 1$ during the time window W_1. All sites (i,j) outside the boundary of the applied pattern will therefore describe identical fixed-point orbits in W_1. PEs or sites for which $I_{ij}{}^v \neq 0$, on the other hand, will describe arbitrary orbits during W_1 that can be fixed-point, period-m, or chaotic. The entries in table 1 for mixed orbits indicate that the first adaptation at the end of W_1 window (n = T) would produce $\delta_{ijk\ell}$ (1) that are nonzero for some of the sites outside the boundaries of input pattern. This means that as the network evolves, coupling weights adaptation is not confined to within the boundaries of the applied input pattern where $I_{ij}{}^{v} \neq 0$ but can selectively spread into the rest of the network, i.e. to sites outside the boundaries of the pattern where $I_{ij}{}^{v} = 0$, in a manner that depends on the nature of $I_{ij}{}^{v}$. It would be natural to adjust parameters of the system then such that the coupling weights matrix, when adaptation halts, be as sparse and as distinctive of the applied input pattern as possible. When achieved, this has the advantage of enabling the learning of additional input patterns to occur with minimum disturbance of the distributed memories formed by earlier input patterns. The adaptation process then acquires some of the traits of holographic storage and would indicate for the first time the viability of the holographic model for brain (cortical) function that goes beyond the usual holographic metaphors for brain function put forth by several investigators in the past few decades.

It is relatively easy at this stage, to appreciate that the postulated adaptation algorithms contains elements that could lead to satisfying requirements (d) to (h) of Section 3.

4. DISCUSSION

Brain activity can be described by occasional purposeful episodes of periodic activity emersed in long temporally irregular chaotic activity. Evidence for such a view is provided in the work of Freeman and colleagues [14]-[16] who have extensively studied the role of chaos in olfaction and higher-level brain function. They find that EEG traces from the olfactory bulb of a rabbit in an unstimulated state appear to be chaotic. However when the rabbit is made to inhale a familiar odor, a burst of activity appears in the EEG recording which they argued to be manifestation of a periodic (limit-cycle) attractor characteristic of that odor. Following the burst the system resets back to resting state. They propose that the olfactory bulb carries a repertoire of learned periodic attractors one for each odorant previously learned by the animal. Each of the periodic attractors has a given basin of attraction and access to it is facilitated when its odor is present by a process, they call, "chaotic basal state".

Ding and Kelso [17], commenting on Freeman's work, suggest that "the brain (the olfactory cortex) may associate specific ordor information with specific unstable periodic orbits embedded in the resting chaotic attractor. Then being chaotic at rest renders the brain dynamical access to any of these unstable periodic orbits for retrieving information. More specifically when a familiar odor is presented, the brain may utilize chaotic motion to select the corresponding periodic orbit and then stabilize this orbit until perception of that given odor is achieved. This simple conceptual picture is intuitively appealing.

The outline of autonomous learning and adaptation in dynamical network we have presented here has elements that match Freeman's and Kelso's views on the role of chaos in brain function. It describes a dynamical network which responds (learns or adapts) only to structured input patterns, i.e. the net becomes fluid then is frozen (is not fluid) with structureless (uniform) input. The fluidity continues until a "loss region" of the state-space is reached where the net collapses into a period-m attractor, hopefully with clustering and therefore with low dimension, as we observed in earlier experiments with parametrically coupled nets of logistic elements and fixed uniform (constant) coupling [17].

5. ACKNOWLEDGEMENT

This work was supported by the Office of Naval Research, grant no. N001-94-093P01.

6. REFERENCES

1. E.M. Harth, et. al., J. Theor. Biol., vol. 26, pp. 93-120, (1970).

2. P.A. Anninos, et. al., J. Theor. Biol., vol. 26, pp. 121-148, (1970).

3. G.M. Edelman, Neural Darwinsim: The Theory of Neuronal Group Selection, Basic Books Inc., Publishers, New York, (1987).

4. M. Usher, H.G. Schuster and E. Neibur, Neural Computation, vol. 5, pp. 570-586, July 1993.

5. C. van Vreeswijk and H. Sompolinski, Science, vol. 274, pp. 1724-1726, Dec. 1976.

6. R.C. Hilborn, Chaos and Nonlinear Dynamics, Oxford Univ. Press, New York (1994).

7. N. Farhat and E. Del Moral Hernandez, SPIE, vol. 2324, SPIE, Bellingham, Wash. (1996), pp. 158-170.

8. K. Kaneko, in Theory and Applications of Coupled Map Lattices, K. Kaneko (Ed.), J. Wiley, New York, 1993, pp. 1-49.

9. J.C. Perez and J.M. Bertille, Proc. INNS First Annual Meeting, Boston, Sept. 1988, p. 121.

10. J.M. Bertille and J.C. Perez, Proc. IJCNN'90, vol. 1, L. Erlbaum Assoc. Pub., Hillsdale, NJ, 1990, pp. 361-364.

11. N. Farhat, S-Y Lin and M. Eldefrawy, Adaptive Computing, S. Chen and J. Caulfield (Eds.), SPIE, vol. CR55, SPIE, Bellingham, Wash. (1994), pp. 77-88.

12. W. Li, J. of Statistical Physics, vol. 60, pp. 823-837, (1990).

13. R.M. Gray, Entropy and Information Theory, Springer-Verlag, Berlin, (1990).

14. W.J. Freeman, in Chaos in brain function, E. Basar (Ed.), Springer-Verlag, Berlin, 1990.

15. W.J. Freeman, Societies of Brains, Lawrence Erlbaum Associates, Hillsdale, N.J. (1995).

16. C.A. Skarda and W.J. Freeman, Behavioral and Brain Sciences, vol. 10, pp. 161-195, (1987).

17. M. Ding and J. Scott Kelso, in Measuring Chaos in the Human Brain, D.W. Duke and W.S. Pritchard (Eds.), World Scientific, Singapore, (1991), pp. 17-31.

Modeling the Parallel Development of Multiple Featuremaps and Topography in Visual Cortex

Winfried A. Fellenz

TU Berlin, FB Informatik, FR-3/11, Franklinstr. 28-29, 10587 Berlin, Germany

Abstract. We study the development of a high dimensional feature map in a retinotopical organised visual cortex. The presented model is an extension of the elastic-net approach employing two basic terms: a stimulus dependent term and a "regularising" constraint term. Interpreting each cortical element as a feature coding vector with selectivity across multiple dimensions, the parallel development of an optimised map with multiple features can be studied. The feature distributions of two feature maps show a structure similar to spatial frequency columns in cat visual cortex supporting the concept of a columnar architecture.

1   Introduction

The "optimal" representation of information is an important aspect of neural network computation. In general, cortical design principles seem to be the minimising of the required connectivity, allowing only few long range connections, and a columnar organisation. The performed computations are still under debate, but the underlying computational structure is anatomically and physiologically observable. Since the goal of early vision computation is the fast and reliable extraction of features in the scene and their coding using a representation which maximises the information content, the underlying computations should additionally constrain the development of cortical maps. The observed computations impose two constraints. The representation should be globally and locally complete, allowing all features to be represented, and the feature selectivity and retinotopy should vary smoothly, allowing local interactions between similar preferences and receptive field positions.

The ontogenetic organisation of the visual cortex could be genetically determined by a simple organisation principle which is governed by some order parameters, and sharpened during postnatal development by activity dependent interactions, which could even lead to further specifity. This view accords to the observed vanishing and reordering of genetically determined connections between cortical neurons during development. The initial selforganization process of a retinotopic structure employing the orientation continuity constraint has been described by some models using different formulations and parameters, to match physiological measured distributions of feature selectivity in early visual cortex. However, the dependencies of the feature dimensions are less clear, since some of them may be derived or well correlated to each other. Another unsolved question regards to the precise topography of the receptive field centres and their correlation to the observed distribution of selectivity in the feature maps. A general assumption about cortical organisation is the early forming of columns, which group sets of related features in close vicinity. If the conception of cortical columns is valid a modular organisation would be achieved with high connectivity within each module and viewer connections to neighbouring columns. Examples are the columns of ocular dominance, vertical columns with similar orientation preference and the hypothetical "hypercolumn" which groups complete sets of features in a spatial neighbourhood. Although the synaptic trees of cortical neurons resemble this connectivity pattern, physiological studies seldom show clustering into anatomically observable modular units. This lack of anatomical evidence for a modular organisation could be interpreted as the result of a parallel and independent development of multiple feature dimensions with different connectivity. The single order parameter of the first scenario will be replaced by multiple parameters to organise a cortical mapping satisfying the imposed constraints on the characteristic wavelength of each "columnar" system.

2   The Elastic-Net Approach

The Model proposed by Durbin and Mitchison [1] develops a retinotopic mapping of oriented cortical units by applying random stimuli to an initially isotropic cortex. Each cortex element is parameterised by a 4-D vector $y_i = (x, y, r, \phi)$, where x, y represents the coordinates of the receptive field centre, and the polar coordinates (r, ϕ) encode feature selectivity and the degree of tuning. The abstract cortical unit could in principle encode multiple features and receptive fields since some cortical neurons have binocular input or represent combinations of feature dimensions. Therefore the higher dimensional feature vector now reads $Y_i = (x_l, y_l, r_m, \phi_m) \in \mathcal{Y}$, with $l \in L$ the number of receptive fields, and $m \in M$, the number of feature dimensions. A further extension would be the introduction of tensor coordinates for multiple receptive field parameters like orientation tuning, selectivity, size, spatial frequency, phase, twist, directionality, curvature, end-stopping and even colour. The equations for the elastic net now become

$$\Delta Y_j = \alpha w_j (S_x - Y_j) + \beta \sum_{k \in N} (Y_k - Y_j), \qquad (1)$$

$$w_j = W_j / \sum W_k, \qquad (2)$$

$$W_j = exp(-(S_x - Y_j)^2 / 2K_m^2). \qquad (3)$$

where ΔY_j stands for the rate of change of each cortical unit in response to the presentation of stimulus $S_x \in \mathcal{Y}$ and the weighted contribution of the neighbouring units Y_k which tend to make response properties similar. The hebbian learning term corresponds to the euclidian distance of stimulus and cortical unit, weighted by a Gaussian receptive field W_j with deviation K_m, which our main order parameter governing the characteristic wavelength. The process equations 1, which define a dimension reduction framework for the mapping of any higher dimensional feature space to a low dimensional (2D) map are iterated using a Gauss-Seidel procedure and uncorrelated random stimuli to produce patterns of feature selectivity across multiple dimensions. The boundary conditions of the quadratic cortical map are cyclic and the cortical units are initialised to an orderly topographic map with a fixed feature preference. Each abstract cortex element comparable to an assembly of feature detecting neurons now represents multiple features which develop their feature specifity according the process

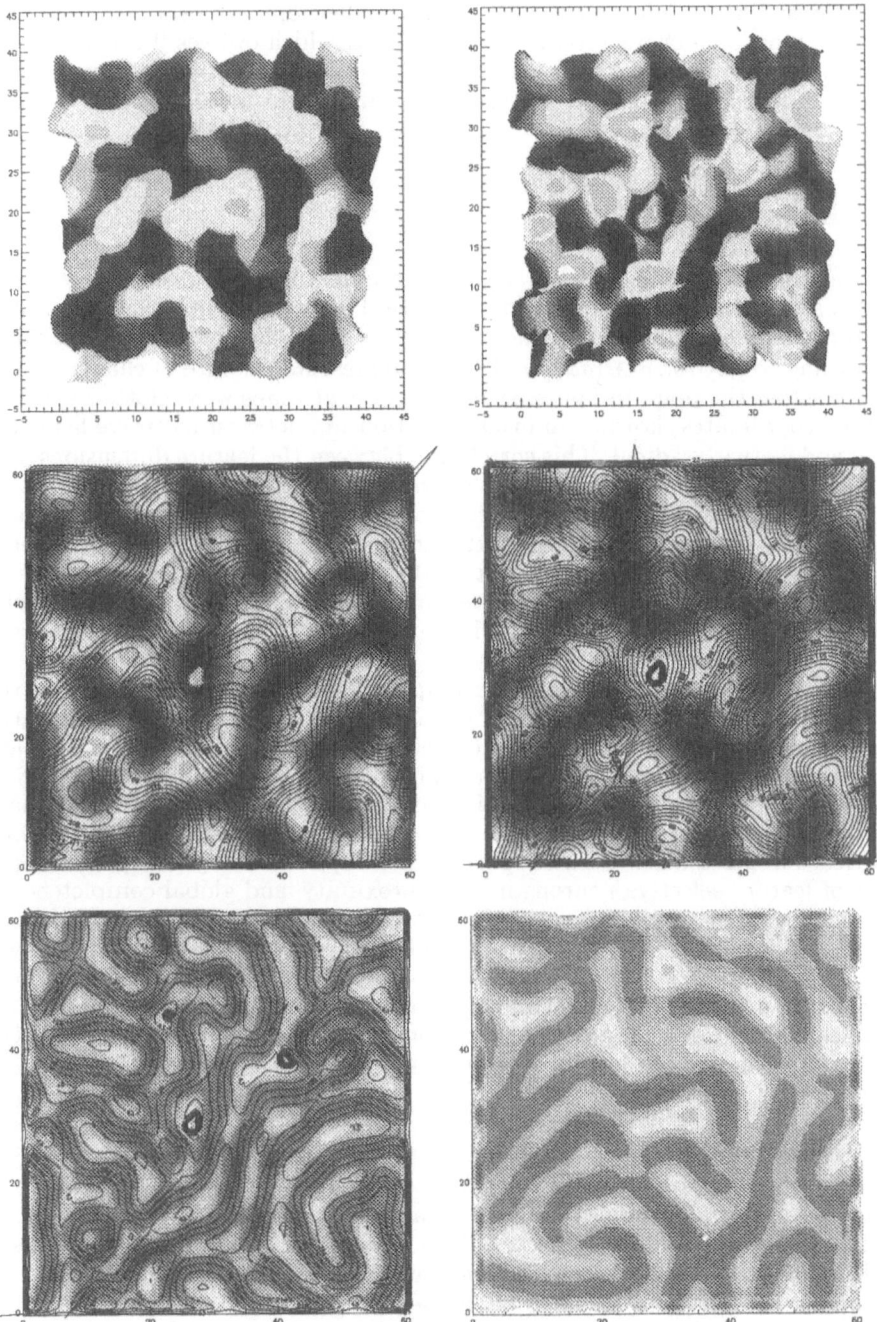

Fig. 1. *(a,b) Retinotopic position plot of feature values in both maps. (c,d) Feature distribution of both feature maps in cortical coordinates; the contour plot in each map indicates iso-orientation lines in the other plot. (e,f) Feature selectivity gradient maps.*

equations. The learning term guarantees the development of a complete representation owing a characteristic wavelength λ_m which evolves if the appropriate parameters are chosen. For the simulations with one receptive field centre (x, y) and two features $(r_{1,2}, \phi_{1,2})$ we used the parameters $\alpha = 0.4$, $beta = 0.0001$, $K_1 = 0.06$, $K_2 = 0.08$, and $\phi_1 = \phi_2 = 0.2$ on a 64 by 64 lattice.

3   Results

To analyse the evolving columnar structure in feature space, the retinotopically organised feature maps shown in figure 1a-d were differentiated to produce a feature gradient map which is shown in figure 1c,d. The gradient was calculated according to $[(\delta\phi/\delta x)^2 + (\delta\phi/\delta y)^2]^{1/2}$ and represents the rate of change in selectivity across the map. In figure 3a-b) the gradient maps were plotted with their retinal coordinates showing an inverse relationship between receptive field variation and feature gradient. This correlation between the feature dimensions is also apparent in plot 3c, showing a double contour plot of both feature distributions. A salient organisation principle is the crossed relationship between both feature dimensions which intersect with orthogonal angels, allowing complete sets of both features to be represented in a local spatial neighbourhood. The fact, that both dimensions have developed independent characteristic wavelengths, which can be evaluated from the autocorrelation plots 2, does not "prevent" the system to reveal a local columnar structure.

A comparison of the retinotopic map to the feature gradient maps reveals the correlation of low receptive field variation with high feature gradients, giving further evidence for a columnar structure in feature space. However, it should be noted that although all dimensions are highly correlated in a functional manner, the measured autocorrelation functions do not match and the emerging map is governed by two order parameters. The columnar organisation of higher order cortical receptive fields both support local computations by grouping complete sets of feature selective neurons in close proximity, and global completeness by introducing areas of high receptive field variation and low feature gradients (compare with figs. 3 a-c. The spatial extent of these areas is obviously bounded by the receptive field in eq. 3 to guarantee completeness of the representation. The organisation principle of a stepwise continuous representation keeps the combined feature gradient of all dimensions at a constant value, allowing cyclic variations in feature gradients to locally optimise the mapping for information processing.

4   Discussion

The spatial organisation of multiple feature selective cells into columns and patches is a well studied subject, but all underlying organisation principles are still unknown. Some features like orientation tuning [5], directionality [3], end-stopping and curvature seem to be derived properties, or develop after completion of the initial map [1]. Whether this is the case for spatial frequency is not clear, since the organisation of spatial frequency columns, observed in cats may

---

[1] Ocular dominance was not treated in this paper, obviously needing two receptive field centres, which is apparent in the shift back in RF-position when an electrode penetrates an ocular dominance border

underly a different columnar organisation principle as in monkeys [4]. The intersections of iso-specifity curves of both simulated maps are orthogonal at most crossings thereby producing patches where all features are locally represented, similar to the maps of orientation and spatial frequency in cat cortex. This organisation favours a spatial frequency filtering system with computations carried out in parallel in a spatially restricted region, whose relaxed responses could be used in a hypercolumnar computational architecture [2].

Although modules, columns and patches seem to be common structures in visual cortex, different ordering principles may govern the development of inter- and intra-columnar connections. Structures like singularities, fractures, roofs and linear zones can be found in the maps, but the formation of an interwoven string-like structure seems to be the most apparent organisation principle. The observed variation of receptive field position is another aspect which is highly correlated to the feature gradients.

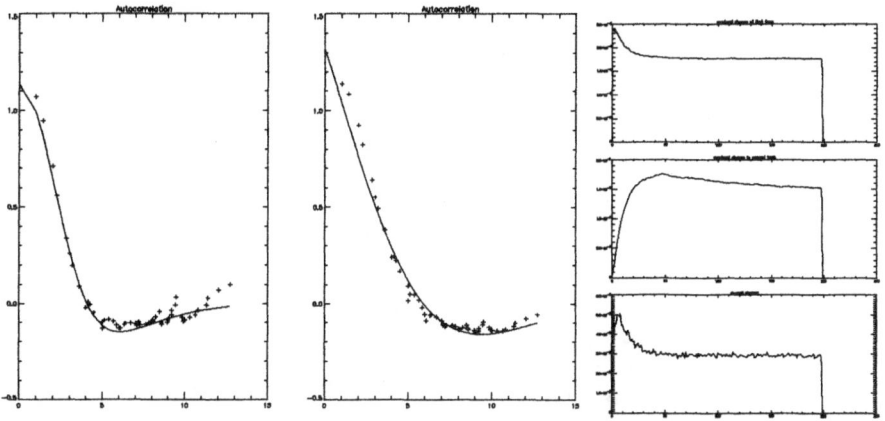

Fig. 2. *Autocorrelation functions of both feature maps, left for $k_m = 0.6$, middle for $k_m = 0.8$. The functions have been fitted by a DoG to resemble the discrete measurements. Maximal change of first and second term of equation 1 and overall change with time (right, top to bottom).*

References

1. R. Durbin and G. Mitchison. A dimension reduction framework for understanding cortical maps. *Nature*, 343:644 – 647, 1990.
2. W. A. Fellenz and G. Hartmann. Image segmentation by phase label diffusion. In *Proceedings of the International Conference on Artificial Neural Networks, ICANN-95*, volume 2, pages 309–314, Paris, 1995.
3. N. V. Swindale, J. A. Matsubara, and M. S. Cynader. Surface organization and direction selectivity in cat area 18. *Journal of Neuroscience*, 7(5):1414–1427, 1987.
4. R. L. De Valois and K. K. Valois. *Spatial Vision.* Oxford University Press, 1988.
5. F. Wörgötter and E. Niebur. Cortical column design: a link between the maps of prefered orientation and orientation tuning strength? *Biological Cybernetics*, 70:1–13, 1993.

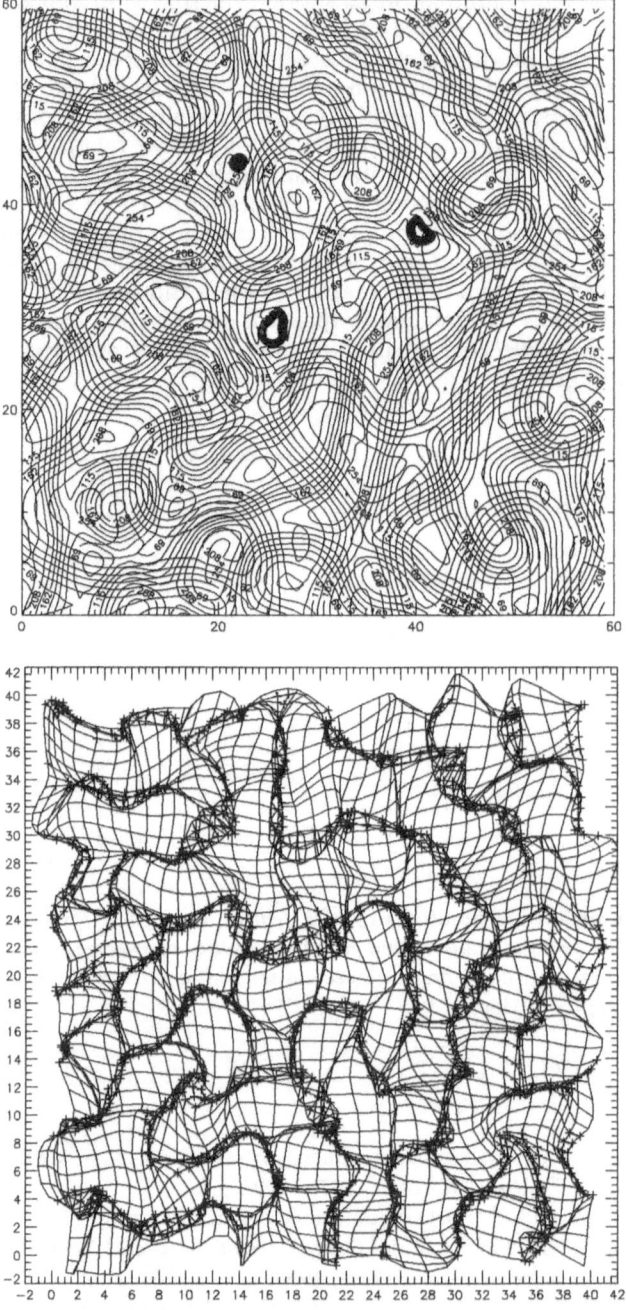

Fig. 3. *Overlaid contour plot of iso-orientation domains of both feature maps in cortical coordinates (upper). Plot of receptive field centres with high feature gradients overplotted for the first feature map (lower). The regions with high RF variation in the second plot correspond to regions with low feature gradients of both dimensions in the first plot.*

Stability and Hebbian Learning in Populations of Probabilistic Neurons

Francisco B. Rodríguez
Vicente López
Instituto de Ingeniería del Conocimiento, Universidad Autónoma de Madrid,
Canto Blanco, Mod. C-XVI, P.2, 28049 Madrid, Spain.

Abstract. The effect of a hebbian learning process in an isolated population of neurons is investigated using numerical simulations on a probabilistic neural network model. An increase of regularity in spike production is observed as a result of exposure to messages received by connections of adapting strength. The simple mechanism of synaptic adaptation that uses local information available at synapses is capable of driving the population towards a stable firing rhythm and does so by selecting a stable set of synaptic weights. The observed stable limit is coherent with a low firing profile in the activity of the isolated population model.

Introduction

Neural activity and brain information processing can be viewed as the emerging property that results from some kind of "organization" occurring in the population of neurons that interact through message interchange. As far as we know messages are transmitted using membrane potentials and, if such is the case, the content should be encoded in the traffic of neural spikes. A great understanding of neural functions will arise from the discovery of the spike coding scheme used by neurons to build and to process information. However, basic properties of neural populations need to be understood before success arrives in solving the decoding problem. For instance, the origin and role of the stability in the neural processing. Information processing performed by neurons is stable in spite of the variable scenario in which a given information is repeatedly acquired. Moreover, it is stable against the instability of the electric activity in isolated units and the imprecise character of the chemical machinery involved in spike transmission.

The work presented in this article is aimed to advance towards the understanding of these "stability" questions and we present the analysis of results measured during numerical simulations of basic neural models. Work has been directed to measure the effect of a Hebbian learning process [1] in the stability of a neural population. Changes induced by the learning process are monitored measuring regularity in the firing activity of units and in the spike production of the ensemble, and measuring the stabilization of synaptic strength in unit connections.

Models used in the work presented here do not include the effect of external stimulus. Results for that kind of models will be analyzed in a paper that will be published elsewhere. In the first section the model used for every unit in the population is presented. The way in which units communicate in the studied ensembles is explained in this section, and the learning process in section 2. A description of magnitudes measured over the units and the ensemble is given in section 3 and results are presented in section 4. A customary discussion of communicated results will end the article.

1 Neural Model

The basic units of our system follow a nondeterministic integrate and fire model [2]. These units interact among themselves, by means of an adequate connection or synapse. We have chosen a neuronal architecture in which we have applied a stochastic Hebbian rule for self-adaptation of synaptic weights.

The isolated neuron

Neurons are modeled [3, 4] as stochastic units with a discrete number of states. Time is also considered to be discrete and only two parameters are relevant for the description of the dynamic behavior of each unit: the number of states and the probability of incrementing the state at every time step.

In what follows we use $a_i(t)$ to represent the state of unit i at time t. Possible states of each unit are in the range from 1 to N_i. In absence of interaction with other neurons, the transition between states is governed by the probabilistic rule:

– For $a_i(t) < N_i$:

$$a_i(t+1) = \begin{cases} a_i(t) + 1 \text{ with probability } p_i \\ a_i(t) \qquad \text{ otherwise} \end{cases} \qquad (1)$$

– And if $a_i(t) = N_i$ then $a_i(t+1) = 1$, being the *neuron spike* represented by this transition to state 1, from which the cycle starts again.

According to this model, an isolated neuron behaves like a stochastic oscillator. The elapsed time between consecutive spikes, T_i, follows a probability law given by the Negative Binomial distribution $P_{N_i}(T_i)$:

$$P_{N_i}(T_i) = \begin{pmatrix} T_i - 1 \\ T_i - N_i \end{pmatrix} p_i^{N_i} (1 - p_i)^{T_i - N_i} \qquad (2)$$

For such a unit the inter-spike interval has an expected value τ_i with standard deviation σ_i. The value of both statistic parameters are easily derived from $P_{N_i}(T_i)$, and they are:

$$\tau_i = 1 + \frac{N_i - 1}{p_i}, \quad \sigma_i = \frac{\sqrt{(N_i - 1)(1 - p_i)}}{p_i} \qquad (3)$$

respectively [5].

The state of a unit at a given time only depends upon the state at a previous time and therefore the dynamics in this model, including the interaction presented below, is Markovian.

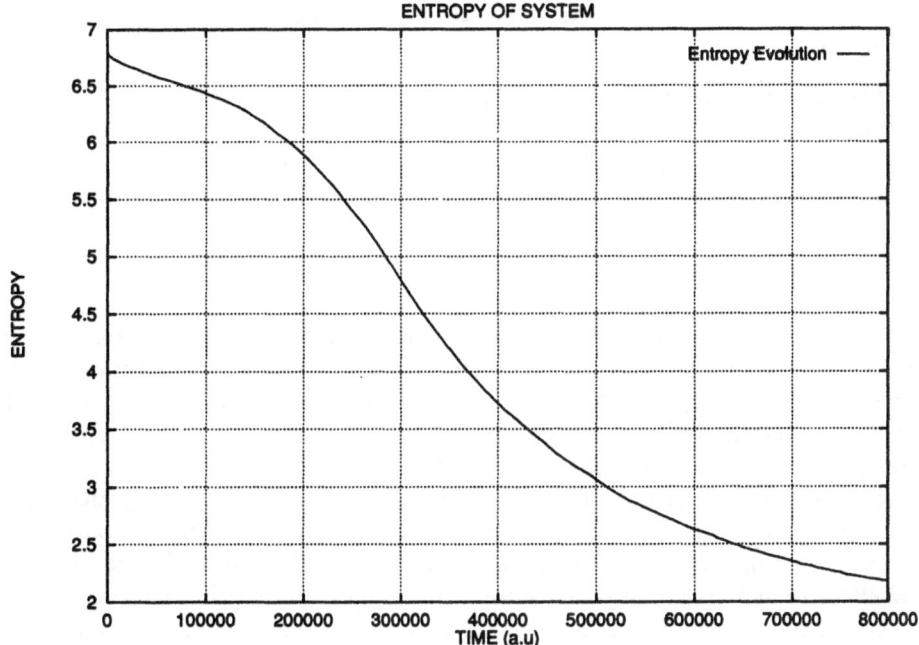

Fig. 1. *Time evolution of connections entropy content during the learning process.*

Units Interaction

The interaction between units is included in the model as a delayed spike transmission through the neural connection, which produces a change in the state of the receiving unit. More explicitly, neuron j in time t will be affected by a spike of neuron i in time $t-1$ according to:

$$a_j(t) = a_j(t-1) + \epsilon_{ij}(t) \tag{4}$$

where the magnitude $\epsilon_{ij}(t)$ refers to the connection strength. If the unit activity is enough to reach the threshold N_j due to spike arrival, then unit will discharge and reset its state to 1, regardless of the exact value $a_j(t)$ before discharging. Each synapse can take values in the range $\epsilon_{ij} \in \{\epsilon_{ij_{min}}, \epsilon_{ij_{max}}\}$ and for each value we have a corresponding probability $Q_{\epsilon_{ij}}$. This is the plasticity range for every connection.

The synapses strength follows a certain dynamics in the range of possible values $\epsilon_{ij}(t) \in \{\epsilon_{ij_{min}}, \epsilon_{ij_{max}}\}$ governed by the probabilistic rule:

– For $\epsilon_{ij}(t) \neq \epsilon_{ij_{min}}$ and $\epsilon_{ij}(t) \neq \epsilon_{ij_{max}}$ the rule is:

$$\epsilon_{ij}(t+1) = \begin{cases} \epsilon_{ij}(t) + 1 & \text{with probability } \frac{1}{2}(1 - Q_{\epsilon_{ij}}(t)) \\ \epsilon_{ij}(t) - 1 & \text{with probability } \frac{1}{2}(1 - Q_{\epsilon_{ij}}(t)) \\ \epsilon_{ij}(t) & \text{otherwise} \end{cases} \tag{5}$$

– And for $\epsilon_{ij}(t) = \epsilon_{ij_{min}}$ and $\epsilon_{ij}(t) = \epsilon_{ij_{max}}$ changes are inside the allowed range.

The function probability $Q_{\epsilon_{ij}}(t)$ is a measure of the probability is finding at time t a synapse with a value $\epsilon_{ij}(t)$. The probability can take values in the range $Q_{\epsilon_{ij}}(t) \in \{Q_{\epsilon min}, Q_{\epsilon max}\}$, being $Q_{\epsilon min} \approx 0$ and $Q_{\epsilon max} \approx 1$. Therefore the synaptic plasticity of the network is always changing during the time.

2 Learning process

Learning in neural networks is the mechanism that adjusts the connections or weights for the network to do a specific task. There are two fundamental paradigms of learning. The first is supervised learning (done adjusting the output of the network with known correct answers). The second is the unsupervised learning, in which learning means self-organization. The model of learning we present falls in the second category. The basic idea of our learning algorithm is based in the Hebb's law of unsupervised learning [6], and there is only reinforcement for optimal effective synapses (later we will define what is an optimal effective synapses). In other words, the synapse connecting two neurons is strengthened when activities are correlated and the connection is not reinforced if the synapse is not effective. The result of learning is encoded in the changes of probabilities $Q_{\epsilon_{ij}}$ associated to synapses of the network (ϵ_{ij}).

In our modified Hebbian rule we need to define which are the optimal effective synapses during the learning process. So, an Optimal Effective Synapse (OES) $\epsilon_{ij}(t)$, in time t, is defined to be the one for which:

1. the neuron i fires at time $t - 1$,
2. the neuron j fires at time t, and
3. the amount of signal that receives the neuron j due to spike transmission from unit i is less than $(N_j - a_j(t)) + \Theta$.

being the parameter Θ, the maximum value that can exceed the exact amount of signal $(N_j - a_j(t))$ needed by the unit j to fire. This parameter is the same for all neurons. Whenever one of these conditions is not satisfied, the synapse $\epsilon_{ij}(t)$ is noneffective.

The learning rule modifies the values of probability $Q_{\epsilon_{ij}}(t)$ using two positive parameters being called the *learning rates*. These parameters are:

1. η_+ for optimal effective synapses (OES).
2. η_- for noneffective synapses.

For the learning process, the modifications in $Q_{\epsilon_{ij}}(t) \in \{Q_{\epsilon min}, Q_{\epsilon max}\}$ associated to the present value $\epsilon_{ij}(t)$ follows the dynamic:

– When $Q_{\epsilon_{ij}}(t) \neq Q_{\epsilon min}, Q_{\epsilon max}$ and unit j fires at time t then:

$$Q_{\epsilon_{ij}}(t+1) = \begin{cases} Q_{\epsilon_{ij}}(t) + \eta_+ & \text{if } \epsilon_{ij}(t) \text{ is an OES} \\ Q_{\epsilon_{ij}}(t) - \eta_- & \text{if } \epsilon_{ij}(t) \text{ only satisfies conditions 2, 3 of OES} \\ Q_{\epsilon_{ij}}(t) & \text{if } \epsilon_{ij}(t) \text{ not satisfies condition 3 of OES} \end{cases}$$

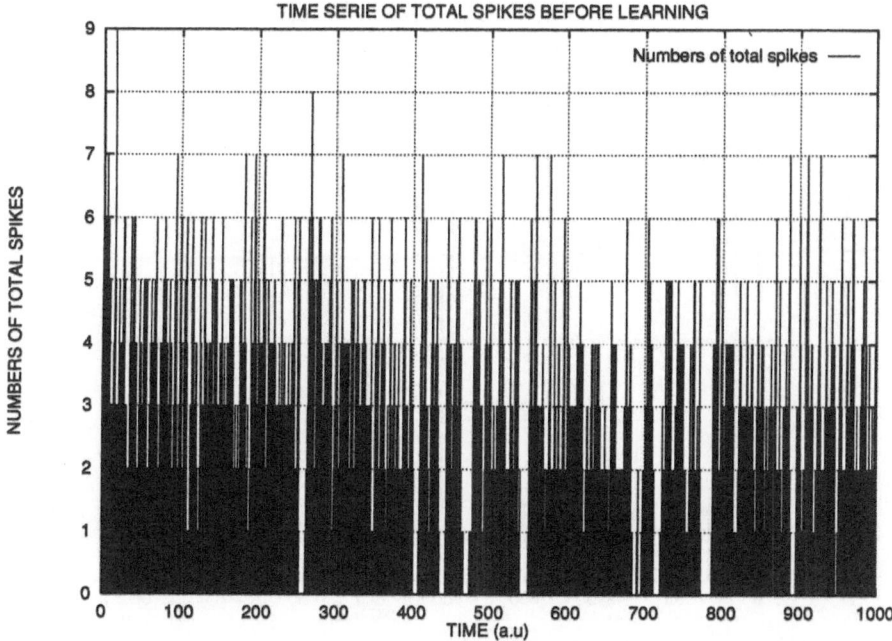

Fig. 2. *Time series of total spikes in the neural model before starting the learning process.*

Only in the third case, when the amount of signals that receives the neuron j due to spike arrival of unit i exceeds $(N_j - a_j(t)) + \Theta$ in a certain amount Δ, then we decrease synapse strength $\epsilon_{ij}(t)$ that contributed to this excess in a quantity proportional to Δ.
- For the limit values with $Q_{\epsilon_{ij}}(t) = Q_{\epsilon min}$, $Q_{\epsilon max}$ modifications are only in the allowed range.

The rule presented is local because every synaptic site is treated like a unit adapting to changes around site. The only information needed for adaptation is the state of the receiving unit at the time when the spike arrives.

3 Learning Process and Monitoring

We have considered a basic network architecture in which each unit is connected to every other unit, except to itself. The initial state of neuron i is selected randomly in the interval from 1 to N_i. The value of initial connections are chosen randomly in the range $\epsilon_{ij}(t) \in \{\epsilon_{ij_{min}}, \epsilon_{ij_{max}}\}$. The probability $Q_{\epsilon_{ij}}$ associated to each ϵ_{ij} is initialized with the value 0.5. Therefore the network will not have preferential synapses before learning. In these conditions we are ready to apply the learning rule explained above.

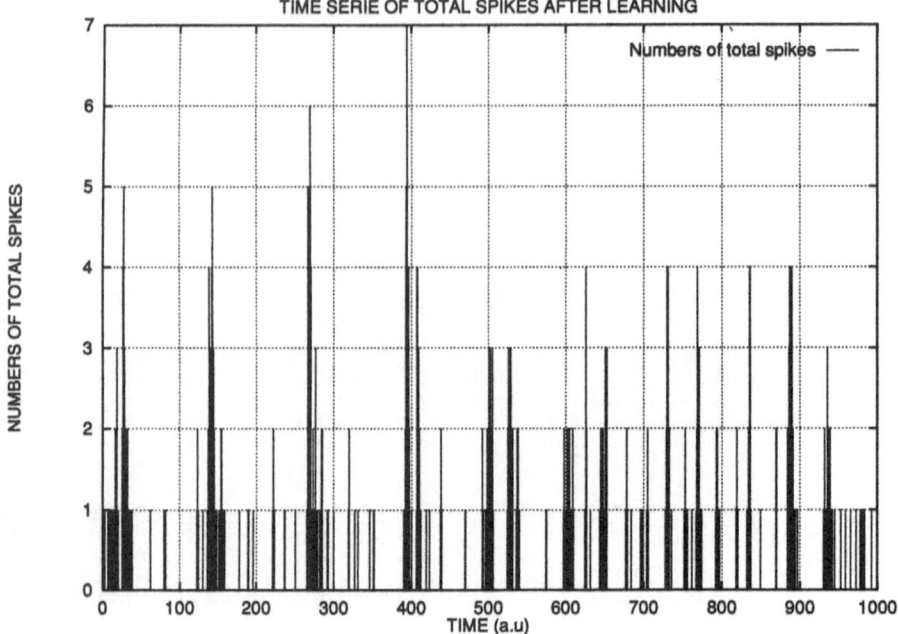

Fig. 3. *Time series of total spikes in the neural model after the learning process finishes.*

In the learning process we monitor the evolution of synaptic strengths by a measure of information entropy [7] in the connections as:

$$S(t) = -K \sum_{\epsilon_{ij}} Q_{\epsilon_{ij}}(t) \ln Q_{\epsilon_{ij}}(t) \tag{6}$$

being n the number of the units in the network, and K a positive constant. The learning process creates an internal organization of the network which is encoded in the synapses. The information entropy measures this organization of the synapses and, at time t we can quantify the level of synaptic organization by means of magnitude (6). A maximum of information entropy corresponds to the case where probabilities $Q_{\epsilon_{ij}}$ associated to ϵ_{ij} are uniform. A minimum of information entropy corresponds to case where probabilities $Q_{\epsilon_{ij}}$ associated to ϵ_{ij} are peaked around a specific value of ϵ_{ij}. Therefore at the beginning of the learning process the value of information entropy is maximum because probabilities $Q_{\epsilon_{ij}}$ are initialized with the values 0.5, and during the learning process the information entropy should decrease.

4 Simulation Results

We have done several simulations, and a case representative is given by the model of fifty neurons with the following parameters:

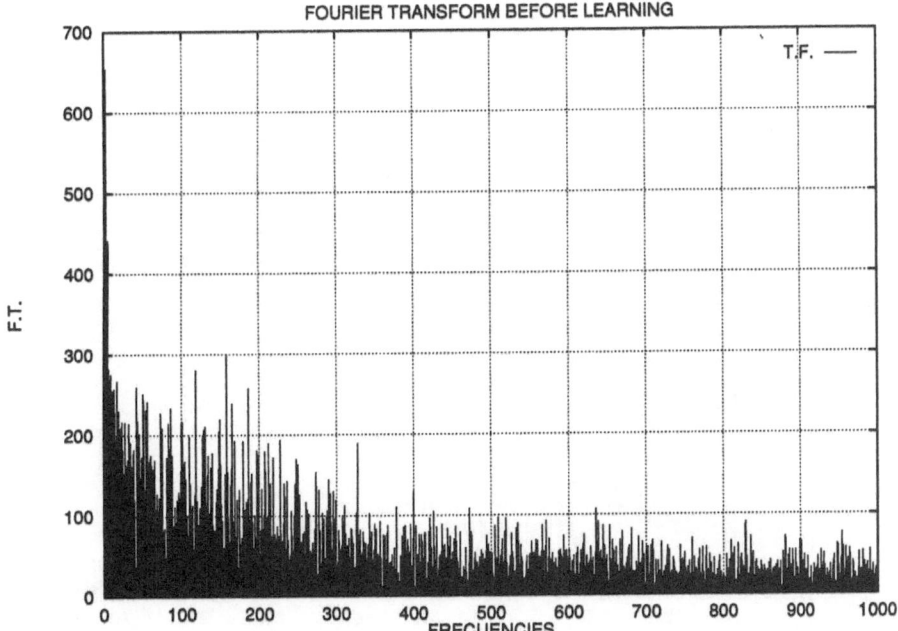

Fig. 4. *Fourier transform of the total spikes measured in the model before learning (Figure 2).*

$$- N_i = 400,\ p_i = 0.75,\ \Theta = 5,\ \eta_+ = 0.1,\ \eta_- = 0.001,\ \epsilon_{ij} \in [1, 20].$$

Results presented for this model are also representative of those obtained for the set of cases we have studied. We have applied the learning algorithm explained above to this network, without an external stimulus. We present in Figure 1 the information entropy evolution during the learning process. The decrease of the entropy means an increase of the internal organization of the neural network which is adapting its synaptic weights.

Neural firing is fundamental in the neural processing of information. We have analyzed the time series of total spikes in this network before and after the learning process. We define the magnitude of *total spikes* at time t, as the total number of spikes measured for the neural ensemble at time t. We have measured the time series of total spikes in the neural network , before and after the learning process, and the results are shown in Figures 2 and 3. It can be seen that spikes distribution is more regular after learning. A quantitative measure can be given with the use of the Fourier Transform.

We present in Figures 4 and 5 the Fourier Transform of time series of total spikes before (Figure 2) and after the learning process (Figure 3) respectively. Low frequency peaks in Figure 5 are the proof of some regularity that was not present before learning (Figure 4).

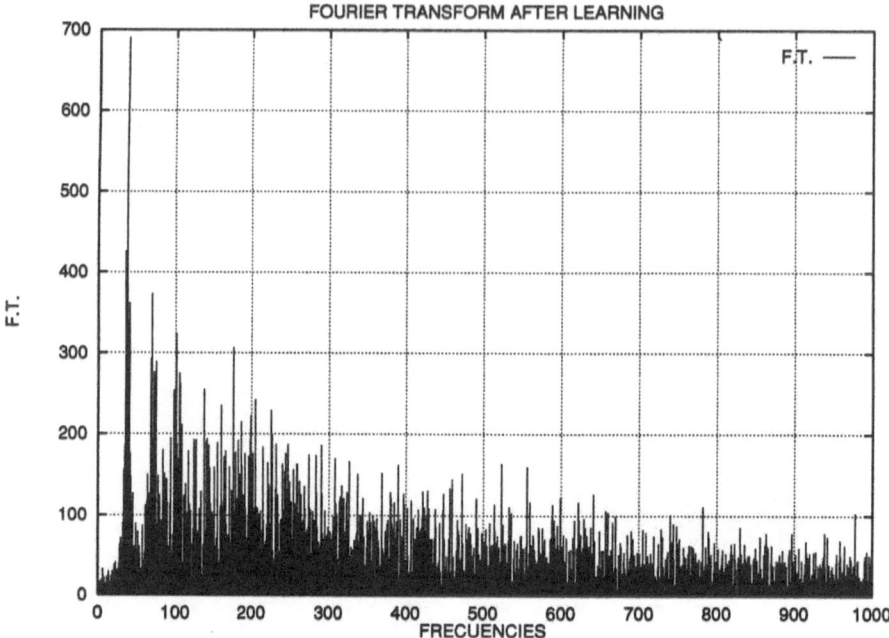

Fig. 5. *Fourier transform of the total spikes measured in the model after learning (Figure 3).*

An interesting result is the decreasing of the total number of spikes in the neural ensemble during the learning process. This result is shown in Figure 6.

5 Discussion

We have studied the effect of Hebbian learning in global properties of a neuronal population. From a general point of view a neural network can be understood as a population of units in which it takes place an adapting process of message interchange. Several are the cases in nature with the same type of interactions. For instance, cricket messages are sounds and the intensity is modulated as a result of individuals adaptation to the group singing pattern. A general property of these populations is that the ensemble behaviour is stable against internal and external sources of noise. For instance, a rhythmic singing pattern is rapidly achieved and long sustained in cricket populations (rhythmic firing is the analog in the case of fireflies ensembles) [8]. Isolated neuronal groups share several characteristics with the insect populations mentioned above. If adaptation of messages follows a hebbian law the result of our simulations indicates that during learning the system evolves towards a stable state. Moreover, during evolution : a) units fire with increasing regularity, b) unit production of messages decreases, c) the information needed for synchronization of activities is encoded in synaptic strengths.

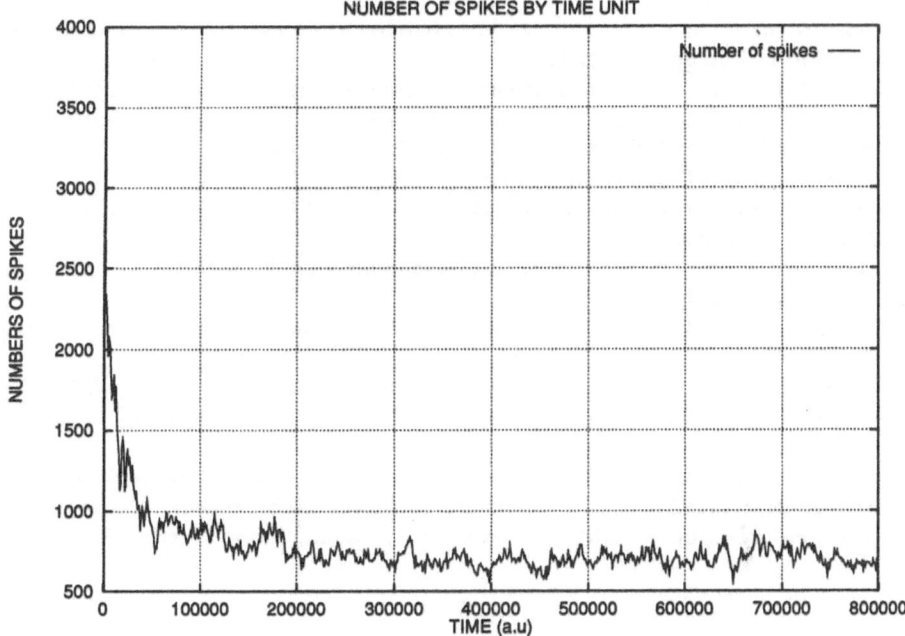

Fig. 6. *Evolution of number of total spikes during to learning process.*

These results are obtained for every model parameters for which we have performed simulations[1]. On the other hand, the model range is quite broad since the large amount of stochasticity included erases the relevance of several model details. General and relevant properties in the model are the noisy evolution of neural activities and synaptic strengths. Units are quite randomly firing in the absence of interactions and a random walk among possible synaptic strengths is only blocked by the efficiency of the hebbian learning process.

The neural ensemble we have studied does not include an "external" driving message. In this sense, it would be the model for an sleeping neural collectivity. None of the properties found for the model, synchronization, stability and economization, are distant from properties assumed in sleeping brains. Results for the case in which an external message drives the population activity will be published elsewhere.

References

1. D. O. Hebb. *"The Organization of Behavior"*. Wiley, New York, 1949.
2. MacGregor R. J. *"Neural and Brain Modeling"*. Academic Press (1989).

---

[1] An implementation of this model to easily study the parameter space is available for XSim Neural Network Simulator [9] (http://www.iic.uam.es/xsim)

3. V. López, J.A. Sigüenza, J.R. Dorronsoro y S. Carrillo-Menendez. *"Stochastic specificity in neural interaction"*. Proceedings ICANN 93 pp 196–199. Eds. S. Gielen and B. Kappen Springer Verlag. 1993.

4. Paul M. Hofman, Francisco B. Rodríguez, Juan A. Sigüenza, Vicente López, Santiago Carrillo-Menendez. *"A Simple Probabilistic Neural Model Producing Multimodal ISHs"* in LNCS 930 pp. 166-173 Mira-Sandoval (Eds.) From Natural to Artificial Neural Computation, Springer Verlag, IWANN95.

5. W. Feller *"An Introduction to Probability Theory and Its Applications"*. Vol. I, Ed. J. Wiley and Sons.

6. Maureen Caudill, Charles Butler. *"Undertanding Neural Networks: Computer Explorations"*. Vol. I, Ed: MIT Press.

7. Shannon, C.E. *"A Mathematical Theory of Communication"*, Bell System Tech. J. 27, 379, 623.

8. Renato E. Mirollo y Steven H. Strogatz. *"Synchronization of Pulse–Coupled Biological Oscillators"*. SIAM Journal on Applied Mathematics, Vol. 50, 6, pp. 1645-1662; December 1990.

9. Varona P., Sigüenza J. A., (1995). *"Introducing XSim: A Neural Network Simulator That Incorporates Biological Parameters"* in LNCS 930 pp. 650-657 Mira-Sandoval (Eds.) From Natural to Artificial Neural Computation, Springer Verlag, IWANN95.

Stochastic Approximation Techniques and Circuits and Systems Associated Tools for Neural Network Optimization

H. Dedieu*, A. Flanagan**, A. Robert*

*E.P.F.L., Electrical Eng. Department, Circuits and Systems Group - CIRC
**E.P.F.L., Micro-Eng. Department, Institute of Micro-Systems - IMS
herve.dedieu@circ.de.epfl.ch, fax : + 41 21 693 67 00

Abstract

This paper is devoted to the optimization of feedforward and feedback Artificial Neural Networks (ANN) working in supervised learning mode. We describe in a general way how it is possible to derive first and second order stochastic approximation methods that provide learning capabilities. We show how certain variables, the sensitivities of the ANN outputs, play a key role in the ANN optimization process. Then we describe how some useful and elementary tools known in circuit theory can be used to compute these sensitivities with a low computational cost. We show by example how to apply these two sets of complementary tools, *i.e.* stochastic approximation and sensitivity theory.

Keywords: Artificial Neural Networks, Stochastic Approximation, Sequential Parameter Estimation, Adaptive Systems, Sensitivity Theory

1   Introduction

This paper is mainly devoted to the optimization of feedforward and feedback Artificial Neural Networks (ANN) working in supervised learning mode.

Although there are many kinds of networks and a huge collection of learning methods we will focus our attention on the *stochastic approximation method* which provides a powerful tool for the on line optimization of most of the ANNs used in practice.

As we shall see in the sequel there is one particular set of parameters which is of paramount importance for the stochastic approximation approach. This particular set is given by the sensitivities of the ANN outputs, *i.e.* the derivatives of the ANN outputs with respect to the ANN weights. The *theory of sensitivity* in dynamical systems has been studied for many years particularly by circuit theorists for different purposes ranging from identification of worst case situations to standard optimization of systems. One of the challenging issues which has received a lot of attention is the low-cost computation of the sensitivities. Powerful tools such as Tellegen's theorem have allowed the circuit theorists to perform efficient circuit optimization. Although well known in the circuit community the 'sensitivity tools' have not been clearly identified by the ANN and the signal processing communities (systematic tools for sensitivity computation are totally absent from ANN textbooks).

It is our intention in this paper to merge the stochastic approximation approach (which requires sensitivity computation) and some elementary sensitivity theory in order to show how these two kinds of tools are complementary and offer a theoretical framework for the optimization process of most of the well known ANN algorithms. Although this view is not completely partial since other different methods could be used, it is our experience that in many practical situations an ANN problem can be solved in this framework. Moreover it is also our experience that this framework can be successfully used for optimization of adaptive systems[6], on-line system identification and on-line optimization of circuits[5].

This paper is divided in three main parts. We will show how to derive first and second order stochastic approximation methods. Next we will give some elementary tools for sensitivity computation. Afterwards we will use the two different tools on a particular example of a backpropagation algorithm.

2 Stochastic approximation

2.1 Introduction

Let an ANN defined as

$$y(n) = f\left(u(n),\ x(n),\ \theta(n-1)\right) \tag{1}$$

where $y(n)$ is the M dimensional ANN output vector at time n, $u(n)$ is the ANN input vector, $x(n)$ is the state space of the ANN and $\theta(n-1)$ is the N dimensional parameter vector computed at time $n-1$.

Our goal is to find $\theta(n)$ in such a way that $y(n)$ matches a desired M dimensional vector $d(n)$ in some statistical sense. In general a Least Mean Squares criterion of the form

$$V(\theta) = E\left[\frac{1}{2}e^T(n)e(n)\right] \tag{2}$$

is chosen where

$$e(n) = d(n) - f\left(u(n),\ x(n),\ \theta\right) \tag{3}$$

The goal is therefore to find θ that minimizes (2). A local minimum of $V(\theta)$ is located at θ^* such that

$$\frac{\partial V(\theta)}{\partial \theta} = -E\left[\frac{\partial y(n)}{\partial \theta}e(n)\right] = 0 \tag{4}$$

Unfortunately the solution of (4) would require the probability distributions of all the pairs $(e_i(n), \frac{\partial y_i(n)}{\partial \theta_j})$ for $i = 1 \ldots M$ and $j = 1 \ldots N$, these probability distributions are not known beforehand.

The underlying idea behind stochastic approximation (which has been developed as a special branch of the sequential parameter estimation in the statistical literature) is to build a sequential estimation of $\theta(n)$ in such a way that

$$\lim_{n \to \infty} \theta(n) = \theta^* \tag{5}$$

where θ^* is some local minimum of $V(\theta)$.

2.2 First order methods (gradient methods)

One simple method to find a sequential estimate is to move the current vector of parameters in the opposite direction to the gradient, *i.e.*

$$\boldsymbol{\theta}(n) = \boldsymbol{\theta}(n-1) - \mu(n)\nabla_{\boldsymbol{\theta}}V\left[\boldsymbol{\theta} = \boldsymbol{\theta}(n-1)\right] \qquad (6)$$

where $\mu(n)$ is a small positive gain chosen in a convenient way. Equation (6) can be rewritten

$$\boldsymbol{\theta}(n) = \boldsymbol{\theta}(n-1) + \mu(n)E\left[\frac{\partial \boldsymbol{y}(n)}{\partial \boldsymbol{\theta}}e(n)\right] \qquad (7)$$

How to compute $E\left[\frac{\partial \boldsymbol{y}(n)}{\partial \boldsymbol{\theta}}e(n)\right]$? One solution first proposed by Robbins and Monro is that for a $\mu(n)$ chosen small enough (in the Robbins and Monro scheme $\mu(n)$ is a sequence of positive scalars that tends to zero) the expectation can be removed in (7). In other terms, if $\boldsymbol{\theta}$ is slowly time variant, 'on the average' the adjustments are made in the negative gradient direction. Hence the name stochastic approximation.

The summary of the stochastic gradient algorithm can be given as follows :

1. At the current estimation n compute the ANN outputs $\boldsymbol{y}(n)$ using the former parameters, *i.e.* $\boldsymbol{y}(n) = f\left(\boldsymbol{u}(n),\ \boldsymbol{x}(n),\ \boldsymbol{\theta}(n-1)\right)$
2. Compute the error of estimation $e(n) = \boldsymbol{d}(n) - f\left(\boldsymbol{u}(n),\ \boldsymbol{x}(n),\ \boldsymbol{\theta}\right)$
3. Compute the derivative of $\boldsymbol{y}(n)$ with respect to the N parameters, *i.e*

$$\psi(n) = \frac{\partial \boldsymbol{y}(n)}{\partial \boldsymbol{\theta}} = \begin{bmatrix} \frac{\partial y_1(n)}{\partial \theta_1} & \frac{\partial y_2(n)}{\partial \theta_1} & \cdots & \frac{\partial y_M(n)}{\partial \theta_1} \\ \frac{\partial y_1(n)}{\partial \theta_2} & \frac{\partial y_2(n)}{\partial \theta_2} & \cdots & \frac{\partial y_M(n)}{\partial \theta_2} \\ \cdots & \cdots & \cdots & \cdots \\ \cdots & \cdots & \cdots & \cdots \\ \frac{\partial y_1(n)}{\partial \theta_N} & \frac{\partial y_2(n)}{\partial \theta_N} & \cdots & \frac{\partial y_M(n)}{\partial \theta_N} \end{bmatrix}$$

4. Update the former parameters according to

$$\boldsymbol{\theta}(n) = \boldsymbol{\theta}(n-1) + \mu(n)\psi(n)e(n) \qquad (8)$$

Observe how the output sensitivities (the reaction of the outputs to an infinitesimal small change of the parameters) play a key role in the ANN on-line optimization.

2.3 Second order methods (quasi-Newton methods)

One major drawback with the Stochastic gradient method is its slow convergence rate. The reason for this poor performance is obvious, on average the method gives the direction to the nearest local minimum but nothing is said to the distance (or the average distance) to the nearest local optimum. Second order methods provide fast convergence rates because in essence they compute at each iteration either the direction and the distance to the nearest local optimum.

The underlying idea behind second order methods is to make a second order Taylor expansion around the current vector of parameters $\theta(n-1)$, *i.e.*

$$
\begin{aligned}
V(\theta(n-1)+\Delta\theta) = {} & V(\theta(n-1)) \\
& + \frac{\partial V(\theta=\theta(n-1))}{\partial\theta}\Delta\theta \\
& + \frac{1}{2}\Delta\theta^T \frac{\partial^2 V(\theta=\theta(n-1))}{\partial\theta^2}\Delta\theta
\end{aligned}
\tag{9}
$$

and to choose an update $\Delta\theta^*$ which provides the best descent for the criterion. This best descent is achieved when choosing a $\Delta\theta^*$ which ensures that the derivation of the right side of equation (9) is set to 0. Setting to 0 this derivative one obtains the best update

$$
\Delta\theta^* = \theta(n) - \theta(n-1) = -\left[\frac{\partial^2 V(\theta=\theta(n-1))}{\partial\theta^2}\right]^{-1}\frac{\partial V(\theta=\theta(n-1))}{\partial\theta}
\tag{10}
$$

which gives the so-called *Newton method*

$$
\theta(n) = \theta(n-1) - \left[\frac{\partial^2 V(\theta=\theta(n-1))}{\partial\theta^2}\right]^{-1}\frac{\partial V(\theta=\theta(n-1))}{\partial\theta}
\tag{11}
$$

It is worth noting that taking $\Delta\theta^*$ as an update we obtain for the value of the criterion at iteration n (by inserting (10) into (9))

$$
\begin{aligned}
V(\theta(n)) = {} & V(\theta(n-1)) \\
& - \frac{\partial V(\theta=\theta(n-1))}{\partial\theta}^T\left[\frac{\partial^2 V(\theta=\theta(n-1))}{\partial\theta^2}\right]^{-1}\frac{\partial V(\theta=\theta(n-1))}{\partial\theta}
\end{aligned}
\tag{12}
$$

Equation (12) has a great consequence in practice since it shows us that if $H(n) = \frac{\partial^2 V(\theta=\theta(n-1))}{\partial\theta^2}$ (H is the so-called Hessian matrix) is a positive-definite matrix we are sure that $V(\theta(n))$ will point downhill with respect to $V(\theta(n-1))$. Close to the optimum the quadratic approximation (9) is generally a good model of V and the Hessian is therefore a positive definite matrix near the optimum since V has a bowl shaped form (upward concavity). The quadratic model is in general not valid far from the local minimum, this is why it is often preferred to replace the Hessian by a *guaranteed semi positive definite approximation*. The methods in which the Hessian is forced to be semi definite positive are referred to as *quasi-Newton methods*.

Quasi-Newton methods ensure that even far from the optimum the parameter search is carried out 'downhill'.

$$
H(n) = E\left[\frac{1}{2}\frac{\partial^2 e(n)e^T(n)}{\partial\theta^2}\right] = E\left[\frac{\partial e(n)}{\partial\theta}\frac{\partial^T e(n)}{\partial\theta} + e(n)\frac{\partial^2 e(n)}{\partial\theta^2}\right]
\tag{13}
$$

is usually replaced by the semi positive definite approximation

$$\hat{H}(n) = E\left[\frac{\partial e(n)}{\partial \theta}\frac{\partial^T e(n)}{\partial \theta}\right] = E\left[\psi(n)\psi^T(n)\right] \tag{14}$$

This choice has also the advantage that only first order derivatives (*i.e.* sensitivities) are required. The so-called *Gauss-Newton method* obeys the following equation

$$\theta(n) = \theta(n-1) + \hat{H}^{-1}(n)E\left[e(n)\psi(n)\right] \tag{15}$$

Computing equation (15) would require two kinds of expectations for which the correspondent probability distributions are not available. This is why as in the Robbins and Monro scheme it is proposed to remove the expectations in (15). The stochastic equivalent of equation (15) is then given by

$$\theta(n) = \theta(n-1) + \hat{H}^{-1}(n)e(n)\psi(n) \tag{16}$$

If $\theta(n)$ is slowly time variant the recursivity of the algorithm will provide in average a good estimate of $E\left[e(n)\psi(n)\right]$. One problem left is to find an estimate of $\hat{H}(n)$ since this quantity is also an expectation. One choice often made is to average on a time window the product $\psi(n)\psi^T(n)$ giving more weight to the current and recent products and discarding (forgetting) the older estimate. For instance the estimation

$$\hat{H}(n) = \lambda\hat{H}(n-1) + \psi(n)\psi^T(n) \tag{17}$$

where λ is a constant close to 1 but less than 1 is often chosen as a good and convenient recursive estimate. The factor λ is referred to as *the forgetting factor*. This factor fixes the tradeoff between the rate of convergence and the precision of the algorithm during the steady state. The closer is λ to 1 the slower is the convergence but the better is the precision (variance of θ during the steady state). In summary second order methods work as follows:

1. At the current estimation n compute the ANN outputs $y(n)$ using the former parameters, *i.e.* $y(n) = f\left(u(n),\ x(n),\ \theta(n-1)\right)$
2. Compute the error of estimation $e(n) = d(n) - f\left(u(n),\ x(n),\ \theta\right)$
3. Compute the derivative of $y(n)$ with respect to the N parameters, *i.e*

$$\psi(n) = \frac{\partial y(n)}{\partial \theta} = \begin{bmatrix} \frac{\partial y_1(n)}{\partial \theta_1} & \frac{\partial y_2(n)}{\partial \theta_1} & \cdots & \frac{\partial y_M(n)}{\partial \theta_1} \\ \frac{\partial y_1(n)}{\partial \theta_2} & \frac{\partial y_2(n)}{\partial \theta_2} & \cdots & \frac{\partial y_M(n)}{\partial \theta_2} \\ \cdots & \cdots & \cdots & \cdots \\ \cdots & \cdots & \cdots & \cdots \\ \frac{\partial y_1(n)}{\partial \theta_N} & \frac{\partial y_2(n)}{\partial \theta_N} & \cdots & \frac{\partial y_M(n)}{\partial \theta_N} \end{bmatrix}$$

4. Update the semi definite Hessian estimate, i.e.

$$\hat{H}(n) = \lambda\hat{H}(n-1) + \psi(n)\psi^T(n) \qquad \hat{H}(0) = \Gamma I \tag{18}$$

5. Update the former parameter according to

$$\theta(n) = \theta(n-1) + \hat{H}^{-1}(n)\psi(n)e(n) \tag{19}$$

Once again observe how the output sensitivities (vector $\psi(n)$ play a key role in the ANN on-line optimization.

3 Elementary sensitivity tools

The aim of this part is to precisely describe the tools which allow the computation of the sensitivities with a low computational cost. When dealing with sensitivities the first idea to come to mind is their calculation using small incremental parameter changes i.e. if the ith parameter of the network θ_i is one of the parameters the influence of which is to be measured with respect t o the output $y_k(n)$ then compute $y_k(n)$ using θ_i, then compute $\hat{y}_k(n)$ using a duplicate net in which the ith parameter θ_i has been incremented by ϵ. The desired sensitivity is then approximated by,

$$\frac{\partial y_k(n)}{\partial \theta_k} = \frac{\hat{y}_k(n) - y_k(n)}{\epsilon} \tag{20}$$

This mode of operation has two considerable drawbacks.

- We obtain only approximate sensitivities and the choice of the correct value for ϵ could be parameter dependent.
- The sensitivities are obtained by duplicating the ANN network as many times as we have parameters. This second drawback rules out this approach for ANNs.

We now present in the following the sensitivity tools which allow the computation of the parameter sensitivities using only one auxilliary network which is called the adjoint network. Furthermore this adjoint network is linearised. Therefore the sensitivity computation is greatly simplified.

3.1 General definitions

Definition 3.1.1 A linear network is one made up of a combination of adders, multipliers and delays. Except source nodes (which are fed with input signals) and sink nodes (which deliver output signals), any node in the network has at least one ingoing branch and at least one outgoing branch.

Definition 3.1.2 A node is either a summation node or a branch node. A **signal at a node** is a signal which is the sum of the signals of the ingoing branches to this node.

Definition 3.1.3 A transposed network of a linear network is built by reversing the signal flow of the original linear network and by replacing, each branch node with a summation node, each summation node with a branch node. Thus inputs and outputs of the transposed network correspond respectively to the outputs and inputs of the original network.

3.2 Linear Network sensitivity properties

Property 3.2.1 *For a linear network which has P input signals $u_k(n)$ $k = 1 \ldots P$ and M output signals $y_j(n)$ $j = 1 \ldots M$, the sensitivity $\frac{\partial Y_j(z)}{\partial w_i}$ of the jth output is given by*

$$\frac{\partial Y_j(z)}{\partial w_i} = T_{y'_j x'_i}^{Adj}(z) \, X_i(z) \tag{21}$$

where $X_i(z)$ is the z transfer of the signal available at the input node of the multiplier w_i. $T_{y'_j x'_i}^{Adj}(z)$ describes the transfer function of the transposed network between the network input node y'_j (source node) and the input node of the multiplier w_i.

3.3 Nonlinear Network sensitivity properties

We will restrict our analysis to Nonlinear Networks which posses memoryless nonlinear functions. These nonlinear functions are furthermore supposed to be differentiable with respect to their inputs. In this case the adjoint of the Nonlinear Network is simply the adjoint of the linearized network at the operating point. We will now state the alter ego property to property 3.2.1.

Property 3.3.1 *For a nonlinear network which has P input signals $u_k(n)$ $k = 1 \ldots P$ and M output signals $y_j(n)$ $j = 1 \ldots M$, the sensivity $\frac{\partial Y_j(z)}{\partial w_i}$ of the jth output is given by*

$$\frac{\partial Y_j(z)}{\partial w_i} = T_{y'_j x'_i}^{Adj}(z) \, X_i(z) \tag{22}$$

where $X_i(z)$ is the z transfer of the signal available at the input node of the multiplier w_i. $T_{y'_j x'_i}^{Adj}(z)$ describes the transfer function of the transpose of the linearized network around the operating point between the network input node y'_j (source node) and the input node of the multiplier w_i. Practically this adjoint is built by reversing the signal flow of the original linear network and by replacing, each branch node with a summation node, each summation node with a branch node, furthermore (in the adjoint) the nonlinear functions are replaced by multipliers whose values are set to the derivative of the nonlinear functions at their operating points.

4 Application to ANN optimization

4.1 General concept

Property 3.3.1 shows that the sensitivities can be computed using only two networks, the original nonlinear network and the adjoint of the linearized network around the operating point. Property 3.3.1 is then applied to compute the sensitivities ($\psi(n)$ matrix) at low computational cost. Once $\psi(n)$ has been computed it can be injected either in (8) (first order stochastic methods) or in (19) (second order stochastic methods) to perform the ANN on-line optimization.

4.2 Example

As an example we show how to derive efficiently the celebrated backpropagation algorithm for the Multilayer perceptron (MLP). In its general form the MLP with n_s hidden layers obey the following equations

$$u_j^{[s]} = \sum_{i=0}^{n_s-1} w_{ji}^{[s]} o_i^{[s-1]} \qquad j = 1 \ldots N^{[s]} \tag{23}$$

$$o_j^{[s]} = \psi_j^{[s]} \left(u_j^{[s]} \right) \qquad s > 0, \quad j = 1 \ldots N^{[s]} \tag{24}$$

$$o_j^{[0]} = u_j \quad j = 1 \ldots N^{[0]} \qquad and \qquad y_j = o_j^{[n_s]} \quad j = 1 \ldots N^{[n_s]} \tag{25}$$

We want to derive the stochastic gradient method which governs the MLP optimization and therefore we have to be able to compute the product $\psi(n)e(n)$ in equation (8), *i.e.*

$$w_{ji}^{[s]}(n) = w_{ji}^{[s]}(n-1) + \mu \sum_{k=1}^{M} \frac{\partial y_k(n)}{\partial w_{ji}} e_k(n) \tag{26}$$

Let us denote

$$\Delta w_{ji}^{[s]}(n) = \mu \sum_{k=1}^{M} \frac{\partial y_k(n)}{\partial w_{ji}} e_k(n) \tag{27}$$

By passing in the z domain and taking into account the property 3.3.1 we obtain

$$\Delta W_{ji}^{[s]}(z) = \mu \sum_{k=1}^{M} T^{Adj}_{y'_k w_{ji}^{[s]}} O_i^{[s-1]}(z) * E_k(z) \tag{28}$$

where the dependence of $T^{Adj}_{y'_k w_{ji}^{[s]}}$ with respect to z has been dropped (since the MLP is a static network, $T^{Adj}_{y'_k w_{ji}^{[s]}}$ is a simple constant). The equation (28) can be then rewritten

$$\Delta W_{ji}^{[s]}(z) = \mu O_i^{[s-1]}(z) * \sum_{k=1}^{M} T^{Adj}_{y'_k w_{ji}^{[s]}} E_k(z) \tag{29}$$

By going back in the time domain we find that

$$\Delta W_{ji}^{[s]}(z) = \mu o_i^{[s-1]}(n) \delta_i^{[s]}(n) \tag{30}$$

where $\delta_i^{[s]}(n)$ is simply the ingoing signal of the w_{ji} coefficient of the adjoint network whose inputs are fed with the error signal $e_1(n), \ldots e_M(n)$. As an example the network architecture of a two layers MLP with its adjoint network is given in figure 1.

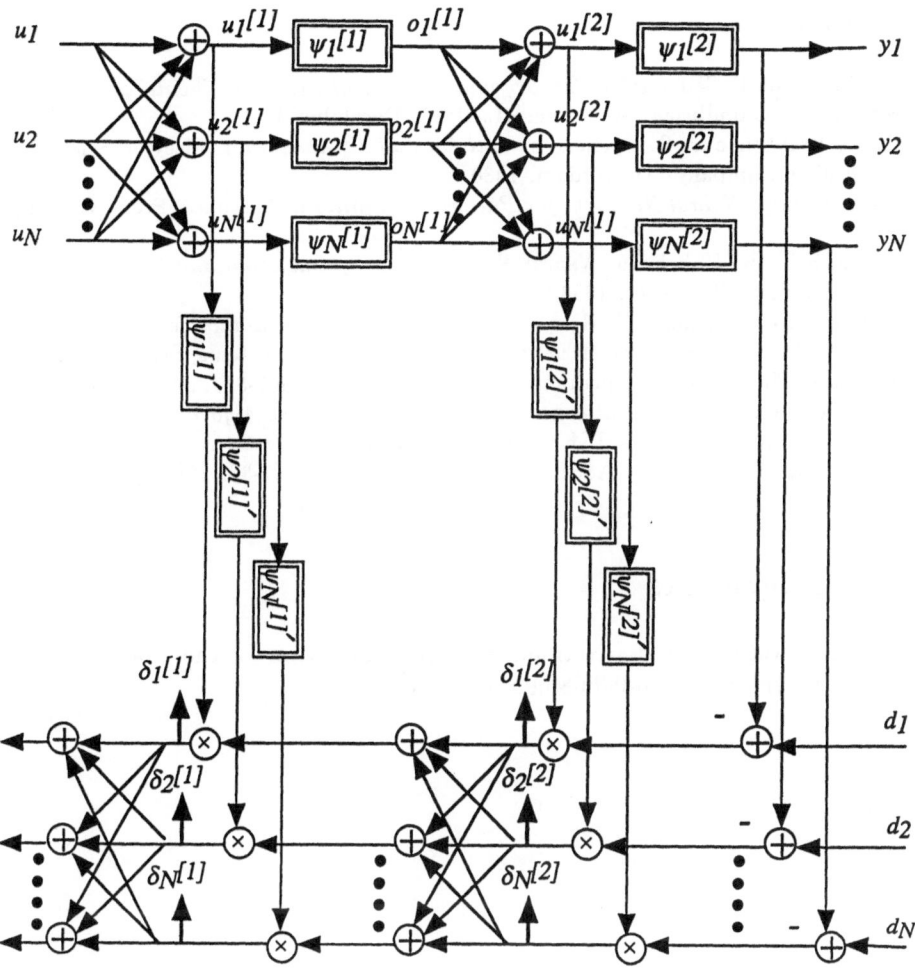

Fig. 1. A two layers MLP with its transpose network computing the MLP updates

5   Conclusion

In this paper we propose a systematic and efficient method to derive learning algorithms for a wide class of ANN. This method is particularly suited for the stochastic approximation technique in which it is of paramount importance to compute the 'senstivities'. It is shown how to use conventional Circuits and Systems tools for the computation of the sensitivities at low computational cost.

References

1. L. Ljung and T. Söderström, *Theory and Practice of Recursive Identification.* The MIT Press, Camdbrige, Massachusetts, 1987. December 1968.
2. R. K. Brayton and R. Spence, *Sensitivity and Optimization.* Elsevier Scientific Publishing Company, Amsterdam, 1980.
3. C. M. Bishop, *Neural Networks for Pattern Recognition.* Clarendon Press, Oxford, 1995.
4. A. Cichocki, R. Unbehaen, *Neural Networks for Optimization and Signal Processing.* J. Wiley, Chichester, 1993.
5. H. Dedieu, C. Dehollain, J. Neirynck, G. Rhodes, *A New Method for Solving Broadband Matching Problems.* IEEE Trans. on Circuits and Systems, Systems-I: Fundamental Theory and Applications, Vol. 41, NO. 9, pp. 561-571, Sept. 1994.
6. H. Dedieu, O. Chételat, *Automatic Derivation of Adaptive Algorithms for a Large Class of Filter Structures.* IEEE International Conference on Acoustics, Speech, and Signal Processing, ICASSP'93, Minneapolis, April 27-30, 1993, pp. III.476-III.479

Acknowledgements

Thanks are due to the Swiss National Science Foundation for the financial support given by the FN 2150-045689.95/1 project.

Recursive Hetero-Associative Memories for Translation

Mikel L. Forcada and Ramón P. Ñeco

Departament de Llenguatges i Sistemes Informàtics,
Universitat d'Alacant,
E-03071 Alacant, Spain.
E-mail: {mlf,neco}@dlsi.ua.es

Abstract. This paper presents a modification of Pollack's RAAM (Recursive Auto-Associative Memory), called a Recursive Hetero-Associative Memory (RHAM), and shows that it is capable of learning simple translation tasks, by building a state-space representation of each input string and unfolding it to obtain the corresponding output string. RHAM-based translators are computationally more powerful and easier to train than their corresponding double-RAAM counterparts in the literature.

1 Introduction

Recently, Pollack (1990) introduced a new connectionist model called the recursive auto-associative memory (RAAM). This model is capable of obtaining compact representations for compositional structures such as trees and lists, *i.e.*, is capable of representing variable-sized recursive data structures. This architecture can indeed be viewed simply as a distributed memory for storing compositional data structures. In particular, it was shown that properly trained RAAM are capable of building representations for strings acting much in the same way as a stack (strings are pushed and may then be popped).

More recently, Chalmers (1990) and Chrisman (1991) used the RAAM formalism to learn simple syntactic transformations and translations. They show how RAAM learn to perform computations directly with the distributed representations obtained by them, without accessing their compositional structure (*holistic* computations). Chrisman (1991) used a (double, stack-like) RAAM architecture in the domain of natural language translation (a small English ↔ Spanish translation task). Chrisman distinguishes two possible ways to perform the translations: *transformational* and *confluent*. In the transformational translations, the auto-association of the sentence and its translation are used to learn separate representations *before* learning the translation function between these representations. In the confluent approach, the network is trained to obtain the *same* internal representation for the sentence and its translation, *i.e.*, the original and final sentences should have identical internal representations, and they may have two different decodings, one corresponding to the original sentence, and the other to its translation. This implies that the network must be trained to auto-associate the two sentences by learning two different encoding-decoding mechanisms.

The translations performed by these models have some limitations. In particular, the tasks used by Chrisman (1991) are limited to one-to-one translations; that is, no two different sentences in one language can have the same translation in the other language. Chalmers' (1991) translators may learn many-to-one translations, but the fact that each input sentence has a unique representation forces the translation function to perform the many-to-one mapping on its own.

We propose a new model, called the RHAM (Recursive Hetero-Associative Memory), which has a computational power which is equivalent or superior to that of the RAAMs used by Chalmers (1990) and Chrisman (1991), and may be applied to learn general translations from examples in which different sentences may have the same translation. The main difference between the RAAM models used so far for translation and the new model is that in the new learning algorithm the network does not need to learn to auto-associate the input and output strings separately; we only train the network to obtain a suitable internal representation of the input, and to decode this representation to obtain the corresponding output string.

2 Definitions

We will study input strings over an input alphabet Σ; the set of all finite-length strings over Σ will be denoted Σ^*. The output strings (translations) are strings over an output alphabet Δ, that is, strings in Δ^*. The class of translations we will consider are those that may be represented by a function (an application)

$$\tau : \Sigma^* \to \Delta^*, \tag{1}$$

so that there is a single translation for each string in Σ^* but two different strings in Σ^* may have the same translation.

In this paper we consider translations that may be performed by *Mealy machines* and more general *deterministic generalized sequential machines* (DGSM) (Hopcroft and Ullman 1979, Salomaa 1987). A Mealy machine is a six-tuple $M = (Q, \Sigma, \Delta, \delta, \lambda, q_1)$, where Q is a finite set of states, Σ a finite set of input symbols (*input alphabet*), Δ a finite set of output symbols (*output alphabet*), $\delta : Q \times \Sigma \to Q$ the next-state function, $\lambda : Q \times \Sigma \to \Delta$ the output function, and $q_1 \in Q$ the initial state. The class of translations that a Mealy machine may perform is limited: for example, input strings and their translations always have the same length (Salomaa (1987) calls these translations length-preserving). A deterministic generalized sequential machine may be formulated as a Mealy machine in which the next-state function is $\delta : Q \times (\Sigma \cup \{\epsilon\}) \to Q$, and the output function is $\lambda : Q \times (\Sigma \cup \{\epsilon\}) \to \Delta \cup \{\epsilon\}$, where ϵ represents the empty string. In this machine, states can have transitions defined either on the empty string or on input symbols but not on both; in addition, for the machine to halt always, it is required that no state be reachable from itself using only transitions with ϵ as input.

3   The RHAM Architecture

A RHAM, as a RAAM, consists basically of two feedforward neural networks, which we will call the *encoder* and the *decoder*. The original RAAM model can encode arbitrary tree structures, but we only need the sequential or "stack" version of Pollack's (1990) RAAM: that is, strings are encoded symbol by symbol. However, the experiments shown in this paper could be generalized to perform translations using the general RAAM version between trees (in this case we would have the restriction that the number of units used to represent a terminal symbol must be equal to the number of hidden neurons used to represent the tree). In the case of the sequential version, where we are basically encoding structures such as lists and stacks, this restriction is not necessary.

In our model each input string $w \in \Sigma^*$ has a vector representation in $[0, 1]^N$ where N is the number of hidden units in the hidden layer. A function

$$\mathbf{r} : \Sigma^* \rightarrow [0, 1]^N \tag{2}$$

assigns a (not necessarily distinct) vector representation to each string.

Let $\mathbf{r}_0$ be the representation of the empty string ϵ or "nil":

$$\mathbf{r}(\epsilon) = \mathbf{r}_0; \tag{3}$$

then, the representation of any string wa ($w \in \Sigma^*$, $a \in \Sigma$) is

$$\mathbf{r}(wa) = \text{encode}\,(\mathbf{r}(w), \mathbf{u}_a) \tag{4}$$

where $\mathbf{u}_a$ is a vector representing symbol a, and

$$\text{encode} : [0, 1]^N \times [0, 1]^K \rightarrow [0, 1]^N, \tag{5}$$

is the encoding function which has a simple connectionist implementation; $\mathbf{r}[t] = \text{encode}(\mathbf{r}[t-1], \mathbf{u}[t])$ is realized as a simple perceptron:

$$r_i[t] = g\left(\sum_{j=1}^{N} W_{ij}^{rr} r_j[t-1] + \sum_{k=1}^{K} W_{ik}^{ru} u_k[t] + W_i^r \right) \tag{6}$$

where the W's represent weights and biases and $g(x) = 1/(1 + \exp(-x))$; that is, a single-layer feedforward neural network having $N + K$ input units and N output units, and therefore $(N + K + 1)N$ different parameters.

With the recursive function $\mathbf{r}$, a string w of length L_w can be encoded by placing a representation $\mathbf{u}[1]$ for the first symbol of the string in the left-hand K units of the input, and a representation for the empty string, $\mathbf{r}(\epsilon) = \mathbf{r}[0]$, on the right-hand N units of the input. Then, the encoder produces a internal representation for a string consisting of the first symbol on the N hidden units. The activations of this hidden units are then copied recursively to the rightmost N units of the input and the representation $\mathbf{u}[2]$ of the second symbol of the string is placed on the left-hand K units; the process is repeated until all the symbols of the string are processed and a representation $\mathbf{r}[L_w] = \mathbf{r}(w)$ for the whole string is obtained.

The decoder, when properly trained, unfolds the representations of input strings to obtain the corresponding output strings. States in $[0,1]^N$ are also interpreted as representations of output strings; and the input representation $\mathbf{r}(w)$ of string w is taken to be the output representation $\mathbf{s}(x)$ of its translation $x = \tau(w)$.

The output representation function $\mathbf{s}$ is analogous to the input representation function $\mathbf{r}$:

$$\mathbf{s} : \Delta^* \to [0,1]^N. \tag{7}$$

The decoder works as follows: the representation $\mathbf{y}_b$ of a symbol b from the output alphabet Δ is popped from the representation $\mathbf{s}(xb)$ of string xb and the representation $\mathbf{s}(x)$ of string x is simultaneously produced. Popping of a complete string ends when a special representation $\mathbf{s}_0$, representing the empty string over the output alphabet, appears in the hidden layer.

The implementation of the decoder is also a perceptron with N input units and $N + M$ output units, having therefore $(N + M)(N + 1)$ different parameters (weights and biases). The first N outputs correspond to the representation of the rest of the output:

$$s_j[t - 1] = g \left(\sum_{i=1}^{N} W_{ji}^{ss} s_i[t] + W_j^s \right), \tag{8}$$

and the remaining M outputs to the representation y of the popped output symbol

$$y_k[t] = g \left(\sum_{i=1}^{N} W_{ki}^{ys} s_i[t] + W_j^y \right). \tag{9}$$

If we have the internal representation of an input string encoded in the hidden units, then we can obtain the translation by repeating the decoding process until the end of the string is detected (the representation $\mathbf{s}_0$). In the first iteration, the leftmost M units of the output return the local representation $\mathbf{y}[L_x]$ for the last symbol of the string x, while the rightmost N units return the representation for the remainder of the string.

The representation $\mathbf{y}_b \in [0,1]^M$ of each output symbol b and the representation of the empty output string $\mathbf{s}_0 \in [0,1]$ may be chosen in advance but might also be learned during the learning. The representation of the empty input string $\mathbf{r}[0] \in [0,1]^N$ may also be chosen or learned.

Figure 1 shows schematically the architecture of a RHAM. In the example the encoder reads the input "1011", symbol by symbol, to obtain a representation of the input sentence $\mathbf{r}(\text{"1011"})$. This representation is also the representation of the translation of this sentence, $\mathbf{s}(\text{"001"})$. The representations of the symbols of the translated sentence are popped by the decoder: first symbol '1', second, symbol '0', and then, symbol '0' and in the right side of the decoder's output, the

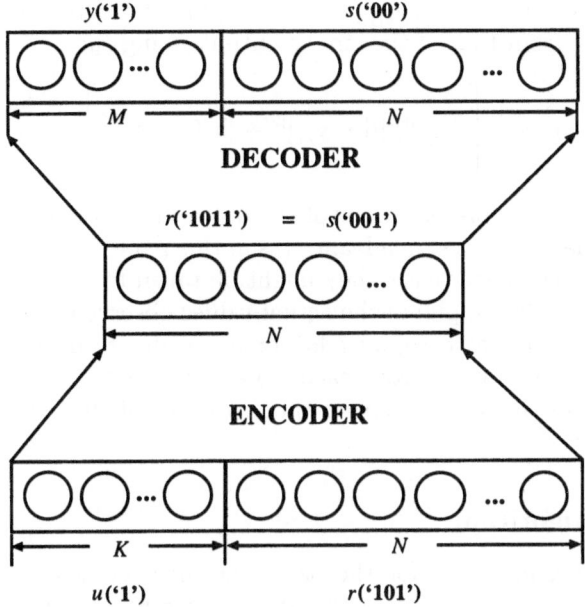

Fig. 1. Architecture of a RHAM performing one step of the translation of sentence "1011" into sentence "001".

representation for the empty output string (s_0), indicating that the translation has finished. Note that if $\tau(x) = x$, $\forall x$, $\Delta = \Sigma$ and $r_0 = s_0$, this is Pollack's (1990) RAAM acting as a stack.

4 Training

4.1 Error Function

We train the network by minimizing an error function for a training sample S containing translations t from $\Sigma^* \times \Delta^*$ which are compatible with the existence of a single-valued function $\tau : \Sigma^* \to \Delta^*$. This error function reflects the difference between the output obtained by the net and the desired output. The total error E for the sample is

$$E = \sum_{t \in S} e_t \tag{10}$$

where e_t is the error for each translation t. In the rest of this section, we will drop the subscript t to alleviate the notation. Assume that the sample translation is (w, z). First, the string w is "pushed" into the RHAM by recursive application of the encoder, to obtain a representation $\mathbf{r}(w)$ which is taken to be the representation $\mathbf{s}(z)$ of its translation $z = b_1 b_2 \ldots b_{|z|}$. The total error in the translation is the departure of the successive output symbol vectors $\mathbf{y}[|z|], \mathbf{y}[|z| - 1], \ldots, \mathbf{y}[1]$

from their desired values $\mathbf{y}_{b_{|z|}}, \mathbf{y}_{b_{|z|-1}}, \ldots, \mathbf{y}_{b_1}$ plus the departure of the final representation $\mathbf{s}[0]$ from that of the empty output string $\mathbf{s}_0$:

$$e = \frac{1}{2} \left[\sum_{l=|z|}^{1} \|\mathbf{y}[l] - \mathbf{y}_{z_l}\|^2 + \|\mathbf{s}[0] - \mathbf{s}_0\|^2 \right] \tag{11}$$

The values of the local representations of the output symbols $\mathbf{y}_b$ and the empty output string $\mathbf{s}_0$ may be chosen in advance (as in this paper) or allowed to change during learning. In the latter case, they might be taken to be the instantaneous average (over the whole sample) of the actual values observed where they should have appeared, but in that case, an additional penalty term would be needed to keep the $\mathbf{y}_b$'s as far apart from each other as possible, in order to avoid the collapse of all of them into a single value. Details of this procedure will be reported elsewhere.

4.2 Training Algorithm

The above formulation allows for the use of gradient-descent-based methods, such as a modified version of RTRL (real-time recurrent learning, Williams and Zipser 1989) or an adaptation of backpropagation through time (Rumelhart *et al.* 1986). This approach to training RHAMs will be reported elsewhere. In the preliminary experiments reported here we have chosen a non-gradient method, Alopex (Unnikrishnan and Venugopal 1994), which is related to simulated annealing but instead of being driven by changes in the error function it is driven by correlations between changes in error and changes in each particular weight. The method relies only on successive evaluations of the error function and needs no information about the particular structure of the system being trained. In a related task (Ñeco and Forcada 1996) where discrete-time recurrent neural networks were trained to perform simple formal translations similar to those shown here, the method proved to be a very attractive alternative to gradient descent, specially when the architecture is new (as in Ñeco and Forcada 1996) or being used in a new way for the first time (as here).

5 Experiments

We have made experiments in order to explore the translation capability of this new architecture. This experiments include simple translation tasks performed by Mealy and deterministic generalized sequential machines.

We train the network with the translations performed by the three automata shown in figure 2. These automata perform translations between strings over the alphabets $\Sigma = \Delta = \{0, 1\}$. We trained networks with $N = 3, \ldots, 12$ hidden neurons and encoded both input and output symbols with the unary (one-hot) encoding, using Alopex (Unnikrishnan and Venugopal 1994) with a step-size $\delta = 0.01$, and an initial temperature $T = 1000$. The training set contained the translations of all strings from length 1 to 8, and we check the generalization for the whole set of strings from length 9 to 10. We stopped training the net when it correctly translated all the sentences in the learning set. A successful translation is reported when all the values of the output neurons depart less than a tolerance $\xi = 0.2$ from their desired values. We consider that a network has failed to learn the task when it takes more than 500,000 epochs.

Table 1 shows the results for the first automaton, which is a Mealy machine. In this table we present the number of epochs needed to learn the task and the generalization performance of the net, for $N = 2, \ldots, 12$ neurons. With bigger values of N, we observed that the net either fails to learn the translation task, or has worse generalization performance. The best generalization result was obtained for $N = 6$ hidden neurons, with 156,000 epochs of Alopex.

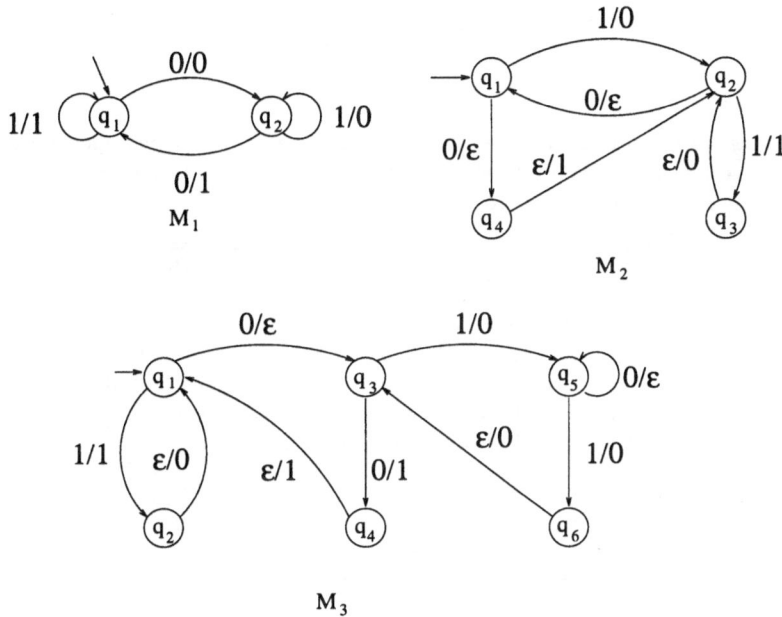

Fig. 2. Automata used in the experiments

In the second experiment we used a DGSM (automaton M_2 shown in figure 2). Note that this automaton performs a many-to-one translation in the sense that two different input strings may have the same output string as translation: for example, $\tau(\text{``0''}) = \tau(\text{``00''}) = \text{``1''}$. The results obtained for this task are shown in table 2. In this case we obtained the best generalization result (87%) for $N = 9$ neurons, with 158,000 epochs, but for $N > 9$ we observed that the generalization results are worse for this task (60 % for $N = 10$, 55 % for $N = 11$, 63% for $N = 12$).

In the third experiment we used automaton M_3 shown in figure 2. This automaton also performs a many-to-one translation. For example, $\tau(\text{``10''}) = \tau(\text{``1''})$

Table 1. Results for automaton M_1. We show the number of Alopex epochs (in thousands) used to learn the training set, and the generalization results (percentages) for all input strings from length 9 to 10.

Neurons	kepochs	Generalization
3	failed, failed, 254, failed	–, –, 65%, –
4	215, 250, failed, failed	68%, 62%, –, –
5	115, 105, 98, 117	67%, 77%, 60%, 78%
6	160, 177, 156, 189	90%, 85%, 91%, 82%
7	98, 126, 117, 108	85%, 87%, 82%, 90%
8	177, 155, 165, 140	75%, 73%, 81%, 79%
9	77, 90, 86, 105	75%, 74%, 55%, 76%
10	155, 284, failed, failed	30%, 47%, –, –
11	failed, 430, 350, failed	–, 53%, 42%, –
12	442, 270, failed, failed	59%, 46%, –, –

Table 2. Results for automaton M_2. We show the number of Alopex epochs (in thousands) used to learn the training set, and the generalization results (percentages) for all input strings from length 9 to 10.

Neurons	kepochs	Generalization
3	failed, 350, failed, failed	–, 53%, –, –
4	270, failed, 362, 388	60%, –, 42%, 31%
5	226, 270, 231, 273	70%, 68%, 55%, 80%
6	210, 224, 230, 205	74%, 73%, 52%, 75%
7	175, 170, 215, 265	70%, 75%, 80%, 65%
8	230, 180, 221, 192	70%, 76%, 71%, 75%
9	150, 124, 290, 158	67%, 84%, 82%, 87%
10	253, 306, 379, 366	45%, 50%, 60%, 48%
11	failed, 340, 365, failed	–, 45%, 55%, –
12	220, 166, 252, 330	50%, 55%, 63%, 54%

= "10". The results obtained for this automaton are shown in table 3. In this case, we obtain the best generalization result (89%) for $N = 8$ neurons. In this experiment, we consider that a network has failed when it takes more than 800,000 epochs.

6 Discussion and Conclusions

The results obtained for examples of the widest class of deterministic finite-state translations (deterministic generalized sequential maps, Salomaa 1987) show that recursive hetero-associative memories (RHAM) are a promising approach to the neural induction of translators from examples. The RHAM architecture has

Table 3. Results for automaton M_3. We show the number of Alopex epochs (in thousands) used to learn the training set, and the generalization results (percentages) for all input strings from length 9 to 10.

Neurons	kepochs	Generalization
3	failed, failed, failed, failed	$-, -, -, -$
4	failed, failed, failed, failed	$-, -, -, -$
5	450, failed, 572, failed	78%, $-$, 65%, $-$
6	failed, 550, 309, 674	$-$, 45%, 63%, 75%
7	420, 580, 331, 427	65%, 60%, 53%, 63%
8	failed, 494, 537, failed	$-$, 75%, 89%, $-$
9	610, 537, 540, 593	85%, 72%, 87%, 63%
10	failed, 510, 638, failed	$-$, 70%, 35%, $-$
11	failed, failed, 705, failed	$-$, $-$, 62%, $-$
12	failed, failed, failed, failed	$-, -, -, -$

been shown to be computationally superior (they may deal with many-to-one translations) to RAAMS as used by Chrisman (1991) and theoretically equivalent to RAAMS as used by Chalmers (1990), but architecturally simpler and easier to train (there is no need to train the network to auto-associate input and output strings separately to learn the translation tasks).

Generalization results are good if we take into account that they refer to strings that are always longer than those in the training set[1]. We are currently studying a new training strategy which forces representations of *all* prefixes of *all* strings in the training set to be *decodable* into prefixes of output strings; we believe that this will improve the generalization performance as well as yield partial translations as a byproduct. The results of this approach and a detailed study of the representations learned by the network will be reported elsewhere.

The approach shown here is complementary to the one involving a new recurrent neural network architecture presented by us (Ñeco and Forcada 1996), which has also been trained to learn deterministic generalized sequential machines. We are currently studying the relationships between both approaches, with a possible hybrid approach in mind.

Acknowledgments: This work has been supported through grant TIC95-0984-C02-01 of the Spanish Comisión Interministerial de Ciencia y Tecnología (CICyT). Ramón P. Ñeco is supported by the Generalitat Valenciana (Spain).

---

[1] Chrisman (1991) and Chalmers (1990) always tested generalization on strings in the same length range.

References

Chalmers, D.J. (1990) "Syntactic Transformations on Distributed Representations", *Connection Science* **2**, 53–62.

Chrisman, L. (1991) "Learning Recursive Distributed Representations for Holistic Computation", *Connection Science* **3**:4, 345–366.

Hopcroft, J.E., Ullman, J.D. (1979) *Introduction to automata theory, languages and computation.* Reading, Massachussets: Addison-Wesley.

Ñeco, R., Forcada, M.L. (1996) "Beyond Mealy machines: Learning translators with recurrent neural networks", *Proc. World Congress on Neural Networks* (San Diego, Calif., Sept. 1996), p. 408–411.

Pollack, J.B. (1990) "Recursive distributed representations", *Artificial Intelligence* **46**, 77–105.

Rumelhart, D.E., Hinton, G.E., Williams, R.J. (1986) "Learning internal representations by error propagation". In *Parallel Distributed Processing: Explorations in the Microstructure of Cognition* (D.E. Rumelhart and J.L. McClelland, eds.), Vol. 1, Chapter 8, Cambridge, MA: MIT Press.

Salomaa, A. (1987) *Formal Languages*, Boston, Massachusetts: Academic Press.

Unnikrishnan, K.P., Venugopal, K.P. (1994) "Alopex: A Correlation-Based Algorithm for Feedforward and Recurrent Neural Networks", *Neural Computation* **6**, 469–490.

Williams, R.J., Zipser, D. (1989) "A learning algorithm for continually running fully recurrent neural networks", *Neural Comp.* **1**, 270–280.

Universal Binary and Multi-valued Neurons Paradigm: Conception, Learning, Applications

Naum N.Aizenberg[1], Igor N.Aizenberg[2]

1- Professor of the Department of Cybernetics of the State University of Uzhgorod (Ukraine); Minaiskaya 28, kv. 49, Uzhgorod, 294015, Ukraine; Phone: (38 03122) 2-39-08; e-mail: igor@pgd.uzhgorod.ua

2 - Visiting researcher in ESAT - KU Leuven (Belgium), Dr. of sc. math.; K.U.LEUVEN, Departement Elektrotechniek ESAT/SISTA, Kardinal Mercierlaan 94, 3001 Heverlee, Belgium; Phone: (32 16) 32-18-42; Fax: (32 16) 32-19-70; e-mail: Igor.Aizenberg@esat.kuleuven.ac.be or igor@pgd.uzhgorod.ua

Abstract.

Futheron development of the Multi-Valued and Universal Binary Neurons conception with basic arithmetic over Complex Numbers Field is presented in this paper. Lot of attention is devoted to Universal Binary Neurons. New high-effective fast convergenced learning algorithm based on Error-correction rule is considered. It is shown that any non-threshold Boolean function can be implemented on the single Universal Binary Neuron. Example for solution of the XOR-problem on the single neuron is considered. Applications of the UBN for solution of the important problems of Image Processing (impulsive noise detection and filtering and edge detection with extraction of the smallest details based on representation of these operations through non-threshold Boolean functions) are also considered.

1. Introduction

McCulloch's neural conception declared in 1943 [1] is of course one of the most revolutionary conception of the present cybernetics. It will be difficult to count how many important problems were solved owing to ideas firstly presented in [1]. Ideas which are developed in this paper are based first of all on the general idea of [1]. This is idea of representation of Boolean function of n variables by n+1 weights and comparison of the weighted sum of variables $W_1 X_1 + \ldots + W_n X_n$ with threshold W_0 for evaluation of the value of corresponding function (or in the terms of Neural Networks - of the neuron's output). This approach has been developed and described in many fundamental papers and books ([2] is one of recent). Disadvantage of this approach is possibility of implementation only of the threshold functions, number of which is so small in comparison with number of all Boolean functions (there are only about 2000 threshold functions of 4 variables, but number of all functions of 4 variables is equal 65536). To implement any other function all authors propose designing of different kinds of networks. One of the most expanded review of such approaches is presented in [2]. Also in [2] review of approaches, which reduce evaluation of multiple-valued output functions of neuron to evaluation of the logistic functions on the interval [0, 1], is presented.

Here we would like to present alternative approach to solution of the both of problems (increasing of the neuron's functionality and implementation of the multiple-valued functions) which has been developed recently. The main idea developed here is consideration of the neuron with the Complex weights and thus with the basic arithmetic over the Complex numbers Field. Of course, central moment of this approach - original definition of the neuron's activation function which makes it possible to design unique fast convergenced learning algorithm and to expand neuron's functionality. Idea of approach presented here has been introduced in [3] and then developed [4-6]. Result of such an developing is carrying out of the models of two type of neurons - Multi-Valued Neuron (MVN) [4,6] which performs multiple-valued functions (functions of k-valued logic) and Universal Binary Neuron (UBN) [5] which performs arbitrary (not only threshold !) Boolean functions. Should be noted that some authors began to work with complex-weight neurons [7-8], but they use absolutely another activation functions, thus they obtain another learning algorithms, another estimates for functionality and so on.

Taking to account that MVN and their applications have been considered recently in detailed form [4, 6] we will concentrate here in majority on UBN and on general advantages of the both type of neurons.

2. Mathematical model and conception

As has been noted above we will consider here two type of neurons : Universal Binary Neuron (UBN) which performs in general arbitrary Boolean function of n variables and Multi-Valued Neuron (MVN) which performs full-defined threshold or partial-defined k-valued function (function of k-valued logic), where k is in general arbitrary integer>0 (there are no problems with k of order 5000, but this restriction based only on power of conventional computer).

So, our basic idea is representation of Boolean function of n variables or of k-valued function of n variables by $n+1$ complex-valued weights $W_0, W_1, \ldots, W_n$

$$f(x_1, \ldots, x_n) = P(w_0 + w_1 x_1 + \ldots + w_n x_n) \qquad (1)$$

where $X_1, \ldots, X_n$-variables of which performed function depends and P - output function which is defined by the next way.

1) For Multi-Valued Neuron:

$$P(z) = \exp(i * 2\pi * j / k),$$
$$\text{if } 2\pi * (j+1) / k > \arg(z) \geq 2\pi * j / k \qquad (2a)$$

where $j=0,1,\ldots,k-1$ - values of the k-valued logic, i- is an imaginary unit, $z = w_0 + w_1 x_1 + w_n x_n$ - weighted sum , $\arg(z)$ is the argument of the complex number z. (values of function and of variables are also coded by complex numbers which are k-th power roots of a unit: $\varepsilon^j = \exp(i2\pi j / k)$, $j \in [0, k-1]$, i- is an imaginary unit, in another words values of the k-valued logic are represented as k-th power roots of a unit: $j \rightarrow \varepsilon^j$).

2) For Universal Binary Neuron

$$P(z) = (-1)^j$$
$$\text{if } (2\pi (j+1)/m) > \arg(z) \geq (2\pi * j/m) \qquad (2b)$$

where m is the some positive integer $4n \geq m > 0$, j- non-negative $0 \leq j < m$, $z = w_0 + w_1 x_1 + w_n x_n$ - weighted sum , $\arg(z)$ is the argument of the complex number z. Variables and implemented function in this case are Boolean and takes values from the set $\{1,-1\}$. For the first time idea of such an representation of Multiple-valued

465

functions has been proposed in [3]. Then it has been developed and the similar approach to Boolean case has been proposed [4-6]. Definition of the functions (2a) and (2b) are illustrated by Fig.1a and 1b respectively:

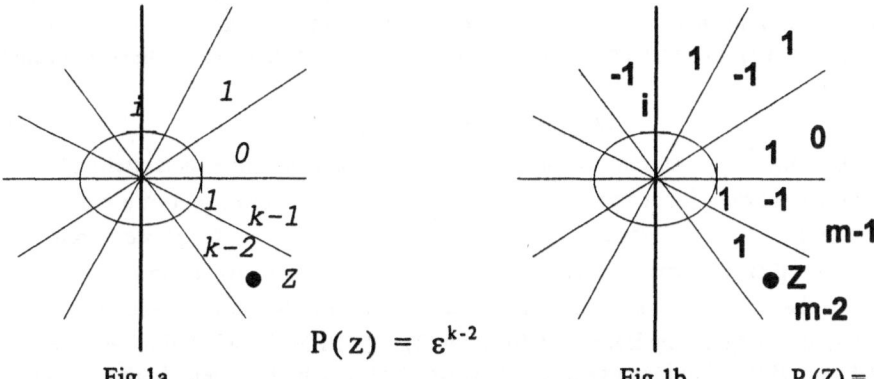

$$P(z) = \varepsilon^{k-2}$$

Fig.1a   Fig.1b   P (Z) = 1

It should be noted that if m=2 in (2b), UBN is restricted to Complex-threshold element [3], which is the same that obvious threshold element with real weights and *sign* as output function.

From the first point of view someone may put a question: "Why it is necessary to pass to the Complex domain ?" Answer is very simple. First of all by such a way it is possible to extend neuron's functionality. Let's consider example. Let n=2 and we would like to implement on the single neuron non-threshold XOR-function $f(x_1, x_2) = x_1 \oplus x_2$. Let m=4 in (2b), therefore we separate Complex plane into 4 sectors (Fig.2). Our XOR function of 2 variables is implemented on the single Universal binary neuron by weighted vector **W**=(0, 1, i), where i is an imaginary unit. Indeed

x_1	x_2	$z = w_0 + w_1 x_1 + w_2 x_2$	$P(z) = f(x_1, x_2)$
1	1	1 + i	1
1	-1	1 - i	-1
-1	1	-1 + i	-1
-1	-1	-1 - i	1

and any network is not need and necessary for solution of this popular problem.

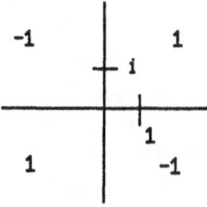

Fig.2

We will return below to general solution of XOR-problem for arbitrary number of variables.

3. Learning

It is evident that learning algorithm both for UBN and MVN have to "direct" weighted sum into one of the sectors on the Complex plane, which corresponds to the correct output (for UBN) or to the sector which corresponds to the desired output (for MVN), if weighted sum falls into sector which corresponds to the incorrect output. Learning algorithm based on this idea has been proposed in [5-6]. It reduced to the next procedure of correction:

$$W_{m+1} = W_m + \omega \varepsilon^q X , \qquad (3)$$

where W_m and W_{m+1} - current and next weighting vectors, ω - correction coefficient, X-vector of the neuron's input signals, ε - is the primitive m-th power root of a unit, q - is number of the "correct" sector $(0, 1, 2, ..., m-1)$. ω have to be chosen from the consideration that weighted sum $Z = W_0 + W_1 X_1 + W_n X_n$ have to be closer as soon as possible to the desired sector (or have immediately get to the desired sector) (sector number q) after step of the learning defined by (3). For UBN choosing of the ω based on the next rule. If weighted sum gets into "incorrect" sector, both of neighborhood sectors are "correct" in such a case, therefore, if Z gets into "incorrect" sector, weights must be corrected by (3) to direct Z into one of the neighborhood sectors (left or right depending what of them current value of Z is closer to). This determine value of the coefficient ω in (3) for learning of the UBN (i - is an imaginary unit):

$$\begin{cases} 1, \text{ if } Z \text{ must be "moved" to the right sector} \\ i, \text{ if } Z \text{ musr be "moved" to the left sector} \end{cases} \qquad (4)$$

ε^q in such a case corresponds to the right bound of the sector on the complex plane to which weighted sum Z must gets (Fig. 1). Such learning process is convergenced pretty fast when m in (2) is so small (if m<10). Despite that many non-threshold functions it is possible to represent by (1)-(2) when m<10, for many other ones greater values of m are needed. But in such a case efficiency of the learning procedure (3)-(4) is law. Way out of such an situation is modification of the correction rule in the algorithm (3)-(4) which we would like to propose here.

Instead of coefficient $\omega \varepsilon^q$, which demands execution of some evaluations for choosing of ω and additional complex multiplication on each step of the learning, we will use error-correction learning rule:

$$W_{m+1} = W_m + \frac{C(\varepsilon^q - \varepsilon^s)}{(n+1)} X , \qquad (5)$$

where W_m and W_{m+1} - current and next weighting vectors, X- vector of the neuron's input signals, ε - is primitive m - th power root of a unit (m is chosen from (2)), C is the scale coefficient, q is the number of "correct" sector on the Complex plane, s is number of the sector on the Complex plane to which actual value of the weighted sum is fallen). For MVN q has to be equal to the desired neuron's output. For UBN q has to be chosen on each step from the same consideration as in algorithm (3): if weighted sum Z gets into "incorrect" sector number s, both of the neighborhood to s - th sectors will be "correct", thus

$$q = s - 1 \ (\text{mod } m) , \text{ if } Z \text{ is closer to } (s-1)\text{-th sector} \qquad (6\ a)$$

$$q = s + 1 \ (\text{mod } m) , \text{ if } Z \text{ is closer to } (s+1)\text{-th sector} \qquad (6\ b)$$

So, (5)-(6) present learning algorithm for UBN with error-correction rule.

In comparison with (3) algorithm (5)-(6) is independent of the value of **m** in (2b) and of **n** (number of neuron's inputs) and leads weighted sum into the necessary sector by 1-2 steps always. As to global convergence of the algorithm (5)-(6) it should be noted that it is especially effective for self-dual functions (it is not restriction because any function of **n** variables may be reduced to self-dual function of **n+1** variables).

Precise proof of the convergence of algorithm (5)-(6) takes lot of place and it is impossible to reproduce it here, but it is very similar to proof of the algorithm (3)-(4) given in [3].

Another modifications which we would like to propose is rejection of the floating-point evaluations in (1) - (2) and (5) - (6) which will give high acceleration of the learning and processing time for the UBN. In [6] we made some similar modifications for algorithm (3)-(4) applied to MVN.

All weights and roots of a unit will be presented by integer 32-bit numbers per real and imaginary part respectively and weighted sum will be presented by integer 64-bit numbers per real and imaginary part respectively. For transformation of the real values into integer ones we will use scale coefficient **D** which may equal to some power of 2 (from 8 till 16384. Also it should be noted that for some complicate cases value of **D** may changed during learning process - sometimes it is useful to begin learning from the greater value of **D** and then low it to 2048, 1024 or lower.). In another words, if **a** is a real number which presents real or imaginary part of the complex number, we will use integer **b=Round(Da)** instead of **a** in all of evaluations. This modification gives 15-17- times acceleration of the evaluations in (2) and (5)-(6), and also is a good mean to break errors connected with evaluations with irrational numbers. After finishing of the learning process all weights always may be reduced to 16-bit per real and imaginary part respectively and even to 12- or 8- bit in some cases.

Learning algorithm (5) - (6) opens unique possibilities for implementation of the non-threshold Boolean functions on the single UBN which may be a good alternative for Multi-Layer Perceptrons [2] and from our point of view may initiate some investigations in the search of Biological analogues with such a type of neurons as UBN.

4. Solution of the XOR-problem on the single UBN

This popular problem is used by lot of authors as example of the necessity of networks designing to implement not-threshold functions. Review of different solutions of this problem on the different kinds of networks presented in [2] (also in many other books). Here we would like to present alternative approach - implementation of the XOR - function of the arbitrary number of variables on the single (!) UBN, without any network. Simple example for two variables (or neuron with two inputs) has been presented above in the section 2.

We will consider here two approaches: one, based on the recoding of the Boolean function to multiple-valued one and then application of the learning algorithm (5) to obtained multiple-valued function (or MVN) and another one, based on the direct application of the learning algorithm (5)-(6) to the Boolean function. Both of approaches may be generalized on the case of arbitrary not-threshold Boolean function.

Let's consider approach based on the recoding. First of all some important mathematical remarks. Let's consider Conjunctive transformation [9] which can be defined for order **n** as Kronecker's **n**-th power of matrix $K_1 = \begin{pmatrix} 1 & 0 \\ 1 & 1 \end{pmatrix}$: $K_n = K_1^{\otimes n}$. It should be noted that over the field GF(2) $K_n^{-1} = K_n$. It is evident that matrix K_n consists (or contains) of all conjunctions of all ranks from 0 to n. We will use Conjunctive transformation for recoding of the Boolean function (especially for recoding of the XOR function) to multiple-valued one. Then we will apply learning algorithm (5) to the obtained multiple-valued function. Recoding consists of the next steps which forms recoding algorithm:

1. Let **f** - vector of the values of Boolean function (in alphabet {0, 1}). We have to evaluate matrix product $g = K_n^{-1}f$ over field GF(2). It is evident that vector **g** represents decomposition of the function **f** over conjunctions.

2. Then we have to evaluate matrix product $h = K_n g$ over the Real numbers field. Vector **h** will contain values of multiple-valued function to which input function has been recoded.

Now it is possible to apply learning algorithm (5) to multiple-valued function **h**. If **f** is XOR function, function **h** has be considered in **2n**-valued logic, or, more precise, for so great values of **n** in logic of value of order **2n** (precise value in such a case must be of order **3n** or lower). It is evident that weighting vector obtained by learning algorithm (5) for multiple-valued function **h** will also satisfy (1)-(2b) for Boolean function **f** , thus functions **f** and **h** will certainly have the same weighting vectors.

Let's consider **f** as XOR function of **n** variables: $f(x_1, \ldots, x_n) = x_1 \oplus x_2 \oplus \ldots \oplus x_n$. Not too complicate reasoning provides the conclusion that in such a case **f**=(0,1,1,0,1,0,0,1,1,0,0,1,0,1,1,0,...) and **h**=(0,1,1,2,1,2,2,3,1,2,2,3,2,3,3,4,..., n-3,n-2,n-2,n-1,n-2,n-1,n-1,n). Function **h** considered in **2n**-valued logic (or for great values of **n** in (2n+s)-valued logic, where s<3n) is partial- defined multiple-valued threshold function [3] and learning algorithm (5) for this function is always convergence with very high speed. Estimate for number of iterations is n^3, which is satisfactory because such umber of iterations needs very small computing time. Therefore we have to make important **CONCLUSION: XOR function of n variables may be implemented on the single Universal Binary Neuron.**

Let's consider example for n=3. $f=(0,1,1,0,1,0,0,1)=x_1 \oplus x_2 \oplus x_3$, where $\oplus$ - mudulo 2 (**XOR**) addition. Then over GF(2) : $g = K_3 f = (0, 1, 1, 0, 1, 0, 0, 0)$ and over Real numbers field: $h = K_3 g = (0, 1, 1, 2, 1, 2, 2, 3)$.

Let now **k=6** in (2a) which is the same that **m=6** in (2b) and therefore **k=m=6=2n** .If now one will take arbitrary vector W_0 as starting vector in (3) only 2-4 iterarions will be necessary for convergence of the learning process. The result of learning process will be weighting vector **W** for multiple-valued function **h**. It is evident that the same weighting vector will implement the original XOR function **f**. By the absolutely same way XOR problem is solved for arbitrary **n** on the single Universal Binary Neuron.

Now we will consider direct application of the learning algorithm (5)-(6) to the Boolean function XOR of the different number of variables. The next weighting vectors for different **n** (number of inputs (variables)) have been carried out during few iterations (3 - for n=3, ... 74 - for n=9) ($W = (w_0, w_1, \ldots, w_n)$, m is number of sectors from (2b)):

n = 3, m=6,   W=((0.00, 0.00), (-20.465, 0.00), (0.045, -35.47), (20.51, 0.00))
n = 8, m=16,   W=((99.72, -165.92), (3.4, 57.32.), (-0.99, 66.16), (-1.39, -58.76),

(-92.39, -1.32), (-1.42, -60.99), (-94.38, -3.36), (90.15, 2.52), (1.20, 57.16))
n = 9, m=22, W=((8.19, 67.86), (-0.54, -16.7), (0.064, -16.83), (-0.053, -16.72), (-0.39, -17.60), (-0.27 , -17.13), (-0.66, -16.59), (21.88, 1.32), (70.11, 1.61), (-0.35, -16.61)).

This example demonstrates power of the UBN and MVN also as efficiency of the learning algorithm (5) (with supplement (6) for UBN), carried out above.

5. Application of the networks based on UBN to image processing and recognition

Application of the networks based on MVN to image recognition, to designing of the associative memory and time-series prediction has been considered in [6]. To confirm efficiency and power of the Universal Binary Neurons we will consider below implementation on them of the some non-threshold functions, which are important for image processing and recognition. Also technology of real-time image processing based on union of the UBNs to Cellular Neural Network (CNN) [10, 5] will be proposed.

Using of the UBN as basic elements of the Cellular Neural Networks [10] and implementation on the UBN of the edge detection algorithm has been presented by us in [5]. But limited possibilities of the learning algorithm (3)-(4) for Boolean functions did not make it possible to expand applications of the UBN in such a field. Elaboration of the learning algorithm (5)-(6) breaks these restrictions.

Let's consider CNN of the dimension N * M (we will process images of the same dimension) based on UBN (fragment of the network is presented on the Fig.3). Each (ij-th) neuron in the CNN is connected only with limited number of other ones (only with neurons from nearest r * r neighborhood). We will consider only 3*3 neighborhood which is sufficient for problems will be solved below. On other hand all neurons of the CNN implements the same processing function.

One of the most important problems in Image Processing, Recognition and Understanding are: 1) noise filtering with minimal blurring of the image and 2) extraction

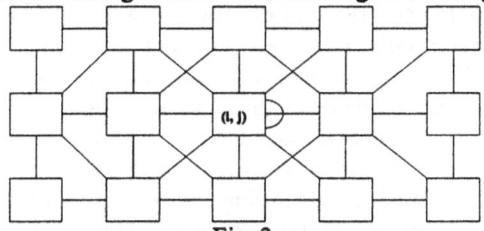

Fig. 3

of the important details against the complex image background. Here we will represent both of problems through Boolean functions which will be implemented on the UBN by learning algorithm (5)-(6) and then weights obtained by learning will be used for realization of the proposed operations on the CNN based on UBN. Also it should be noted that algorithms represented by Boolean functions are oriented first of all (or precisely) on the binary Image Processing. But we will also apply them to gray-scale images by separation of the image with 256 gray levels into 8 binary planes, separate processing of these planes as binary images and final union of the processed plains to the resulting gray-scale image. Such an technique was proposed by us in [5].

Let's consider problem of the impulsive noise filtering. Boolean function proposed here for detection of the impulsive noise and it's filtering is oriented on the complicate noise - combination of the single impulses and horizontal, vertical and diagonal "scratches"(function for detection of the single impulses only also is non-threshold and can be implemented on the single UBN, but we would like to present here more complicate case). Our function is function of 9 variables. These variables are values of brightness in local 3*3 window around it's central pixel. So, function for detection and filtering of the combination of single impulses, vertical, horizontal and diagonal "scratches" may be defined by the next formula:

$$Y\begin{bmatrix} x_1 & x_2 & x_3 \\ x_4 & x_5 & x_6 \\ x_7 & x_8 & x_9 \end{bmatrix} = \begin{cases} -x_5, & \text{if lot of } x_1,x_3,x_7,x_9=-x_5 \text{ or lot of } x_2,x_4,x_6,x_8=-x_5 \\ x_5, & \text{otherwise} \end{cases} \quad (7)$$

This is non-threshold function, but it can be implemented on the single UBN. Learning algorithm (5)-(6) needs nearly 8000 iterations (they request few minutes even on the 33 Mhz computer) and gives the next weighting template (bias W_0 and other weights in the matrix) for function (7) ($m = 18$ in (2b)) :

$$W = (106.416 , 205.537) \begin{bmatrix} (24.015 , -13.113) & (-9.167 , 33.299) & (6.222 , -32.004) \\ (-31.648 , -25.418) & (22.269 , -8.172) & (-0.023 , 1.112) \\ (-1.355 , 1.138) & (0.237 , 1.704) & (-0.621 , 0.91) \end{bmatrix}. \quad (8)$$

Knowing dynamic range of the noise, it is possible to process only binary planes, which on noise exactly will be found. Main advantage of this method from the point of view of Image Processing is the minimal blurring (smoothing) of the image during filtering in comparison with other filters (median filtering, rank-order filtering, etc.). On the Fig.4b one can see result of the noised gray-scale image (Fig. 4a) processing (probability of distortion - 0.07, range of the noise [40, 240]) by the template (8) which corresponds to function (7) and on the Fig.4d - result of the processing of the same image by the rank-order filter. Fig. 4c presents difference between input noised image and image filtered by template (8) on the CNN-UBN software simulator. Fig. 4e presents difference between input noised image and image filtered by the rank-order filter. It is so evident that method of the noise filtering proposed here is much more powerful than rank filter (last gives best results for impulsive noise filtering in comparison with other classical filters). First of all our method with the same quality of detection and smoothing of the noise gives minimal smoothing of the output image. Difference on the Fig. 4c corresponding to method presented here contains practically only noise component, but on the image from Fig. 4d corresponding to the rank-order filter edges of image are seen so clearly, which confirms smoothing of the useful information on the output image by rank-order filtering.

Very important for image understanding and recognition is also extraction of details which may be got by different ways. One of them - edge detection with high precise. High effective method for solution of this problem has been proposed by us in [5]. It is edge detection by the same way as impulsive noise filtering above - on the same cellular neural network based on UBN. Joint using of the technique presented in [5] and here gives unique possibilities for image recognition and understanding. Thus, our task is extraction of the smallest details on the noised image. On the Fig. 5a one can see noised input satellite picture. On the Fig. 5b - result of it's processing on the CNN with weighting template (8). On the Fig. 5c - result of processing of the filtered image on the same CNN

Fig. 4a

Fig. 4b

Fig. 4c

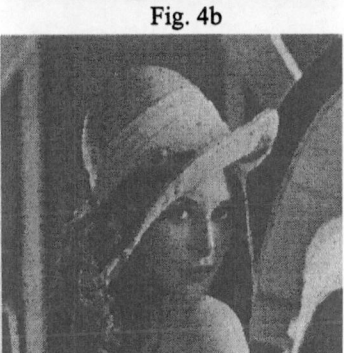

Fig. 4d

Fig. 4e

with weighting template obtained in [5] for edge detection. Edged image obtained by this method looks like to map. Fig. 5d presents combination of images from Fig.5b and 5c (with coefficients 0.8 and 0.2 respectively) - all even smallest details are sharpened.

6. Conclusions

The main conclusion which we have to do is high efficiency of the conception of MVN and UBN for solution of different kinds of problems: image processing and recognition, designing of associative memory, time-series prediction, implementation of arbitrary functions on the single neuron without any network. Effective leaning algorithm elaborated above makes it possible to solve lot of different and important problems on the MVN and UBN also as on networks based on them.

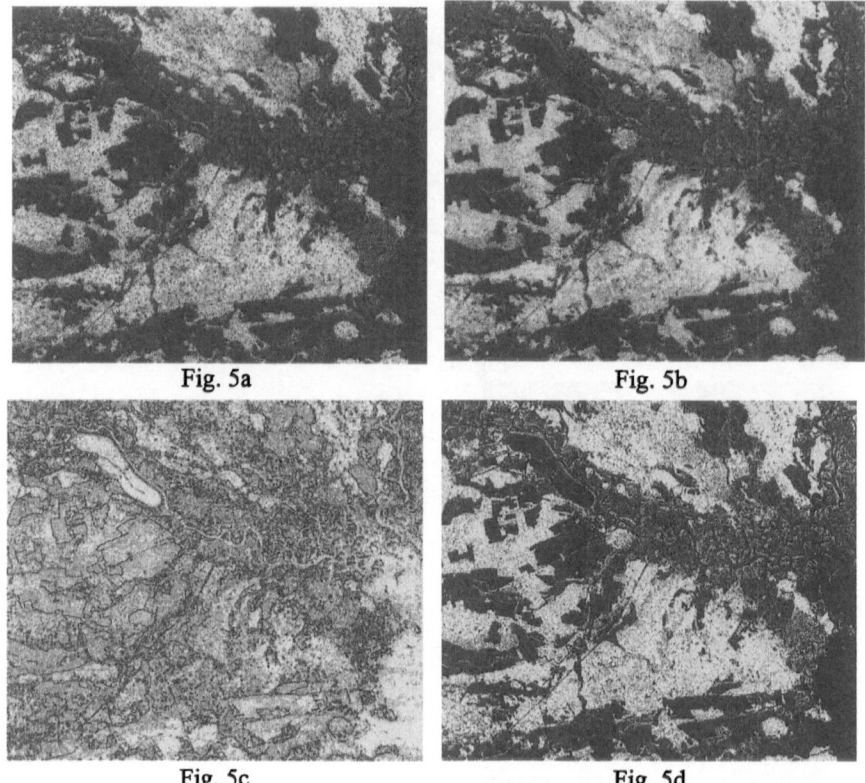

Fig. 5a Fig. 5b

Fig. 5c Fig. 5d

References

[1] W.S.McCuloch, W.A.Pitts "Logical Caculus of the Ideas Immanent in Nervous Activity", Bul. Math. Biophys., 5, pp. 115-133 (1943).

[2] S.Haykin "Neural Networks. A Comprehensive Foundation". Macmillan College Publishing Company, New York, 1994.

[3] N.N.Aizenberg, Yu.L.Ivaskiv " Multiple-Valued Threshold Logic". Kiev: Naukova Dumka Publisher House (1977) (in Russian).

[4] N.N.Aizenberg, I.N.Aizenberg "CNN based on Multi-Valued neuron as a model of Associative Memory for Grey-Scale Images", Proc. of the 2-d International Workshop CNNA-92, Munich, 1992, IEEE Catalog No. 92TH0498-6, pp. 36-41.

[5] N.N.Aizenberg , I.N.Aizenberg "Fast Convergenced Learning Algorithms for Multi-Level and Universal Binary Neurons and Solving of the some Image Processing Problems", Lecture Notes in Computer Science,Ed.-J.Mira,J.Cabestany,A.Prieto,v.686, Shpringer-Verlag, (1993),pp.230-236.

[6] Aizenberg N.N., AizenbergI.N., Krivosheev G.A. "Multi-Valued Neurons: Learning, Networks, Application to Image Recognition and Extrapolation of Temporal Series","Lect. Notes in Computer Science",v. 930, (J.Mira, F.Sandoval - Eds.), Shpringer-Verlag, 1995,pp.389-395.

[7] G.M.Georgiou and C.Koutsougeras "Complex Domain Backpropagation", IEEE Trans. CAS- II.Analog and Digital Signal Processing, vol.39, No 5, 1992, pp. 330-334

[8] H.Leung, S.Haykin "The Complex Backpropagation Algorithm", IEEE Trans. on Signal Processing, vol.39, No 9, September, 1991, pp. 2101-2104.

[9] N.N.Aizenberg "Spectrum of the Convolution of Discrete Signals in the Arbitrary Basis", Dokladi Academii Nauk SSSR, vol.247,No 3, pp.551-554, 1978 (in Russian).

[10] L.O. Chua and L.Yang, "Cellular neural networks: Theory", IEEE Trans.Circuits Syst.vol. 35, pp. 1257-1290, Oct. 1988.

Learning a Markov Process with a Synchronous Boltzmann Machine

by

Ursula Iturrarán[1] and Antonia J. Jones[2]

Abstract. In this paper we present the simulations of a synchronous Boltzmann machine which learns to model a Markov process. The advantage in using synchronous updating is lies in the parallelism that the model offers. Learning with synchronous Boltzmann machines provides an attractive alternative to asynchronous learning provided that one can establish a suitable theoretical framework. The dynamics of synchronous Boltzmann machines were first studied by W.A. Little [Little 1974], [Little 1978] and [Perreto 1984]. The aim of the present study is to present the results generated by a new local learning algorithm for synchronous Boltzmann machines. The algorithm uses *Gradient-Descent* to update weights and thresholds. Three different Markov processes were set to be modelled with a three unit network.

Keywords: Boltzmann machine, Synchronous, Learning, Markov process, Gradient-descent.

INTRODUCTION.

That we learn by imitating our environment is perhaps the first and simplest model that occurs to us when we think about the nature of learning. We obtain an evaluation of our performance by producing a variety of actions that try to mimic environmental transitions and then comparing the consequences with reality. In order to achieve goals progressively we tend to favour those actions that will most probably give the desired outcome.

In the present paper the goal chosen to illustrate a theoretical model of learning for a synchronous stochastic network is to train such a network to produce the same *temporal stochastic behaviour* exhibited in the environment. During the process of training the system selects an environmental state with a uniform random probability and then observes the next environmental state. The neural network attempts to

---

[1] *Department of Computing, Imperial College of Science Technology and Medicine, University of London, 180 Queen's Gate, London, SW7 2BZ, England UK.*

[2] *Department of Computer Science, University of Wales, Cardiff, PO Box 916, CF2 3XF, Wales UK.*

reproduce this state transition in a sequence of *rehearsals*. When a rehearsal is successful the system supplies the network with a *global reinforcement* signal and after a number of such rehearsals the network then updates its weights and thresholds using the theoretically derived local learning rule based on *local gradient descent*.

We suppose that the temporal behaviour of the environment is described by a Markov process (where the next state only depends on the current state). A synchronous Boltzmann machine updates all units at every clock tick. The *activation* of unit i is given by

$$net_i(x) = \sum_{j=1}^{n} w_{ij} x_i - \theta_i \qquad (1)$$

where w_{ij} represents the weight from unit j to unit i, and θ_i is the threshold of unit i. The probability that the $i th$ unit outputs x'_i on updating, when the network is in global state x, is defined as

$$p(x'_i) = \frac{1}{1 + e^{-(2x'_i - 1)\ net_i(x)/T}} \qquad (2)$$

where T is a global parameter, called *temperature*. When T is large all units are equally likely to output a one or a zero. At low temperature the update rule represents an almost deterministic choice of output: one if $net_i(x) > \theta_i$ and zero otherwise.

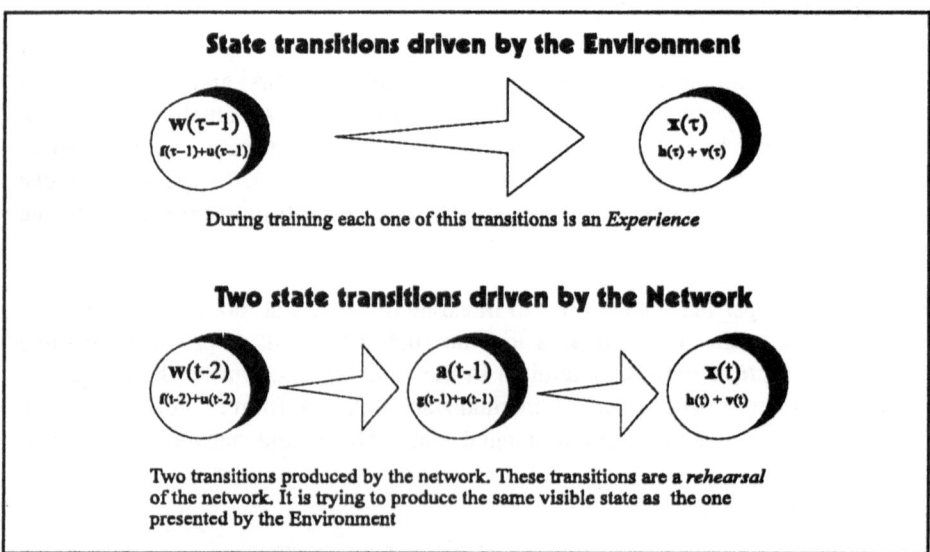

State transitions driven by the Environment

During training each one of this transitions is an *Experience*

Two state transitions driven by the Network

Two transitions produced by the network. These transitions are a *rehearsal* of the network. It is trying to produce the same visible state as the one presented by the Environment

Figure 1 Difference in the timing for the environment and for the network

The set of updated outputs is a new global state for the whole machine. Thus the network update process can itself be naturally regarded as a Markov process. In other words Markov chains are generated in a synchronous Boltzmann machine by considering the sequence of global state transitions. Thus it is natural to apply such a machine to the problem of mimicking an environment described in terms of a Markov process.

In contrast to the model in [Hinton et. al 1986], the model described here has the property that in free-running mode the output is generated *two* clock cycles after the presentation of the input. Thus the model supposes that the network runs twice as fast as the environment, see Figure 1. The reason for taking two transitions in the network whilst the environment produces only one is to take advantage of the hidden units. Hidden units offer an additional and extendable capability for modifying the output of the system on the basis of particular features observed in the input.

MATHEMATICAL PRELIMINARIES.

In order to define the error function to apply *gradient-descent* we need some preliminaries.

The state space $X = \{0, 1\}^n$, where n is the number of units, can be decomposed into the direct sum $X = H \oplus V$, therefore all $x \in X$ can be expressed uniquely as a sum $x = h + v$ ($h \in H$, $v \in V$) of orthogonal hidden and visible vectors. The notation is conveniently summarised in Table 1 where $w(t-2) = f + u \rightarrow a(t-1) = g + s \rightarrow x(t) = h + v$.

Table 1 Summary of notation.

State space	State Transitions
X	w → a → x
H	f → g → h
V	u → s → v

The probability of the two-step *global* state transition w(t-2) -> x(t) in the free running network, where the intermediate state is arbitrary, is given by

$$P\big(x(t)|w(t-2)\big) = \sum_{a \in X} P\big(x(t)|a(t-1)\big) \cdot P\big(a(t-1)|w(t-2)\big) \qquad (3)$$

Next we obtain the two-step transition probabilities for the *visible units* of the network starting from some visible state $u(t-2)$. Let $P_V(v|u)$ be the probability of seeing the transition u(t-2) -> v(t) $\in$ V on the visible units of the network (passing through some arbitrary global state a(t-1) $\in$ X at time t-1). Then

$$P_V(v(t)|u(t-2)) = \sum_{a(t-1) \in X} P(v(t)|a(t-1)) \cdot P(a(t-1)|u(t-2)) \qquad (4)$$

Now

$$P(v(t)|a(t-1)) = \sum_{h(t) \in H} P(h + v|a) \qquad (5)$$

and

$$P(a(t-1)|u(t-2)) = \sum_{f(t-2) \in H} P(a|f + u) \cdot \frac{G(f + u)}{\sum_{f^* \in H} G(f^* + u)} \qquad (6)$$

where $G(f + u)$ is the probability of seeing $f + u \in X$. From (4), (5) and (6) we obtain

$$P_V(v(t)|u(t-2)) = \sum_{h(t), f(t-2) \in H} \sum_{a(t-1) \in X} M(x, a, w) \qquad (7)$$

where $w = f + u$, $x = h + v$ and the auxiliary function

$$M(x, a, w) = P(h + v|a) \cdot P(a|f + u) \cdot \frac{G(f + u)}{\sum_{f^* \in H} G(f^* + u)} \qquad (8)$$

is the probability of observing the global state transitions $w(t-2) \rightarrow a(t-1) \rightarrow x(t)$ *given the particular circumstances defining G* on global states.

In learning ideally the training procedure would be driven by the environment. However, this training regime leads to significant problems:

  • If the environment drives the training process the mathematical analysis shows that a derivation based *directly* on *gradient descent* leads to a learning rule which is non-local.

This problem can be circumvented by altering the distribution G in the training procedure.

  • In what follows we shall assume that during training the initial visible states **u** and hidden states **f** (for a two step transition) are selected with a *uniform random distribution G.*

This assumption eliminates any dependence on the equilibrium distribution of the environmental Markov process. On this basis a locally computed learning rule based directly on *gradient descent* can indeed be constructed [Iturrarán 1996].

An estimate of the difference of the (one-step) environmental transition probabilities and the (two-step) transition probabilities of the visible units of the network whilst training can be constructed using the *Kullback measure* [Kullback 1959]. We denote the difference between two distributions $Q(v \mid u)$ (distribution produced by the environment) and $P_V(v \mid u)$ (distribution produced by the visible units of the network) by $D(Q(v \mid u), P_V(v \mid u))$.

THE GRADIENT THEOREM

In this section we present the theorem from which we have extracted an algorithm to train synchronous networks. The basic idea is to derive the Kullback measure with respect to the weights and use *gradient descent* to minimize the error between the transition probability matrix produced by the network and the one given by the environment. A proof of this theorem can be found in [Iturrarán 1996].

Theorem 1.1. *(Gradient-Descent for Synchronous Boltzmann machines).* If training exemplars $w = f + u$ are presented with a uniform random distribution and the environmental transition is $u \rightarrow v$ then the partial derivatives of the Kullback measure $D(Q(v \mid u), P_V(v \mid u))$ with respect to a connection strength w_{ij} and threshold θ_i are

$$\frac{\partial D}{\partial w_{ij}} = \frac{1}{T} \sum_{u,v \in V} \left(\sum_{h, f \in H} \sum_{a \in X} A(i,j,x,a,w) \frac{M(x,a,w)}{\sum\limits_{h^* f^* \in H} \sum\limits_{a^* \in X} M(x^*,a^*,w^*)} \right) Q(v \mid u) \tag{9}$$

$$\frac{\partial D}{\partial \theta_i} = \frac{1}{T} \sum_{u,v \in V} \left(\sum_{h, f \in H} \sum_{a \in X} B(i,x,a,w) \frac{M(x,a,w)}{\sum\limits_{h^* f^* \in H} \sum\limits_{a^* \in X} M(x^*,a^*,w^*)} \right) Q(v \mid u)$$

where $x = h + v$, $w = f + u$, $x^* = h^* + v$, $w^* = f^* + u$,

$$A(i, j, x, a, w) = \frac{(1-2x_i)a_j}{1 + \exp\left(-\varphi_i(x_i, a)\right)} + \frac{(1-2a_i)w_j}{1 + \exp\left(-\varphi_i(a_i, w)\right)} \tag{10}$$

$$B(i, x, a, w) = \frac{1-2x_i}{1 + \exp\left(-\varphi_i(x_i, a)\right)} + \frac{1-2a_i}{1 + \exp\left(-\varphi_i(a_i, w)\right)}$$

and φ_i is given by

$$\varphi_i(x', x) = \frac{(1 - 2x')}{T} \left[\sum_{\substack{j=1 \\ j \neq i}}^{n} w_{ij} x_i - \theta_i \right] \tag{11}$$

Note. Observe that the computation of $A(i, j, x, a, w)$ can be done by a combination of the ith and jth units by combining information available to these units at times t-2,

t-1, for the second term in (10), and times t-1, t for the first term in (10). Hence we have to assume that a small short-term memory is available for each unit when learning takes place. Similarly to compute function $B(i, \mathbf{x}, \mathbf{a}, \mathbf{w})$ all computations are local for each unit.

EXTRACTING THE LEARNING ALGORITHM

In order to update weights and thresholds we need a rule that applies the formula in the gradient theorem during the training. The inner sum in (9) can be regarded as averaging with respect to the event $\mathbf{w} = \mathbf{f} + \mathbf{u} \to \mathbf{a} \to \mathbf{x} = \mathbf{h} + \mathbf{v}$. If we perform the weight adjustment

$$\Delta w_{ij} = -\frac{\eta}{T} A[i, j, x = h + v, a, w = f + u] \tag{12}$$

and the threshold adjustment

$$\Delta \theta_i = -\frac{\eta}{T} B[i, x = h + v, a, w = f + u] \tag{13}$$

each weighted by the correct probability of seeing the event $\mathbf{w} \to \mathbf{a} \to \mathbf{x}$, the cumulative effect of such adjustments will create the required descent. The correct weighting is obtained using the summation of rewards for successful *rehersals*.

EXPERIMENTAL RESULTS

The aim is to present the results of modelling a Markov process using a synchronous network. The results were generated using a combination of C and *Mathematica*™ programs. We consider a small network with two hidden units and one visible unit trained applying (12) and (13). We have set two different problems to be solved by a synchronous network. The first one is simulate *an environmental Markov process which is itself generated by two-step transitions of a synchronous neural network*, an *archetypal* network. Thus we specify a network with a given architecture, weights and thresholds and hence derive a set of two-step transition probabilities which we will identify with transition probabilities for the environment. This ensures the existence of a network with a known architecture whose two-step transition probabilities can model the environment exactly. The second problem is to model an arbitrary given Markov process.

The following two experiments are the results of training a synchronous network to model a Markov process which has been generated by two-step transitions of another synchronous network (an *archetypal* network) with temperature $T = 2$. The transition probability matrix produced with this archetypal network it is called the *Target*. The initial weights for the learning network are randomly selected within the range [-0.1, 0.1]. The adjustment of the learning rate and momentum parameters (η, α) to obtain the reasonable convergence is done by trial and error - choosing suitable values is rather similar to the problems encountered in Backpropagation.

EXPERIMENT 1: NN-ENVIRONMENT #1, η = 0.03.

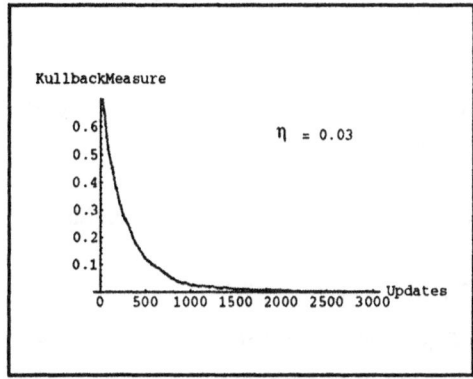

Figure 2 Experiment 1: decreasing Kullback measure as learning progresses.

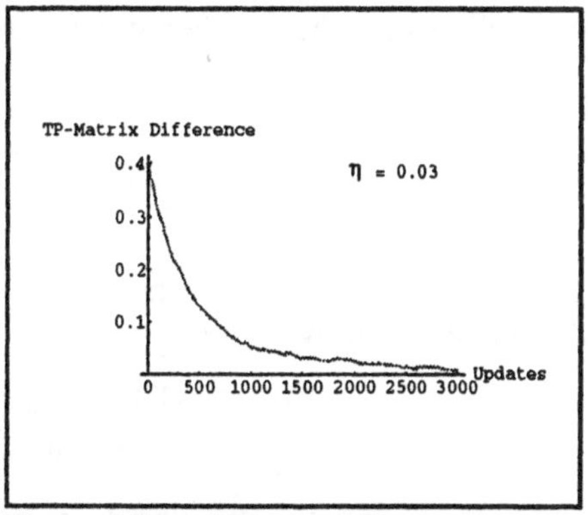

Figure 3 Experiment 1: difference between the transition probability matrices.

In this experiment we use an environment generated from the two-step transitions of a synchronous network. The target transition probabilities for the visible unit in this experiment are

$$
TP_{TAR-1} = \begin{array}{c} \\ 0 \\ 1 \end{array} \begin{array}{cc} 0 & 1 \\ \left[\begin{array}{cc} 0.08650 & 0.91349 \\ 0.08017 & 0.91982 \end{array}\right] \end{array} \tag{14}
$$

generated by a three unit network with weights and thresholds given by

$$
\begin{array}{ccc}
\theta_1 = -1 & w_{12} = 3 & w_{13} = 3 \\
w_{21} = 1 & \theta_2 = 1 & w_{23} = 2 \\
w_{31} = -1 & w_{32} = 2 & \theta_3 = -3
\end{array} \tag{15}
$$

Figure 2 illustrates the Kullback graph over several hundred updates. Similarly, Figure 3 shows

$$
Diff(TP_{TAR-1}, TP_{NET}) = \frac{1}{|V|^2} \sum_{u, v \in V} \left| TP_{TAR-1_{u, v}} - TP_{NET_{u, v}} \right| \tag{16}
$$

computed from the current weight/thresholds.

EXPERIMENT 2: NN-ENVIRONMENT #2, $\eta = 0.7$ $\alpha = 0.9$

In this experiment we have a different environment and we can observe a rather unstable behaviour. Compared with the smooth decrease of the Kullback measure in Experiment 1, here we have a very stochastic descent. The target transition probabilities for the visible unit here are

$$TP_{TAR\text{-}2} = \begin{array}{cc} & \begin{array}{cc} 0 & 1 \end{array} \\ \begin{array}{c} 0 \\ 1 \end{array} & \begin{bmatrix} 0.80934 & 0.19065 \\ 0.64530 & 0.35469 \end{bmatrix} \end{array} \qquad (17)$$

generated by the two-step transitions of the three unit network with weights and thresholds given by

$$\theta_1 = -1.24261 \quad w_{12} = -2.26075 \quad w_{13} = -3.27131$$
$$w_{21} = -3.95804 \quad \theta_2 = -308560 \quad w_{23} = -1.63622 \qquad (18)$$
$$w_{31} = -2.96794 \quad w_{32} = 1.45293 \quad \theta_3 = -3.17911$$

Figure 4 illustrates the Kullback graph over several hundred updates. Figure 5 shows the difference between the transition probability matrices given by (16).

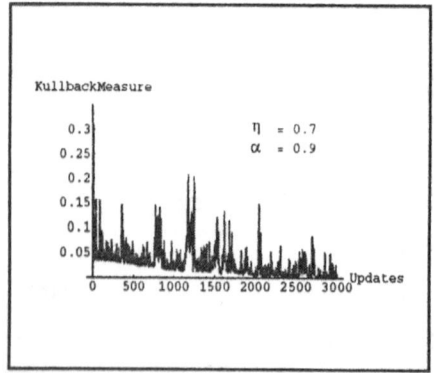

Figure 4 Experiment 2: decreasing Kullback measure as learning progresses.

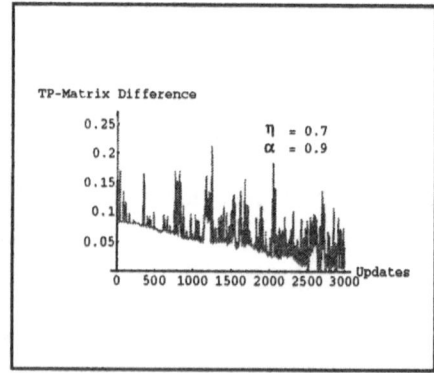

Figure 5 Experiment 2: difference between the transition probability matrices.

EXPERIMENT 3: ARBITRARY ENVIRONMENT #3, $\eta = 0.8$

The target transition probabilities for the visible unit here are

$$TP_{TAR\text{-}3} = \begin{array}{cc} & \begin{array}{cc} 0 & 1 \end{array} \\ \begin{array}{c} 0 \\ 1 \end{array} & \begin{bmatrix} 0.2 & 0.8 \\ 0.7 & 0.3 \end{bmatrix} \end{array} \qquad (19)$$

which is arbitrary and has not been generated by a synchronous network. Therefore in this example we do not know weights and thresholds that produce the model. Figure 6 illustrates the Kullback graph over several hundred updates. Similarly, Figure 7 shows the difference between the transition probability matrices given by (16).

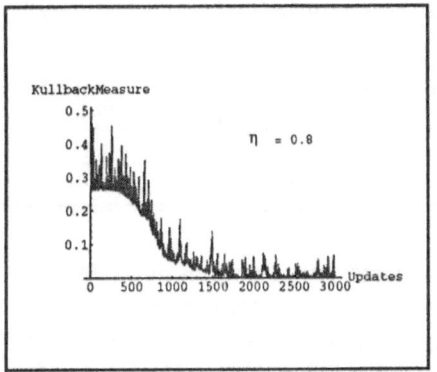

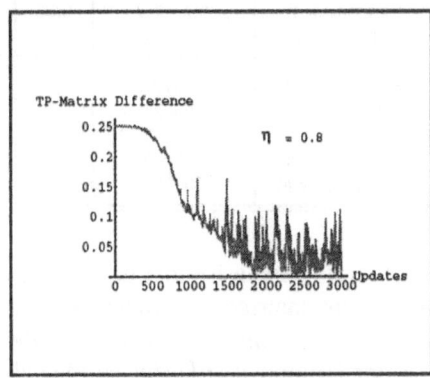

Figure 6 Experiment 3: decreasing Kullback measure as learning progresses.

Figure 7 Experiment 3: difference between the transition probability matrices.

CONCLUSIONS

Various experiments have shown that synchronous learning is feasible using local information, short term local memory, a global reward signal and a Hebbian rule. The main result of the present work is that utilising *gradient descent* based on a measure of the difference between the two sets of transition probabilities, a synchronous Boltzmann machine can be trained to model a Markov process. We can see in the Kullback measure graphs (Figure 2, Figure 4, Figure 6) a progressive decrease, illustrating that the synchronous Boltzmann machine can learn to mimic a Markov process.

We were able to reduce an error function for three different environments. One notable result is that for the arbitrary environment of Experiment 3 the algorithm is successfully learnt a model. This suggests that given any Markov process the algorithm can be applied to train a network to model it. To give a formal proof of this hypothesis will be one of the subjects for future work.

We summarise the experimental results in Table 2 with a synopsis of each experiment including their learning rates and momentum. Here we also compare the final transition probability matrix (generated by the trained network) with the target.

Table 2 Summary of results.

Experiments	Initial / Final Kullback measure	η	α	Final TP- matrix		Target	
Experiment 1: NN-Env #1	0.77226/ 0.00159	0.03	0	0.09352 0.90647	0.09193 0.90806	0.08650 0.91349	0.08017 0.91982
Experiment 2: NN-Env #2	0.23727/ 0.00029	0.7	0.9	0.80164 0.19835	0.65300 0.34699	0.80934 0.19065	0.64530 0.35469
Experiment 3: Arbitrary-Env #3	0.28099/ 0.00031	0.8	0	0.20952 0.79047	0.70475 0.29524	0.2 0.8	0.7 0.3

Finally, we emphasise that during these experiments and as a result of the learning, the weight matrices generated are not symmetric: this contrasts sharply with the restriction to symmetric matrices in the theory of asynchronous Boltzmann machine learning [Hinton 1986]. The learning algorithm presented in this paper is independent of any hypothesis regarding weight symmetry and will in general produce a network with non-symmetric weights.

REFERENCES

[Hinton et. al, 1986] Hinton G.E, and T.J Sejnowski. *Learning and relearning in Boltzmann machines*. In Parallel Distributed Processing: Explorations in Microstucture of Cognition (D.E. Rumelhart and J.L. McClelland eds.) Cambridge MA: MIT Press.

[Iturrarán 1996] Ursula Iturrarán and Antonia J. Jones. *Learning a Markov process with a synchronous Boltzmann machine*. Research Report. Available from: Department of Computer Science, University of Wales, Cardiff, PO Box 916, CF2 3XF, UK.

[Kullback 1959] *Information theory and statistics*, Wiley, N.Y, 1959.

[Little 1974] W. A. Little. *The existence of persistent states in the brain*. Math. Biosci. 19:101, 1974.

[Little 1978] W. A. Little. and G. L. Shaw. *Analytic study of the memory storage capability of a neural network*. Mathematical Biosciences **39**:281-290, 1978.

[Perreto 1984] P. Perreto. *Collective properties of neural networks: A statistical physics approach*. Biological Cybernetics. 50, 51-62.

The α-EM Algorithm: A Block Connectable Generalized Leaning Tool for Neural Networks

Yasuo MATSUYAMA[†, ‡]

† Department of Electrical, Electronics & Computer Engineering,
Waseda University, Tokyo, 169 Japan
‡ Sympat Committee of the RWCP, Japan

Abstract   The α-divergence is utilized to derive a generalized expectation and maximization algorithm (EM algorithm). This algorithm has a wide range of applications. In this paper, neural network learning for mixture probabilities is focused. The α-EM algorithm includes the existing EM algorithm as a special case since that corresponds to $\alpha = -1$. The parameter α specifies a probability weight for the learning. This number affects learning speed and local optimality. In the discussions of update equations of neural nets, extensions of basic statistics such as Fisher's efficient score, his measure of information and Cramér-Rao's inequality are also given. Besides, this paper unveils another new idea. It is found that the cyclic EM structure can be used as a building block to generate a learning systolic array. Attaching monitors to this systolic array makes it possible to create a functionally distributed learning system.

1 Introduction

Computing an expectation and maximization is a set of powerful tools in statistical data processing. Dempster, Laird and Rubin [1] collected examples from diverse areas, built a unified theory, and then created the name "EM algorithm." This algorithm has diverse areas of applications. Among them, learning on mixtures of probability density showed a strong power of this algorithm. Jordan and Jacobs [2] connected their leaning strategy on hierarchical mixtures of experts with this EM algorithm. Since then, many learning strategies related to the EM algorithm appeared by bringing new ideas [3], [4], [5], [6].

The core part of the EM algorithm is a repeated computation of the following two steps.
[E-step] Computing the expectation of the log-likelihood.
[M-step] Computing an argument that maximizes this expectation.
Thus, this algorithm heavily depends on the logarithm and the non-negativity of the Kullback-Leibler's divergence. But, we claim the following fact. There is a wider class for this logarithm/non-negativity pair. This comes from the divergence of order α [7], [8], [9].

There is one more contribution in this paper. The α-EM structure can be used as a building block to construct a total system. Thus, the purpose of this paper is stated as follows. (i) Deriving a general EM algorithm based upon the above divergence. (ii) Obtaining connection update algorithms for neural networks on mixtures of experts. (iii) Presenting a hierarchical structure of the α-EM algorithm. The item (ii) requires extensions of existing statistical theories; the Fisher's efficient score, his measure of information and the Cramér-Rao's bound.

The parameter α is found in probability weights for the learning. Thus, the α-EM algorithm can also be called the probability weighted EM algorithm (WEM).

2 Order α Divergences

2.1 Rényi's divergence and α-divergence

Rényi [7] is the first who presented the order α divergence. But, it is better to replace his α by $(1+\alpha)/2$ and normalize the total amount. Then, this version of the Rényi's divergence between two probability measures $p = \{p_i\}$ and $q = \{q_i\}$ (actually, the total mass need not be the unity) is

$$D_R^{(\alpha)}(p\|q) = -4/(1 - \alpha^2) \log\{\textstyle\sum_i p_i(q_i/p_i)^{(1+\alpha)/2}\}. \tag{1}$$

If p and q are continuous, the summation is replaced by an integration. Thus, readers are requested to switch the summation and the integration appropriately.

The α-divergence [8], [9] has the same kernel.

$$D^{(\alpha)}(p\|q) = 4/(1 - \alpha^2)\{1 - \textstyle\sum_i p_i(q_i/p_i)^{(1+\alpha)/2}\} \tag{2}$$

There is a monotone relationship between these two measures. Therefore, $D_R^{(\alpha)}(p\|q)$ and $D^{(\alpha)}(p\|q)$ are equivalent in the sense of optimization. Our idea started from $D_R^{(\alpha)}(p\|q)$ [11]. In the following sections, however, $D^{(\alpha)}(p\|q)$ is used for the derivation of the α-EM algorithm.

2.2 More general divergence and extended logarithm

Csiszár [10] (its early version is given by Rényi [7]) presented a general divergence measure:

$$D_C(p\|q) = \textstyle\sum_i p_i f(q_i/p_i) = \sum_i q_i g(p_i/q_i),$$

where f and g are twice differentiable convex function with $f(1) = g(1) = 0$. Thus, the α-divergence $D^{(\alpha)}(p\|q)$ is the case of $f(x) = 4\{x - x^{(1+\alpha)/2}\}/(1 - \alpha^2)$. If $\alpha = -1$, then $D^{(\alpha)}(p\|q)$ is reduced to the well-known Kullback-Leibler divergence:

$$K(p\|q) = D^{(-1)}(p\|q) = \textstyle\sum_i p_i \log(p_i/q_i). \tag{3}$$

Note that both $D_R^{(\alpha)}(p\|q)$ and $D^{(\alpha)}(p\|q)$ are symmetric for $\pm\alpha$.

One finds that $\log x$ corresponds to $\{2/(1+\alpha)\}\{x^{(1+\alpha)/2} + c(x, \alpha)\}$. The term $c(x, \alpha)$ is not unique as long as it satisfy a couple of necessary conditions. We select $c(x, \alpha) = -1$. Then,

$$L^{(\alpha)}(x) = \tfrac{2}{1+\alpha}(x^{\frac{1+\alpha}{2}} - 1) \tag{4}$$

is regarded as an extension of $\log x$. Note that the constant "-1" is often cancelled out when $L^{(\alpha)}(x)$ is differentiated by a parameter or compared with other $L^{(\alpha)}(y)$.

3 Extended EM Algorithms based Upon the α-Divergence

3.1 Derivation of the α-EM

Let $p_{Y|\phi}(y|\phi)$ be the probability density of observed data y, Let x be the complete data which contains unknown parts for the observer. ϕ and ψ denote structures defining the probability densities. The simplest case is that ϕ and ψ are parameters. Thus,

$$p_{Y|\phi}(y|\phi) = \int_{\mathcal{X}(y)} p_{X|\phi}(x|\phi)dx$$

is the relationship describing the observation. Let

$$R_Y^{(\alpha)}(\psi|\phi) = \{p_{Y|\psi}(y|\psi)/p_{Y|\phi}(y|\phi)\}^{(1+\alpha)/2},$$

and

$$L_Y^{(\alpha)}(\psi|\phi) = 2\{R_Y^{(\alpha)}(\psi|\phi) - 1\}/(1+\alpha). \tag{5}$$

Let the conditional probability be

$$p_{X|Y,\phi}(x|y,\phi) = p_{X|\phi}(x|\phi)/p_{Y|\phi}(y|\phi).$$

Then, we use the following convention:

$$\phi \leftrightarrow p_{X|Y,\phi}(x|y,\phi); \qquad \psi \leftrightarrow p_{X|Y,\psi}(x|y,\psi).$$

Then, $D^{(\alpha)}(\phi\|\psi) \geq 0$ and the definition of the conditional probability give the following basic inequality.

$$\frac{4}{1-\alpha^2}\left\{\frac{p_{Y|\psi}(y|\psi)}{p_{Y|\phi}(y|\phi)}\right\}^{\frac{1+\alpha}{2}} \geq \frac{4}{1-\alpha^2}\int_{\mathcal{X}(y)} p_{X|Y,\phi}(x|y,\phi)\left\{\frac{p_{Y|X,\psi}(y|x,\psi)}{p_{Y|X,\phi}(y|x,\phi)}\right\}^{\frac{1+\alpha}{2}} dx$$

Let

$$S_{Y|X}^{(\alpha)}(\psi|\phi) \stackrel{\text{def}}{=} \int_{\mathcal{X}(y)} p_{X|Y,\phi}(x|y,\phi)\left\{\frac{p_{Y|X,\psi}(y|x,\psi)}{p_{Y|X,\phi}p(y|x,\phi)}\right\}^{\frac{1+\alpha}{2}} dx$$

$$\stackrel{\text{def}}{=} E_{p_{X|Y,\phi}}\left[\left\{\frac{p_{Y|X,\psi}}{p_{Y|X,\phi}}\right\}^{\frac{1+\alpha}{2}}\right] \stackrel{\text{def}}{=} E_{p_{X|Y,\phi}}\left[R_{Y|X}^{(\alpha)}(\psi|\phi)\right]. \tag{6}$$

Let

$$Q_{Y|X}^{(\alpha)}(\psi|\phi) = \frac{2}{1+\alpha}\{S_{Y|X}^{(\alpha)}(\psi|\phi) - 1\}. \tag{7}$$

Then, one obtains

$$\frac{2}{1-\alpha}L_Y^{(\alpha)}(\psi|\phi) \geq \frac{2}{1-\alpha}Q_{Y|X}^{(\alpha)}(\psi|\phi). \tag{8}$$

Thus, the following theorem and corollary are obtained.

[**Theorem 3.1**] The α-EM algorithm is a series of applications of E-step and M-step.

E-step: Compute $Q_{Y|X}^{(\alpha)}(\psi|\phi)$.

M-step: Compute $\psi^* = \arg\max_{\psi} Q_{Y|X}^{(\alpha)}(\psi|\phi)$.

[**Corollary 3.2**] The α-EM algorithm is classified into the following three cases depending on the number α.

E-step: Compute $S_{Y|X}^{(\alpha)}(\psi|\phi)$.

M-step: Compute the following.

1. $\alpha < -1$: $\psi^* = \arg\min_{\psi} S_{Y|X}^{(\alpha)}(\psi|\phi)$,

2. $\alpha = -1$: $\psi^* = \arg\max_{\psi} E_{p_{X|Y,\phi}}\left[\log p_{Y|X,\psi}\right]$,

3. $\alpha > -1$: $\psi^* = \arg\max_{\psi} S_{Y|X}^{(\alpha)}(\psi|\phi)$.

There are two comments on the above theorem and corollary. First, the convergence is claimed on $Q_{Y|X}^{(\alpha)}(\psi|\phi)$, but not necessarily on each parameter. Secondly, for $\alpha < -1$, the M-step is not the maximization but the minimization.

3.2 Cyclic α-EM and change of conditioning

In Section 3.1, the α-divergence between $p_{X|Y,\phi}$ and $p_{X|Y,\psi}$ was computed in the world of $p_{X|Y,\phi}$. The same holds by replacing the conditioning. That is, the α-divergence between $p_{Y|X,\zeta}$ and $p_{Y|X,\xi}$ is computed in the world of $p_{Y|X,\zeta}$. Then, for $Q_{X|Y}^{(\alpha)}(\xi|\zeta)$ and $S_{X|Y}^{(\alpha)}(\xi|\zeta)$, equivalent versions of Theorem 3.1 and Corollary 3.2 hold. Thus, the pair $(Q_{Y|X}^{(\alpha)}(\psi|\phi), Q_{X|Y}^{(\alpha)}(\xi|\zeta))$, or equivalently, $(S_{Y|X}^{(\alpha)}(\psi|\phi), S_{X|Y}^{(\alpha)}(\xi|\zeta))$ generates a *cyclic* α-EM. Different philosophies concerning to this are given in [5] and [6] for the case of $\alpha = -1$. It is important to point out here that the cyclic α-EM will be just a building block of the systolic and monitoring α-EM [11], [12], [13] of Section 6.

There is one more different version of the conditioning in the α-EM. The non-negativity of $D^{(\alpha)}(\phi\|\psi) \geq 0$ and Bayes' equations give

$$\tfrac{4}{1-\alpha^2}\left\{\tfrac{p_{\psi|Y}(\psi|y)}{p_{\phi|Y}(\phi|y)}\right\}^{\frac{1+\alpha}{2}} \geq \tfrac{4}{1-\alpha^2}\int_{\mathcal{X}(y)} p_{X|Y,\phi}(x|y,\phi)\left\{\tfrac{p_{\psi|X,Y}(\psi|x,y)}{p_{\phi|X,Y}(\phi|x,y)}\right\}^{\frac{1+\alpha}{2}}dy$$

when ϕ and ψ are considered as random variables. By allowing the above notations on probabilities, one obtains

$$S_\phi^{(\alpha)}(\psi|\phi) = E_{p_{X|Y,\phi}}\left[\left\{\tfrac{p_{\psi|X,Y}(\psi|x,y)}{p_{\phi|X,Y}(\phi|x,y)}\right\}^{\frac{1+\alpha}{2}}\right] \tag{9}$$

and

$$Q_\phi^{(\alpha)}(\psi|\phi) = \tfrac{2}{1+\alpha}\left\{S_\phi^{(\alpha)}(\psi|\phi) - 1\right\} \tag{10}$$

for the α-EM on parameter estimation.

4 The α-EM Algorithm for Neural Networks on Mixtures
4.1 The α-EM for neural networks via experts

In the neural network via hierarchical experts [2], random variables have the following correspondences.

$$X \leftrightarrow X; \qquad Y \leftrightarrow (Y, Z).$$

The random variable X stands for an input, Y is the teacher, and Z gives a path in a mixture. The mixture model is

$$p(y|x,\phi) = \sum_{i=1}^K p(y, z_i|x,\phi) = \sum_{i=1}^K g_i \sum_{j=1}^k g_{j|i} p_{ij}(y)$$

with

$$p_{ij}(y) = \exp\{\tfrac{-(y-\mu_{ij})^T(y-\mu_{ij})}{2\sigma_{ij}^2}\}/(2\pi)^{2/n}\sigma_{ij}^n.$$

Then,

$$S_{YZ|X}^{(\alpha)}(\psi|\phi) = E_{p_{Z|XY,\phi}}\left[\left\{\tfrac{p_{YZ|X,\psi}(y,z|x,\psi)}{p_{YZ|X,\phi}(y,z|x,\phi)}\right\}^{\frac{1+\alpha}{2}}\right], \tag{11}$$

$$Q_{YZ|X}^{(\alpha)}(\psi|\phi) = \tfrac{2}{1+\alpha}\left\{S_{YZ|X}^{(\alpha)}(\psi|\phi) - 1\right\}, \tag{12}$$

and

$$\psi^* = \arg\max_\psi Q_{YZ|X}^{(\alpha)}(\psi|\phi). \tag{13}$$

4.2 Gradient ascent learning based upon the α-EM

Let $U_{ij}^{(\alpha)}(\phi)$ be a neural weight connecting element j to i in the probability world of ϕ. Let ψ be a post-learning world. Then $\psi = \phi + \Delta\phi$ corresponds to

$$\begin{cases} U_{ij}^{(\alpha)}(\psi) = U_{ij}^{(\alpha)}(\phi) + \Delta U_{ij}^{(\alpha)}(\phi), \\ \Delta U_{ij}^{(\alpha)}(\phi) = \rho \frac{\partial}{\partial U_{ij}} \left[\frac{2}{(1+\alpha)} \left\{ p_{Y|X,\phi}(y|x,\phi) \right\}^{\frac{1+\alpha}{2}} \right], \end{cases}$$

where ρ is a small constant. First, we consider a successive version of the learning. The increment is as follows.

$$\Delta U_{ij}^{(\alpha)}(\phi) = \rho p_{Y|X,\phi}(y|x,\phi)^{\frac{-1+\alpha}{2}} \frac{\partial}{\partial U_{ij}} \left\{ p_{Y|X,\phi}(y|x,\phi) \right\}$$

The case of $\alpha = -1$ is the traditional "log" version.

$$\Delta U_{ij}^{(-1)}(\phi) = \rho p_{Y|X,\phi}(y|x,\phi)^{-1} \frac{\partial}{\partial U_{ij}} \left\{ p_{Y|X,\phi}(y|x,\phi) \right\}$$

Thus,

$$\Delta U_{ij}^{(\alpha)}(\phi) = p_{Y|X,\phi}(y|x,\phi)^{\frac{1+\alpha}{2}} \Delta U_{ij}^{(-1)}(\phi)$$

holds. Therefore, a large α emphasizes learning at a high probability density. Smaller α's watch errors rigorously at low density parts. For $\alpha < -1$, the increment $\Delta U_{ij}^{(\alpha)}(\phi)$ is actually a decrement for $\{p_{Y|X,\psi}(y|x,\psi)/p_{Y|X,\phi}(y|x,\phi)\}^{\frac{1+\alpha}{2}}$. This is a gradient descent. The effect of α is understood as follows: The number α chooses a path or a world to achieve $\max_\psi \ p_{Y|X,\psi}(y|x,\psi)$.

Next, we give a batch learning version. In this case, $\prod_t p_{Y|X,\psi}(y(t)|x(t),\psi)$ and $\prod_t p_{Y|X,\phi}(y(t)|x(t),\phi)$ are considered. Then, one obtains

$$\begin{aligned} \Delta U_{ij}^{(\alpha)}(\phi) &= \rho \textstyle\prod_t p_{Y|X,\phi}(y(t)|x(t),\phi)^{\frac{-1+\alpha}{2}} \frac{\partial}{\partial U_{ij}} \left\{ \textstyle\prod_t p_{Y|X,\phi}(y(t)|x(t),\phi) \right\} \\ &= \rho \textstyle\prod_t p_{Y|X,\phi}(y(t)|x(t),\phi)^{\frac{1+\alpha}{2}} \Delta U_{ij}^{(-1)}(\phi). \end{aligned} \tag{14}$$

as the batch version.

5 Expectation Learning and Various Statistics Measures for the α-EM

5.1 The α-versions of important statistics measures

The probability density $\{\prod_t p_{Y|X,\psi}(y(t)|x(t),\psi)\}^{(1+\alpha)/2}$ and its extended logarithm

$$L_{Y|X}^{(\alpha)}(\psi) = \frac{2}{1+\alpha} \left[\left\{ \textstyle\prod_t p_{Y|X,\psi}(y(t)|x(t),\psi) \right\}^{\frac{1+\alpha}{2}} - 1 \right] \tag{15}$$

defines an efficient score from the following equation.

$$L_{Y|X}^{(\alpha)}(\psi) = L_{Y|X}^{(\alpha)}(\phi) + \Delta L_{Y|X}^{(\alpha)}(\phi).$$

That is, the α-efficient score is

$$\begin{aligned} \frac{\partial L_{Y|X}^{(\alpha)}(\phi)}{\partial \phi} &= \left\{ \textstyle\prod_t p_{Y|X,\phi}(y(t)|x(t),\phi) \right\}^{\frac{-1+\alpha}{2}} \frac{\partial}{\partial \phi} \left\{ \textstyle\prod_t p_{Y|X,\phi}(y(t)|x(t),\phi) \right\} \\ &= \left\{ \textstyle\prod_t p_{Y|X,\phi}(y(t)|x(t),\phi) \right\}^{\frac{1+\alpha}{2}} \frac{\partial \ell_{Y|X}(\phi)}{\partial \phi}, \end{aligned} \tag{16}$$

Here, $\ell_{Y|X}(\phi)$ is the logarithm of $\prod_t p_{Y|X,\phi}(y(t)|x(t),\phi)$. The number $\partial \ell_{Y|X}(\phi)/\partial\phi$ is the Fisher's efficient score. Hereafter, $L_{Y|X}^{(\alpha)}(\phi)$ is denoted simply by $L^{(\alpha)}$ or L depending on emphases.

There can be many α-information measures corresponding to the Fisher's information measure. We list up the following two versions.

1. Exponential expectation:

$$E_{\exp(L)}\left[\left(\frac{\partial L^{(\alpha)}}{\partial \phi}\right)\left(\frac{\partial L^{(\alpha)}}{\partial \phi^T}\right)\right] = -E_{\exp(L)}\left[\frac{\partial^2 L^{(\alpha)}}{\partial \phi \partial \phi^T}\right] \stackrel{\text{def}}{=} M_{\exp}^{(\alpha)}(\phi). \qquad (17)$$

2. Plain expectation:

$$E\left[\frac{1-\alpha}{2}p^{-(1+\alpha)}\left(\frac{\partial L^{(\alpha)}}{\partial \phi}\right)\left(\frac{\partial L^{(\alpha)}}{\partial \phi^T}\right)\right] = -E\left[p^{-\frac{1+\alpha}{2}}\left(\frac{\partial^2 L^{(\alpha)}}{\partial \phi \partial \phi^T}\right)\right] \stackrel{\text{def}}{=} M^{(\alpha)}(\phi)$$
$$(18)$$

Both cases are reduced to Fisher's information measure when $\alpha = -1$. But, $M_{\exp(L)}^{(\alpha)}(\phi)$ is complicated to compute since an exponential of the probability density itself is the base of the expectation. Therefore, we adopt the second case; $M^{(\alpha)}(\phi)$.

Related to $M^{(\alpha)}(\phi)$ is the Cramér-Rao's bound for parameter estimation:

$$V(\hat{\phi} - \phi) \geq 1/V(\partial \ell/\partial \phi) = 1/M^{(-1)}(\phi).$$

Here, V is the variance, and $\hat{\phi}$ is an estimate of ϕ satisfying $E[\hat{\phi}] = \phi$. This Cramér-Rao bound can be derived from the α-efficient score:

$$V(\hat{\phi} - \phi) \geq 1/V(p^{-(1+\alpha)/2}\partial L^{(\alpha)}/\partial \phi). \qquad (19)$$

The righthand side is reduced to

$$1/V(p^{-(1+\alpha)/2}\partial L^{(\alpha)}/\partial \phi) = 1/V(p^{-1}\partial p/\partial \phi) = 1/V(\partial \ell/\partial \phi).$$

Therefore, one obtains

$$m^{(\alpha)} \stackrel{\text{def}}{=} \frac{M^{(\alpha)}(\phi)}{M^{(-1)}(\phi)} = \frac{1-\alpha}{2}. \qquad (20)$$

Thus, this number $m^{(\alpha)}$ reflects the speed that the learning system acquires knowledge from the training. This $m^{(\alpha)}$ can be called the aptitude number. If $\alpha = -1$, i.e., $M^{(\alpha)}(\phi) = M^{(-1)}(\phi)$, then this is the case of the logarithm. If $\alpha=1$, then $M^{(\alpha)}(\phi) = M^{(0)}(\phi) = 0$. $M^{(\alpha)}$ becomes negative for $\alpha > 1$. Then, the system has a reverse learning aptitude.

5.2 The α-information measure used in the learning

Let the τ-th iteration value of the extended logarithm of p_τ be

$$L_\tau^{(\alpha)}(\phi) = \frac{2}{1+\alpha}\left\{p_\tau^{\frac{1+\alpha}{2}} - 1\right\}.$$

The Newton-Raphson method using $L_\tau^{(\alpha)}(\phi)$ is

$$\phi_{\tau+1} = \phi_\tau - \left[\frac{\partial^2 L_\tau^{(\alpha)}}{\partial \phi \partial \phi^T}\right]^{-1}\frac{\partial L_\tau^{(\alpha)}}{\partial \phi}.$$

Therefore, a simple expectation gives

$$\phi_{\tau+1} = \phi_\tau - \left[E\left\{\frac{\partial^2 L_\tau^{(\alpha)}}{\partial \phi \partial \phi^T}\right\}\right]^{-1}\frac{\partial L_\tau^{(\alpha)}}{\partial \phi}. \qquad (21)$$

This is one possibility. But, the update matrix does not have exact relationship to the α-information measure. Consider the relationship

$$\frac{\partial L_\tau^{(\alpha)}}{\partial \phi} = p_\tau^{\frac{1+\alpha}{2}}\frac{\partial \ell_\tau(\phi)}{\partial \phi}.$$

Then, we have

$$\phi_{\tau+1} = \phi_\tau - \left[\frac{\partial^2 L_\tau^{(\alpha)}}{\partial\phi\partial\phi^T}\right]^{-1} p_\tau^{-\frac{1+\alpha}{2}} \frac{\partial \ell_\tau}{\partial\phi} = \phi_\tau - \left[p_\tau^{-\frac{1+\alpha}{2}}\left(\frac{\partial^2 L_\tau^{(\alpha)}}{\partial\phi\partial\phi^T}\right)\right]^{-1} \frac{\partial \ell_\tau}{\partial\phi}.$$

Thus, the expectation version is

$$\begin{aligned}
\phi_{\tau+1} &= \phi_\tau - \left[E\left\{p_\tau^{-\frac{1+\alpha}{2}}\left(\frac{\partial^2 L_\tau^{(\alpha)}}{\partial\phi\partial\phi^T}\right)\right\}\right]^{-1}\frac{\partial \ell_\tau}{\partial\phi} \\
&= \phi_\tau + \left[M^{(\alpha)}(\phi)\right]^{-1}\frac{\partial \ell_\tau}{\partial\phi} = \phi_\tau + \frac{2}{1-\alpha}\left[M^{(-1)}(\phi)\right]^{-1}\frac{\partial \ell_\tau}{\partial\phi}. \quad (22)
\end{aligned}$$

This case is effective for $\alpha \neq 1$.

5.3 Update equations for the expert networks

Here, we give examples of update equations for expert networks. Let $\mathcal{I} = \{(x(t), y(t)); \ t = 1, \ldots, N\}$ be the incomplete data set. Let $\mathcal{Z} = \{z_j(t); \ j = 1, \ldots, K, \ t = 1, \ldots, N\}$ be the missing or hidden variables. Then, $\mathcal{C} = \{\mathcal{I}, \mathcal{Z}\}$ is regarded as the complete data set. Then, the incomplete probabilities are described by

$$\begin{cases}
P(\mathcal{C}|\Psi) = \prod_{t=1}^N \prod_{j=1}^K g_j(x(t), \psi_0)p(y(t)|x(t), \psi_j), \\
P(\mathcal{C}|\Phi) = \prod_{t=1}^N \prod_{j=1}^K g_j(x(t), \phi_0)p(y(t)|x(t), \phi_j).
\end{cases}$$

Therefore,

$$R_{\mathcal{C}}^{(\alpha)}(\Psi|\Phi) = \left[\prod_{t=1}^N \prod_{j=1}^K \left\{\frac{g_j(x(t), \psi_0)p(y(t)|x(t), \psi_j)}{g_j(x(t), \phi_0)p(y(t)|x(t), \phi_j)}^{I_j^{(t)}}\right\}\right]^{\frac{1+\alpha}{2}}$$

and

$$L_{\mathcal{C}}^{(\alpha)}(\Psi|\Phi) = \frac{2}{1+\alpha}\left\{R_{\mathcal{C}}^{(\alpha)}(\Psi|\Phi) - 1\right\}$$

hold. Then, one obtains

$$\arg\max_\Psi Q^{(\alpha)}(\Psi|\Phi) = \arg\max \frac{2}{1+\alpha}S^{(\alpha)}(\Psi|\Phi)$$

with

$$S^{(\alpha)} = E_{[\mathcal{Z}|\mathcal{I}]}R^{(\alpha)},$$

and

$$Q^{(\alpha)}(\Psi|\Phi) = \frac{2}{1+\alpha}\left[\prod_{t=1}^N \sum_{j=1}^K h_j^{(t)}\left\{\frac{g_j(x(t)|\psi_0)p(y(t)|x(t), \psi_j)}{g_j(x(t)|\phi_0)p(y(t)|x(t), \phi_j)}\right\}^{\frac{1+\alpha}{2}} - 1\right],$$

where

$$h_j^{(t)} = p(z_j(t) = 1|x(t), y(t), \phi_0).$$

For the update equations, properties of the following derivatives are used:

$$\frac{\partial Q^{(\alpha)}}{\partial\psi_0}, \quad \frac{\partial Q^{(\alpha)}}{\partial\psi_j}, \quad \frac{\partial Q^{(\alpha)}}{\partial\Sigma_{\psi_j}}, \quad j = 1 \ldots, K.$$

Here, Σ_{ψ_j} is a covariance matrix $(j = 1, \ldots, K)$. Since each optimization affects others, we have to use the following update method if strict increase of the cost $Q^{(\alpha)}$ is requested.

[Step k+1/3] Given $\phi_0^{[k]}$, $\psi_j = \phi_j^{[k]}$, $\Sigma_{\psi_j} = \Sigma_{\phi_j}^{[k]}$, $j = 1, \ldots, K$, compute $\phi_0^{[k+1/3]} = \psi_0^*$. Then, set $\phi_j^{[k+1/3]} = \phi_j^{[k]}$ and set $\Sigma_{\phi_j}^{[k+1/3]} = \Sigma_{\phi_j}^{[k]}$.

[Step k+2/3] Given $\psi_0 = \phi_0^{[k+1/3]}$, $\phi_j^{[k+1/3]}$, $\Sigma_{\psi_j} = \Sigma_{\phi_j}^{[k+1/3]}$, $j = 1, \ldots, K$, compute $\phi_j^{[k+2/3]} = \psi_j^*$. Then, set $\phi_0^{[k+2/3]} = \phi_0^{[k+1/3]}$ and set $\Sigma_{\phi_j}^{[k+2/3]} = \Sigma_{\phi_j}^{[k+1/3]}$.

[Step k+1] Given $\psi_0 = \phi_0^{[k+2/3]}$, $\psi_j = \phi_j^{[k+2/3]}$, $\Sigma_{\phi_j}^{[k+2/3]}$, $j = 1, \ldots, K$, compute $\Sigma_{\phi_j}^{[k+1]} = \Sigma_{\psi_j}^*$. Then, set $\phi_0^{[k+1]} = \phi_0^{[k+2/3]}$ and set $\phi_j^{[k+1]} = \phi_j^{[k+2/3]}$.
Note that the scheduling

$$\psi_0 \to \psi_j \to \Sigma_{\psi_j} \to \psi_0 \to \psi_j \to \cdots$$

is not the only one.

[Update of ψ_0] $\partial Q^{(\alpha)}/\partial \psi_0$ is usede as an update vector. The indexing is as follows.

$$\psi_0 = \phi_0^{[k]} = \phi_0, \quad \psi_j = \phi_j^{[k]} = \phi_j, \quad \Sigma_{\psi_j} = \Sigma_{\phi_j} = \Sigma_{\phi_j}^{[k]}.$$

Therefore, this step is the same as the case of logarithm.

$$\phi_0^{[k+1/3]} = \phi_0^{[k]} - \gamma_g^{(\alpha)} \left(R_g^{(\alpha)}[k] \right)^{-1} \mathbf{e}_g^{(\alpha)}[k],$$

where

$$\mathbf{e}_g^{(\alpha)}[k] = \sum_{t=1}^{N} \sum_{j=1}^{K} \left\{ h_j^{(t)}[k] - g_j^{(t)}[k] \left(\frac{\partial s_j}{\partial \psi_0} \right) \right\}$$

and

$$R_g^{(\alpha)}[k] = -\sum_{t=1}^{N} \sum_{j=1}^{K} g_j^{(t)} (1 - g_j^{(t)}) \left(\frac{\partial s_j}{\partial \psi_0} \frac{\partial s_j}{\partial \psi_0^T} \right)^{[k]}.$$

[Update of ψ_j] From

$$\frac{\partial Q^{(\alpha)}}{\partial \psi_j} = 0,$$

one obtains

$$\phi_j^{[k+2/3]} = \left\{ R_j^{[k+1/3]}(\alpha) \right\}^{-1} \mathbf{c}_j^{[k+1/3]},$$

where

$$\mathbf{c}_j^{[k+1/3]} = \sum_{t=1}^{N} \prod_{\tau=1}^{N} \{W^{(\tau)}\}^{1-\delta_{t\tau}} \{h_j^{(\alpha)}(\tau) X^{(\tau)} \Sigma_{\phi_j}^{-1} y(\tau)\}^{\delta_{t\tau}}$$

and

$$R_j^{[k+1/3]}(\alpha) = \sum_{t=1}^{N} \prod_{\tau=1}^{N} \{W^{(\tau)}\}^{1-\delta_{t\tau}} \{h_j^{(\alpha)}(\tau) X^{(\tau)} \Sigma_{\phi_j}^{-1} X^{(\tau)T}\}^{\delta_{t\tau}},$$

where

$$W^{(\tau)} = \sum_{j=1}^{K} h_j^{(\alpha)}(\tau) = \sum_{j=1}^{K} h_j^{(\alpha)} \left\{ \frac{g_j(x(t),\psi_0)p(y(t)|x(t),\psi_j)}{g_j(x(t),\phi_0)p(y(t)|x(t),\phi_j)} \right\}^{\frac{1+\alpha}{2}}.$$

[Update of Σ_{ψ_j}] From

$$\frac{\partial Q^{(\alpha)}}{\partial \Sigma_{\psi_J}} = 0,$$

the following update equation is obtained.

$$\begin{aligned}
\Sigma_{\phi_j}^{[k+1]} &= \sum_{t=1}^{N} \prod_{\tau=1}^{N} \{W^{(\tau)}\}^{1-\delta_{t\tau}} \{h_j^{(\alpha)}(\tau)(y(\tau) - \mu_j)(y(\tau) - \mu_j)^T\}^{\delta_{t\tau}} \\
&\quad / \sum_{t=1}^{N} \prod_{\tau=1}^{N} \{W^{(\tau)}\}^{1-\delta_{t\tau}} \{h_j^{(\alpha)}(\tau)\}^{\delta_{t\tau}}.
\end{aligned}$$

6 Monitoring α-EM

6.1 Systolic layer of α-EM

In Section 3.2, we considered the cyclic α-EM. But, this structure is too monolithic to model complex systems such as brains. Therefore, it is at least necessary to have block connection structures. The building block of the α-EM has multiple input/output pairs passing each other. Therefore, block connections can make a systolic layer. Fig. 6.1 illustrates such an example.

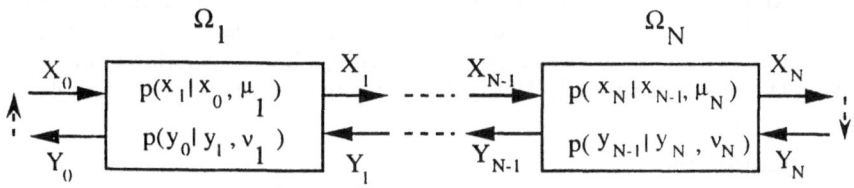

Fig. 6.1 A systolic layer of the α-EM.

In this figure, processing of X_{i-1} to generate X_i in the sub-world Ω_i becomes active at an even clock. On the other hand, processing of Y_i to generate Y_{i-1} in the sub-world Ω_i becomes active at an odd clock. A more concrete version of this figure is obtained by the following specifications:

$$\mu_i = \nu_i, \quad Y_i = X_{i-1}, \quad X_N = Y_0 = \Lambda, \quad p(x_N|x_{N-1}, \mu_N) = p(y_0|y_1, \nu_1) = \lambda.$$

Here, Λ and λ stand for null outputs or null structures.

6.2 Asynchronous monitoring α-EM

The systolic structure of Fig. 6.1 is a base for the next sophistication. The main stream of the systolic layer can fork. This can be regarded as attaching a monitor such as in Fig. 6.2. In this case, even/odd clocking is generalized to mutual exclusion. That is, multiple directions of the streams can not be active at the same time in the same sub-world. There can be a learning scheduler, say $\mathcal{A}$, that orders the activity of modules. The scheduler $\mathcal{A}$ is proper if it requests every subsystem to work infinitely often.

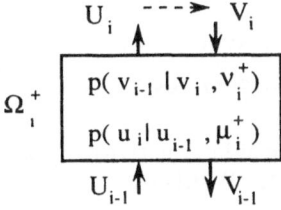

Fig. 6.2 Monitoring α-EM.

7 Concluding Remarks

In this paper, the α-EM algorithm was presented first. Then, this algorithm was used to build up a block connected system. Main findings are as follows.

1. The α-EM algorithm is derived based upon the generalized divergences. The number α corresponds to a probability weight in learning.

2. The case of $\alpha = -1$ is reduced to the traditional log-likelihood EM.

3. The efficient score, the information measure matrix, the Cramér-Rao bound, and the expectation of the Hessian matrix are derived for the general α.

4. The α-EM can be blockwisely connected to build up the total system.

The traditional case of $\alpha = -1$ gives an elegant structure, especially for the exponential family. But, discussions from the general case are meritorious. It was understood that α chooses a path of learning. Thus, switching this number α during the learning can help the system to converge to a more desirable state.

Acknowledgment

The author is grateful to his graduate student, Mr. S. Furukawa, for his discussions.

References

1. Dempster, A.P., Laird, N.M., Rubin, D.B.: Maximum likelihood from incomplete data via the EM algorithm, J. R. Stat. Soc. Sr. B. **39** (1978) 1-38

2. Jordan, M.J., Jacobs, R.A.: Hierarchical mixtures of experts and the EM algorithm, Neural Computation **6** (1994) 181-214

3. Jordan, M.J., Xu, L.: Convergence results for the EM approach to mixtures of experts architecture, Neural Networks **8** (1995) 1409-1431

4. Amari, S.: Information geometry of the EM and em algorithms for neural nets, Neural Networks **8** (1995) 1379-1408

5. Amari, S.: Information geometry of neural networks - New Bayesian duality theory -, Proc. ICONIP'96 **1** (1996) 3-6

6. Xu, L.: Bayesian-Kullback Ying-Yang Machine: Reviews and new results, Proc. ICONIP'96, **1** (1996) 59-67

7. Rényi, A.: On measures of entropy and information, Proc. 4th Berkeley Symp. Math. Stat. and Pr., **1** (1960) 547-561

8. Havrda, J.H., Chavat, F.: Qualification methods of classification processes: Concepts of structural α entropy, Kybernetica, **3** (1967) 30-35

9. Amari, S., Nagaoka, H.: *Methods of Information Geometry* (in Japanese), Iwanami Publishing Co. (1993)

10. Csiszár, I: A class of measure of informativity of observation channels, Periodica Mathematica Hungarica **2** (1972) 191-213

11. Matsuyama, Y.: The α-EM algorithm and its block connections, Tech. Report, **CML-96-1** (1996) Computation Mechanism Lab., Dept. of EECE, Waseda Univ.

12. Matsuyama, Y.: The weighted EM learning and monitoring structure, Info. Proc. Soc. of Japan, 54th Convention Record (1997) 6G-04

13. Matsuyama, Y.: The Weighted EM algorithm and Block Monitoring, Proc. ICNN, (1997)

Training Simple Recurrent Networks Through Gradient Descent Algorithms*

M.A. Castaño[†], F. Casacuberta[††], A. Bonet[††]

[†] Dpto. de Informática. Universitat Jaume I de Castellón. Spain.
[††] Dpto. Sistemas Informáticos y Computación. Universidad Politécnica de Valencia. Spain.
e-mail: castano@inf.uji.es, fcn@iti.upv.es

Abstract

In the literature Simple Recurrent Networks have been successfully trained through both Exact and Truncated Gradient Descent algorithms. This paper empirically compares these two learning methods, training an Elman architecture with first-order connections in order to approach a simple Language Understanding task.

Keywords: Language Understanding, Simple Recurrent Neural Networks, Training Algorithms, Gradient Descent.

1. Introduction

Simple Recurrent Networks (SRNs), as Elman architectures [Elman,90], have been usually trained through an adequate modification of the standard *Backward-Error-Propagation algorithm* [Rumelhart,86]. While this procedure does not take into account the recurrent connections in order to compute the weight changes, it is not guaranteed to follow the precise negative gradient of the total error throughout the trajectory (i.e., *truncates the gradient*), although it works well in practice [Cleeremans,89] [Stolcke,90] [Castaño,93] [Castaño,95].

On the other hand, SRNs (with first and/or higher-order connections) can be considered as a particular architecture of the general Recurrent Networks; according to this, adequate modifications of the (corresponding first or higher-order version of the) well-known *Real Time Recurrent Learning algorithm* [Williams,89] [Giles,92] have been also recently employed to train the recurrent layer of Elman SRNs [Sopena,94] [Alquézar,95] [Giles,95]. Faced to the above learning method, the batch version of these last algorithms actually follows the gradient (*exact gradient*).

---

* Work supported in part by the Spanish CICYT, under grant TIC-95-084-CO2

Anyway, empirical comparisons between both the exact gradient descent method and the truncated algorithm have not been previously carried out. Some experiments on this issue are presented in this paper assuming a first-order Elman SRN. To do so, a simple Language Understanding task is considered [Castaño,93].

The paper is organized as follows. First, both the exact and the truncated descent methodology are briefly and formally described. Later, Section 2 presents comparative experimental results of these two training algorithms, taking into account the associated on-line and batch versions. The conclusions of all this experimental process are finally discussed in Section 3.

2. Learning algorithms for Simple Recurrent Networks

2.1. Exact Gradient Descent algorithms

Let us assume a first-order Elman SRN with D inputs, M outputs and H hidden units (see Figure 1). A general architecture with V preceding and V following context of the input signal is considered. According to this features, the functional behaviour of the connectionist model is formally presented as:

$$s_k^1(t) = f(net_k^1(t)) = f\left[\sum_{v=-V}^{V} \sum_{l=1}^{D} w_{kvl}^1 x_l(t+v) + \sum_{l=1}^{H} W_{kl}^1 s_l^1(t-1) \right] \quad 1 \le k \le H \tag{1}$$

$$s_p^2(t) = f(net_p^2(t)) = f\left[\sum_{l=1}^{H} w_{pl}^2 s_l^1(t-1) \right] \quad 1 \le p \le M. \tag{2}$$

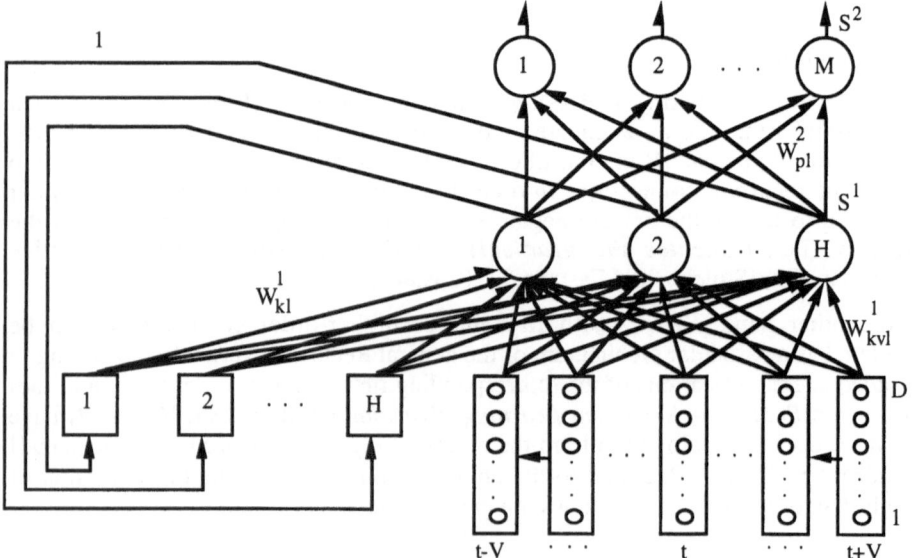

Figure 1. Elman Simple Recurrent Network.

In order to obtain an exact gradient descent learning algorithm for this (firs-order) Elman net, the same procedure as that employed for Real Time Recurrent Networks [Williams,89] can be applied. To do so, a training sequence (x,y) should be considered, where $x(t) \in \Re^D$ denotes the D-tuple of external input signals to the net at time t, $1 \leq t \leq |x|$; and $y(t) \in \Re^M$ the corresponding M-tuple of target values that the outputs should match. The objective function to be minimized is the squared error between the expected output in t and the associated output actually provided by the net. Assuming that the model is run starting at time t_0 up to some final time t_1, the total error over this trajectory is:

$$E(t_0, t_1) = \sum_{t=t_0}^{t_1} E(t) = \sum_{t=t_0}^{t_1} \frac{1}{2} \sum_{p=1}^{M} (y_p(t) - s_p^2(t))^2. \tag{3}$$

The application of the gradient descent method to this total error measure in the weight space can be done as follows:

- The <u>weights of the second layer</u> w_{pl}^2 $1 \leq p \leq M$ $1 \leq l \leq H$ should be updated according to:

$$\Delta w_{pl}^2 = \rho \sum_{t=t_0}^{t_1} (y_p(t) - s_p^2(t)) s_l^1(t). \tag{4}$$

- The <u>recurrent connections of the first layer</u> $w_{kl}^1 1 \leq k,l \leq H$ should be altered by:

$$\Delta w_{kl}^1 = \rho \sum_{t=t_0}^{t_1} \sum_{q=1}^{H} \left[\sum_{p=1}^{M} (y_p(t) - s_p^2(t)) w_{pq}^2 \right] \frac{\partial s_q^1(t)}{\partial w_{kl}^1}; \tag{5}$$

for each $1 \leq k,l,q \leq H$, equation (5) must be computed by an iterative procedure through:

$$\frac{\partial s_q^1(t)}{\partial w_{kl}^1} = f'(net_q^1(t)) \left[\delta(k,q) s_l^1(t-1) + \sum_{r=1}^{H} w_{qr}^1 \frac{\partial s_r^1(t-1)}{\partial w_{kl}^1} \right] \tag{6}$$

$$\left. \frac{\partial s_q^1(t)}{\partial w_{kl}^1} \right|_{t=t_0} = 0 \tag{7}$$

where δ denotes the Kronecker delta function.

- For the <u>non-recurrent connections of the first layer</u> w_{kvl}^1 $1 \leq k \leq H$, $-V \leq v \leq V$ $1 \leq l \leq D$

$$\Delta w_{kvl}^1 = \rho \sum_{t=t_0}^{t_1} \sum_{r=1}^{H} \left[\sum_{p=1}^{M} (y_p(t) - s_p^2(t)) w_{pr}^2 \right] \frac{\partial s_r^1(t)}{\partial w_{kvl}^1}; \tag{8}$$

for each $1 \leq k,r \leq H$, $-V \leq v \leq V$ and $1 \leq l \leq D$ equation (8) must be computed by an iterative procedure through:

$$\frac{\partial s_r^l(t)}{\partial w_{kvl}^l} = f'(net_r^l(t)) \left[x_1(t+v)\delta(k,r) + \sum_{q=1}^{H} w_{rq}^l \frac{\partial s_q^l(t-1)}{\partial w_{kvl}^l} \right] \tag{9}$$

$$\left. \frac{\partial s_r^l(t)}{\partial w_{kvl}^l} \right|_{t=t_0} = 0. \tag{10}$$

The above method was derived under the assumption that the weights remained fixed throughout the trajectory from t_0 to t_1 (batch learning). In order to allow on-line training, the weights should be updated at every time step, t. Thus, the resulting algorithm does not truly lead to weight changes along the negative gradient of $E(t_0,t_1)$, since the weights themselves are actually altered over the course of the trajectory.

The overall updating of the net weights according to these formulae has a computational cost of $\Theta((M+H)H^2(DV+H))$ for each t, $t_0 \leq t \leq t_1$; and for the sake of simplicity, assuming that $M \in O(H)$ and $DV \in O(H)$, this expression is stated as $\Theta(H^4)$.

Similar learning formulae can be obtained for another SRN introduced by Jordan [Jordan,88], although the associated computational cost is $\Theta(H^5)$ in that case.

2.2. Truncated Gradient Descent algorithms

The truncated gradient descent algorithm usually employed to train Elman SRNs consists of a modification of the Backward-Error-Propagation algorithm [Rumelhart,86]. This procedure updates the connections of the output layer according to the same equations (4) used for the previous exact gradient descent method. However, both algorithms differ in the computations for modifying the weights of the first layer; that is, the equations (5) and (8) become respectively:

$$\Delta w_{kl}^l = \rho \sum_{t=t_0}^{t_1} \left[\sum_{p=1}^{M} (y_p(t) - s_p^2(t)) f'(net_p^2(t)) w_{pk}^2 \right] f'(net_k^l(t)) s_l^l(t-1) \tag{15}$$

$$\Delta w_{vkl}^l = \rho \sum_{t=t_0}^{t_1} \left[\sum_{p=1}^{M} (y_p(t) - s_p^2(t)) f'(net_p^2(t)) w_{pk}^2 \right] f'(net_k^l(t)) x_l(t+v). \tag{16}$$

The computational cost to modify all weights of the architecture according to this truncated gradient algorithm is $\Theta(H^2)$ for each t, $t_0 \leq t \leq t_1$.

3. Experimental results

3.1. The task

The previous gradient descent training method for SRNs was applied to a simple Language Understanding task which had been previously and successfully approached with Elman and hybrid Elman-Jordan nets trained through the above truncated algorithm [Castaño,93] [Castaño,95]. The task consists of recognizing the Spanish text numbers in the one-million range, so that orthographic representations of (Spanish) numbers have to be converted into convenient semantic output representations; for example, /setecientosveintemiltreintaicuatro/ (720034, in Spanish) is to be translated into (+7*100+2*10)*1000+(+3*10+4).

The set of Semantic Units for this language and the corresponding orthographic realizations are shown in Table 1, for a total of 16 items. This choice very directly allows for a perfect sequentially with the input, even in the cases where the orthographic representation would not be in sequential correspondence with the usual decimal transcription. It should be noted that, given the irregularities and exceptions of the Spanish numbers language, this task is significantly more difficult than the corresponding task in English.

SEMANTIC CATEGORY	ORTHOGRAPHIC REALIZATIONS	SEMANTIC CATEGORY	ORTHOGRAPHIC REALIZATIONS
+0	cero	+8	ocho, oche
+1	un, uno, on	+9	nueve, nove
+2	dos, do, vei	+10	diez, dieci, ce
+3	tres, tre, trei	*10	nti, nte, nta, ntai
+4	cuatro, cator, cuare	+100	cien, ciento
+5	cinco,quin,cincue,quini	*100	cientos, entos
+6	seis, sese	+1000	mil
+7	siete, sete	)*1000+(	mil

Table 1. Semantic Units and their orthographic representations.

3.2. Features of the network

Attending to the results previously obtained on this task [Castaño,93] [Castaño,95], a (first-order) Elman network with a single hidden layer of 20 units is adopted.

In both the input and output layers, a *local representation* of the input and output language alphabet is respectively adopted. This means that input orthographic labels are encoded by orthogonal vectors. In a similar way, every output unit is dedicated to the representation of one of the possible semantic categories. Consequently, 19 binary inputs and 16 output units are adopted for the architecture.

Initial weights are set randomly in the range -0,5 to +0,5. A sigmoid (0,1) is assumed as the non-linear activation function and, consequently, context activations are initialized to 0,5 at the beginning of every message.

3.3. Training procedures

For comparison purposes the above described Elman network was trained through both the previous gradient descent and the truncated algorithm usually employed. For each learning method, both the corresponding on-line updating version and a pseudo-batch updating version were tried. On the other hand, input labels of the orthographic messages were presented sequentially at the input layer of the net, while the corresponding target semantic category was held constant for the duration of such a category. Consequently, in order to train the Elman SRN through the on-line learning algorithm, its weights should be updated after each input label is processed. However, when the net was trained through the pseudo-batch algorithm, connections should be updated only after the end of the sentence is reached.

Before each training began, a systematic scan in the 2-dimensional space defined by the *learning rate* and the *momentum* was carried out. This suggested (0.1,0.9) and (0.15,0.9) as adequate pairs for the on-line and pseudo-bath gradient descent training, respectively; and (0.3,0.5) and (0.5,0.3) for the on-line and pseudo-batch truncated training, respectively.

For any training algorithm or version, the net is evaluated on a different (validation) set of numbers randomly generated, after a certain block of arbitrary training samples is learnt. And learning stops when some established criterion is verified.

3.4. Training, and validation set

The training set adopted in the experiments, consisted of 5,000 input-output pairs, each consisting of the orthographic representation of a Spanish number randomly chosen in the one-million range and its associated semantic translation. Performance was evaluated on a different corpus of 5,000 pairs corresponding to the validation set.

3.5. Criterion assessing correct understanding

Attending to the local representation of the target semantic categories, only one of the output neurons should be activated at a time. On the other hand, the net continuously generates a semantic output: the category associated to the maximum activation, which should be also higher than a established threshold 0.85. In this case, identical consecutive categories are assumed to correspond to a single unit.

After a semantic output message is provided by the net, the input orthographic *string* is assumed to be *correctly understood* if it coincides with the expected translation. In order to determine a *semantic category recognition rate*, the obtained and expected semantic categories corresponding to every string in the validation sample are compared and the right categories are computed.

3.6. Results

The orthographic language was mapped into the semantic one, by training the SRN previously described through both the on-line and pseudo-batch versions of the gradient descent and truncated algorithms. The performance evolution of each trained net was repeatedly evaluated on the 5,000 validation strings, after each successive random block of 500 training strings was presented to the net. Learning stopped when 10 of these blocks were processed. For each of the four learning algorithms, this process was repeated for 5 different training seeds and the averaged (validation) performances are shown in Figure 2.

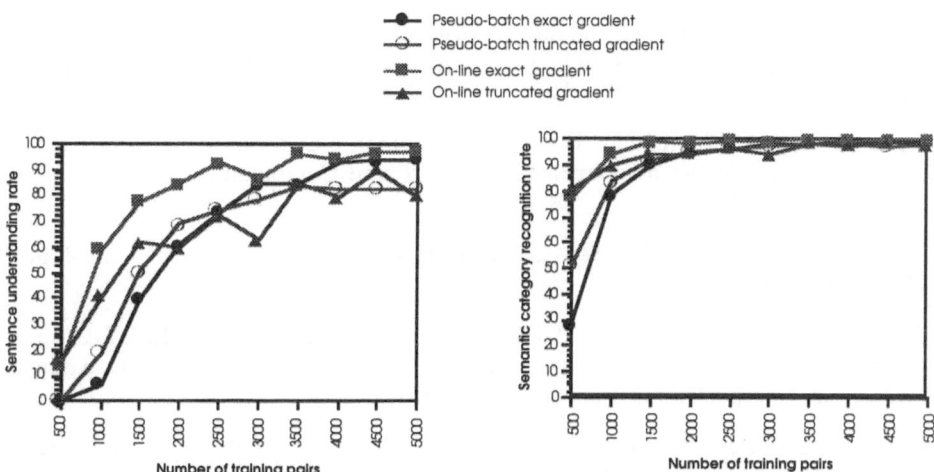

Figure 2. Sentence understanding rates and semantic category recognition rates for an Elman net trained through exact and truncated gradient descent algorithms.

Several conclusions can be drawn from these results: First, all training methods tend to converge, although on-line versions seem to be faster. Of course, since the on-line truncated method carries out the most heuristic weight updating, it leads to some oscillations. On the other hand, the behaviour in the long run for exact gradient descent methods is better than for truncated gradient algorithms, as it was expected. Summarizing, the best sentence understanding rates are achieved by training the net through the on-line exact gradient descent algorithm.

4. Conclusions and future work

In order to compare training SRNs through both Exact and Truncated Gradient Descent methodologies, an Elman architecture with first-order connections was considered. The results obtained on the considered (Language Understanding) task reveal that, as it was expected, exact descent learning algorithms provide the best performances. Nevertheless, good results are also achieved by using truncated descent procedures with which the computational cost saving is substantial.

Anyway, new tasks should be approached by using both exact and truncated gradient learning methods so that more decisive results could be established.

REFERENCES

[Alquézar,95] R. Alquézar, A. Sanfeliú. *An Algebraic Framework to Represent Finite-State Machines in Single-Layer Recurrent Neural Networks*. Neural Computation, vol. 7, no. 5, pp. 931--949, 1995.

[Castaño,93] M.A. Castaño, E. Vidal, F. Casacuberta. *Learning Direct Acoustic-to-Semantic Mapping through Simple Recurrent Networks*. Procs. EUROSPEECH-93, vol. 2, pp. 1017--1020, 1993.

[Castaño,95] M.A. Castaño, E. Vidal, F. Casacuberta. *Preliminary Experiments for Automatic Speech Understanding through Simple Recurrent Networks*. Procs. EUROSPEECH-95, vol. 3, pp. 1673--1676, 1995.

[Cleeremans,89] A. Cleeremans, D. Servan-Schreiber, J.L. McClelland. *Finite State Automata and Simple Recurrent Networks*. Neural Computation, no. 1, pp. 372-381, 1989.

[Elman,90] J.L. Elman. *Finding Structure in Time*. Cognitive Science, vol. 2, no. 4, pp. 279-311, 1990.

[Giles,92] C.L. Giles, C.B. Miller, D. Chen, H.H. Chen, G.Z. Sun, Y.C. Lee. *Learning and Extracting Finite State Automata with Second-Order Recurrent Neural Networks*. Neural Computation, no. 4, pp. 393-405, 1992.

[Giles,95] C.L. Giles, D. Chen, G.Z. Sun, H.H. Chen, Y.C. Lee, M.W. Goudreau. *Constructive Learning of Recurrent Neural Networks: Limitations of Recurrent Cascade Correlation and a Simple Solution*. IEEE Trans. on Neural Networks, vol. 6, pp. 829-836, 1995.

[Jordan,88] M.I. Jordan. *Serial Order: A Parallel Distributed Processing Approach*. Technical Report no. 8604, Institute of Cognitive Science, University of California, San Diego, 1988.

[Rumelhart,86] D.E. Rumelhart, G. Hinton, R. Williams. *Learning Sequential sSructure in Simple Recurrent Networks*. In Parallel distributed processing: Experiments in the microstructure of cognition, vol. 1. Rumelhart D.E., McClelland J.L. and the PDP Research Group (Eds),. MIT Press, Cambridge, 1986.

[Sopena,94] J.M. Sopena, R. Alquézar. *Improvement of Learning in Recurrent Networks by Substituting the Sigmoid Activation Function*. Procs. ot the Intenational Conference on Artificial Neural Networks, Springer Verlag, vol. 1, pp. 417-420, 1994.

[Williams,89] R.J. Williams, D. Zipser. *Experimental Analysis of the Real-time Recurrent Learning Algorithm*. Connection Science, vol. 1, no. 1, pp. 87--111, 1989.

On Simultaneous Weight and Architecture Learning

Santiago Rementeria [1,2]   Xabier Olabe [2]

[1] *European Software Institute, Bilbao, Spain*

[2] *Dpto. Ingeniería de Sistemas y Automática, Universidad del País Vasco - Euskal Herriko Unibertsitatea, Bilbao, Spain*

Abstract

Neural network learning is most often understood in the sense of automatic parameter adaptation. Connection strengths between units are typically updated after successive presentations of exemplar data so that the system is able to generalize to previously unseen cases the underlying function or pattern classification rule. Despite the impact of other operational, architectural and analysis aspects, only a minority of the algorithms following this inductive approach focus on parameters others than synaptic weights. In this paper we discuss a pruning method to automatically determine not only the weights but also the topology of a class of learning systems. A procedure to adapt dynamically the pruning strength is also discussed.

1 Introduction

Selecting the right structure (number and type of units, as well as their connectivity) is still one of the main difficulties faced by neural network designers. A common approach when the optimum network size is ignored consists of training an array of networks of different sizes and then keeping the smallest one learning the data. This parsimonious approach is guided by the empirical observation that smaller systems, i.e. those with a smaller number of degrees of freedom, often generalize better to previously unseen cases. Moreover, small networks are faster and cheaper to build and both their internal workings and the solutions provided are easier to interpret. A way to alleviate the burden and uncertainties of this trial-and-error, *ad-hoc* process, is to automate architecture selection.

Several structural optimization methods have been proposed to automatically come up with a network that has neither too many units (over-determined system that generalizes poorly) nor too few (under-determined one unable to learn properly). Constructive methods (Kwok and Yeung, 1995) build networks starting from simple architectures. They progressively add nodes until a (quasi-)optimal structure is obtained. In contrast to this generative approach, stabilization techniques start with an overparameterized network and then cut out redundant weights or nodes that do not contribute to the model. There are also hybrid approaches that combine both alternatives.

In this paper we present a general method to automatically determine the topology of a class of learning systems. As opposed to other methods that are based on some measure of parameter saliency, we discuss an alternative approach that includes gating factors as explicit components of synaptic connections. The following section provides some context by discussing the strong and weaker aspects of conventional network pruning techniques. Section 3 presents the fundamentals of the proposed approach and shows how it can be integrated within traditional gradient descent-based learning algorithms. Section 4 illustrates the operation of the algorithm on a simple classification task and with fixed pruning strength. In section 5 a dynamic schedule to adapt the pruning strength is introduced and its application on a classical regression benchmarking problem is commented. The paper concludes with a summary of the main ideas discussed and our plans for future research.

2 Strengths and limitations of traditional network trimming

Stabilization methods can be broadly divided into two groups. Formal stabilization, also known as regularization, adds penalty terms to the cost function so that efficient solutions are favored. The modified cost function makes the learning algorithm drive non-essential weights to a null value. As a consequence, even if the weights are not physically removed, the network functions as if it was smaller. Typical structural stabilization or pruning algorithms progressively trim connections based on estimates of the effect of weight removal on the error function. The parameters with the least impact can be effectively ignored. One could also think of hybrid methods in which impact or sensitivity calculations are included in the cost function. (Reed, 1993) provides more details on the main types of stabilization methods.

Traditional pruning techniques based on sensitivity calculations require that the network be trained to the error minimum before weights can be removed (LeCun, Denker and Solla, 1990). The pruned network can be further trained and pruned if required. Having the test variables calculated after convergence of the training process may bring in undesired overfitting that is difficult to repair. Regularization methods, on the other hand, implicitly provide dynamical weight "elimination" while training takes place (Plaut ,1986) (Chauvin, 1990). In this way, the network can adapt and decrease the error as "pruning" happens. Moreover, the training time is not increased by having to retrain the network after successive weight removals.

Another acknowledged limitation of sensitivity-based methods is their inability to deal with correlated parameters. The effect on the cost function of removing each weight, for example, is estimated assuming that no other weight would be eliminated. In most cases, once a node is effectively deleted, the sensitivities of the others may also easily result affected.

The method presented in the next section tries to overcome such shortcomings by combining the most attractive features of both formal and structural stabilization techniques. As it is the case with regularization methods, training and pruning will take place simultaneously and without the need of external variables requiring subjective interpretation. Weights will not be physically, but functionally eliminated. In contrast to such methods, however, the cost function will not be required to include any structural complexity penalty term.

3 Algorithm description

A slightly varied form of the common neural network additive model will be assumed. Synapses include a multiplicative gating factor π_{ji} whose value is learnt at the same time as the weight w_{ji} connecting node i to node j, as shown in Figure 1.

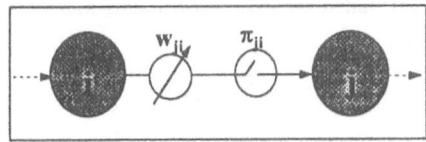

Figure 1: Modified connection model

If $o_j(k)$ is the output of node j, then:

$$o_j(k+1) = f\left(\sum_{i=0}^{N+M} w_{ji} o_i(k) \pi_{ji} \right) \tag{1}$$

where k is a time index, f is a non-linear squashing function, say the logistic one $f(x, \alpha) = (1 + e^{-\alpha x})^{-1}$, M is the size of the input vector, N is the number of non-input nodes in the network, and $o_0(k) = 1$ is defined to account for the bias weights.

The gating factor effectively determines whether the corresponding connection is active or not. It is defined in a way such that it will approach a value of one or zero as learning proceeds, thus freeing the designer from the need to decide which connections are worthwhile pruning. If all output connections from a node are deactivated, then the node itself can be considered to be non-existent. Similarly, when all the incoming connections to a node are effectively removed, it will have a constant output corresponding to its bias unit's contribution. Updating the bias of subsequent nodes accordingly will then allow us to delete that node.

The difference with the parameter originally introduced in (Mozer and Smolensky, 1989) to estimate the sensitivity of the error function to the elimination of connections is that their gating term is merely a notation convenience, a mathematical trick that is not implemented in the actual network. In contrast, not only to them, but also to (Karnin, 1990) and (Wynne-Jones, 1993), for example, we introduce a parameter that constitutes part of the network architecture. The gating factor associated to each connection is "real" and it is learnt during the training stage. In this way, some of the drawbacks of pruning methods based on sensitivity measures that were mentioned in the previous section are avoided. The same as there are lots of possible ways to update connection weights, one could think of a number of methods to learn the value of gating factors, the only constraint being that they must converge to 0 or 1. The first case implies a non-effective connection and the network will act as if there was no link between units i and j. The second situation is interpreted as a full connection indistinguishable from the classical case.

These ideas are in principle not restricted to any specific type of initial network configuration or learning algorithm. In this paper we make a conceptually simple choice by adopting the same type of gradient descent approach for connection weights w_{ji} and gating terms π_{ji}. In the remainder of this section a learning algorithm with the proposed alteration in node connections is derived for gradient descent searches in the general case of fully connected, discrete time recurrent neural networks. The gradient is calculated recursively (Williams and Zipser, 1989) to minimize the general case criterion function:

$$J(w) = \sum_{p=1}^{P} J_p(w) \tag{2}$$

where P is the number of training sequences, w is the weight vector and $J_p(w)$ is the usual total squared error of the pth sequence:

$$J_p(w) = \frac{1}{2} \sum_{k=1}^{K_p} \sum_{j \in \Omega} (d_j(k) - o_j(k))^2 \tag{3}$$

Here K_p is the length of the pth training sequence and Ω represents the set of output nodes in the network. Defining an iteration index n for the weights, the weight update formula is as follows:

$$\Delta w_{ji}(n) = w_{ji}(n+1) - w_p(n) = -\mu_w \sum_{p=1}^{P} \frac{\partial J_p(w)}{\partial w_{ji}} \bigg|_{w(n)} \tag{4}$$

where the partial derivative can be written as:

$$\frac{\partial J_p(w)}{\partial w_{ji}} = -\sum_{k=1}^{K_p} \sum_{h \in \Omega} (d_h(k) - o_h(k)) p_{ji}^h(k), \text{ with } p_{ji}^h(k) = \frac{\partial o_h(k)}{\partial w_{ji}} \tag{5}$$

Elaborating further we have the following:

$$p_{ji}^h(k) = f'(\sum_{\alpha=0}^{N+M} w_{h\alpha} o_\alpha(k-1) \pi_{h\alpha}) \left[\delta_{hj} \pi_{ji} o_i(k-1) + \sum_{\beta=1}^{N} \pi_{h\beta} w_{h\beta} p_{ji}^\beta(k-1) \right] \tag{6}$$

where $f'(x) = df(x)/dx$, and δ_{hj} is the Kronecker delta function. The partial derivatives indicating how each weight of the network influences each unit activation are kept in memory at the regular forward pass.

It has been seen above that π_{ji} must be bound between the values 0 and 1. If, in the general case, the gating factor is a function of an independent variable, $\pi_{ji} = \pi(z_{ji})$, then a similar procedure can be followed to update that variable (which we will call *gating term*) z_{ji}:

$$z_{ji}(n+1) = z_{ji}(n) + \mu_z \sum_{p=1}^{P} \sum_{k=1}^{K_p} \sum_{h \in \Omega} (d_h(k) - o_h(k)) q_{ji}^h(k) \tag{7}$$

where $q_{ji}^h(k) = \partial o_h(k)/\partial z_{ji}$, and its value can be derived analogously:

$$q_{ji}^h(k) = f'\left(\sum_{\alpha=0}^{N+M} w_{h\alpha} o_\alpha (k-1) \pi_{h\alpha}\right)\left[\delta_{hj} w_{ji} \pi'_{ji} o_i(k-1) + \sum_{\beta=1}^{N} \pi_{h\beta} w_{h\beta} q_{ji}^\beta (k-1)\right]$$ (8)

with $\pi'_{ji} = d\pi(z_{ji})/dz_{ji}$.

4 Application example with constant pruning strength

Preliminary experiments were done by instantiating the previous general gradient descent algorithm for non-recurrent multilayer networks. Classical three-layer perceptrons were used to allow an easier interpretation of the results and to avoid the longer training times required by the most general types of recurrent architectures.

The equations thus obtained are a generalization of the well-known backpropagation algorithm:

$$\Delta w_{ji}(n) = \mu_w \delta_j o_i \pi_{ji}$$ (9)

$$\Delta z_{ji}(n) = \mu_z \delta_j o_i w_{ji} \pi'_{ji}$$ (10)

where $\delta_j = (d_j - o_j) f'_j (net_j)$ if j is an output unit, and $\delta_j = f'_j (net_j) \sum_{h \in \Omega} \delta_h w_{hj} \pi_{hj}$ if j is a unit in an arbitrary hidden layer. When $\pi_{ji} = 1$, these are the usual backpropagation formulas, and if $\pi_{ji} = 0$, then $\Delta w_{ji} = 0$. This result is compatible with (KrishnaKumar, 1993). The connection between units i and j stabilizes $(\Delta z_{ji} = 0)$ when the gating function saturates, i.e. in both previous cases.

Although it is not required by the algorithm, an asymmetric sigmoid function $\pi_{ji} = f(z_{ji}, \alpha_z)$ similar to the one used to compute node outputs in equation (1) was selected for the derivations and experiments reported in this section. This arbitrary choice is only due to simplicity reasons as, in that case, $\pi'_{ji} = \alpha_z \pi(z_{ji})(1 - \pi(z_{ji}))$.

It is also observed that when the magnitude of w_{ji} increases, then z_{ji} and π_{ji} also increase, the opposite being true when $|w_{ji}|$ decreases. Therefore, this method can be partly viewed as providing an automatic, context-sensitive means to determine the threshold below which weight magnitudes can be considered small enough to be ignored.

A sample experiment of network configuration and weight learning using the algorithm instantiation for three-layer perceptrons is discussed below. This problem, which has been termed the "rule-plus-exception" one, is adopted from (Mozer and Smolensky, 1989) and (Karnin, 1990). It is an artificial and conceptually simple classification task that can nevertheless result helpful to make clear how the described method works. The goal is to have a four input, two hidden units and one output network learn the Boolean function $AB + \overline{A}\,\overline{B}\,\overline{C}\,\overline{D}$. The output unit should have a value of 1 either when the "rule" ($A = 1$ and $B = 1$) or the "exception" $\overline{A}\,\overline{B}\,\overline{C}\,\overline{D}$ occur. A proper pruning scheme should account for the fact that the former is more important than the latter, since by itself it explains 15 out of the 16 possible combinations.

Figure 2 shows the initial, potentially oversized structure of the network. It was initialized with small random weights, network inputs were either on (1) or off (0) and unit outputs remained in the range (0,1). In this example we had $\mu_w = 0.2$ and $\alpha_z = 1$. Both connection weights and gating terms were updated stochastically, i.e. after the presentation of every training pattern.

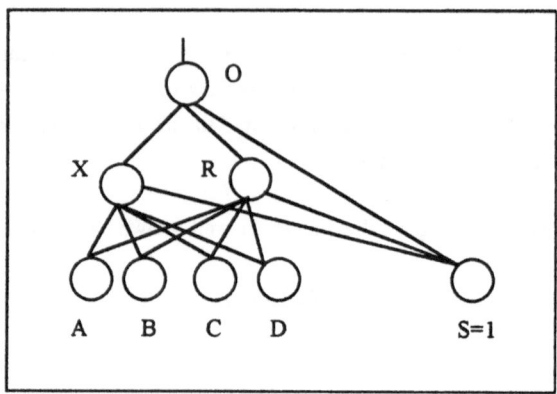

Figure 2: *Starting configuration for the "rule-plus-exception" problem*

The algorithm was run several times with values of μ_z ranging from 0.01 to 10. Despite the influence of the starting set of random weights, the initial architecture was able to configure itself into stable and reasonable configurations relatively quickly. The following is a typical set ($\mu_z = 3$) of resulting weights and gating terms (these are given in parentheses and italics):

First layer weights and gating terms

	S	A	B	C	D
X	-1.98 (0.68)	4.67 (-1.84)	0.00 (-0.01)	-2.67 (0.47)	0.17 (-0.03)
R	4.76 (2.87)	-3.59 (3.17)	-4.61 (1.45)	3.46 (-2.02)	3.07 (-1.95)

Second layer weights and gating terms

	S	X	R
O	1.69 (2.86)	-1.63 (-2.40)	-4.30 (3.27)

The squashing shape selected for π_{ji} in this example allows one to roughly interpret a gating term z_{ji} just by looking at its sign. In effect, those connections for which the gating term is positive will have $\pi_{ji} \to 1$, i.e. full throughput, whereas those with $z_{ji} < 0$ will have $\pi_{ji} \to 0$, that is, they will have been effectively pruned. It can be observed that the output node O is only connected to the bias unit and to node R, which approximately computes the "rule" AB. The connection between nodes O and X is not in place, so any possible effect of the network inputs on the output through X is neglected. The system has come up, then, with a good two input approximation that sacrifices the "exception".

5 Adaptive pruning strength

During the experiments with the previous example it was observed that the magnitude of the gating term learning coefficient μ_z affects the rate of convergence to the final architecture. In addition, too big values of this parameter force a quick pruning of connections before the weight space has been explored in the early stages of the learning phase. As a consequence, the final network configuration may result non-optimal. This fact, which agrees with the observations reported in (Weigend, Rumelhart and Huberman, 1990) and (Prechelt, 1996) for different methods, raises the issue of how to control the amount of pruning. The performance of regularization techniques, for example, depends on a delicate balance of the objective function terms scaling factors. Since it is often not possible to devise a good enough schedule for those factors before training begins, the best (and least trivial) choice is to have the scaling factors adapt dynamically. In our case it is μ_z that controls the pruning strength, and extensive experimentation has shown that a fixed value can sometimes get to degrade the generalization ability of the network. The same that happens in alternative stabilization methods with other types of parameters, the gating factor option is sensitive to the choice of the learning coefficient μ_z. The variability in the initial set of weights as well as application-specific characteristics impede the existence of a universal value for this coefficient. The problem is more severe in function approximation or regression tasks. In forecasting applications, for example, the objective is to approximate a continuous-valued target function, and so the precision in quantitative outputs is crucial. It is not enough to look at the relative values of outputs or to check whether some threshold is passed. An adaptive pruning strength schedule is required to control the connection trimming process at a finer level.

The approach followed in our experiments is based on a variation of the RPROP algorithm for local adaptation of weight-updates (Riedmiller and Braun, 1993). That algorithm extends the heuristics given in (Jacobs, 1988), and already applied in, e.g. (Tollenaere, 1990). The original adaptation rule is:

$$\Delta w_{ji}^{(t)} = \begin{cases} -\Delta_{ji}^{(t)}, & \text{if } \dfrac{\partial E^{(t)}}{\partial w_{ji}} > 0 \\ +\Delta_{ji}^{(t)}, & \text{if } \dfrac{\partial E^{(t)}}{\partial w_{ji}} < 0 \\ 0, & \text{else} \end{cases}, \text{ where } \Delta_{ji}^{(t)} = \begin{cases} \eta^+ * \Delta_{ji}^{(t-1)}, & \text{if } \dfrac{\partial E^{(t-1)}}{\partial w_{ji}} * \dfrac{\partial E^{(t)}}{\partial w_{ji}} > 0 \\ \eta^- * \Delta_{ji}^{(t-1)}, & \text{if } \dfrac{\partial E^{(t-1)}}{\partial w_{ji}} * \dfrac{\partial E^{(t)}}{\partial w_{ji}} < 0 \\ \Delta_{ji}^{(t-1)}, & \text{else} \end{cases} \qquad (11)$$

and $0 < \eta^- < 1 < \eta^+$. The idea is to decrease the weight by an update value Δ_{ji} if the error increases, and to add the update value when the error decreases. Every time the partial derivative corresponding to the weight w_{ji} changes its sign, the update value is decreased by a factor η^-. If the slope retains its sign, then the update-value is increased to accelerate convergence.

Due to the gating term introduced in equation (1), instead of considering the simple weight w_{ji}, in our case an *effective* weight $W_{ji}^{eff} = w_{ji}\pi_{ji}$ is defined and used. The following expression can be derived for the *effective* weight increment:

$$\Delta W_{ji}^{\textit{eff}} = \Delta w_{ji} \pi_{ji} (1 + \frac{\mu_z}{\mu_w} w_{ji}^2 \alpha_z^2 (1 - \pi_{ji})^2) \qquad (12)$$

Note in the previous equation that when π_{ji} tends to 1, then $\Delta W_{ji}^{\textit{eff}} \rightarrow \Delta w_{ji}$, whilst, if π_{ji} approaches 0, then $\Delta W_{ji}^{\textit{eff}} \rightarrow 0$, as expected. It is by applying (11) to the *effective* weights (and starting from $\Delta_{ji}^0 = 0$) that the learning coefficients μ_z of gating terms are updated epoch-wise. *Effective* weights are not network parameters, but a product of two real parameters that is forced to satisfy the constraints imposed by the RPROP variant we have described. The coefficient μ_z is the only available variable and it is updated to ensure that those constraints are met. The actualization of μ_z must not be mixed up with the one of weights w_{ji} and gating factors π_{ji}. These are learnt independently and according to the equations of sections 3 and 4.

This method was applied to the classical sunspot series forecasting problem. A three-layer perceptron with eight hidden nodes was trained with the data from 1700 to 1920. The results obtained with $\eta^- = 0.5$ and $\eta^+ = 1.2$ for the test set are much better than what can be achieved with regular backpropagation and also comparable to the ones reported in (Weigend et al., 1990) without refinements like having the value of the weight learning coefficient μ_w reduced at the end of the training phase. The connections between the output node and the nodes in the hidden layer are progressively eliminated until only three of them are left. Given the space limitations, the details of this and other experiments will be published elsewhere.

6 Conclusion

The present paper has reviewed the main features of traditional stabilization techniques that, starting from an overparameterized network, delete connections progressively until a satisfactory solution is found. Including a gating factor that is actually implemented in every connection can be a mechanism to retain the advantages of both penalty-term techniques and pruning methods. The gating factor is learned by the network in parallel to conventional weights during the training stage.

It has been shown how this idea can be naturally integrated into a general gradient descent learning algorithm for fully connected, discrete time recurrent neural networks and also into the particular case of backpropagation. Experience shows that demanding applications require pruning severity not to be fixed beforehand. Problem dependency, sensitivity to the initial weights and the need for numerically accurate outputs are all aspects that justify the investigation of adaptive schedules for the pruning strength that allow the parameter space to be searched more carefully. Initial experiments seem to support the validity of the ideas presented in the paper. Ongoing work aims at further validation in additional real-world applications and at an extension of the methods presented to more complex architectures, using different types of gating terms and pruning strength updating procedures.

References

Chauvin, Y., "Dynamic behavior of constrained back-propagation networks", *in Advances in Neural Information Processing (2)*, Touretzky, D.S., Ed., pp. 642-649, 1990.

Jacobs, R.A., "Increased rates of convergence through learning rate adaptation", *Neural Networks*, vol. 1, pp. 295-307, 1988.

Karnin, E.D., "A simple procedure for pruning back-propagation trained neural networks", *IEEE Trans. Neural Networks*, vol. 1, no. 2, pp. 239-242, 1990.

KrishnaKumar, K., "Optimization of the neural net connectivity pattern using a back-propagation algorithm", *Neurocomputing*, vol. 5, no. 6, pp. 273-286, 1993.

Kwok, T-Y. and Yeung, D-Y., "Constructive feedforward neural networks for regression problems: A survey", Tech. Rep. HKUST-CS95-43, The Hong Kong Univ. of Science & Technology, 1995.

Le Cun, Y., Denker, J.S. and Solla, S.A., "Optimal brain damage", in *Advances in Neural Information Processing (2)*, Touretzky, D.S., Ed., pp. 598-605, 1990.

Mozer, M.C. and Smolensky, P., "Skeletonization: A technique for trimming the fat from a network via relevance assessment", in *Advances in Neural Information Processing (1)*, Touretzky, D.S., Ed., pp. 107-115, 1989.

Plaut, D.C., Nowlan, S.J. and Hinton, G.E., "Experiments on learning by back propagation", Tech. Rep. CMU-CS-86-126, Carnegie Mellon Univ., 1986.

Prechelt, L., "Adaptive parameter pruning in neural networks", Tech. Rep. 95-009, International Computer Science Institute, Berkeley, CA, 1995.

Reed, R., "Pruning algorithms-A survey", *IEEE Trans. Neural Networks*, vol. 4, no. 5, pp. 740-747, 1993.

Riedmiller, M. and Braun, H., "A direct adaptive method for faster backpropagation learning: The RPROP algorithm", in *Proc. of the IEEE International Conference on Neural Networks.*, pp. 586-591, 1993.

Tollenaere, T., "SuperSAB: Fast adaptive backpropagation with good scaling properties", *Neural Networks*, vol. 3, pp. 561-573, 1990.

Weigend, A.S., Rumelhart, D.E. and Huberman, B.A., "Back-propagation, weight-elimination and time series prediction", in *Proc. 1990 Connectionist Models Summer School*, Touretzky, D., Elman, J., Sejnowski, T. and Hinton, J., Eds., pp. 105-116, 1990.

Williams, R. and Zipser, D., "A learning algorithm for continually running fully recurrent neural networks", *Neural Computation*, vol. 1, no. 2, pp. 270-280, 1989.

Wynne-Jones, M., "Node splitting: A constructive algorithm for feedforward neural networks", *NeuralComputing & Applications*, vol. 1 no. 1, pp. 17-22, 1993.

Evolution of Structure and Learning – A GP Approach

K. Govinda Char

Block-72, Department of Electronics and Electrical Engineering,
University of Glasgow, Glasgow G12 8LT, UK.

E-mail:kchar@elec.gla.ac.uk

Abstract

Recently evolutionary algorithms have been shown to be successful in evolving optimal neural network topologies and also novel learning rules. Genetic programming is a new paradigm that has proved to solve a number of complex problems in various domains. In this paper, I have suggested a novel approach to show how genetic programming can be an effective tool in evolving neural networks that work on the principles of interaction, competition, self-organization and adaptation, that is a self-organizing neural network. Can we evolve new learning algorithms with this approach? Can we extend this approach to evolve complex self-organizing systems? Can we employ this approach to evolve and simulate the mechanisms that are found in various sub-systems in the brain and hence for biological modelling? In this work, I have attempted to answer some of these questions.

Keywords

Genetic programming, connectionist learning rules, micro-macro dynamics, self-organizing feature maps, quantization error.

1. Introduction

Artificial neural networks are computational paradigms that mimic the information processing mechanisms in natural systems through the implementation of brain-like structures and learning algorithms. Evolutionary algorithms have proved to be quite successful in evolving a variety of connectionist architectures and learning rules [1]. In the present work, I have chosen a self-organizing neural network [2] as a framework for my experiments and discussions. I view a learning rule as a sequence of interacting concepts. I have illustrated how a novel learning rule can be evolved for a known structure. In addition, I have briefly discussed how the structures can dynamically evolve with a mechanism for morphogenesis, such as cellular encoding (CE) [3]. Genetic programming (GP) [4] may prove to be a potential paradigm for implementing such notions. The key aspect of such an approach is that the environment which is an essential part of a learning process can be integrated within the evolutionary mechanism.

This paper is organised as follows. Section 2 provides the background necessary for the work. Section 3 explains the framework that I have chosen for the simulation work. Section 4 will discuss the GP approach. Section 5 suggests a developmental mechanism for structure evolution that can be effectively employed in this kind of simulation. Section 6 will discuss the theme of the research and provide the concluding remarks.

2. The background

This section will discuss an evolutionary approach wherein a genetic algorithm was used for evolving a number of neural network learning rules by encoding the dynamic parameters of the neural network in the genome and subjecting these to selection pressures, assuming a known topology. Two such cases are discussed. In the first case, a supervised learning rule is evolved and in the second case an unsupervised learning rule is evolved.

2.1. The supervised learning rule

Chalmers [5] evolved a number of potential learning rules for a feedforward neural network using the genetic algorithm. The genome encodes a function F given by:

$$\Delta w_{ij} = F (a_j , o_i , t_i , w_{ij}) \tag{1}$$

where:

a_j - the activation of the input unit j;
o_i - the activation of the output unit i;
t_i - the training signal on the output unit i;
w_{ij} - the current value of the connection weight.

Thus the change in the weight of a given connection should be a function of only the information local to that connection and the same function should be employed for every connection. Chalmers used a genome of 35- bits assuming that the function F to be a linear function of the four dependent variables and their six pair-wise products. The genome specifies the ten coefficients with the help of an eleventh scaling parameter. With this approach Chalmers succeeded in evolving a number of potential learning rules that included the well-known delta rule. The rules that evolved were evaluated for their fitness by testing them on a number of various learnable tasks on different networks. The fitness of the network that used the learning rule reflect the fitness of the learning rule.

2.2. The unsupervised learning rule

Dasdan [6] employs a similar approach to evolve an unsupervised learning rule such as a Kohonen learning rule. In this case the target value of the exemplars is not known. The equation for the weight adaptation is:

$$\Delta w_{ij} = F (w_{ij}, x_j, t, y_j)$$
(2)

where:

w_{ij} - the current value of the connection weight;
x_j - the signal on the input node;
t - the training iteration number;
y_j - the correlation between the signal x and m, m being the weight associated with the output neuron.

The final equation was in terms of a scaling parameter and fifteen other coefficients. In both the cases the learning rule first evolves and then is subsequently adapted to the assumed structure. The rules are thus evaluated in their success in tackling the given tasks.

3. The framework

I have chosen a self-organizing neural network as a framework for my experiments. The reasons for my choice are that it has shown to mimic some of the biological functions of the brain. Also, it works on the principles of interaction, competition, and co-ordination. Genetic programming may be a potential paradigm for evolving these principles or concepts.

3.1. Self-organizing neural networks

Kohonen's self-organizing feature map is motivated by biological evidence. The map is based on the interaction between a number of neurons, typically arranged on a two-dimensional grid structure. Each of the neurons is connected to the n-dimensional signal through a n-dimensional reference vector. The adaptation structure essentially characterizes a learning rule. The structure is adapted by an unsupervised learning rule such as Kohonen's learning rule which is based on the concept of a *winning* neuron. Each neuron on the grid is sensitive to a particular component of the input signal space and will be maximally excited as compared to the rest of the neighboring neurons. The learning rule consists of the following steps:

- Apply the exemplars for a number of epochs;
- Find the *winning* neuron;
- Evolve a strategy to adapt the *winning* neuron maximally and the neighboring neurons to a lesser extent.

Repeat these steps until a topological map is evolved wherein the adaptive structure faithfully models the input signal distribution. This point is identified in terms of a quantization error that reflects the closeness of the weights to the signals. This can be observed graphically. Figure-1 illustrates a self-organizing feature map.

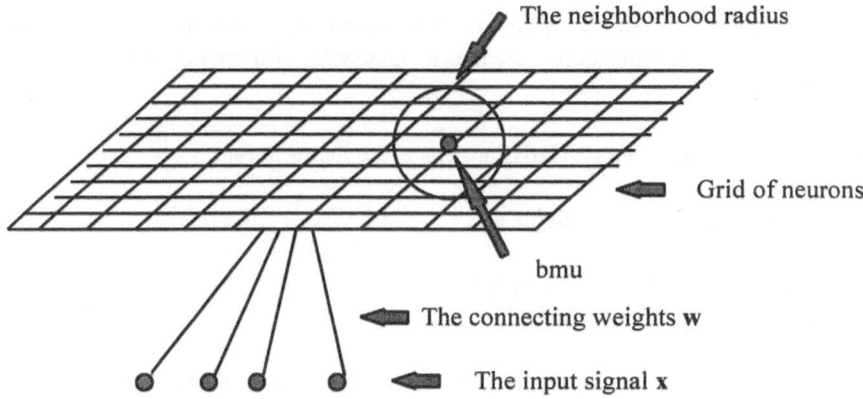

Figure-1: The self-organizing feature map

The location of each of the neurons is vital. The best matching unit (bmu) refers to the *winning* neuron. Thus the topological order evolves as a result of interaction between the neurons. Now the Kohonen learning rule will be explained. Assume a fixed number of neurons nx *ny = 10*10. Assuming a two-dimensional input signal, the learning rule looks like:

Init-synapses(random); // Initialize the two-dimensional synaptic weights for random values;

For a number of iterations:

{ Apply-signals(x,y);// Apply two-dimensional signals;

Find the *winning* neuron as:

For the number of neurons:

mismatch += ABS(x-wix) + ABS(y-wiy); //The metric used is the Euclidean distance;

Repeat these steps until the *winning* neuron is obtained, which gives the smallest mismatch;

For the number of neurons:

Find the distance of each neuron from the bmu;

$$\text{distance} = SQR(ix - ix_{bmu}) + SQR(iy - iy_{bmu}); \text{ // ix, iy are the co-ordinates of any neuron}$$

Update the synaptic weights based on the distance through a neighborhood function that can reflect lateral inhibition   such as a   Gaussian with the neighboring radius and the distance.

Form a function: temp = eps * Gaussian; eps is the learning rate;

Adapt the two-dimensional weights as:

w[0][ix][iy]+= temp * (x - w[0][ix][iy]);
w[1][ix][iy]+ = temp * (y - w[1][ix][iy]);

Decrease the learning rate and the neighboring radius by a small fraction;
}

Display-net (epochs, mismatch); // Observe whether the input distribution is faithfully modelled; In the next section I will describe my approach.

4. The genetic programming approach

By viewing a learning rule as a sequence of interacting concepts, I tried to implement the above learning rule. First of all,   GP, through the primitives should be able to evolve all the concepts such as the bmu, a Gaussian (or a similar neighborhood function) and the right direction for the weight adaptation so as to enforce the topological order.   Also it is essential that GP needs to evolve this sequence correctly. I assumed a known structure, a 2-D grid, initially. The input stimuli are provided by a two-dimensional vector, with components distributed in a chosen subset of a square $[-1,+1]^2$ (such as   1/3 quadrants).   Each neuron is associated with the input stimuli via the two-dimensional synaptic weights. The function set and terminal sets of GP are as follows:

F[s] = { +, -, *, %, IFLTE, ABST, Adaptx, Adapty};
T[s] = { x, y, wix, wiy, ix, iy, F0, F1, F2};

The functions:

ABST returns an absolute value; // If this primitive evolves the experiments will be more interesting.

Adaptx adapts the weights that are associated with the x-signals;
Adapty adapts the weights that are associated with the y-signals;

As an example:

Adaptx looks like: { w[0][ix][iy] = }; // for x-signals;
Adapty looks like: { w[1][ix][iy] = }; // for y-signals;

No other information is provided. GP has to come up with the right parameters so as to minimize the quantization error. The learning rule that evolves should find the "direction" for the weight adaptation employing the right set of primitives. GP will have to evolve the concept of "bmu" via the function IFLTE along with the other primitives and the *ephemeral* constants, to get the crucial information of the "location" of each of the neurons. Here it should be noted that IFLTE is a micro concept that GP employs to evolve a macro concept such as a *winning* neuron. Vaario [7] has emphasised on the concept of "micro-macro" dynamics for evolving intelligent behaviour.

The terminals:

x- the x-component of the two-dimensional signal;
y- the y-component of the two-dimensional signal;
wix, wiy are variables for accessing weights;
ix, iy are variables for accessing the location of neuron. GP will need to assign the evolved values from the *ephemeral* constants and assign to these variables. Few more terminals for implementing the learning rate and the variance may be additionally defined. It is also essential to have a large number of epochs for the adaptation phase.

The Fitness:

The fitness is a quality function G(x,y):

$$Error = \sum ABS(x - wix) + ABS(y - wiy) ; // \text{ over a number}$$
of iterations;

$$Quantization\ Error = Error / (number\ of\ neurons);$$

$$Fitness = 1/ (Quantization\ Error)^2 \qquad (3)$$

The preliminary simulation was done on a PC. A large population (which is absolutely essential for this problem) could not be used due to memory limitations. Two sample programs are shown:

((IFLTE (IFLTE(Adaptx (y) (+ (+ (* (F0 (wix) (F0 (Adapty (y)) (- (Adapt (y) (veps=0.1) (F0 * (F0 (wix)));

The Fitness: 20;
Structural Complexity: 22;

In the above program it should be noted that any combination such as (Adaptx (y)) or (Adapty (x)) are both detrimental for a self-organization. A correct learning rule will push the wix- weights towards the x signals and wiy weights towards the y signals to reduce the quantization error. The next program shows how GP has managed to induce this information.

((IFLTE (IFLTE (IFLTE (Adaptx(x) (+ (+ (* F0 (wix) (x) (F0 (Adapty (y))) (- (Adapty(y) (veps=0.002) (F0 (* (wix).........

The Fitness: 80;
Structural Complexity: 30;

GP was not given any bias to form such a combination. This simulations will be tried with a large population to obtain a complete picture on GP's performance.

5. Embedding a developmental mechanism

In my paper [8] I used GP and cellular encoding to evolve the adaptation structure. It seems CE will be very effective in these simulations as the neurons can be added or removed dynamically to optimize the network topology. Also, the cellular operators can easily be defined as GP primitives.

6. Discussions and conclusions

Preliminary results suggest that GP can be an efficient tool for evolving and combining concepts yielding powerful learning rules. The search space is the space of potential concepts. With this approach GP does not directly solve a problem but provides the combination of potential macro concepts to form a learning rule that can solve a problem. Another important aspect of this simulation is that it is likely that GP may employ the *same* primitive under different contexts. Basically GP has to evolve and invoke the right combination of concepts (in the form of a learning rule) that can be effective in solving a problem. The learning is incremental. Undoubtedly, such an approach is computationally expensive. Nevertheless it may lead to a rich landscape of learning mechanisms justifying the cost. Furthermore, an optimum network topology can be evolved by including a mechanism for morphogenesis. The experiments were simulated on a PC where large population size could not be employed due to the memory limitations. These simulations will continue on a UNIX workstation. A large population size is vital for this kind of simulation work. It is hoped that many potential learning rules including the Kohonen learning rule are likely to evolve. If so, the approach can be used for biological modelling. Based on the success with further simulations, the implications could be that cognitive processes might be simulated and hypotheses proven without having to act directly upon living beings. In addition, these simulations, in principle, can be extended further to evolve complex self-organizing systems. GP may prove to be the potential paradigm for such an endeavour.

References

[1] J. Branke, Evolutionary algorithms for Neural Network Design and Training. In: Proceedings of the First workshop on Genetic Algorithms and its Applications, Vaasa, Finland, 1995.

[2] T. Kohonen, *Self-organization and Associative Memory, volume* 8 of Springer Series in: Information Sciences. Springer-Verlag, Berlin, Heidelberg, New York, third edition, May 1989.

[3] Frederic Gruau, Efficient Computer Morphogenesis: A Pictorial Demonstration, Report No. 94-04-027, Santa Fe Institute, April 29, 1994.

[4] John. R. Koza, Genetic Programming: On the Programming of Computers by Means of Natural Selection. The MIT Press, Massachusetts Institute of Technology, Cambridge, Massachusetts, 1992.

[5] D.J. Chalmers, The Evolution of Learning: An experiment on Genetic Connectionism. In: Proceedings of the 1990 Connectionist Models Summer School, CA: Morgan Kaufmann.

[6] A. Dasdan and K. Oflazar, Genetic Synthesis of Unsupervised Learning Algorithms. In: Proceedings of the Second Turkish Symposium on Artificial Intelligence and Artificial Neural Networks, Istanbul, June 1993.

[7] J. Vaario, Modelling Adaptive Self-Organization, ATR Laboratories, Kyoto, Japan.

[8] K.G. Char, Emergence of Structures With Genetic Programming and Cellular Encoding, Tainn96, Turkey.

Self-Organizing Formation of Receptive Fields and Competitive Systems

Satoshi Maekawa and Hidefumi Sawai

Communications Research Laboratory,
588-2 Iwaoka, Iwaoka-cho, Nishi-ku, Kobe 651-24, Japan

Abstract. In previous work[1] it was shown that a mechanism competing for a presynaptic factor enables the self-organizing formation of local receptive fields with orientation selectivity, even though the synapses between the input and output layers are all nonlocal, i.e., fully connected. The previous model, however, assumed *a priori* competitive systems, called hypercolumns, that may not appropriately represent the inherent structure of the input, which is a hierarchy of low- to high-level features. In this paper we propose to use a self-organizing competitive system, rather than an *a priori* determined system. Self-organization is implemented by including Földiák's anti-Hebbian learning rule in out system. Computer simulations show that this model allows for the formation of local receptive fields with orientation selectivity, and that self-organization successfully structures the competitive system.

1 Introduction

Complex patterns usually have a hierarchical structure and can be decomposed into subpatterns. Such subpatterns are used as building blocks, and the combination of subpatterns makes many representations of complex information possible. For examples, in the case of character images, a combination of pixels forms a line segment, a combination of line segments forms a curve, a corner or a cross, and a combination of these simple patterns forms a character.

Receptive fields created by self-organizing neural networks such as Fukushima's neocognitron[2] represent subpatterns. However, for the formation of receptive fields some *a priori* knowledge and constraints are needed, such as the topology of the physical arrangement of the neurons and the predetermination of connective regions and so on. To have a more constraint-free and biologically plausible formation of receptive fields, it is important to have a self-organizing scheme that does not need *a priori* knowledge about the hierarchical structure of the information source. Such a self-organizing formation of receptive fields is achieved in [1] by employing synaptic competitive learning that competes for a presynaptic factor. Using this learning model, local receptive fields with orientation selectivity can be formed in a self-organizing fashion, even though synapses between the input and output layers are fully connected.

A drawback of this model is its *ad hoc* fixed structure consisting of hypercolumns, a drawback that is also found in the neocognitron. The conventional hypercolumn structure of a fixed competitive system has preset *a priori* parameters for the number of hypercolumns and the number of neurons in a hypercolumn. In a hypercolumn, neurons compete with each other, and only one neuron can fire at a time. So, the number of hypercolumns determines how many partial features a hierarchical input information source can be divided into, and the number of neurons in a hypercolumn determines the kind of partial features that emerge nonsimultaneously.

The fixed predetermined structure of this model is its main problem: the input patterns are always divided into the same number of partial features. As there are sometimes complicated and sometimes simple input patterns, the number of partial features is not necessarily always the same.

In this paper, we employ Földiák's anti-Hebbian rule[3], which develops laterally inhibitory connections and is used for self-organization in competitive systems. The resulting learning rule has as its *a priori* parameters the total number of neurons, and the expectation for the number of neurons firing simultaneously. Since the number of neurons firing simultaneously is given on average and not strictly set, a more flexible structure is obtained.

2 Synaptic Competitive Learning

We assume that each synapse needs a synaptic factor to maintain it when its corresponding postsynaptic neuron is firing, and only then. The necessary amount of synaptic factor C_{jk} is proportional to the synaptic weight:

$$C_{jk} = a\, y_j w_{jk}, \tag{1}$$

where, a is a constant, y_j is the activation potential of postsynaptic neuron j, and w_{jk} is the synaptic weight between presynaptic neuron k and postsynaptic neuron j.

The total amount of synaptic factor emitted by a presynaptic neuron is limited and is proportional to the action potential of the presynaptic neuron k:

$$S_k = \sum_j S_{jk} \propto x_k. \tag{2}$$

Synaptic factor is supplied only to the synapses that connect to the firing postsynaptic neurons; it is supplied via axonal transportation and paracrine (see Fig.1). The amount of synaptic factor supplied through axonal transportation is proportional to the synaptic weight, and the amount through paracrine is constant and denoted by β. Taking into account the total amount of synaptic factor in Eq.(2), the supply for each synapse, S_{jk}, is given as follows:

$$S_{jk} = \frac{b\,(w_{jk} + \beta) y_j x_k}{\sum_i y_i (w_{ik} + \beta)}. \tag{3}$$

Change in the synaptic weights depends on the balance between the demand and the supply of synaptic factor. If the supply exceeds the demand, a synaptic

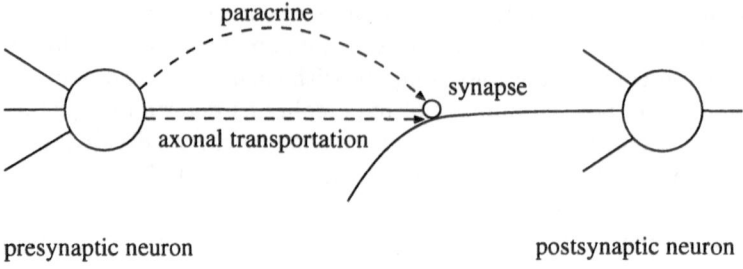

paracrine

synapse

axonal transportation

presynaptic neuron   postsynaptic neuron

Fig. 1. The presynaptic factor is supplied via axonal transportation and paracrine.

weight is strengthened, otherwise it is weakened. The dynamics of change in the synaptic weights is described by the following equation:

$$\Delta w_{jk} = \varepsilon(S_{jk} - C_{jk})$$

$$= \varepsilon \left\{ \frac{b(w_{jk} + \beta)y_j x_k}{\sum_i y_i(w_{ik} + \beta)} - a\,y_j w_{jk} \right\}. \tag{4}$$

Now suppose that we have one presynaptic neuron and two postsynaptic neurons, that β is adequately small, and the firing probabilities of the two postsynaptic neurons are equal. Then the equilibrium states of the corresponding synaptic weights are given by[4]:

$$w_1 = \frac{1}{1 - \delta^2} \{P(X|Y_1) - \delta P(X|Y_2)\} \tag{5}$$

$$w_2 = \frac{1}{1 - \delta^2} \{P(X|Y_2) - \delta P(X|Y_1)\} \tag{6}$$

$$\delta = \frac{P(XY_1Y_2)}{P(XY_1) + P(XY_2) - P(XY_1Y_2) + 2\beta P(Y)}. \tag{7}$$

Here, X and Y_i ($i = 1, 2$) are random variables, taking on values 0 or 1, which denote the activation potentials of pre- and postsynaptic neurons, respectively. $P(XY_i)$ denotes the joint probability of the event $X = 1$ and $Y_i = 1$, and $P(X|Y_i)$ denotes the conditional probability of the event $X = 1$ under the condition $Y_i = 1$.

These equations show that if the conditional probabilities $P(X|Y_1)$ and $P(X|Y_2)$ are different, the difference is reflected in the synaptic weights. This effect is called synaptic competition.

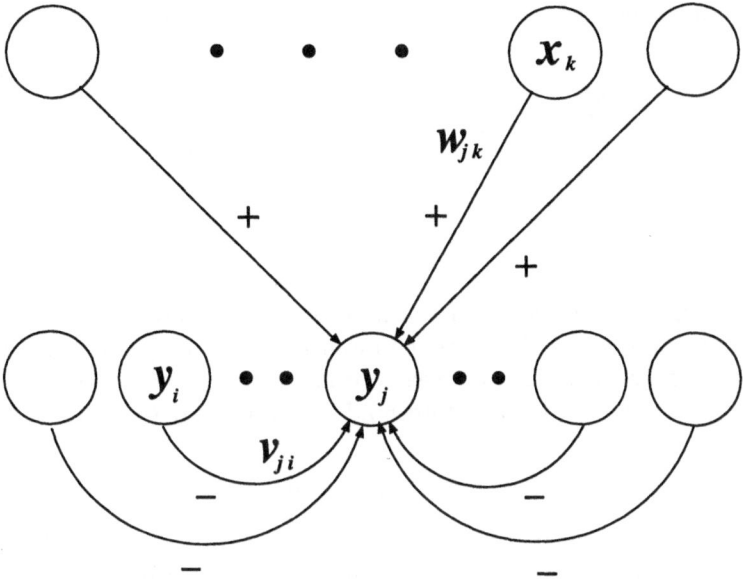

Fig. 2. Network Architecture

3 Self-organization of the Competitive System

For a successful self-organizing formation of receptive fields by synaptic competitive learning, it is necessary to simultaneously fire more than one neuron. Conventional competitive systems with hypercolumns, such as in [1, 2], have an *a priori* predetermined structure and permit the firing of only one neuron in each hypercolumn. Since the number of simultaneously firing neurons is constrained to be always constant and independent of the complexity of the input patterns, conventional systems fall short. So, a mechanism which can self-organize a competitive system should be introduced.

To realize such a competitive system we propose to use the same network architecture and learning rule as Földiák's models, except for the learning of inter-layer connections. The network architecture is a two-layer network with an input and an output layer (see Fig. 2). Neurons in the output layer receive input through excitatory connections from the neurons in the input layer, and the neurons in the output layer are connected with each other through inhibitory intra-layer connections. However, there are no self-connections. The self-organizing learning of the inter-layer connections is performed by the synaptic competitive learning mechanism described in Sect. 2, rather than the Hebbian learning mechanism used by Földiák.

The activation potential y_i of the i-th neuron in the output layer is determined by the converged value of y_i^* in the following differential equation:

$$\frac{dy_i^*}{dt} = f\left(\sum_{k=1}^{m} w_{ik}x_k + \sum_{j=1}^{n} v_{ij}y_j^* - t_i\right) - y_i^* \tag{8}$$

$$y_i = \begin{cases} 1 & \text{if } y_i^* > 0.5 \\ 0 & \text{otherwise} \end{cases} \tag{9}$$

where, x_k denotes a activation potential of the k-th neuron in the input layer, w_{ik} the inter-layer connection between input neuron k and output neuron i, and v_{ij} the intra-layer connection between output neuron i and j. The function $f(\cdot)$ is a sigmoid function and t_i is a threshold determined by the following learning rule:

$$\Delta t_i = \gamma(y_i - p), \tag{10}$$

where p is a constant denoting the average firing rate of the output neurons. Since y_i is 0 or 1, the above equation implies that the threshold converges so that the firing rate of each neuron becomes p. When the total number of output neurons is n, on average pn neurons fire simultaneously.

As the learning rule for the laterally inhibitory intra-layer connections v_{ij}, the anti-Hebbian learning rule of Földiák is used:

$$\Delta v_{ij} = -\alpha(y_i y_j - p^2), \tag{11}$$

where no self-connections exist ($v_{ii} = 0$). Variable v_{ij} is hard-limited from above: if it becomes positive in accordance to Eq.(11), it is automatically set to $v_{ij} = 0$. According to this equation, the simultaneous firing probability for each neuron becomes equal to or less than p^2. As the firing probability for each neuron is p, the neurons' firing signals are independent or negatively correlated.

On the other hand, the inter-layer connections (w_{ij}) are not trained based on the Hebbian learning which Földiák used; i.e.,

$$\Delta w_{ik} = \varepsilon\, y_i(x_k - w_{ik}), \tag{12}$$

but based on the synaptic competitive learning[1] by pre-synaptic factors:

$$\Delta w_{ik} = \varepsilon\, y_i \left\{ \frac{(w_{ik} + \beta)x_k}{\sum\limits_{j=1}^{n} y_j(w_{jk} + \beta)} - w_{ik} \right\}. \tag{13}$$

This is identical to (4), except that constants a and b are set to 1. In the above equations, α, β, γ and ε are positive constants.

4 Simulation

Computer simulations were performed using a network with an input layer of 24×24 neurons and an output layer of $n = 100$ neurons. As input images, Japanese Hiragana, Japanese Katakana, alphabetic characters, and numerals were used that were randomly translated and rotated. The parameters used were set as $p = 0.07, \beta = 0.005$ and $\alpha = \gamma = \varepsilon = 0.01$. In this case, the number of simultaneously firing neurons was 7 on average. The inter-layer connections between the input and the output layers were fully connected and their weights were initialized as uniformly random values.

We first use the proposed model, which employs synaptic competitive learning in Eq.(13), and then compare the obtained results with those obtained with Földiák's model, which employs Hebbian learning in Eq.(12).

Synaptic Competitive Learning

Use of a model that employs synaptic competition among its inter-layer connections gives the results in Fig.3. In this case, the formed receptive fields are clearly local and orientation selective in spite of the full connections between the input and the output layers.

The positions in the input layer where the neurons' receptive fields are formed depend on the connections' initial values and the statistics of the input images. The self-organizing mechanism employed here consists of synaptic competition and lateral inhibition and does not rely on any topological relationship among the neurons, i.e. there is no meaning in the arrangement of neurons. For this reason, in this model a functional map as observed in the primary visual cortex can not be observed, and each neuron forms its receptive field in a random position. In case synaptic competitive learning is adopted as the learning rule for a self-organizing multi-layered neural network, such a deficit of the functional map is not a substantial weakness, because full connections between the inter-layers can be trained according to synaptic competitive learning, resulting in local receptive fields. So, compared to conventional competitive learning methods, the network using synaptic competitive learning does not need to impose *a priori* hard-wired connections on the receptive fields. Because of the full inter-layer connectivity, it is also unnecessary to have *a priori* knowledge of what feature each neuron in the preceding layer can extract.

Though functional map formation is not strictly necessary in case synaptic competitive learning is employed, it does improve the efficiency of learning. It allows for the initial synaptic connections to be roughly determined in advance. As neurons with high firing correlations gather into each other's physical neighborhood when forming of a functional map, the initial synaptic connections between a layer and the next higher layer in a multi-layer system can be genetically biased to form local regions using information about the physical arrangement of the neurons.

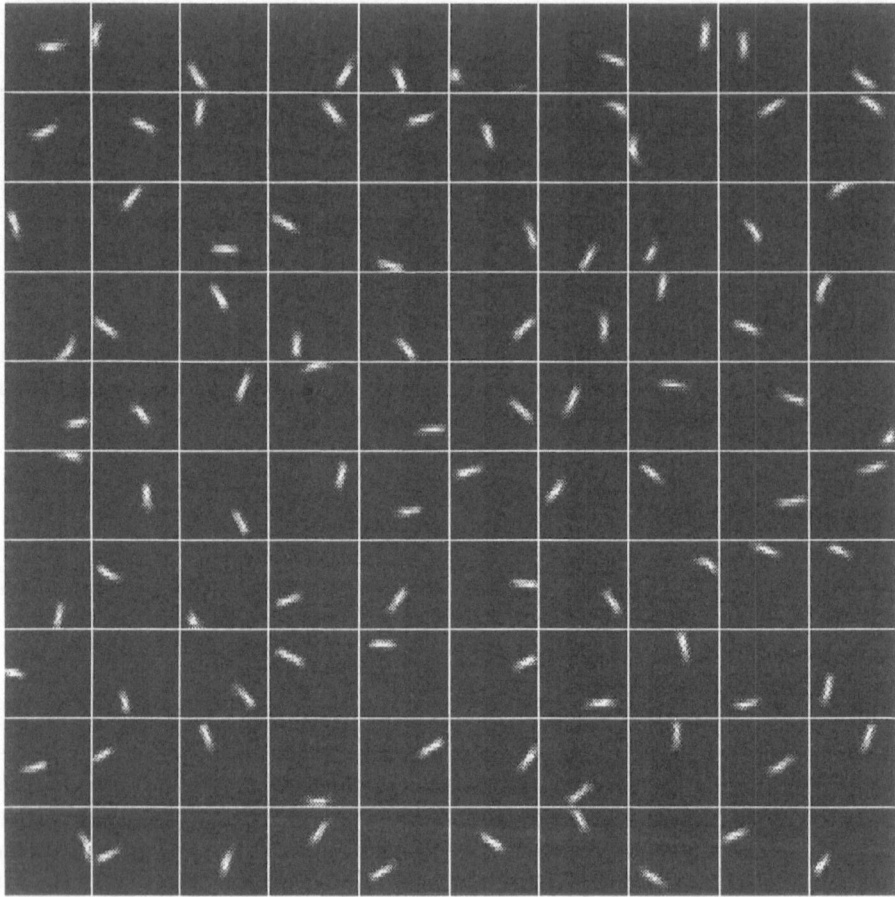

Fig. 3. Receptive fields formed by a synaptic competitive learning, p = 0.07, number of neurons is 100

Hebbian Learning

Földiák's learning model, which employs a Hebbian learning rule for the inter-layer connections, gives the results in Fig.4. In this case, there is no locality in the receptive fields and it is found that connections are formed in the whole input layer. As the input images are character images consisting of segments in this simulation, these receptive fields do not represent any partial features of the input patterns.

Using Hebbian learning, the receptive field of a neuron is formed with just the average of input patterns that are presented when the neuron becomes winner and fires. Since the presented patterns are not local in the input layer, the averaged pattern neither has locality. This is the reason for the formation of non-local receptive fields in the case of using Hebbian learning.

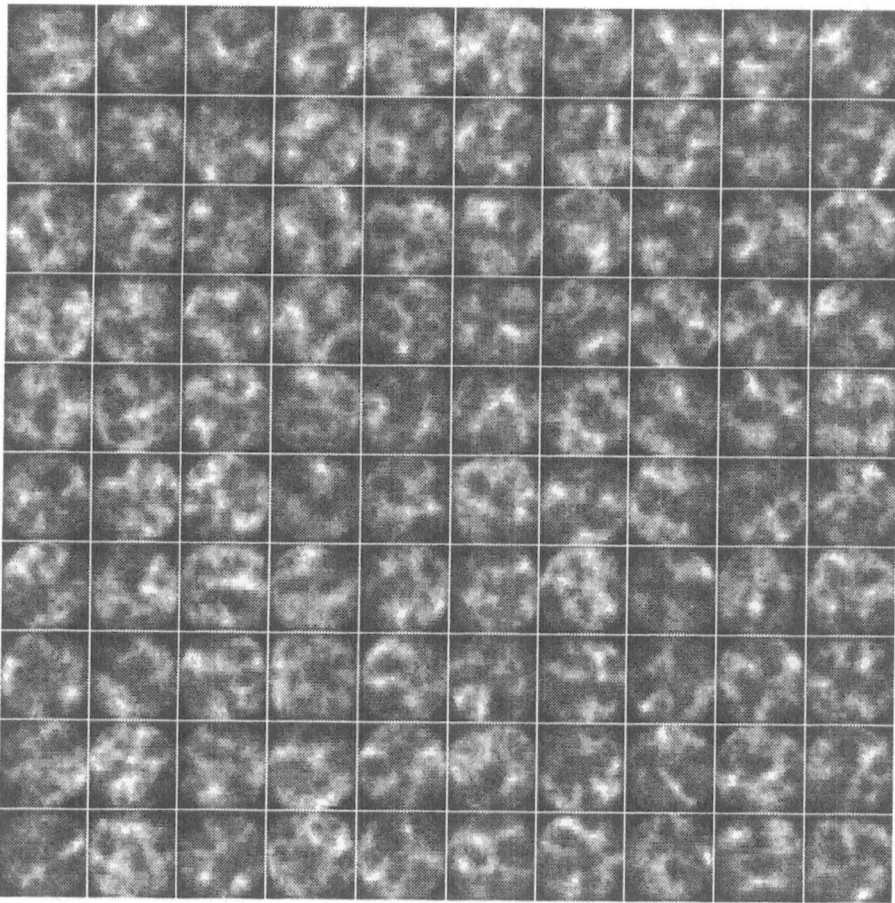

Fig. 4. Receptive fields formed by a Hebbian learning, p = 0.07, number of neurons is 100

Though the receptive fields exhibit no locality, as a result of lateral inhibition, each neuron possibly fires to a line in some specific orientation. However, according to the experiments by Ferster et.al[5], the orientation selectivity of simple cells in the primary visual cortex is not due to laterally inhibitory connections, but due to the features that receptive fields themselves express. The receptive fields shown in Fig.4 are thus implausible in view of Ferster et al.'s results.

5 Conclusion

The conventional hypercolumn structure of a fixed competitive system has pre-set *a priori* parameters for the number of hypercolumns and number of neurons in a hypercolumn. The number of hypercolumns determines how many partial features a hierarchical input information source can be divided into, and the number of neurons in a hypercolumn determines the kind of partial features that emerge in exclusion of each other. One of the problems of the conventional model is that the input patterns are always divided into the same number of partial features. As input patterns have various degree of complexity, the number of partial features is not necessarily the same. To improve this situation, we used the anti-Hebbian learning rule of Földiák to make such the fixed hypercolumn structure self-organizing. The *a priori* parameters accordingly introduced are the total number of neurons, the firing probabilities, and the upper limits on the simultaneously firing probabilities of the neurons. As the resulting number of divided partial features is not always constant but constant on average, the model allows the firing of comparatively more neurons for complicated input patterns and less neurons for simple input patterns. Computer simulations showed that local receptive fields with orientation selectivity can be formed without introducing *a priori* predetermined connective regions, but instead by making the competitive system self-organizing.

Acknowledgments

We thank Dr. Peper Ferdinand and Dr. Mahdad N. Shirazi at Communications Research Laboratory (CRL) for the helpful discussions. This work was partially financed by the Council for the Promotion of Advanced Information and Communications Technology.

References

1. Maekawa, S., Kita, H., Nishikawa, Y.: Self-organizing extraction of hierarchical information with use of synaptic competition for the presynaptic factor. Proceedings of the World Conference on Neural Networks, Washington D. C. **2** (1995) 555–558
2. Fukushima, K.: Neocognitron : A self-organizing neural network model for a mechanism of pattern recognition unaffected by shift in position. Biol. Cybern. **36** (1980) 192–202
3. Földiák, P.: Forming sparse representations by local anti-hebbian learning. Biol. Cybern. **64** (1990) 165–170
4. Maekawa, S., Kita, H., Nishikawa, Y.: Self-organizing learning rule with use of synaptic competition for the synaptic factor. SICE Joint Symposium (In Japanese), Okinawa (1994) 293–296
5. Ferster, D., Chung, S., Wheat, H.: Orientation selectivity of thalamic input to simple cells of cat visual cortex. Nature **380** (1996) 249–252

Optimizing a Neural Network Architecture with an Adaptive Parameter Genetic Algorithm

Arnaud RIBERT, Emmanuel STOCKER, Yves LECOURTIER, Abdel ENNAJI

Université de Rouen, PSI/La3i, UFR des Sciences et Techniques
F-76821 Mont Saint Aignan Cedex, France
E-mail : Arnaud.Ribert@univ-rouen.fr

Abstract : This article deals with the use of genetic algorithms to optimize the architecture of a neural network. After a brief recall of our original neural network (named Yprel network), we show that a simulated-annealing-like technique has been advantageously replaced by genetic operators. Indeed, tests on character recognition (NIST handwritten database) have shown that the generalization rate has been improved, the mean network size has been reduced by a factor 3 and the learning speed has been significantly increased. Moreover, a portable adaptive mutation probability has been introduced which enables a parameter-free learning.

I. Introduction

Among the various ways to perform an automatic classification, one of the most popular is the neural network. However, it is often very difficult for a user to find a well-suited architecture for his problem. In 1993, a new type of neural network - Yprel network - was introduced which could find its own architecture [LEC93][STO96]. At this time, the building heuristic was based on a simulated annealing-like procedure. It required two parameters, which could have a great impact on the network size, learning speed and generalization performances. In order to optimize the network structure and to avoid parameters finding, genetic algorithm tools : reproduction and mutation have been tested.

Some authors have proposed to use genetic algorithms with neural networks. They can be used to determine weights of an MLP [SCH95][BEL91][LIS95][WHI90] or to perform feature selection [KUS95], but the main use of them is to determine the network architecture [ROB95] [MIL89] [HAR89] [DOB95] [MIC95] [GRU93] [GRU94] [KIT90] [MER91]. Indeed, they are of great interest to avoid local minima and are attractive when the process to be optimized is not known as a mathematical function. Moreover, the fitness function of genetic algorithms can be designed to optimize several parameters a time.

This paper gives an example of successful use of genetic algorithms for a non-MLP neural network architecture fitting. Moreover, it proposes a way to obtain an adaptive mutation probability for a genetic algorithm.

II. A brief recall of Yprel network building

Our aim is not to introduce Yprel neural networks. Details can be found in [LEC93][STO96]. Consequently, we just give elements to understand genetic tool use.

An Yprel network is associated to one class C and is able to decide whether a prototype belongs to C or not. An Yprel (Y because it presents 2 inputs and 1 output, and PRocessing ELement) is a particular neuron which is able to classify some prototypes, depending on their position in the feature space. An Yprel network building is an incremental process. At the beginning, there are only neurons associated to features which constitute the only layer of the network. Each input yprel is able to classify a certain number of the learning database prototypes. A new yprel is added until all prototypes are classified.

An example of Yprel network building is given on figure 1.

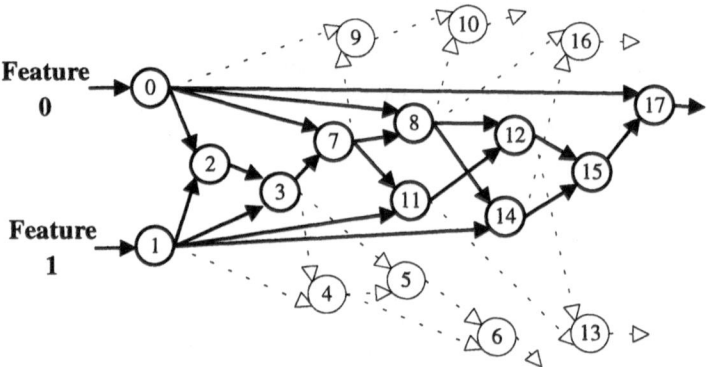

Figure 1 : Building of an yprel network : a population of neurons is generated. When an yprel is able to classify every prototypes, it is kept with all its ancestors to form the winner network, while non-contributing neurons are discarded.

When adding a new yprel, the problem is to find its two parents. Two criteria should be taken into account to choose them : we want a small network and a fast convergence of the learning. So, the basic heuristic favoured yprels with few ancestors and being very efficient (classifying a maximal number of prototypes). However, if best yprels are to be favoured, less efficient ones are not completely discarded because a combinaison of them may lead to a good classifier.

The first way to implement such a heuristic has been to use a simulated annealing-like technique. To favour small size networks, the idea is to sort yprels according to their number of ancestors. Yprels needing approximately the same ancestor number N are grouped in a list. Lists are sorted according to N. Then a second stage illustrated on figure 2 consisted in a biased draw following an exponential law to designate a list of yprels L.

Efficient yprels are favoured using the same principle : yprels of the list L are sorted according to their efficiency (number of classified prototypes). Then a second biased draw is performed, which designates an yprel.

For a user, the main problem was to find a good curvature for the exponential, which is named the temperature in a simulated annealing algorithm.

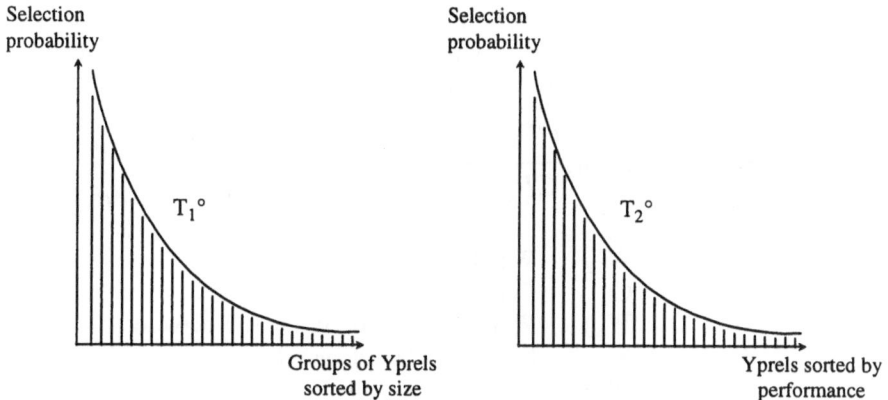

Figure 2 : Original yprel selection rules

In fact, two temperatures had to be determined. The choice was obviously critical, because it influenced the network architecture, that is to say learning and generalisation capabilities. It can be noticed that strictly speaking, since the temperatures were fixed, this technique could not be called simulated annealing.

This selection rule enabled the learning to converge for primitive-based problems. Unfortunately, when the network had to learn images of character, the convergence was very hard - not to say impossible.

Moreover, the main criticism that can be made is that even if yprels have very close performances, the sort will impose that the very best of them are selected although an equiprobability would be preferable.

III. Using genetic algorithms

Genetic algorithms are used to optimize a certain function, called "fitness". The aim is to begin with a random population of points and to improve its global performance, according to the fitness function. A basic genetic algorithm applies to individuals three operators : a reproduction, a crossover and a mutation. The first one enables a selection of best individuals of the population. The second one enables an exchange of capabilities between two individuals, while the mutation may introduce new capabilities in the population.

When using genetic algorithms, and whatever the application is, the first and biggest problem to solve is the encoding technique. As it will be shown, in our case, a particular encoding - without any string - was easy to employ. The second problem is to define the fitness function : how to quantify an individual performance.

When dealing with neural networks, the first worry is to obtain an efficient learning. An yprel is considered as efficient if it classifies a lot of prototypes. So it seems natural to imply the number of classified prototypes in the fitness function. Moreover, generalization performances are usually better when the network size is low. So, the size of the network has to appear in our fitness function. Eventually, the fitness function used can be expressed by the following equation.

$$F(Yprel) = 1 + W_{CP} \cdot \frac{CP_{Yprel} - CP_{Min}}{CP_{Max} - CP_{Min}} + W_S \cdot \frac{S_{Yprel} - S_{Min}}{S_{Max} - S_{Min}}$$

With :
- W_{CP} : Weight of Classified Prototypes
- CP : Classified Prototypes
- W_S : Weight of the Size of the sub-network
- S : Size = number of ancestors
- $W_S + W_{CP} = 99$

It is clear that the fitness of an yprel varies from 1 to 100. The selection probability of an yprel to be a parent is given by :

$$P_S(Yprel) = \frac{Fitness(Yprel)}{\sum_i Fitness(Yprel_i)}$$

A lot of tests have been carried out to determine good weight values on a handwritten character recognition problem. When the size weight was more important than the weight of classified prototypes, the network was clearly less efficient in learning, which was significantly longer than with the previous selection rule. Moreover, the final network size (the number of yprels) was not smaller, and involved more features. Eventually, the best weight configuration was $W_{CP} = 99$ and $W_S = 0$.

This result is quite surprising, but can be interpreted considering that it is more efficient to progress quickly in the problem solving. As a matter of fact, the learning time is significantly improved compared with the original selection rule. Although the final network size and generalization performances are equivalent for both selection rules, the genetic selection is more advantageous because it is parameter-free, whereas the simulated-annealing selection rule requires two parameters.

Mutation is known by genetic algorithm users to be very powerful to find good solutions. Indeed, with a standard reproduction operator, no parent selection could be done without considering yprel characteristics. The mutation we use enables to choose one of the parents of an yprel whatever its performances are.

In our case, the population size is not fixed but is only kept under a pre-defined threshold : 1000 is frequently used. This feature enables to use high mutation rates, which would completely degenerate a standard population. So, we could test mutation rates up to 90%. Our mutation operator randomly selects an yprel, creates its twin and randomly changes one of its parents. The twin of an yprel is in fact a new yprel which have the same parents, but no child. For high mutation rates, the maximum population size had to be increased to 2500 to enable convergence. The learning time was then very long (multiplied by a factor 25), but the neuron number in a network was divided by a factor 2. This result is easy to interpret : a high mutation rate enables a wider search in the configuration space, and leads to a more efficient final architecture.

In order to reduce the learning time, an adaptive mutation rate has been introduced. A second motivation was that the optimal mutation rate depended on the problem to learn. Indeed, sometimes the mutation is desirable because it gives some "oxygen" to the population, but sometimes it is only a waste of time. When designing an adaptive parameter, the usual way is to begin with a certain value and to modify it according to the process evolution. For instance, it is possible to begin a learning stage with $P_m = 0$ (P_m : Probability of Mutation), and to increase it of 5% if the learning does not progress over the 10 last added yprels. On the contrary, P_m can be decreased of 5% if the progression has been good over the same period. It can be noticed that 5 and 10 are a priori parameters which cannot be found easily.

A second problem appears when a mutation rate is fixed. The classical mutation operator guarantees that the rate will be respected after a certain number of draws. In our application, experiences have shown that it was more efficient to have a deterministic sequence of mutation. For instance, a 50% rate can be obtained doing N mutations followed by N standard selections. N is a new parameter difficult to adjust.

The retained solution uses a statistical parameter measured on the complete population which enables to determine if the next yprel will be a mutant or a standard one. Such a criterion is very difficult to find if a given value has to indicate directly the next yprel type. However, it is easier to find a criterion which indicates if it is necessary to change the operator. Our operator is based on the standard deviation σ and the mean μ of the classified prototype number of yprels in the population. Experiments have shown that σ/μ had to decrease while σ increased to get an efficient architecture.

So, an adaptive mutation rate can be obtained using the following algorithm :

If ((σ/μ has decreased) and (σ has increased)) then
        Keep the same selection rule
Else
        Change the selection rule
End If

For the first drawing, the selection rule is set to "standard reproduction operator". So, when the problem is easy to solve, no mutation is performed. On the contrary, when the problem is harder, the mutation rate increases. It is worth noticing that such an algorithm could be adapted to a classical genetic algorithm. The statistical measures give an image of the population evolution between two generations. So, if the criterion leads to keep the same selection rule, it could be interpreted like "no mutation is needed". On the contrary, the classical mutation operator could be used, with a pre-defined probability P. This one would stay adaptive over several generation, P only being a maximum mutation rate.

This algorithm gave good results concerning the final architectures, but the learning time was still important. So a third selection rule has been introduced to guide the learning. Indeed, no operator in the building algorithm takes into account the possibility to choose two complementary yprels to be the parents of the next added one. So, when a new yprel is to be added, a search is performed. The aim is to find an yprel (Y_1) classifying the maximum unclassified prototypes by the best yprel (Y_b). If this maximum is greater than 0, Y_1 and Y_b are the parents of the new yprel. On the contrary, the previous algorithm is applied. This operator is very efficient, because it leads to shorter learning times than initial ones while keeping efficient final architectures.

IV. Results

The final version of our building rule has been tested on an optical handwritten-digit recognition problem. The learning database was constituted of 10 000 figures of the NIST database while the test database contained 10 000 others NIST prototypes. The initial building rule leaded to some results published in 1995 [LEC95] which are reproduced in table 1. A classifier is composed of 10 yprel networks, each of them being associated to one class. A comparison between the initial and the current rules is given in the following table.

	Original Version	Genetic Version
Recognition	76,60%	86,63%
Rejection	19,55%	12,28%
Confusion	3,85%	1,09%
Nb Yprels	4111	1235

Table 1:Comparative results for 10 000 NIST characters

It appears that the genetic network building leads to better architectures, improving recognition, rejection and confusion rates. Networks are smaller : from 4111 yprels for the original version to 1235 yprels for the entire classifier (10 networks). The learning time has also been improved, being divided by a factor 15.

Results are therefore very promising, but a last worry could subsist : is our building rule stable? Indeed, it is a semi-random process and the final solution may not be always the same. So, stability tests have been carried out. A typical example for a network size is given on the following figure. It is clear that the genetic selection rule is by far more stable, thus more reliable.

Yprel Number

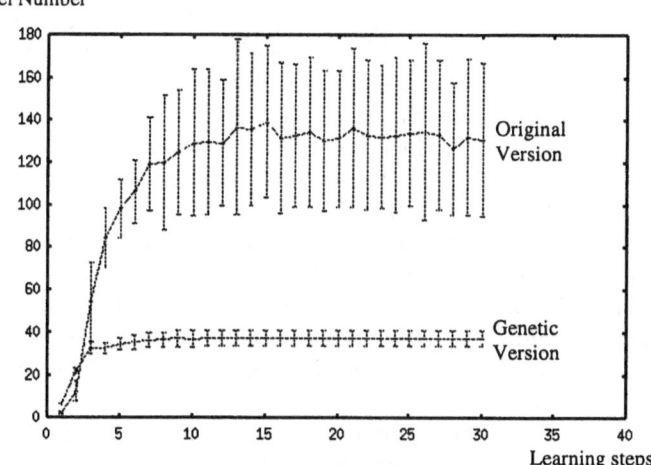

Figure 3 : Yprel-number evolution for a typical learning stage

Another important point is that bitmap characters can now be learned without any difficulty, whereas it was impossible with the initial version.

V. Conclusion

This paper shows the interest of genetic tools in the building of a neural network. We have adapted a reproduction operator which has saved 2 user-defined parameters. A portable adaptive mutation operator has been successfully implemented, which has led to a significant network size reduction. Eventually, compared with the original building strategy, new architectures are at least three times smaller, generalize better, while the learning is more stable and 15 times faster.

In order to improve our classification methodology, a different network distribution strategy has been studied in parallel to this work. Building a classifier is then viewed as the adaptation of a network population to an evolving learning database. This new frame will integrate the genetic network building to improve the evolving abilities of the population in a dynamic environment, thus making it close to works on artificial life.

534

Bibliography

[BEL91]   Belew R.K., Mc Inerney J., Schraudolph N.N. (1991). *Evolving network :
          using the genetic algorithm with connectionnist learning.* In Langton C.G.,
          Taylor C., Farmer J.D., Rasmussen (Eds.), Artificial life II, SFI studies in
          the sciences of complexity, Vol X (pp.511-547). Reading, Massachusetts :
          Addison-Wesley.

[DOB95]   Dobnikar A. (1995) Genetic synthesis of task oriented neural networks. In
          D.W. Pearson, N.C. Steele, R.F. Albrecht (Eds.), *Artificial neural nets and
          genetic algorithms* (pp.329-332). New-York : Springer-Verlag.

[GRU93]   Gruau F. (1993). Genetic synthesis of modular neural networks. In
          S.Forrest (Eds.), *Proceedings of the fifth International Conference on
          Genetic Algorithms* (pp. 318-325). San Mateo, CA : M.Kaufmann.

[GRU94]   Gruau F. (1994). Synthèse de réseaux de neurones par codage cellulaire et
          algorithmes génétiques, Ph. D. Thesis, France.

[HAR89]   Harp S.A., Samad T., Guha A. (1989). Towards the genetic synthesis of
          neural networks. In D.J. Schaffer (Eds.), *3rd International Conference on
          Genetic Algorithms* (pp.360-369). San Mateo, CA : M.Kaufmann.

[KIT90]   Kitano H. (1990). Designing neural networks using genetic algorithms with
          graph generation system, In *Complex Systems*, **4**, 461-476.

[KUS95]   Kussul E.M., Baidyk T.N. (1995). Genetic algorithm for neurocomputer
          image recognition. In D.W. Pearson, N.C. Steele, R.F. Albrecht (Eds.),
          Artificial Neural Nets and Genetic Algorithms (pp.120-123). New-York :
          Springer-Verlag.

[LEC93]   Lecourtier Y., Ennaji A., Gilles F., Chavy P. (1993). Yprel networks and
          classification. In *Proceedings of the 1993 IEEE International Conference
          on Systems, Man and Cybernetics Vol 3* (pp.463-468). New-York : IEEE.

[LEC95]   Lecourtier Y., Ennaji A., Stocker E., Gilles F. (1995). Yprel networks,
          classification and incremental learning, Traitement du signal, Vol. 12, **6**,
          597-607.

[LIS95]   Lis J. (1995). The synthesis of the ranked neural networks applying genetic
          algorithm with the dynamic probability of mutation. In J. Mira, F. Sandoval
          (Eds.), *Proceedings of the 1995 International Workshop on Artificial
          Neural Networks : From Natural to Artificial Neural Computation* (pp.
          498-504). New-York : Springer-Verlag.

[MER91]   Merrill J.W.L., Port R.F. (1991). Fractally configured neural networks,
          Neural Networks, **4**, 53-60.

[MIC95]   Michel O., Biondi J. (1995). From the chromosome to the neural network.
          In D.W. Pearson, N.C. Steele, R.F. Albrecht (Eds.), *Artificial Neural Nets
          and Genetic Algorithms* (pp.80-83). New-York : Springer-Verlag.

[MIL89]   Miller G.F., Todd P.M., Hedge S.U. (1989). Designing neural networks
          using genetic algorithms. In J.D. Schaffer (Eds.), *Proceedings of the Third
          International Conference on Genetic Algorithms* (pp. 379-384). San Mateo
          : M.Kaufmann

[ROB95] Roberts S.G., Turega M. (1995). Evolving neural network structures : an evaluation of encoding techniques. In D.W. Pearson, N.C. Steele, R.F. Albrecht (Eds.), *Artificial Neural Nets and Genetic Algorithms* (pp.96-99). New-York : Springer-Verlag.

[SHA95] Schaffer J., Braun H. (1995). Optimizing classifiers for handwritten digits by genetic algorithms. In D.W. Pearson, N.C. Steele, R.F. Albrecht (Eds.), *Artificial Neural Nets and Genetic Algorithms* (pp.10-13). New-York : Springer-Verlag.

[STO96] Stocker E., Ribert A., Lecourtier Y., Ennaji A., (1996). An incremental distributed classifier building. In *13th International Conference on Pattern Recognition (ICPR'96) Vol IV*, (pp. 128-132). Washington : IEEE Computer Society Press.

[WHI90] Whitley D., Starkweaker T., Bogart C. (1990). Genetic algorithms and neural networks : optimizing connections and connectivity, *Parallel Computing*, **14**, 347-361.

Self-Organizing Symbolic Learned Rules

Antonio Bahamonde, Enrique A. de la Cal, José Ranilla, Jaime Alonso
Centro de Inteligencia Artificial. Universidad de Oviedo at Gijón
Campus de Viesques. E-33271 Gijón, Spain
email: {antonio, delacal, ranilla, jalonso}@aic.uniovi.es
http://ntserver.aic.uniovi.es

ABSTRACT

In this paper we present a self-organizing process for rules obtained from a machine learning system. The resulting map can be interpreted back into the symbolic field in an attempt to make the logical representation of the original rules reflect the relationships codified by map distances. Thus, we improve the quality of the starting set of rules both in classification accuracy and in conceptual clarity.

INTRODUCTION

In this paper we deal with **explicitly set knowledge** about a classification problem. The representation unit used will be one of the most popular formalisms in symbolic artificial intelligence: productions **rules** written in a plain logic language. We assume a set of rules induced from a set of training examples, and our goal is to improve the quality of our learned knowledge in a general sense.

To fix the ideas and denotational conventions used throughout the paper, our training **examples** will be described by a set of **attributes** or features like *color*, *size*, *with-red-spots*, etc.. Additionally, the examples will provide us with a singular attribute, usually called class. Our intended goal is find out the relations of attributes and their values that guarantee the presence of each class. These relations, will be the explicit knowledge pieces called rules and look like this:

$$1 \leftarrow \text{Att-2} = 2 \wedge \text{Att-5} = 3 \wedge \text{Att-6} = 2,$$

where **1** codifies a class, and **2** and **3** represent values of the corresponding attributes.

The **quality** of a set of rules can be measured in a quantitative way by the proportion of success in their classification task. However, since we stress the role of explicitness, there is a qualitative dimension worthy of being taken into account: the **conceptual clarity** of rules.

The approach followed in this paper tries to improve both senses of quality in rule sets: we obtain more accurate classifications with fewer and more reliable rules. With this aim in mind, we have devised a process by which the rules suffer **exposure** to a **neural** treatment to then return back to the symbolic field. The idea is to endow our rules with a flexible metric that allows their self-organization in a kind of **map of**

rules and attribute values following a process similar to Kohonen's SOM [Kohonen, 95].

Once the map has been obtained, we read it trying to make the external face of rules (their description in their logical language) look similar to the internal relationships of rules and attribute values codified by the map distances. Thus, close values can be added to logical expressions of rules. At the same time, grown rules can overlap other rules which will then become unnecessary.

The final step in representing neural results back into the symbolic world is aggregation. Here we try to produce a more compact version of our set of rules. The aim is to group together rules alluding to nearby values of attributes.

In [Bahamonde, 91], and then in SHAPE [Botana, 95], a restricted version of aggregation was introduced as a **syntactic** process where no induction was carried out. The rules with a similar description (the same class and set of attributes in their conditions) were interpreted as a regular expression which was minimized by means of non-deterministic finite automata techniques. Therefore, training examples were not considered at all, and rules were dealt with as algebraic objects.

Here, we are going to present a **semantic** version since our rules are now endowed with patiently worked out data: the tables of differences. In a few words, we extend the limits of rules in an attempt to cover their decision regions more explicitly. The result is a simplification filter where the number of rules can be drastically reduced. In this way, the set of rules obtained is clearer than the original one, and an improvement in conceptual quality is thus achieved.

In this paper we will only deal with symbolic valued attributes, but the results could easily be extended to a general situation; that is to say, attributes whose values are numbers or symbols.

The last section of the paper is devoted to presenting some **experimental results**. There we show that machine learning systems like C4.5 [Quinlan, 93] or our Abanico [Ranilla, Bahamonde, 95] can be improved in noisy well-known problems. In Monk's 3, starting with Abanico rules, we obtained 43 times of 50 rules with 100% success and a conceptual clarity similar to the logical formula that defines the classification problem. In noise free problems we can significantly reduce the number of rules without penalizing the classification performance.

METRICS FOR SYMBOLIC RULES

A set of rules must be endowed with an **application algorithm** in order to implement a classification procedure. Sometimes a case fulfills more than one rule set of conditions; if the conclusions do not agree, the result is a conflict that must be solved by some kind of priority order. At the same time, we might have cases that do not fulfill any rule, but some decision must be taken in any case.

We will use a distance-based criterion to decide which rule to apply to any case. The **nearest rule** will provide the decision about the class of a given case. In a draw state, we prefer the most specific rule; that is to say, the rule with more antecedents. Even if we have more than one possibility we will use the best rule according to a priority given by the supplier of the rule set.

Additionally, we need metrics in the rule spaces, because we want to arrange them in such a way that similar properties will be placed in contiguous places. When dealing with numbers, distances can be measured with the usual metrics; so, for instance, we are tempted to aggregate weights of people of *65341* and *65342* grams into a decision rule. However, although the distance between *yes* and *no* is usually taken to be 1; this number does not mean the same as the above weight difference. In general, the metrics should take into account the relevance of differences in order to take a good decision with a case and a given set of rules.

In this section, we present the formulas of distances used throughout the paper. We start by setting the distance between an example **ex** and a **rule**. It is worthy noting that here *example* or *case* are used as synonymous; we do not care if we are in training phase (we usually say *examples* then) or in testing phase (where we prefer *case* instead).

$$\text{example\_distance(rule,ex)} = \frac{\sqrt{\sum\left(\text{difference(rule, att, value\_ex(ex, att))}^2 : \text{att in rule}\right)}}{\sqrt{\text{length (rule)}}}$$

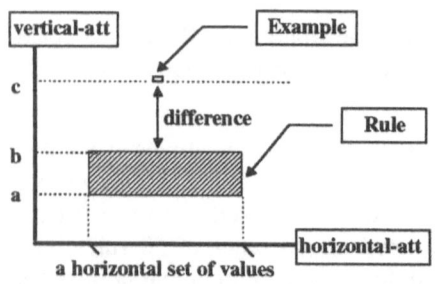

Let us point out that our distances refer to differences between rules and individual values of attributes; in the above formula the value in the example. Thus, each rule will be endowed with a table of differences in every attribute. In the adjoining figure, the difference table of the rule, in the vertical attribute, should show difference zero with a and b, and a not zero number for value c. Now, we can compute the

$$\text{rule\_distance(r1, r2)} = \frac{\sqrt{\sum\left(\text{min\_edges\_difference(r1, r2, att)}^2 : \text{att in Set\_of\_attributes}\right)}}{\sqrt{\text{length (r1)}}}$$

where **min_edges_difference** stands for the minimum distance (difference, to be precise) from the first rule to the edges of the second one in a given attribute. In formulas, when dealing with symbolic valued attributes, we define

min_edges_difference(r1, r2, att) =
    minimum{difference(r1,att, v): v ∈ values(r2, att)}   **If att appears in r2**
    0   **else**

It is important to point out that rule distance it is not a symmetrical relationship given that the differences between rules and individual values do not exhibit any reciprocity behavior. Additionally, we wish to emphasize the fact that all distances can be computed if tables of differences of all rules are available. So our aim in this paper is to provide an effective way to compute realistic and useful measures for these differences.

THE SELF-ORGANIZING PROCEDURE

The SOM algorithm has been introduced by Kohonen in [Kohonen, 95; p. 77] as a kind of entities adaptation in a hypothetical elastic network according to the interaction of a set of examples and the topological arrangement of the entities. The self-organizing procedure that we are going to present in this section follows this spirit along with the ideas involved in LVQ algorithms; but it is not a simple application of them.

Taking into account the remarks of the last section, our goal is to compute a table of differences for each rule. Thus, we start by considering trivial tables for each rule; that is to say, in every attribute the difference will be 0 if the value is part of the rule condition and 1 otherwise. Then for all the examples **ex** (randomly sorted) we find the nearest rule and a neighborhood of it. Every one of these rules update their table of differences for each value mentioned in the example ex and for each *relevant attribute* according to the following formulas:

difference := difference - h(t,d) * difference {when ej belongs to the same class as the rule}
difference := difference + h(t,d) * difference {else}

Where **h** is called (following [Kohonen, 95; p. 79]) the *neighborhood function* and is given by

h(t,d) := α(t) * σ(d),   α(t) := 0.9 * (1 - (t / upper_limit_examples)),   σ(x) := 1/ (1 + exp(10x-5))

The intention of h is to weight the influence of examples over rules. So, at the beginning of the process, h outputs bigger values according to a *learning-rate factor* quantified by α; its argument, **t**, is usually interpreted as the time or, in simpler words, is just the ordinal of the example being dealt with (its upper bound is **upper_limit_examples**). However, the final value returned by h is α(t) filtered by the sigmoid σ that tries to weight the results according to the distance (**d**) from the example to the rule. Its intention is to favor influences of near examples and to punish those of far ones.

To determine the neighborhood of rules for an example **ex** we first compute the

radius := minimum_distance(ex) + h(t, minimum_distance(ex)),

where **minimun_distance(ex)** is the smallest distance from ex to any rule. Then we order the rules whose distance from ex is smaller than the radius, from the nearest to the farthest. Rules at the same distance are ordered according to the preference policy given with the set of rules. The list thus obtained is **pruned** if the class of the first rule of the list appears later with rules of different classes in the middle; we say that these middle rules *ban* the influence of the example. Finally, we only allow lists of rules of n elements at most (usually n = 3); thus we separate the first n elements as the neighborhood of our example.

To end the description of our self-organizing process, we only need to spell out what we mean by a ***relevant attribute*** for an example ex and its neighborhood N(ex). These attributes are those appearing in the nearest rules in N(ex) for each class. The idea is to fix the attention of the process only on the really important features of the examples instead of the arbitrary ones.

SYMBOLIC SEMANTICS OF NEURAL RESULTS

Once the rules have been self-organized, we must return to the symbolic field. So we read the results to try to give a conventional logic semantics to rule tables of differences. In order to do this, we only need to fix a **zero threshold** λ. So, values under λ (we use $\lambda = 0.1$) will be assumed as zero and will then be incorporated into the external face of the rule. For instance, the rule

$1 \leftarrow$ Att-2 = 2 $\wedge$ Att-5 = 3 $\wedge$ Att-6 = 2,

after interaction with Monk's 3 training examples, produced the following table:

attributes	Att-2			Att-5				Att-6	
value	1	2	3	1	2	3	4	1	2
difference	1.68e-243	0	1.55	4.46e-213	1.05e-202	0	1.56	1.68e-243	0

In other words, the difference between the rule and value 1 in attribute number 2 is almost insignificant, therefore, we are forced to include it in the final symbolic version of our rule. In fact, the new rule should look very much like this:

$1 \leftarrow$ Att-2 = {1, 2} $\wedge$ Att-5 = {1, 2, 3} $\wedge$ Att-6 = {1, 2}.

The next step is to skip the attributes like attribute number 6 in the above rule. It has become meaningless since the set of all possible values for said attribute is {1, 2}, so the condition described in the rule is now a redundancy. Hence, the final version is

$1 \leftarrow$ Att-2 = {1, 2} $\wedge$ Att-5 = {1, 2, 3}.

But if we are going to consider this rule, there are some others that will no longer be necessary: the rules included in it. So, rules like

$1 \leftarrow$ Att-2 = 1 $\wedge$ Att-5 = 1
$1 \leftarrow$ Att-2 = 2 $\wedge$ Att-5 = 3 $\wedge$ Att-3 = 2

are now particular cases of the final release of our original rule. However, inclusion can be more subtle. Let us consider the rule

class $\leftarrow$ A : {c1, c2}, B : {b1, b2}.

This is not included in any of the rules

class $\leftarrow$ A : {a1, c1, c2}, B : b2
class $\leftarrow$ A : {a2, c1, c2}, B : b1

but it is a part of the union of both.

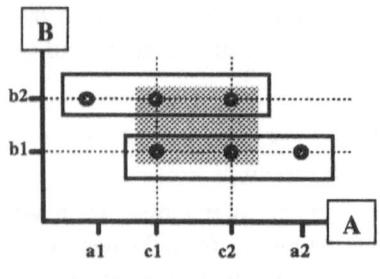

Multiple inclusion

AGGREGATION

The basic operation in the aggregation process is the **extension** of one rule **r1** over another one **r2** through an attribute **at**. If we represent **at** in the horizontal axis, the process can be graphically sketched out as in the following figure:

Extension of r1 over r2 through the horizontal attribute

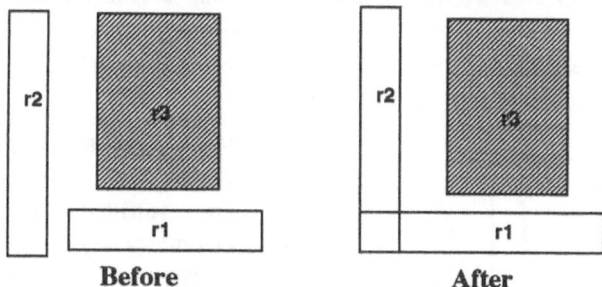

Before   After

To allow the extension, we must ensure that rules belong to the same class and their distance is small. Fortunately, our metrics is not a symmetric one, but in any case we have to set conditions that reject the extension of r2 over r1 in the above figure; the new r2 would become too wide: it would include r1 and r3, a rule of another class.

However, before going into details, let us review the kind of outputs that we can expect from the aggregation process. To illustrate this process let us assume that we have the 4 rules of the following figure (a). In order to avoid cumbersome numerical peculiarities, we represent distances by units in a symmetric scenario: horizontal and vertical differences for all rules are 1 unit. After two consecutive extensions we reduce the rules to those shown in figure (b). Finally, we will aggregate the original 4 rules into one only rule (see figure (c)).

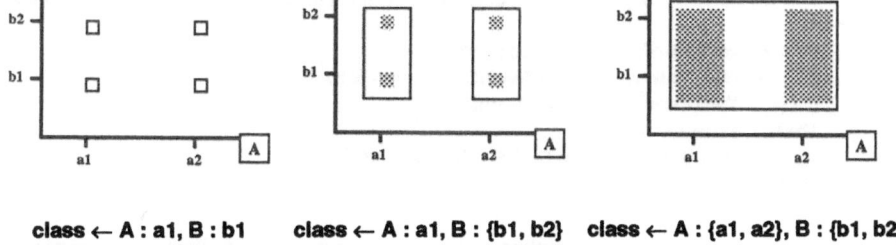

class ← A : a1, B : b1	class ← A : a1, B : {b1, b2}	class ← A : {a1, a2}, B : {b1, b2}
class ← A : a1, B : b2	class ← A : a2, B : {b1, b2}	
class ← A : a2, B : b1		
class ← A : a2, B : b2		

To specify these operations we need the **degree of coincidence** as a decision parameter to apply the extension of rules. Formally, we define

$$\text{degree\_of\_coincidence}(r1, r2, att) = \frac{\text{measure}\big[(\text{values}(r1, att)) \cap (\text{values}(r2, att))\big]}{\text{measure}\big[\text{values}(r1, att)\big]}$$

where *measure* of a set of values can be understood here as the sum of all their membership degrees given by

membership_degree (rule, attribute, value) = 1 - difference (rule, attribute, value)

and, as usual, for intersection

**membership_degree (r1 ∩ r2, att, value) =
membership_degree (r1, att, value) * membership_degree (r2, att, value).**

With this definition, we only have to take care with rules of different sets of attributes in the description of their conditions. The next figure shows what kind of extension we would like to provide in such cases.

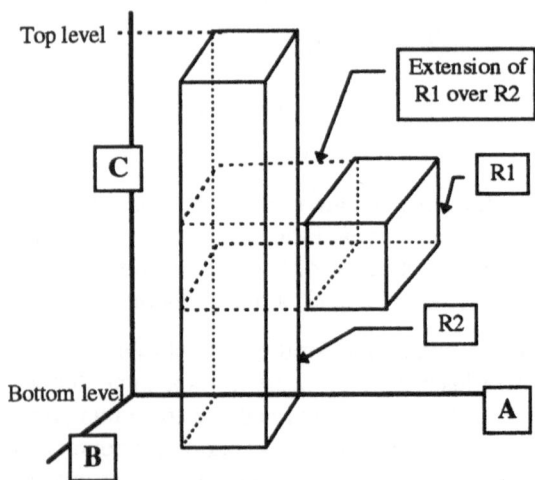

Here, r2 is described by attributes A and B; so attribute C conditions are fulfilled by any value. Geometrically, this means that the extent of r2 in axis C goes from their bottom level to its top. On the other hand, r1 (rule of the same class as r2) needs the 3 attributes to be specified. If the condition of attribute B is the same in both rules, then the extension of r1 over r2 should be in the address indicated by axis A as depicted in the adjoining figure.

Therefore, the aggregation process should be fired according to the following rule

```
If   rule_distance (r1, r2) is small                              AND
     conclusion(r1) = conclusion(r2)                              AND
     For all attribute att but AT      degree_of_coincidence(rule-1, rule-2, att) is large
then
     r3:= extension of r1 over r2 through AT
     If   r3  cuts with rules of different classes are included in
             the union of cuts of r1 and r2
     then    r1 := r3
             If r2 is included in r1 then delete r2 end_If
     end_If
end_If
```

Finally, we only need to point out that extension through an attribute represents the addition to the first rule of values used in the description of the second rule. The table of differences of the thus extended rule should now be updated in the attribute that points to the extension address. This is done by taking the minimum of the difference between each value and any of the involved rules.

EXPERIMENTAL RESULTS

The experiments reported here were carried out with 10 iterations of training examples. To avoid the influence of the initial random ordering of the examples, we repeated the proof, for each problem, 50 times. Thus, the scores that follow will provide average values.

The learning problems were taken from the UCI repository [Murphy, Aha, 96] and all are well known bench marks for learning systems. The learning algorithms used here to produce rules to be self-organized by our algorithm were C4.5 [Quinlan, 93] and Abanico [Ranilla, Bahamonde, 95].

Quinlan's system is probably the best known machine learning system, and produces ordered sets of rules with a final default class, so that default should be used just in case the other rules do not have an explicit answer. Since we use a nearest rule approach, we skipped the default class as a rule in our experiments with release 8 of C4.5.

On the other hand, Abanico (a Spanish acronym for learning based on continuous numbers aggregation into intervals), in its actual release, is based in part on SHAPE [Botana, Bahamonde, 95] and [Alguero, 96]. It produces good results both for symbolic and numerical valued attributes with or without noise.

The first problem that we are going to review is a typical symbolic **noisy** problem: **Monk's 3**. It is part of a family that relies on an artificial robot domain, in which robots are described by six different attributes with symbolic values [Wnek, Sarma, Wahab and Michalski, 1991]. An exhaustive performance comparison study of different learning systems can be found in [Thrun et al., 91]. The learning task is a binary classification and each problem is given by logical description of a class. Monk's 3 is defined by

a robot belongs to the *good* class Iff (Att-5 = 3 $\wedge$ Att-4 = 1) or (Att-2 $\neq$ 3 and Att-5 $\neq$ 4)

From 432 possible examples (the test set), the training set are 122 examples selected randomly, and among them there were 5% misclassifications. To conclude this general presentation, let us recall that this is a hard problem to be learned by typical neural systems, for instance, backpropagation only reaches 93.1% of classification success, and with weight decay, the results only rise to 97.2%.

Monk's #3		Original scores		Average Final Scores	
learning system	#rules	%success	#rules	%success	
C4.5	15	96.3	5.68	97.5	
Abanico	10	99.07	3.72	99.83[1]	

Abanico scores [1] were found 41 times out of 50 with 3 rules and 100% of success and the other 9 times we obtained 7 rules without any improvement in classification tasks; that is, 99.07%. In other words, most of the times we improve the rules from Abanico both in classification success and in conceptual clarity. The few times that the original scores are not worsened, the same level of success is maintained while the number of rules is improved (the worse time we obtained 7 rules). It is worth noting here that, as far as we know ([Thrun et *al.*, 91] and [Murphy, Aha, 96]), there is no other explicit knowledge learning system with such scores.

Next we will present results with **noiseless problems**. The first one is **Monk's 1**. Here the story is a short one since our Abanico (as Alguero's system also does [Alguero, 96]) obtains 100% of classification success with the least number of rules possible: 5. The self-organizing process respects the original results.

Now let us deal with a famous chess end game problem originally proposed by Quinlan in [Quinlan, 83]: white King-Rook versus black King-Knight. The aim is to decide whether knight's side is lost in at most 2-ply in a black-to-move situation. The learning corpus are 647 examples described by 7 symbolic valued attributes. The goal is to compress the set of classification rules as much as possible since it is not a problem, in general, to reach 100% success. The following table shows the results obtained.

A Chess Problem	Original scores		Final Scores	
learning system	#rules	%success	#rules	%success
C4.5	15	100	6	100
Abanico	14	100	9	100
Abanico weighted	6	99.38	5	100
Shape without automata aggregation	17	100	9.02	100

Shape, endowed with its original automata aggregation, only needs 9 rules to cope 100% classification success. We obtained 49 out of 50 times 9 rules and the other time 10 rules.

The final problem faced in this paper was proposed in [Bahamonde et *al.*, 91] to illustrate automata aggregation. Starting from Monk's data, we modified the attribute ranges and the decision rule slightly in order to obtain a model for the following 11 rules concluding the class codified by **1**:

$1 \leftarrow$ att-1 = 1 $\wedge$ att-2 = 1 $\wedge$ att-5 = 1 $1 \leftarrow$ att-1 = 1 $\wedge$ att-2 = 3 $\wedge$ att-5 = 2
$1 \leftarrow$ att-1 = 1 $\wedge$ att-2 = 1 $\wedge$ att-5 = 2 $1 \leftarrow$ att-1 = 1 $\wedge$ att-2 = 3 $\wedge$ att-5 = 3
$1 \leftarrow$ att-1 = 2 $\wedge$ att-2 = 1 $\wedge$ att-5 = 1 $1 \leftarrow$ att-1 = 1 $\wedge$ att-2 = 3 $\wedge$ att-5 = 4
$1 \leftarrow$ att-1 = 2 $\wedge$ att-2 = 1 $\wedge$ att-5 = 2 $1 \leftarrow$ att-1 = 2 $\wedge$ att-2 = 3 $\wedge$ att-5 = 3
$1 \leftarrow$ att-1 = 1 $\wedge$ att-2 = 2 $\wedge$ att-5 = 2 $1 \leftarrow$ att-1 = 2 $\wedge$ att-2 = 3 $\wedge$ att-5 = 4
$1 \leftarrow$ att-1 = 1 $\wedge$ att-2 = 2 $\wedge$ att-5 = 3

any other possibilies conclude **0**. In [Bahamonde et *al.*, 91] we showed that these rules can be condensed into 3 rules:

$1 \leftarrow$ att-1 = {1,2} $\wedge$ att-2 = 1 $\wedge$ att-5 = {1,2} $1 \leftarrow$ att-1 = {1,2} $\wedge$ att-2 = 3 $\wedge$ att-5 = {3,4}
$1 \leftarrow$ att-1 = 1 $\wedge$ att-2 = {2,3} $\wedge$ att-5 = {2,3}

We provoked a self-organization of the above rules with the examples from which Abanico induced them. The results obtained were a set of rules with, of course 100%

success in classification tasks, and an average number of rules concluding class 1 of 3.68. Specifically, 30 out of 50 times we found 3 rules, 10 out of 50 times 4 rules appeared, and the other 10 times we got 5 (6 times) or 6 (4 times) rules.

BIBLIOGRAPHY

ALGUERO GARCIA, A., Algoritmos para el tratamiento de reglas aprendidas a partir de ejemplos. Ph. D. Dissertation. Centro de Inteligencia Artificial, Universidad de Oviedo at Gijón, November 1996.

BAHAMONDE, A., VELA, C. R., BOTANA, F.: Generalización de reglas de clasificación. *Proceedings of IV Reunión Técnica de la Asociación Española para la Inteligencia Artificial, AEPIA-91*, Madrid, (1991), 231-242.

BOTANA F., BAHAMONDE, A., SHAPE: A Machine Learning System from Examples. *International Journal of Human-Computer Studies*, **42** (1995), 137-155.

KOHONEN, T., Self-Organizing Maps, p. xv+362. Springer Series in Information Sciencies, Vol. 30, 1995.

MURPHY, P., AHA, D.W., UCI repository of machine learning databases -a machine-readable data repository. Department of Information and Computer Science, University of California, Irvine. Anonymous *ftp* from ics.uci.edu in the directory pub/machine-learning-databases, 1996.

QUINLAN, J. R., Learning efficient classification procedures and their application to chess end games. In R. S. MICHALSKI, J. G. CARBONELL & T. MITCHELL (Eds.), *Machine learning: An artificial intelligence approach*, Palo Alto, Tioga (1983), 463-482.

QUINLAN, J. R.: C4.5, Programs for Machine Learning, p. x+302, Morgan Kaufmann Publishers, San Mateo (California), 1993.

RANILLA, J., BAHAMONDE, A., Segmentación de valores numéricos para el aprendizaje a partir de ejemplos. *Proceedings of VI Conferencia de la Asociación Española para la Inteligencia Artificial, CAEPIA-95*, Alicante (1995), 225-234

THRUN, S. B., BALA, J., BLOEDORN, E., *et al.*, The MONK's Problems - A Performance Comparison of Different Learning algorithms, Technical Report CS-CMU-91-197, Carnegie Mellon University, 1991.

WNEK, J., SARMA, J., WAHAB, A, and MICHALSKI, R., Comparison learning paradigms via diagrammatic visualization: A case study in single concept learning using symbolic, neural net and genetic algorithm methods. Technical Report, George Mason University, Computer Science Department, 1990.

Viewing a Class of Neurodynamics on Parameter Space

Jianfeng Feng and David Brown

Biomathematics Laboratory, The Babraham Institute, Cambridge CB2 4AT, UK

Abstract. Nearly all models in neural networks start from the assumption that the input-output characteristic is a sigmoidal function. On parameter space we present a systematic and feasible method for analyzing the whole spectrum of attractors–all saturated, all-but-one saturated, all-but-two saturated, etc. – of a neurodynamical system with a *saturated* sigmoidal function as its input-output characteristic. We present an argument which claims, under a mild condition, that only all saturated or all-but-one saturated attractors are observable for the neurodynamics. For any given all saturated configuration ξ (all-but-one saturated configuration η) the paper shows how to construct an exact parameter region $R(\xi)$ ($\bar{R}(\eta)$) such that if and only if the parameters fall within $R(\xi)$ ($\bar{R}(\eta)$), then ξ (η) is an attractor (a fixed point) of the dynamics. The parameter region for an all saturated fixed point attractor is independent of the specific choice of a saturated sigmoidal function, whereas for an all-but-one saturated fixed point it is sensitive to the input-output characteristic.

1 Introduction

Neural networks provide a systematic approach to massively parallel computation, as well as a possibly better understanding of brain function. Interestingly, to date, sigmoidal functions have been utilized in the vast majority of neural networks as the input-output characteristic [18], either on the state space or on the weight space. The sigmoidal function is nearly saturated outside a region, by suitably adjusting some parameters of the function. In the present paper we first report our recent work on neurodynamics defined by

$$y_i(t+1) = f(y_i(t) + \sum_{j=1}^{N}(a_{ij} + k_2)r_j y_j(t) + k_1), \ i = 1, \cdots, N, \ t = 1, \cdots, \quad (1)$$

where $\mathbf{y}(t) = (y_i(t), i = 1, \cdots, N) \in \mathbb{R}^N$, $A = (a_{ij}, i, j = 1, \cdots, N)$ an $N \times N$ matrix representing interaction between units (either weights or states), (k_1, k_2) are two key parameters of the dynamics, N can be thought of either as the number of neurons or the number of synaptic efficacies connected to the i-th neuron, $R = (r_i \delta_{ij}, i, j = 1, \cdots, N)$ plays the role of normalization, f is a *saturated sigmoidal function* which is continuous and defined by[1]

$$f(x) = \begin{cases} y\text{max} & \text{if } x > y\text{max} \\ f(x) & f(x) \text{ is strictly increasing for } x \in [y_{\min}, y_{\max}] \\ y_{\min} & \text{if } x < y_{\min} \end{cases} \quad (2)$$

As has been amply shown before [1, 2, 6, 9, 11] and as further demonstrated here, the dynamical system (1) reflects the main features of dynamics using *sigmoidal functions.*

We introduce the 'Nemytskij operator' [5] $F : I\!\!R^N \to I\!\!R^N$ acting coordinate-wise

$$(F(\mathbf{y}))_j = f(y_j), \quad \mathbf{y} = (y_1, \cdots, y_N), \quad j = 1, \cdots, N \quad (3)$$

Dynamics (1) reads

$$\mathbf{y}(t+1) = F(\mathbf{y}(t) + (A + K_2)R(\mathbf{y}(t))' + \mathbf{k}_1) \quad (4)$$

for $\mathbf{k}_1 = (k_1, \cdots, k_1)$, $(\cdot)'$ representing the transpose of a vector and

$$K_2 = k_2 \begin{pmatrix} 1 \cdots 1 \\ \cdots \\ 1 \cdots 1 \end{pmatrix}$$

Dynamics (4) is characterized by the following three properties

P1 There is a linear operation given by the matrix $(A + K_2)R$;

P2 There is a 'simple' nonlinear function, saturated outside $[y_{\min}, y_{\max}]^N$, making dynamics (4) a nontrivial one;

P3 And there is monotonicity inside the region $[y_{\min}, y_{\max}]^N$ with respect to a partial ordering of the underlying space.

A variety of methods (see [3, 5, 17]) have been developed in recent years for exploring different aspects of the properties of dynamics (4) with some further assumptions on f or A. In [1, 6, 9, 11] the behavior of a neurodynamical system with a *limiter function* as its input-output characteristic is analyzed. Nevertheless there are two severe restrictions which prevent a wide application of the approach presented in the latter set of papers.

1. Limiter functions have been applied in a few models and, because of this, linear analysis can be carried out and informative results obtained as in [17]. But nearly all models of neural networks start from the assumption that the firing rate is a *sigmoidal* function of the summed inputs rather than a *linear* function, although both allow single neurons to make linear discriminations in the space of input features.

---

[1] Some biological implications of dynamics (1) can be further found in [1, 5, 6, 10, 11], [HKP],[Swi].

2. All results obtained in [1, 6, 9, 11] are based upon an assumption that we exclusively consider the set of all saturated attractors and tacitly suppose that this is a generic case. This assumption is partly confirmed by Linsker [16] for the limiter function, using both numerical simulation and theoretical proof. But he also points out the possibility of the emergence of all-but-one saturated attractors, even in the limiter function case. When we consider the (more general) sigmoidal function case (1), we are not certain, at least at a first glance, what form of attractor occurs in general.

With the aim of providing a systematic and feasible tool for grasping some informative properties of dynamics (4), here we generalize the *saturated* fixed point attractor analysis on parameter space with *limiter* functions developed in [1, 6, 9, 11] to fixed point attractor analysis with saturated sigmoidal functions. By this we mean the following.

- We perform an analysis on parameter space for the whole set of fixed point attractors rather than only the set of saturated fixed point attractors. We provide an argument to claim that we are only likely to observe the set of all saturated or all-but-one saturated attractors for dynamics (1). In other words, the generic outcome of dynamics (1) is an all saturated or all-but-one saturated attractor.
- We derive a necessary and sufficient condition to test whether a given all saturated (all-but-one saturated) state is an attractor (fixed point) or not for any given set of system parameters. This result in turn enables us to study Linsker's model in a more general setting and consider the set of all saturated attractors, as well as the set of all-but-one fixed points. Using extreme value theory in statistics, we give an exact parameter region for the threshold of the Hopfield model within which a stored pattern is an attractor.

The first property **P1** of dynamics (4) is the linearity of the operation $(A + K_2)R$, which includes many learning rules in neural networks. For a learning rule of this kind, Miller and MacKay [17] have carried out a detailed discussion directed at elucidating the effect of different constraints–subtractive and divisive. They show that divisive enforcement causes the weight pattern to tend to the principal eigenvector of the synaptic development operator(matrix AR), whereas subtractive enforcement causes almost all weights to reach either their minimum or maximum values. Our results on dynamics (1) partly serve as a complement of their results: under an exactly given condition we assert that only all saturated or all-but-one saturated attractors are possible outcomes of the dynamics, similar to the situation of subtractive enforcement.

Applications of our results to competitive learning[10, 14], the Hopfield model, Linsker's model and a continuous time model can be found in our full paper [4].

2 Fixed Point Attractors Of Dynamics (1)

2.1 All Saturated Fixed Point Attractors

The following definition and theorem are keys for our further development of the present paper. Without loss of generality we suppose that $y_{\min} = -1$ and $y_{\max} = 1$. As we already pointed out in [1, 2, 6, 11, 9] saturated states in the space $\{-1, 1\}^N$ represent the most common outcome of many learning and retrieval models of neural networks and so we address the following definition.

Definition 1. A fixed point attractor $y \in \{-1, 1\}^N$ is called a saturated attractor if

$$k_1 + \sum_{j=1}^{N}(a_{ij} + k_2)r_j y_j \neq 0 \quad \text{for all } i \tag{5}$$

Restriction (5) will be relaxed gradually in the following subsection. The case that one unit violates condition (5) is dealt with in the next subsection; the case of more than one unit not satisfying (5) is discussed in subsection 2.3.

Motivated by the Hopfield model(see [4]) we introduce the following definition.

Definition 2. The quantity

$$h_i(\mathbf{y}) := \sum_{j \in J^+(\mathbf{y})} a_{ij}r_j - \sum_{j \in J^-(\mathbf{y})} a_{ij}r_j \tag{6}$$

is called the local field of the i-th neuron where $J^+(\mathbf{y}) = \{i, y_i = 1\}$, $\quad J^-(\mathbf{y}) = \{i, y_i = -1\}$. And we say that there is a local field gap between neurons in $J^+(\mathbf{y})$ and $J^-(\mathbf{y})$ if and only if

$$\min_{i \in J^+(\mathbf{y})} h_i(\mathbf{y}) > \max_{i \in J^-(\mathbf{y})} h_i(\mathbf{y}) \tag{7}$$

In spite of the fact that dynamics (4) is a generalization of what we consider in [6], the proof of the following theorem is similar to that of Theorem 2 in [6]. However we prefer to sketch a proof of it here (Appendix A) since it is essential for understanding the rest of this paper.

Theorem see also theorem 2 in [6] *y is a saturated attractor of dynamics (4) if and only if*

$$d_1(\mathbf{y}) < k_1 + k_2 c(\mathbf{y}) < d_2(\mathbf{y}) \tag{8}$$

where the slope function $c(\mathbf{y}) = [\sum_{j \in J^-(\mathbf{y})} r_j - \sum_{j \in J^+(\mathbf{y})} r_j]$ and two intercept functions

$$d_1(\mathbf{y}) = \begin{cases} \max_{i \in J^+(\mathbf{y})}[-h_i(\mathbf{y})] & \text{if } J^+(\mathbf{y}) \neq \phi \\ -\infty & \text{otherwise} \end{cases} \tag{9}$$

and

$$d_2(\mathbf{y}) = \begin{cases} \min_{i \in J^-(\mathbf{y})}[-h_i(\mathbf{y})] & \text{if } J^-(\mathbf{y}) \neq \phi \\ \infty & \text{otherwise} \end{cases} \tag{10}$$

In other words, a saturated state $\mathbf{y}$ is a saturated fixed point attractor of dynamics (4) if and only if there is a local field gap between neurons in $J^+(\mathbf{y})$ and $J^-(\mathbf{y})$.

These two functions d_2 and d_1 were introduced in 1993(see [6] and references therein) but their physical meaning, extremes of local fields, is clear only after we apply Theorem see also theorem 2 in [6] to the Hopfield model [4].

2.2 All-But-One Saturated Attractors

Now we consider the set of all-but-one saturated attractors. Without loss of generality we assume that $y_1 \in (-1, 1)$ is the only unsaturated state and $y_i \in \{-1, 1\}$ with

$$k_1 + \sum_{j=1}^{N} (a_{ij} + k_2) r_j y_j \neq 0 \tag{11}$$

for $i \neq 1$.

Since y_i, $i \neq 1$, are saturated fulfilling condition (11) our arguments of the previous subsection hold which imply that

$$\bar{d}_1(\mathbf{y}) < k_1 + \bar{c}(\mathbf{y}) k_2 < \bar{d}_2(\mathbf{y}). \tag{12}$$

for

$$
\begin{cases}
\bar{c}(\mathbf{y}) = \sum_{j \in J^-(\mathbf{y})} r_j - \sum_{j \in J^+(\mathbf{y})} r_j + y_1 r_1 \\
\bar{d}_2(\mathbf{y}) = \min_{i \in J^-(\mathbf{y})} \left[\sum_{j \in J^-(\mathbf{y})} a_{ij} r_j - \sum_{j \in J^+(\mathbf{y})} a_{ij} r_j + a_{i1} y_1 r_1 \right] \\
\bar{d}_1(\mathbf{y}) = \max_{i \in J^+(\mathbf{y})} \left[\sum_{j \in J^-(\mathbf{y})} a_{ij} r_j - \sum_{j \in J^+(\mathbf{y})} a_{ij} r_j + a_{i1} y_1 r_1 \right]
\end{cases}
\tag{13}
$$

Note that there is a slight difference between the definition of $\bar{d}$'s and d's: the maximum and minimum for d_2 and d_1 is taken over a set of N elements, but for $\bar{d}_2$ and $\bar{d}_1$ it is over a set of $N-1$ elements. For y_1 we have the following identity

$$y_1 = f\left(y_1 + \sum_{j=1}^{N} (a_{1j} + k_2) r_j y_j + k_1\right)$$

or equivalently

$$d^{(1)}(\mathbf{y}) = k_1 + \bar{c}(\mathbf{y}) k_2 = f^{-1}(y_1) - y_1 - \sum_{j \in J^+(\mathbf{y})} a_{1j} r_j + \sum_{j \in J^-(\mathbf{y})} a_{1j} r_j - a_{11} y_1 r_1 \tag{14}$$

Hence the parameter region of (k_1, k_2) in which $\mathbf{y}$, an all-but-one saturated state, is a fixed point of dynamics (4) is not empty if and only if

$$\bar{d}_1(\mathbf{y}) < d^{(1)}(\mathbf{y}) < \bar{d}_2(\mathbf{y}) \tag{15}$$

Under condition (15) the parameter region for $\mathbf{y}$ to be a fixed point of dynamics (4) is the line (see Fig. 1) given by

$$\{(k_1, k_2) : k_1 + \bar{c}(\mathbf{y}) k_2 = d^{(1)}(\mathbf{y})\}$$

Theorem 3. *Under condition (11) an all-but-one saturated configuration* $\mathbf{y}$ *is a fixed point of dynamics (4) if and only if* (k_1, k_2) *is in the set*

$$\{(k_1, k_2) : \bar{d}_1(\mathbf{y}) < k_1 + \bar{c}(\mathbf{y})k_2 = d^{(1)}(\mathbf{y}) < \bar{d}_2(\mathbf{y})\} \tag{16}$$

Remark 1 For an all saturated configuration $\mathbf{y}$ except one unit say y_1 which violates restriction (5)

$$k_1 + \sum_{j=1}^{N}(a_{1j} + k_1)r_j y_j = 0 \tag{17}$$

we have a similar conclusion as Theorem 3, namely the parameter region in which $\mathbf{y}$ is a fixed point is line (16) inside a band.

2.3 Other Forms Of Attractors

For concreteness of expression we assume that $y_1, y_2 \in (-1, 1)$, the only two unsaturated states, and $y_i \in \{-1, 1\}$ with the property

$$k_1 + \sum_{j}(a_{ij} + k_2)r_j y_j \neq 0, \tag{18}$$

for $i \neq 1, 2$. After processing similarly as done above for all-but-one saturated configuration, we readily see that a necessary and sufficient condition for $\mathbf{y}$ to be a fixed point of dynamics (1) is

$$\tilde{d}_2(\mathbf{y}) > k_1 + k_2\tilde{c}(\mathbf{y}) > \tilde{d}_1(\mathbf{y}) \tag{19}$$

and

$$\left\{\begin{aligned}
d_1^{(2)}(\mathbf{y}) &:= k_1 + k_2\tilde{c}(\mathbf{y}) = \sum_{j\in J^-(\mathbf{y})} a_{1j}r_j - \sum_{j\in J^+(\mathbf{y})} a_{1j}r_j - a_{11}y_1 r_1 \\
&\qquad\qquad -a_{12}y_2 r_2 + f^{-1}(y_1) - y_1 \\
d_2^{(2)}(\mathbf{y}) &:= k_1 + k_2\tilde{c}(\mathbf{y}) = \sum_{j\in J^-(\mathbf{y})} a_{2j}r_j - \sum_{j\in J^+(\mathbf{y})} a_{1j}r_j - a_{21}y_1 r_1 \\
&\qquad\qquad -a_{22}y_2 r_2 + f^{-1}(y_2) - y_2
\end{aligned}\right. \tag{20}$$

where

$$\left\{\begin{aligned}
\tilde{c}(\mathbf{y}) &= \sum_{j\in J^-(\mathbf{y})} r_j - \sum_{j\in J^+(\mathbf{y})} r_j + y_1 r_1 + y_2 r_2 \\
\tilde{d}_2(\mathbf{y}) &= \min_{i\in J^-(\mathbf{y})} [\sum_{j\in J^-(\mathbf{y})} a_{ij}r_j - \sum_{j\in J^+(\mathbf{y})} a_{ij}r_j + a_{i1}y_1 r_1 + a_{i2}y_2 r_2] \\
\tilde{d}_1(\mathbf{y}) &= \max_{i\in J^+(\mathbf{y})} [\sum_{j\in J^-(\mathbf{y})} a_{ij}r_j - \sum_{j\in J^+(\mathbf{y})} a_{ij}r_j + a_{i1}y_1 r_1 + a_{i2}y_2 r_2]
\end{aligned}\right. \tag{21}$$

An interesting new phenomenon occurs: the two lines corresponding to the two unsaturated states defined by Eq. (20) are *parallel*, which indicates that as long as

$$d_1^{(2)}(\mathbf{y}) \neq d_2^{(2)}(\mathbf{y}) \tag{22}$$

then the parameter region in which $\mathbf{y}$ is a fixed point of dynamics (1) is *empty*. The fulfillment of Eq. (22) is a generic situation essentially depending on the property of the matrix A. When $f(x) = x$, $x \in I\!\!R$ this conclusion has been confirmed by Linsker[16] in his numerical simulation and theoretical proof.

Theorem 4. *If and only if (k_1, k_2) is in the following set*

$$\{(k_1, k_2); \tilde{d}_1(\mathbf{y}) < k_1 + \tilde{c}(\mathbf{y})k_2 = d_1^{(2)}(\mathbf{y}) = d_2^{(2)}(\mathbf{y}) < \tilde{d}_2(\mathbf{y})\} \tag{23}$$

an all-but-two saturated configuration $\mathbf{y}$ is a fixed point of dynamics (4).

Remark 2 If for an all saturated configuration there are two saturated units not satisfying Eq. (5) or an all-but-one saturated configuration with one saturated unit violating Eq. (11) we have a similar conclusion as in Theorem 4.

We are able to carry out a cascade study, continuing to consider three unsaturated units and so on. The situation to ensure the existence of a nonempty parameter region in which $\mathbf{y}$ is a fixed point of dynamics (1) becomes more and more difficult when the number of unsaturated units is larger and larger since it requires all parallel lines corresponding to unsaturated units to intersect. In general two parallel lines $k_1 + \tilde{c}(\mathbf{y})k_2 = d_2^{(2)}(\mathbf{y})$ and $k_1 + \tilde{c}(\mathbf{y})k_2 = d_1^{(2)}(\mathbf{y})$ do not coincide. Hence we stop here and believe that the general outcomes of dynamics (4) are all saturated and all-but-one saturated attractors.

In conclusion for dynamics (4) the full spectrum of its outcomes is summarized the following table.

3 Conclusions

We have studied the dynamics of neural network models with saturated sigmoidal functions as their input-output characteristics. A complete spectrum on the parameter space for all possible outcomes of dynamics (4) is obtained. Under a stated condition we have shown that the possible outcomes of dynamics (1) are all saturated or all-but-one saturated fixed point attractors. An exact parameter region is given for all saturated attractors and all-but-one saturated fixed points.

In a single theoretical framework we have managed to treat diverse models in neural networks. The significance of this unified treatment lies in that, in additional to some novel discoveries after revisiting these models, we have exposed some common mechanisms behind them (for example we have understood the physical meaning of d_2 and d_1 from the study of the Hopfield model [13]) which will provide useful guidance in further designing and understanding new models,

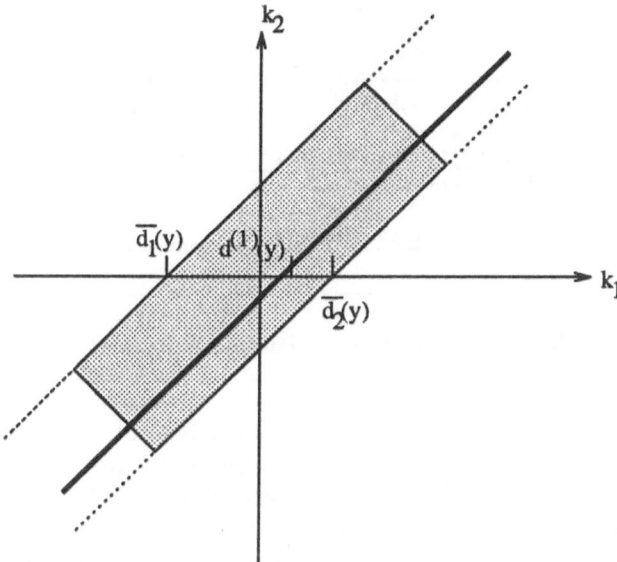

Fig. 1. The parameter region in which **y**, an all-but-one saturated configuration, is a fixed point of dynamics (4) is the line(dark line) inside the band(filled region).

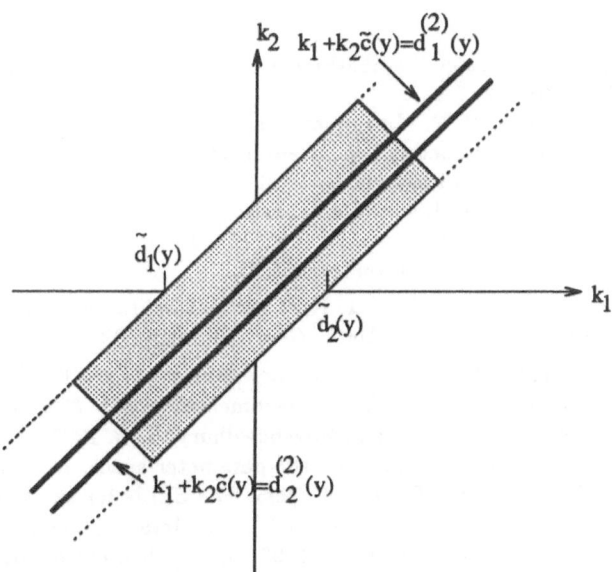

Fig. 2. The parameter region in which **y**, an all-but-two saturated configuration, is a fixed point of dynamics (4) is not empty only when the two parallel lines (dark lines) inside the band (filled region) coincide.

554

ATTRACTOR TYPE	PARAMETER REGION	POSSIBILITY
y: all saturated	A band determined by $d_2(y) > k_1 + c(y)k_2 > d_1(y)$ (Theorem see also theorem 2 in [6]), independent of f	Most Possible
y: all-but-one saturated	A line $k_1 + \bar{c}(y)k_2 = d^{(1)}(y)$ inside a band $\bar{d}_1(y) < k_1 + \bar{c}(y)k_2 < \bar{d}_2(y)$ (Theorem 3), dependent on f	Less Possible
y: all but two saturated	Intersection of two parallel lines inside a band (Theorem 4), dependent on f	Hardly possible
⋮		

Table 1: The General Parameter Region

both for learning and retrieving. In [5] we have carried out a detailed investigation for a dynamical system possessing the property **P3** listed in Section 1 on the uniqueness and existence of fixed point attractors. In [1] we have risen to the challenge that the property **P1** is violated, i.e. there is a nonlinear operation given by the matrix $(A + K_2)R$ (like competitive learning considered in this research), which marks important new dimensions into which our approach may grow.

References

1. Feng, J. 1995. Establishment of topological maps–a model study. *Neural Processing Letters* **2**, 1-4.
2. Feng, J. 1997. Lyapunov functions for neural nets with nondifferentiable input-output characteristics. *Neural Computation* **9**, 45-51.
3. Feng, J., and Brown, D. 1996. A novel approach for analyzing dynamics in neural networks with saturated characteristics. *Neural Processing Letters* **4**, 9-16.
4. Feng, J., and Brown, D. 1997. Fixed point attractors for a class of neurodynamics. *Neural Computation* (accepted).
5. Feng, J., and Hadeler, K.P. 1996. Qualitative behavior of some simple networks. *Jour. of Phys. A: Math. Gen.* **29**, 5019-5033.
6. Feng, J., Pan, H., and Roychowdhury, V. P. 1996. On neurodynamics with limiter function and Linsker's developmental model. *Neural Computation* **8**, 1003-1019.
7. Feng, J., Pan, H., and Roychowdhury, V. P. 1997. Linsker-type Hebbian learning: a qualitative analysis on the parameter space. *Neural Networks* (in press).
8. Feng, J., and Tirozzi, B. 1995. The SLLN for the free-energy of the Hopfield and spin glass model. *Helvetica Physica Acta* **68**, 365-379.
9. Feng, J., and Tirozzi, B. 1995. An application of the saturated attractor analysis to three typical models. *Lecture notes in computer science* **930**, 353-360.
10. Feng, J., and Tirozzi, B. 1996. Convergence theorems for Kohonen feature mapping with VLRPs. *Computers and Mathematics with Applications* **32**, (in press).
11. Feng, J., and Tirozzi, B. 1997. A discrete version of the dynamic link network. *Neurocomputing* **14**, (in press).

12. Feng, J., and Tirozzi, B. 1997. Convergence of learning processes, stability of attractors and critical capacity of neural networks. in Bovier, A.(ed.) Springer-Verlag, (in press).

13. Feng, J., and Tirozzi, B. 1997. Capacity of the Hopfield model. *Jour. of Phys. A: Math. Gen.* (accepted).

14. Goodhill, G., and Barrow, H.G. 1994. The role of weight normalization in competitive learning. *Neural Computation* 6, 255-269.

15. Hertz, J., Krogh, A., and Palmer, R. 1991. *Introduction to the Theory of Neural Computation.* Addison-Wesley Publishing Company.

16. Linsker, R. 1986. From basic network principle to neural architecture (series). *Proc. Natl. Acad. Sci. USA* 83, 7508-7512, 8390-8394, 8779-8783.

17. Miller, K., and MacKay, D. 1994. The role of constraints in Hebbian learning. *Neural Computation* 6, 100-126.

18. Sejnowski, T.J. 1995. Time for a new neural code? *Nature* 376, 21-22.

19. Swindale, N.V. 1996. The development of topography in the visual cortex: a review of models. *Network: Computation in Neural Systems* 7, 161-247.

Hopfield Neural Network Applied to Optimization Problems: Some Theoretical and Simulation Results

G. Joya, M. A. Atencia, and F. Sandoval
Dpto. Tecnología Electrónica, E.T.S.I. Telecomunicación.
Universidad de Málaga. Campus de Teatinos, 29071 Málaga (Spain)

ABSTRACT

This paper is devoted to the study of continuous Hopfield-like neural networks, either in its original version [Hopfield,1984] or in its high order generalization [Samad, 1990], [Kobuchi, 1991], applied to the solution of optimization problems. Main problems affecting the practical application of these networks are brought to light : a) Incoherence between the network dynamics and the associated energy function ; b) Error due to discretization of the continuous dynamical equations caused by simulation on a digital computer ; c) Existence of local minima. The behavior of this kind of neural networks with respect to these problems is analyzed and simulated, indicating possible mechanisms to avoid them. The last part of the paper we shown that the integral term in the energy function is bounded, in contrast with Hopfield's statement. Using this result, a new local minima avoidance strategy is proposed with an enhanced efficiency.

1. INTRODUCTION

The neural paradigm initially proposed by Hopfield as an associative memory, [Hopfield, 1982], [Hopfield, 1984], has been later widely used in optimization problems, either its original version (first order Hopfield networks [Hopfield, 1985]) or its high order generalized version [Samad, 1990], [Kobuchi, 1991], [Joya, 1997]. The definition of this kind of networks implies fixing two key characteristics : its activation dynamics and an associated energy function which decreases as the network spontaneously evolves. These characteristics allow to use these networks to solve optimization problems. The applied methodology may be summarized in the following way [Hertz, 1991]: given an optimization problem, finding the cost function that describes it ; designing a Hopfield neural network whose energy function must reach its minima in the same points as the cost function, so that the stability configurations of the network correspond to solutions of the problem.

This method faces some application problems that bound both the network that can be used and the results that may be obtained. We think that these problems must be studied either analytically or experimentally, to avoid the lack of rigor in the application of Hopfield networks that is present in many published papers about this subject, which do not explicitly take into account these limitations. These problems may be grouped in three kinds :

a) Many described applications do not coherently make a correspondence between the network dynamics and the energy function associated to that network. A common situation is using an analog neuron network with a continuous dynamics and associating it an energy function corresponding to a discrete neuron network [Hopfield, 1985].

b) The energy function is forced to decrease only if the network evolves according to its dynamical equations. If these equations are continuous (differential equations), they can not be strictly represented by means of a computer simulation. That is, the simulation implies the discretization of these equations (difference equations), so that the bigger is the simulation step the more different are the real and the theoretical network behavior.

c) The energy function of a Hopfield network has many local minima. Consequently, the network probably will reach an equilibrium state that does not correspond to a problem solution. The search of evolution strategies to move the network out of local minima and take it to a global minimum is a main task in this field.

In this paper we realize a study of these problems refered to the high order generalized version of the Hopfield neural network with continuous activation function and continuous dynamics, originally described in [Hopfield,1984]. Starting from the dynamics equations of the system and its associated energy function, we study the mathematical properties of the integral term appearing in this energy function and we proof that it is bounded in contradiction to Hopfield's statements ; this result provides us with a new strategy for the network evolution, based upon a non heuristic criterion.

Our study is restricted to those optimization problems that fulfill the following requirements (most optimization problems solved with the Hopfield paradigm fulfill them [Hopfield,1985][Takefuji,1991][Kim,1991][Joya,1991] [Metha,1993]):

1. The solution may be expressed by a set of n variables $\mathbf{x} = (x_1, x_2, ..., x_n)$ each taking one of two discrete values. That is, $x_i \in \{-1,1\} \forall i$ or $x_i \in \{0,1\} \forall i$.

2. The cost function may be described by an arbitrary order multinomial expression in variables $\mathbf{x}$:

$$f(\mathbf{x}) = \sum_{j=0}^{r} \sum_{(i_1, i_2 ... i_{j+1}) \in C_{j+1}^n} a_{i_1, i_2 ... i_{j+1}} x_{i_1} x_{i_2} ... x_{i_{j+1}} + C \qquad (1)$$

These conditions force the following properties of the neural networks :

1. As variables x_i must be discrete, neurons must have a discrete activation function ($s_i \in \{-1,1\} \forall i$ or $s_i \in \{0,1\} \forall i$) or a continuous activation function ($s_i \in [-1,1] \forall i$) provided that the equilibrium state is reached at one of the extremes of the interval.

2. The cost function has not any term raised to a power greater than one in each x_i because $x_i^k = x_i$ if $s_i \in \{0,1\}$ or $x_i^k = 1$ (even k) and $x_i^k = x_i$ (odd k) if $x_i \in [-1,1]$. Consequently, the network has no self connections.

3. Each term of the cost function contains a different combination of variables, so that the symmetrical connections in the associated neural network are the same at every order.

The rest of this paper has the following structure : in section 2, the dynamics of the system and its energy function are presented, and its behavior analyzed with respect to the problems mentioned above ; we show that the integral term appearing in the energy function is bounded, in contradiction to Hopfield's statements. This result is used in Section 3 to design a non heuristic evolution strategy for the avoidance of local minima, which is compared to the traditional one, showing a greater efficiency with regard to the number of simulation steps. In section 4 main results are summarized. All simulations have been applied to an example problem : solution of a diophantine equation [Garey, 1979] [Joya, 1991], which is described in the Appendix.

2. HOPFIELD NEURAL NETWORK WITH CONTINUOUS ACTIVATION FUNCTION AND CONTINUOUS DYNAMICS.

An arbitrary order Hopfield feedback neural network with continuous activation function $g(u_i/\beta)$ (being g a sigmoid or hyperbolic tangent function) whose evolution dynamics is described by equations (2) and (3), has the energy function shown in equation (4) :

$$s_i(t) = g(\frac{u_i(t)}{\beta}) \tag{2}$$

$$\frac{du_i}{dt} = -u_i + \sum_{j=1}^{q} \sum_{\substack{(i_1,\ldots i_j) \in C_j^n \\ i_1,\ldots,i_j \neq i}} T_{i,i_1\ldots i_j} s_{i_1} \ldots s_{i_j} - I_i \tag{3}$$

$$E = -\sum_{j=1}^{q} \sum_{\substack{(i_1\ldots i_j) \in C_j^n \\ i_1,\ldots,i_j \neq i}} T_{i,i_1\ldots i_j} s_i s_{i_1} \ldots s_{i_j} + \sum_i I_i s_i + \beta \sum_i \int_0^{s_i} g^{-1}(s)\, ds \tag{4}$$

The proof for a first order network is found in [Hopfield, 1984] and the high order generalization is found in immediate. The parameter that describes the steepness of the activation function in [Hopfield, 1984] is $\lambda = 1/\beta$. In this paper, the inverse parameter, β, is used instead because the application of this dynamics to optimization problems described in Section I requires the evolution of the activation functions towards a step-like function which is obtained by moving λ towards ∞ or β towards 0. The later case has an easier simulation on a computer.

Due to the existence of the integral term, the energy function can not be equated to the cost function f(x). This dynamics is often applied without considering this fact, that is, removing the integral term (problem *a)* in section 1). This is an erroneous procedure, because in this case the minima in the two functions are no longer the same. Just in the case $\beta=0$ ($\lambda =\infty$) functions E and f are the same, but this is the discrete neuron case described in [Hopfield, 1982].

Unlike the other possible dynamics ([Hopfield, 1982] [Abe, 1989]), the integral term in eq. (4) causes the existence of stable states (minima of the energy function) of the network at points inside the hypercube in the state space. Moreover, Hopfield states that in this kind of networks no neuron can reach its extreme values ±1, which means that the network can not reach a stable state in vertexes, edges or sides of the hypercube. This fact that, in principle, would make these networks useless to solve optimization problems as those in Section 1, is presented as a consequence of the integral term unboundly increasing when s_i approaches ±1. Literally: "The integral is zero for $V_i = 0$ and positive otherwise, getting very large as V_i approaches ±1 because of the slowness with which g(V) approaches its asymptotes " [Hopfield,1984].

We think that Hopfield's statement is erroneous and show that using the hyperbolic tangent function the integral term keeps bounded .

Theorem I : Given an arbitrary order Hopfield network, whose neurons have an activation function $s_i = g(u_i/\beta)= \tanh(u_i /\beta)$, with $\beta > 0$, every integral term in its energy function remains bounded by the value $\beta \ln2$, that is :

$$0 \le \beta \int_{0}^{s_i} g^{-1}(s)\,ds < \beta \ln 2 \qquad (5)$$

Proof :
Hyperbolic tangent function may be expressed as :

$$g(x) = \frac{e^x - e^{-x}}{e^x + e^{-x}} = \frac{e^{2x} - 1}{e^{2x} + 1} \qquad (6)$$

finding x, we have:

$$x = \tfrac{1}{2}[\ln(1 + g(x)) - \ln(1 - g(x))] \qquad (7)$$

so, the inverse function g^{-1}, may be expressed as:

$$g^{-1}(x) = \tfrac{1}{2}[\ln(1 + x) - \ln(1 - x)] \qquad (8)$$

The term $\int \ln(1 + ax)\,dx$ may be integrated in parts by making

$$u = \ln(1 + ax); v = \frac{1 + ax}{a}, \text{ obtaining}$$

$$\beta \int_0^{s_i} g^{-1}(s)\,ds = \frac{\beta}{2}[(1+s_i)\ln(1+s_i)+(1-s_i)\ln(1-s_i)] \qquad (9)$$

This expression is zero when $s_i = 0$; when s_i tends to 1 we have:

$$\lim_{x\to 1}\beta \int_0^{s_i} g^{-1}(s)\,ds = \frac{\beta}{2}[\lim_{x\to 1}(1+s_i)\ln(1+s_i)+\lim_{x\to 1}(1-s_i)\ln(1-s_i)] \qquad (10)$$

The former limit of the second member of (10) is 2 ln2, and the latter may be calculated using the L'Hopital rule:

$$\lim_{x\to 1}(1-s_i)\ln(1-s_i)] = \lim_{x\to 1}\frac{\ln(1-s_i)}{1/(1-s_i)} = \lim_{x\to 1}\frac{-1/(1-s_i)}{1/(1-s_i)^2} = \lim_{x\to 1}-(1-s_i)=0 \qquad (11)$$

Doing the same when s_i tends to -1, we obtain :

$$\lim_{x\to \pm 1}\beta \int_0^{s_i} g^{-1}(s)\,ds = \beta\ln 2 \qquad (12)$$

q.e.d.

Corollary : If β is the same for every neuron, the sum I_β of all integral terms in E is bounded : $0 \le I_\beta < n\cdot\beta\cdot\ln 2$.

Figure 1 shows the graphic representation of the integral term as presented by Hopfield and the representation obtained from expression (9).

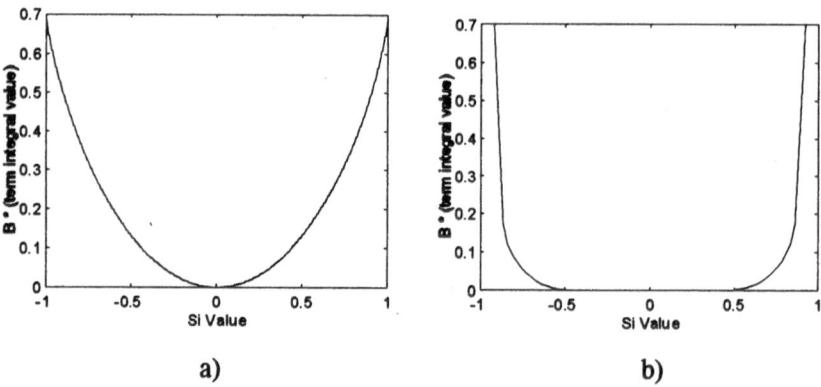

a)   b)

Figure 1. a) Representation of integral term obtained from eq. (9). b) Representation of integral term as presented by Hopfield.

It is worth noting that although Hopfield's statement is erroneously based, it is strictly true, because, given any neuron i, there is always a contour in the state space around the value $s_i=1$ (or $s_i=-1$) where the energy is lower than that of $s_i=1$ (or $s_i=-1$). The explanation of this fact is based on the integral term in E always increasing when s_i approaches 1 (or -1) ; on the other hand the discrete sum term in

E varies linearly with s_i; even if this variation is negative, there is always a contour around $s_i=1$ (or $s_i=-1$) where the increasing gradient of the integral term is greater than the lineal decreasing of the discrete sum term. In spite of this limitation, in practical implementations values ±1 are reached by some neurons due to the radius of this contour being smaller than the simulation precision, depending on the connection weights and the value of β.

Problem b) in section 1, produced by time discretization in a computer simulation, affects also this kind of networks, resulting a practical discrete dynamics (13) that is not required to produce a decreasing of the energy function (4).

$$u_i(k+1) = (1-\Delta t)u_i(k) + (\sum_{j=1}^{q} \sum_{(i_1,\cdots,i_j)\in C_j^n} T_{i,i_1,\cdots,i_j} s_{i_1} \cdots s_{i_j} - I_i)\Delta t \ (13)$$

The effect produced by Δt value is difficult to quantify due to the mentioned above existence of local minima into the hypercube (problem c) in section 1). Besides, the evolution strategies [Samad,1990][Mehta,1993] that must be used in optimization problems to move the network out of these minima mask the effect of time discretization. So, $\Delta t=1$ is used in our simulations. The following section is centered on these strategies and presents a new one by using the property of boundness of the integral term that has been shown.

3. EVOLUTION STRATEGY IN A HOPFIELD NETWORK WITH CONTINUOUS DYNAMICS.

Once the bound of the integral term has been obtained, its influence over a network with continuous neurons and dynamics of eq. (3) applied to optimization problems is analyzed. In this case, while $\beta > 0$, function cost f that describes the problem and must be minimized, can not be associated to the energy function of the network (eq. (4)) but with just a part of it : the part that includes the discrete sum term. Consequently, equilibrium states of the network, that are minima of the whole energy function, are not solutions to the problem. Moreover, as every s_i of the solution must be discrete, a continuous network can not find a solution due to the existence of minima inside the hypercube. The way to avoid these problems and make the network to converge up to a solution state is to decrease β along the simulation, theoretically until $\beta = 0$; at this moment, the integral term will become zero, and E will be exactly the same as f. This strategy reminds of the Simulated Annealing (S.A.) used in discrete dynamics : both modify the parameter that determines the steepness of a sigmoid-like function up to it is converted into a step-like function. The difference consists in while the function in S.A. is the probability of a neuron changing from -1 to 1 (or vice-versa) even if the energy increases, in our case the modified function is the own activation function of the neuron. In other words, this strategy is, actually, equivalent to using many different neural networks, one for each used β.

As far as we know, there is no described algorithm for the variation of β along the simulation with a justification other than heuristic. In general, the used procedure may be described as follows : starting from a value for β high enough, let the network evolve up to reaching a stable state -which is not a solution-, then multiply β by a factor less than 1, let the network evolve again up to a new stable state, and so on ; the process ends when β becomes zero and, at this moment, the reached stable state should be a global minimum of the cost function [Samad, 1990] [Metha, 1993].

A new evolution strategy is proposed below, based upon the property of the integral term being bounded by the quantity $CI_\beta = n\ \beta\ \ln2$ (as shown in Theorem I). This strategy is oriented to reducing the number of evolution steps, and may be justified by the following reasoning : the cost function is associated just to the discrete sum terms of the energy function (eq. 4), thus, the problem solution corresponds to a state where these terms have a minimum value E_m; on the other hand, the integral term of the energy function reaches its maximum value CI_β in the vertexes of the hypercube, i.e., for $s_i = \pm1$ $\forall i$; then, if for a particular β, the network evolves until reaching an energy value E_β, fulfilling

$$\left|E_\beta - E_m\right| < CI_\beta \tag{14}$$

it means that the network is in a state with a smaller energy than that of a vertex ; consequently, if now the network is left to evolve, it will move farther from that vertex , i.e., from a possible solution. For the network to evolve in the direction of approaching the vertex, β must be decreased so that the state falls out of the hypersphere centered at E_m and with radius CI_β (Figure 2).

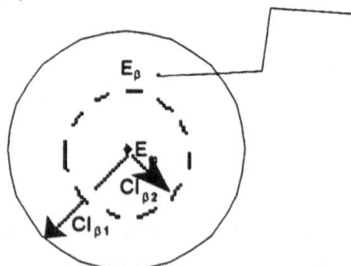

Figure 2. Schematic representation of the state space. Each circle represents the set of states whose energy is between the minimal value E_m (center) and the value corresponding to a vertex $E_m + CI_\beta$.

To sum up, this strategy does not need to wait for the network to reach a local minimum inside the hypercube to carry out the change of β, but this change will be realized just when the evolution is proved to tend to that minimum.

Using the environment for design and simulation of high order Ann developed in [Atencia, 1996], two kinds of experiments have been carried out for the example problem :

a) Simulation of a network of continuous neurons and evolution strategy based on the bound of the integral term (Experiment I)

In this case, neurons have tanh() as an activation function, that is $s_i \in [-1,1]$. Starting from β high enough (2000), the network is left to evolve either until it reaches a stable state or the condition in eq. (14) is fulfilled. Then, β is decreased (multiplying it by 0.8), and the evolution starts again. See Table I.

b) Simulation of a network with continuous neurons and traditional evolution strategy (Experiment II).

The neural network has the same characteristics as in the former case, but now, the network must evolve until it reaches an equilibrium state for each β. In this moment, β is decreased (multiplying it by 0.8), and the evolution starts again. See Table I.

	Experiment I	Experiment II
Initial β	2000	2000
Decreasing factor of β	0.8	0.8
Percentage of correct solutions (Global minima)	100 %	100 %
Average number of simulation steps	68	750

Table I. Simulation results for the evolution of a continuous high order Hopfield ANN oriented to solving a diophantine equation. In Experiment I the network evolves according to the new strategy based on the bound of the integral term of the energy function. In Experiment II the network evolves according to the traditional strategy.

4. CONCLUSIONS

This paper has been addressed to studying Hopfield-like neural networks when applied to optimization problems. We think that this was a necessary task, due to the large number of published papers about this subject, that consist in a practical application of this model to a particular problem, without explicitly considering the problems due to the divergence from the theoretical principles imposed by a simulated realization. These problems may be classified in three categories :

a) Incoherence between the network dynamics and the associated energy function.

b) Error due to discretization of the continuous dynamical equations caused by simulation on a digital computer.

c) Existence of local minima.

The behavior of our neural systems is analyzed by means of simulations for a particular optimization problem : solution of a diophantine equation. This kind of neural networks are mainly affected by problems a), due to the existence of the integral term in the energy function and c), due to the existence of local minima inside the hypercube. Theorem I proves that this integral term is bounded :

$0 \le I_\beta < n \cdot \beta \cdot \ln 2$, in contrast with Hopfield's statement that predicts an asymptotic increasing of this term. This result supports the development of a new strategy for the avoidance of local minima, which is compared to the traditional one, showing a greater efficiency whit regards to the number of simulation steps. The effectiveness of this strategy mask the effect of time discretization.

ACKNOWLEDGEMENTS

This work has been partially supported by the Spanish Comisión Interministerial de Ciencia y Tecnología (CICYT), Project No. TIC95-0589.

REFERENCES

Abe, S., (1989), "Theories on the Hopfield Neural Networks", *Int. Joint Conf. on Neural Network*, Vol. 1, pp. 557-564.

Atencia, M. A., (1996), "Entorno de diseño y simulación de redes neuronales artificiales de alto orden", Internal Report, Dpto Tecnología Electrónica, Univ.Málaga, Spain.

Garey, M. R., y Johnson,D.S., (1979), *Computers and intractabily. A guide to the Theory of NP-Completeness*, W.H. Freeman and Company, p. 245.

Hertz, J., Krogh, A., Palmer, R.G.,(1991), *Introduction to the theory of neural computation*, Addison-Wesley.

Hopfield, J.J., (1982), "Neural networks and physical systems with emergent collective computational abilities", in *Proc. National Academic Sciences U.S.A.*, Vol. 79, April, pp. 2554-2558.

Hopfield, J.J., (1984), "Neurons with graded response have collective computational properties like those of two-state neurons", in *Proc. National Academic Sciences U.S.A.*, Vol. 81, May, pp. 3088-3092.

Hopfield, J. J. y Tank, D. W., (1985), "Neural computation of decisions in optimization problems", *Biological Cybernetics*, No. 52, pp. 141-152.

Joya, G., Atencia M. A. and Sandoval, F., (1991), "Application of high-order Hopfield neural networks to the solution of diophantine equations", in *Artificial Neural Networks*, A. Prieto (Ed.), LNCS 540, Springer-Verlag, pp. 395-400.

Joya, G., Atencia, M. A., Sandoval, F., (1997), "Associating arbitrary-order energy functions to an artificial neural network. Implications concerning the resolution of optimization problems", *accepted to be published in Neurocomputing.*.

Kim, K.H., Lee C.H., Kim, B.Y. y Hwang, H.Y., (1991), " Neural optimization network for minimum-via layer assignment", *Neurocomputing*, Vol. 3, pp. 15-27.

Kobuchi, Y., (1991), "State Evaluation Functions and Lyapunov Functions for Neural Networks", *Neural Networks*, Vol. 4, pp. 505-510.

Metha, S., y Fulop, L., (1993), "An analog Neural Network to Solve the Hamiltonian Cycle Problem", *Neural Networks*, Vol. 6, pp. 869-881.

Samad, T., y Harper, P., (1990), "High-order Hopfield and Tank optimization networks", *Parallel Computing*, Vol.16, pp. 287-292.

Takefuji, Y. y Lee, K.C., (1991), "Artificial Neural Networks for Four-Coloring Map Problems and K-Colorability Problems", *IEEE Transactions on Circuits and Systems*, Vol. 38, No. 3, pp. 326-333.

APPENDIX.

Our example problem consists in the solution of the following diophantine equation [Joya, 1991], [Garey, 1979] :

$$a x^2 + b y = c \, ; (a = 1, b = 3, c = 37) \tag{A.1}$$

It may be solved with a third order Hopfield network with 10 neurons. The cost function is :

$$f(x,y) = (a x^2 + b y - c)^2 \tag{A.2}$$

Solutions are obtained as :

$$x = \sum_{i=0}^{m-1} 2^i \frac{s_i+1}{2} \, ; y = \sum_{i=m}^{n-1} 2^{i-m} \frac{s_i+1}{2} \tag{A.3}$$

Values of the connection weights are:

$$
\begin{aligned}
I_i = & -[(1-\delta_i)(a^2(-8 K_1 2^{3(i-l)} + 12 K_1 K_2 2^{i-l} + 4 K_1^3 2^{i-l}) \\
& -4ac K_1 2^{i-l} + 4abC_1 K_1 2^{i-l}) \\
& +\delta_i (2 b^2 C_1 2^{i-n-l} - 2bc\, 2^{i-n-l} + 2ab(K_2 + K_1^2) 2^{i-n-l})]
\end{aligned} \tag{A.4}
$$

$$
\begin{aligned}
T_{ij} = & -2![(1-\delta_i)(1-\delta_j)(a^2(-4 \cdot 2^{3(i-l)+j-1} - 4 \cdot 2^{3(j-l)+i-1} + 6 K_2 2^{i+j-2} + 6 K_1^2 2^{i+j-2}) \\
& -2ac\, 2^{i+j-2} + 2ab C_1 2^{i+j-2}) \\
& +\delta_i \cdot \delta_j b^2 2^{i+j-2n-2} + \frac{1}{2}[((1-\delta_i)\delta_j + (1-\delta_j)\delta_i)4ab K_1 2^{i+j-n-2}]
\end{aligned} \tag{A.5}
$$

$$
\begin{aligned}
T_{ijk} = & -3![(1-\delta_i)(1-\delta_j)(1-\delta_k)(4 a^2 K_1 2^{i+j+k-3}) \\
& +\frac{1}{3}((1-\delta_i)(1-\delta_j)\delta_k + (1-\delta_i)(1-\delta_k)\delta_j + (1-\delta_j)(1-\delta_k)\delta_i)2ab\, 2^{i+j+k-n-3}]
\end{aligned} \tag{A.6}
$$

$$T_{ijkl} = -4![(1-\delta_i)(1-\delta_j)(1-\delta_k)(1-\delta_l)a^2 2^{i+j+k+l-4}] \tag{A.7}$$

$$\delta_i = \begin{cases} 0 \; si \; i < m \\ 1 \; si \; i \geq m \end{cases} ; K_p = \sum_{i=0}^{m-1} 2^{p(i-1)} \; ; \; C_p = \sum_{i=m}^{n-1} 2^{p(i-m-1)} \tag{A.8}$$

A Genetic Approach to Computing Independent AND Parallelism in Logic Programs[*]

Camino R. Vela, Cesar Alonso, Ramiro Varela, Jorge Puente
Artificial Intelligence Centre. University of Oviedo at Gijón
Campus of Viesques. E-33271 Gijón. Spain
Tel. +34-8-5182032. FAX +34-8-5182125.
e-mail:{camino, calonso, ramiro, puente}@trasgu.aic.uniovi.es

ABSTRACT

In this paper we face the problem of determining the best partial order among the subgoals of a query in order for this query to be evaluated under Independent AND Parallelism. This is the most common source of parallelism exploited by the different models that have been proposed to evaluate logic programs in parallel. This problem is proved to be NP-hard, so every model utilises its own heuristic strategy in order to estimate the best ordering. Here, a Genetic approach is proposed and compared to conventional heuristic ones. The experimental results show that the Genetic Algorithm produces better solutions, as well as comparable execution times for reasonably sized problems.

KEYWORDS: Parallel Logic Programming, Independent AND Parallelism, Genetic Algorithms, Evolutive Optimization.

1 INTRODUCTION

Over the last few years, Parallel Logic Programming has emerged as an intensive field of research, and accordingly a great number of interpretation models have been proposed to exploit the different sources of parallelism that the language of the logic offers. In this context, *Independent AND Parallelism (IAP)* became one of the most promising approaches and subsequently was incorporated in many of the proposed models [Con83, DeG84, Kal91, Mut90, Var95b]. *IAP* consists of the ordered evaluation of subgoals of a query in such a way that two of them are not evaluated in parallel if they are dependent, i.e., if they share some free variable. In order to utilise *IAP*, several problems have first to be solved. The first is that of determining the best ordering among the literals in order to obtain as much parallelism as possible. This is a NP-hard problem, so several heuristic strategies have been proposed to solve it. In this paper, we propose a new approach within the framework of Genetic Algorithms, and give experimental results to compare with other classic approaches. These results show that, in general, the proposed Genetic Algorithm discovers better solutions, and that the computation time is comparable in reasonably sized problems.

The rest of the paper is organised as follows. Section 2 introduces our proposed model for evaluating logic programs, and in particular how *IAP* is represented by

---

[*] This work has been partially supported by the University of Oviedo under the project DF-96-513-10.

means of an ordered structure called the *Data Flow Lattice (DFL)*. In Section 3, we describe the classic approach to determining *IAP* which is used in our model. Section 4 describes the genetic approach proposed in this paper. Section 5 analyses certain experimental results. And finally, we present our conclusions in Section 6.

2 INDEPENDENT *AND* PARALLELISM

The classic strategy for evaluating logic programs is *SLD resolution*. This is the one utilised by the Prolog systems and is inherently sequential. *SLD* resolution is a simple, elegant strategy, but is inefficient in certain situations. Hence, in order to improve efficiency and as a result of the ability of the logic to express parallel computations, a lot of models for evaluating logic programs in parallel have been proposed over the last few years. Most of them exploit one or both of the main sources of parallelism of logic programs: *AND parallelism* and *OR parallelism*. The first one stands for the simultaneous evaluation of several literals within a query; and the second one permits several clauses with the same conclusion to be exploited in parallel.

In order to avoid the generation and management of a high number of incompatible solutions to the dependent subgoals, a variant of AND parallelism is commonly used, the so called *Independent AND Parallelism (IAP)* proposed in [Con83] and widely accepted [DeG84, Kal91, Mut90]. *IAP* consists of the ordered evaluation of literals in such a way that two literals are not evaluated in parallel if they share some free variable. Therefore, if a variable is shared by several literals, one of them is chosen to be evaluated first, in a producer way, and then the others can be evaluated as consumers with the shared variable bounded to the value computed by the producer. This produces a partial ordering among the literals of a given query. This order is usually represented by ordered structures such as the *Data Flow Graph (DFG)* [Con83] or the *Data Join Graph (DJG)* [Kal91]. In our model, the partial ordering is codified into a graph called the *Data Flow Lattice (DFL)* where the nodes are labelled with literals (or they are null nodes) and the arcs are unlabelled. In order to achieve an efficient strategy for managing partial solutions, as shown in [Var95b], we impose the following properties onto the *DFL*.

- *P1. It has degree two.*
- *P2. Dependence of consumed variables: every variable of a literal either appears in an immediate predecessor or it does not appear in any predecessor.*
- *P3. It is a semi-lattice with respect to the infimum.*
- *P4. It is a structured graph: given two nodes with a common child and with supremum, every path between a preceeding node of the supremum and one of these nodes passes necessarily through the supremum.*

Figure1 shows two examples of *DFLs*. As we can see, the null nodes are necessary in order to guarantee the above properties.

To measure the amount of parallelism expressed by a partial ordering, several evaluation functions are commonly used. For instance the *number of steps (ns)*, a

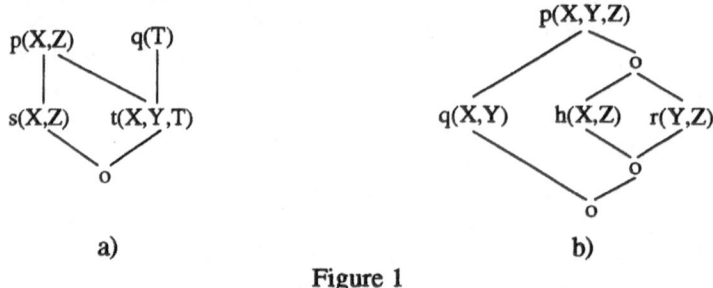

$$p(X,Z) \quad q(T)$$

$$s(X,Z) \quad t(X,Y,T)$$

$$o$$

$$p(X,Y,Z)$$

$$q(X,Y) \quad h(X,Z) \quad r(Y,Z)$$

$$o$$

$$o$$

a)

b)

Figure 1

step being a subset of literals that can be evaluated in parallel after the evaluation of the literals in the previous steps. And the *waiting factor (wf)* [Var95b], defined as

$$\sum_{i=1}^{n} number\_of\_predecessors(p_i) \Big/ \sum_{i=1}^{n-1} i \, ,$$

n being the number of literals of the query, which measures how much the literals have to wait before being evaluated.

Partial solutions management in our model is carried out by means of another ordered structure: the *Process and Solutions Net (PSN)*. This is constructed by an AND process from the *DFL*, and represents the partial solutions as well as the OR process identifiers that computed them. Figure 2 shows the *DFL* (a) and the *PSN* (b) generated from the query

$$q(X,Y),p(X,Z),r(Y,T),s(Z,T,U),$$

with respect to the logic program defined by the set of facts

$$q(a,b),q(d,b),q(a,c),p(a,i),p(d,j),r(b,k),r(c,l),s(i,k,m),s(j,k,n).$$

Additionally, Figure 2c shows the application of the *inference function, INF*, to the solution nodes *(U/n)* and *(U/m)* of the *PSN*. These nodes represent partial solutions computed by the OR processes labelled with $^{proc}s(i,k,U)$ and $^{proc}s(j,k,U)$ respectively. The function *INF* makes solutions to the query explicit by joining the partial solutions spread over the *PSN*. The correctness and efficiency of this function is founded on the properties of the *DFL*. This sophisticated model for representing

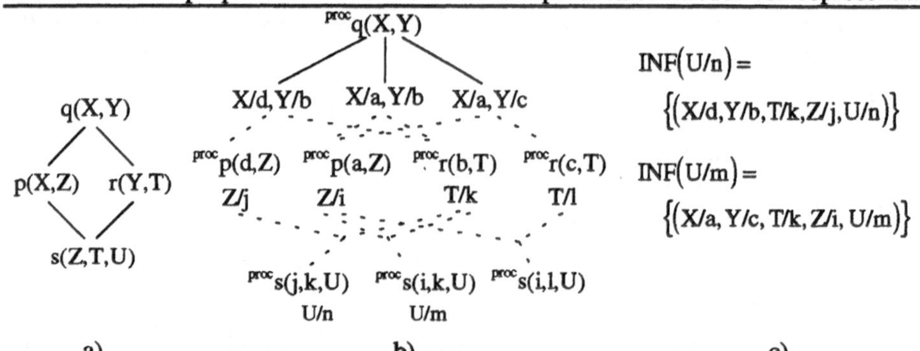

$$^{proc}q(X,Y)$$

$$q(X,Y)$$

$$p(X,Z) \quad r(Y,T)$$

$$s(Z,T,U)$$

X/d,Y/b X/a,Y/b X/a,Y/c

$^{proc}p(d,Z) \quad ^{proc}p(a,Z) \quad ^{proc}r(b,T) \quad ^{proc}r(c,T)$

Z/j Z/i T/k T/l

$^{proc}s(j,k,U) \quad ^{proc}s(i,k,U) \quad ^{proc}s(i,l,U)$

U/n U/m

$$INF(U/n) = \left\{ (X/d,Y/b,T/k,Z/j,U/n) \right\}$$

$$INF(U/m) = \left\{ (X/a,Y/c,T/k,Z/i,U/m) \right\}$$

a)

b)

c)

Figure 2

partial solutions permits us to exploit the *IAP* and *full OR parallelism* at the same time, as well as other secondary sources of parallelism, such as the *producer/consumer* one, in such a way that the duplication of processes is avoided. For example, the processes $^{Proc}p(a,Z)$ and $^{Proc}r(b,T)$ of Figure 2b are generated twice with other models [Kal91]. We do not explain the model in more detail here but refer the interested reader to [Var94, Var95b]. This interpretation model for logic programs proved to be good in problems with many literals with a high number of answers; as for example in the *map-colouring* problem [Var96a]. There, we compared our model with the *Reduce-Or Process Model (RPM)* [Kal91] in the resolution of several instances of this problem.

3 COMPUTING THE DATA FLOW LATTICE

In this section, we introduce the strategy used to determine the *DFL* from a query. The algorithm is proposed in [Var95b] and consists of three main stages. In the first one, the steps for evaluating the query are computed. This is a NP-hard problem as proven in [Del89], so an heuristic strategy becomes necessary. The second determines an intermediate structure named the *relation graph*. In the third stage, the relation graph is transformed, by introducing null nodes and reorganising the predecessors of some nodes, in order to guaranty the four properties of the *DFL*. Figure 3 shows the application of the algorithm to the query

$$q(X,Y),p(Z,T),r(Y,T),h(X,Z,U),s(U,X).$$

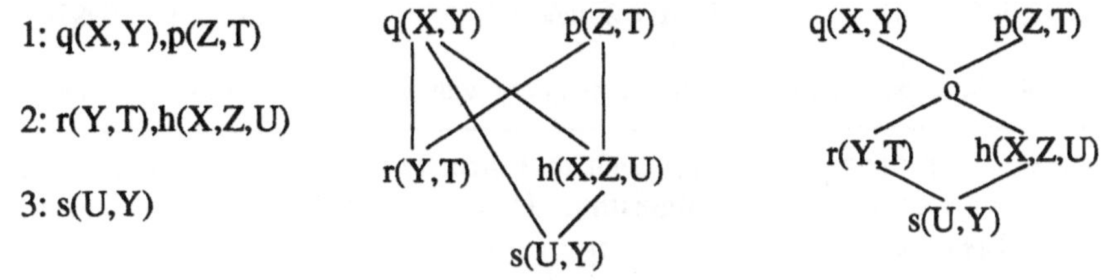

1: q(X,Y),p(Z,T)

2: r(Y,T),h(X,Z,U)

3: s(U,Y)

a) steps   b) relation graph (*wf=0.7*)   c) DFL (*wf=0.8*)

Figure 3

The waiting factor of the relation graph is the minimum possible for the computed steps, but in the final *DFL* this parameter can worsen with respect to the relation graph as we can see in the example. But the number of steps remains constant from the first stage.

To compute the number of steps, several heuristic strategies can be used, for example the following ones which are proposed in [Del89]. The first one computes the next step, starting from the set of literals not included in the previous ones, by applying the following actions until no more literals can be chosen

- i) *Select a literal with the least number of dependent literals among the rest of the literals of the set (here two literals are dependent when they share some variable which does not appear in any literal of the previous steps).*
- ii) *Clear the selected literal and the dependent ones from the set.*

Another strategy consist of substituting the former step *i)* by the following

- *i') Select a literal with the largest number of variables not included in any literal within the former computed steps.*

From here onwards, we refer to these strategies as *heuristic-1* and *heuristic-2* respectively. In order to make these heuristics clear, let us consider the query

$$p(W, X), q(Y, Z), r(Q, W, X, Z), s(Q, X, Z).$$

Figure 4 shows three possible partial orderings. If we use *heuristic-1*, we get the partial order of Figure 4a, whereas with *heuristic-2* that of Figure 4b is computed. This is the best ordering, as is indicated by the *ns* and *wf* values. Although it would be easy to find another example for which *heuristic-1* is better than *heuristic-2* strategy. This is due to the fact that the result depends on the initial linear order of the literals. For example, if the initial linear ordering was the opposite, the order of Figure 4c would be produced by *heuristic-1*.

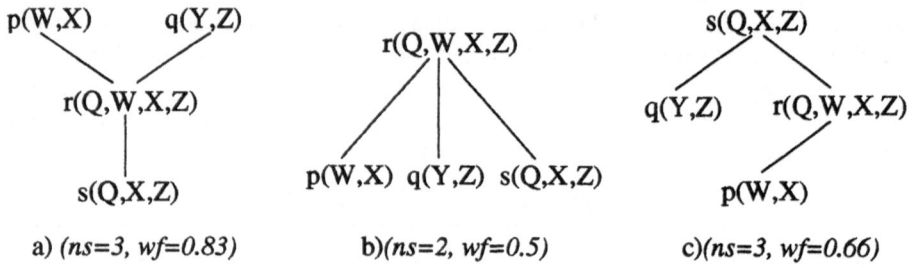

a) *(ns=3, wf=0.83)*   b)*(ns=2, wf=0.5)*   c)*(ns=3, wf=0.66)*

Figure 4. Three partial orders for the same query with different values of *ns* and *wf*.

In this paper, we propose a new approach that directly computes the relation graph from the query by using a genetic algorithm. This is in fact the main contribution of this paper and is described in the following section.

4 THE GENETIC APPROACH TO COMPUTING RELATION GRAPHS

As we have pointed out, we confront our problem within the Genetic Algorithms framework. In particular, we use a *Hybrid Genetic Algorithm (HGA)* which is inspired by the ones proposed in [Mic94, Mar95, Rie96] used to solve similar problems such as the *map-colouring* problem, the design of a Neural Net and the *TSP*.

The encoding technique is as follows. First, we define a mapping from the *n* literals of the query to the subset *{1..n}*, in accordance with the initial order of the literals in the query. Thus, an individual (the phenotype) is a permutation of the list *(1...n)*. At the same time, the genotype, i.e. the relation graph, is computed by applying the following actions to get the next step, starting from the set of literals of the query ordered according to the individual encoding

- *Search the set of literals from left to right, and move which are independent from the literals already included in this step to the current step.*

The initial population is a set of random permutations of the list *(1...n)*. One important fact is that we incorporate the initial order of the query into our initial population. This guarantees that our algorithm does not work worse than the algorithm we are hybridising.

To evaluate the individual fitness, we build the relation graph from the individual by means of the above heuristic, and then compute its waiting factor and number of steps. The individuals are ordered in the population by their *wf* from low to high values, and in the case of equality by the *ns* value.

The genetic operators that we use are the *order-based crossover* and *order-based mutation* as proposed in [Law91]. The first one preserves part of the first parent while incorporating information from the second one. The information encoded here is not a fixed value associated to a position on the individual, but the relative orderings among elements of the individuals. In this sense what is being passed back and forth is not information of the form "literal 3 is in fifth position" but "literal 3 is before literal 2 and after literal 5". An example of how order-based crossover works is the following. Let the parents be

$$paren\text{-}1: (1\ 2\ 3\ 4\ 5) \qquad parent\text{-}2: (4\ 2\ 3\ 5\ 1).$$

First, a random bit string with the same length as the parents is generated; for example *(0 0 1 0 1)*. After that, we fill in some positions on *child-1* copying from *parent-1* wherever the bit string contains a "1" and copying to *child-2* from *parent-2* wherever the bit string contains a "0", thus obtaining

$$child\text{-}1: (\_\ \_\ 3\ \_\ 5) \qquad child\text{-}2: (4\ 2\ \_\ 5\ \_)$$

Now, the remaining elements from *parent-1* are put in *child-1*, maintaining the relative order from *parent-2*. Likewise for *child-2*. So, we obtain

$$child\text{-}1: (4\ 2\ 3\ 1\ 5) \qquad child\text{-}2: (4\ 2\ 1\ 5\ 3)$$

On the other hand, the mutation selects a sublist of the *parent* and permutes it into the *child*, keeping the rest of the *child* equal to the parent. For example

$$parent: (1\mid 2\ 3\ 4\mid 5) \qquad child: (1\mid 3\ 2\ 4\mid 5).$$

It is worth noting that all of these operations are computed with a complexity of $O(n)$.

The probabilities of crossover and mutation are not constant across generations, but vary according to an interpolation linear function. The variation rates are controlled by parameters introduced by the user.

Now, we are ready to explain how the algorithm works. It starts from a set of inputs such as the query that fixes the individual size, the number of iterations (*max_gen*), the population size (*pop_size*) and the initial and final probabilities for crossover and mutation (*Pc_ini, Pm_ini, Pc_fin, Pm_fin*). Then, an initial population is randomly generated, but which include the individual *(1 2...n)*. In each iteration, the crossover and mutation probabilities, *Pm* and *Pc* respectively, are adjusted by the interpolation function, according to the respective parameters. Then, the genetic operation to be applied is selected from the relative *Pm* and *Pc* values. If the operation is mutation, one individual is selected, otherwise two individuals are chosen to apply crossover. In either case, the selection of individuals is by the roulette-wheel rule with a probability that is proportional to their order in the population rate. A produced individual is merged with the population only if it is not yet a member. After merging, the worst individual of the population is removed. A top level description is the following:

```
Algorithm HGA
initialise_population;
for i from 1 to max_gen do
    begin  Pc := interpolation(Pc_ini, Pc_fin, i);
           Pm := interpolation(Pm_ini, Pm_fin, i);
           Op := operator_selection(Pc, Pm);
           parent-1 := individual_selection(pop);
           if Op = Pc then   begin parent-2 := individual_selection(pop);
                                    list_childs := crossover(parent-1,parent-2)
                      end
                 else    list_childs := mutation(parent-1);
           for each child in list_childs do
               if not member(child,pop) then begin insert_in_pop(child);
                                                    delete_last(pop)
                                             end
    end
end.
```

5 EXPERIMENTAL RESULTS

In this section, we present some experimental results. On the one hand, results from different applications of the proposed *HGA* applied to various examples with different values of the initial parameters; and on the other hand, results comparing the algorithm to the ones based in the *heuristic-1* and *heuristic-2* strategies mentioned in Section 3. The first example is the following *QUERY-1*.

QUERY-1: $p(X_1,...,X_9)$, $q(X_1,X_2,X_{10},X_{11},X_{12},X_{13},X_{14})$, $r(X_{15},X_3,X_{16},X_{17})$, $s(X_{18},X_4,X_{19},X_{20},X_5,X_{21})$, $q(X_1,X_{10},X_{11},X_{21},X_{22},X_{23},X_{24})$, $s(X_1,X_{10},X_{12},X_{25},X_{26},X_{27})$, $r(X_1,X_{10},X_{11},X_{12})$, $s(X_{15},X_3,X_{17},X_{28},X_{29},X_{30})$, $s(X_5,X_{15},X_{19},X_{31},X_{32},X_{33})$, $t(X_{11},X_{12},X_3,X_{16},X_{17})$, $r(X_{17},X_{18},X_4,X_{12})$, $r(X_{18},X_4,X_{10},X_{19})$, $u(X_{13})$

We have done several runs varying the input parameters. Figure 5 summarises the results obtained with two different number of generations, 20 and 100 respectively, and with different initial weights for mutation and crossover operations, in particular, with (*Pc_ini, Pc-fin*) taking different values from (80, 20) to (20, 80), as indicated in the graphs, in every run (*Pm_ini, Pm-fin*) being equal to (100-*Pc_ini*, 100-*Pc_fin*). In all cases, a population size of six individuals is used. We ran the algorithm 10 times for each set of initial values, and show the average results of *wf*, *ns* and the number of evaluated individuals in Figure 5.

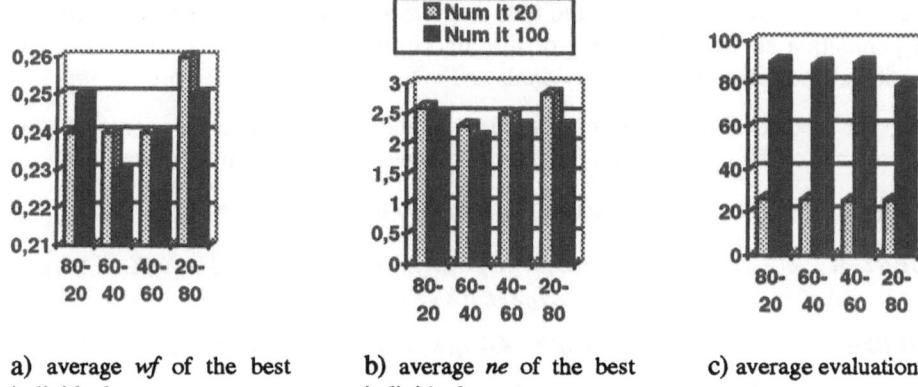

a) average *wf* of the best individual.

b) average *ne* of the best individual

c) average evaluations.

Figure 5

As we can see in Figures 5a and 5b, the solutions are not much better with 100 iterations than with 20, whereas the number of individuals which are finally evaluated is approximately three times higher with 100 iterations. So, in this example a number of 20 iterations appears to be better than 100. On the other hand, the values (60, 40) for (*Pc_ini*, *Pc_fin*) appear to be the best ones in the example. Furthermore, we proved that the best solution is found more frequently than with other values. A population size of six individuals is good enough in this example, since after many runs with higher sizes, we observed that the best individual is often found in the population. Of course, a different problem might require a greater size of population.

We also ran examples by applying only crossover or mutation. In the first case, the number of evaluations is notably higher, in the second one it is a little bit lower. But

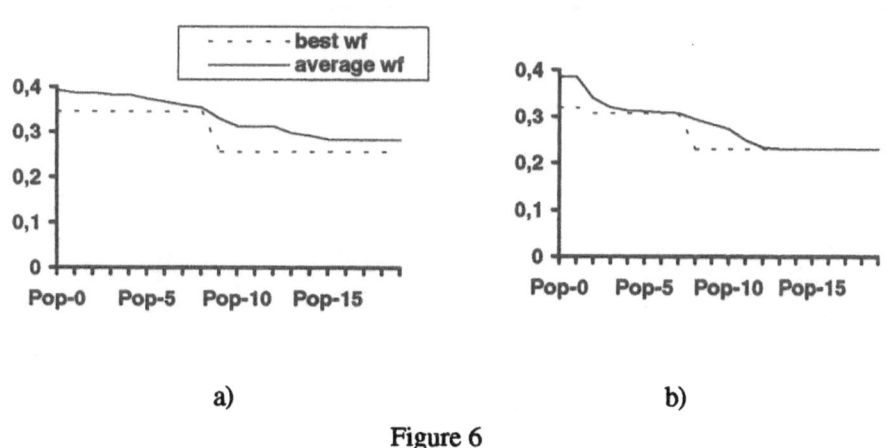

a)

b)

Figure 6

in either case the quality of the solutions clearly worsens. Therefore, it is worth remarking that the conjunction of the two operations is necessary.

Figure 6 shows the evolution of the best and average values of the waiting factor of the population in two executions of the algorithm. Here, we have *QUERY-1*, 20 iterations, a population of six individuals and values (60, 40) for (*Pc_ini, Pc_fin*)

Now, we consider three more examples of queries, in order to compare the performance of the *HGA* to both the *heuristic-1* and *heuristic-2* approaches.

QUERY-2: $\qquad\qquad\qquad p(W,X),\ q(Y,Z),\ r(Q,W,X,Z),\ s(Q,X,Z)$

QUERY-3: $\quad p(X_1,...,X_8),\ q(X_6),\ r(X_1,X_2,X_3,X_6,X_9),\ s(X_4,X_{10},X_{11},X_{12}),\ s(X_7,X_8,X_{13},X_{14}),$
$\qquad\qquad t(X_4,X_{11},X_9),\ t(X_7,X_{13},X_{12}),\ t(X_1,X_3,X_{14})$

QUERY-4: $\qquad\qquad q(X_1,X_2),\ q(X_3,X_4),\ q(X_2,X_4),\ p(X_1,X_3,X_5),\ r(X_5,X_1)$

As we can see in Figure 7, the *HGA* always produces better results than the *heuristic-1* and *heuristic-2* approaches, *Alg-1* and *Alg-2* respectively. Even the execution time is shorter with the *HGA* for small examples. Although the time is longer in the biggest one (but less than twice) with the *HGA*, the difference in the quality of the solutions clearly overcomes the higher time.

6 CONCLUSSIONS

In this paper a *HGA* for computing Independent AND Parallelism is proposed that produces, in general, better results than previous ones based on various heuristic strategies. This is validated by the experimental results and we expect that the genetic approach to be even more powerful in bigger examples than the others in finding good solutions. However, the execution time generally increases with the size of the problem because of the number of evaluations. Nevertheless, in problems with a reasonable size the genetic approach appears to be good enough. In order to improve the performance of the *HGA*, we will try to hybridise the strategy even more by introducing specific knowledge from the problem domain into the genetic operators.

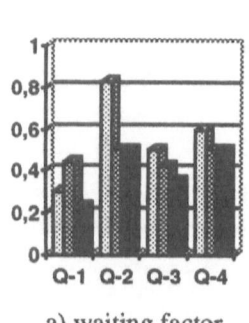

a) waiting factor

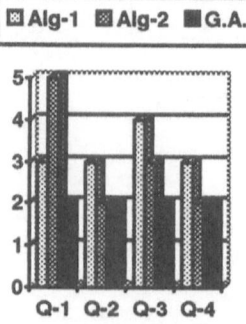

b) number of steps

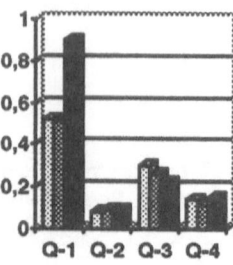

c) execution time

Figure 7

7 REFERENCES

[Con83] Conery, J. S. *The AND/OR Process Model for Parallel Interpretation of Logic Programs*. Ph. D. Th.. Dpto. Inf. and Comp. Science. Univ. California. Irvine. 1983

[DeG84] DeGroot, D. *Restricted AND-Parallelism*. Procs. of the Int. Conf. on Fifth Generation Comp. Systems. North Holland. pp. 471-478. 1984.

[Del89] Delcher, A. and Kasif, S. *Some Result on the Complexity of Exploiting Data Dependency in Parallel Logic Programs*. JLP, Vol. 6, pp. 229-241. 1989.

[Gol89] Goldberg, D. E. *Genetic Algorithms in Search, Optimization & Machine Learning*. Addison-Wesley, 1989.

[Kal91] Kalé, L. V. *The REDUCE-OR Process Model for Parallel Interpretation of Logic Programs*. The Journal of Logic Prog.. Vol 11, pp. 55-84. 1991

[Law91] Lawrence, D. *Handbook of Genetic Algorithms*. Van Nostrand Reinhold, 1991.

[Mar95] Marín, F. J.; Sandoval, F. *Diseño de Redes Neuronales Artificiales Mediante Algoritmos Genéticos*. Computación Neuronal. Ed: Senén Barro y José Mira, Universidad de Santiago, pp 385-424. 1995.

[Mic94] Michalewicz, Z. *Genetic Algorithms + Data Structures = Evolution Programs*. Second, Etended Edition, Springer-Verlag. 1994.

[Mut90] Muthukumar, K. and Hermenegildo, M. *The CDG, UDG, and MEL Methods for Automatic Compile-Time Parallelization of Logic Programs for Independent And-Parallelism*. ICLP'90, pp. 221-237. MIT Press, Jun. 1990.

[Rie96] Riesco Peláez, F. *Optimization of the Initial Generation In the Solving of the Travelling Salesman Problem with a Genetic Algorithm*. (ITHURS' 96), pp. 161-167. 1996.

[Var94] Varela, R. *El Modelo RPS para la Gestión del Paralelismo AND Independiente en Programas Lógicos*. Proccedings of the 1994 Joint Conf. on Declarative Prog.ng GULP_PRODE'94, pp. 251-265. 1994.

[Var95a] Varela, R., Sierra, E., Jiménez L. y Vela, C. R. *Combinación de Soluciones Parciales en Programación Lógica Paralela*. C-AEPIA'95. Alicante. 1995.

[Var95b] Varela, R. *Un Modelo para el Cálculo Paralelo de Deducciones en Lógica de Predicados*. Tesis Doctoral. Dpto. de Matemáticas, Univ. Oviedo. 1995.

[Var96a] Varela, R. and Vela, C. R. *AND/OR Trees for Parallel Deductions*, ITHURS'96. León, Spain. July 1996.

[Var96b] Varela, R.; Vela, C. R. and Puente, J. *Efficient Producer/Consumer Parallelism in Logic Programming*. APPIA-GULP-PRODE'96. San Sebastian.. July 1996.

Predicting Toxicity of Complex Mixtures by Artificial Neural Networks

F. Gagné and C. Blaise. St. Lawrence Center, Environnement Canada, 105 McGill, Montréal, Québec, Canada H2Y 2E7

Abstract

Industrial and municipal wastewaters constitute major sources of contamination of the aquatic compartment and represent a threat to the aquatic life. Artificial neural networks, based on three different learning paradigms, were studied to predict the acute toxicity of trouts (5 days exposure to wastewaters) with the inputs of two simple microbiotests which requires only 5 and 15 min incubation time. These microbiotests were 1) the chemoluminescent peroxidase (Cl-Per) assay which can detect radical scavengers and enzyme-inhibiting substances, and 2) the bacterial luminescent toxicity test (Microtox[TM]) which is responsive to toxic substances affecting the vital function of bacteria. The responses obtained with the the trout bioassay, the Cl-Per and the Microtox[TM] microbiotests were analyzed for statistical correlations (Pearson-moment correlation), unsupervised learning by self-organizing network, and assisted learning by the backpropagation and the Hopfield (probabilistic) paradigms. The results showed that no significant correlation (p<0.05) was obtained between either the responses obtained with Cl-Per (p = 0.121) or the Microtox[TM] (p = 0.061) microbiotests with the ones obtained with the trout bioassay. The self-organizing network was able to identify by itself a maximum number of 5 classes that were more or less related to fish toxicity: class 1 contained 2 samples that were toxic to fish, class 2 contained 2/3 samples that were toxic to fish, class 3 showed 6/8 samples that were not toxic, class 4 contained 5/6 samples that were non-toxic and class 5 identified one sample that was toxic. Supervised learning with backpropagation analysis yielded 2 kinds of networks that proved promising. The first one, was able to predict the actual toxic wastewater concentration with an overall performance of 65 % with unseen data while the second one, which was designed to identify toxic effluents from those that were not, had a much better performance (90 %). However, the probabilistic network proved also a very good prediction model to predict toxicity to fish with an overall performance of 90 %. Although more data is needed, the network based on the backpropagation paradigm seems a better predictor or classifier of trout toxicity with the Cl-Per and the Microtox[TM] microbiotests.

Introduction

Industrial and municipal wastewaters represent major sources of contamination of aquatic biota, accounting for several thousand types of chemicals released into the environment. While some of these chemicals are persistent, others are susceptible to biotransformation or biodegration, thereby augmenting the number of chemicals released (Babich and Borenfreund, 1991). The toxic properties of these complex mixtures can be assessed, in part, by a standardized rainbow trout acute lethality bioassay (Environment Canada, 1990; Bergman et al., 1986), which is the method used for compliance monitoring of industrial (or municipal) wastewaters. However, this assay requires about one hundred fingerling rainbow trouts per sample, along with several liters of wastewater, and is labor intensive. In order to reduce the number of fishes resulting from these screening studies, several alternatives can be proposed. It would be worthwile to seek alternatives to the fish bioassay which requires very small incubation time and would reduce further the use of animals. For example, the bacterial luminescent bacteria, *Vibrio fisheri*, commercially named the Microtox[TM], and the chemoluminescent peroxydase (Cl per) assay are tests requiring only minutes of incubation time with the test sample (Gagné and Blaise, 1997 ; Whitehead et al., 1993, Environment Canada, 1992). However, it remains to be confirmed

577

whether such simple and rapid tests have predictive value toward fishes after a 5 day exposure period to industrial and municipal wastewaters.

Artificial neural networks (ANN) are well known for their capacity to learn from real-life examples and to predict the outcome of events when only exposed to a partial sequence of data (Baghat, 1990 ; Collins, 1993). ANN are mathematical simulations of different learning processes of neural networks in brain tissues (figure 1). Learning is sometimes achieved by showing the ANN to real input data and the corresponding output data, hence by supervised training (Clark, 1991). In this case, the ANN adapt their interconnection weights so the input leads to the desired output neurons. The backpropagation network is a well known system for supervised or assisted learning (Baghat, 1990). Each neurons cumulates the signal from the input data, transfer the data to another neuron, and adjust its interconnection weights according to the error in respect to output data so that, at the end, the input data is transformed into the output data with the least possible error. The connecting weights, where generalization during learning takes place, are adjusted according to the error in respect to the output data. An interesting variation of supervised learning model is the Boltzmann machine, which is a probabilistic ANN (Clark, 1991). This ANN is different from the backpropagation netwwork in that the neurons are fully interconnected and each neuron has a probability of being in a relaxed or an excitated state. Therefore, the Boltzmann machine introduce a probability factor in the outcome of some events. However, unsupervised learning is also possible by a special type of ANN : competitive learning with self-organizing nets (Clark, 1991). In this case, the ANN can be adapted to classify by itself large complex data sets according to their «vectorial» direction in n-dimensionnal space where n corresponds to the number of input data. Competitive learning appears to be a usefull tool to explore classifications in a multivariable data set. Therefore, ANN modelling is an interesting tool to model the behavior of complex systems because of their capacity to learn and to adapt to (non)linear variations. ANN may be of interest for the prediction of toxicity to rainbow trouts exposed to wastewaters according to the Cl Per and Microtox^{TM} responses. The purpose of this study was therefore to study the predictibility of the Cl-Per and the Microtox^{TM} responses towards fish toxicity exposed to complex mixtures by ANN methodology. The data obtained with the bioassays will be studied with statistical correlation analysis, unsupervised competitive learning, supervised learning with the backpropagation and the Boltzmann machine algorhythms. The performance of each of the above-mentionned models will be studied as a means of evaluating the predictive properties of these simple and very rapid microbiotests toward fish mortality resulting from a 96 h exposure.

Method
Toxicity assays on wastewaters and data preparation
In this study, industrial and municipal wastewaters were chosen for toxicity screening. A 24 h composite was brought back to the laboratory and stored at 4°C before analysis. The static 96 h rainbow trout acute lethality test was performed according to a standard methodology (Environment Canada, 1990). Briefly, 10 fingerling trouts were placed in 60 L polyethylene-lined containers and exposed to several concentrations of the wastewaters at 15°C under aeration. After the exposure period, mortality was counted and the data is expressed as the first concentration were mortality significantly occur. Toxicity was assayed with the luminescent bacteria, *Vibrio fisheri* (Awong et al., 1989). Luminescent

bacteria (1 x 10^6 cells/ml) were exposed to several concentrations of wastewaters for 15 min at 15°C and light emission was recorded with a luminometer. The data are expressed as the lowest concentration where light inhibition occurs (LOEC or lowest observable effect concentration) and the last concentration were no effect is observed (NOEC or no observed effect concentration). The LOEC was determined by analysis of variance followed by a *post hoc* test (Dunnett *t* test) to confirm significant difference from controls (i.e unexposed bacteria). The data are expressed as toxicity threshold (TT) : TT = (LOEC x NOEC)$^{1/2}$. The chemoluminescent peroxydase assay was also chosen to evaluate the potential toxicity of wastewaters (Gagné and Blaise, 1997). Briefly, the assay conditions were 1 ng/ml of horseradish peroxydase (53 units per mg solid), 1 ng/ml serum bovine albumin, (fraction V), 1 mM hydrogen peroxide, 0.5 mM luminol, and 0.34 mM p-iodophenol as a photon emission enhancer. The reaction was started by the addition of the enzyme. The enzyme mixture was exposed to concentrations (0, 0.1, 1, 10, 25 and 50%) of wastewaters for 5 min at 20°C and luminescence was measured in 96-well microplates using a microluminometer (ML-1000, Dynatech). The data were expressed as LOEC and NOEC as described above. Twenty wastewater samples were considered in this study.

Correlation analysis
The data obtained from the rainbow trout assay, the Microtox^{TM} and the Cl Per assay were analyzed for (non)parametric correlations (Pearson-moment or Spearman rank). Significance was set at p<0.05. The power of correlation were also calculated for confirming absence of correlation, thus considering the possibility of making type 1 (presence of a trend or false-positives) or type 2 errors (absence of any trend or false negatives).

Unsupervised learning
The data was subjected to unsupervised learning by analysis with self-organizing nets (Neuroclassifier, Advanced Technology Transfer Group, Canada). The network was designed so as to give the maximum number of classes . The Cl Per and the Microtox^{TM} input neurons were linked to 6 competitive neurons (fig. 1b). After 50 to 70 trials with the whole data set (Cl Per assay and Microtox^{TM}), the net yielded a maximum of 5 classes. Learning vector quantitation was also tempted with our data set. In this case, supervised classification was based either on the 5 classes identified earlier or on the absence/presence of toxic effects to fish. In both cases, weights were randomly changed and the gain was manually changed as to accelerate the convergence to satisfactory local minima, if any.

Backpropagation network
The data was also subjected to backpropagation neural network analysis (Neuronet Net Tutor, Advanced Technology Transfer Group, Canada). The input data (Cl Per and Microtox^{TM} assays) were linked to one hidden layer containing 3 neurons and 1 bias neuron and these were connected to the output neuron (trout bioassay data). The learning coefficient was dynamically changed and the weights were initially randomized by the software. The error function and the transfer function were linear and sigmoïdal respectively. The momentum was set at 0.7 and Falhman's derivatives at the hidden and output layers was applied. Higher momentum values resulted in overflow of data. Two neural nets were produced by the backpropagation engine. The first model was designed to

calculate the actual toxic concentration to fish while the second was designed to classify toxic effluents (value = 1) from those that were not (value = 0). The input data were the same in each case. The performance of the networks was defined by the following rule : the number of correct predicted values that fall within an interval of ± 20 % of the real value with previously unseen samples by the ANN. For this performance evaluation, 10 wastewater samples were considered for learning/training while the remaining 10 were used for testing the network performance.

Boltzmann machine (probabilistic network)
The data was finally analyzed using a probabilistic learning paradigm of the Boltzmann machine (Attrasoft, Georgia, USA). The neurons are either in a relaxed (0) or excitated state (1). When an effect is encountered the value 10 is encoded to two neurons and where absence of effect is observed the value 01 is then encoded. For example, a positive response observed with Cl Per assay, a negative response observed the Microtox[TM] and a positive response with the rainbow trout assay translate into 100110 (6 neurons). Therefore this network acts as a probabilistic classifier under supervised training which differ, in this respect, from the unsupervised learning of competitive nets.

Results and discussions
Over the 20 wastewater samples tested, 45 % of them were toxic to fish and 35 % of them were toxic to bacteria (Table 1). The Cl Per assay was able to detect light-inhibiting effects in 45 % of the samples. It appears that when effects were observed with both of the Cl Per and the Microtox[TM] assays, a positive effect was observed to fishes exposed to the same effluent. Moreover, when both microbiotests did not measure an effect, no toxicity was also obtained to fish. However, some samples showed toxicity to fish and Cl Per assay while none was measured with the Microtox[TM]. Therefore the results suggest that some trends exist qualitatively between the responses of the microbiotests and the toxic properties of the wastewater to rainbow trout. Correlation analysis of the toxicity responses obtained with the three tests are shown (Table 2). We obtained no significant correlation between the responses obtained with the Microtox[TM] and the CL Per assay or the trout assay, albeit a marginal correlation (p= 0.061) between the Microtox[TM] and the rainbow trout assay. If we consider samples that were concordant (i.e correspondance between absence/presence of toxicity with both assays), the correlation coefficient improved only slightly and remained marginally significant (p=0.055). The bacterial luminescent assay is known to be qualitatively concordant with the toxicity of effluents to rainbow trouts (Blaise et al., 1987). Thus, the toxic effects displayed by the Microtox[TM] should follow the ones displayed by the rainbow trout assay. However, if the data are transformed in ranks before analysis, the Microtox[TM] is now significantly correlated with the trout bioassay. Therefore, correlative analysis is not able to display any clear significant trends at the $0.05 < p < 0.001$ interval. Because marginal effects ($0.1 < p < 0.05$) are encountered, we cannot dismiss without any doubt that these two microbiotests have no predictive information in respect to the toxicity of rainbow trout assay exposed to environmental wastewaters. Perhaps, if the data were treated together then more relevant trends would be observed in respect to the prediction of fish toxicity.

When the data (Cl Per and Microtox[TM] assays only) were subjected to unsupervised competitive learning (table 3), the ANN identified, by itself, 5 classes having some relevance towards fish toxicity. The classification results were more relevant from our viewpoint (i.e. trout toxicity) when at least 5 classes were imposed on the network (i.e. at least 5 competitive neurons). Classes 1, 2 and 5 contained wastewaters that proved to be generally toxic to rainbow trout (5 / 6 samples, combined classes) while classes 3, and 4 contained samples that were not toxic most of the times (10/13 samples, combined classes). Therefore, unsupervised learning can classify toxic effluents from those that are not, 80 % of times. It is therefore conceivable by using learning vector quantitation (LVQ) analysis, which is a cross between backpropagtion and Kohonen nets but lacking hidden layer(s), that we can classify wastewaters according to fish toxicity. However, no satsifactory LVQ model was obtained with our data set with an error lower than approximately 25 %. Morevoer, no LVQ model was obtained with even a presence/absence of toxicity classification.

Data analysis using backpropagation supervised learning was studied for prediction of trout toxicity (Figure 2). The first network, which was designed to calculate the actual concentration toxic to trouts with unseen data, had a performance of 60 % (figure 2A) while the second network, which was designed to classify the toxic effluents from those that were not, had a performance of 90 %. Therefore it appears that backpropagation supervised learning is more efficient to classify samples than to calculate the precise toxic wastewater concentration to trouts. This observation is consistent with ANN technology (Baghat, 1991). Indeed, ANN more effectively predict the general outcome of output data than calculate the actual output data with high precision. Although our data size can be considered small (n= 20), the supervised learning methodology is effective in classifying toxic samples to trouts (96 h exposure) based of the responses obtained with the Cl Per assay (5 min exposure) and the Microtox^{TM} (15 min exposure). It is also probable, that increasing the number wasterwater sample the ANN performances would be increased without excluding the possibility of the inverse situation (loss of predictive properties). Nevertheless, learning the interconnecting weights with 10 samples (while the remaining 10 samples served for testing) leads to generalizations that have predictive value. Eventhough the Cl Per assay and the Microtox^{TM} does not appear to be relevant toward rainbow trout toxicity, the results showed that indeed, they can be used along with ANN to predict the outcome of trout viability after a 96 h exposure time. The Cl Per assay is sensitive, in part, to radical scanvengers thereby limiting the avaibility of oxygen electrons to the organism (Whitehead et al., 1993), and this effect could likely affect trouts. The Microtox^{TM} responds generally to (in)organic or metallic compounds that are also likely to be toxic to trouts (Ribo and Kaiser, 1987). Others have found that macroinvertebrates toxicity tests (4 toxicity tests) and chemical analysis (38 compounds) may be of predictive value towards human toxicity when ANN methodology was applied (Calleja et al., 1994). In another study, the chemicals parameters of a refinery stream were found to possess predictive properties on rat teratology (fetal weights and the number of resorption) when the effluents were applied dermally to gestating rats (Feuston et al., 1994). Therefore, backpropagation analysis can be used in a variety of ways to explore trends in the toxic effects of complex mixture's. The Boltzmann machine was also able to predict the toxic outcome to fish according to Cl-Per and Microtox^{TM} data when trained with trout toxicity data (figure 3). The network was able to predict the presence or the absence of fish mortality 90 % of times. Only one effluent which displayed toxicity to fish but was not toxic according to the ANN. This effluent was toxic with the Cl Per assay while it was not with the Microtox™. This type of pattern leads to samples that are generally not toxic to trouts (sample 2, 3, 7, and 10). Therefore, this network generalize data in a probabilistic way; the response leads to absence of toxicity because the pattern generally correlates with no fish mortality.

We have compared the performance of three ANN learning paradigms for the prediction of fish mortality with the results obtained with the Cl Per and the luminescent bacterial assays. Eventhough no significant correlations ($p < 0.05$) were obtained, the three learning paradigms yielded generalizations that were more or less relevant toward prediction of trout toxicity when exposed to chemically-complex mixtures. Finally, the use of ANN appears to be a powerfull and interesting tool in the area of predictive environmental toxicology. Because of their learning capability of inter-relations within a data set, a predictive model can be sometimes produced. The inter-relationships between multitrophic toxicty tests can be explored with this methodology with interesting outcomes. For example, microbiotests along with ANN methodology constitute a potential alternative testing procedure for screening the toxic properties of industrial and municipal wastewaters. This would greatly reduce the number organisms required for testing, augment considerably the rapidity of testing while reducing the cost. It would have been interesting to study the performance of genetic algorhythms in finding nets with predictive value.

References
Awong, J. G. Bitton, B. Koopman and J.L. Morel. 1989 Evaluation of ATP photometer for toxicity testing using Microtox™ luminescent bacterial reagent. Bull. Environ. Contam. Toxicol. 43, 118-122.

Babich, H., and E. Borenfreund. 1991. Cytotoxicity/genotoxicity assays with cultured fish cells: a review. Toxicol. *In vitro* 5, 91-100.

Bergman, H.L., R.A. Kimerle, and A.W. Maki. 1986. Environmental hazard assessment of effluents. Pergamon Press, New York, USA.

Blaise, C., Van Coillie, R., Bermingham, N., Coulombe, G. 1987. Comparaison des réponses toxiques de trois indicateurs biologiques (nactéries, algues, poissons) exposés à des effluents de fabriques de pâtes et papiers. Revue Int. Sci. Eaux 3, 9-17.

Bhagat, P. 1990 An introduction to neural nets. *Chem.Eng.Prog.* August, 55-60

Clark, J.W. 1991 Neural network modelling. *Phys.Med.Biol.* 36, 1259-1317.

Calleja, M.C., Geladi, P. and Persoone, G. 1994 Modelling of human acute toxicity from physicochemical properties and non-verterbrate acute toxicity of the 38 organic chemicals of the MEIC priority list by PLS regression and neural network. *Food Chemistry and toxicity* 32, 923-941.

Collins, M. 1993 Empiricism strikes back: neural networks in biotechnology. *Biotechnology* 11, 163-166.

Environment Canada. 1990. Biological test method, acute lethality test method using rainbow trout. Conservation and Protection, Environment Canada, Ottawa. Report No. EPS I/RM/9.

Environment Canada. 1992 Biological test method : Toxicity test using luminescent bacteria *(Photobacterium phosphoreum)*. Report EPS 1/RM/24, Environmental protection Series. Ottawa, Canada.

Feuston-MH; Feuston-BP; Hamilton-CE; Mackerer-CR 1994 An improved neural net model for predicting developmental toxicity of refinery streams. Teratology 49, 416.

Gagné, F. and Blaise, C. 1997. Evaluation of industrial wastewater quality with a chemiluminescent peroxidase activity assay. Environ. Toxic. Wat. Qual. In press.

Gagné, F. and Blaise, C. 1997 Validation of rainbow trout hepatocyte model for ecotoxicity testing of industrial wastewaters. Environ, Toxic. Wat. Qual. In press.

Ribo, J.M., and Kaiser, K.L.E. 1987 *Photobacterium phosphoreum* Toxicity bioassay. 1. Test procedures and applications. Toxic. Assessm. 2, 305-323.

Whitehead, T.P., Thorpe, G., Lane, M., Watson, A., and Billings, C. 1993. A rapid and simple chemiluminescent assay for water quality monitoring. *Biology Perspect.,* 377-381.

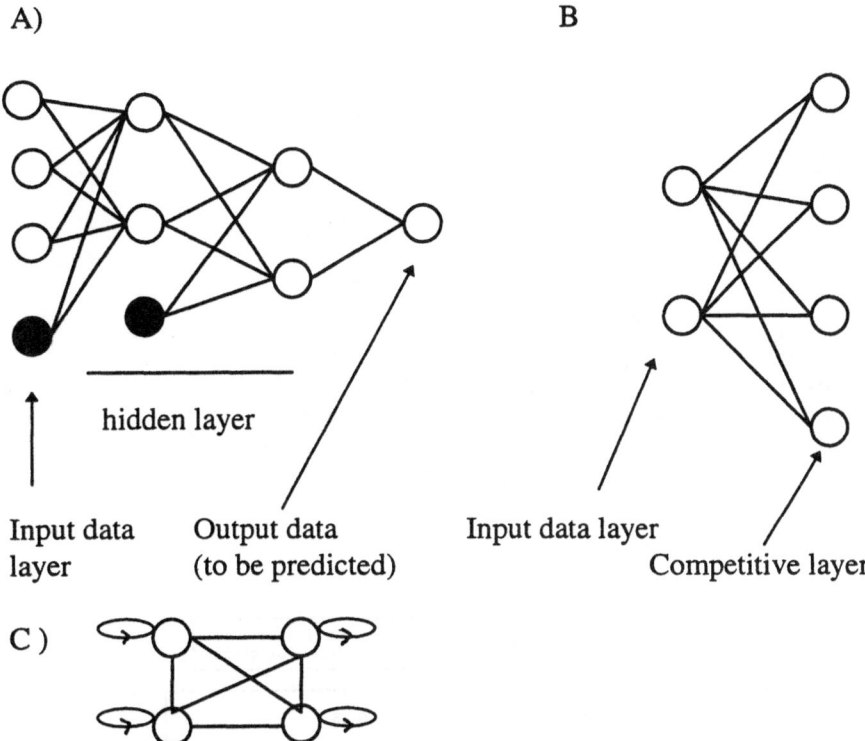

A)

B

hidden layer

Input data   Output data   Input data layer
layer   (to be predicted)   Competitive layer

C)

Figure 1 : Artificial neural networks.
Different topology are shown : a) backpropagation topology with 3 input neurons, 2 hidden layers and one output neuron, b) competitive learning topology (unsupervised learning) with input data neurons connecting to classification neurons, and c) a fully connected Boltzmann machine network. Note that in a) filled neurons are bias neurons which are invariant, in b) no hidden layers are present nor are bias neurons and in c) all neurons are interconnected. Each line represent a mathematical connection aw_{ij} between neuron i to neuron j. The input neurons are the data directly inputed to the hidden neurons: aw where a is the normalized value of the data and w the weight. The hidden neuron in a) cumulates the data : $S = a_1w_1 + a_2w_2 + a_3w_3 + ...$ and transfer the function to the following neuron by a sigmoid equation : $f(S) = 1/(1+e^{-s})$ or an exponential function : $f(s) = e^{-S}$ most of the times. The weights are ajusted according to the error in respect to the output data : $\Delta w = \eta\,(t_{ij} - O_{ij})f'(I_{ij})a_i$ where η is a learning coefficient, t_{ij} is the target value and O_{ij} is the output value. The network error is calculated : error $= (\Sigma\Sigma(T_{ij}-a_{ij})^2/n_pn_i$ where n_p is the number of patterns, n_i the number of neurons in the output and T_{ij} is the target value and a_{ij} is the activation value. In the case of competitive learning b), the classification neurons cumulate the data has above ($S = a_1w_1 + a_2w_2 + a_3w_3 + ...$) and the resulting vectors are classified in relation to their respective distance : distance $= (\Sigma(a_i - w_{ij})^2)^{1/2}$. The number of classes corresponds to the number of competitive neurons. In the case c), the Boltzmann machine can be described in terms of a simple rule: let a synaptic connection be $M_{i,j,t}$ from neuron i to neuron j, at time t, then the connection at the next time t+1, $M_{i, j, t+1}$, is $M_{i, j, t+1} = M_{i, j, t}\,a\,(q_{i, j} - p_{i, j} + ...)$ where a is a real number, $q_{i, j}$ represents the probability of neuron i and neuron j being excited together in the training data, and $p_{i, j}$ is an ANN-generated probability between neuron i and neuron j. The neurons are either in an excitated (1) or relaxed state (0).

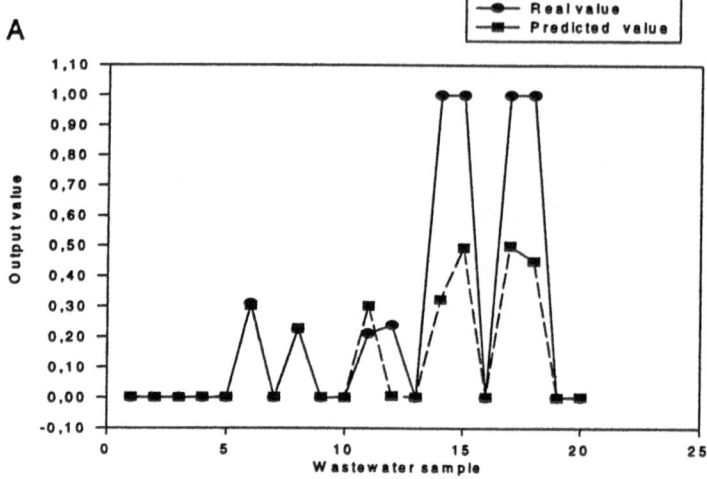

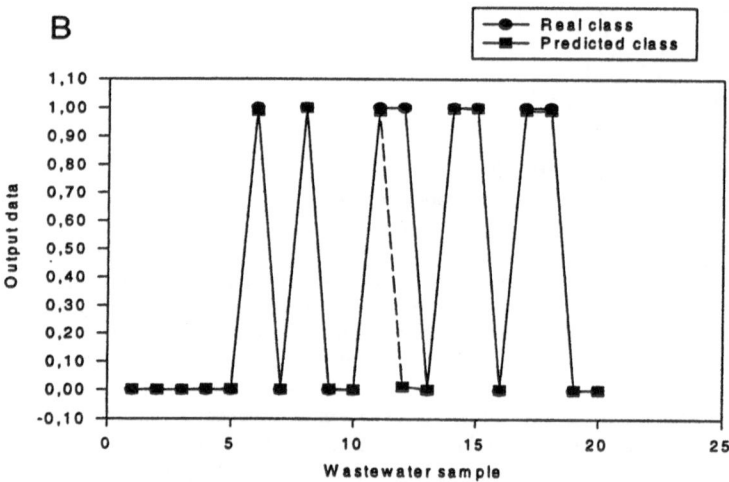

Figure 2 : Backpropagation analysis for the prediction of trout mortality.

The data reported in table 1 were analyzed to predict trout mortality (output neuron) according to the Cl Per and the Microtox™ data (input data). One hidden layer made of 3 neurons and 1 bias neuron gave the best results. The network was designed to a) calculate the actual toxic concentration to the effluent or b) to classify the toxic wastewaters toward exposed trouts. A value of 1 designs toxicity while the value 0 designs absence of toxicity.

Table 1 : Toxicity of industrial and municipal wastewaters.

Wastewater No.	Chemoluminescent peroxydase activity (% v/v)[1]	Microtox[TM] (% v/v)	Trout Bioassay (% v/v)
1	0[2]	0	0
2	16	0	0
3	36	0	0
4	0	0	0
5	0	0	0
6	3	2.2	22
7	3	0	0
8	3	17.5	16
9	0	0	0
10	36	0	22
11	16	0	0
12	36	0	32
13	3	2.2	15
14	0.3	0.3	17
15	3	0	0
16	3	7.5	71
17	0.3	4.2	71
18	36	4.4	71
19	0	0	0
20	0	0	0

1) % v/v is the toxic threshold concentration of the wastewater.
2) 0 means that no effect was detected at the highest concentration used (not detected).

Table 2. (Non-)Parametric correlation analysis.

	Microtox™	Chemoluminescent assay
Rainbow trout assay	R = 0.427 (R$_k$=0.760)[1] p = 0.061 (p<0.001)[1] Power = 0.470 n = 20	R = 0.236 p = 0.316 Power = 0.166 n = 20
Microtox™	-----	R = - 0.110 p = 0.645 Power = 0.07 n = 20

1) Spearman rank correlation analysis where the data are transformed in ranks before analysis.

Table 3 : Toxicity classification of wastewaters according to an unsupervised learning paradigm network.

Class	Number of toxic sample[1]	Non-toxic sample	Total
1	2	---	2
2	2	1	3
3	2	6	8
4	1	5	6
5	1	--	1

1) the number of samples that proved to be toxic to rainbow trouts after a 96 h exposure period.

Table 4 : Predictive analysis of trout toxicity with a probabilistic network.

Unseen samples[1]	Actual value	Predicted value
1	-[2]	-
2	-	+**
3	+	+
4	+	+
5	-	-
6	+	+
7	+	+
8	+	+
9	-	-
10	-	-

1) The probalistic neural network (Boltzmann Machine) was trained with the fisrt 10 data rows (table 1) and the remaining 10 samples were used to test the performance of the network. Note an errror in the prediction of sample 2 (**).

2) The symbol + indicates trout mortality while - indicates no apparent mortality.

·Regularisation by Convolution in Symmetric-α-Stable Function Networks

Christophe G. Molina*, William J. Fitzgerald & Peter J. W. Rayner

Signal Processing and Communications Laboratory
Cambridge University Engineering Department
Trumpington Street
Cambridge CB2 1PZ
email cm3/wjf@eng.cam.ac.uk

Abstract. In previous work, *Regularisation by Convolution* was proposed to improve the generalisation on regression of Gaussian Radial Basis Function Networks [Molina and Niranjan, 1997]. In this paper, we demonstrate that the same technique can be applied to a more general family of RBF networks called *Symmetric-α-Stable function networks* (*SαS* networks) which contains the Gaussian and Cauchy functions as particular cases. We also demonstrate that *Regularisation by Convolution* can be applied to *sigmoidal-like function networks* obtained by integration of *SαS* kernels. We illustrate the performance of *Regularisation by Convolution* on *Wahba's toy problem* and the probability density estimation of ink in ancient manuscript letters (*British library Beowulf manuscript*).

1 Introduction

Regularisation theory is now a mature tool in the fields of signal processing, statistics and neural networks (see [Girosi et al., 1995, Bishop, 1995, Ripley, 1996] and cited papers therein for an in-depth review of regularisation). Most of the regularisation carried out on neural networks concerns models based on a mixture of Gaussian or sigmoidal kernels that are regularised during training. In [Molina and Niranjan, 1997], it was shown that regularisation can be achieved independently of the training stage by convolving a trained network with Gaussian filters.

In this paper, we extend the application of this technique to the *Symmetric−α − Stable (SαS)* family of kernels [Nikias and Shao, 1995]. This technique does not need retraining and consists of the following steps: a) Training a *SαS* network with a large number of kernels; b) Convolving the network with *SαS* filters of varying dispersion and normalising the network to maintain a constant L_1-norm (this implicitly leads to a reduction in the energy of the second derivative

---

* This research is sponsored by grant RDD/G/228 from the British Library.

of $F(\mathbf{x})$); c) Verifying the generalisation performance of the convolved $S\alpha S$ networks by cross-validation, and retaining the best solution.

In section 2 we present the $S\alpha S$ family of Kernels and its integral family, the sigmoidal-like kernels. Section 3 presents the mathematical principles of *Regularisation by Convolution* for these two families. Section 4 describes the main regularisation algorithm and a binary technique used to search for the best $S\alpha S$ filter. Section 5 gives an experimental evaluation of the regularisation technique by regression and probability density estimation of synthetic and real problems.

2 Symmetric-α-Stable Family of Kernels and Networks

Symmetric-α-Stable distributions are a family of unimodal symmetric and bell-shaped distribution functions used in statistics and signal processing to model non-Gaussian processes[2] [Nikias and Shao, 1995]. The family is fully described by its d-dimensional multivariate characteristic function

$$\varphi(\mathbf{z}|\theta) = \exp\{\sqrt{-1}\,\mu\,\mathbf{z} - \gamma\mid\mathbf{z}\mid^\alpha\} \tag{1}$$

where $\theta = (\mu, \gamma, \alpha)$. The characteristic exponent α, $0 < \alpha \leq 2$, determines the "thickness" of the tails of the distribution. Values of α equal to 1 and 2 correspond to Cauchy and Gaussian distributions respectively. The central location of the distribution is given by parameter μ (the *mean* and *median* for Gaussian and Cauchy distributions respectively). Finally, the dispersion parameter γ is a measure of the deviation from the mean of the distribution and is equal to half the variance in the Gaussian case. Realisations other than Gaussian or Cauchy do not have explicit expressions but can be approximated using absolutely convergent series [Nikias and Shao, 1995]. The normalised $S\alpha S$ distributions are transformed into general kernels, as follows

$$f(\mathbf{x}|\theta) = (\gamma\,\pi)^{d/\alpha}\,\mathcal{F}^{-1}\{\varphi(\mathbf{z}|\theta)\} \tag{2}$$

where $(\gamma\,\pi)^{d/\alpha}$ is the L_1-norm and $\mathcal{F}\{\cdot\}$ the Fourier transform. The integration of the $S\alpha S$ kernel family leads to the *sigmoidal-like family*. Figure (1) illustrates both families for different values of α. An $\alpha = 2$ corresponds to the error function erf$(\cdot)$ and $\alpha = 1$ to the arc-tangent function atan$(\cdot)$. The whole sigmoidal-like family, as just defined, is similar in shape to the well known sigmoidal function in one dimension, but becomes orthogonal to the axes in higher dimensions since it originates from the fully centre symmetric $S\alpha S$ family.

$S\alpha S$ networks are defined as a finite linear mixture of $S\alpha S$ kernels. Let this be denoted by

$$F(\mathbf{x}) = \sum_{i=1}^{N} \lambda_i\,f_i(\mathbf{x}|\theta_i), \tag{3}$$

---

[2] Symmetric-α-Stables belong to the more general family of α-stable distributions.

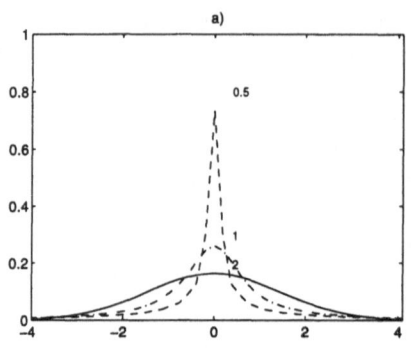 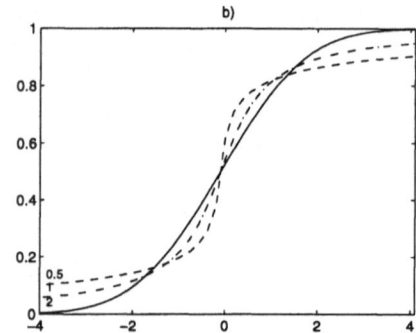

Fig. 1. *a) SαS distributions for α = 0.5 (dash), Cauchy α = 1 (dash-dot) and Gaussian α = 2 (solid). b) Sigmoidal-like kernels obtained from the SαS distribution integrals. Note that for this figure the original kernels have been normalised so that their integrals tend to value one.*

where N is the number of $S\alpha S$ kernels as given in general expression (2), and λ_i the mixing weights. In the particular case of probability density estimation by $S\alpha S$ networks, mixture weights are constrained to be positive and of unit sum, and kernels to be normalised (as originally stated in equation (1)). In general there is no requirement that the component kernels appearing in (2) should all have the same characteristic exponent α, but it is the case for this application since we require the kernels to be closed under convolution. A sufficient condition for this requirement is to have a characteristic function $\varphi(\cdot)$ closed under multiplication which is the case for kernels of the same α (closeness under multiplication of $\varphi(\cdot)$ can easily be demonstrated from equation 1). We also consider for this study the linear mixture of sigmoidal-like kernels as defined above for its application to non-linear regression.

3 $S\alpha S$ Network convolution with $S\alpha S$ filters

As stated in the introduction, *Regularisation by Convolution* is based on the convolution of a $S\alpha S$ network with $S\alpha S$ filters of different dispersion factors γ and equal characteristic exponent α. Let the filter $g_j(\mathbf{x}|\theta_j)$ be of general form (2) with centre $\mu = 0$, and a kernel $f(\mathbf{x}|\theta_i)$. We first define the integral convolution in the Hilbert space $\mathcal{H}$ of these two functions as

$$h_{i,j}(\tau) = \int_{-\infty}^{\infty} f_i(\mathbf{x}|\theta_i)\, g_j(\tau - \mathbf{x}|\theta_j)\, d\mathbf{x}. \tag{4}$$

Since $S\alpha S$ are bounded and absolutely integrable on $\mathcal{H}$, the computation of their convolution may more easily be done by means of the convolution theorem

$$\mathcal{F}\{h_{i,j}\}(\mathbf{z}) = \sqrt{2\pi}\, \mathcal{F}\{f_i\}(\mathbf{z})\, \mathcal{F}\{g_j\}(\mathbf{z}), \tag{5}$$

where $\mathcal{F}\{\cdot\}$ is the Fourier transform of the $S\alpha S$ kernel given in (2). From this convolution product, one has

$$\mathcal{F}\{h_{i,j}\}(\mathbf{z}) = \sqrt{2\pi}\ (\gamma_i\ \gamma_j\ \pi)^{d/\alpha}\ \exp\{\sqrt{-1}\ \mu_i\ \mathbf{z}\ -(\gamma_i+\gamma_j)\ |\mathbf{z}|^\alpha\}. \tag{6}$$

From the inverse Fourier transform of the above equation, a new $S\alpha S$ kernel of the same location but different amplitude and dispersion is obtained[3]

$$h_{i,j}(\tau) = \left(\frac{\gamma_i\ \gamma_j}{\gamma_i+\gamma_j}\pi\right)^{d/\alpha} f_{i,j}(\tau|\theta_{i,j}), \tag{7}$$

where $\theta_{i,j} = (\mu_i, \gamma_i+\gamma_j)$. The integral convolution of a $S\alpha S$ network $F(\mathbf{x})$ as defined in (3) and a $S\alpha S$ filter $g_j(\mathbf{x}|\theta_j)$ is a straightforward consequence of equation (7), since

$$\int_{-\infty}^{\infty} F(\mathbf{x})\cdot g_j(\tau-\mathbf{x}|\theta_j)d\mathbf{x} = \sum_{i=1}^{N}\lambda_i \int_{-\infty}^{\infty} f_i(\mathbf{x}|\theta_i)\cdot g_j(\tau-\mathbf{x}|\theta_j)d\mathbf{x} \tag{8}$$

and is equal to a new $S\alpha S$ network with the same number of kernels, located at the same positions but with different amplitude and dispersion.

In the particular context of *Regularisation by Convolution*, it is of great importance to keep the convolved network as close as possible to the original $F(\mathbf{x})$ except at those points where high frequencies have been filtered. This may be achieved by requiring the $S\alpha S$ filter to be normalised. Thus, the L_1-normalised convolution $F(\mathbf{x}) \otimes g_j(\mathbf{x}|\theta_j)$ of a $S\alpha S$ network $F(\mathbf{x})$ and a filter $g_j(\mathbf{x}|\theta_j)$ which retains the same norm of the original network may then be stated as

$$F(\mathbf{x}) \otimes g_j(\mathbf{x}|\theta_j) \doteq (\gamma_j\ \pi)^{-\frac{d}{\alpha}} \int_{-\infty}^{\infty} F(\mathbf{x})g_j(\tau-\mathbf{x},\theta_j)d\mathbf{x}. \tag{9}$$

Moreover, since the derivative operator $\mathcal{D}_\mathcal{X}$ of a function is also linear, both can be combined to apply *Regularisation by Convolution* on sigmoidal-like functions as defined in the previous section. In this case, the convolution operator $\otimes$ is applied on the derivative $\mathcal{D}_\mathcal{X}\{F(\mathbf{x})\}$ of the sigmoidal-like network and performance of generalisation is tested after integration of the filtered network.

---

[3] In the particular case of normalised kernels the term $\left(\frac{\gamma_i\ \gamma_j}{\gamma_i+\gamma_j}\pi\right)^{d/\alpha}$ vanishes (leaving the kernels normalised by their new dispersion) and therefore the amplitudes remain the same.

4 The Regularisation Technique

This section describes the regularising technique based on the convolution of $S\alpha S$ kernels with $S\alpha S$ filters, developed in the previous section. The technique is to be applied to a pre-trained $S\alpha S$ network as defined in equation (3). In non-linear regression, we use the $Mean - p - Error$ (MPE) to measure the error obtained on a cross-validation data set, $CV = \{\mathbf{x}_k, y_k\} \ k = 1, \dots, n$, after regularisation, as a performance criterion, given by

$$E = \frac{1}{n} \cdot \sum_{k=1}^{n} \mid y_k - F(\mathbf{x}_k) \mid^p, \tag{10}$$

where the best regularising $S\alpha S$ filter is found by binary search under the assumption that the underlying noise is of $S\alpha S$ nature with characteristic exponent $\alpha = p$ and then has no local minima in the search space. In p.d.f. estimation, we use the negative log likelihood of a cross-validation data set, $CV = \{\mathbf{x}_k\} \ k = 1, \dots, n$, as the performance criterion, given by

$$E = - \sum_{k=1}^{n} \ln F(\mathbf{x}_k). \tag{11}$$

Reasonable bounds for the binary searching are easily obtained from the bounded space on which the underlying function F_* lies. Thus the minimal bound for the filter dispersion is equal to zero and the maximal bound is equal to the square of the maximal distance between two points in the bounded space $\mathcal{H}$.

The structure of the related regularisation algorithm and its initialisation is shown below as pseudo-code,

- **Initial parameters and data**
  - $S\alpha S$ network : $F(\mathbf{x}) = \sum_{i=1}^{N} \lambda_i f_i(\mathbf{x}|\theta_i)$
  - Cross Validation set : CV (as defined previously)
  - $S\alpha S$ filter : $g_j(\mathbf{x}|\mu = 0, \gamma_j)$
  - Bounds of $S\alpha S$ filter dispersion and their average:
    $\gamma_{min} = 0$
    $\gamma_{max} = \max \left(\sum_{k=1}^{d} \left(x_i^k - x_j^k \right)^2 \right) \ \forall \, \mathbf{x} \in CV$
    $\gamma_{avg} = (\gamma_{max} + \gamma_{min})/2$
  - E errors obtained on the CV data set after convolving $F(\mathbf{x})$ with $g(\mathbf{x}|\gamma_{min})$, $g(\mathbf{x}|\gamma_{avg})$ and $g(\mathbf{x}|\gamma_{max})$:
    E_{min}, E_{avg} and E_{max}
  - Stop threshold for the binary search : ϵ

– **Regularising Algorithm**

   Loop until $\mid E_{max} - E_{min} \mid < \epsilon$

   1. Calculate the E_1 and E_2 errors obtained after convolution of $F(\mathbf{x})$ with $g(\mathbf{x}, \gamma)$, for dispersions γ_1 and γ_2, where,

   $$\gamma_1 = (\gamma_{min} + \gamma_{avg})/2$$
   $$\gamma_2 = (\gamma_{avg} + \gamma_{max})/2$$

   2. If $E_1 < E_2$ then,

   $$\gamma_{max} = \gamma_{avg} \ \& \quad \gamma_{avg} = \gamma_1$$
   $$E_{max} = E_{avg} \ \& \ E_{avg} = E_1$$

   else

   $$\gamma_{min} = \gamma_{avg} \ \& \quad \gamma_{avg} = \gamma_2$$
   $$E_{min} = E_{avg} \ \& \ E_{avg} = E_2$$

   end loop

5 Numerical Results

5.1 Wahba's Toy Problem

Wahba's problem consists of the regression of a noisy function generated synthetically according to the model $F_*(x) = 4.26\left(e^{-2x} - 4e^{-x} + 3e^{-3x}\right) + \nu$, where ν is normally distributed random noise with zero mean and standard deviation 0.2. In the original problem, 100 noisy observations were generated and used as training set for the training of a sigmoidal network. Wahba performed regularisation during training and the regulariser resulting in the lowest Root Mean Squared error was obtained by *leaving-one-out* cross-validation. Thus retraining was needed for each different choice of regularisation factor.

In our case we applied *Regularisation by Convolution* to the same problem after training a sigmoidal-like network (200 kernels) with over-fitting of the noisy data as shown in figure (2.a). We used an extra set of 100 samples as the Cross-Validation set CV to calculate the best regularising filter. The stop threshold ϵ was fixed at 10^{-4}. Figure (2.b) illustrates the final result after binary search for the best regulariser. Similar results obtained from a Gaussian RBF network on the same problem are shown in [Molina and Niranjan, 1997].

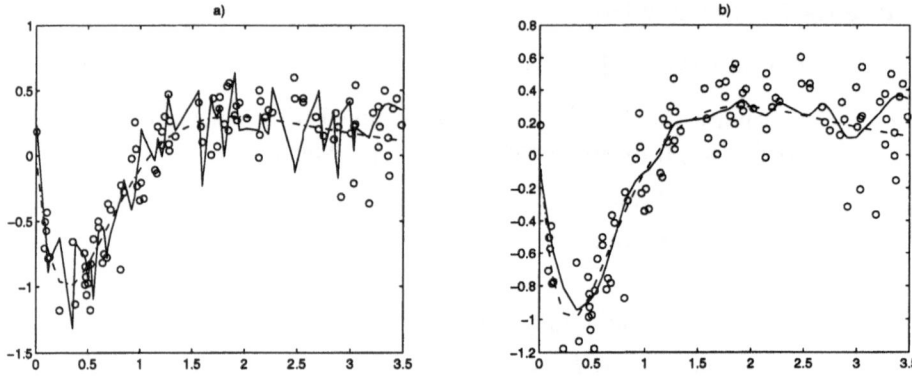

Fig. 2. *Original function (dash-dot). Sigmoidal-like ($\alpha = 1$) network regression (solid). Cross-validation observations (circles). a) Network regression after training on Wahba's synthetic problem with $E_{p=2}$ error equal to 0.276 on the CV data set. b) The same network after regularisation by convolution with a Cauchy filter ($\alpha = 1, \gamma_{avg} = 0.0135$) with $E_{p=2}$ error equal to 0.204.*

5.2 Beowulf Manuscript Letters

The *British Library Beowulf* poem is an English literary masterpiece dating from the eleventh century. The unique copy was damaged by a fire in 1731 and now suffers from a process of severe degradation. To preserve the original and make it available to researchers requiring regular access to the manuscript, each leaf was digitised under "the Electronic Beowulf Project"[4]. The digital version of *Beowulf* accumulates both the degradations of the damaged original and those of the digitisation. Among other ancient manuscripts, we are using *Beowulf* to develop image processing techniques which will facilitate the reading of the manuscript and highlight its hidden features. Our project is model based and consists on the following stages : a) Foreground leaf modeling and subtraction, b) Paragraph detection and sequential ink density estimation of letters by mixtures of Symmetric-α-Stables, c) Regularisation of $S\alpha S$ networks, and d) Divergence measure between estimated letters and a set of networks representing the corpus of symbols.

Results presented in this section concern stage c) and illustrate how *Regularisation by Convolution* increases the likelihood of the digitised data given a $S\alpha S$ network. Figure (3.a) shows a portion of image extracted from *Beowulf* containing the word "pas." We have considered the image to be a bi-dimensional histogram representation of letter ink density, where each pixel corresponds to a bin with frequency equal to the grey-level. Grey-levels have been normalised so that they sum to one. From these pixels, a *training*, a *Cross-validation CV* and

---

[4] A report about "the Electronic Beowulf Project" is available by ftp at:
http://portico.bl.uk/access/beowulf/electronic-beowulf.html.

a *test* set of respectively 1680, 840 and 840 samples were constructed. Each sample, characterised by its normalised position and gray-level, was taken randomly from the image. A network of 500 Gaussian kernels was trained using Supervised Expectation Maximisation (SEM) on the training set. SEM uses the position as input data and the frequency as desired response of the network. Figure (3.b) shows the contour of the density estimation produced by the network trained on the training set. Figure (3.c) shows the contour after regularisation of the network using the cross-validation set as reference. We used equation (11) as a performance criterion with $E_{train} = -1380$, $E_{CV} = 99$ and $E_{test} = 70$, before regularisation and $E_{train} = -963$, $E_{CV} = -238$ and $E_{test} = -0.11$, after convolution with the best $S\alpha S$ filter. Extended numerical results highlighting the improvement in performance achieved after regularisation are given in table (1).

 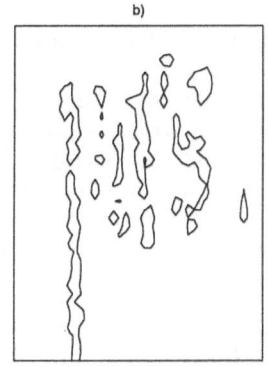 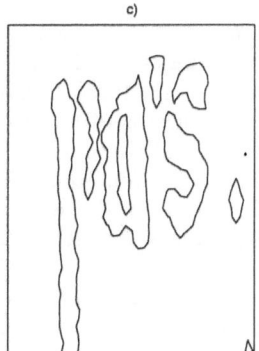

Fig. 3. . *a) Original image with word "pas." from the Beowulf manuscript. b) SαS network estimation of ink density before regularisation with final negativeloglikelihood of $E_{train} = -1380$, $E_{CV} = 99$ and $E_{test} = 70$. c) Regularised SαS network estimation using the CV set with final negativeloglikelihood of $E_{train} = -963$, $E_{CV} = -238$ and $E_{test} = -0.11$.*

	σ_{min}	σavg	σmax	E_{CV}		σ_{min}	σavg	σmax	E_{CV}
1	0.0000	0.0397	0.0794	−39.6	5	0.0099	0.0124	0.0149	−237.3
2	0.0000	0.0199	0.0397	−205.9	6	0.0124	0.0136	0.0149	−237.0
3	0.0000	0.0099	0.0199	−224.8	7	0.0124	0.0130	0.0136	−237.6
4	0.0099	0.0149	0.0199	−233.8					

Table 1. *Standard deviation bounds (σ_{min} & σ_{max}, where $\gamma = (2\sigma)^2$), and likelihood obtained for Beowulf letters by a 500 kernel Gaussian network after regularisation with gaussian filters of different standard deviation σ_{avg}.*

6 Conclusion

We have extended *Regularisation by Convolution* to $S\alpha S$ networks with the following main features:

- *Regularisation by Convolution* is suitable for $S\alpha S$ networks and sigmoidal-like networks as defined in section 2,
- the technique is independent of the training stage and algorithm,
- it does not require retraining for the calculation of the best regulariser,
- it improves the generalisation of overfitted $S\alpha S$ networks as shown by means of synthetic and real problems.

References

[Bishop, 1995] Bishop, C. M. (1995). *Neural network for Pattern Recognition*. Clarendon Press-Oxford, Oxford.

[Girosi et al., 1995] Girosi, F., Jones, M., and Poggio, T. (1995). Regularization theory and neural networks architectures. *Neural Computation*, 7(2):219–269.

[Molina and Niranjan, 1997] Molina, C. and Niranjan, M. (1997). *Generalisation and Regularisation by Gaussian Filter Convolution of Radial Basis Function Networks*. To appear in Mathematics of Neural Networks: Models, Algorithms and Applications. Operations Research/Computer Science Interface Series, Kluwer Academic Publishers.

[Nikias and Shao, 1995] Nikias, C. L. and Shao, M. (1995). *Signal Processing with α-stable Distributions and Applications*. John Wiley & Sons.

[Ripley, 1996] Ripley, B. D. (1996). *Pattern Recognition and Neural Networks*. Cambridge University Press, Oxford.

Continuation of Chaotic Fields by RBFNN

Igor Grabec, Simon Mandelj

Faculty of Mechanical Engineering,
University of Ljubljana, pob 394,
SI-1001 Ljubljana, Slovenia

Abstract. A chaotic field generator is represented by a non-linear equation. Its generating function is modeled empirically by a statistical non-parametric estimator. The estimator corresponds to a radial basis function neural network which learns from a record of a field given in some initial domain to predict the field distribution elsewhere. The performance of the generator is demonstrated by prediction of a chaotic series and a regular as well as a chaotic surface.

1 Introduction

Modeling of stochastic field generators stems from the statistical description of natural phenomena. As typical examples of stochastic fields we can mention rough surfaces produced by chaotic technological processes [8] and the density of an inhomogeneous material. The modeling can be carried out either analytically or empirically. A synergetic description of physical processes which generate chaotic fields often requires a cumbersome analytical treatment which cannot be completely carried out. [4] Consequently an empirical treatment is needed in which the field generator is modeled based on recorded samples of the field. Formation of surfaces and structures of inhomogeneous materials in various technological processes, generally exhibit a stochastic character [4,2] therefore the records of the corresponding field can be considered as statistical samples of a stochastic process. The empirical modeling represents a mathematical procedure by which the information hidden in the field records is transformed into functions describing the field generator. The most well known procedure is the ARMA algorithm which is based on a linear recursive equation whose parameters are estimated from records of the field. [6,8,9] The weak points of this method are: 1) the application of a linear model, because it is very restrictive with respect to the generality of the description, and 2) the utilization of a random disturbance for the description of stochastic properties of the field, because it is arbitrary specified. Processes which generate chaotic fields generally exhibit a non-linear and unstable dynamic behavior, [1,9] and therefore we want to provide for a non-linear description of unstable field generators. As the stochastic character of a chaotic field stems from the generating process instability the utilization of an arbitrary random disturbance is then superfluous. [3,7] We want to avoid both deficiencies by introducing a new method based on non-parametric, non-linear empirical model of field generators which can also be inherently unstable. A similar method was previously developed for the modeling of chaotic time series as are for example generated by manufacturing processes [2,3] and consequently we expect that it generalization could provide for a proper empirical modeling of chaotic fields.

2 Theoretical Background

Let us consider a scalar field $z(\vec{r})$ and utilize the parameter vector $\vec{r} = (x, y, u)$ to denote the point of field observation. We assume discrete description of field on a lattice of points spaced for $\triangle x = 1$, $\triangle y = 1$, $\triangle u = 1$. When the field represents a surface the variable z measures the height above the reference $(x, y, 0)$ plane. We further assume that a record of field on some compact domain $\mathcal{D}$ is given and that the statistical properties of the field z are invariant with respect to translation. Our task is to define a procedure by which the field values could be generated outside the domain $\mathcal{D}$ consistently with the statistical properties of the given record.

In accordance with the theory of deterministic chaos [7] we suppose that the stochastic process is self-generating. Therefore, the value of the field z at point $\vec{r}$ is a function of the values in the surrounding points. At the analytical description of this property we first define *a shell of i-th order* $\vec{s}_i$ by the set of points which are for i steps away form $\vec{r}$. The generator of the field is then described by :

$$z(\vec{r}) = G(z(\vec{r} + \vec{s}_1), \dots, z(\vec{r} + \vec{s}_{D-1})) \qquad (1)$$

In the generating function $G(\dots)$ it is not indicated which points from the shell contribute to the generation of the field inside it. The corresponding information must still be provided. For this purpose the statistical description of the field will be utilized. In the case of stochastic generator we expect that the function $G(\dots)$ is non-linear and yields unstable generation of field distribution. [3,7] Index $D - 1$ describes the number of shells cooperating in the field generation. Consequently D is called *a self-generating dimension*. It can be determined from the observation of the generator performance as will be explained later, but let us first assume that D is given. Using the generating equation (1) the field values along the boundary of the domain, where the field is initially specified, can be calculated. By a recursive repetition of this procedure the field can be continued from some domain of initially specified field.

At the empirical modeling of generating function $G(\dots)$ we suppose that the given field values make feasible creation of a state vector :

$$v(\vec{r}) = \{z(\vec{r}), z(\vec{r} + \vec{s}_1), \dots, z(\vec{r} + \vec{s}_{D-1})\} \qquad (2)$$

for a set of points $\{\vec{r}_n; n = 1, \dots, N\}$ inside $\mathcal{D}$. We assume that the vector $v(\vec{r}_n) = v_n$ represents a sample of a continuous stochastic variable whose properties can be statistically described by the empirical data basis:

$$\{v_n; n = 1, \ldots, N\} \tag{3}$$

The probability density function (*pdf*) of v is described by the Parzen's estimator [3]:

$$f(v) = \frac{1}{N} \sum_{n=1}^{N} w(v - v_n) \tag{4}$$

in which $w(\ldots)$ denotes a smooth approximation of the delta function. The Gaussian function with the width σ corresponding to a distance between samples v_n is applicable for this purpose. The invariant description of field properties with respect to translation is assured because the samples $\vec{v}_n$ are utilized in Eq. 4 irrespectively of the sample position determined by $\vec{r}_n$.

Eq. 1 is a relation between $z(\vec{r})$ and the remaining components of vector v which are represented by the vector

$$g(\vec{r}) = \{z(\vec{r} + \vec{s}_1), \ldots, z(\vec{r} + \vec{s}_{D-1})\} \tag{5}$$

From the *pdf* in Eq. 4 the conditional average is obtained as an optimal, non-parametric estimator of this relation. [2,3] It is expressed by N empirical in the form:

$$z(\vec{r}) = \sum_{n=1}^{N} B_n(g) \, z(\vec{r}_n) \tag{6}$$

The coefficient or a *basis function*

$$B_n(g) = \frac{w(g - g_n)}{\sum\limits_{k=1}^{N} w(g - g_k)} \tag{7}$$

corresponds to a normalized non-linear measure of similarity between the vector g, representing the field around the point $\vec{r}$, and the data basis vectors g_n. The estimated value $z(\vec{r})$ is therefore similar to the values which were recorded in similar structures of field around the other points of the basic domain $\mathcal{D}$. The structure of vector g can be arbitrary selected which is very convenient for calculation of the field values in various configurations of domain boundaries. Therefore particular points of shells are not explicitly denoted in the Eq. 1. This equation can be treated just as a support for a statistical description of the filed generator which is represented by the estimator in Eq. 6. This makes feasible very flexible statistical estimation of relations between field values in various configurations. The estimator is defined in terms of statistical samples only, therefore it represents a non-parametric regression. The non-linear character of basis function $B_n(g)$ renders simple empirical description of a non-linear field generator without any *a priori* supposition about its non-linear properties. The same approach is applicable also for modeling of one-dimensional generators like chaotic time series. In this case the parameter vector is $\vec{r} = (x, 0, 0)$.

The field value z, as estimated by Eq. 6, can be interpreted as a response of a radial basis function neural network to excitation g. [3] In this interpretation the sample vector g_n denotes the synaptic weights of n-th neuron. At an empirical characterization of the field the number N of acquired field samples can surpass the number of memory cells K determined by the capacity of the applied storage device. In this case it is convenient not to stop the acquisition of data, but to modify the existing samples so that the empirical information is best preserved. This leads to a self-organized learning of neural network by which a set of K optimal prototype vectors

$$\{q_k; \ k = 1, \ldots, K\} \tag{8}$$

is formed in its memory. The learning algorithm, which describes the adaptive changes of prototype vectors at $N > K$ was derived elsewhere [3] and attains in the first approximation the form

$$\triangle q_k = \frac{K}{N} \left\{ (v_N - q_k) \exp\left[\frac{-\|v_N - q_k\|^2}{4\,\sigma^2} \right] - \frac{1}{K} \sum_{l=1}^{K} (q_l - q_k) \exp\left[\frac{-\|q_l - q_k\|^2}{4\,\sigma^2} \right] \right\} \tag{9}$$

Let us briefly explain the theoretical concept of inherent instability of stochastic field generators. Consider a generation of field by similar shells in two different regions. A small difference between the field values in both regions can grow exponentially with the number of generating steps. Such a divergence indicates inherent instability of the generator and yields uncertain predictions in many steps which is further reflected in the chaotic character of the field. [3]

The quality of the modeling can be described based on a comparison of generated and given field values in some testing domain $\mathcal{D}_t$. The prediction error is defined by the mean square difference between both values. Its behavior at the increasing self-generating dimension leads to the estimation of a proper value D. At this value the error is either minimal or below some prescribed limit. The increment of the exponential growth of the prediction error with increasing number of steps corresponds to the maximal Lyapunov coefficient of the generator by which the generator instability can be characterized. [2,3]

2 Examples of modeling

The method was tested on 1- and 2-dimensional, regular and chaotic records which were generated either analytically or experimentally. [5] Here we demonstrate the generator modeling for a chaotic time series, a regular and a chaotic surface.

The basic record of the chaotic time series was generated by the logistic map:

$$z_{x+1} = 3.8\, z_x\, (1 - z_x) \tag{10}$$

We describe the results obtained with two different sets of modeling parameters. In the first set we used $N = 596$ and $\sigma = 0.01$. Fig. 1 shows the test domain with the original and empirically generated record. The empirical generation based on 4 shells ($D = 5$) started on the left side of the domain and was recursively repeated to the left. Fig. 2 shows the difference of both records. Because the generator is not exactly modeled both record diverge with the number of forecasting steps.

A question we have tried to answer in the next experiment was: **how good forecasting can be achieved by empirically modeled generator.** With this aim we have been repeating the modeling with increasing number of samples and shells and decreasing value of σ. And surprisingly, we have found that it is possible to model the generator **exactly**. Fig. 3 shows the initial and final section of the overlapped original and forecast records in the testing domain. The modeling was performed using $N = 5979$, $D = 22$, $\sigma = 0.0001$ and the observed difference between original and empirically generated test record was 0 calculated in double precision inside the complete testing domain of 10.000 points. This is in contradiction with the theoretical expectations for modeling of unstable chaotic generators, but the computer experiment has confirmed it also for the chaotic time series generated by the Henon map. [5] According to the theory of deterministic chaos a small discrepancy in the original and modeled chaos generator should in any case cause a small difference between both series which should grow exponentially with the number of steps, similarly as in Fig. 2. Why this is not the case in our computation is still not clear.

The second example demonstrates applicability of empirical method to modeling of a regular surface generator. From the $(N_x = 27) \cdot (N_x = 54) = 1458$ data points $N = 1378$ basic vectors were formed. In the modeling two shells ($D = 3$) and $\sigma = 0.045$ were used. The surface in the testing domain $\mathcal{D}_t$ was specified separately from the basic domain and is shown in Fig. 4. Empirical generation of the surface started in the middle of $\mathcal{D}_t$ from two initial shells. Fig. 5 shows the topography of the generated wave-like surface. The square difference between original and the generated surface is shown in Fig. 6. Agreement between both surfaces is good inside the testing domain. A reason for a small discrepancy which appears at the domain boundary is presumably accumulation of numerical errors.

The third example stems from the description of chaotic surfaces that resemble profiles of glass-paper.[5] Fig. 7 shows a surface topography in the basic domain $\mathcal{D}$ comprised of 900 points. The correlation function R_z of the profiles at all four sides of the domain are shown in Fig. 8a,b in x and y direction respectively. Its amplitude is above statistical variations inside the characteristic length l_c of about three steps. The surface generator was empirically modeled using two shells, $N = 841$, and $\sigma = 0.055$. Figs. 9a, b, c show the topography of original and generated surface in the testing domain as well as their square difference. The generation started from two given shells in the middle of $\mathcal{D}_t$. The surface generation appears acceptable over the region that is for $\sim l_c$ from the starting shells. Similar examples of surfaces which are produced in various manufacturing processes were modeled elsewhere. [5]

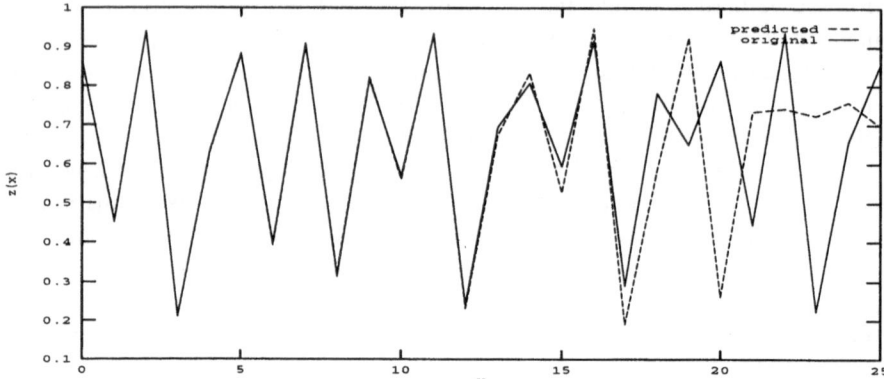

Fig. 1 Original and forecast records of chaotic time series

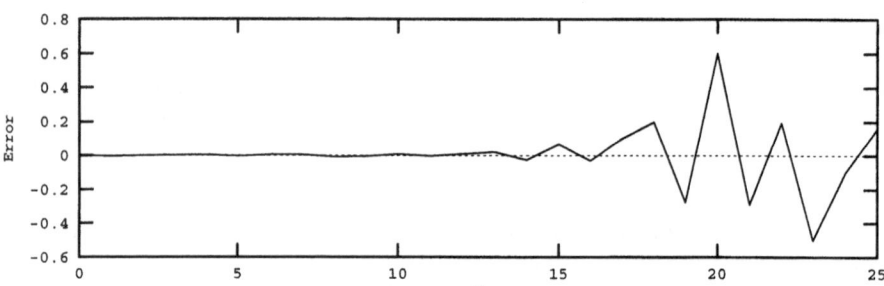

Fig. 2 Divergence of the original and the forecast time series

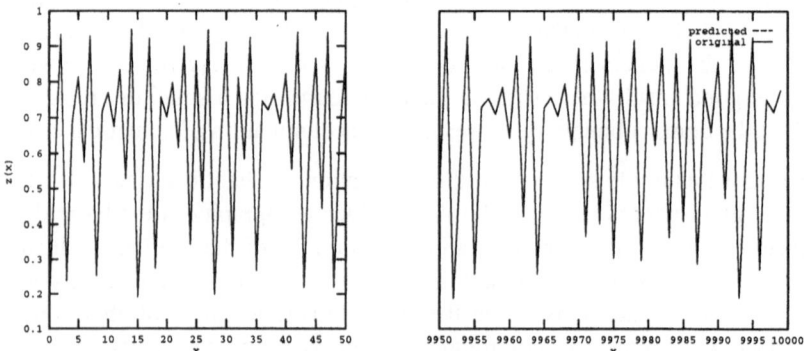

Fig. 3. Accurate forecasting of time series based on 6000 samples

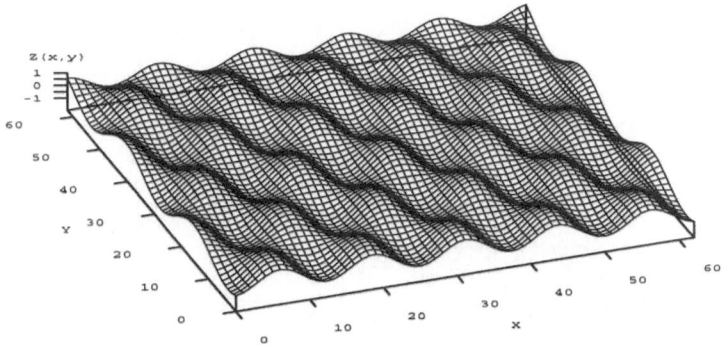

Fig. 4. Regular wave-like surface topography in $\mathcal{D}_t$.

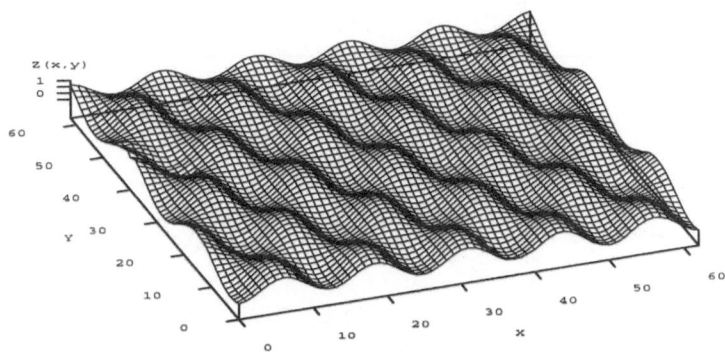

Fig. 5. Predicted regular surface in $\mathcal{D}_t$.

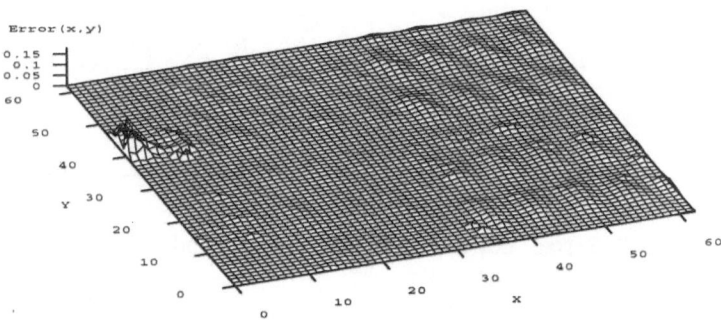

Fig. 6. Distribution of square difference between given and generated regular surface in $\mathcal{D}_t$.

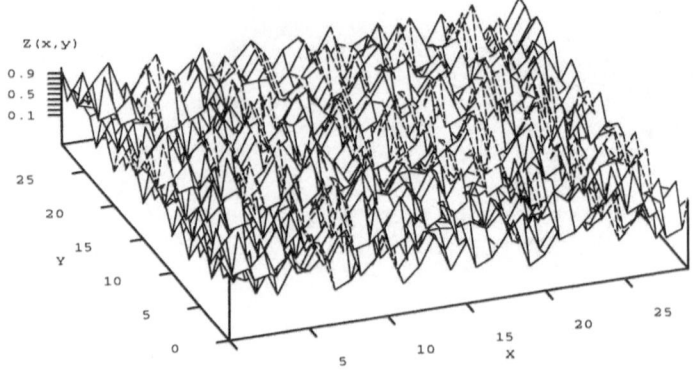

Fig. 7. Chaotic surface topography in $\mathcal{D}$.

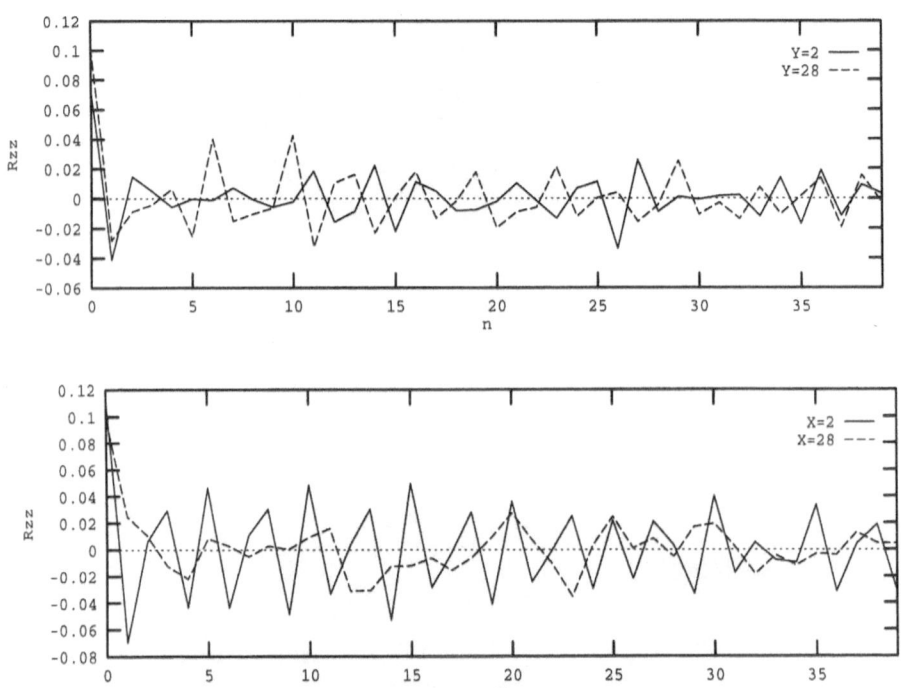

Fig. 8. Surface correlation functions $R_z(x)$ and $R_z(y)$.

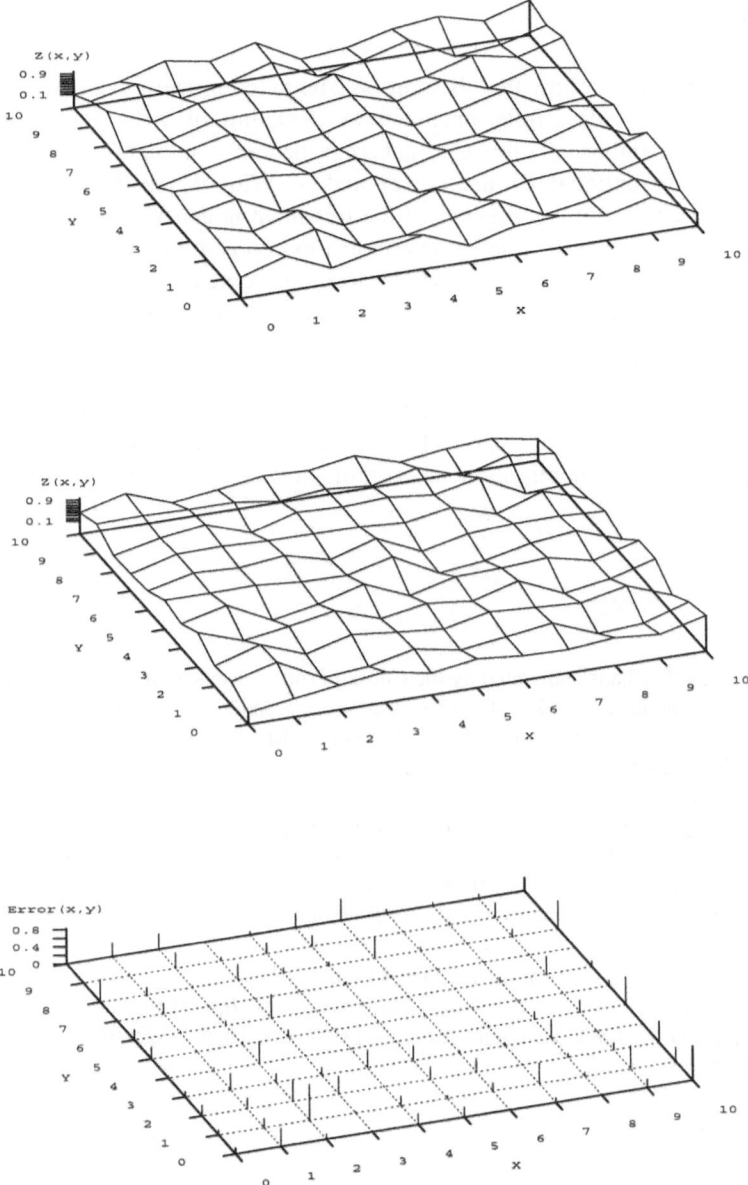

Fig. 9. Original (a) and predicted (b) surface and their square difference (c) in $\mathcal{D}_t$.

3 Conclusions

The presented examples show that the non-parametric empirical modeling of chaotic time series can be successfully extended to modeling of field generators. The performance of an empirically modeled generator depends on the number of samples N, as well as on optional parameters D and σ. With increasing N and D the performance, but also the complexity, of the treatment is generally increased. Their acceptable values can in principle be determined by a repetition of the modeling and analysis of prediction error. This is generally a cumbersome task and we expect that it could be avoided by some optimization procedure based on measure of complexity. A complementary method for the estimation of the proper dimension D was developed in relation to forecasting of chaotic time series and corresponds to growth of a hierarchical structure in the corresponding radial basis function neural network. [3] The method presented here is applicable also for empirical modeling of dynamic vector fields, however in this case the complexity of calculations is significantly increased.

References

[1]. Grabec, I., 1986, Chaos Generated by the Cutting Process, Phys. Lett., 117: 384-386

[2]. Grabec, I., Kuljanic', E., 1994, Characterization of Manufacturing Processes Based upon Acoustic Emission Analysis by Neural Networks, Ann. CIRP, 43 (1): 77-80

[3]. Grabec, I., Sachse, W., Synergetics of Measurement, Prediction and Control, Springer-Verlag, Heidelberg, 1997

[4]. Haken, H., Synergetics, Springer-Verlag, Berlin, 1983

[5]. Mandelj, S., 1996, Modeling of Surface Properties by Neural Networks, Dipl. Eng. Thesis, Fac. of Mech. Eng., University of Ljubljana, Slovenia, (slovenian)

[6] Patir, N., 1978, A Numerical Procedure for Random Generation of Rough Surfaces, Wear, 47: 263-277

[7] Schuster, H. G., 1989, Deterministic Chaos, Physik-Verlag

[8]. Whitehouse, D. J., 1994, Handbook of Surface Metrology, Inst. Phys. Publ., Bristol

[9]. Whitehouse, D. J., 1983, The Generation of 2D Surfaces Having Specified Function, Ann. CIRP, 32 (1)

Improving the Performance of Piecewise Linear Separation Incremental Algorithms for Practical Hardware Implementations

A. Chinea, J.M. Moreno, J. Madrenas, J. Cabestany

Universitat Politécnica de Catalunya, Departament d'Enginyeria Electrònica, c/ Gran Capità s/n, 08034, Barcelona, Spain

Abstract. In this paper we shall review the common problems associated with Piecewise Linear Separation incremental algorithms. This kind of neural models yield poor performances when dealing with some classification problems, due to the evolving schemes used to construct the resulting networks. So as to avoid this undesirable behavior we shall propose a modification criterion. It is based upon the definition of a function which will provide information about the quality of the network growth process during the learning phase. This function is evaluated periodically as the network structure evolves, and will permit, as we shall show through exhaustive benchmarks, to considerably improve the performance (measured in terms of network complexity and generalization capabilities) offered by the networks generated by these incremental models.

1. INTRODUCTION

In the last few years a substantial effort in the field of Artificial Neural Networks' theory has been devoted to the study and development of incremental Neural Networks models [1], [2]. The most important feature of this kind of neural models is their ability to determine the proper network structure (i.e., the number of neurons and connections between neurons) to handle a particular task.

As was pointed out in [1], there are several types of incremental algorithms, and we shall concentrate on the Piecewise Linear Separation (PLS) models, due to the fact that they present a low computational complexity for both learning and recall phases, thus being well suited for VLSI implementations. PLS models are used mainly for classification tasks, and, starting from a network composed of just one neuron, they try to find in an incremental way the discriminant function able to separate the categories defined in the input space. This discriminant function is obtained by combining the linear discriminant functions associated with the perceptron-like units generated during the training process.

In this paper we shall first briefly review the methods which have been used for training the individual units generated by the PLS models, as well as the usual drawbacks posed by such algorithms. Then we shall present a novel modification criterion which can be applied to this incremental models. This method will permit, given the desired generalization error as a starting parameter, to construct an appropriate network structure to meet such specification, allowing, at the same time a substantial reduction in the size of the generated network structures. Afterwards, we shall present a comparative and exhaustive simulation study on classification performance of the method proposed in this paper when is applied to a particular PLS model: The *Neural Trees* algorithm [3]. Finally, the conclusions and future work related to the proposed criterion will be outlined

2. PROBLEM STATEMENT

Perceptron [4] and Pocket [5] are the most common training algorithms used for the units generated by this PLS incremental models. When the learning process is completed, each unit has the weight vector which ideally yields the best correct classification rate in accordance with its input training set.

However, optimizing a function which accounts for the number of patterns correctly classified may impose an erroneous scheme for the incremental algorithm, so that, as stated in [6], a network structure of infinite size may be generated by the evolving process. Several methods have already been proposed in order to allow for a correct network evolving process [6], [7]. Among them, we have initially adopted the improvement method proposed in [6], which consists in running the Pocket algorithm with the modification that the two following conditions have to be met before a weight vector is stored as a new best weight vector: Both sides of the separating hyperplane defined by the weight vector are not empty, and the correct classification rate provided by this weight vector is larger than that provided by the best weight vector stored previously.

Nevertheless, this method generates quite complex network structures with rather poor generalization capabilities. Furthermore, the network structure depends on the order the inputs patterns are presented to the network during the training process. On the other hand, empirical observations carried out in artificial as well as in real databases have demonstrated that a large percentage of the units generated by the algorithm are used in the precise establishment of the boundaries used to separate the categories defined in the input space (causing therefore overfitting problems). This is due to the fact that PLS models present serious difficulties for solving problems in which there exists a high degree of overlapping between classes, and as a consequence produce a large amount of units trying to separate distributions of patterns very close to each other.

Taking into account the considerations stated above, we shall define a function which is calculated periodically during the training process and which detects when the network begins to have problems in determining the separation borders. This will permit to stop the growth of the network before it generates units which hardly provide information about the problem to be solved.

3. PROPOSED SOLUTION

Without loss of generality, let us consider a problem which consists in separating two classes, denoted hereafter as class 0 and class 1. We define the class-i (i=0,1) centroid, c_i, as the vector whose components are obtained by calculating the mean value from the components of the vectors which represent the patterns belonging to this class.

Let us define the following functions:

$$J_1 = \frac{1}{N_{10} + N_{11}} \left[\sum_{i=1}^{N_{10}} \sum_{j=1}^{d} \frac{1}{\left|x_{ij}^{10} - c_j^{11}\right|} + \sum_{i=1}^{N_{11}} \sum_{j=1}^{d} \frac{1}{\left|x_{ij}^{11} - c_j^{10}\right|} \right] \tag{1}$$

$$J_0 = \frac{1}{N_{00} + N_{01}} \left[\sum_{i=1}^{N_{00}} \sum_{j=1}^{d} \frac{1}{\left|x_{ij}^{00} - c_j^{01}\right|} + \sum_{i=1}^{N_{01}} \sum_{j=1}^{d} \frac{1}{\left|x_{ij}^{01} - c_j^{00}\right|} \right] \tag{2}$$

$$J_c = J_0 + J_1 \tag{3}$$

Where:

• d: Dimension of the input data space.
• N_{kl} : Number of patterns belonging to class k the network classifies as belonging to class l (k, l = 0, 1).
• x_{ij}^{kl} : j-th component of the i-th pattern belonging to class l and classified by the network as class k.
• c_{ij}^{kl} : j-th component of the i-th centroid determined for the patterns which belong to the class k and are classified by the network as class l.

As it can be deduced, this function will present maximum values when the network begins to have problems in determining the separation borders between the two categories. This is the reason for calculating in the auxiliary functions J_1 and J_0 the inverse of the distance from the components of the patterns to the components of the centroids, so as to emphasize this fact. On the other hand, these auxiliary functions are normalized, so that they do not depend on the number of patterns. Moreover, in the case any of the c^{kl} centroids do not exist, the corresponding function J_k will have value zero.

The initial value of J_c, J_{ci}, before the learning process is started, can be thus defined, using the same notation as for (1)-(3), as follows:

$$J_{ci} = \frac{1}{N_0 + N_1} \left[\sum_{i=1}^{N_0} \sum_{j=1}^{d} \frac{1}{\left|x_{ij}^0 - c_j^1\right|} + \sum_{i=1}^{N_1} \sum_{j=1}^{d} \frac{1}{\left|x_{ij}^1 - c_j^0\right|} \right] \tag{4}$$

Figure 1 depicts the evolution of the J_c function when trying to solve the classification problem stated in the phoneme database, provided by the ROARS

Esprit Project. The aim of this database is to distinguish between nasal and oral vowels (therefore, two different categories are defined in the input space) coming from 1809 isolated syllabes. Each vector constituting this database is characterized by five features, corresponding to the first harmonics, normalized by the total energy. The database is composed of 5404 vectors, 3818 of class 0 (nasal vowel) and 1586 of class 1 (oral vowel).

The algorithm used to train the network is the Neural Trees algorithm [3], which belongs to the category of PLS models. This algorithm tries to solve a certain classification task by dividing iteratively the initial problem in subsequent reduced versions, which are used as training sets for the units generated during the network construction process. As a consequence, this algorithm produces finally a network structure which resembles a binary decision tree. In this way, Fig.1 represents the value of the J_c function for each level of the tree generated by the incremental algorithm. The unit training principle has been the Pocket algorithm with the modification criterion mentioned previously. The number of iterations (i.e., the number of weights updates) for each unit was set to 10000.

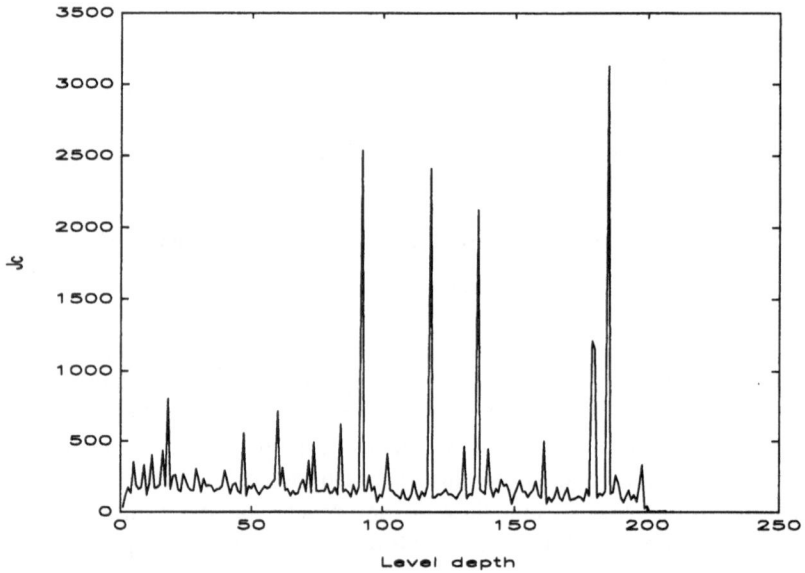

Fig.1. Evolution of the J_c function for the *phoneme* database

The evolution of the J_c function depicted in Fig.1 suggests a stopping criterion for the network growth process. It consists in freezing the network construction procedure when a relative peak (with respect to a given reference) of the J_c function arises. We have set this reference as the value for the J_c function before the training process is started. Thus, if we further use some penalty/merit functions in order to maximize the quotient generalization / network complexity, we shall improve the overall performance provided by the network constructed by PLS incremental models.

A suitable penalty/merit function, I, could be defined as follows:

$$I = \begin{cases} \left[\dfrac{G}{1+\left(\dfrac{G-G_{max}}{G_{max}}\right)^2} \right]^2 \dfrac{1}{1+\left(\dfrac{C}{C_{max}}\right)} & G \le G_{max} \\[2em] \dfrac{G^2}{1+\left(\dfrac{C}{C_{max}}\right)} & G > G_{max} \end{cases} \qquad (5)$$

Where G is the generalization given by the network (i.e, the correct classification rate provided for the test set) and C is the complexity of the network (measured as number of units). The constant parameters G_{max} and C_{max} can be determined using the theoretical results obtained by Baum and Haussler [8] concerning the probability of poor generalization. More specifically, they showed that at least about W/ε training examples are needed to obtain a generalization error less than ε, being W the number of weights in the network. In the case of the Neural Trees algorithm the previously described lower bound on the number of patterns leads to the following expression:

$$C_{max} = \frac{p\varepsilon}{d+1} \qquad (6)$$

Where p is the number of patterns on the training set, ε the generalization error and d the dimension of the input space. Thus, we have completely characterized expression (5) since we have related the maximum complexity of the network C_{max} with the generalization error ε. Hence, once the value of the generalization error is fixed, the parameters G_{max} (since $G_{max} = 1 - \varepsilon$) and C_{max} in function I are determined automatically.

Taking into account the considerations stated above we can summarize the proposed network evolving scheme as follows:

1. Given a desired generalization error calculate the value of the function J_c for the entire distribution.
2. Start the network evolving scheme imposed by the particular PLS model. During the network construction process, evaluate periodically the function J_c in order to detect peaks of amplitude λ times greater than the initial value.
3. If the function J_c presents a peak value, then calculate the current value of the function I. If this value is greater than that calculated in the previous maximum, it

is updated and the network construction process is still allowed. Otherwise, the network evolution is stopped at this point.

It is important to note, however, that the election of the parameter λ may be critical. If we choose a small value for λ, the network growth process will be stopped too early, even when the network size will not be enough for providing a satisfactory correct classification rate, leading to misclassifications and a resulting network structure with rather poor generalization capabilities.

On the other hand, if we choose an arbitrary large value for the parameter λ the complexity of the generated network structures will increase due to the fact that maximum values given by the function J_c never reach the value given by the parameter λ. Furthermore, even in the case we could choose an optimal value for λ the standard deviation of the complexity of the structures generated during the training phase, would be quite large as a consequence of the behaviour of the function J_c. The maximum values presented by the function J_c are highly dependent on the particular evolution followed by the network during the learning phase, specifically on the order the training patterns are presented to the network. Therefore, in order to avoid the problems previously addressed we finally defined the parameter λ as a linear decreasing function over the complexity (number of units) followed by the network during the course of training.

Expression (7) shows the above mentioned relationship between the parameter λ and the network complexity C. It is simple to see how the parameter λ will take large values while the complexity of the network remains small, decreasing its value at the extent network complexity grows up. Thus, the proposed evolving scheme will generate structures with a maximum complexity around the optimal value given by C_{max}.

$$
\lambda = \begin{cases} C_{max}\left(1 - \dfrac{C}{C_{max}}\right) & C \le C_{max} \\ 0 & C > C_{max} \end{cases}
\tag{7}
$$

In the next section, a meaningful comparison on classification performance will be carried out between the Neural Trees algorithm and the same algorithm with the proposed improvement method.

4. EXPERIMENTAL RESULTS

The classification tasks used to check the proposed modification criterion are contained in the following artificial and real databases:

• *gauss2, gauss4, gauss6* and *gauss8*: These four databases are composed by 5000 vectors belonging to two normal distributed classes (2500 vectors for each category) in dimensions 2, 4, 6 and 8, respectively. Both distributions have the same mean but different variance.

- *rectangular*: It is composed of 2500 bidimensional vectors, 1248 belonging to class 0, and 1252 belonging to class 1, which are distributed following bidimensional distributions in the square (0,0)-(1,1) with overlapping.
- *clouds*: This database consists of 5000 vectors belonging to two different classes, with 2500 vectors in each class. The first class is obtained by the sum of three different gaussian distributions, while the second class corresponds to a single normal distribution.
- *phoneme*: The main features of this database have been explained previously in section 3.

The experiments were carried out as using the *leave-k-out* cross-validation procedure [9]. In this way, the original database is divided in ten equal sized parts, and then the network is trained with nine parts and tested with the remaining part, so that a total of ten training-test sets are obtained for each database. This process is repeated six times for each partition, so that finally a total amount of sixty evolving processes are performed for each database.

Table 1 shows the results, indicated as mean number of units generated and generalization percentage (together with the corresponding standard deviations of both values) provided by the original Neural Trees algorithm (which evolves a network structure providing a 100 % correct classification rate for the training set) for the classification tasks stated previously.

Problem	Number of units	Std. dev.	Generalization (%)	Std. dev.
Gauss2	1748.0	24.26	63.66	1.91
gauss4	1117.1	21.74	68.6	2.09
gauss6	891.0	25.29	72.2	2.14
gauss8	752.77	26.05	73.78	1.95
rectangular	626.43	12.77	76.62	2.34
clouds	762.13	19.84	84.6	2.55
phoneme	744.0	22.4	84.9	1.51

Table1. Results for the *Neural Trees* algorithm

Table 2 reproduces the results provided by the Neural Trees algorithm when the network evolving process is modified by the criterion proposed in this paper. In our experiments, the desired generalization error, which is represented in the last column of the table, was set to that given by the theoretical Bayes limit. In the case of the phoneme database we tested different values for the generalization error since Bayesian limit was not known.

As can be deduced from the comparison of these tables, the results provided by the proposed method not only represents a substantial reduction on the number of units generated by the algorithm, but also a meaningful improvement on the generalization capability. However, simulation results corresponding to the clouds database and the gaussian database (except for the gauss2 problem) illustrate the difficulty in obtaining generalization errors close to the Bayesian limit. For the case of the gaussian database this behaviour can be explained due to inherent sparseness of high dimensional training data (known as the curse of dimensionality).

To have a better understanding about this effect it is important to note that for the gaussian database the number of training patterns is the same in all dimensions. Hence, from statistical point of view it does not exist an amount of enough training samples in order to estimate a probability density distribution. Moreover, linear smoothers (like the PLS models) generally have insufficient data in high dimensional spaces to reliably estimate a probability density distribution.

Problem	Number of units	Std. dev.	Generalization(%)	Std. dev.	Error (%)
gauss2	428.56	11.26	71.36	1.94	26.37
gauss4	200.12	11.25	76.49	1.98	17.64
gauss6	120.8	10.35	77.89	1.89	12.44
gauss8	80.31	9.46	76.56	1.9	9.0
rectangular	122.33	5.26	82.55	2.2	15.51
clouds	157.2	6.22	78.25	1.89	9.66
phoneme	133.36	6.16	83.72	1.39	15.0
phoneme	97.53	8.38	83.23	1.5	10.0
phoneme	61.68	7.43	83.05	1.28	5.0

Table 2. Results for the *Neural Trees* algorithm modified with the proposed criterion

In the case of the clouds database not only the previously commented difficulties related to the Bayes limit can be observed, but also a substantial looseness of the generalization error with respect to that given in table 1. At this point, it is helpful to note that the lower bound given in the previous section (see expression (6)) was made under the assumption of a network with a large number of weights. Nevertheless, this fact constitutes the main reason of such discouraging results, since working conditions of the mentioned hypothesis fail.

A natural way to overcome this limitation is to use the upper bound on the number of patterns given by Baum and Haussler. More specifically, they showed that if the error on the training set was less than $\varepsilon/2$ at most of the order of $(W/\varepsilon)log(M/\varepsilon)$ examples are needed to obtain a generalization error less than ε. Where W is the number of weights on the network and M the number of threshold units. For the case of the *Neural Trees* algorithm the following expression is straightforward:

$$ p = \frac{C_{max}(d+1)}{\varepsilon} \log\left(\frac{C_{max}}{\varepsilon}\right) \tag{8} $$

Where p is the number of patterns, d the dimension of the input space and C_{max} is the complexity of the network. However, the value of the parameter C_{max} is very difficult to compute due to the non-linearity of the equation derived above.

A heuristic for avoiding such a problem can be summarized as follows:

1. Compute the parameter C_{max} from expression (6).
2. Given the value of C_{max} compute p from expression (8).
3. Re-compute C_{max} from expression (6) using the new value of p.

Table 3 shows the results provided by the Neural Trees algorithm when the network evolving process is modified by the criterion expounded in section 3 together with the previously described heuristic procedure.
From the observation of these results, it is clear to see how the proposed heuristic overcome the aforementioned difficulties. On the one hand, for the clouds database, as was expected, it is appreciated a meaningful increase of the generalization capability.

Problem	Number of units	Std. Dev.	Generalization(%)	Std. dev.	Error (%)
gauss2	1259.5	4.26	67.32	2.13	26.37
gauss4	486.13	6.3	75.87	1.77	17.64
gauss6	255.98	11.53	77.88	1.7	12.44
gauss8	161.46	10.57	77.71	2.2	9.0
rectangular	335.78	8.51	80.1	2.08	15.51
clouds	486.31	9.82	84.0	2.09	9.66
phoneme	353.15	15.42	84.54	1.52	15.0
phoneme	239.25	4.25	84.2	1.5	10.0
phoneme	129.73	6.25	83.91	1.38	5.0

Table 3. Results for the *Neural Trees* algorithm with the proposed criterion and the heuristic

As a consequence of the results presented previously, two main conclusions can be derived:

• The proposed criterion is able to reduce the network complexity generated by the incremental algorithm, thus facilitating an eventual hardware implementation of the classifier evolved by the PLS incremental model. Furthermore, the standard deviation of the number of units generated is considerably reduced, minimizing the influence of the vector presentation order during the training phase.

• The generalization performance provided by the final network structure is held or even improved, as was expected for less complex networks.

5. CONCLUSIONS AND FUTURE WORK

We have shown in this paper that the linear discriminant solutions provided by the usual training algorithms are not always useful for evolving the network structures generated by PLS incremental models.
After reviewing some modification proposal for these training algorithms and the problems posed by the network evolving scheme associated with PLS models, we have introduced a novel approach for improving the performance of these models. This modification criterion is able to stop the network construction process when no further improvement is obtained by adding new units to the network. Furthermore, the method permits to construct automatically the proper network structure for a

given classification task with only one input parameter, the generalization capability expected for the resulting classifier

As a consequence, the proposed method produces very compact network structures for a given problem to be handled, thus facilitating an eventual hardware implementation (or software emulation). Together with the expected improvement in the generalization capabilities of the resulting networks, this method is able to reduce the influence of the vector presentation order during the training phase of the network evolving process.

Our current work is concentrated in comparing the proposed method with some criterions proposed recently [10] and aimed also at selecting the proper network structure able to handle a given classification task. We contemplate also the possibility to apply this method to incremental models aimed at solving regression tasks.

6. REFERENCES

[1] J.M. Moreno, "VLSI Architectures for Evolutive Neural Models", Ph. D. thesis. Universitat Politécnica de Catalunya, 1994.
[2] Y. Kwok, D.- Y. Yeung, "Constructive Feedforward Neural Networks for Regression Problems: A Survey", Technical Report HKUST-CS95-43, Hong Kong University of Science and Technology , 1995.
[3] J.A., Sirat, J.P. Nadal, "Neural Trees: A New Tool for Classification", Technical Report, Laboratoires d'Electronique Philips,1990.
[4] M. Rosenblatt, "Principles of Neurodynamics", Spartan, New York, 1962.
[5] S.I. Gallant, "Optimal Linear Discriminants". Proc. of the 8th Intl. Conf. on Pattern Recognition, pps. 849-854, Paris, 1988.
[6] J.M. Moreno, F. Castillo, J. Cabestany, "Optimized Learning for Improving the Evolution of Piecewise Linear Separation Incremental Algorithms", New Trends in Neural Computation, J. Mira, J. Cabestany, A. Prieto (eds.), pps. 272-277, Springer-Verlag,1993.
[7] J.M. Moreno, F. Castillo, J. Cabestany, "Improving Piecewise Linear Separation Incremental Algorithms Using Complexity Reduction Methods", Proc. of the European Symposium on Artificial Neural Networks, ESANN'94, pps. 141-146, 1994.
[8] E.B. Baum, D. Hausler, "What Size Net Gives Valid Generalization", Neural Computation, Vol. 1, pps. 151-160, 1989.
[9] B. Efron, "Bootstrap Methods: Another Look at the Jacknife", The Annals of Statistics, Vol. 7, No. 1, pps. 1-26, 1979.
[10] Murata, S. Yoshizawa, S.-I. Amari, "Network Information Criterion. Determining the Number of Hidden Units for an Artificial Neural Network Model", IEEE Trans. on Neural Networks, Vol. 5, No 6, pps. 865-872, November 1994.

Accurate Decomposition of Standard MLP Classification Responses into Symbolic Rules

Guido Bologna and Christian Pellegrini

Université de Genève, Centre Universitaire d'Informatique,
24, Rue Général Dufour
CH-1211 Genève, Suisse

Abstract. In this work we determine hyper-plane equations from three MLP models. The first one is the standard MLP model, the second one is called OMLP (oblique MLP) and the last one is called IMLP (Interpretable MLP). From OMLP and IMLP, hyper-plane equations are determined easily, whereas for MLP we just give a sufficient condition for the detection of potential hyper-plane discriminators.
Our goal is to justify MLP classification responses in terms of symbolic rules. For this, we use a standard MLP network for classification and an IMLP network for justification of MLP responses. The system consists in the training of the IMLP network with MLP responses and in the extraction of symbolic rules from IMLP. The approach is sufficiently general to work even when input variables are continuous. Moreover, the justification provided by IMLP is accurate because if MLP and IMLP responses are contradictory with respect to a new unknown example, IMLP is retrained with the addition of the new example until the system becomes coherent. Finally, we show results given by a medical diagnosis application with continuous input variables.

1 Introduction

MLP (Multi Layer Perceptron) [1] and inductive decision trees (IDT) [2], [3] are widely used in classification problems. The training phase of the latter is faster than those of neural networks. Moreover, IDT models are interpretable through symbolic rules. From the accuracy point of view, some classification problems give better results with MLP networks, and other ones give better results with inductive decision trees [4], [5], [6], [7], [8], [9]. One limitation of MLP is that symbolic rules are difficult to extract, especially when input neurons have continuous input variables. In a real world diagnosis application, justification of responses can be given through symbolic rule extraction. In fact, especially for critical diseases, the physician using a diagnosis help system strictly needs to understand the decision making mechanism.

Several rule extraction techniques applied to neural networks are developed from binary input variables (as an example see [10], [11]). The rule extraction becomes particularly difficult when input data contain continuous variables [12]. Andrews and Jeva introduced the Rulex model that is close to a dynamic RBF model in which rules are extracted in a natural manner even with continuous input neurons [13].

In the context of a medical diagnosis application we compare three MLP models and the C4.5 algorithm. The first neural network model we consider is the standard MLP model, the second one is OMLP (Oblique MLP) [14], and the last one is IMLP (Interpretable MLP) [15]. We show experimentally that standard MLP gives better accuracy results on our medical application. The interesting property of OMLP and IMLP is that symbolic rules are determined in a natural manner, even with continuous input variables. Concerning standard MLP, we determine a sufficient condition from which a hyper-plane equation is a potential discriminator. We show that directly extracting symbolic rules from MLP that exactly mimic MLP responses is very hard. However, we propose to give justification of MLP classifications as symbolic rules that accurately mimic MLP diagnosis. This is carried out by using an IMLP network that learns MLP responses. The rulebase extracted from IMLP is not used to perform inferences on unknown examples; the diagnosis is performed by MLP and the justification of MLP responses by the rules extracted from IMLP.

2 Neural Network Models

For each of the neural network models we describe the architecture and the related learning algorithm.

During the training phase, each model minimizes the error function E given by:

$$E = \sum_p \sum_i (t_i - o_i)^2 \tag{1}$$

Symbols p and i represent respectively the index over all training patterns and the index over all output units; t_i represents a teaching value and o_i represents an output value.

2.1 The Standard MLP Model

This last decade the MLP model has been widely studied, so we just give a small description. For more details see [1].

The Architecture. As shown by figure 1, we define an MLP with one input layer, one hidden layer, and one output layer[1]. We represent the output values

---

[1] Bias are included in weight vectors.

of the three layers respectively by symbols x, h, and o. The output value h_j of the jth hidden unit of the hidden layer is given by the sigmoid function σ:

$$h_j = \sigma(\sum_k w_{jk} \cdot x_k) = 1/(1 + exp(-\sum_k w_{jk} \cdot x_k)) \qquad (2)$$

where w_{jk} represents the weight associated with the connection from input neuron k to hidden neuron j. The output value o_i of the ith output unit is given by the sigmoid function σ:

$$o_i = \sigma(\sum_j v_{ij} \cdot h_j) = 1/(1 + exp(-\sum_j v_{ij} \cdot h_j)) \qquad (3)$$

where v_{ij} represents the weight associated with the connection from hidden neuron j to output neuron i.

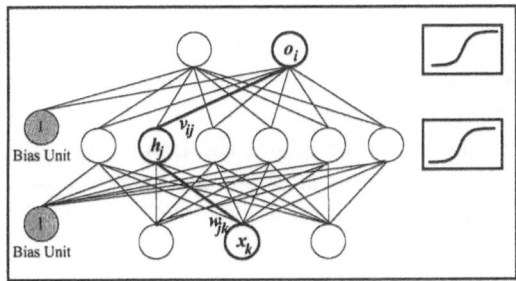

Fig. 1. An MLP network with one hidden layer. Symbols x, h, and o represent respectively input units, hidden units, and output units. Weights going from the first to the second layer are denoted by symbol w, and weights going from the second to the third layer by symbol v.

The Learning Algorithm. We use the backpropagation algorithm with momentum and flat spot elimination.

2.2 The OMLP Model

The Architecture. The only difference between the standard MLP model and OMLP [14] resides in the output of the units of the hidden layer. They are given by the threshold function instead of sigmoid function. So, expression (2) is replaced by the following one:

$$h_j = \begin{cases} 1 \text{ if } \sum_k w_{jk}.x_k > 0 \\ 0 \text{ otherwise} \end{cases} \qquad (4)$$

The Learning Algorithm. The training phase begins with backpropagation and continues by simulated annealing [16].

For the backpropagation algorithm, since (1) is not differentiable with threshold units, we approximate threshold functions with a sigmoid in which the slope has a suitable large value[2]. During learning, weight variations are calculated with the backpropagation equations associated to sigmoid functions, whereas the error is calculated with threshold units. Typically, this method can learn almost the whole training set of examples [14]. Nevertheless, near the end of the training phase the method may be stuck for a very long time in local minima[3], so at this stage we use simulated annealing.

2.3 The IMLP Model

The Architecture. IMLP [15] differs from standard MLP in three main aspects:

1. each hidden unit in the hidden layer[4] is connected to only one input unit;
2. the activation function used for the hidden layer units is the threshold function instead of sigmoid function.
3. The hidden layer size is a multiple of the input layer size. This guarantees that during the training phase all input neurons will have the same opportunity of affecting network responses.

Learning Algorithm. As in the OMLP model, we start the training phase with backpropagation and we finish with simulated annealing.

3 Discriminant Hyper-Planes Determination from MLP, OMLP, and IMLP

We define a hyper-plane as the linear combination of weights and input variables associated to (4). We define a hyper-plane discriminator as a hyper-plane that separate examples of different classes. At the beginning of discriminant hyper-plane determination each hidden unit in the hidden layer represents a potential discriminator, that may become a true discriminator. The reason is, that weights going from hidden units to output units ultimately control its visibility (see sect. 3.2 and fig. 3).

---

[2] the larger the slope is, the close the sigmoid is to the threshold function.

[3] The error calculated with sigmoid function in the hidden layer decreases whereas the error calculated with threshold functions stays at the same level.

[4] If the model has two hidden layers, the second one is fully connected to the first one and the related outputs are given by the sigmoid function.

3.1 MLP Hyper-Plane Determination

Without loss of generality, we consider the case of an MLP with n input neurons, m hidden neurons and two output neurons coding for two classes. We assume that when $o_1 > o_2$ then a related example belongs to class 1 and to class 2 otherwise. Thus, we assume that the frontier between the two classes is determined by the condition $o_1 = o_2$. This can be rewritten as:

$$\sum_{l=0}^{m} \Delta V_l \cdot h_l = 0 \tag{5}$$

where $\Delta V_l = v_{1l} - v_{2l}$ (the difference between weights going from hidden unit l to output units one and two). From (5) we obtain:

$$\sum_{k=0}^{n} w_{jk} x_k + ln|\Delta V_j / (\sum_{l=0,l\neq j}^{m} \Delta V_l \cdot h_l) + 1| = 0 \tag{6}$$

Thus, if the logarithmic term is constant, then h_j is a potential hyper-plane

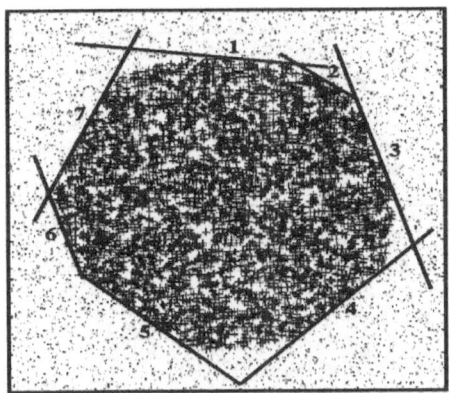

Fig. 2. Plot of classified points provided by a standard MLP with 6 hidden units. The classification problem corresponds to the discrimination of data inside and outside a circle (with only 6 hidden units many points are misclassified). Lines represent MLP discriminant hyper-planes. Note that one hidden unit has constructed two parallel lines (3 and 6). Finally, sometimes at the intersection of lines the frontier has been curved by the interaction of the related hidden units.

discriminator with slope determined by weights going from input layer to h_j. In other terms, the constant logarithmic term shifts the hyper-plane defined by weights going from input units to hidden units.

We have found a sufficient condition for finding potential hyper-plane discriminators. The question now is to know if this is also a necessary condition. Unfortunately, this is not the case. The reason is explained in the following example. From (5), with two hidden neurons and assuming that $\Delta V_1 = \Delta V_2$, it follows that the slope of hyper-plane associated to h_1 is added to the slope of hyper-plane associated to h_2. In order words, the slopes of the two hyper-planes are mixed together. This situation is observed in other cases, but work is still in progress. We illustrate discriminant hyper-planes in figure 2. Note that in some line intersection regions, the interaction of hidden units gives rise to a curvature. Our point of view is that from (6), when the logarithmic term is no longer constant, the curvature appears.

For the time being we can determine hyper-planes equations in simple cases, but we are not able to express the curvature process in terms of parameters. Moreover, we are not able to determine exactly when the slope of several hyper-planes are mixed together. Thus, for standard MLP discriminating boundary characterization, we cannot give an algorithm.

3.2 OMLP and IMLP Hyper-Plane Determination

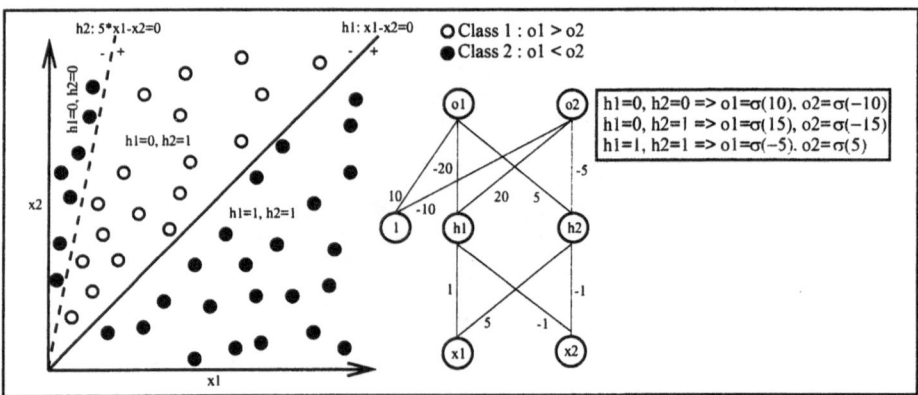

Fig. 3. An example of OMLP network that misclassifies data of two classes (data on the left are misclassified). The reason is that h_2 discrimination is not activated by weight values going from h_1, h_2 to o_1, o_2.

Since each hidden unit output value is binary, each hidden unit defines a potential discriminator. The weights going from the hidden layer to the output layer will determine if each hidden unit is a real discriminator, as shown by figure 3. If the input layer and the hidden layer are fully connected as in the OMLP network, hyper-planes are in general oblique with respect to the axis defined by input variables. When only one input variable (and the bias) is connected to one hidden unit, as in the IMLP model, hyper-planes are parallel to the axis of input variables. For the details of the hyper-plane determination see [15].

3.3 Discussion on the Expressivness Capability

MLP is more adaptive than the OMLP and IMLP models. Indeed, MLP is able to put into the hyper-space discriminating hyper-planes and discriminating hyper-curves by smoothing hyper-planes. Moreover, hyper-planes can be shifted by weights going from the hidden layer to the output layer (cf. (6)).

OMLP is only able to make discriminations by putting oblique hyper-planes into the hyper-space. Moreover, weights going from the hidden layer to the output layer can only decide if a potential discriminator is visible or not. However, when the discrimination between two classes is curved, many oblique hyper-planes can approximate the curvature.

IMLP is like OMLP with the difference that hyper-planes are always parallel to the axis. In a certain sense, this model has less "degrees of freedom" for adaptability with respect to MLP and OMLP.

The OMLP model is interpretable by inspecting the oblique hyper-plane equations. However, if many equations characterize an application, the expressions given by IMLP could be more comprehensible.

4 Justification of MLP Responses in Terms of Symbolic Rules

The goal is to justify MLP responses in terms of symbolic rules. The idea is to use the IMLP network to mimic MLP responses. Note that for MLP response justification, inductive decision tree models are not excluded in the future. Work is still in progress on this subject.

Our classification system is composed by an MLP network and an IMLP network. Classification responses are given by MLP and justification is provided by IMLP in terms of symbolic rules. Typically, for the classification of an unknown example the detailed steps in the process of classification and justification are as follows:

1. MLP is trained on the training set.
2. The MLP output values of the training set are used as targets for training IMLP.
3. Symbolic rules are extracted from IMLP using an exact rule extraction method.
4. An unknown new example belonging to the testing set is fed to MLP and classified.
5. The same example is fed to IMLP; if the response is close to that of MLP return to step 4 (the rules activated by that example justify the MLP classification); else go to step 6.

6. As MLP and IMLP are contradictory, the new example is added to the training set of IMLP with target values given by MLP output neuron values.
7. IMLP is retrained with the new training set until it gives a response close to the MLP response; return to step 3.

The classification and justification process shows that the rulebase extracted from IMLP is not static but is evolving as long as new unknown examples generate contradictions. It could be a limitation for the understandability of justification responses. Moreover, at some step the current IMLP architecture could not be able to learn new examples.

5 Results

Two series of experiments have been carried out. The first one compares the predictive accuracies obtained by each model. The second one illustrates IMLP rulebase evolution as long as unknown examples belonging to a testing set are presented to the classification system.

5.1 Application Description

The Coronary Heart Disease application data was supplied by the Institute of Cardiology of the University of Pisa (Italy) [17].
The total number of patients, symptomatic of CHD (Coronary Heart Disease) included in the study is 884; each patient had a complete angiographic study of the coronary tree. The results of angiography are considered the "gold standard" for correct classification. The patients were thus classified as not having (314 cases) or having (570 cases) CHD. A subject was included in the CHD group when his coronary angiography showed the presence of at least one stenosis greater than 75% of the lumen. For each patient the following 16 variables were recorded: age, sex, height, weight, smoking habits (yes/no, where yes denotes smokers or ex-smokers with less than 10 years of abstinence), family history (yes/no, where yes denotes the presence of CHD in at least one relative), presence of diabetes (according to blood glucose level values and the use of anti-diabetic drugs), arterial systolic and diastolic pressures measured at the admission, serum cholesterol level, total High Density Lipoprotein, α-High Density Lipoprotein, triglycerides, β/α ratio (defined as the ratio between the β-High Density Lipoprotein and the α-High Density Lipoprotein), low density lipoprotein, and fibrinogen.

5.2 Model's Configuration

For each network model we give its architecture: MLP: 16-8-2; OMLP: 16-12-2; IMLP: 16-240-8-2. Note that we use two hidden layers with the IMLP network. All results are obtained by randomly selecting half of the examples for training and remaining ones for testing. This procedure is repeated 100 times for each

model, including the C4.5 algorithm.[5]
For all neural network models the training phase is stopped when the accuracy
on training examples goes over 97%, except for IMLP networks of the second
series in which 100% of examples are learned.
Concerning C4.5 configurations, we tried 360 choice of parameters. We show the
most accurate results that have been obtained.

5.3 First Series of Experiments

Figure 4 shows that the best mean predictive accuracy is obtained by standard
MLP model. However, OMLP is very close. The number of rules obtained from
IMLP is greater compared to C4.5. The number of conditions per rule is slightly
smaller for IMLP.

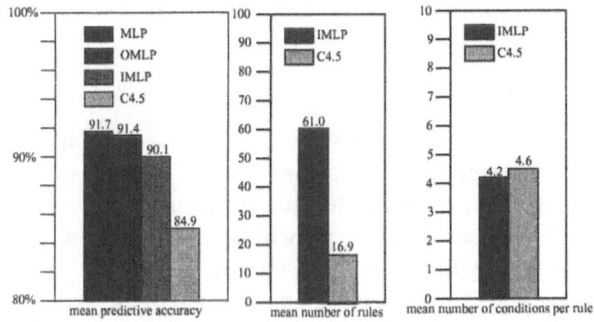

Fig. 4. From the left: mean predictive accuracies (MLP, OMLP, IMLP, and
C4.5); mean number of rules (IMLP, C4.5); mean number of conditions per rule
(IMLP, C4.5).

5.4 Second Series of Experiments

Here the target values of IMLP are the MLP responses. We show in figure 5
the evolution of the rulebase before feeding unknown examples belonging to the
testing set and after having fed all these examples.

Below, we show an example of a rule extracted from an IMLP network that
is the most activated for class Non_CHD. The rule is shown before and after
unknown examples are tested.

1. **If** (*Woman*) **and** (*Diastolic_Pressure* < 0.325) **and**
   (*Low_Density_Lipo_Protein* < 0.446) **and** (*Fibrinogen* < 0.364) **then**
   Non_CHD

---

[5] In this case rules are extracted by *C4.5rules* algorithm.

626

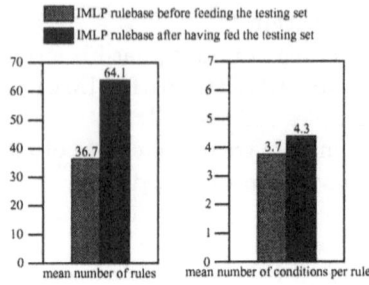

Fig. 5. Evolution of the rulebase extracted from IMLP.

2. **If** (*Woman*) **and** (*Diastolic_Pressure* < 0.246) **and** (*Low_Density_Lipo_Protein* < 0.433) **and** (*Fibrinogen* < 0.399) **and** (*Age* < 0.734) **and** **not** (*Family_History*) **and** (*High_Density_Lipo_Protein* < 0.600) **then** *Non_CHD*

5.5 Discussion

The first series of experiments have shown that for our diagnosis application IMLP is less accurate than MLP. For that reason, in the second series MLP is used for classification and IMLP for justification. If we consult IMLP for classification rather than justification, we will be in the case of the first series of experiments and thus, we will loss some accuracy.

Now, even if the number of rules extracted from IMLP is greater compared to C4.5, it will not represent an irreparable drawback. When an unknown example is presented (MLP and IMLP responses are supposed coherent), only few rules are activated for justification. From this set of rules the user will judge if the justification is plausible with respect to its knowledge.

6 Conclusion

We have examined three MLP models and inductive decision trees with C4.5 for justification of their responses in the context of a medical diagnosis application. The most accurate was the standard MLP model.

We have shown that if a condition is satisfied hyper-planes discriminators can be found from standard MLP model. However, this is only a sufficient condition; besides the MLP model is also characterized by discriminant curved boundaries whose parameterization is not straightforward. Thus, we decided to give justification of MLP responses indirectly by the use of IMLP model.

The final system is composed of one MLP network that makes the decision and one IMLP network that is consulted for the justification. When an unknown new example is presented to MLP and the IMLP response is contradictory, the latter is retrained with the new example added to its training set. Of course, the quality of the evolving rulebase requires verification by physicians.

References

1. White H.: Connectionist non-parametric regression. multi-layer feedforward networks can learn arbitrary mappings. Neural Networks. **3** (1990) 535–551.
2. Breimann L., Friedmann J.H., Olshen R.A., Stone J.: Classification and Regression Trees. Wadsworth and Brooks, Monterey, Calif (1984).
3. Quinlan J.R.: C4.5: Programs for Machine Learning. Morgan Kaufmann (1993).
4. Mooney R., Shavlik J., Towell G., Gove A.: An experimental Comparison of Symbolic and Connectionist Learning Algorithms. Proc. IJCAI-89 Morgan Kaufmann Los Altos. (1989) 775–780.
5. Atlas L., Cole R., Connor J., El-Sharkawi M., Marks R.J., Muthusumi Y., Barnard E.: Performance Comparison Between Backpropagation Networks and Classification Trees on Three Real-World Applications. NIPS 2, Morgan Kaufmann, San Mateo, CA. (1990) 622–629.
6. Tsoi A.C., Pearson R.A.: Comparison of Three Classification Techniques, CART, C4.5 and MLP. NIPS 3, Morgan Kaufmann, San Mateo CA. (1991) 963–969.
7. Mitchell T.M., Thsun S.B.: Explanation Based Learning. A comparison of symbolic and connectionist Learning Algorithms. Proc. 10th Int. Conf. on Machine Learning, Morgan Kaufmann San Mateo CA. (1993) 197–204.
8. Feng G., Sutherland A., King R., Muggleton S., Henery R.: Comparison of Machine Learning Classifiers to Statistics and Neural Networks. Proc. 4th Int. Workshop on Artificial Intelligence and Statistics, Florida (1993).
9. Quinlan J.R.: Comparing Connectionist and Symbolic Learning Methods. Hanson et al, (1994) 445–456.
10. Towell G.G., Shawlik J.W.: Extracting Refined Rules from Knowledge-Based Neural Networks. Machine Learning. **13** (1993) (1).
11. Setiono R., Liu H.: Understanding Neural Networks via Rule Extraction. Proc. IJCAI. **1** (1995) 480–485.
12. Gorman R.P., Sejnowski T.J.: Analysis of Hidden Units in a Layered Network Trained to Classify Sonar Targets. Neural Networks. **1** (1988) 75–88.
13. Andrews R., Geva S.: Extracting Rules From a Constrained Error Backpropagation Network. Proc. of the 5th Australian Conference on Neural Networks, Brisbane (1994).
14. Bologna G., Pellegrini C.: Extraction de Règles d'un réseau PMC à Valeurs d'Entrée Continues utilisant des Unités Cachées à Seuil dans la Couche Cachée. Proc. des Huitièmes journées Neurosciences et Sciences de l'Ingénieur, (1996) 199–204.
15. Bologna G.: Rule Extraction from the IMLP Neural Network: a Comparative Study. Proc. of the NIPS-96 Workshop of Rule Extraction from Trained Artificial Neural Network, (1996) 13–19.
16. Huang H.H., Zhang C., Lee S.: Implementation and Comparison of Neural Network Learning Paradigms: Back Propagation, Simulated Annealing and Tabu Search. Artificial Neural Networks in Engineering 1991: ASME Press, New York. (1991) 95–100.
17. Amendolia S.R., Bertolucci E., Biadi O., Bottigli U., Caravelli P., Fantacci M.E., Fidecaro E., Mariani M., Messineo A., Rosso V., Stefanini A.: Neural Network Expert System for Screening Coronary Heart Disease. Physica Medica 1993: **9** (1); 13–17.

A Hybrid Intelligent System for the Pre-processing of Fetal Heart Rate Signals in Antenatal Testing

Bertha Guijarro-Berdiñas (1), Amparo Alonso-Betanzos(1), Soledad Prados-Méndez(1), Olga Fernández-Chaves(2), Miguel Alvarez-Seoane(2), Francisco Ucieda-Pardinas(2)

(1) Dep. Computación. Facultad de Informática. Universidad de A Coruña. Campus de Elviña s/n 15071 A Coruña. Spain.
(2) Hospital Materno-Infantil Teresa Herrera de A Coruña

Abstract. The Non-Stress Test (NST) is a non-invasive test widely used in obstetrics to assess the fetal state by means of the analysis of Fetal Heart Rate (FHR) and Uterine Pressure (UP) signals. NST-EXPERT is an expert system designed for the diagnosis, therapy and prognosis of fetal state using the NST. NST-EXPERT uses the information extracted from the FHR/UP signals together with information regarding the maternal-fetal context. However, the acquisition of the parameters related to the signals was not automated yet. In achieving this task, the poor quality that the FHR signal presents very often, due to the presence of artifacts and maternal heart rate (MHR) coupling, needs to be faced. In this paper, a hybrid system capable of distinguish between artifacts, MHR and fetal signal to eliminate incorrect data that interfere in the analysis is described.

INTRODUCTION

The evaluation of intrauterine fetal condition is a standard of care in clinical obstetrics. The antepartum tests most often used to screen fetal status are based on continuous observations of the fetal heart rate (FHR) baseline [1] and the empirical associations between intrapartum FHR patterns and fetal outcomes [2]. Among the methods currently employed for antepartum fetal assessment, the nonstress test (NST), has become the most commonly used [3] for assessing fetal well-being, as it appears to be simple and reliable. To realize the NST the FHR and the Uterine Activity (UA) signals have to be obtained by an external FHR monitor, usually in the Doppler mode. The rationale of the NST results from the observation that reactive FHR accelerations associated with fetal movements are good prognostic signs for fetal well-being, but improving the overall diagnostic usefulness of the NST also requires consideration of other FHR parameters, such as FHR baseline, FHR variability, and number and type of FHR decelerations [4]. Although, apparently, there seem to exist clear rules to determine whether, from the signals registered by the cardiotocograph, a NST shows a normal or an abnormal result, in practice the decision is not so easy: there is a big amount of contextual information that need to be taken into account for the diagnosis. Also, the interpretation of the test is subject to intra and interobserver variation, which can lead to incorrect classification of fetal status [5]. Thus, a reliable and consistent method for the diagnosis of the NST is needed.

There have been different systems that tried to approach the clinical domain of the nonstress testing [6-7]. Among those, NST-EXPERT performs a diagnosis of the test and formulates therapeutic plans, while taking into account different aspects of the maternal-fetal context, and incorporating these analysis into a model for predicting fetal outcome. To evaluate the test, NST-EXPERT employs the FHR parameters mentioned above, and later interprets the results taking into account the specific maternal-fetal context. The acquisition of these inputs, however, is not automated yet, and it is the user the responsible of entering the required data. The maternal-fetal information can be retrieved by the system from the hospital databases, and the FHR parameters need to be extracted directly from the cardiotocograph. There have been previous attempts to automate the process of feature extraction using empirical methods[8-9], but detailed studies have not been published to demonstrate their performance compared to that of the experienced perinatologist. In contrast, artificial neural networks (ANNs) are trained using examples that are representative of the problem, not depending on empirically derived means, and are known to recognize nonlinear patterns. For that reason, they seemed appropriate to recognize the FHR-UP features of the cardiotocographic records. In this paper, the module developed to analyze the signal and extract the relevant parameters, using an ANN approach is described. This module is later interfaced with NST-EXPERT so as to supply the parameters needed by the expert system.

MATERIALS AND METHODS

Fetal monitors chart the FHR and UP against time, in a paper printout called cardiotocogram (CTG). In antenatal monitoring, the FHR signal (in beats per minute, bpm) is obtained by Doppler ultrasound, and the UP (in mmHg) signal is measured by an external tocodynamometer (Figure 1). The cardiotocographs used in this research are Hewlett Packard M1315A. This fetal monitors are equipped with a serial communication port RS422A, and employs a sampling rate of 0.25Hz for both FHR and UP signals. Using the RS232 port, the cardiotocograph is connected to a PC, which sends the corresponding request to the fetal monitor, so as to obtain the FHR and UP signals. The information packages sent by the monitor also contain two bits regarding the quality of the FHR signal. In this way, FHR samples with bad quality can be eliminated prior to the analysis. As mentioned above, the FHR signal is obtained using a non-invasive technique, the Doppler ultrasound. However, this method causes frequently several artifacts, due to signal loss, maternal heart rate (MHR) coupling and other factors that are not taken into account by the quality bits of the data package. Still, these artifacts must be detected and later an appropriate action must be taken, such as eliminate them (if for example, MHR coupling is detected) or reconstruct the signal (if, for example, it is part of an acceleration or deceleration), before the analysis of the FHR parameters can be done.

 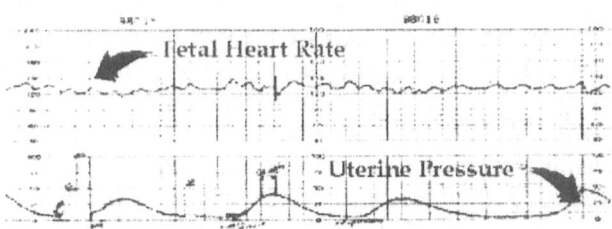

Figure 1. Cardiotocograph, transducers and an example of a cardiotocogram.

In this work, a hybrid intelligent system that combines a small rule-based system and an ANN is presented. The rule-based system is implemented in C language and is encharged of detecting the cases of signal loss, and also of solving the simple ones. The ANN is called in when the rule-based system is not capable of solving the problem using its knowledge, being necessary to take into account the morphological characteristics of the already detected pattern to arrive to a decision. The ANN judges whether the segment belongs to an artifact, it is MHR coupling or it belongs to a fetal deceleration/acceleration. Based on this decision, an algorithmic module reconstructs the FHR signal.

A HYBRID SOLUTION

As previously mentioned, this approach combines several modules. First, the FHR signal is continuously analyzed by an algorithmic module that calculates the FHR baseline and variability. If a loss or sudden change of the signal is detected [10], the module tries to delimit the conflictive segment of signal looking for the next sample that is around the prior baseline, taking into account the FHR variability . Once the segment is located between two samples belonging to the regular tracing (that is, samples located in signal baseline), a rule-based system analyzes it. This system starts by checking the appropriateness of one of the elimination criteria, which are the following:

1.- If in the instant in which the baseline changes abruptly there is no signal loss but a sudden change in the signal, and the difference between both values is higher than the maximum difference between heart beats allowed (0.12 s) [10], then the system assumes that the sampling is incorrect or an artifact has occurred, and the segment is eliminated. The signal is reconstructed between the last point before the abrupt change, and the next valid sample.

2.- If the segment between the baseline sample prior to the abrupt change and the next valid sample (also in the baseline) contains less than 50% of the signal corresponding to that period of time, then it is eliminated and the signal is reconstructed as in the previous case.

3.- If the reconstruction between both valid samples will cause the appearance of incorrect samples (i.e. the difference in heart beats is higher than the maximum allowed), then the segment between both valid samples is also eliminated.

4.- If the temporal distance between the sample prior to the abrupt change and the next valid sample is less than 15s for the descendent segments, (that is, possible decelerations) or 10s for the ascendent segments (that is, possible accelerations), then the signal is reconstructed using both valid samples directly, without further analysis. This criterium is employed due to the fact that only the patterns above this temporal limits are taken into account for the NST diagnosis.

All the criteria described are useful to eliminate part of the artifacts and MHR coupling already mentioned in the second group of signal loss described in the previous part of this paper. If none of this criteria can be applied it will be necessary to determine whether the signal segments between valid FHR samples are MHR coupling, and artifact, or part of a FHR pattern (acceleration or deceleration).

In order to accomplish this task, two groups of signal loss were considered:
1.- Cases with signal loss in which the FHR sample following the loss belonged clearly to the FHR signal. This membership is determined if a) the amplitude of the signal is similar in both sides of the loss (figure 2a), or b) even if the amplitude range is different, the samples after the loss present some degree of continuity with the rest of the FHR signal (figures 2b and 2c). In these cases, the reconstruction of the signal is simple, and it is practically reduced to an interpolation between samples.

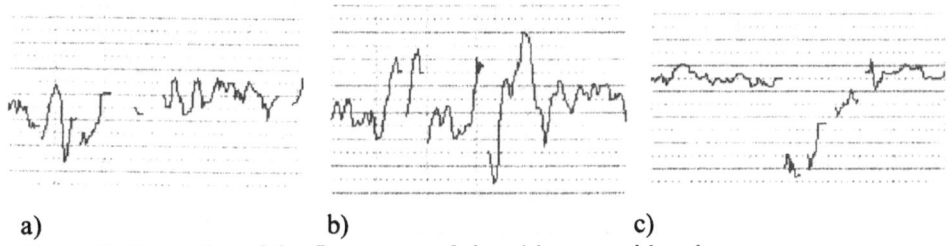

a)   b)   c)

Figure 2. Examples of the first group of signal loss considered.

2.- More complex cases, in which the signal loss delimitate abrupt transitions (falls or recoveries) in the FHR signal (Figure 3). The problem with these cases is that the signal section that appears between the losses can belong to an MHR coupling, can be an artifact or can belong to a FHR acceleration or deceleration. The first two problems are solved eliminating the corresponding section (MHR or artifact), and reconstructing the signal as in the first group cases, while the third problem implies a more detailed processing, maintaining that section and reconstructing the complete pattern. As the presence, number and type of decelerations and accelerations is decisive for the diagnosis of the NST, the method used to distinguish among the different problems, specially in this second group, must be very reliable.

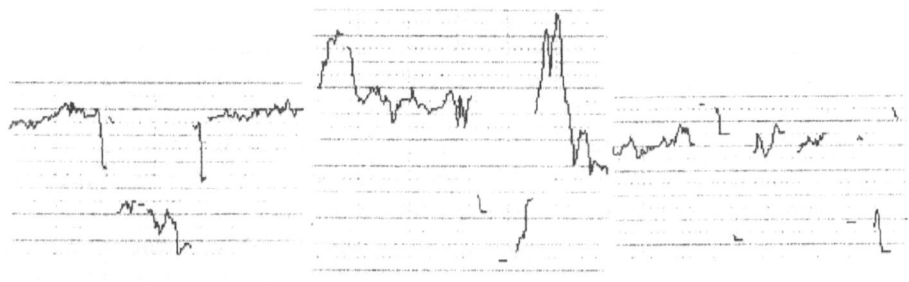

a) MHR coupling.   b) Fetal pattern   c) Other artifacts

Figure 3. Examples of the second group of signal loss considered.

The characteristics and morphology of this second type of patterns were very different, thus limiting the approaches that could be used. The only restriction to process the signals is that at least 50% of the pattern must be present in order to reconstruct the signal [10].

Normally, the perinatologists distinguish both cases using the morphology of the segment. For that reason, certain structural characteristics of the signals were calculated, and later used as inputs of an ANN which is encharged of the classification of the segment, in order to eliminate it or not. The structural characteristics chosen were: a) the proportion between the amplitude and the duration of the segment, b) the slope of the regression line that adjusts the points between the initial point of the segment and the maximum (in the case of a possible acceleration), or the mimimum (in the case of a possible deceleration), c) the slope of the regression line between this inflexion point and the last point of the segment, d) the errors associated to both regression lines, and e) the curvature parameter of a regression second degree polynomium. All these characteristics reflect the morphological differences between MHR and FHR events.

Once all the characteristics are calculated by the algorithmic module, the rule-based system checks them, in order to corroborate that the signs of the slopes and the curvature parameter match with the type of possible FHR pattern detected (acceleration or deceleration). For example, if the pattern is a possible deceleration, because the sample values of the segment are below the baseline, but the signs indicate that the samples have a convex form, then the rule-based system decides that the segment is not part of the FHR signal, and eliminates it.

In order to build the test and training sets, 672 minutes of cardiotocographic recordings belonging to 39 different patients were used. This 672 minutes were analyzed separately by three obstetricians that indicated the fetal patterns (accelerations and decelerations), the MHR coupling segments, artifacts and the parts of the recording in which it was impossible to diagnose anything due to an excessive signal loss. Taking into account the opinion of the clinicians the patterns needed for the training and test sets were chosen and the characteristics were calculated. All cases in which there was not complete agreement between the classification made by the obstetricians were eliminated.

RESULTS AND FUTURE WORK

First, three groups of examples were constructed: a) fetal patterns with signal loss, b) maternal coupling and c) other artifacts. Later on, the 672 minutes of signals were analyzed by the system. In the execution phase previous to the activation of the ANN 174 from 197 (88.3%) artifacts were detected by the rule-based system. Those artifacts not detected by the rule-based system were added to the test set of the ANN together with the maternal and fetal patterns already selected.

The final training set contained 101 examples, 45 of them were MHR couplings, and 56 were fetal patterns to be reconstructed, and which were susceptible of being mistaken with MHR coupling. To develop the neural networks the NeuroSolutions software package was used [11]. The ANN constructed is a two layer backpropagation with six input elements (one per characteristic) and two output elements, indicating if the segment belongs to a maternal heart rate or to a fetal heart rate signal. After trying different architectures, an optimum 6-3-2 ANN using hyperbolic tangent as transfer functions was selected. When testing the network, an example were considered to be assigned to one class (maternal/fetal signal) when the

633

output of the corresponding neuron exceeds the threshold of 0.5. If the example is not assigned to any of the two previous classes, then it is classified as an artifact. The final results are as follows: 93.4% of the artifacts and 70% of MHR coupling were properly detected, while the fetal signal was correctly recognized in all cases.

At present, a second approach that consists on the direct input of the FHR signal to an ANN trained in a supervised mode, using the reconstructed FHR signal as the desired output is being considered. This second approach implies the design and implementation of an ANN to accomplish the tasks that were previously done by the hybrid system (algorithmic module, rule-based system, and static ANN) of the first approach. The input of the ANN is the FHR signal, and the desired output is the reconstructed FHR signal, so the neural net must decide elimination/reconstruction in a unique step.

Finally, both approaches are to be compared, so as to select the optimum method to automate the acquisition of the FHR parameters to be fed to NST-EXPERT.

ACKNOWLEDGEMENTS

This work has been supported in part by project XUGA-10501B96.

REFERENCES

1.- K. Hammacher. "The clinical significance of cardiotocography". In P. Huntingford, M. Hunter and E. Saling, eds. Perinatal Medicine, Academic Press, pp 80-93, 1970.
2.- M.L. Cabaniss. "Fetal monitoring . Interpretation". Ed. J.B. Lippincott Co, 1995.
3.- J.P. Lavery. "Nonstress fetal heart rate testing". Clinical Obstetrics & Gynecology, Vol. 25, pp 689-705, 1982.
4.- L.D. Devoe. "The nonstress test". In R.D. Eden, F.H. Boehm Eds. Assessment and Care of the fetus: Physiologic, clinical and medicolegal principles, Appleton and Lange, pp 365-384, 1990.
5.- F.K. Lotgering, H.C.S. Wallenburg and H.J.A. Schouten. "Inter-observer and intra-observer variation in the assessment of antepartum cardiotocograms". American Journal of Obstetrics & Gynecology, Vol. 144, pp 701-705, 1982.
6.- A. Alonso-Betanzos, V. Moret-Bonillo, C. Hernández-Sande. "FOETOS: An expert system for fetal assessment". IEEE Trans. on Biomed. Eng., Vol 38, pp 199-211, 1991.
7.- A. Alonso-Betanzos, B. Guijarro-Berdiñas, V. Moret-Bonillo, S. López-González. "The NST-EXPERT project: The need to evolve". Artificial Intelligence in Medicine, Vol 7, No. 4, pp 297-314, 1995.
8.- G.S. Dawes, M. Moulden, C.W.G. Redman. "System 8000: Computerized antenatal FHR analysis". J. Perinatal Medicine, Vol. 19, pp 47-51, 1991.
9.- K. Maeda. "Computerized analysis of cardiotocograms and fetal movements". Lilford Eds., In Balliere's clinical obstetrics and gynaecology, Vol. 4, pp 797-810, 1990.
10.- B.R.L. Barlett, A. Murray and W. Dunlop. "Properties of fetal heartbeat intervals during labour". J. Biomed. Eng. Vol. 13, pp 169-172, 1991.
11.- NeuroSolutions v.2.11. Neuro Dimension Inc., Gainesville, FL, USA, 1994.

The Pattern Extraction Architecture: A Connectionist Alternative to the Von Neumann Architecture

L. Andrew Coward

Nortel, 140-13551 Commerce Parkway, Richmond BC V6V 2L1, Canada

Abstract

A detailed connectionist architecture is described which is capable of relating psychological behavior to the functioning of neurons and neurochemicals. The need to be able to build, repair and modify current electronic systems with billions of hardware components has been met through a seldom appreciated aspect of the von Neumann architecture: the hardware architecture is compatible with a simple functional architecture which can support precise translation between functional descriptions at many levels of detail down to the individual hardware components, through the use of a common functional element, the instruction. Existing neural network models have been developed to simulate different aspects of cognition, but do not offer a behavioral architecture analogous with the von Neumann architecture. The brain has experienced intense evolutionary selection pressures analogous with the requirements to build, repair and modify electronic systems. These pressures have resulted in a neural architecture which is compatible with a simple functional architecture based on the use of the common functional element, the pattern extraction. The pattern extraction functional architecture is the basis for intellectual understanding of brain functioning. Physiological structures can be understood as an efficient partitioning of function within the constraints imposed by the properties of neurons. Behavioral phenomena such as declarative memory can be understood in neuron terms, including the reasons for distribution of memory traces. A range of physiological and psychological evidence is discussed. Electronic simulation demonstrates that key psychological functions can be emulated by the architecture.

Introduction

A range of neural networks have been proposed as models for different aspects of biological brain operation. These models employ the perceptron to model the neuron. Information is coded in the connection weights of inputs to the perceptrons, and learning proceeds by gradual adjustment of the weights.

There are several objections to these models. Firstly they require biologically implausible control signals or prior knowledge of the learning task for successful learning. Even Kohonen's topological map scheme for unsupervised learning (Kohonen 1988) has been criticized on the basis that the patterns which are presented to the model must be preselected (Pfeiffer 1996). Secondly, gradual adjustment of weights has difficulty in accounting for permanent, instantaneously created declarative memory.

The most fundamental objection is the gap between high level behavioral descriptions and the problems addressed by neural networks, which tend to be the modeling of sensory or motor functions etc. There has been little work addressing the constraints imposed on the functional partitioning of behavior by the use of perceptron type models. Such work would lead to functional separations which could be compared with physiological structure. Kohonen (1995) has written "Some researchers [have] the goal ... to develop autonomous robots; accordingly, the main functions to be implemented ... are sensory functions, motor functions, decision making, verbal behavior." This list of functions may or may not be a plausible partitioning of the operation of an "autonomous robot" given the use of perceptron models of neurons.

Constraints on Possible System Architectures

Electronic systems are operational today which contain billions of individual hardware elements. The largest telecommunications switches, able to handle 100 thousand telephone subscribers, contain close to 5 billion transistors.

In such systems, the relationship between a high level function (for example, creating a conference connection between three telephones) and the operation of individual transistors can be very complex. However, there are three factors which require a relatively simple hierarchy of relationship between descriptions in terms of function and descriptions in terms of component states. These factors are the need to build many copies of a system, the need to repair component failures, and the need to add features.

To illustrate the requirement, imagine a system which had been created by repeated cycles of random connection of randomly selected transistors, followed by test, until a system was found which performed as desired. The only way to build a copy of such a system would be to duplicate it component by component, connection by connection, simplification like "do the following x times" would be rare. Furthermore, slight differences between the original component and a copy might have large, unpredictable effects on system function. Failures, experienced as a loss of a system function, could not be easily related to the individual component which failed. Adding a new feature (for example, making it possible to have conference connections between more than three telephones) would not be achievable with a simple set of component changes.

The von Neumann architecture is the basis for almost all commercial systems. It has a capability to support the construction of a hierarchy of functional descriptions at many different levels of detail. These descriptions are precisely translatable between levels, down to the level of the individual hardware components used. They are capable of being partitioned in different ways within a level to allow different system functions to depend in different ways on a common set of more detailed building blocks. In addition, these descriptions can be mapped into the physical partitioning of the hardware functionality and into the logical partitioning of the data used by the system.

There are in fact three interlocking architectures in a complex system. The hardware architecture partitions the system into physically separate functions. The data architecture partitions representations of the external world into elements which can be considered separately. The functional architecture is the hierarchy of functional descriptions which makes it possible to identify the individual components associated with system functions, and vice versa. Failure to maintain a simple functional architecture results in systems which are very difficult to build, test, repair, or modify.

The requirement for precise translation and flexible partitioning is severe, and a major factor in the ability of von Neumann based systems to meet the requirement is the use of the instruction concept. Functional descriptions are built at every level from instances of this concept. This common base allows translatability and permits flexible partitioning. A limitation is that the implicit time sequence of descriptions leads to considerable difficulty in designing parallel processing systems.

Analogous pressures to those which lead to simple functional architectures in electronic systems apply to the brain. There are requirements to construct many copies from DNA 'blueprints', to recover from damage, and to add features by a process which only allows simple mutations. These requirements have given an immense evolutionary advantage to brains with a simple functional architecture.

What is the neural architecture of which this functional architecture is a key component? The first step in answering this question is to recognize that a condition for the existence of life is the existence of repetition in the environment. A behavior found to be successful in a given combination of environmental conditions is repeated when the combination of conditions is perceived to repeat. A combination of environmental conditions which repeats is a pattern, and the fundamental role of repetition has resulted in a neural architecture in which pattern extraction plays a similar role to that of instruction in von Neumann systems. The simple functional architecture based upon pattern extraction, although it has emerged as a result of natural pressures, can be the basis for intellectual understanding of the brain, because it provides a simple way to relate psychological behavior to functional descriptions in terms of neurons and neurochemicals. The neural architecture was originally described in Coward 1990. The work reported here is a refinement of that architecture.

Neuron Models

Standard neural networks use neuron models of the perceptron type. Such models have a range of states (not firing, firing at a range of rates) which can be communicated to the rest of the network via connections to other neurons. These connections stimulate or inhibit the firing of their targets to different degrees (or weights). The next firing state of a neuron is determined by its current state and by the ratio of the weighted sum of its active inputs to a fixed, internally determined threshold.

The information content of such networks is coded in the input weights, and learning proceeds by slight modification of weights. There is a layering of neurons, with a layer receiving external inputs separated from a layer generating external outputs by one or more 'hidden' layers. A layer receives the bulk of its inputs from the immediately preceding layer.

An additional neuron model introduced in Coward 1990 has inputs of differing weights from large numbers of other neurons. Neurons are in a passive (or 'virgin') state until a single, unique learning event. This learning event is triggered by a high level of firing of regular neurons in the neighborhood of the virgin neuron but no output firing from the neighborhood. The combination of these conditions lowers thresholds in all local virgin neurons until enough virgin neurons fire to generate an output from the neighborhood. Learning occurs by a large, instantaneous, permanent weakening of the inactive input connections to any virgin neurons which fired, and the setting of the threshold of the neuron at a level which will cause it to fire in the future if a similar combination of active inputs occurs. This mechanism provides a permanent record of a pattern extracted at a particular instant from a perceived object.

New patterns become part of the set which can trigger learning in the same neighborhood in the future. Category learning by similarity starting from a single example is therefore possible as described in more detail later. At that point the neighborhoods will be referred to as modules. While the relative weights of the imprinted inputs remains constant, it is possible for some level of addition of relevant inputs to occur later as described in Coward 1996. This addition plays a role in the management of learning.

Learning is successful with random inputs to virgin neurons, provided the number of inputs is large enough (Coward 1996). Fewer resources are used if there is a statistical bias in favor of inputs which frequently fire regular neurons at the same level in the same module. Coward (1990) therefore proposed that the function of REM sleep is to impose this statistical bias. A rerun of past neuron firing, with a bias towards the recent past as the best available indicator of the immediate future, allows virgin neurons to accept inputs from axons which frequently fire regular neurons at their level in the module. The learning process is thus a variant of the Hebb mechanism, but taking place in two functionally separate stages.

One additional feature of these neurons is that global parameters can lower the threshold of all regular neurons within a particular set of modules. This lowering increases the strength of outputs from modules within the set and, as discussed later, plays a role in global modulation of behavior. Some perceptron type neurons exhibit the same mechanism for similar reasons.

Pattern Extraction Hierarchy Architectural Description

Data Architecture
A data architecture defines the partitioning and flow of information about the external world within a system. In general the data architecture for the pattern extraction hierarchy architecture is similar to the data architecture implied in standard neural networks and is illustrated in figure 1. The primary data connectivity is from one neuron layer to the next, but lateral connections play major functional roles (see below). A neuron can be regarded as being programmed with a pattern which is the combination of the set of patterns

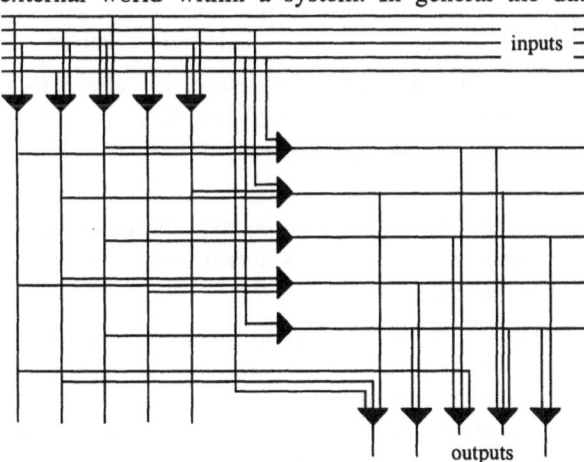

Figure 1. Data Architecture

extracted at the previous level by its inputs. The input weights indicate the relative importance of each input pattern. The firing of a neuron indicates the presence of a high enough proportion of its currently programmed pattern to require attention under current conditions, with the rate of firing indicating the level of attention required. The proportion is the current neuron threshold.

Patterns close to sensory input can be functionally interpreted as simple sensory characteristics of objects or internal body conditions, while deeper in the hierarchy the patterns are complex combinations of characteristics, or relationships between objects. Deeper still the patterns acquire the functional role of behavioral recommendations (' *this* is present, therefore do *that* '). Patterns evolve by weight adjustment or by imprinting; the mechanism depends on the functional domain in which the neuron is located.

Functional Architecture
The functional architecture is shown in figure 2. Sensory preprocessing extracts patterns from raw input. These sensory patterns are stable with respect to external objects (an example would be object color independent of illumination) and are generally established by the imprinting mechanism in the developmental phase. The massive cell death in that period is associated with the elimination of imprinted patterns which are determined not to repeat over long periods and are therefore not object specific.

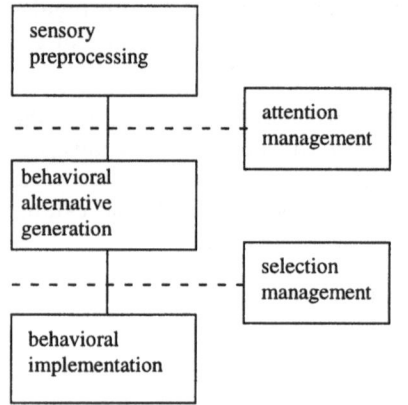

Figure 2. Pattern Extraction High Level Functional Architecture

Attention management allows the set of sensory pattern associated with a single object to proceed to behavioral alternative generation. This function generates a set of alternative responses to the object. Selection management selects a consistent set of behaviors to proceed to behavioral implementation, where a portfolio of muscle movements are managed to produce the selected behavior.

As discussed below, the imprinting model applies in behavioral alternative generation, and in sensory preprocessing particularly during early development. A more perceptron like model applies in behavioral implementation and in the management functions.

The power of the architecture will now be illustrated by descending through several layers of descriptive detail to the neuron operational levels in two areas: behavioral alternative generation and selection management.

The next level of detail for behavioral alternative generation is shown in figure 3. The function contains a number of regions, each of which generates behavioral recommendations of a different type. Thus in response to perceiving a dog, alternative recommendations might be 'kick the dog' from the aggressive region, 'avoid the dog' from a fearful region, 'pat the dog' from a cooperative region, 'say "dog" ' from a speech generating region, and 'focus attention on the dog' from

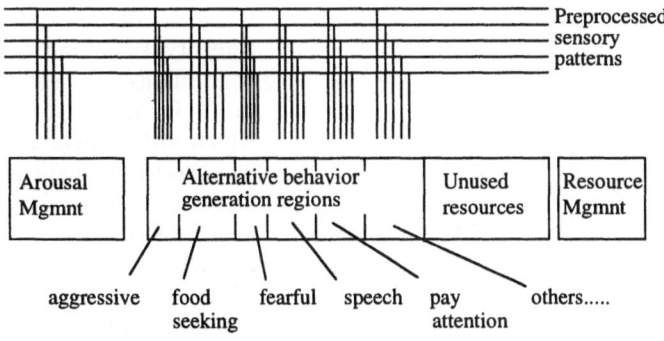

aggressive   food   fearful speech pay   others.....
           seeking   attention

Figure 3. Functional Model for Behavior Generation

an attention management region. These alternatives can be visualized as parallel cascades of neuron firing all driven by the same set of sensory patterns. Arousal management uses a subset of the preprocessed sensory patterns, including patterns extracted from internal physiological states, to control the probability of a strong cascade emerging from different regions. Thus anger lowers the threshold of all neurons in the aggressive region and increases the strength of any cascade. Hunger lowers neuron thresholds in a food seeking region. One feedback route is that an aggressive recommendation including a recommendation to become more angry. Generation of behavioral recommendations consumes neuron resources. Resource management assigns more resources in response to the extraction of a resource-depleted pattern . This management function takes the form of a resource map with an active boundary, and resources are assigned from the unused resources corresponding with the current position of the active boundary (Coward 1990).

category 1	category 2	category 3	category 4	category 5	category 6	category 7	More Categories

Figure 4. Functional Model: Behavior Generation Region

The next level of detail is generation of behavioral alternatives within a region as shown in figure 4. A category module extracts the presence of an object similar to an existing category, or can be created in response to an object which has no such similarity. Detailed category output can be unique to an individual object in a particular situation since patterns extracted from the individual object and situation will be included in the neurons imprinted to generate the output. This uniqueness allows control of the particular type of (for example) aggressive action to be taken. Some categories may also input to other categories: a leg may be an object/action recommendation in its own right or a component pattern to a dog category.

The next level of detail is the internal structure of a category module as shown in figure 5. The memory trace of, for example, a new dog, consists of patterns extracted from previously perceived dogs plus a small set extracted from the new dog. This trace exists within all the behavioral generation regions which generated a recommendation, each such region therefore contains an independent module which can be regarded as a 'dog recognizing' module. The new patterns in each module add to that module's definition of 'dog'. An unknown animal would not create a cascade in any existing module strong enough to produce an output or trigger imprinting. This condition triggers assignment of a new, randomly connected module in which imprinting occurs until an output results. This animal thus becomes the initial

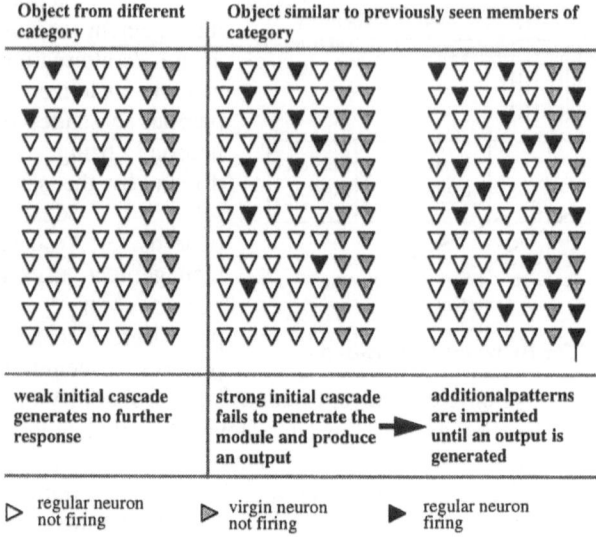

Object from different category	Object similar to previously seen members of category	
weak initial cascade generates no further response	strong initial cascade fails to penetrate the module and produce an output ➤	additionalpatterns are imprinted until an output is generated

▷ regular neuron not firing ▶ virgin neuron not firing ▶ regular neuron firing

Figure 5. Learning within a Category Module

member of a new category. Experimental connections are made to implementation, which will be evolved by pleasure and pain as discussed later. The new category will be evolved by perception of any similar animals in the future. In practice provision must be made for merging duplicate categories within a region when, for example, the two animals of a particular type first seen are extremes within the type (Coward 1996). The same reference describes electronic simulation demonstrating that imprinting within the proposed architecture is an effective means of sorting and recording perceived objects in categories on the basis of similarity, with no guidance or feedback of any kind, and that dream sleep significantly improves the efficiency of the process.

Turning now to the more detailed descriptions of the selection management function as shown in figure 6, the alternative recommendations are directed to the selection management function as well as towards behavioral implementation, but are blocked in the latter direction. Each alternative enters a pipe within selection management which is specific to the type of behavior. Competition between the active pipes in general results in an output from one (or no) pipe. An output opens

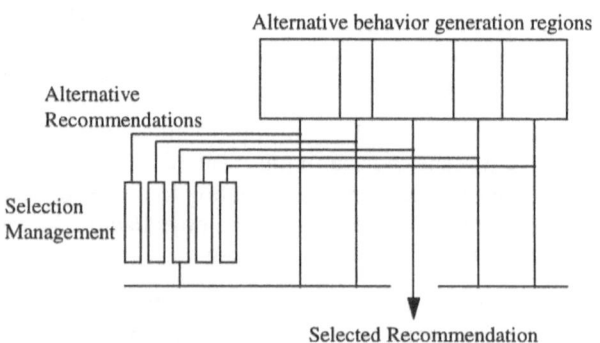

Figure 6. Selection Management High Level Functional Architecture

the gate to allow the recommendation which provided the successful input to proceed towards behavioral implementation. At the next level of detail shown in figure 7, there is cross inhibition between all pipes. A recommendation in one pipe inhibits all other pipes. There are two control functions. One modulates total activity: if the cross inhibition process is resulting in more than one recommendation getting close to output, all recommendations are reduced in strength until a single clear recommendation emerges, or none. Faults in   this   mechanism would lead to rapid, erratic shifts in behavior as conflicts are resolved at the muscle control level. The other   control function is   pleasure/pain. If patterns of success are

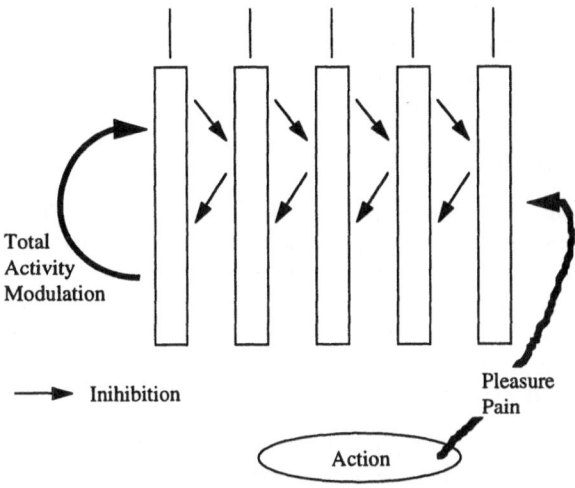

extracted from the result of a behavior, pleasure is triggered which weakens all recently active inhibitive connections in the successful pipe. Pain strengthens such connections. The effect is to modulate the probability of future acceptance of similar behavioral recommendations.

Figure 7. Selection Management Detailed Functional Architecture

At a detailed level, perceptron type neurons are appropriate for the function, except that their thresholds must be increased or decreased in accordance with the activity modulation function, and the learning mechanism does not need a detailed feedback of an expected result. Connections from layer to layer within a pipe are stimulative.

A simple simulation of this function has demonstrated that because output from a category in the behavioral generation function is generally unique to the individual object, a simple pain mechanism acting on provisional connections to behavioral implementation eliminates categorization errors (Coward 1996). Such categorization errors could be of the type identifying a "dog-like cat" incorrectly as a dog.

Physiological Architecture
The mapping of functional architecture into brain physiology is shown in figure 8.

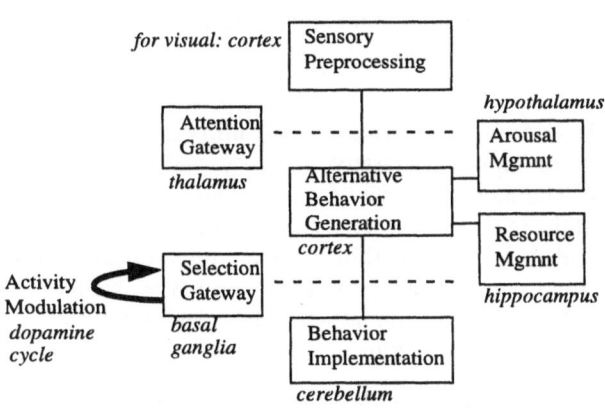

Figure 8. Physiological Architecture

The thalamus attention gateway uses the primal sketches described   by Marr (1980) to compete for access of all the sensory input within the domain   defined   by the primal sketch to the cortex   generating recommendations   for bahavior   (Coward 1997). The mechanisms described by Taylor et al (1993)   form   a plausible   detailed description of this function. The role of the hippocampus as a map of cortex resources assigning them to cortex modules as required results in the development of an implicit time sequence of memory. Damage to the hippocampus or to structures linking it to the cortex can therefore result in time related deficits such as the loss of a time slice of memory and the inability to create new memories observed in Korsakov's syndrome (Coward 1990). Defective operation of the dopamine feedback loop in the basal ganglia results in the rapid, erratic behavioral shifts observed in Tourette's syndrome (Coward 1990).

Physiological and Psychological Evidence

Some of the most persuasive evidence is the way in which the pattern extraction hierarchy architecture makes it possible to create a systematic, integrated description of an extensive range of psychological and physiological evidence (Coward 1990). A wide range of detailed pattern management mechanisms can be expected at the neuron level, but there is a clear requirement for an imprinting mechanism triggered by firing of other neurons in the neighborhood for neurons in the cortex and the hippocampus to account for declarative memory.

Permanent, instantaneous weakening of the weights of all inactive inputs could be followed by elimination of the synapses. There is some evidence that learning by initial excess and subsequent reduction of neuron connectivity occurs in the human cortex (Huttenlocher et alii. 1982). Synaptic density has been observed to be greater in rats raised in a stimulating environment (Turner et al. 1988) and there is evidence of a higher proportion of synapses in cortex regions undergoing active learning (Greenough et alii 1988). There is also evidence that synaptic plasticity depends on the activity of nearby neurons (Fazeli 1992).

At a structural level, the observation of a predominance of local connectivity resulting in a column like appearance in the cortex in some areas is as expected for category modules. The observation that long range connectivity in the cortex is stimulative but a proportion of local connectivity is inhibitive (Douglas and Martin 1991) is also as expected. One role of the local inhibitive connectivity is inhibition of imprinting if a module output is being produced. Observation of "multiple, parallel, segregated circuits ... [which] receive inputs from several separate areas, traverse specific portions of the basal ganglia and thalamus and project back upon one of the cortical areas providing input" (Parent and Hazrati 1995) is the structure predicted by the architecture.

At a phenomenological level, the architecture provides a straightforward explanation for the inability of local damage to remove declarative memory, and its impact on motor abilities and personality. The memory trace of a particular encounter with, say, a dog, is distributed over all the regions in which a category module produced a behavioral recommendation and distributed within each module. Local cortex damage would generally be within one region of one behavioral type, leading to personality shift and/or motor deficits, but to eliminate the total memory trace of an event, damage would have to occur to similar modules in multiple regions.

The proposed role of REM sleep in memory would predict (Coward 1990) that REM sleep deprivation coupled with major activity in a particular domain should result in behavior atypical of that domain. This prediction has not been tested experimentally, but anecdotal evidence that intractable labor or political disputes are sometimes settled under conditions of protracted negotiation with sleep deprivation is in agreement with the prediction.

Conclusions

Evolutionary pressure has resulted in a neural architecture with a relatively simple functional architecture based upon the concept of pattern extraction. This functional architecture can be the basis for intellectual understanding of the brain, relating psychological phenomena to neural physiology. The same architecture can also be the basis for construction of an electronic system which would exhibit equivalent phenomenology to the human brain.

References

Coward, L. A. (1990), Pattern Thinking, New York: Praeger.

Coward, L.A.(1996), 'Understanding of Consciousness through Application of Techniques for Design of Extremely Complex Electronic Systems' Towards a Science of Consciousness , Tucson, Arizona.

Coward, L.A.(1997), 'An Integrated Theory of the Phenomenal Content of Consciousness', to be published.

Douglas, R.J. and Martin, K.A. (1991), 'A Functional Microcircuit for Cat Visual Cortex', Journal of Physiology, 440, 735-769.

Fazeli, M.S. (1992), Synaptic Plasticity: on the trail of the retrograde messenger, Trends in Neuroscience 15, 4, 115-7.

Greenough, W.T., Hwang, H.M. and Gorman C. (1988), 'Evidence for Active Synapse Formation, or Altered Postsynaptic Metabolism, in Visual Cortex of Rats raised in Complex Environments', Proceedings of the National Academy of Sciences USA .

Huttenlocher, P.R., de Courtin, C., Gary, L.J. and van der Loos, H (1982), 'Synaptogenesis in the Human Visual Cortex - evidence for synapse elimination during ordinary development', Neurosciences Letters 33, 247-52.

Kohonen, T. (1988), Self Organization and Associative Memory, Berlin: Springer-Verlag.

Kohonen, T (1995), Self-Organizing Maps, Berlin: Springer-Verlag.

Marr, D. (1982), Vision , New York: W.H. Freeman.

Pfeifer (1996), Symbols, Patterns and Behavior: Beyond the Information Processing Metaphor, in Encyclopedia of Microcomputers, 17, 253-75, New York: Marcel Dekker.

Taylor, J.G. and Alavi, F.N., (1993), 'Mathematical Analysis of a Competitive Network for Attention' in Mathematical Approaches to Neural Networks ed. J.G. Taylor (Elsevier).

Turner, A.M. and Greenough, W.T. (1985), 'Differential Rearing Effects on Rat Visual Cortex Synapses I: Synaptic and Neuronal Density and Synapses per Neuron', Brain Research 329, 195-203.

A Two-level Heterogeneous Hybrid Model

Nicolae B. Szirbik
Department of Computer Science
"Politehnica" University of Timisoara
V. Parvan 2, 1900 Timisoara, Romania

Abstract

A new type of architecture for hybrid symbolic-connectionist systems named the multimodular heterogeneous model is proposed. In this model the reasoning process is driven by the symbolic computation, and the connectionist computations are only secondary processes, that intervene to unlock the symbolic process when it is stuck. The proposed architecture is inspired from the logic-based compositional model, formed by collections of heterogeneous modules, communicating via a unique, "esperanto-like" language. Conclusions about experiments described and the possible enhancements are drawn.

1 INTRODUCTION

There are two major trends in modern theorizing about reasoning: one symbolic and another subsymbolic. Each is usually described by its tendencies rather than any definitive property. Symbolic reasoning is generally characterized by hard-coded, explicit rules operating on discrete, static tokens, while subsymbolic reasoning is associated with learned, fuzzy constraints affecting continuous, distributed representations.

Hybrid systems, which amalgamate both paradigms, are a promising avenue towards developing models for intelligent systems. Even though some recent papers [Sun et al., 1995] propose models and functional systems, a lot of issues are yet unclear, especially those involving the theoretical objectives and the application domains.

Starting from the idea that the symbolic part of the system is leading the reasoning process, this paper is presenting a model for an automated reasoner which has a secondary connectionist subsystem. In contrast, most of the models that have been published up to this point, consist of multiple differentially specialized and heterogeneously represented expert modules each of which constitutes an equal partner in accomplishing the general task.

The proposed model uses analog processing (implemented by connectionist modules) only for unlocking the symbolic reasoning process. The results have no certainty values; they are plausible or at least, make sense. Chapter 2 renews an old idea, that is, the reasoning process can be simulated with a relaxation network; the idea is used to implement the connectionist part. Chapter 3 describes the architecture of the model, and the meta-processing, subprocessing and co-processing issues. Chapter 4 examines the structure and implementation of a connectionist module. Chapter 5 concludes about the possible enhancements and the auto-learning features.

2 REASONING AS A TRAJECTORY OF A DYNAMIC SYSTEM

In [Smolensky, 1986] it is demonstrated that a symbolic reasoning chain (at a conceptual level) may be interpreted as a trajectory of a linear dynamic system (for example, a neural network with linear units). The initial premises can be viewed as a point in the multidimensional state space of such a system, and the final conclusions are "compressed" in another point. If the weights are computed in such a way that starting from a point in the state space (representing premises), another point can be reached (representing conclusions) by letting the system evolve on a trajectory, an isomorphism between the conceptual world and the neural world can be stated.

Smolensky also shows that this isomorphism does not hold for nonlinear dynamic systems. Speaking in connectionist terms, for implementing a nonlinear dynamic system, a fully connected network with nonlinear units is needed. If linear units are used instead, the state space will be the entire euclidian space R^N (the network has N units). For nonlinear units, using the hyperbolic tangent as unit transfer function, the trajectories will be limited to the a hypercube with the edge length = 2.0, centred in the origin. The network can be trained to learn contiguous trajectories. If the network is used as a heteroassociator [Pineda, 1988], it has to learn a pair of trajectories, one for input, and one for output, and these can be in different dimensional hypercubes. In this case, a small part of the units are used for input (they are forced to follow the input trajectory), a small part for output (they give the output trajectory) and the rest are hidden units. The important fact is that the network is fully connected. To establish an overall image, it must be mentioned that an "internal trajectory" is followed by the system in a N dimensional hypercube (where N=I+O+H, I-the number of input units, O-the number of output units, and H-the number of hidden units).

Although the isomorphism does not hold for nonlinear units, for very small trajectories within the hypercube it can be considered that these trajectories are almost similar with the trajectories of linear systems. If the trajectories are in the centre of the hypercube, this hypothesis is stronger, because the transfer functions are quasi-linear in this region. With this approximation, the presented ideas have led to the very attractive possibility of reasoning simulation using neurodynamic networks. There are three main problems which have to be solved:

    i. - how the network is trained?

    ii. - how the network is used?

    iii. - which is the interface between the symbolic world and the connectionist representation to use?

i. Training needs patterns and a learning algorithm. If the algorithm can be picked from the class of learning algorithms for dynamic recurrent nets, the training base must be built from the conceptual world. The simplest method is to associate a sequence of symbols (sentences) to a trajectory. A text (sequence of sentences) presented to the input units, may be interpreted as a trajectory in the input hypercube, from which a trajectory in the output hypercube is obtained. This can be associated with another text. Some points of the trajectories are associated with symbols of the domain of discourse, and if these points are reached, the symbols are activated. If the input and the output texts have the same domain of discourse, then the input and output hypercubes will have the same dimensionality and the same associations between points (real vectors) and symbols. Finally, the training base is formed by pairs of texts which have each an associated trajectory.

ii. The neurodynamic network can be used in two ways. In the first one, the system is stable in a point, and a symbol is appearing at the input. The input units are forced to move to the associated point (which must be near the stable point) and the output units are moving to another stable point (in an optimistic scenario, it will be a point on a learned trajectory). This is associated with a symbol (if possible) which will be the result of this kind of computation. This mode is named ONE-MOVE-on-TRAJECTORY (OMT). In the second mode, an entire text is presented to the system, and the output pair is generated. This mode is named ALL-MOVES. In both modes, the choice of the initial stable point is very important.

iii. The author of this paper has studied the possibilities of translating the symbolic structures into activation vectors for input and output units [Szirbik, 1995a]. Tensorial translation schemes [Smolensky 1990], [Smolensky, 1992] and their derivatives have been used. The symbols were structured in *conceptual graphs* to capture the sentence semantics in an efficient way. The tensorial scheme is very useful, due to possibility of inverting the tensorial function used for direct translation (a *vectors-to-symbols* translation scheme is also necessary).

If the number of units is small, the trajectories can be easily trained, and atomic symbols (A, B, C,...) without a semantic are associated to the points on these trajectories. It is more complicated to upgrade the symbols into sentences with semantics, and transform the sentences into activation vectors. In order to solve this, the sentences were internally represented as conceptual graphs. The tensorial function used [Szirbik, 1995a] results in a fifth rank tensor, with a high dimensionality (10^3) and the resulting network was very large (10^4 units). The learning problem becomes incommensurate in this case (it is important to know that these learning algorithms have $o(n^4)$ complexity - n being the number of units) and the needed computing power for training reaches the order of 10^{20} FLOPS. Also, the memory for the weight matrices reaches the GigaByte magnitude (that can be feasible). All this theoretic and experimental conclusions have led the researcher to abandon this type of architecture (known as the *monomodular model*).

An essential problem is the binary vectors learning. Tensorial translation schemes linked with linear independent representations yield binary vectors. If the sentences are associated with these type of vectors, the resulting trajectory for a text is not in the centre of the hypercube, and is not small. Actually, it has the form illustrated in fig.1.

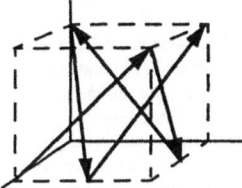

Figure 1: Hopping trajectory

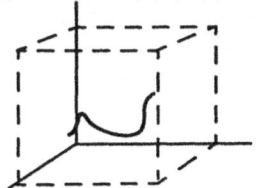

Figure 2: Smooth trajectory

Some papers [Pearlmutter, 1990] clearly indicate that learning very discontinuous trajectories is almost impossible. But is very easy to learn a trajectory as the one depicted in fig.2. The main conclusion of this stage of research was: use small networks, grouped in some modular architecture. Also, is important to transform the hopping trajectories into smooth trajectories.

3   THE HETEROGENOUS MULTIMODULAR MODEL

One of our goals was to define an architecture which benefits from symbolic evaluation processes and all advantages that can be derived (clarity, consistency, soundness). The proposed architecture performs the reasoning process mainly through purely symbolic evaluations, and only when these processes are stuck, a connectionist part is activated. In order to make a closer link with the symbolic level, the connectionist part is "looking" at every step of the symbolic part. In this case, the neural part of the system is represented by a collection of connectionist modules (a *homogeneous multimodule* structure).

For the symbolic part, an existing cognitive framework is used. A multimodular heterogeneous reasoning systems is DESIRE [Treur, 1990]. It can comprise modules with different evaluation algorithms, knowledge representation schemes, even different logic values (two-value, three-value, or multi-value logic). In DESIRE the exchange of information between different modules is solved by a special module (the "generic" module) using predicative structures (PSs). A predicative structure is basically a simplified conceptual graph, with a rigid structure:

<predicate>(<first_argument>,<second_argument>).

A PS can represent a simple sentence as "a1 pred a2". More complicated sentences can be transmitted as sets of PSs, for example the set {p1(a1,a2);p2(a2,a3)} represent the sentence "a1 p1 a2 p2 a3".

Although the framework allows different knowledge representations and inference engines, for the hybrid architecture, all symbolic processing modules have the same rule based, forward-chaining implementation scheme. This provides both simple structure and more control. The "generic" module, which performs synchronization and communication (the metaprocessing, [Hilario, 1995]), is changed to a simplified blackboard with step based refreshing algorithm (BB1 - in fig.3). The "generic" module of DESIRE provides goal-driven reasoning schemes, based on a compositional formalism, but a simplified productive, data-driven reasoning is sufficient for hybrid experiments. To use the system in this fashion, requires the user intervention for termination and occasional hints. The hybrid system has the structure illustrated in fig.3.

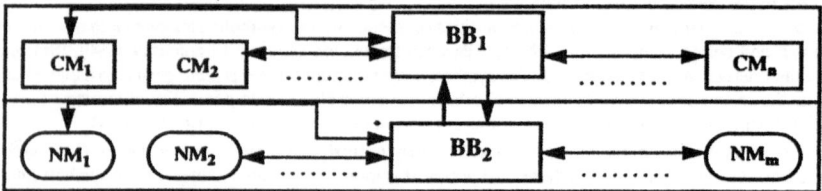

Figure 3: The compositional hybrid architecture

The CM$_i$ are the "classical" reasoning modules, which have all the same external behavior. They accept sets of PSs as input and in a one step reasoning, generate a set of PSs as output. All modules are synchronized to accept the input and generate the output in a ONE-MOVE-REASONING-STEP (OMRS). The input is taken from the blackboard (where the user inserts the starting PSs and hints during the reasoning process). Not all modules can make an OMRS, because the blackboard usually contains a limited set of PSs, which can be considered as inputs. For the overall functioning, the management of the blackboard is the most important aspect. This can be described as a deletion strategy of the blackboard. Two strategies can be used:

    Ia. After an OMRS, the blackboard is written with the resulting PS's from the modules, and a new OMRS is started with this information. After the modules have read the blackboard, all PS's are deleted.

    IIa. Only those PS's which are "consumed" by modules are deleted. The rest are remaining on the blackboard as a "history" and can be used in the following steps.

Both strategies can induce two undesired phenomena. The first is *cycling*. The same module, or a ring of modules repeats infinitely the same sequence of PS's. It is demonstrated [Van Langevelde, 1991] that the cause of cycling is not the productive reasoning scheme or the structure of the model, but one or more redundant reasoning chains within one or more modules. This phenomenon can be avoided by constructing the knowledge bases carefully. The second problem is *starvation*. This happens when after an OMRS, the blackboard is empty, or the existing PS's can not be used by existing modules. If the number of modules is small, the problem can be easily fixed by adding some missing rules. If the number of modules is great, the problem becomes increasingly difficult, and a hint from the user, which is following the reasoning process step by step, can unlock the situation. This can be also difficult, because for a long process (hundreds of steps) and for a large amount of knowledge, the user may be totally confused, and even the designers of the system may not be able to offer the appropriate hints.

The proposed model is intended to resolve the starvation problem. On a second level, the system is enlarged with a set of connectionist modules (NM$_i$ - "neural modules"), which are working in the same rhythm with the CMs.

The internal behaviour of the NMs is based on neurodynamic systems, which follow, during the reasoning process, a trajectory in a multidimensional hypercube. At the beginning the recurrent nets are fixed in starting points, which can be randomly selected, or to be the first point of the most frequent trajectory. All modules (CMs and NM's) are synchronized to work simultaneously step by step (OMRS for CMs and OMT for NMs). Every module can accept as input only a limited set of sentences from the domain of discourse. This restriction is more evident for NMs where the accepted sets of sentences depend on the limited number of learned trajectories, due to the learning capacity of recurrent networks. From the entire set of NMs, only a little part will do an OMT, because they can read the inputs from another blackboard (BB2 on fig.3.). This blackboard is dedicated for NMs, and contains PSs imported from BB1.

After an OMRS, all the generated PSs written on BB1 are transmitted to BB2, to which NMs have access. In the same step (a step can be divided in two phases, one for CMs, and another for NMs), all "input activated" NMs will generate results, which are written on BB2. Some protocols to combine the information on both blackboards can be used, along with a deletion strategy:

I. If a starvation situation is NOT encountered on the symbolic level, all PSs from BB2 are erased.

II. The existing PSs are used in the next OMT (only by NMs), or like in BB1 strategy (IIa.), only those which are not used are kept.

III. All PSs from BB2 are transmitted to BB1, to be used in the next OMRS, where they are kept according to the deletion strategy previously imposed for BB1.

IV. NM level generated PSs are used only on the symbolic level, that is, transmitted to BB1 and erased from BB2.

Note. In all cases, if a starvation situation appears, the PSs generated on the connectionist level, will be used to try to unlock the reasoning process.

These strategies can be classified in two classes: one is conceived to not make heavy interferences between the module levels (I. and II.) and the other (III. and IV.) implies a lot of information exchange between CMs and NMs. As our basic goal is to unlock symbolic computation, we have selected the strategy I. The experiments revealed some advantages of this "Occam's razor" approach. However, a definitive solution cannot be given yet, due to insufficient experimentation with strategies II., III and VI.

The advantage of this type of structure is that hints can be offered by the connectionist level. For a very simple system, the hints can be programmed by the designer. For a purely symbolic multimodular system, the starvation points can be easily identified, and NM's can be inserted to provide the needed hints. In a very simple and illustrative example, for a two CMs system, the following scenario can take place:

```
p1(a1,a2) -> CM1 -> p2(a3,a4)
p2(a3,a4) -> CM2 -> p3(a5,a6)
p3(a5,a6) -> CM1 -> p4(a7,a8)
** starvation
** is needed: p5(a9,a10) -> CM2 -> final_conclusion
```

A neural module must generate the needed PS in the right step. The NM is to be trained with the following trajectory:

```
move1( input == p1() / output == don't_care )
move2( input == p2() / output == don't_care )
move3( input == p3() / output == don't_care )
move4( input == p4() / output == p5(a9,a10) )
```

As mentioned before, it is very important to have the associated trajectory (with the text p1-p2-p3-p4) as a quasi-linear succession of points in the middle of the hypercube. The output trajectory must be quasi-linear too. This restriction imposed a supplementary translation, from binary vectors (resulted via tensorial methods) to real vectors. Initially this mapping was made by hand.

Some design aspects have emerged from the experiments. The points in the hypercube which are associated with PSs lined in a reasoning sequence must be close one to another. That is important from the semantic point of view, because sentences within a semantic cluster associate with clusters of points. A developing trajectory can leave the learned pattern to move on another learned pattern, but only when the patterns are intersecting or have a very close section.

If the number of modules is large, and the system has a lot of starvation moments, the learning example presented above is not feasible to develop the system with NMs. A more general learning methodology is needed. The reader can observe that it is not necessary to use such a intricate scheme with neural networks. The PSs sequences generated by NMs can be easily implemented with a very simple module, based on an identification/generation algorithm. In [Van Langevelde, 1991] a module architecture for hint generation has a formal description for DESIRE. But the benefit of using neural networks is more subtle and lies in the nondeterministic behavior of the system.

Recurrent networks tend to be nondeterministic when the input is somewhat different from the learned trajectory. Atractor dynamics can even yield chaotic or oscillatory behavior. The bifurcation phenomena can be encountered too [Pearlmutter, 1990]. OMT tactics can generate unexpected results. This is a strong reason to not transmit information from BB2 to BB1 at every step (as stated in communication protocols III. and IV.). In the our view the nondeterminism is the interesting aspect to be studied. This is supported by some experimental results which show that trajectories followed by NMs during a reasoning process, can be very close to points which have associated PS's managed in BB1 in the same time at the conceptual level. That is, the NMs are "looking" at CMs and tend to remain in the same semantic clusters. If the CMs are "moving" in another semantic cluster, the NMs are following them. This is a very interesting property, and has to be theoretically studied. If the number of NMs is enlarged very much, it is plausible to believe that the NMs level can be useful to the CMs level. Our experiments involved only a small number of NMs and CMs, but it was clear that the necessary hints are appearing on BB2.

The first experiments involved two or three CMs and only one or most two NMs. Preparing and learning the NMs and especially mapping the points to PSs by hand is a very difficult task, and a new, somewhat holistic learning strategy was used when the increase of the number of NMs was seen as a solution to enhance the system performance (measured in "right hint at the right moment"). If the CM's are working alone, without the NMs level, with hints offered by the user, the current contents of BB1 can be traced in a history table. The resulting PSs are in a temporal sequence. For repeated experiments a set of these sequences can be built. The sequences are used to construct the learning trajectories, dividing them between a population of nontrained NMs. No other rules for constructing the learning patterns were applied yet. Only the semantic and necessary sense preserving of the texts impose some structure, and as a result, the sentences from input and output are representing a sort of dialogue.

Using these training bases for a a larger set of NMs (a dozen) with limited vocabulary (15 words/NM) and two or three learned trajectories per NM, the overall behavior and performance of the system was almost the same, and in some cases better. This means more hints were generated in the right moments. But the majority of the useful hints were generated a couple of steps too late and in a few cases earlier than needed. Apparently, some changes in the blackboard communication protocols and deletion strategies could resolve this problem. But this is not so straightforward. For example, if the BB2 content is copied in BB1 and kept until is consumed, it often produces undesired cycling at the NMs level, because at every step, the content of BB1 is transmitted back to BB2. If the NM generated PSs are kept separately in a special blackboard (BB3) the protocols for the entire system will became very intricate, while our intention was to define a protocol as simple as possible. A complicated scheme with a greater number of blackboards and access priorities and restrictions, needs a strong formal description, and the initial goal of the research (the hybridization of "classic" and "neural" modules) is lost.

4 THE INTERNAL STRUCTURE OF A NEURAL MODULE (NM)

A NM must have a very simple architecture, centred around a recurrent neural network. The dimensional restriction is necessary, because the intended developments of the system, require the enlargement of the NMs level with new modules. The modules must be small, fast and easy to train. That considerations lead to a RNN with a small number of units. The generic architecture is illustrated in fig.4.

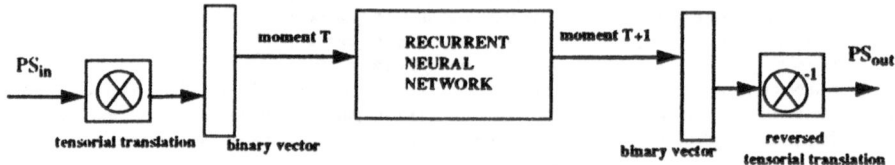

Figure 4: The NM's first structure

Two problems appear in the above presented structure:

i. -the trajectories are not quasi-contiguous

ii. -the dimensionality of the binary vectors is too high. For example for PSs with pred(a1,a2) structure, where the cardinalities of the vocabulary are:

```
card{pred|pred in Voc(NMi)}=5
card(a1|a1 in Voc(NMi))=5
card(a2|a2 in Voc(NMi))=5 (--Voc(NMi) is the vocabulary of NMi)
```

The domain of discourse with 5 verbs, 5 nouns and 5 adjectives is very limited (10-20 sentences) restricting the number of trajectories to 2-3. Using a linear independent vector representation:

```
pred1 -> [1,0,0,0,0]
pred2 -> [0,1,0,0,0]
    ....
pred5 -> [0,0,0,0,1] (similarly for a1 and a2)
```

and a tensorial translation:

$$\bar{r} = r_p(pred) \otimes r_1(a1) \otimes r_2(a2)$$

the dimesionality of the vector $\bar{r}$ is 125. A RNN with 125 units for input and other 125 units for output needs H=200+500 "hidden" units (depending on the number of trajectories to be learned and their forms). This RNN needs 4+12 hours for 500 epochs on a distributed system consisting of 80 SPARC processors, using a parallel version of the learning algorithm. But it is even more difficult is to train the RNN directly with the binary vectors resulted from the tensorial product. An improved architecture resolved both problems, reducing the dimensionality and making the trajectories quasi-contiguous (see fig.5.)

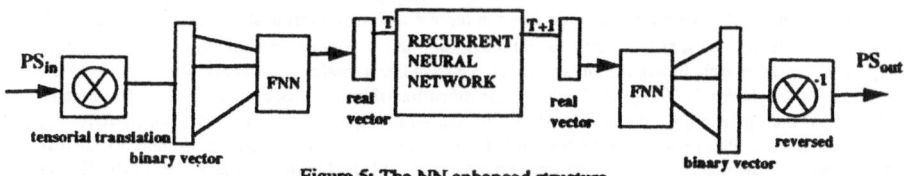

Figure 5: The NN enhanced structure

Two new neural nets are inserted, one in the input channel and another in the output channel. Their role is to transform the "hopping" binary trajectories into "smooth" trajectories evolving in hypercubes with a smaller dimensionality. The selection of mappings between the binary vector space and the reduced space remains a major problem, but in experiments, a 3-dimensional hypercube was used, and the points were selected by hand, trying to construct semantic clusters and "smooth" trajectories in the same time. An automation of this process is not very difficult, using clustering techniques for sentence semantics [Memmi, 1992] and imposing a restriction due to the form of the trajectories. The technique used for constructing these feedforward networks was cascade-correlation [Fahlman, 1991], which permits additions to the training base, without starting the learning from scratch. Cascor trains fast, and the dimensions of the FNNs are optimal. No major problems were encountered learning the mapping between the binary vectors and the real points in 3D space. The inverse mapping was more difficult to train, but applying a threshold technique for the output units (output=1 when the sigmoidal activation is >0.65 and output=0 when the sigmoidal activation is<0.35) the training produced a perfect mapping.

Dividing the training tasks between NMs is produced by dividing the universe of discourse in equal sections, with vocabularies having approximately equal cardinalities. These vocabularies must have overlapping zones. For example, in the case of three NMs the overall domain of the discourse can be illustrated as in fig.6.

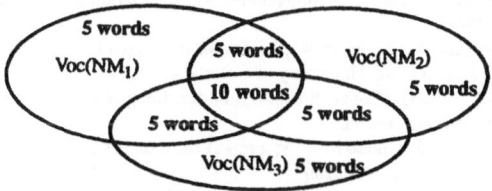

Figure 6: NMs-vocabulary overlapping

If a single NM has a vocabulary formed of 20 items (6 verbs, 7 nouns, 7 adjectives) and the intended number of NMs is 400 (memory needed: 20-40 kilobytes for each, which means 8-16 Mbytes for all) the common vocabulary for all NMs can be 10 (the most used words). If the rest of the overlapping is 5 words for groups averaging 8 NMs, every NMs can have 5 individual words, unknown by the rest of NMs (but useful for constructing sentences for CMs). This dividing scheme leads to a total vocabulary of:

400x5(individual words)+50x5(small overlappings)+10(common for all)=2260 words

For a population of 400 NMs, approximately a thousand learning patterns are needed, each consisting of a pair of texts, every text formed by a sequence of 10-20 sentences. The training is not very difficult. In our experiments, one day was sufficient for training all the networks of an NM on a single SPARC processor. The tremendous task is the construction of the entire training base. The author considers that the selection and sequences of sentences can be made without looking to the CMs level and the problem of hint generation. Forming texts with sense may be enough, and a NMs level with a great number of modules can play an important role in the functioning of the system.

5 CONCLUSIONS

Our experiments with an implementation of the model revealed interesting phenomena, and the most striking is the semantic dependence between the levels and the stability of the connectionist level. However, the system still lacks a strong formal description and presents some degree of nondeterminism. The model has some advantages:

-it is flexible and has a great plasticity. New knowledge can be inserted or deleted from CMs and NMs, but it is even more simple to add new modules.

-implementing the synchronization and the functioning scheme is very straightforward.

-the communicating language based on PSs can represent sentences from the natural language.

Among future enhancements, two deserve special attention. First, a better communication protocol and deletion strategy to enhance the right hint appearance; second, a possibility to perform automatic learning (via new trajectories and even new NMs). In [Sun, 1996], where a classification of hybrid models is presented, this model can be classified as a "third generation" model, but it lacks the auto-learning facilities. This enhancement requires a new kind of modules (an intermediate level between "classic" and connectionist), which are able to identify new needed trajectories and capable to select (or even create a new NM) and train the appropriate NMs. This may be the key to obtain the desired behavior of the system, without explicitly train the NMs.

References

Fahlman, S.B., Lebiere, C., (1991), The Cascade-Correlation Learning Architecture, TR-CMU-CS-90-100, Carnegie-Mellon University, Pittsburgh, PA.
Hilario, M., (1995), An overview of strategies for neurosymbolic integration, in R. Sun and F. Alexandre (eds.), *Connectionist-Symbolic Integration: From Unified to Integrated Approaches*, IJCAI'95, Montreal, Canada.
Memmi, D., (1990), Connectionism and AI as Cognitive Models, *AI & Society*, p.115-136, 4/90.
Pearlmutter, B.A., (1990) Dynamic Recurrent Neural Networks, TR-CMU-CS-88-191, Carnegie-Mellon University, Pittsburgh, PA.
Pineda, F.J., (1988), Dynamics and architecture for neural computation, *Journal of Complexity* 4/88.
Smolensky, P., (1986), Neural and Conceptual Interpretation of PDP models, in Rumelhart & McClelland (eds.), *Parallel Distributed Processing (vol.2): Explorations in the Microstructure of Cognition*, p.390-431, Bradford Books, Cambridge, MA.
Smolensky, P., (1990), Tensor Product Variable Binding and the Representation of Symbolic Structures, *Artificial Intelligence*, p.159-216, 46/90.
Smolensky, P., Legendre, G., Miyata, Y.,(1992), Principles for an integrated Connectionist/Symbolic Theory of Higher Cognition, TR-CU-CS-600-92, University of Colorado, Boulder, CO.
Sun, R., Alexandre, F., (eds.), (1995), Connectionist-Symbolic Integration: From Unified to Integrated Approaches, IJCAI'95, Montreal, Canada.
Sun, R., (1996), Hybrid Connectionist-Symbolic Models; a report from the IJCAI'95 workshop on connectionist-symbolic integration, to appear in *AI-Magazine*.
Szirbik, N.B., Somlo, G.L., Buliga, D.L., (1995a), Using the Conceptual Graph Model as Intermediate Representation for Knowledge Translation in Hybrid Systems, in L. Niklasson and M. Boden (eds.), *Current Trends in Connectionism*, p.141-152, Lawrence Erlbaum Publishers, Hove, United Kingdom.
Szirbik, N.B., (1995b), Towards a Model of Multi-Agent Connectionist Hybrid System, in *Proceedings of 1995 Second New Zealand International Two-Stream Conference on Artificial Neural Networks and Expert Systems*, p.xiii+397, 273-6, IEEE Computer.Soc.Press, Los Alamitos, CA.
Treur, J., (1990), Modelling Non-Classical Reasoning Patterns by Interacting Reasoning Modules, TR-IR-263, The Free University, Amsterdam, The Netherlands.
Van Langevelde, I.,Treur, J., (1991), Tackling the Incompletness of Chaining, TR-IR-274, The Free University, Amsterdam, The Netherlands.

Interpretation of a Hierarchical Neural Network

Jürgen Rahmel, Christian Blum
University of Kaiserslautern
Centre for Learning Systems and Applications
PO Box 3049, 67653 Kaiserslautern, Germany
e-mail:*rahmel@informatik.uni-kl.de*

Peter Hahn
Neustadt Hand Centre
Salzburger Leite 1
97616 Bad Neustadt, Germany
e-mail:*hahn@hand.franken.de*

Abstract

In this paper, we concentrate on the expressive power of hierarchical structures in data analysis. Recently, the so-called SplitNet model was introduced. It develops a dynamic, growing network structure and belongs to the class of topology preserving networks. We briefly introduce the basics of this model and explain the different sources of information built up during the training phase. Our focus then lies on the interpretation of the hierarchy produced by the training algorithm and we relate our findings to existing data analysis methods. We illustrate the results with an example from a real medical diagnosis and monitoring task.

1   Introduction

Existing approaches to hierarchical clustering and classification neglect the spatial relations of clusters or partial solution spaces. A particular class of neural network models has the potential to overcome this problem. Regarding the mapping from the input space onto the space spanned by the neighborhood relations of the neurons in the network, the property of certain neural network models to keep track of neighborhood relationships of clusters of data even in cases of reduction of dimensionality is called *topology preservation*. The degree of topology preservation can be determined by the observation, how well neighborhood relationships in one space are preserved by the mapping onto the other space. Thus, one question is: for two input vectors that are close in input space, are their best matching units close[1] in the network topology? The other question is: for two neurons that are neighbors in the network topology, are their associated weight vectors close in the input space? These questions led to the development of the topographic function [VDHM96], that effectively quantifies the topology preservation in topographic maps. Topology preserving models are, among others, the Self-Organizing Map (SOM) [Koh90], the Growing Cell Structures (GCS) [Fri93] and the Topology Representing Network (TRN) [MS94] as well as several descendants of these examples. But all those models lack the ability of hierarchically structuring the training set.

The SplitNet model for the first time succeeded in developing a hierarchical structure over the training set. It is an unsupervised learning method and in some sense comparable to the hierarchical cluster analysis (see e.g. [DH73]). For static models, like e.g. the Self-Organizing Map, the task of the network application determines the desired interpretability of the network and thus controls such parameters of network design as number and connectivity of neurons. In dynamically growing networks, the approach is necessarily different. The incremental construction of the network up to its final size and topology is in general controlled by specific performance criteria. The training result is a network, where not only the weights contain relevant information. Additionally, the size of the network, the distribution of neurons and the emerged connectivity inside the network implicitly code information on the trained sample set. Compared to the above-mentioned topology preserving models, the tree-structured organization of topologically connected parts of the SplitNet network adds a completely new dimension to the interpretability of the network model. The hierarchy offers structured knowledge on various levels of abstraction and generalization as well as optimized access to samples of the training population.

The rest of the paper is organized as follows. In the next section, we will outline the basic methods related to the neural model used in our approach. Section 3 will present the principle of the SplitNet model and in Sec. 4 the role of the emerging hierarchical structure is explained. We then present results of the application of the SplitNet model in the medical domain of finger movement pattern analysis. A summary and final remarks will conclude the paper.

2 Related Work

The hierarchical cluster analysis, either the divisive or the agglomerative approaches, are methods that progressively split or link clusters of data. The result of the analysis can be visualized as a dendrogram, which is a two-dimensional tree structure that shows the order of linkage (for the agglomerative case) and the distance or similarity at which this linkage of clusters was performed. Thus,

---

[1] As different input vectors may be mapped onto the same neuron, in this informal explanation, the *closeness* of neurons includes also the identity of those neurons.

this method is able to display the clustering of data, whereby the result depends on the distance measure that determines the closeness of clusters and/or samples. Specific variants of agglomerative versions of the hierarchical cluster analysis are e.g. the *single linkage* and *complete linkage* methods, which minimize the minimum or maximum distance of cluster elements, respectively, during each merger of two clusters. For the comparison intended in this paper, we will use the *centroid method*, which selects those clusters with the minimum distance between their means.

Because of the explicit distance information contained in the dendrogram for each linkage of clusters, hierarchical clustering is a flexible way of detecting the resulting number of clusters given a certain threshold value. However, it is not possible to reason about the real spatial relationship of the observed pattern. There is no similarity information other than the one for linked clusters. Similar statements are true of course for divisive methods. So the hierarchical clustering methods are useful tools for a preliminary analysis of the data, but they do not provide additional ways for explanation of the clustering results and do not enable reasoning on alternative solutions based on neighborhood observations.

Such inspection of neighboring clusters and samples can be performed by topology preserving networks. As indicated above, several models like the GCS or the TRN already exist for topology preserving representation of a training set. The Growing Cell Structures (GCS) are a dynamic vector quantization model. Different criteria, e.g. the quantization error, determine the insertion position of a new neuron. Removal strategies yield an adaptive quantizer that is superior to the original SOM, but the GCS model also uses an a priori specified dimensionality (through the choice of simplices) and still involves computations that are performed for all the neurons in the network. The TRN algorithm also approximates the distribution of input data and constructs topology preserving connections between its neurons. In the limit, it is able to find the Delaunay triangulation of a data set, thus it generates, by virtue of not being fixed to a given dimensionality, a nearly perfectly topology preserving map. But those neural models only provide data analysis on a flat level. They cannot provide views on the data at different granularities, and thus lack the advantages of methods like the hierarchical cluster analysis.

3 The SplitNet Model

SplitNet is a topology preserving, dynamically growing model for unsupervised learning and hierarchical structuring of data [Rah96b]. Starting with a single, small Kohonen chain [Koh90], localized insertion and deletion criteria enable an efficient quantization of the data space. The hierarchy in the architecture grows, if one of the following splitting criteria is satisfied: (i) detection of topological defects, (ii) deletion of neurons by an aging mechanism, (iii) significant local variances in quantization errors and (iv) significant local variances in edge lengths. Those criteria are checked several times during progress of training. If a criterion is satisfied, the affected chain is split into two or more subchains which are added to the network at one level lower in the hierarchy. The node in the hierarchy that formerly represented the unsplit chain now serves as a generalized description and access structure for the new son nodes. Figure 1 illustrates this basic mechanism. If a topological defect is found (e.g. between neurons 1 and 7, which are close in input space but distant in the chain of neurons), the chain is split and nodes representing the fragments are added as descendants to the tree. Path decisions in the so constructed hierarchy will be drawn according to the mean of the weight vectors of the neurons in the chain, therefore the mean is also indicated in the figure. The topology preserving construction of the network structure provides local neighborhood information that is necessary for incremental retrieval of nearest neighbor to a given input vector. The dashed

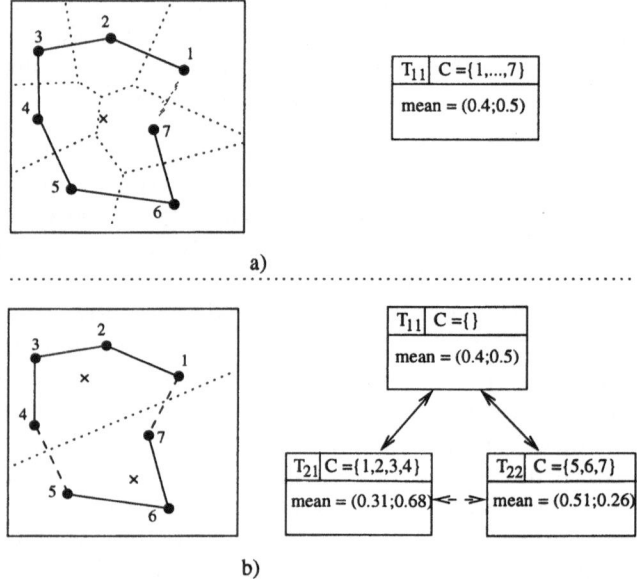

Figure 1: Example of splitting a chain because of a topological defect

lines in Fig. 1 indicate this type of knowledge. The neighborhood relations are kept as lateral connections defining the topology of the network space. They are responsible for the high degree of topology preservation in SplitNet and enable a fast and incremental search for a set of nearest neighbors. A more rigorous and exhaustive treatment of these aspects and retrieval results can be found in [RV96]. The purpose of such a set of nearest neighbors is e.g. to serve as the basis for the k-nearest-neighbor rule in decision making processes (cf. Sec. 5).

Since unsupervised learning methods provide no direct classification, the training result has to be interpreted in the context given by the training data. For the SplitNet model, we observe three containers of knowledge that can be used for the tasks in applications like the one described in this paper:

Neuron distribution: The insertion criterion determines the error function to be minimized. Quantization of the data set allows local estimation of sample density.

Topology: The connections between neighboring neurons provide information on where to find similar cases. Measuring topological defects yields the search depth for incremental retrieval of nearest neighbors to a given query.

Hierarchy: The hierarchical structure of the network contains different levels of generalization and abstraction. It allows a fast tree search for best matches and insightful visualization of the data structure for the domain expert.

The utility of the neuron distribution is comparable to reference vector placement in quantization algorithms [Gra84]. Interpretation of the network topology is described e.g. in [Rah96a]. In the following, we illustrate the semantics of the hierarchical structure developed by methods like SplitNet.

4   Interpretation of Hierarchy

Hierarchical organizations offer additional properties to make use of in data analysis and structure utilization. One rather general aspect is the fact that a hierarchical structure – like any search tree – provides fast access to the terminal nodes. For neural networks like SplitNet, this results in accelerated training runs, since the search for the best matching unit is supported by the hierarchical network structure.

The hierarchies produced by classification or decision trees [BFOS84] [Qui93] yield simple, crisp, and explicit tests as path decisions in nodes at the expense of flexibility of the decision regions. The orientation of hyperplanes generated by those tests is limited to the dimensionality of the respective test. Unfortunately, higher dimensionality of the test yields both higher flexibility and massively growing computational effort for determination of optimal tests. Therefore, in practical applications, test are often one- or two- dimensional and the corresponding hyperplanes separating subspaces of the sample space are orthogonal to the coordinate axes or depending on only two of the possible vector components. In contrast to this, the SplitNet structure offers implicit tests that cover the whole information contained in the description of a sample. The decision regions of SplitNet approximate the Voronoi regions given by the sample population and thus minimize quantization errors imposed by generalization inside the regions.

Classification or decision trees select the tests for path decisions according to the gain criterion for classification of samples. In an unsupervised setting where class information is not available, efficient sample localization plays the most important role. In order to minimize the search effort, we need a test that maximizes the information about the location of the nearest sample. The Kohonen model provides a solution for this task. As demonstrated in [RMS92], the weight adaptation of the algorithm leads to a discrete approximation of principal curves by Kohonen chains. If we divide the sample population according to the placement of neurons and recursively repeat this subset construction, we get a hierarchic structure that on every level optimizes the information on sample locations. Thus, for an average test sample, we have an efficient access to the best match of the training population. In this respect, the tree is interpretable as a decision tree, regarding the optimization of spatial information, and still trainable and adpatable to slight changes in the training data set.

The interpretation of a hierarchical structure like the one generated by Split-Net thus combines the knowledge on the above-mentioned property of the Kohonen algorithm with the splitting reasons of the SplitNet model which deviate from this principle. Path decisions in the SplitNet tree have thus definite semantics to be used when descending the tree and relating accessed clusters with others that are reachable through the topology of the network.

5   Diagnosis and Monitoring of Ulnaris lesions

We now briefly present an application of the SplitNet model in the domain of nerve lesions of the human hand. We will outline the general problem and describe the results obtained by using the hierarchical neural model.

The human hand is provided with the radial, median and ulnar nerve. The ulnar nerve provides sensory function for the small and ring finger and innervates the intrinsic muscles of the hand. These muscles are crucial in balancing and coordinating the flexor and extensor muscles, rendering possible fine movement such as grip and pinch. While assessing sensory function is feasible, objective analysis of motor function is quite difficult. Clinical investigation includes grip force measurement and recording of active and passive range of motion. Besides these factors, ulnar nerve dysfunction causes changes in coordination of the movement which cannot be measured by instruments. In contrast to a normal, physiological movement pattern (Fig. 2(a)), the dynamic disorder 'rolling' describes the pathological flexion of the finger. This movement resembles the rolling of a carpet (Fig. 2(b)). As an effect, patients are not able to grasp an object because their fingers push it out of the palm. The dynamic disorder 'clawing' describes the hyperextension of the MP joint with flexion of the PIP and DIP joint[2] while the finger is in resting position (Fig. 2(c)). These descriptions are based on the experience of the examiner. Changes in quality and especially improvement of fine motor activities after nerve repair are difficult to detect and

---

[2] The MP, PIP and DIP joints are the three finger joints ordered from the base joint between hand and finger to the tip.

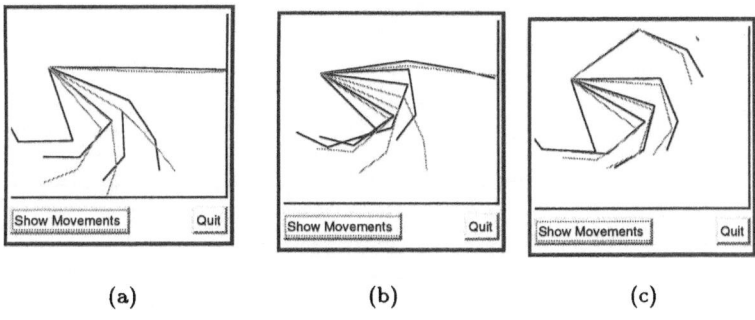

(a)   (b)   (c)

Figure 2: Different forms of finger movement pattern: (a) normal, physiological movement, (b) rolling, and (c) clawing (see text). Each picture shows nine steps of finger movement during one cycle of closing (black lines) and opening (gray lines) the fist.

to quantify. If nerve repair fails, there are different operations to rebuild the movement pattern. In these cases, the outcome of surgery also cannot be quantified. Until now, there was no convenient measurement system to distinguish finger movement patterns.

Based on kinematic research we established a measurement system to get real-time data of human finger movement. Attempts to analyze these data with classical mathematical methods like discriminant analysis failed to distinguish between normal and pathological movement. Statistical clustering provides a good first insight into the structuring of the data but is not able to support the specific needs in this application like, for example, retrieval of samples and their comparison to a group of similar data, as it is required for diagnostic applications.

Figure 3 shows an example of a tree structure generated with a small fraction of the available data. Despite the fact that our preprocessing generates high-dimensional training vectors, we used no further dimension reduction method. The reason is the necessity to display the hierarchy with the neuron weights re-translated into finger positions. By this, physicians can evaluate the position of a newly encountered data vector in the tree. In the above figure, the upper part of the tree contains roughly the patterns for 'clawing', while the bottom part corresponds to physiological movement and 'rolling'. Further analysis of branches and subtrees reveals the full organization of the hierarchy. The sequence of generalizing non-terminal nodes supports the physician in understanding relative positions of different patient vectors. At each non-terminal node, the reason for branching – outlier splitting, topological defect, etc. – is accessible, so a reasonable interpretation of the emerging hierarchy, supported by the local topological connections (not shown in the figure), is rendered feasible.

658

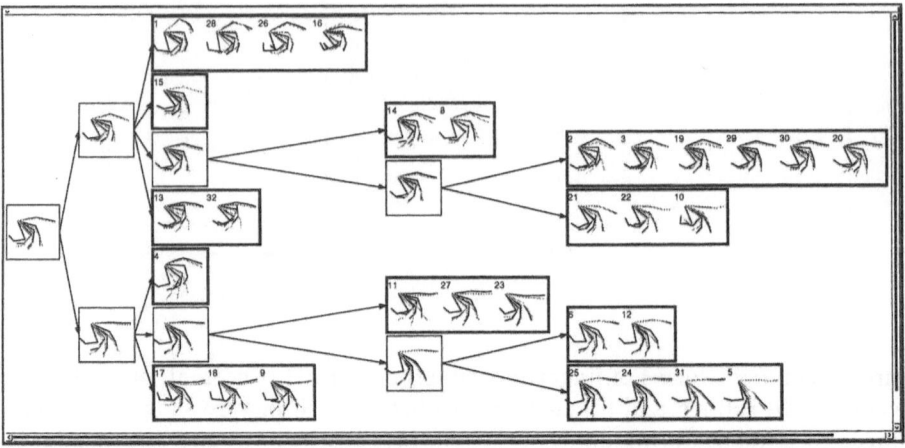

Figure 3: Hierarchical representations generated by SplitNet. The retranslation of neuron weights allows the display of interpretable finger movements in the learned hierarchical arrangement.

A comparison of the results obtained by SplitNet with those of a hierarchical cluster analysis clarifies the strength of the neural model. We performed a run of the clustering process and examined a subgraph of the dendrogram with about as many terminal nodes as the SplitNet tree described above. The result was not surprising. The clustering produced nearly the same groups of data represented by leaf nodes, thus supporting the clustering abilities of the SplitNet model. However, despite the distance information that is available for the merging level of two clusters, interpretation of the dendrogram from a medical point of view was possible only in a very limited way. Whereas the neuron chains representing the terminal nodes in SplitNet arrange themselves in a direction that best reflects the largest variation in the associated movement pattern (the intrinsic property of the underlying Kohonen model), such information is not available in the cluster analysis. Moreover, the dendrogram provides information on the order of the cluster linkages, yet it does not contain explicit or implicit information on the spatial relationships of clusters. This is a crucial property for a reliable diagnosis of new cases which are not contained in the center of existing clusters. In order to compare those with nearest neighbors, reliable information on cluster connectivity is necessary. The lateral connections between neurons in the SplitNet model facilitate reasoning for class assignment based on neighborhood considerations. We can use the retrieval properties of the topology preserving network structure for the enumeration of the nearest neighbors and application of the k-nearest-neighbor rule [DH73] yields a majority vote, if training samples contain classification information.

6 Conclusion

Our approach for the first time applies pattern recognition by a neural net approach to human finger movement. Besides simple clustering tasks, the applied SplitNet model provides support for the interpretation of the learning processes which have occurred and the emerged hierarchical structure. Thus, in our case, interpretation of the images (which are retranslations of neuron weights into the semantics of training vectors) enhances our knowledge of the finger movement pattern. So far, from the medical point of view, we do not know if we portray the whole spectrum of ulnar nerve dysfunction. More data have to be recorded and our aim is to build up a neural net containing all types of normal and pathological movement. Then we are able to represent all ulnar nerve lesions by recording finger movement and classify the new movement pattern by observing the mapping performed by the neural net onto a certain location in the tree, for which clinical diagnosis is already accessible.

References

[BFOS84] L. Breiman, J.H. Friedman, R.A. Olsen, and C.J. Stone. *Classification and Regression Trees.* Belmont, CA, Wadsworth, 1984.

[DH73] R.O. Duda and P.E. Hart. *Pattern Classification and Scene Analysis.* Wiley, 1973.

[Fri93] B. Fritzke. Growing cell structures - a self-organizing network for unsupervised and supervised learning. Technical Report TR-93-026, ICSI, 1993.

[Gra84] R. M. Gray. Vector quantization. *IEEE ASSP Magazine*, pages 4–29, April 1984.

[Koh90] T. Kohonen. The self-organizing map. *Proceedings of the IEEE*, 78(9):1464–1480, 1990.

[MS94] Thomas Martinetz and Klaus Schulten. Topology representing networks. *Neural Networks*, 7(2), 1994.

[Qui93] J.R. Quinlan. *C4.5: Programs for Machine Learning.* Morgan Kaufman, 1993.

[Rah96a] J. Rahmel. On the Role of Topology for Neural Network Interpretation. In W. Wahlster, editor, *Proc. of the ECAI*, 1996.

[Rah96b] J. Rahmel. SplitNet: Learning of Hierarchical Kohonen Chains. In *Proc. of the ICNN '96*, Washington, 1996.

[RMS92] H. Ritter, Th. Martinetz, and K. Schulten. *Neural Computation and Self-Organizing Maps.* Addison Wesley, 2nd edition, 1992.

[RV96] J. Rahmel and T. Villmann. Interpreting Topology Preserving Networks. Technical Report LSA-96-01E, University of Kaiserslautern, 1996.

[VDHM96] Th. Villmann, R. Der, M. Herrmann, and Th. Martinetz. Topology preservation in self-organizing feature maps: Exact definition and measurement. *IEEE Transactions on Neural Networks*, 1996. To appear.

Cognitive Processes in Social Interactions - A Neural Networks' Approach[*]

J. Barahona da Fonseca[1], I. Barahona da Fonseca[2] and J. Simões da Fonseca[3]

[1]FCT/UNL, Department of Electrical Engineering
QUINTA DA TORRE, 2825 MONTE DA CAPARICA, PORTUGAL,
Phone: 351-1-295 44 64, Fax: 351-1-2957786,
E-mail:jbf@fct.unl.pt
[2]Faculty of Psychology of Lisbon,
Alameda da universidade, 1600 Lisboa, Portugal,
Fax: 351-1- 793 34 08
[3]Faculty of Medicine of Lisbon,
Av Prof Egas Moniz, 1600 Lisboa, Portugal,
Fax: 351-1-796 40 59
E-mail:j.s.da.fonseca@ip.pt

Abstract:

In a well controlled experimental situation, interpersonal relationships were studied using a double script model representation of the social situation that subjects were required to identify.

Results obtained in a group of normal volunteers subjects were analyzed. It is used an abductive model to describe the process of concept attainment by experimental subjects.

The interaction between partial attribute identifications and global characterizations of the simulated situations is described in the context of other psychological processes evoked by the interactions. A block diagram specifies the operations which are performed by a neural network model which implements the cognitive operations of a subject involved in the social interaction under study.

Key Words: social interaction, cognitive processes, script representation

I. Introduction

Cognitive processes which occur during social interactions involve knowledge structures of great complexity.

Nevertheless if it is high the number of characteristics belonging to the situation which have to be taken into account they impose restrictions which allow the identification of some simple but not trivial cognitive processes. These cognitions are required for the effectiveness of adjustment of experimental subjects to the social interactions under consideration. Our aim is to report data obtained during a study of interpersonal relationships in which subjects were required to interact verbally with a group of experimenters that performed a strictly ritualized play. This representation simulated a real situation that subjects were asked to identify.

The experimenters performed their play according with a script in which concepts and relationships were characterized verbally using partial attribute variables that experimental subjects should use to attain and recognize the social meaning of the situation. Ostensive denotation of clues which might allow an immediate solution to the problem was completely avoided. The flow of information in the ritualized play was kept constant as far as the number of attributes for each sequence of utterances produced during each stage of the performance was concerned.

---

[*] This work was partially supported by a grant from BIAL Foundation, OPorto, Portugal and by JNICT PBIC/TIT/2527/95

Every eight verbal interventions of the participants in the simulation the performance was interrupted and experimental subjects were asked a set of questions which aimed to render ostensive the knowledge they had acquired. Those data were recorded and treated statistically to study cognitive differences between normal volunteers and subjects suffering from distinct psychopathological disturbances [J. Simões da Fonseca et al, 1979a, 1979b, 1982, 1984].

The aim of the present work is to show the adequation of a double script model to represent the interaction and to build a neural and dendritic network representation of the cognitive processes of subjects under study. Hyperincursive computation [Dubois et al, 1992] is used to represent anticipative and control processes in concept attainment.

Cognition appears in this approach as a component of a complex script representation of the performance of an individual when a structured social environment acts upon him. There are then evoked motivational emotional, perceptive, attentional reactions, normative interactions together with some specific cognitive processes which must be understood taking into account the global context of the interaction. This context imposes restrictions on the freedom to react but simultaneously provides cues which are required for the conceptual identification.

II. Material and Methods

Experiments involving social interaction were performed in eleven groups of fifteen subjects, one of normal volunteers and ten other groups each one including subjects which did suffer from well characterized psychopathological disturbances. The initial motivation was to study patterns of interpersonal relationships characteristic for each type of psychopathological disturbance.

The experimental situation involved the simulation of (1) a psychopharmacological experiment in which subjects cooperation was asked (2) interactions in the context of a nuclear family structure; (3) admission for treatment in a Hospital ward; (4) a situation of trial.

These situations occurred in this sequence without any explicit marker for the change from one to the next of the four successive simulations. A group of four experimenters performed the simulations according with a strictly ritualized script .

No change in the simulation would occur as a result of the verbal intervention of subjects submitted to the experiment. Every eight verbal interventions produced by the members of the group, which performed the simulated action, the situation was interrupted and a fifth experimenter posed a set of questions which allowed the characterization of the state of knowledge which was this way rendered ostensive. Then the simulation would start again for another sequence of eight interventions by the members of the group of experimenters.

Audio and video records were obtained and a content analysis of the verbal communication of the experimental subjects was performed. The cognitive strategies used in subjects identifications were analyzed aiming to characterize the script of the action produced by experimental subjects when they tried to solve the problems posed to them.

A model for the abductive processes and cognitive identifications was then constructed considering those processes as components of a second script which the experimenters' communication did evoke in experimental subjects. This model was then implemented as an analog computation in a neural and dendritic network for which we present the information flow diagram.

III. Results

Table 1 summarizes results obtained in the group of normal subjects concerning the mean value of the number of characterizations produced by the members of this group for each one of twenty types of attributes. As the characteristic distribution of the number of attributes of each type was small and stable along subjects interaction in successive interruptions only global data for each situation are reported.

	1st situation		2nd situation		3th situation		4th situation	
A	M	SD	M	SD	M	SD	M	SD
A1	8.07	7.986	9.6	11.153	7.20	6.774	5.33	5.972
A2	5.27	2.915	6.40	4.421	7.13	4.121	6.93	4.832
A3	4.93	4.496	7.00	8.177	5.13	5.357	4.73	5.7
A4	3.33	1.839	3.87	2.066	3.47	1.922	3.27	2.738
A5	1.87	1.807	4.13	2.696	3.07	2.187	3.0	2.236
A6	5.13	2.696	4.2	4.004	4.20	2.933	4.40	3.719
A7	1.33	1.718	1.13	1.807	1.0	1.363	1.33	1.988
A8	1.07	1.335	1.20	1.474	0.93	1.280	2.2	3.550
A9	1.93	1.387	1.47	1.642	0.8	1.207	0.53	1.060
A10	0.67	1.113	0.67	0.900	0.13	0.352	1.2	1.612
A11	0.47	0.915	0.67	1.234	0.27	0.458	0.33	1.047
A12	3.8	4.678	6.20	7.608	4.47	7.303	7.40	10.716
A13	1.27	1.223	1.60	2.197	2.07	3.262	0.73	1.335
A14	12.07	7.905	13.40	6.588	12.13	9.094	11.87	8.314
A15	1.2	1.424	1.0	2.035	1.8	3.075	1.8	2.783
A16	3.4	3.334	4.46	4.533	2.6	2.923	2.0	2.204
A17	0.0	0.0	0.27	0.799	0.2	0.561	0.27	0.799
A18	0.33	0.816	0.20	0.561	1.33	2.944	0.8	1.320
A19	0.0	0.0	0.40	1.298	0.0	0.0	0.0	0.0
A20	0.40	0.910	0.06	0.258	0.20	0.561	0.0	0.0

Table 1- 1st Situation- Psychopharmacological experiment; 2nd situation- Nuclear family interaction; 3 rd Situation- Admission in a medical ward; 4th situation- Trial. M-Mean; SD- standard deviation; A1-A20-Attributes identified in content analysis. A1-Elementary references; A2- Aggregation of references with one level of inference; A3- Complex referential structure with hypothesis construction; A4- Definition of individual standpoint from the viewpoint of personality attributes; A5- Definition of the personal standpoint of other persons; A6- Statements about private subjective experiences; A7- Positive Affective and Valorative statements about the other; A8- Negative Affective and Valorative statements about the other; A9- Positive affective and valorative statements about the subject himself; A10- Negative affective and valorative statements about the subject himself; A11- Normative statements; A12- Semantic, affective or valorative accentuated qualifications; A13- Statement of hypothesis together with their alternatives; A14- Holophrastic statements; A15- Rigidity; A16- Statement of incapacity; A17- Justifications on the basis of illness; A18- Statement about health, medical care and medical testing; A19- Semantic drifting and distortion; A20- Mixed comments, both referential and about private experiences.

The content analysis of verbal statements produced by experimental subjects during the first simulation was centered on the role of the four experimenters expressing a set of referential attributes as well as normative or moral justifications for their personal participation in the experiment. Statements about their own personality, or else the personality of the intervenients in the simulation, valorative statements about the verbal interventions of the experimenters as well as justifications for their statements based in past private subjective experiences were used as well as hypotheses not based in referential data which were used together with a very small number of identifications of the global meaning of the simulation which the experimenters were performing.

Note that the meaning of the global situation could be inferred from the statements of the member which played a role with leadership attributes in this first simulation, but it was not ostensively denoted. As the transition to the next situation was not identified by any marker, experimental subjects had great difficulty to grasp the global meaning of this second situation. They characterized mainly their own affective standpoint or else the affective characteristics of the interventions based on the valorative attributes produced by the experimenters. Four of the fifteen normal volunteers were unable to discover that the situation involved the representation of two parents and two brothers in a family in which one of the two brothers was being criticized by some misdeed. Although referential attributes were produced in a higher number comparatively to the first situation most subjects were unable to identify the global script before the third interruption. Again normative statements, statements justified on the basis of past private experiences, production of hypothesis dissociated from referential contents of the play or else mentions of characteristics of their own personality or of the personality of the intervenients in the play were made and the global meaning of participating in a medical experiment was maintained until the third interruption. The third simulation was also identified with difficulty. Subjects returned to a concrete reference to a medical experiment. The fourth simulation which represented a trial was nevertheless easily identified as an examination in an educational institution or else effectively as a trial without any further mention of a medical experiment.

3.3

Normal volunteers identified two types of scripts besides those which were intentionally simulated, namely one in which they stated that members of the group of experimenters performed a medical experiment or else invoked a medical action.

In the case of subjects suffering from psychopathological disturbances it was furthermore recorded that the situation was part of a treatment process, or else their personal suffering or disturbances were used to justify a situation of therapeutical intervention in which experimenters did participate. These interpretations were due to the concrete personal context of characteristics of subjects under observation.

The transition from concrete attributions to global interpretations was mediated by partial attributes referentially related to the participation of experimenters during the simulation. In the transition from one situation to the next one a bias due to preceding interpretations did interfere with the identification of later simulations.

It should be noted that in the experiment the global interaction contained two types of scripts-(1) the script that is played by the experimenters which is rigidly maintained without any variation and (2) the script that provides a structure that organizes personal interventions produced by experimental subjects which included (a) referential cognitive characterizations of the simulated script, but also (b) valorative statements, (c) normative statements, (d) statements based in past subjective experiences (e) statements about the personality of the subject or (f) about the personality characteristics attributed to one or another of the experimenters in the context of the simulation, respectively, (g) production of hypotheses on the basis of not referential data.

In our approach cognition corresponds to a restricted subset of the integral pattern of adaptation of subjects where references, as well as hypothesis even those not based on objective data, are the sources of attributions and interpretations which are used to construct cognitive representations as they are commonly understood in psychological theory. The process of abduction did not present particular difficulties in its modeling because it was based in high level symbolic representations and not in elementary mappings of the environment which would require an interpretation for their high level significant to significate relationship.

3.4 The Abductive Model

In our characterization of cognitive processes which occur in the paradigm of four social interactions we have chosen, subjects perform a process of concept attainment using the information which is rendered available to them. In this context, their success or failure must be defined according with a reference which was established during the design of the experiment.

$\overline{R}_i$, the reference function, is a vectorial function already available which is invoked using the global or partial characterization of $\overline{F}_i$ according with the choice and acceptance criteria that will be specified. We may define the degree of attainment in the cognitive process by $dist(\overline{R}_i, \overline{F}_i)$ in which $\overline{R}_i$ and $\overline{F}_i$ correspond respectively to relationships specified over a field of lower order functions $r_i(a_j)$ and $f_i(a_j)$ which may only assume scalar values depending on their arguments. $\overline{R}_i$ specifies a global characterization of the situation as it is proposed by an experimental subject. $r_i(a_j)$ and $f_i(a_j)$ are entries of $\overline{R}_i$ and $\overline{F}_i$ respectively. For each pair of entries $r_i(a_j)$ and $f_i(a_j)$ we define an elementar component of the global distance between $\overline{R}_i$ and $\overline{F}_i$.

Furthermore, we associate to each elementary distance $dist(r_i(a_j), f_i(a_j))$ an elementary cost value $c_i = dist(r_i(a_j), f_i(a_j))\, \alpha_i$ in which α_i is a scalar that expresses cost for each unit of the $dist(r_i(a_j), f_i(a_j))$. Distinct α_i represent the congruence that the corresponding $f_i(a_j)$ possess for the specification of the degree of concept attainment..

The relevance of $f_i(a_j)$ to specify the $\overline{R}_i$ which best fits $\overline{F}_i$ is expressed by $k_i = e\text{-}c_i$ with $\alpha_i \geq 0$ and $dist(r_i(a_j), f_i(a_j)) \geq 0$. Then k is included in the closed interval $[0,1]$. The best fit corresponds to k=1 for $dist(r_i(a_j), f_i(a_j)) = 0$. The contribution β_i of $f_i(a_j)$ to identify $\overline{R}_i$ is defined by $\beta_i = v_i\, e^{-\alpha_i\, dist(r_i(a_j), f_i(a_j))}$ with v_i, value for identification, $v_i \geq 0$. Cognitive experience acquisition and learning may be expressed at the level of α_i and v_i which change with passed success or failure in identifications based on their values.

Then, given a set of alternative $\overline{F}_k \subset \overline{F}_x$ where adequacy may be evaluated under $\overline{R}_k$ the selection procedure will take place. The concept which will be chosen is represented by some $\overline{F}_k$ which maximizes the value of S_k, $S_k = \sum_i k_i v_i$ for a fixed number of terms i, for all the $\overline{R}$ and $\overline{F}$ contained in the set. In case S_k exceeds a threshold ω the k_i which contribute most for the S_k are used to accept $\overline{R}_k$. If not, the procedure is repeated to criteria ω is met through another combination of k_j. The procedure may be described as one in which subjects may essay successive $\overline{F}_m, \overline{F}_n, \ldots$ distinct from $\overline{F}_k$ chosen according with information about particular k_i and evaluate if the corresponding $S_m, S_n, \ldots$ are greater than the S_k associated to $\overline{F}_k$. It is then chosen some $\overline{R}_x$ which maximize S_x, $S_x > \omega$.

The process we have characterized operates on a set $\{\overline{F}_x\}$ whose members have been chosen according with hypothesis, or knowledge about the value of criteria k_i and S_k. In a reverse sense the choice of any $\overline{F}_x$ will imply an anticipation for the remaining $f_x(a_j)$ which produce the best choice for a predicted $\overline{R}_x$ This is the process which seems adequate to represent the essay of hypotheses as they were produced by subjects while they were submitted to the situation of social interaction. In those situations it was observed that the choice of $f_i(a_j)$ preceded the choice of $\overline{F}$ and at least three $\overline{F}$ were proposed - one corresponding to a concrete medical characterization, the second to the identification proposed - one corresponding to a concrete medical characterization, the second to the identification of an experiment and a third to the meaning intended in the design of the experiment. Furthermore patients proposed global interpretations in terms of their illness or of a supposed treatment.

3.5 Block Diagram which represents the social interaction

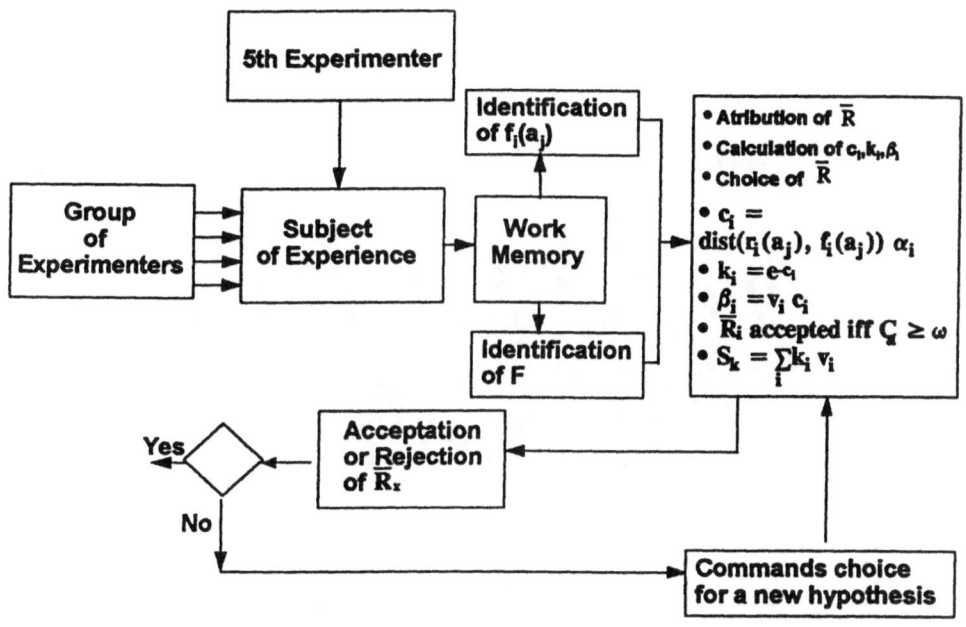

IV. Discussion and Conclusions

Data obtained with 150 subjects (see Figure 1) belonging to ten diagnostic groups provided cross-validation for the cogency of the categories used in the content analysis of verbal communications of experimental subjects - the twenty categories we did chose and their frequency of use provided by themselves a cross validation of the classification of subjects in the ten diagnostic groups which had been obtained previously on the basis of clinical criteria.

When a further group of adolescents [M. P. Horta and J. Simões Fonseca, 1983] with adjustment problems and without further psychopathological disturbances was submitted to the same social interactions the use of the discriminant functions previously obtained allowed an adequate separation of this group from anyone of the other ten diagnostic groups.

Results were confirmed by two further independent experiments concerning Paranoid Schizophrenic patients and Normal volunteers and independent samples of Dysthimic Disorder (Neurotic Depression). The double script paradigm provided an insight into the genesis of delusional concepts as they were produced independently of affective or normative attributes [J. L. Simões da Fonseca et al, 1984; M. L. Figueira, 1986].

They rather did occur as a special type of reference which is dealt with as if it had the status of a subjective experience. Paranoid Schizophrenic subjects would require only an internal evidence with the exclusion of the external validation which is usually necessary to accept judgments about the external reality. As far as the abductive model is concerned, the design of the experiment implied a straight forward diagnosis of the global situations on the basis of lower order attributes. In the circumstances under which the experiment was performed, the experimental setup together with the content of the verbal interventions produced by the experimenters did render unlikely other concepts distinct from those which were predicted. The only exception was observed in the group of Paranoid Schizophrenic patients who did not require any validation for many of their statements about external reality.

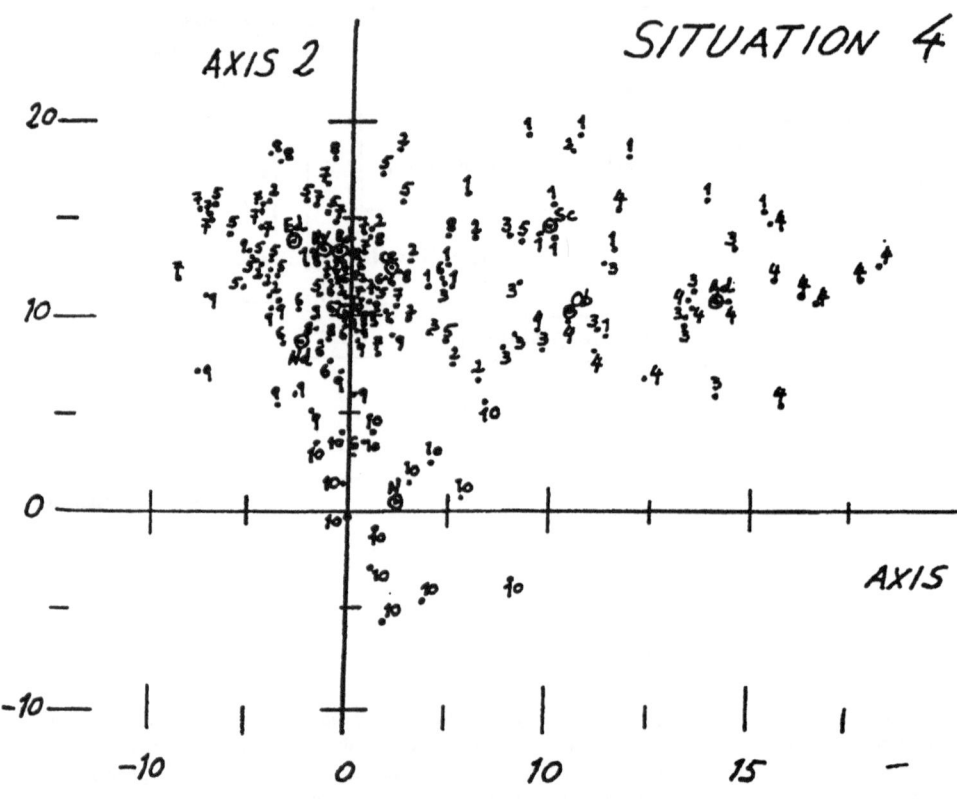

Figure 1. Results from Multivariate Discriminant Analysis; 1-Sc-Paranoid Schizophrenic subjects; 2-Ce-Generalized Epilepsy; 3-Ob-Obsessive Neurotic; 4-Ad-Adolescent; 5-Hy-Hysteric Neurotic; 6-Te-Temporal Lobe Epilepsy; 7-Ed-Major Depression; 8-Rd-Major Reactive Depression; 9-Nd-Disthymic Disorder; 10-N-Normal Volunteers. In the plot, numbers 1-10 indicate the position of subjects of each group. Letters near to circles signal group centroids.

The cognitive task was not to find a new higher order concept using induction in a pure form, but rather to chose among cognitive relationships already available and easily attainable those which would produce the best match with the functions empirically defined on lower order attributes. The neural network model for this process that we propose has two stages - (1) a lower order associacionist stage in which lower order attributes are identified and associated to a criteria α_i and v_i of congruence and relevance concerning the acceptation of these variables to define a higher order vectorial function; (2) a higher order decision procedure which involves (a) the imposition of a limited choice among the members of a set of vectorial functions and (b) the use of the Euclidean distance between the reference function on one hand and the empirically obtained function on the other to decide if the match between both was close enough.

Our block diagram represents the principles which are used to construct a neural network capable of performing abduction.

Finally, note that the model of cognitive adaptation implied in this process should be understood as one in which attribution relevance is evaluated on the basis of a process of cognitive congruence involved in the reinforcement concerning the choice of the global concept which is judged as the more adequate for the situation. The variables $\alpha_i, c_i, v_i, \beta_i$ and ω may have values dependent on passed experiences.

References

DUBOIS D, RESCONI G (1992) *Hyperincursivity- a new mathematical theory*. First Edition, Presses Universitaires de Liège.

FIGUEIRA ML (1986): *Relações Interpessoais na Esquizofrenia Paranoide*. Doctoral Thesis.

HORTA MP, SIMÕES DA FONSECA JL (1983): Characteristics of Interpersonal Relationships of Normal Adolescence and the Obsessional Cognitive Set. *Acta Psiq. Port.*(29-1,2) pp17-25.

SIMÕES DA FONSECA JL (1996).Representation of psychic events in neural networks in R. Moreno-Diaz and J. Mira y Mira (Eds) *Brain Processes, Theories and Models* An International Conference in Honor of WS McCulloch 25 Years after His Death, Mit Press, pp173-183.

SIMÕES DA FONSECA JL, FIGUEIRA ML,GIL MT, ARRIAGA F, LARA E, HORTA MP (1979a): Interpersonal Relationships. A Rigorous Mathematical Model from a System's Theoretical Viewpoint, *Soc. Port. Psicol*, Feb.

SIMÕES DA FONSECA JL, FIGUEIRA ML,GIL MT, ARRIAGA F, LARA E, HORTA MP (1979b): Relações interpessoais: um modelo estrutural e dinâmico baseado na Teoria dos Sistemas. *Revista Port. Psic. (14/15/16)* pp167-182.

SIMÕES DA FONSECA JL, FIGUEIRA ML,GIL MT, ARRIAGA F, LARA E, HORTA MP, SILVA A (1982): When they interact socially do normal subjects know what is happening and is it true what they think? *Acta Psiq. Port.* 27, pp. 11-22.

Adding Phase to Recurrent Backpropagation Networks: An Application to Binding Tasks in Vision

Hanna Majewski and Janet Wiles
School of Information Technology,
The University of Queensland, Queensland 4072
Australia

ABSTRACT

Visual representations differ in their capacity to encode binding information: in this paper we present a sequential binding task which requires a recurrent neural network to translate from a feature-based to a combinatorial scene-based representation. The mechanisms for binding information also depend on representational capacity. Binding information is easily carried by phase, but is not usually a component of neural network models. We propose a complex version of backpropagation for use with complex domain recurrent networks and assess the resources and requirements of the Simple Recurrent Network (SRN) and the Complex Domain Recurrent Network (CDRN) in simulations of the sequential binding task. Simulations demonstrate the improved performance and capacity of the CDRN.

1. INTRODUCTION

A scene is a composition of objects, related to each other in terms of position. However, the scene is initially perceived in terms of primary features such as a colour and movement. To construct a whole scene, a binding process is required which involves grouping sets of features together. The information for grouping features is based on spatial and timing cues: the eye fixates on different positions in turn where each position in the sequence provides space-time information for binding groups of co-occurring features and objects. The term, "sequential binding task", is used to refer to a task which requires binding several groups of features into distinct objects.

Representations differ in their ability to maintain distinct bindings of features to objects, with more powerful representations requiring more computational resources.

In this project we have been studying biological, psychological and computational aspects of visual binding tasks, primarily from a artificial neural network perspective. In this paper, we present a simple task analogous to a sequence of successive eye fixations, as a benchmark sequential binding task. In an initial simulation, we assess the representations and resources required for a Simple Recurrent Network (SRN) to learn the task. This simulation illustrates how the representational space is required to grow mutiplicatively with the number of items in a sequence, which is clearly a substantial problem for a conventional recurrent network. To deal with the combinatorial problem, we extended the conventional recurrent backprop network by adding phase to all the units, weights and biases (which we call a Complex Domain Recurrent Network - CDRN). The extension to the complex domain allows different phases to be used as "binding parameters" for different objects in the scene. With this extension, the CDRN requires twice as many parameters as the SRN, but the representational space requirements increase only linearly with the number of objects in a sequence, rather than multiplicatively.

In the section below, we first describe a novel sequential binding (SB) task, briefly place it in the context of biological and psychological studies of vision, then present the SRN and CDRN simulations and finally close with a discussion of the results.

1.1. NOVEL SEQUENTIAL BINDING TASK ANALOGOUS TO A SEQUENCE OF EYE FIXATIONS.

The task is a simplified analog of the task of building a representation of a scene from its component features. In our task, a "scene" comprises a set of coloured shapes, which are presented one at a time to a neural network. For example, the network is presented with a "red square", then "green triangle". Each time step comprises the presentation of the input units and updating each layer in turn in the SRN. The input representation is given in two fields, one for colour, and the other for shape. The task of the network is to associate each colour-shape pair together. The output representation is given in terms of a single colour-shape field. Thus, if there are m colours and n shapes, then the network requires $m+n$ input units, and $m \times n$ output units.

As mentioned above, representations differ in the binding information they can contain. The output representation clearly requires more resources than the input space (in the case of the neural networks, $m \times n$ outputs compared to only $m+n$ input units). However, it can represent composite scenes in a way that the input space cannot: the input space, using one field for colour, and another for shape, can represent at most one object (colour-shape pair) per time step; the output space can represent any possible combination of objects (however, note that it cannot represent the order in which they are observed - for that additional representational power is required). The SB task is interesting in that it requires the network to convert a representation of a scene given as a sequence of objects, into a static representation of the whole scene. The neural network must remember the early objects from the sequence as it adds later items. For this task, recurrent connections are required on the hidden layers so that the network can update its internal representation of the emerging scene. If there are k coloured shapes in a scene, then k time steps are required. For the network to learn this processing task, it must include a learning algorithm. For the first set of simulations, we used Elman's Simple Recurrent Network (Elman, 1990), which is a multi-layer network with recurrent connections only on the hidden layer. The method of training is backpropagation.

1.2. PSYCHOLOGICAL AND PHYSIOLOGICAL STUDIES IN BIOLOGICAL VISION

To what extend should we expect to find the computational concerns raised in the section above reflected in human vision? There is considerable evidence from psychological and neurophysiological studies of early vision as to the structure of the visual cortex. One of the most significant discoveries in recent years is the existence of multiple visual areas in the cortex, each optimized for the processing of different visual attributes of the stimuli (Wilson, H., R., 1995). Each of these areas consists a number of local groups of neurons sharing the same receptive field preferences. It has been reported (Eckhorn et al., 1988) that the coherence of neuron responses to specific stimuli occur not only for nearby neurons which belong to the same group, but also can occur between the neurons that are separated by several cortical columns (several millimetres of cortex) when these neurons share some receptive field preferences specific to this stimuli. Stimulus position, orientation, movement direction, and velocity are among those properties that were found to be responsible for producing stimulus-evoked coherent responses. It was found that there is synchronization between the responding neurons, between cells within cortical columns (Eckhorn et al., 1988), (Gray and Singer, 1989), in neighbouring hypercolumns, in distant hypercolumns (Gray and Singer, 1989) and between two different cortical areas (Eckhorn et al., 1988).

Visual feature integration could be explained by the concept of directed attention as central to visual binding (Prinzmetal, Presti and Posner, 1986). Some neurophysiological experiments have shown that neurons in the area 7 of the visual cortex can be identified in terms of their physiological properties, particularly the correlation of their responses and activity associated with fixation of gaze (Lynch et al., 1977). It is believed that the behaviour of neurons in area 7 of the visual cortex could help to explain the mechanism of directed visual attention (Lynch et al., 1977).

1.3. MODELLING OF THE BINDING SYSTEM BASED ON THE EVIDENCE FROM PSYCHOLOGY AND PHYSIOLOGY

The neurophysiological and psychological findings influence the modelling of visual processes and specifically those of the binding mechanisms (Andreou and Edwards, 1994; Damasio, 1989; Eckhorn et al., 1990; Frohn et al., 1987; Grossberg and Somers, 1991; Horn et al., 1991; Mozer et al., 1991; Sporns et al., 1989; Treisman, 1988). The initial assumption used in current binding theories is that different sensory features (such as colour, size, orientation, or shape) are registered in some specialized groups of neurons (Treisman, 1988), or are located in geographically separate regions in early sensory cortices (Damasio, 1989). Each group of neurons codes for different features values; for example red, green, or white is coded within the colour group. In Treisman's model, these basic features are coded automatically, in a preattentive stage, without focused attention, and spatially in parallel. In models based on the neurophysiological findings this preattentive stage of vision involves a stimulus-forced synchronization between common stimulus features (event-locked) (Eckhorn et al., 1990). When features must be located and conjoined to specify an object or sequence of objects, attention is required (Treisman, 1988). Based on the evidence from physiology and psychology the attention mechanism appears to be carried out by a network of anatomical areas. Orienting attention to location affects the perception of features and their integration (Treisman and Gelade, 1980; Treisman and Schmidt, 1982; Lynch et al. 1977).

1.4. COMPLEX VARIABLES AND ARTIFICIAL NEURAL NETWORKS

Most of the artificial neural network applications using complex variables are found in the area of signal processing (Widrow et al., 1974; Georgiou, 1992) and solution to Exclusive - Or problem (Kim and Guest, 1988). At present only one paper by Mozer et al. (1991) employs complex numbers to solve the visual binding problem. In this paper a neural network model, "Magic" learns how to group features of objects. Feature units are represented by complex numbers (amplitude and phase).The amplitude indicates presence (1.0) or absence (0.0) of the image feature. The phase indicates orientation of line segments (0, 45, 90, and 135 degrees) from which objects such as rectangles, diamonds, crosses, triangles, hexagons, and octagons are constructed from. Another function of the phase is to determine which image features are part of the same object and which should be segregated into different objects. Visual binding in this model is represented by phase-locking of features that belong to the same object and objects that belong to the same image. This model incorporates spatial feature integration, but does not include spatio-temporal feature integration.

2. EXPERIMENT 1 - SRN APPROACH TO SPATIO-TEMPORAL BINDING REPRESENTATION

2.1 AIMS

The sequence binding (SB) task allows us to investigate the spatio-temporal integration of objects for modelling visual binding. In this first simulation we analyse the behaviour of Multi-Layer Networks (Simple Recurrent Network - SRN) as binding mechanism for multiple components over time studying: convergence performance, level of erratic responses, representation of the hidden units illustrating binding model, and generalization. The above task is also called the Temporal Association Task [Hertz et al., 1991].

2.2 METHOD

For our task an input pattern comprises a sequence of 3 colour-shape objects, selected: 3 colours and 4 shapes. One shape and one colour unit are specified as 1, if the object exists. All input units (colour and shape) are 0s if the object does not exist. The SRN architecture comprises: 7 input units, 10 fully recurrent sigmoid hidden units and 12 sigmoid output units. The first step in the training process involved a pair of features (colour and shape) that is being presented to the network. The corresponding output unit is required to turn on. The hidden unit activations for this step are copied to the

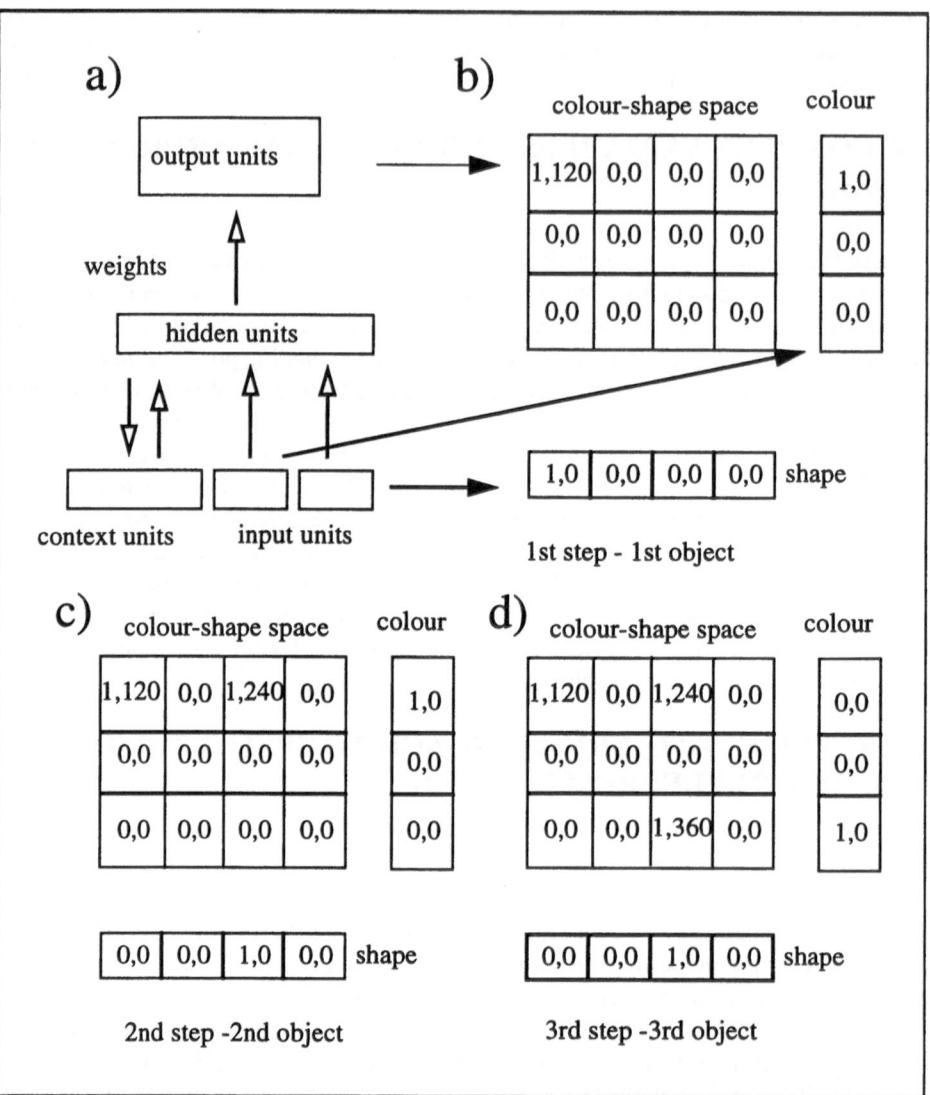

Figure 1. Network diagram and sequence of 3 objects. (a) In both the SRN and CDRN, the neural network consists 7 input units, (3 colour and 4 shape units), 10 context units, 10 hidden units, and 12 output units. (b), (c) and (d) examples of 1st, 2nd and 3rd step of the sequence. For the SRN, all units are real numbers. For the CDRN, absolute and phase values are indicated by two digits separated by comma. The first digit indicates the absolute value and second indicates the phase.

context units for storage. At the second step another pair of features is selected and presented to the network. The hidden units receive the current input information as well as the information stored in the context units from the first step. These two groups of contributing units should provide enough information to the hidden units to specify two objects on the output. At the third step the hidden units receive the third input pattern and the context unit information (consisting of the information from the first and second steps). The output units are expected to identify all 3 objects from the first, second and third steps. These 3 steps are repeated for all patterns in the pairs set. All weights and units (input units, hidden units, output units, context units and biases) are real numbers (see Fig.1). Two sets of input patterns for training and testing were selected randomly each containing 70 sequences of 3 input patterns. Weights were generated randomly. Each training and testing set of patterns represents ~5% of the total input pattern space. The network was trained for 1000 epochs.

2.3 RESULTS

After 1000 epochs the percentage of "bad patterns" was 14% (i.e. correct on a forced choice criterion). All patterns in the first position of the input sequence were correct, 93% in the second, and 66% in third were correctly coded (see Fig. 2)
The structure of the hidden unit space was explored using Principal Components Analysis (PCA) and Canonical Discriminants Analysis (CDA). These statistical methods identify best separation of the logical groups. The analysis has shown that the hidden unit space is maintaining an order of the colour space and the shape space. Also, the hidden unit space maintains groupings colours and shapes for multiple of sequences, showing that the order of the colour and shape spaces is transferred across multiple bindings.

2.4 DISCUSSION

The SRN simulation shows that the network is able to bind a single colour and shape (step 1 of the training process) with minimal error, but has severe capacity limitations for a sequence of three such bindings. To correctly represent *any sequence* of three objects, the hidden layer of the network must be able to discriminate between all sequences. For $m=3$ colours, $n=4$ shapes and $k=3$ items, $(mn)^k = 1728$ input sequences. However, since the output representation does not record the order of items, nor repeated items, its actual representational capacity need only be $\frac{(mn)!}{k!(mn-k)!} = 220$ sets of objects. For the training and the data set provided, the 10 hidden units are clearly having problems discriminating all patterns by step three. Using PCA and CDA to examine the hidden unit space, it was clear that the network was maintaining the position of items in the sequence, even though it was not required in this task. The combinatorial growth of the required hidden unit space is due to the way the SRN uses a single, real-valued output unit for each colour-shape pair. There is no "spare" variable to perform the binding function. In the next simulation we investigate the addition of phase to the SRN to overcome this problem.

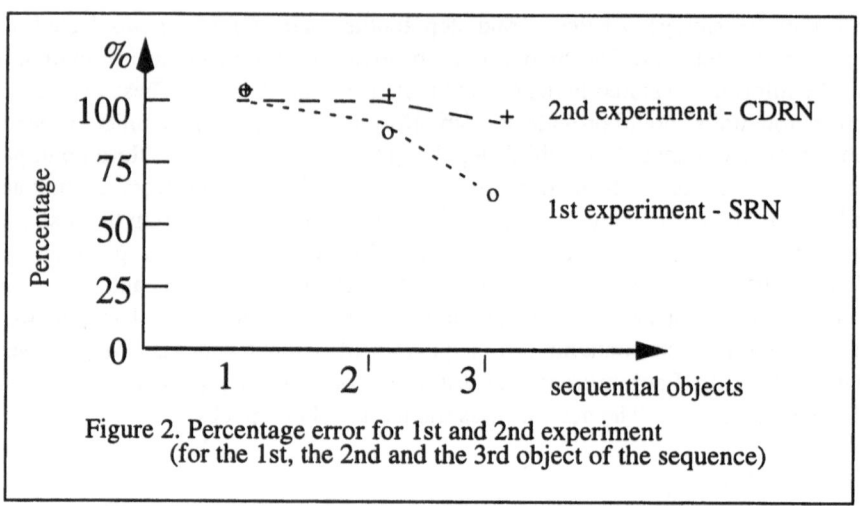

Figure 2. Percentage error for 1st and 2nd experiment
(for the 1st, the 2nd and the 3rd object of the sequence)

3. EXPERIMENT 2 - CDRN APPROACH TO SPATIO-TEMPO-RAL BINDING REPRESENTATION

3.1 CDRN

In this experiment, we added phase to the SRN as a binding mechanism. In contrast to the SRN, in the CDRN all weights and units (input units, hidden units, output units, context units and biases) are complex numbers. In the polar notation they consist of an absolute value, a, and a phase value, θ, (a, θ).

The network input, N_j, to the specific unit j is given by:

$$N_j = \sum_h A_h \cdot W_{jh}$$

where A_h is a complex valued activation, (a_h, α_h), and W_{jh} is a complex valued weight, (w_{jh}, β_{jh}), for unit h to unit j. Now the network input can be expressed as follows:

$$N_j = \sum_h \{a_h w_{jh} [\cos(\alpha_h + \beta_{jh}) + i\sin(\alpha_h + \beta_{jh})]\}$$

The network input N_j is passed through the squashing (activation) function

$$f(N_j) = \frac{N_j}{c + \frac{1}{r}|N_j|}$$

(Georgiou and Koutsougeras, 1992) where c controls steepness of $|f[N_j]|$ and r satisfies the condition $|N_j| < r$. The phase of N_j is the same as that of $|f(N_j)|$. Following Georgiou's discussion this function has required properties to be a suitable activation function. Another useful property of this activation function, particularly for our application, is that $|N_j|$ is squashed to $|f(N_j)|$ in an analogous way the real sigmoid maps point on the real line, from the range $-\infty \to \infty$ to $0 \to 1$. The complex valued output pattern is compared to the complex valued target pattern and the error for an input pattern is calculated $E = \frac{1}{2}\sum_k \varepsilon_k \bar{\varepsilon_k}$ where $\varepsilon_k = T_k - O_k$ and $\bar{\varepsilon}$ indicates the complex conjugate (T_k and O_k are the complex target and output values). The target, $T_k = (t_k, \phi_k)$, is specified by the amplitude t_k and by the phase ϕ_k. The output $O_k = (o_k, \delta_k)$, is specified by the amplitude o_k and by the phase δ_k. For these target and output units the error function is given by:

$$E = \frac{1}{2}\sum_k [t_k^2 + o_k^2 - 2t_k o_k \cos(\phi_k - \delta_k)]$$

As in experiment 1, the CDRN was trained and subsequently tested with 70 sequences of 3 input patterns. The absolute value of the input pattern of an object can be specified: 1 or 0, (depends whether object exists or not). However now the representation requires phase as well as absolute value. Phases of all the inputs in the pattern are set to zero. Weights and biases are generated randomly (absolute and phase values).
The phase space of an output is divided into three equal sections (each 120 degrees) representing the first, second and third objects presented to the network during first, second and third step (see Fig.1). Phase, in this experiment, is a feature-binding agent as well as an identifier of the sequential position of an object.

3.2 RESULTS

After training the network for 1000 epochs, all patterns in the first, second and 98% in the third position were correctly coded (see Fig.2) and the sequential position of correctly coded objects was correctly displayed in the colour-shape output space. The analysis (using PDA and CDA) of the absolute value of the hidden unit space showed grouping of the colour-shape spaces for multiple of sequences and the separation of the logical groups as for SRN. The phase space of the hidden units was also analysed. It has shown that hidden units built explicit representation of consecutive objects of the sequence. Objects that belong to a specific sequence are clearly separated from other sequences and the sequential position of the objects within sequence can easily be traced. As a consequence of this, multiple identical objects are identifiable.

3.3 DISCUSSION

This experiment clearly shows improved performance of the CDRN over the SRN. The main point of comparison between the two types of network is the capacity in the real space to represent the set of items in the sequence. In this task, both networks have 12 output units with real values representing the presence or absence of the items. The CDRN has additional capacity to represent bindings in the phase domain, and clearly this makes a significant different in performing the task. Further simulations are required to test the capacity of the SRN with the same number of parameters as the CDRN. Such a comparison would not be satisfied simply with twice as many hidden units, since that would quadruple the number of recurrent weights. Simulations are currently being designed to test these capacity issues.

In addition to the ability to represent the set of items in the final sequence, the CDRN is also able to represent their order. This increased representational power allows the output values to represent multiple identical items.

4. CONCLUSIONS AND FUTURE WORK

In this paper we have proposed a benchmark sequential binding (SB) task, and developed a complex-domain recurrent network (CDRN), with a complex version of the backpropagation algorithm. Simulations of the SRN and CDRN show the improved representational and learning capacity of the CDRN on a small version of the SB task. In current simulations we are exploring the ability of the CDRN to learn additional feature sets and longer sequences, and in future work intend to explore its applications to more complex visual binding tasks.

5. ACKNOWLEDGEMENTS

We acknowledge the support we were given from School of Information Technology for this research. We thank the Cognitive Science group for stimulating discussions and feedback.

6. REFERENCES

Andreou, A., G.and Edwards, T., G. (1994). VLSI Phase Locking Architecture for Feature Linking in Multiple Target tracking System. In Cowan, J., G. Tesauro, G. and Alspector, J. editors, *Advances in Neural Information Processing Systems* 6, pages 866-873, San Francisco CA: Morgan Kaufmann

Damasio, A. R. (1989). The brain binds entities and events by multi-regional activation from convergence zones. In Churchland, P., editor, *Neural Computation* 1, pages 123–32.

Eckhorn, R., R.Bauer, Jordan, W., Brosch, M., Kruse, W., Munk, M., and Reitboeck, H. J. (1988). Coherent oscillations: A mechanism of feature linking in the visual cortex? In *Biological Cybernetics, 60*, pages 121–130.

Eckhorn, R., Reitboeck, H. J., Arndt, M., and Dicke, P. (1990). Feature linking via synchronization among distributed assemblies: Simulations of results from cat visual cortex. In Koch, C., editor, *Neural Computation* 2, pages 293–307.

Elman, J., L. (1990). Finding Structure in Time. Cognitive Science 14, 179-211.

Frohn, H., Geiger, H., and Singer, W. (1987). A self-organizing neural network sharing features of the mammalian visual system. In *Biological Cybernetics, vol.*55, pages 333–343.

Georgiou, G., M. and Koutsougeras, C. (1992). Complex Domain Backpropagation. In *IEEE Trsansactions on Circuits and Systems II: Analog and Digital Signal Processing, vol. 39(5)*, pages 330-334

Gray, C. M. and Singer, W. (1987). Stimulus-dependent neuronal oscillations in the cat visual cortex area 17. In *Neuroscience [Suppl]* 22, page 1301.

Grossberg, S. and Somers, D.(1991).Synchronized oscillations during cooperative feature linking in a cortical model of visual perception. In *Neural Networks* 4,pages 453–466. Pergamon Press.

Hertz, J. Krogh, A., and Palmer,R., G. (1991) *Introduction to the theory of neural computation;* Addison-Wesley Publishing Company

Horn, D., Sagi, D., and Usher, M. (1991). Segmentation, binding and illusory conjunctions. In Koch, C., editor, *Neural Computation 3*, pages 510–525.

Kim, M., S. and Guest, C., C. .(1988) . Modification of Backpropagation Network for Complex-Valued Signal Processing In *Proceedings of the 1988 International Conference on Neural Networks, vol.*III. pages 27-31. Morgan Kaufmann

Lynch, J., C. Mountcastle, V., B. Talbot, W., H. and Yin, T., C., T. (1977).Parietal lobe mechanisms for directed visual attention. In *Journal of Neurophysiology Vol.*40, *No* 2, pages 362-389.

Mozer, M. Zemel, R., S. and Behrmann, M. (1991). Learning to segment images using dynamic feature binding. Technical Report CU-CS-540-91. University of Colorado, Boulder, USA, Department of Computer Science and Institute of Cognitive Science

Prinzmetal, W., Presti., D.,E., and Posner, M I. (1986). Does attention affect visual feature integration? *Experimental Psychology, Human Perception and Performance,* vol.12, No.3

Sporns, O. Gally, J., A. Reeke, G., N. and Edelman, G., M. (1989) Reentrant signalling among simulated neuronal groups leads to coherency in their oscillatory activity. In *Proceedings of the National Academy of Science USA,* vol. 86. pages 7265-7269

Treisman, A. (1988) . Features and objects : The fourteenth Bartlett memorial lecture *Experimental Psychology, Human Experimental Psychology, 40A(2)*, pages 201-237.

Treisman, A. and Gelade, G. A. (1980). A feature integration theory of attention. *Cognitive Psychology*, 12, pages 97–136.Widrow, B. McCool, J. and Ball, M. (1975) The complex LMX algorithm. In *Proceedings of the IEEE, Proceedings Letters*, pages 719-720.

Wilson, H., R. (1993). Nonlinear process in pattern discrimination and motion perception. In Harris, L. and Jenkins, M., editors, *Spatial vision in humans and robots: the Proceedings of the 1991 York conference on Spatial Vision in Humans and Robots. Cambridge University Press.*

Schema-Based Learning:

Biologically Inspired Principles of Dynamic Organization

Fernando J. Corbacho and Michael A. Arbib

Center for Neural Engineering
University of Southern California
Los Angeles, CA 90089, U.S.A.
corbacho@pollux.usc.edu,   arbib@pollux.usc.edu

ABSTRACT

We propose a generalized framework Schema-based learning (SBL) for the design of complete and integrated adaptive autonomous agents incorporating general principles of adaptive organization e.g., bootstrap coherence and coherence maximization principles. A schema is an evolutionarily or experience-based constructed recurrent pattern of interaction or expectation (perceptual, motor, reactive, and predictive schemas) with the environment, and coherence is a measure of the congruence between the result of an interaction with the environment and the expectations the agent has for that interaction. SBL attempts to provide a general and formal framework independent of the particularities of implementation, thus allowing the design and analysis of a wide variety of agents. SBL allows the growth of increasingly complex patterns of interaction between the agent and its environment from an initially restricted stock of schemas while allowing for efficient learning by confining statistical estimation to a narrow credit assignment space.

1. INTRODUCTION

This paper attempts to provide a theory of organization for the analysis and design of Adaptive Autonomous Agents (*AAA*). By *autonomous* we mean that the system can perform many of its "survivability functions" as well as play a critical role in selecting its own training set while deciding for itself how to relate perceptions to actions. Different disciplines have attempted to develop design theories for AAA: Artificial Intelligence, Control Theory, and Neural Networks, to name a few. They all have provided with partial success in restricted tasks and domains. nevertheless, none has yet provided an overall integrated theory. On the other hand, biological agents are flexible, adaptable, and highly robust; and they are so at all levels of organization and functionality. These agents "designed" by evolution survive in natural (usually very complex) environments where no current artificial agent could.

So we claim that fundamental principles have yet escaped current theories for AAA. We must rethink in a principled way what are the principles of organization, and what are the unit(s) of organization. In this paper we propose a theory of organization for more robust, flexible and adaptive AAA inspired by the organization of their biological siblings. One of the backbones of our theory is based on the general idea that AAA should be designed to maintain *coherence* with their environment. Coherence of a particular agent with respect to a particular environment is a measure of the congruence between the result of an interaction with the environment and the expectations the agent has for that interaction. We will then claim that coherence is very much related to survival. In this respect we will introduce the *Coherence Maximization Principle* (CMP): Given a current interaction with the environment measure the coherence of the expectation with respect to the current results of the interaction, and try to maximize the coherence by adaptation.

We propose that the design of the overall agent (natural or artificial) should be done (has been done by evolution) by aggregation of both units of interaction with the environment, as wells as "dual" units of *expectation* about those interactions (formed from successful experiences in similar interactions). Thus, by allowing the system to compare its expectations with the actual results obtained after the interactions, the system may focus its adaptation mainly on the incoherent units. We, therefore, introduce the *Bootstrap Coherence Principle* (BCP) which defines which initial units of interaction and expectation the agent should be initially provided with to achieve a particular chance of survival in a particular environment. Bootstrap coherence provides the agent with the ability to anticipate the results of some of its "seed" interactions, and, thus, expect certain changes in its internal representations accordingly. All the seed schemas included by BCP can be considered the initial bias for the adaptive system. Yet this bias reflects the regularities in the interactions between the agent and the environment.

Schema-Based Learning (SBL) is the architecture implementing these principles of organization. It provides with a generalized framework incorporating general principles of adaptive organization, e.g., CMP and BCP. The basic unit of organization in SBL is the *schema*. A schema is an evolutionary constructed or experience-based constructed unit of interaction (perceptual, motor and reactive schemas) or expectation (predictive schema). Animal behavior is controlled by significant patterns of interaction which are useful again and again, e.g., grasping. They provide a somehow vague description of a situation and action, and on the other hand are specific enough to be applicable. They represent what is stable and therefore generalizable over variability becoming precise through adaptation (Arbib, 1992).

One of the main goals of SBL is to grow increasingly complex patterns of interactions (schemas) from an initial stock (seed schemas) in an harmonious way which does not destroy the primitive basic functionality but which enhances, and increments it. SBL relies on two related important aspects/assumptions:
i) It is advantageous (efficient) to cope with new experiences based on positive results from previously similar experiences. It is efficient since the agent avoids the statistical estimation required

to form the past successful action-perception correlations. The current stock of schemas allows to perform a broad-brush analysis of the action-perception "scene" e.g. prey detection/catching, rude grasping in babies. SBL does not replace statistical estimation, rather, it confines it to a narrow search space (initial islands of reliability), so that the system learns in the right ballpark with minimum *scene* statistics.
ii) SBL allows for the incremental construction of increasingly more complex schemas by aggregation of schemas (stable units of composition) to reflect increasingly more complex interactions with the environment. These new constructed schemas may then be tuned by experience into a well-adapted integrated unity. Without this, given a new task, the system would start *ab initio* to train a new unstructured network dedicated to the task. Thus, learning is based on the current stock of schemas (seeds). Ultimately SBL must be able to explain the design of the set of "seed" schemas as they are the basis for the construction of more complex ones.

2. CASE STUDIES

Due to space limitations we will focus mainly in *Rana Computatrix*, other case studies are included in (Corbacho and Arbib, 1996ab) (henceforth referred to as C&A96ab). *Rana Computatrix* - the computational frog, Arbib (1987) consists of a set of evolving models of integrated adaptive visuo-motor coordination in the frog. The reasons for using *Rana Computatrix* as one of our test beds for SBL were spelled out in (Corbacho and Arbib, 1995) (henceforth referred to as C&A95; here let us simply recapitulate the most important ones: (i) SBL is constrained by data on a neuroethologically sound system -as to the task, the environment and the agent, (ii) the work on *Rana Computatrix* allows for horizontal integration (across many integrated functionalities) and not just vertical integration (action-perception within one central functionality, e.g. prey catching) thus, giving rise to truly "complete" agents, and (iii) Frogs are quite flexible and adaptable systems (specially if compared to any existing robot), for instance, learning to detour has proved to be a very adaptive process relying on important processes of learning (C&A95; Corbacho et al. 1996).

680

SBL has also been applied to the construction of motor synergies and the formation of cognitive maps for navigation (C&A96b).

2.1. Learning to Detour

A frog or toad, viewing a vertical paling fence barrier (e.g., a row of chopsticks) through which it can see a worm, may either approach directly to snap at the worm, or detour around the barrier (Collett, 1982). We address the questions: "How does the probability of detouring change with experience for a given width of barrier?" and "Can the animal generalize from one configuration to another one?" We present data suggesting that for certain prey-barrier configurations the detour behavior is acquired through refinement and extension of an innate behavioral repertoire. Here we sample a few of our observations of the main capabilities of frogs *Rana pipiens* for detour behavior which set challenges for our learning model. We refer the reader to (Corbacho et al., 1996) for more details.

Experiment: Barrier 20 cm wide

From now on we will refer to a frog which has not been exposed to the barrier paradigm as naive.
Observation 1: If the chopsticks are placed the same distance apart, so that the gaps have the same width, and the barrier is 20 cm wide, then the naive frog tends to go for the gap in the direction of the prey (this was the case for 88% of the trials). The frog starts out approaching the fence trying to make its way through the gaps. During the first trials with the 20 cm barrier the frog goes straight towards the prey thus bumping into the barrier. When the frog is not able to go through a gap towards the prey it backs-up about 2 cm and then reorients towards one of the neighboring gaps with no apparent bias (see Fig. 1A).
Observation 2: After 2 (43%) or 3 (57%) trials, the frog is already detouring around the barrier without bumping into the barrier (see Fig. 1B). The behavior involves a synergy of both forward and lateral body (sidestep) movements in a very smooth and continuous single movement.

3. SBL UNDERPINNINGS

3.1. Efficient Learning in SBL: Bias

SBL will try to address two of the most fundamental problems in adaptive systems, namely, how to decide what substructure(s) is (are) to blame when the system is in error, and how to fix it (them) (cf. credit assignment).

It is advantageous (efficient) to cope with new experiences based on positive results from previously similar experiences.

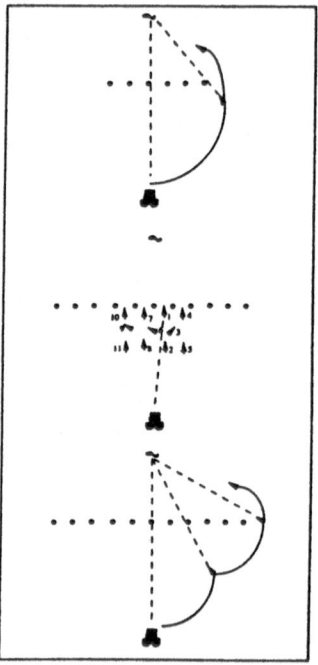

Figure 1. A. Approach to prey with single 20 cm barrier interposed: first trial with frog in front of 20 cm barrier (numbers indicate the succession of the movements). **B.** Approach to prey with single 20 cm barrier interposed: after 3 trials with frog in front of 20 cm barrier. Arrowheads indicate the position and orientation of the frog following a single continuous movement after which the frog pauses. This figures show actual "typical" trajectories from start to finish (traced from video).

This is similar to the learning across tasks proposed by (Thrun, 1996) in his life-long learning approach to machine learning. It is efficient since the agent avoids the statistical estimation required to form again past successful action-perception correlations. The current stock of schemas (bias) allows the performance of a broad-brush analysis of the action-perception "scene" e.g., prey detection/catching, crude grasping in babies, etc. SBL does not replace statistical estimation, rather, it confines it to a narrow search space, so that the system learns in the right ballpark with minimum scene statistics.

When the agent interacts with the environment, many of its component structures will be undergoing major changes in their patterns of activity. We claim that a system initially designed to actually expect many of these changes will have a much reduced space of unexpected changes where the adaptation can be "played", thus, reducing the space for credit assignment. The space of potential "culprits" is mainly reduced to the space of incoherent units - and, hence, will require minimum statistical estimation.

3.2. Bootstrap Coherence Principle

Bootstrap design should include initial "seed" units of interaction to allow good enough chances of survival. In the case of *Rana computatrix* -as many other AAA- the agent needs to feed, thus recognize food, food may be scattered so the agent must be able to navigate -avoid obstacles not to self-destroy- and approach food. It may also incorporate adaptation capabilities if so required by the agent's ecological niche. In other words Bootstrap Coherence corresponds to the necessary initial structure to guarantee initial (partial) coherence with the environment.

Space constraints do not allow the introduction of all the seed schemas for Rana Computatrix. We introduce a few used for the examples in this paper and refer the reader to (C&A96a) for more details. Several perceptual schemas e.g. PREY-REC (food recognition), SO-REC (obstacle recognition), several reactive schemas e.g. SO-MHM (obstacle avoidance), several motor schemas

FORWARD and SIDESTEP (further notation will be introduced when we "formally" define schemas).

The Bootstrap Coherence Principle (BCP) also determines the predictive schemas that should be incorporated in the initial structure to guarantee initial survivability - react fast enough- avoiding huge credit assignment spaces. All predictive schemas could be learned by interactions, nevertheless, the agent would not probably survive. So evolution has built in "pairwise" not letting too much complexity on the interaction side without introducing corresponding complexity on the predictive side. This will allowin better chances of survival by better action selection (e.g. goal regression) and better capability for learning (reduced "search" spaces).

3.2.1. Duality: Schemas Dependencies

As already mentioned the agent comes equipped not only with units of interaction but also with *dual* units to predict the result of the corresponding interaction. Hence, providing the agent with the ability to anticipate the results of its basic actions and, thus, expect certain changes in its internal representations accordingly. We have introduced the corresponding predictive schemas linked to the dual perceptual, motor or reactive schemas to allow the agent to begin its interactions with bootstrap coherence.

Thus, SBL needs to "represent" both the predictive response and the actual observed responses. In general we will define a space as the schema instances space (vector spaces in neural networks). For instance the PREY-REC schema will "instantiate" (represent the particular parameters of the instance) in the parameter space PREY. For every space Q we introduce $\hat{Q}$ as the space of expectations about Q. $\hat{Q}$ can be thought as a "working memory" of Q. A particular instance e.g., $\hat{q}(t)$ represents the expectation for $q(t+1)$. This allows for the comparison/match of both representations, hence, detecting incoherences.

To "start" the agent with partial bootstrap coherence several *dual* predictive schemas should be included. C&A96a included several *dual* predictive schema to reflect upon the changes that a given motor action exerts over the state of the particular perceptual schema. For instance PREY-M-$\hat{PREY}$ where M is a motor action and $\hat{PREY}$ anticipates the resultant pattern in PREY in case that action is taken (analogously for $SO - M - \hat{SO}$).

C&A96a also included several predictive schemas to take into consideration what the system expects when activating the corresponding reactive schema. The predictive schema "checks" whether the reactive schema performs as expected -is the result of the interaction with the environment as expected? For instance SO-MHM (obstacle avoidance) may have to be tuned and the role of its predictive schema is to determine when it should be tuned. $SO - MHM - TAC\hat{T}ILE$ represents a predictive schema related to the reactive schema SO-MHM with expectation space for TACTILE. If the SO-MHM schema works appropriately, then the agent should not detect tactile stimulation due to bumping into the stationary objects perceived, that is, $SO - MHM - TAC\hat{T}ILE$ will be coherent.

4. SBL ARCHITECTURE

We claim that any agent must be analyzed (designed) with respect to its environment. The environment provides the agent with an interaction space. Ultimately the behavior of any agent can only be understood in relation to the synergy agent-environment. We now go beyond the particularities of each case study to provide the more general SBL framework. Our goal includes defining a formal framework that will allow to explicitly define schemas, and the corresponding operations upon them.

Definition. An *Environment* E is a space which includes a collection of entities and their relations (interactions). A particular instance configuration of E at time t will be denoted as $e(t)$. E is defined by a set of statistical regularities including structural relations within $e(t)$ and constraints on transitions from $e(t)$ and $e(t+1)$.

C&A95 introduced a specific E namely, a 150x150 grid where different entities (e.g., barrier, frog, worm) interact. Part of the environment functions is the mapping $F_r: E \rightarrow receptors$ where receptors correspond to agent's "surfaces" "in contact" and receiving input from the external environment. An agent may have several types of receptor surfaces e.g. retina receptor cells, tactile receptor cells. C&A95 introduced a particular example where $receptors^i \subset \Re^{n \times n}$ is the n^2 dimensional space of views from that environment. A particular instance $receptors^i(t) = F_r(e(t))$ corresponds to a particular "view" of $e(t)$. Another environmental function is $F_e: effectors \rightarrow E$ mapping the effect of the agent actuators on a subset of the environment's parameters; where effectors correspond to agent's "surfaces" "in contact" and affecting the external environment. Different environments may have different mappings for the same effectors activity (e.g. ground level vs. under-water).

Definition. An *Adaptive Autonomous Agent* (*AAA*) is defined by a triplet $(receptors, effectors, S(t))$ where $S(t)$ is a collection of interconnected "seed" schemas (defined below). Receptor surfaces define the projection from E upon AAA. Effector surfaces define the projection from AAA upon E. For instance in *Rana Computatrix* we consider the retina receptor cells, tactile receptors, and propioreceptive receptors. Perceptual schemas have their "seed" on the receptor surfaces. Motor schemas end in the effector surfaces. In order to define the structure of the agent we need to define the schemas that directly interact/communicate with the environment as well as the internal schemas for internal interactions (all sorts of internal interactions take place defining in some sense another "internal environment").

Definition. A *schema* S^i is determined by $\left(C^i, P^i, \Pi^i, R^{h,i}\right)$ where Π^i is the schema mapping $\Pi^i : C^i \to P^i$ from $C^i \subset \prod_{\alpha \in D} L_\alpha^\alpha$, that is an arbitrary number of (other schema) parameter spaces (or AA receptors) to $P^i \subset \prod_{\beta \in R} L^\beta$ the schema parameter space (the instance space, typically a vector space in *Brain Theory*). $R^{h,i}$ is the support vector from other schemas S^h that are connected to S^i.

For instance the obstacle avoidance schema contains the mapping $\Pi^{SO-MHM} : SO \to MHM$ where SO is the parameter space for the obstacle recognition schema (SO-REC) and MHM is the parameter space for obstacle avoidance (a Motor Heading Map). The meaning/grounding of a schema comes partially from its relation to the other schemas within the AAA system. A schema may be composed of several (sub)schemas and each component schema may also be composed of further subschemas recursively.

Definition. A *schema instance* (SI) $s_m^i(t)$ consists of a pair $\left(\left(l_m^{i,1}(t), \ldots, l_m^{i,\#R}(t)\right), a_m^i(t)\right)$ where $\left(l_m^{i,1}(t), \ldots, l_m^{i,\#R}(t)\right)$ define the schema instance parameters (e.g. size, location). The instance has also an associated activation variable $a_m^i(t) \in \Re$ reflecting its activation level. Each instance must have its own *activation variable* label since different instances of the same schema may be active at the same time and be part of different schema assemblages. The instance spaces employed in Neural Networks are mainly vector spaces, e.g. neural layers with certain properties. The actual pattern of activity in the layer being an instance -element in the space (C&A96b).

Definition. A *predictive schema* $\hat{S}_j^i$ (dual of the schema S^i) is a schema with the schema mapping $\hat{\Pi}_j^i : L^j \times |S^i\rangle \to \hat{L}^j$ where L^j indicates the "context", $|S^i\rangle$ indicates the intended schema instantiation, i.e. internal simulation not actual

instantiation - the notation $|X\rangle$ means triggering internal processes associated with X (we will later introduce brain operators); and $\hat{L}^j$ corresponds to the expectation parameter space. Many units of interaction have associated dual units of expectation. The dual mapping takes into consideration the effect of the interaction on the internal representations.

5. SBL SHEMA OPERATIONS

5.1. Instantiation/Activation

Schema activations are largely task driven, reflecting the goals of the agent and the physical and functional requirements of the task. Activation of schemas is a distributed dynamic process without the intervention of a central manager. A schema network does not, in general, need a top-level executor since schema instances can combine their effects by distributed processes of competition and cooperation (i.e., interactions which, respectively, decrease and increase the activity levels of these instances -von der Malsburg (1986, 1994), rather than the operation of an inference engine on a passive store of knowledge (e.g. GOFAI). This may lead to apparently *emergent behavior*, due to the absence of global control.

Definition. A schema S^i is *instantiated* with instance $s_m^i(t)$ when $a_m^i(t) > th^i$ where $a_m^i(t)$ depends on the *support* from the other related schema instances through $R^{h,i}$

$$a_m^i(t+1) = a_m^i(t) + \sum_h a_m^h(t) \times R^{h,i}$$

For instance the motor schemas action on the controlled "musculature" is only enabled when their *activation variable* surpasses the threshold th^i. The instance will then correspond to the pattern in the schema parameter space.

5.2. Coherence Maximization

Learning is triggered when predictions made by predictive schemas do not match the observed results and when the goals are not achieved. The dynamics of the system should first try to reduce the error by tuning its current stock of schemas and when that fails it should then construct new schemas to reduce the error. In the long term, the reliability of the predictions will increase, making the system increasingly better in predicting the results of its actions for a given context.

An incoherence is detected iff

$$\left\| l^j(t+1) - \hat{l}^j(t) \right\| > th^k$$

Coherence Maximization reduces the difference by accommodation of the mappings related to both spaces while not increasing other incoherences. This can be a complex process involving in itself large credit assignment in many of our cases this is solved by decomposing this mapping in simpler mappings -along with their respective predictive schemas- . By decomposing in simple mappings credit spaces become smaller (C&A96b).

How is the system to decide which operation to perform? For any incoherence several operations could take place e.g. tuning of the reactive or/and dual predictive schema; construction of a new schema and its corresponding dual predictive schema (e.g. in learning to detour).

In many cases the system will construct both the Reactive and Predictive Schemas - cf. learning the inverse and the forward models simultaneously (Jordan and Rumelhart, 1992). In Learning to Detour both reactive and predictive are learned simultaneously, the reactive reflects a new pattern of interaction and the predictive reflects the results of that interaction in a particular context.

5.3. Schema Tuning

Schema tuning corresponds more to "parametric" changes within a schema (e.g., skill refinement) rather than to any structural change or creation of a new unit of interaction (e.g., skill formation) which corresponds more to schema construction (refer to next section for a complete formal definition). Tuning consists of changes on an already existing mapping.

For a schema to be tunable it must be equipped with yet another mapping, a mapping from some "incoherence space" to the space of the schema's tunable parameters θ^i. For tuning we require the

existence of a predictive schema $\hat{S}^i_j$ (its construction defined later). So that $L^j \rightarrow S^i_\Theta$, that is $\left(l^j(t+1) - \hat{l}^j(t) \right) \mapsto (\theta^i_1, \ldots, \theta^i_N)$

We will provide with an example in section 6.1.

5.4. Schema Construction

SBL is also able to learn "unexpected" things by detecting new correlations of unexpected patterns with their possible causes. What is the context for the unexpected pattern, what was the related action?

Definition. Schema *construction* builds a new mapping $\prod_{\alpha \in D} L^\alpha \rightarrow \prod_{\beta \in R} L^\beta$ where Domain and Range are parameter spaces. The full exposition is beyond the scope of this article (please refer to C&A96b). Here just point out that first a new relation is noticed $\left(l^1_D(t), \ldots, l^N_D(t) \right)$ and then a new mapping is defined

$$\left(l^1_D(t), \ldots, l^N_D(t) \right) \mapsto \left(l^1_R(t), \ldots, l^N_R(t) \right)$$

(C&A96b) discuss the construction of perceptual schemas (construction of complex feature detectors, creation of a new "context"). As well as the construction of motor schemas: new pattern over effectors that proves to produce new successful coherent interaction with the environment. Motor schemas may be activated simultaneously, thus producing emergent motor patterns e.g. FORWARD(t) & SIDESTEP(t) or in series to produce a sequence of actions.

Definition. Predictive schema *construction* builds a new mapping with L^j and $\left| S^i \right\rangle$ in the Domain and $\hat{L}^j$ in the Range. Where $l^j(t+1) \neq \hat{l}^j(t)$ was the incohence caused by the instantiation of S^i. The system can construct predictive schemas given a current *context* state and the current *observed* result, so that the current observed result becomes the predicted result in future applications of the schema in the same context.

6. LEARNING SEQUENCE

The adaptation of the system is reflected as a sequence of schema tuning and/or construction when incoherences arise, thus giving rise to the emergence of transitions.

The present section provides the most relevant results of the simulations on learning that occurs when a naive frog is repeatedly presented with a wide (20 cm) barrier separating it from a worm. All these simulations take place in Rana Computatrix and only reflect a limited subset of the possible learned behaviors.

6.1. Obstacle-Avoidance Tuning

When a barrier is present in the visual field of the agent SO-REC will be instantiated in turn instantiating SO-MHM (obstacle avoidance) which in turn may activate its dual predictive schema

$$SO - MHM - TAC\hat{T}ILE \quad (TAC\hat{T}ILE = 0).$$

The view field of the agent after having approached the barrier (Fig. 2A) contains a gap which triggers an approach (forward) action towards it (Fig. 2B). Upon reaching the gap, the agent bumps into the fence, thus, making the TACTILE schema active (i.e. TACTILE = 1). Since

$$TACTILE(t+1) = 1 \quad \text{yet} \quad TAC\hat{T}ILE(t) = 0$$

thus, $SO - MHM - TAC\hat{T}ILE$ is in incoherence. By the CMP the system "realizes" that the reactive schema did not perform as expected, and thus the reactive schema must be tuned. So for the current schema (see Schema Tuning for more details)

$$T^{SO-MHM} : TACTILE^{INCO} \rightarrow SO - MHM_\theta$$

where $SO - MHM_\theta$ represents the adaptive parameters associated with the SO-MHM schema. By instantiating this function the reactive schema SO-MHM gets tuned. (C&A96a) discuss in detail the tuning of this reactive schema by tuning its schema mapping -implemented by a connectivity kernel that becomes more inhibitory.

After the increase in inhibition the system reaches an *impasse* in MHM (Fig. 2C). Namely the

system has learned that these gaps should not be approached.

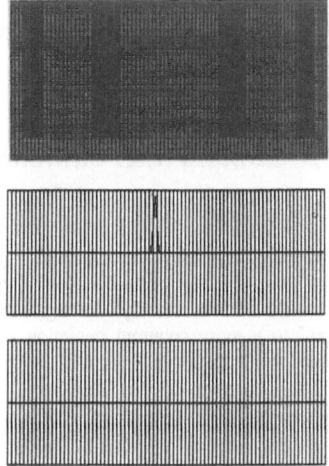

Figure 2. A. The view when the agent is close to the barrier. **B.** The activity in MHM before adaptation reflecting the target heading angle for the gap in front of the agent. **C.** Activity in MHM after adaptation reflecting the "impasse", i.e., no target heading angle. The central gap appears wider since the perceived gaps width decreases with eccentricity.

6.2. Constructing Detour Predictive Schema

Then, the system dynamics give rise to "exploratory" behavior (biased by perceptual gating; see C&A for details) which as a result eventually produces an action to be taken (in our example FORWARD+SIDESTEP motor synergy). After taking several sidesteps the agent happens to be close to the end of the barrier (Fig. 3A). At this point the same strategy of FORWARD+SIDESTEP unexpectedly gets the agent at the end of the barrier. This gives rise to an *incoherence* as the expected result

$(M\hat{H}M(t) = \{0^*\}$ the impasse activity in Fig. 2C) of the action does not match the observed result ($MHM(t+1) = \{0^{\bullet}10^{\bullet}\}$ target heading angle in Fig. 3B).

$$\xi_k(t+1) = \left| MHM(t+1) - M\hat{H}M(t) \right| > th^k$$

The unexpected change triggers the construction of a new predictive schema consisting of

context $q^i(t+1) = MHM(t)$;
action

$$q^j(t+1) = SIDESTEP(t)\ \&\ FORWARD(t)$$

expected result $\hat{q}^k(t+1) = MHM(t+1)$.

This new predictive schema then starts as a gross approximation of the interaction between the agent and the barrier and eventually, through tuning, "converges" to a more coherent depiction of the interaction.

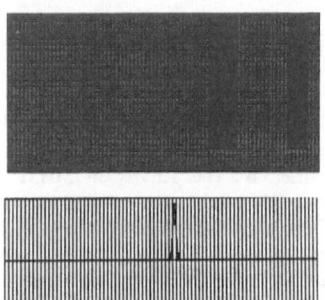

Figure 3. A. The view when the agent is close to the barrier and on the edge. **B.** The activity in MHM reflecting the large gap to the side of the barrier.

7. CONCLUSION

That is, *Rana Computatrix* ends up with its Obstacle Avoidance schema tuned to better discriminate between passable and not passable gaps. The system also ends up with a new predictive schema which allows the system to detour around a barrier without bumping into it. The model generates several predictions at the behavioral and physiological levels (C&A96b).

A general question is how the system "decides" which structures are modulated, and what type of modulation is exerted upon the "chosen" structures. In general the "culprit" may come from different structures, and thus the system faces a *structural credit assignment problem* - the problem of deciding which subset of those structures should be adapted and how. We have shown how to restrict the space for Credit Assignment, and, thus, Reduce Statistical Estimation. Only Incoherent Units define initial Credit Assignment Space. The result is a "almost one-shot" (2-3 trials) learning, system dynamics focus onto the right structures. We also claim that as long as the Bootstrap Coherence is maintained the system scales up without introducing credit assignment problems. We claim that SBL supports more efficient learning and allows for more complex dynamic construction of behaviors. Thus, we demonstrate how predictive schemas under the BCP and CMP can reduce the credit assignment spaces giving rise to efficient learning beyond that of Foner and Maes (1994) who only reduce the credit assignment space by focus of attention. SBL also allows for more flexible and robust learning within the *active learning* paradigm (the agent can select its own training set), thus elaborating on some issues not sufficiently addressed in supervised learning.

7.1. Comparison to related approaches

Learning to detour could be implemented in a backpropagation network (Rumelhart et al., 1986), where the input map would be the same as the current input to our perceptual maps, and the output map would be the motor action the system is to perform. The problem, then, becomes clear: the learning time for the backpropagation network to "converge" to some reasonable solution would be very large; and it would grow exponentially for even larger systems (bad scaling up). This is mainly due to the lack of structure in the weight space of the backpropagation network.

For related approaches to learning in Artificial Intelligence (AI), refer to Carbonell & Gil (1990), Drescher (1991), and Shen (1994). All of them were greatly influenced by Piaget's work on development in the child (Piaget, 1954). They all provide autonomous learning from the environment, as we claim in this paper, as well as using expectations to "help" adapting the system.

The reactive navigation component here described is similar to the potential field method employed on the mobile robotics circle. For instance Arkin (1989) AuRA's path planning model is based on obstacles exerting repulsive forces onto the robot, while targets apply attractive forces to the robot. Our work on the design of autonomous agents also has several points in common with the autonomous robots community. For instance we share with Brooks (1986) the behavioral decomposition and the absence of a single central representation. This is achieved in SBL by the schema decomposition and the assemblage of schema instances. The application of SBL to Robotics follows naturally and is currently under investigation (C&A96b).

Acknowledgments

We acknowledge very illuminating discussions with Dr. Christoph von der Malsburg. Preparation of this article was supported by NSF award no. IBN-9411503 (M. A. Arbib and A. Weerasuriya).

REFERENCES

Arbib, M. A. (1987). Levels of modeling of visually guided behavior. *Behav. Brain Sci.* 10, 407-465.

Arbib, M. A. (1992). Schema Theory. In: The Encyclopedia of Artificial Intelligence. (Shapiro, S., ed.) 2nd Edn pp. 1427-1443. New York, NY: Wiley Interscience.

Arkin, R. (1989). Motor Schema based mobile robot navigation. *Int. J. Robotics. Res.* 8(4), 92-112.

Brooks, R. A. (1986). A robust layered control system for a mobile robot. *IEEE Trans. Rob. Automation*, 2: 14-23.

Carbonell, J. G. & Gil, Y. (1990). Learning by experimentation: The Operator Refinement Method. (In *Machine Learning: An Artificial Intelligence Approach.* Vol. III. Y. Kodratoff & Ryszard Michalski Eds.). Morgan Kaufmann: San Mateo.

Collett, T. (1982). Do toads plan routes? A study of detour behavior of *B. viridis. J. Comp. Physiol. A*, 146:261-271.

Corbacho, F. & Arbib, M. A. (1995). Learning to Detour. *Adaptive Behavior*, 3(4), 419-468.

Corbacho, F. & Arbib, M. A. (1996a). Learning to Detour and Schema-based Learning. *Proceedings of the Fourth International Conference on Simulation of Adaptive Behavior.*

Corbacho, F. & Arbib, M. A. (1996b). Schema-based Learning. (in preparation).

Drescher, G. L. (1991). *Made-Up Minds.* Cambridge: MIT Press.

Duhamel, J. R., Colby, C. L., & Goldberg, M. E. (1992). The Updating of the Representation of Visual Space in Parietal Cortex by Intended Eye Movements. *Science*, 255: 90-92.

Foner, L. N. & Maes. P. (1994) Paying attention to what's important: using focus of attention to improve unsupervised learning. *Proceedings of the Third International Conference on Simulation of Adaptive Behavior.*

Hummel, R. A., & Zucker, S. W. (1983) On the Foundations of Relaxation Labelling Processes, *IEEE Trans. Pattern Analysis and Machine Intelligence*, 5:267-287.

Maes, P. & Brooks, R. A. (1990). Learning to coordinate behaviors. *AAAI-90*, Boston, 796-802.

Piaget, J. (1954). *The Construction of Reality in the Child.* New York: Ballantine.

Rumelhart, D. E., Hinton, G. E., and Williams, R. J. (1986). Learning internal representations by error propagation. In *Parallel distributed processing: Explorations in the microstructure of cognition* (D. E. Rumelhart, and J. McClelland, eds.), vol. 1, pp. 318-362. The MIT Press/Bradford Books.

Shen, W-M. (1994). *Autonomous Learning from the Environment.* New York: W. H. Freeman and Company.

Thrun, S. (1996). *Explanation-based Neural Network Learning: A lifelong learning problem.* Kluwer Academic Publishers: MA.

von der Malsburg, C. (1986). Am I Thinking Assemblies? In *Brain Theory*, G. Palm & A. Aertsen (Eds.). Springer-Berlag, Berlin.

von der Malsburg, C. (1994). The Correlation Theory of Brain Function, (Reprint) In *Models of Neural Networks II* (E. Domany, J. L. van Hemmen, and K. Schulten, eds.), Ch. 2, pp. 95-119. Springer Verlag.